Advanced Materials for Membrane Fabrication and Modification

Advanced Materials for Membrane Fabrication and Modification

Edited by
Stephen Gray
Toshinori Tsuru
Yoram Cohen
Woei-Jye Lau

CRC Press
Taylor & Francis Group
Boca Raton London New York

CRC Press is an imprint of the
Taylor & Francis Group, an **informa** business

CRC Press
Taylor & Francis Group
6000 Broken Sound Parkway NW, Suite 300
Boca Raton, FL 33487-2742

First issued in paperback 2020

ISBN-13: 978-1-138-73905-5 (hbk)
ISBN-13: 978-0-367-65698-0 (pbk)

Library of Congress Cataloging-in-Publication Data

Names: Gray, Stephen (Chemical engineer), editor. | Tsuru, Toshinori, editor. | Cohen, Yoram | Lau, Woei Jye, editor.
Title: Advanced materials for membrane fabrication and modification / editors, Stephen Gray, Toshinori Tsuru, Yoram Cohen, and Woei Jye Lau.
Description: Boca Raton : Taylor & Francis a CRC title, part of the Taylor & Francis imprint, a member of the Taylor & Francis Group, the academic division of T&F Informa, plc, 2018. | Includes bibliographical references and index.
Identifiers: LCCN 2018013173| ISBN 9781138739055 (hardback : acid-free paper) | ISBN 9781315184357 (e-book)
Subjects: LCSH: Membranes (Technology)--Materials.
Classification: LCC TP159.M4 A37 2018 | DDC 660/.28424--dc23
LC record available at https://lccn.loc.gov/2018013173

Visit the Taylor & Francis Web site at
http://www.taylorandfrancis.com

and the CRC Press Web site at
http://www.crcpress.com

Advanced Materials for Membrane Fabrication and Modification

Edited by
Stephen Gray
Toshinori Tsuru
Yoram Cohen
Woei-Jye Lau

CRC Press
Taylor & Francis Group
Boca Raton London New York

CRC Press is an imprint of the
Taylor & Francis Group, an **informa** business

CRC Press
Taylor & Francis Group
6000 Broken Sound Parkway NW, Suite 300
Boca Raton, FL 33487-2742

First issued in paperback 2020

ISBN-13: 978-1-138-73905-5 (hbk)
ISBN-13: 978-0-367-65698-0 (pbk)

Library of Congress Cataloging-in-Publication Data

Names: Gray, Stephen (Chemical engineer), editor. | Tsuru, Toshinori, editor. | Cohen, Yoram | Lau, Woei Jye, editor.
Title: Advanced materials for membrane fabrication and modification / editors, Stephen Gray, Toshinori Tsuru, Yoram Cohen, and Woei Jye Lau.
Description: Boca Raton : Taylor & Francis a CRC title, part of the Taylor & Francis imprint, a member of the Taylor & Francis Group, the academic division of T&F Informa, plc, 2018. | Includes bibliographical references and index.
Identifiers: LCCN 2018013173| ISBN 9781138739055 (hardback : acid-free paper) | ISBN 9781315184357 (e-book)
Subjects: LCSH: Membranes (Technology)--Materials.
Classification: LCC TP159.M4 A37 2018 | DDC 660/.28424--dc23
LC record available at https://lccn.loc.gov/2018013173

Visit the Taylor & Francis Web site at
http://www.taylorandfrancis.com

and the CRC Press Web site at
http://www.crcpress.com

Contents

SECTION II Membranes for Gas Separation Process

SECTION III Membranes for Organic Solvent Applications

SECTION IV Membranes for Energy Applications

SECTION V Membranes for Biomedical Applications

Preface

The first recorded use of a membrane was Jean-Antoine Nollet in 1748, who used a pig's bladder membrane to demonstrate osmosis. It was another 200 years later in the 1950s that synthetic membranes for reverse osmosis (RO) were developed at the University of California, Los Angeles (UCLA). These cellulose acetate membranes were demonstrated in the mid-1950s for seawater desalination by UCLA and the University of Florida, but their low water flux rendered them commercially unviable. Later enhancements to membrane morphology with the development of asymmetric membranes by Sidney Loeb at UCLA and Srinivasa Sourirajan at the National Research Council of Canada, Ottawa rendered RO membranes commercially viable. The thin skin layer on top of a thicker, open porous support layer enabled significantly higher flux to be achieved while retaining the separation efficiency. Further improvements were made to RO membrane design by John Cadotte of FilmTec Corporation in the 1960s, and he was able to produce even higher flux polyamide thin film composite membranes with high salt separation efficiency by using interfacial polymerization. While further improvements in RO membranes have occurred since then, the basic polyamide chemistry developed by Cadotte has remained as the basis for most RO and nanofiltration (NF) membranes used today. These membranes, along with the later development of microfiltration (MF) and ultrafiltration (UF) membranes, have pervaded the water industry and are now ubiquitous.

However, further improvements in membrane materials are still of interest, as such improved materials could overcome operational issues such as fouling via surface modification or increased membrane lifetimes via choice of membrane materials. Additionally, improved membrane materials may also enhance performance by increasing particle and pathogen rejection via smaller/narrower pore size distribution or improving flux, rejection or selectivity. For applications beyond water, such as for gas separation, separation in solvent-based systems, energy production and storage, and biomedical applications, membrane processes have not yet been fully exploited and have potential to provide greater processing efficiencies and improved outcomes. Indeed, the potential benefits of membrane processes in the energy, hydrocarbon, and pharmaceutical industries through low energy processing and greater selectivity is enormous, so there is still significant demand for new membrane materials to greatly impact many applications.

This book provides current information on emerging membrane materials and fabrication approaches for membranes, and spans materials from polymeric, surface modified, nanocomposites, and ceramics. This book aims to serve as a reference for membrane scientists and technologists working on the development of new membrane materials for a wide range of industrial separation processes. We have assembled the contributions from membrane experts from different disciplines and divided the chapters based on the five categories, (1) membranes for water and wastewater applications (5 chapters), (2) membranes for gas separation process (6 chapters), (3) membranes for organic solvent purification and concentration (4 chapters),

(4) membranes for energy applications (3 chapters), (5) and membranes for biomedical applications (2 chapters).

We wish to express our sincere gratitude to all the authors who have contributed to the writing of this book. We very much appreciate the efforts and time they have devoted to deliver high-quality content to make this new book a great success.

Stephen Gray
Victoria University, Melbourne, Australia

Toshinori Tsuru
Hiroshima University, Hiroshima, Japan

Yoram Cohen
University of California, Los Angeles, United States

Woei-Jye Lau
Universiti Teknologi Malaysia, Johor, Malaysia

Editors

Professor Dr. Stephen Gray
Victoria University
Australia

Dr. Stephen Gray is currently executive director of the Institute for Sustainable Industries and Liveable Cities at Victoria University, Australia. He has had a focus on water and membrane research for more than 25 years and was part of the CSIRO teams that developed MIEX™ resin and that conducted the Melbourne Water Climate Change Impact study. He is part of research teams that are working towards commercialization of membrane distillation, have introduced ceramic membranes to Australian water authorities, and is working towards commercialization of membrane integrity monitoring technology for reverse osmosis and ultrafiltration membranes. He also worked with the gas industry to develop silica removal and management technology for coal seam gas brines. In 2013 he was recognized as Thought Leader in Lux Research's international review of water research. He is known for his research on membrane distillation, organic fouling of low pressure membranes, and high recovery reverse osmosis systems. In the field of membrane fabrication, he has projects that span nanocomposite ultrafiltration membranes for increased abrasion resistance, nanocomposite pervaporation membranes for niche applications, ceramic membranes for desalination, and small pore size metal membranes.

Professor Dr. Toshinori Tsuru
Hiroshima University
Japan

Dr. Toshinori Tsuru earned his Dr. Eng. in 1991 from the University of Tokyo, Japan. After working at the University of Tokyo and Hiroshima University as a research associate and an associate professor, respectively, he was promoted to a full professor in 2006. He was a principal investigator of CREST (Core Research for Evolutionary Science and Technology) for development of robust reverse osmosis/nanofiltration membranes for various types of water resources (October 2012–March 2017) and has been appointed as a distinguished professor since 2015 at Hiroshima University. His research interests are preparation of subnano-nanoporous membranes and their applications to gas and liquid phase separation. He has been working on membrane science and technology for more than 30 years from the viewpoints of materials science and chemical engineering. His research experience started with nanofiltration in terms of the preparation of polymeric membranes and the transport mechanism of electrolytes through charged porous membranes. His research on nanofiltration, including the theory and experiments on extended Nernst-Planck equation for permeation of single electrolyte and separation of mixed electrolytes, is one of the pioneering works and has been cited more than several hundred times. After moving to Hiroshima University, his interest has expanded to inorganic membranes, including preparation and characterization of various types of nano/subnanoporous materials, mostly SiO_2-based materials. His research group successfully fabricated high-performance silica-based membranes by sol-gel processing and atmospheric plasma-enhanced chemical vapor deposition. Recently, he has successfully developed a highly permeable hydrogen separation membrane with high selectivity. He also applied them to various separation systems not only in liquid phase such as reverse osmosis and nanofiltration, but also in pervaporation and gas separation. Based on the precise and reproducible measurements of permeation properties in single and mixed systems, he is an expert in the transport mechanism through nano/subnanoprous membranes. In addition, he proposed the permeation mechanism and the modified gas translation model, which enables the determination of pore sizes of subnanometer range.

Professor Dr. Yoram Cohen
University of California
United States

Dr. Yoram Cohen is a distinguished professor of Chemical & Biomolecular Engineering (CBE) at the University of California, Los Angeles (UCLA) where he has been on the faculty since 1981. He is also on the faculty of the Institute of the Environment & Sustainability, Director of the UCLA Younes and Soraya Nazarian Center for Israel Studies, Director of the Water Technology Research (WaTeR), and co-founder of the UCLA/NSF Center for Environmental Implications of Nanotechnology (CEIN). He is a UCLA Luskin Scholar and holds the Rosalinde and Arthur Gilbert Foundation Chair. Dr. Cohen earned his B.A.Sc., M.A.Sc., in 1975 and 1977, respectively, both in Chemical Engineering, from the University of Toronto, and his Ph.D. in 1981 from the University of Delaware. He is a recognized expert in the areas of water technology, water purification and desalination, environmental protection and sustainability, environmental impact assessment, and nanoinformatics. He has over 230 published research papers and book chapters, 400 conference presentations, and 150 invited talks. Dr. Cohen has contributed to policy and regulatory efforts focused on environmental protection and economics of water reuse and has an active program devoted to assisting disadvantaged communities develop safe drinking water resources. He developed and patented water treatment and desalination technologies, new membranes, in addition to software for environmental impact assessment.

Associate Professor Dr. Woei-Jye Lau
Universiti Teknologi Malaysia
Malaysia

Dr. Woei-Jye Lau is currently an associate professor at the Faculty of Chemical and Energy Engineering, Universiti Teknologi Malaysia (UTM) and a research fellow at Advanced Membrane Technology Research Centre (AMTEC). He earned his Bachelor of Engineering in Chemical-Gas Engineering (2006) and Doctor of Philosophy (PhD) in Chemical Engineering (2009) from UTM. Dr. Lau has a very strong research interest in the field of membrane science and technology for water applications. He has published more than 100 scientific papers, 12 reviews, and 10 book chapters and is the author of the book titled *Nanofiltration Membranes: Synthesis, Characterization and Applications* published by CRC Press in January 2017. He is credited with 5 patents in the field of membrane science and technology with his collaborators from academia and industry. He wrote articles on the subjects of water saving and reclamation for newspapers and magazines at both the national and international level. Dr. Lau is currently the co-editor for *Journal of Applied Membrane Science & Technology* and has been appointed as guest editor for several peer-reviewed journals such as *Journal of Engineering Science & Technology*, *The Malaysian Journal of Analytical Sciences* and *Chemical Engineering & Technology*. Dr. Lau has received many national and international awards since he started his career in 2009. These include Australian Endeavour Research Fellowship 2015, UI-RESOLV Program 2016 (Indonesia), Mevlana International Exchange Program 2017/2018 (Turkey), and Sakura Exchange Program 2018 (Japan).

Contributors

Amura, Ida
Department of Chemical Engineering
University of Bath
Claverton Down, United Kingdom

Bastin, Maarten
Membrane Technology Group
Centre for Surface Chemistry and Catalysis
Faculty of Bioscience Engineering
KU Leuven, Belgium

Ben Amar, Raja
Faculty of Science of Sfax
University of Sfax
Sfax, Tunisia

Buekenhoudt, Anita
VITO NV—Flemish Institute for Technological Research NV
Mol, Belgium

Carbonnier, Benjamin
University of Paris-East
ICMPE (UMR7182), CNRS, UPEC
Thiais, France

Diniz da Costa, João C.
Functional and Interfacial Materials and Membranes Laboratory
School of Chemical Engineering
University of Queensland
Brisbane, Australia

Emanuelsson, Emma A.C.
Department of Chemical Engineering
University of Bath
Claverton Down, United Kingdom

Gray, Stephen
Institute for Sustainable Industries and Liveable Cities
Victoria University Research
Victoria University
Melbourne, Australia

He, Yubin
CAS Key Laboratory of Soft Matter Chemistry
Collaborative Innovation Centre of Chemistry for Energy Materials
School of Chemistry and Material Science
University of Science and Technology of China
Hefei, China

Hernández-Fernández, Francisco José
Department of Chemical and Environmental Engineering
Campus Muralla del Mar
Technical University of Cartagena
Cartagena, Spain
and
Department of Chemical Engineering
Campus Espinardo
University of Murcia
Murcia, Spain

Hou, Jianqiu
CAS Key Laboratory of Soft Matter Chemistry
Collaborative Innovation Centre of Chemistry for Energy Materials
School of Chemistry and Material Science
University of Science and Technology of China
Hefei, China

Idris, Ani
Institute of Bioproduct Development
Faculty of Chemical and Energy Engineering
Universiti Teknologi Malaysia
Johor, Malaysia

Irfan, Masooma
Department of Chemistry
Comsats University Islamabad
Lahore Campus
Lahore, Pakistan

Irfan, Muhammad
Interdisciplinary Research Centre in Biomedical Materials
Comsats University Islamabad
Lahore Campus
Lahore, Pakistan

Ismail, Ahmad Fauzi
Advanced Membrane Technology Research Centre
Universiti Teknologi Malaysia
Johor, Malaysia

Jin, Hua
School of Materials Science and Chemical Engineering
Ningbo University
Zhejiang, China

Jin, Wanqin
State Key Laboratory of Materials-oriented Chemical Engineering
College of Chemical Engineering
Nanjing Tech University
Nanjing, China

Kentish, Sandra
Peter Cook Centre for Carbon Capture and Storage Research
Department of Chemical Engineering
University of Melbourne
Melbourne, Australia

Khemakhem, Mouna
Faculty of Science
University of Sfax
Sfax, Tunisia

Kita, Hidetoshi
Graduate School of Science and Technology for Innovation
Yamaguchi University
Yamaguchi, Japan

Lau, Woei-Jye
Advanced Membrane Technology Research Centre
Universiti Teknologi Malaysia
Johor, Malaysia

Lee, Sung-Eun
College of Agriculture & Life Sciences
School of Applied Biosciences
Division of Environmental Life Science
Kyungpook National University
Daegu, Republic of Korea

Li, Yanshuo
School of Materials Science and Chemical Engineering
Ningbo University
Zhejiang, China

Liu, Gongping
State Key Laboratory of Materials-oriented Chemical Engineering
College of Chemical Engineering
Nanjing Tech University
Nanjing, China

Liu, Liang
Peter Cook Centre for Carbon Capture and Storage Research
Department of Chemical Engineering
University of Melbourne
Melbourne, Australia

Mahouche-Chergui, Samia
University of Paris-East
Thiais, France

Matsukata, Masahiko
Department of Applied Chemistry
Advanced Research Institute for Science and Engineering
Waseda University
Tokyo, Japan

Matsuyama, Hideto
Department of Chemical Science and Engineering
Graduate School of Engineering
Kobe University
Kobe, Japan

Muhamad, Mimi Suliza
Advanced Technology Centre
Faculty of Engineering Technology
Universiti Tun Hussein Onn Malaysia
Johor, Malaysia

Mulyati, Sri
Chemical Engineering Department
Faculty of Engineering
Syiah Kuala University
Aceh, Indonesia

Nagasawa, Hiroki
Department of Chemical Engineering
Hiroshima University
Hiroshima, Japan

Nazri, Noor Aina Mohd
Section of Chemical Engineering
International College
Universiti Kuala Lumpur
Melaka, Malaysia

Ong, Chi-Siang
Advanced Membrane Technology Research Centre
Universiti Teknologi Malaysia
Johor, Malaysia

Ortiz-Martínez, Víctor Manuel
Department of Chemical and Environmental Engineering
Campus Muralla del Mar
Technical University of Cartagena
Cartagena, Spain
and
Department of Chemical Engineering
Campus Espinardo
University of Murcia
Murcia, Spain

Oun, Abdallah
University of Paris-East, ICMPE (UMR7182), CNRS, UPEC
Thiais, France
and
University of Sfax, FSS (MESLab)
Sfax, Tunisia

Pérez de los Ríos, Antonia
Department of Chemical and Environmental Engineering
Campus Muralla del Mar
Technical University of Cartagena
Cartagena, Spain
and
Department of Chemical Engineering
Campus Espinardo
University of Murcia
Murcia, Spain

Sakai, Motomu
Research Organization of Nano & Life Innovation
Waseda University
Tokyo, Japan

Salar-García, María Jose
Department of Chemical and Environmental Engineering
Campus Muralla del Mar
Technical University of Cartagena
Cartagena, Spain
and
Department of Chemical Engineering
Campus Espinardo
University of Murcia
Murcia, Spain

Seshimo, Masahiro
Research Organization of Nano & Life Innovation
Waseda University
Tokyo, Japan

Shahid, Salman
Department of Chemical Engineering
University of Bath
Claverton Down, United Kingdom

Shon, Ho Kyong
Centre for Technology in Water and Wastewater
School of Civil and Environmental Engineering
University of Technology Sydney
Sydney, Australia

Takagi, Ryosuke
Department of Science, Technology and Innovation
Graduate School of Science, Technology and Innovation
Kobe University
Kobe, Japan

Tijing, Leonard D.
Centre for Technology in Water and Wastewater
School of Civil and Environmental Engineering
University of Technology Sydney
Sydney, Australia

Tsuru, Toshinori
Department of Chemical Engineering
Hiroshima University
Hiroshima, Japan

ur Rehman, Ghani
Advanced Membrane Technology Research Centre
Universiti Teknologi Malaysia
Johor, Malaysia

Vankelecom, Ivo F.J.
Membrane Technology Group
Centre for Surface Chemistry and Catalysis
Faculty of Bioscience Engineering
KU Leuven, Belgium

Wang, David K.
School of Chemical and Biomolecular Engineering
University of Sydney
Sydney, Australia

Woo, Yun Chul
Department of Land, Water and Environment Research
Korea Institute of Civil Engineering and Building Technology
Gyeonggi-Do, Republic of Korea

Xie, Ming
Institute for Sustainable Industries and Liveable Cities
Victoria University Research
Victoria University
Melbourne, Australia

Xu, Tongwen
CAS Key Laboratory of Soft Matter Chemistry
Collaborative Innovation Centre of Chemistry for Energy Materials
School of Chemistry and Material Science
University of Science and Technology of China
Hefei, China

Yang, Weishen
State Key Laboratory of Catalysis
Dalian Institute of Chemical Physics
Chinese Academy of Sciences
Liaoning, China

Yao, Minwei
Centre for Technology in Water and Wastewater
School of Civil and Environmental Engineering
University of Technology Sydney
Sydney, Australia

Yoshihara, Kei
Department of Applied Chemistry
Waseda University
Tokyo, Japan

Yoshimune, Miki
Research Institute for Chemical Process Technology
National Institute of Advanced Industrial Science and Technology
Tsukuba, Japan

Zhao, Changsheng
College of Polymer Science and Engineering
State Key Laboratory of Polymer Materials Engineering
Sichuan University
Chengdu, China

Zhao, Weifeng
College of Polymer Science and Engineering
State Key Laboratory of Polymer Materials Engineering
Sichuan University
Chengdu, China

Section I

Membranes for Water and Wastewater Applications

1 Recent Progress of Thin Film Composite Membrane Fabrication and Modification for NF/RO/FO Applications

Chi-Siang Ong, Ming Xie, Woei-Jye Lau, Stephen Gray, and Ahmad Fauzi Ismail

CONTENTS

1.1 INTRODUCTION

With a growing world population, global climate change, and intensification of human activities, water availability is becoming one of the most important environmental challenges facing humanity. Membrane-based technologies for water treatment, water reclamation, and desalination are some of the most effective strategies to address the global water quality and scarcity issues.

Thin film composite (TFC) membranes are the state-of-the-art technology used to augment freshwater supply because of their high water permeability, excellent salt selectivity, and wide pH operation range. In addition, the ability to separately optimize two layers, a dense active (selective) layer and a microporous support layer, during membrane fabrication renders TFC membranes superior compared to asymmetric membranes formed by phase inversion technique.

The ultrathin (<500 nm) active layer of TFC membranes is fabricated on top of a microporous support layer through interfacial polymerization. Surface characteristics such as morphology, functional groups, charge, and hydrophilicity of the active layer play a significant role in determining TFC membrane performance, including transport properties and fouling propensity. Thus, numerous efforts have been dedicated to either fabricate TFC membranes with better rejection and flux performance or to modify the membrane active layer to tailor fouling characteristics for various challenging feed streams. An improved understanding between surface properties and the TFC membrane performance will enable and guide next generation of membrane design.

In this chapter, we review and discuss the latest development in the TFC membrane fabrication and techniques for the membrane surface modification. In particular, a comprehensive review of fabrication approaches for microporous substrates as well as polyamide active layer is presented. Inorganic nanofillers as an effective material to enhance membrane performance will be highlighted and discussed. Several surface modification approaches will also be discussed to impart membrane with low fouling propensity, including polyethylene glycol (PEG)-based derivations, graphene oxide (GO) or GO/silver (Ag) nanocomposite and surface patterning.

1.2 BRIEF DESCRIPTION OF TFC MEMBRANE FABRICATION AND SURFACE MODIFICATION METHOD

In this section, the approaches commonly used to synthesize or modify the substrate and polyamide selective layer of TFC membranes will be briefly described in order to provide readers quick access to information about techniques involved in TFC membrane fabrication and modification.

1.2.1 Phase Inversion Process for Microporous Substrate Synthesis

The phase inversion method, also known as the polymer precipitation process, is the process of transforming polymer solution from a liquid state into a solid state (i.e., forming a membrane) using a non-solvent (e.g., water) as the medium. This method is widely used in the preparation of microporous substrate for commercial TFC membranes. A homogenous polymer solution can be prepared by dissolving a pre-weighed amount of the main membrane forming material into suitable organic solvents (e.g., *n*-methyl-2-pyrrolidone (NMP) and dimethylformamide (DMF)) followed by continuous stirring for up to 1 day. In most of the cases, a small amount of polymer additive (e.g., PEG or PVP (polyvinylpyrrolidone)) is added into the polymer/solvent mixture to serve as a pore former for the membrane structure during the phase inversion process.

Although a vast variety of polymers have been successfully used as microporous supports for lab-scale TFC membrane fabrication over the years, polysulfone (PSF) still remains a mainstay in the commercial TFC membrane (Lau et al., 2012). Many works have been previously conducted to optimize the properties of PSF-based substrate for TFC membrane. But, the exact substrate's pore properties (mean pore size, pore size distribution, and porosity) that are ideal for TFC membranes are not easy to define as there are many other factors such as surface roughness and hydrophilicity of substrate and interfacial polymerization conditions that govern the formation of the polyamide layer. Ghosh and Hoek (2009) found that by varying the structure and chemistry of PSF-based substrate, TFC membranes with different water permeability and structure could be produced. They explained four different scenarios based on the observations from support membranes with pore size in the range of 30–70 nm, water contact angles between 60° and 80°, and RMS roughness of 5–10 nm.

PSF is generally well accepted as a support film, but it is not without drawbacks as composite membrane support. Its hydrophobic properties are the main concerns to many particularly when the composite membranes are used in water separation process (Peterson, 1993). In order to achieve higher surface hydrophilicity and better antifouling properties of the resultant substrate, the introduction of appropriate amounts of hydrophilic nanoparticles into the polymer solution is an effective strategy to produce a nanocomposite substrate with synergistic properties without compromising solute rejection (Lau et al., 2015; Ghanbari et al., 2015b). In order for the nanoparticles to disperse well in the polymeric solution without causing severe aggregation, mild modification on the nanoparticle surface is normally performed to improve its compatibility with organic polymers. Such mixed matrix slurry is then subjected to several hours of sonication prior to use in order to ensure the homogeneity of the solution.

It must be noted that a large number of variables can influence the properties of a microporous substrate. These include the characteristics of main membrane forming material (e.g., molecular weight and concentration in polymeric solution), solvent type (e.g., solubility parameter and viscosity), synthesis conditions (e.g., shear rate/take up speed, coagulation medium, evaporation temperature, and humidity), and post treatment (e.g., drying method and period). The variation of these conditions can affect not only the characteristics of the substrate but also its interaction with polyamide layer during interfacial polymerization, leading to production of composite membranes with wide range of properties.

1.2.2 Electrospinning Process for Microporous Substrate Synthesis

In recent years, the use of nanofiber substrate for manufacturing TFC membranes has become more and more popular particularly for the engineered osmosis applications (Li et al., 2016; Tian et al., 2017; Park et al., 2018). The relatively thick, low porosity support materials commonly used in the nanofiltration (NF) and reverse osmosis (RO) membrane substrates are not suitable for forward osmosis (FO) membranes as they tend to create higher resistance to solute mass transfer during FO/pressure retarded osmosis (PRO) applications, leading to the boundary layer phenomenon known as internal concentration polarization (ICP) and lowering the water transport rate (Bui et al., 2011).

Typically, the electrospinning process that is used to fabricate nanofiber substrate involves the application of an electrical field between the polymer solution (with or without nanoparticles) in a plastic syringe and a grounded rotating collector covered with aluminum foil. Under a sufficiently high field, electrical force would overcome the surface tension of the polymer solution, deforming a pendant drop at the tip of the syringe and resulting in ejection of a thin jet. The charged jet undergoes a stable stretching before starting to bend and whip randomly as a result of the combined effects of solvent evaporation and charge repulsion. Post-treatment by subjecting the nanofiber substrate to heating process is normally performed to eliminate the residual solvent from the nanofiber substrate.

Similar to the asymmetric substrate made by phase inversion technique, the characteristics of the nanofiber are dependent not only on the properties of the polymer solution but also the fabrication conditions. Some of the key parameters that can influence the diameter, shape, texture, and surface morphology of the nanofiber substrate are the strength of the electrical field, the electrode geometry, the distance between the syringe and collector, the rotating speed of the collector, and the feeding flow rate of the polymer solution. The specific pore structure that is best for polyamides to interact with is not clear, as too many factors govern the interaction between polyamide layers and nanofiber substrates. Nevertheless, Huang and McCutcheon (2014) reported that besides exhibiting lower water contact angle, the nanofiber should possess excellent tensile strength to make it highly desirable as TFC membrane substrate. Additionally, the overall mechanical property of nanofiber can be enhanced with the use of nonwoven support prior to interfacial polymerization for pressure-driven NF process (Wang et al., 2014).

1.2.3 Interfacial Polymerization for Polyamide Layer Synthesis and Its Surface Modification Techniques

The interfacial polymerization technique that has been known since the 1960s is a type of step-growth polymerization in which polymerization occurs at an interface between an aqueous solution containing amine monomer and an organic solution containing polyacyl chloride monomer. By employing this technique, an ultrathin polyamide layer with a thickness of several hundreds of nanometers is able to form over the surface of a microporous substrate, producing a TFC membrane with a good combination of water flux and solute rejection. This technique is the commercial technique used in the fabrication of TFC membranes for NF, RO, and FO processes.

To establish a very thin polyamide layer on the surface of microporous substrate, an aqueous solution containing amine monomer of 1–3 w/v% (prepared in the unit of w/v%, that is, amine monomer/100 mL pure water) is first poured on the support surface and allowed to come into contact with the substrate surface for several minutes before draining the excess aqueous solution (Lau and Ismail, 2017). The polyamide layer will start to develop right after an organic solution (usually *n*-hexane or cyclohexane) that contains acyl chloride monomer of 0.05–0.2 w/v% is poured on the same substrate surface (Lau and Ismail, 2017). Once the interaction between the two active monomers is complete, the organic solution is drained and the resulting TFC membrane is rinsed using either pure organic solution or pure water. At last, heat treatment is carried out to densify the polymerization properties of the polyamide layer.

Similar to the TFC membrane, the thin film nanocomposite (TFN) membrane that incorporates inorganic nanoparticles within the polyamide layer can also be produced using the same interfacial polymerization technique. Although the nanoparticles with typical concentration of 0.01–0.1 w/v% could be introduced into either aqueous or organic solutions, most research has preferred to introduce nanoparticles that are hydrophilic in nature into the amine aqueous solution (Lau et al., 2015). This could be due to the difficulties in producing a homogenous mixture in the nonpolar organic phase. Nevertheless, dispersing modified nanoparticles into an organic solution is more practical to produce a TFN membrane with better surface integrity and minimum leaching of nanoparticles (Lai et al., 2016a). This is because the soft rubber roller that is used to remove excess amine aqueous solution during the interfacial polymerization process could remove a large amount of nanoparticles together with the amine solution (in the case of nanoparticles-aqueous mixture), leaving only a small amount of nanoparticles in the substrate. Good dispersion of nanofillers into a polyamide layer is not only able to overcome the water flux/salt selectivity trade-off of the typical TFC membrane but also to improve its resistances against organic fouling, biofouling, bacterial adhesion, and chlorination attack (Lau et al., 2015). As a rule of thumb, the polyamide layer produced (with and without incorporation of nanofillers) should be able to reject effectively divalent salt (e.g., Na_2SO_4 and $MgSO_4$) for NF application and monovalent salt (e.g., NaCl) for RO and FO application.

Instead of embedding the nanoparticles within the polyamide layer, post-fabrication functionalization on the top polyamide surface of TFC membrane using inorganic particles or hydrophilic polymers is also one of the approaches to tailor membrane surface chemistry. As the materials are placed specifically at the polyamide surface, the bulk properties of the polyamide layer underneath are not significantly affected. Furthermore, the surface modification approach is more material- and cost-efficient since fewer materials are required. These materials can be immobilized onto the polyamide surface via physical interaction, chemical binding, and layer-by-layer technique, depending on the characteristics of modifiers.

1.3 DEVELOPMENT OF MICROPOROUS SUBSTRATE OF COMPOSITE MEMBRANE

1.3.1 Polymer/Polymer-Polymer Blended Substrate

Most of the relevant works on TFC membranes used PSF or polyethersulfone (PES) for support film fabrication, but these polymeric materials are not without their drawbacks. Their relatively poor thermal properties and solvent resistance are the main concerns, particularly when such membranes are subject to harsh environments. In addition, the relatively high water contact angle of the PSF and PES-based substrates might not suitable for the engineered osmosis process in which the substrate of the TFC membrane is in direct contact with an organic-loaded feed solution in PRO orientation. In light of this, many studies have been carried out to improve the properties of substrates with respect to chemical, thermal, and antifouling resistance without compensating water permeability and solute rejection.

It must be noted that apart from developing substrates with improved resistance, studies on the interaction between a polyamide selective layer and microporous substrate made of new type of polymeric materials are also one of the main research focus areas over the years. A review by Lau et al. (2012) highlighted several excellent candidates that can be used as materials for microporous substrates. These include plasma-modified polyvinylidene fluoride (PVDF) (Kim et al., 2009), chemically-modified polypropylene (PP) (Korikov et al., 2006), polyacrylonitrile (PAN) (Dalwani et al., 2011), poly(pyromellitic dianhydride-co-4,4′-oxydianiline) (PMDA/ODA) (Ba and Economy, 2010) and sulfonated poly(phthalazinone ether sulfone ketone) (SPPESK) (Dai et al., 2002). These substrate materials are made of either commercially available polymers or new laboratory synthesized polymers. In this section, focus will be placed on the latest development of new types of substrates for various applications of TFC membranes over the past 5 years.

Kong et al. (2016) fabricated a series of TFC NF membranes using hollow fiber substrate made of polyvinyl chloride (PVC). Although PVC is one of the commonly used materials for microfiltration (MF) and ultrafiltration (UF) membrane preparation, the use of it as substrate materials for TFC membranes is rarely reported. The findings showed that the differences in the substrate pore size (25.8–13.0 nm) and porosity (32.8–51.6%) had their respective contributions to the changes in the characteristics (cross-linking degree and thickness) of the polyamide layer formed. In general, the substrate with smaller pore size but higher porosity tended to produce the TFC membrane associated with higher salt rejection but lower water flux. The TFC membrane made of the substrate with intermediate properties (17.9 nm pore size and 45.7% porosity) was reported to exhibit $MgSO_4$ and NaCl rejection of 98% and 30%, respectively, with pure water permeability of 7 L/m^2.h.bar. Furthermore, it was found that the performance of the TFC membrane remained unchanged after continuous filtration for 24 h using either 0.5 wt% NaOH (pH 11) or 0.5 wt% citric acid (pH 2.5) solution. This indicated the good stability of the developed membranes for water filtration over a relatively wide pH range.

Polytetrafluoroethylene (PTFE) substrates made by the bi-stretching process were also reported in the work of Tang et al. (2017), in which the impact of bi-stretching ratios (longitudinal stretching ration and transverse stretching ratio) on the PTFE substrate properties (pore size and porosity) and thus on TFC membrane separation performance were studied. Compared to the PSF- and PES-based substrates, the substrates made of PTFE offer several unique features including excellent chemical and thermal stability. The increase in longitudinal stretching ratio increased the length of fibrils, which further increased the pore sizes and porosities of substrate. This led to the production of a TFC membrane with a lower rejection rate but higher permeate flux. The increase of transverse stretching ratio after the longitudinal stretching process, however, caused the pore sizes to decrease but porosities to increase, owing to interlaced fibrils during the bi-stretching process. Accordingly, it improved TFC membrane rejection and flux simultaneously. Besides showing good rejection rates against Na_2SO_4 (98.1%) and methylene blue (98.2%), the developed TFC membrane also exhibited good acid stability (over 30 days) and high stability on long-term filtration process (over 90 days) when tested at constant pressure of 4 bar. No mechanical test was conducted on such membranes in this work, but the stretching process is likely to make the membrane thinner and mechanically weak and might not suitable for high-pressure operation.

Widjojo et al. (2013) self-synthesized sulfonated polyphenylenesulfone (sPPSU) via direct route by varying the content of sulfonated units of monomer (2.5 or 5 mol% 3,30-di-sodiumdisulfate-4,40-dichlorodiphenyl sulfone (sDCDPS)) and used the synthesized polymer as the supporting layer material for the TFC membrane fabrication. When tested using 2 M NaCl as draw solution under PRO mode, the pure water flux (54 L/m^2.h) of the membrane made of sPPSU substrate (2.5 mol% DCDPS) was 5 times higher than that of a TFC membrane made of non-sulfonated substrate with reverse salt flux maintained at 8.8 g/m^2.h. The newly developed membrane also displayed promising water flux (22 L/m^2.h) when it was subjected to synthetic seawater concentration solution that used 3.5 wt% NaCl as feed seawater and 2 M NaCl as draw solution under PRO mode.

To improve the mechanical strength of the TFC membranes, Han et al. (2016) developed a mesh-reinforced substrate by combining the strength of robust polyester (PET) open mesh with hydrophilic sPPSU and used it to develop high-performance TFC membranes. The resultant membrane substrate showed a small membrane thickness (45 μm), small structural parameter (<300 μm), and high mechanical strength (59–80 MPa). In addition, the prepared sPPSU-TFC membranes exhibited very excellent and stable osmotic water fluxes of 69.3–76.5 L m^2.h and 38.5–47.0 L/m^2.h with low reverse salt fluxes of <0.22 g/L using DI water as feed and 2 M NaCl as draw solution under PRO and FO modes, respectively. This is possibly attributed to the super-hydrophilic nature and the relative large pore size of the sPPSU substrate.

Besides using a single type of polymer for substrate making, many research papers reported the effect of two different types of polymer materials on the properties of substrate and evaluated how the changes in a substrate's properties could affect TFC membrane performance. The research on polymer-polymer blended substrates has received much attention in recent years particularly for the FO/PRO application in which the substrate of TFC membrane contacts directly with either a feed solution or a draw solution during osmotic process.

Wang et al. (2012) fabricated a PES/sulfonated polysulfone (SPSF) blended substrate and used it for TFC membrane development intended for FO/PRO process. Compared to the substrate made of single polymer (PES or PVDF), the PES/SPSF blended substrate exhibited much higher water permeability owing to a lower water contact angle and better hydrophilicity upon incorporation of a hydrophilic sulfonated polymer into the membrane matrix. When tested under PRO mode using 5 M NaCl as the draw solution and DI water as the feed, the TFC membrane made of PES/SPSF substrate displayed water flux of 69.8 L/m^2.h with reverse solute flux as low as 25 g/ ^{2}.h. As a comparison, the TFC membrane made of pure PVDF and PES substrates only exhibited 50 L/m^2.h / 23 g/m^2.h and 32.8 L/m^2.h / 44.7 g/m^2.h, respectively, under the same testing conditions.

Using hydrophilic sulfonated polyetheretherketone (SPEEK) as secondary polymer, Corvilain et al. (2017) fabricated a series of PES/SPEEK blended substrates for TFC FO membranes by varying the content of SPEEK in the dope solution. By blending PES with SPEEK, the bottom side of the substrate changed into an inhomogeneous porous structure consisting of areas with a high pore density present in what had previously been a low pore density area. Results showed that polyamide layer became smoother with increasing SPEEK content in the substrate, resulting in a decreased retention and increased water flux of the TFC membranes. The TFC membrane incorporated with optimum SPEEK content (5%) showed water flux of 6 L/m^2.h, which was superior compared to the commercial cellulose triacetate (CTA) membranes when tested with 0.5 M NaCl as draw solution and DI water as feed solution.

Instead of using sulfonated polymer, Sun et al. (2014) blended PES substrate with PSF and studied the effect of PSF/PES ratio on the membrane performance. Their results showed that by using an optimized PSF/PES ratio of 2/3 (total polymer weight in dope solution: 18 wt%), the TFC membrane made of such substrate could reach FO flux of 27.6 L/m^2.h, which is 5.5 times higher than that of substrate made without PES (5 L/m^2.h). The remarkable improvement in the FO water flux is due to the greater surface roughness coupled with highly microporous structure of substrate, thus resulting in increase in contact surface area with water. Nevertheless, rejection capability of the developed membranes was not reported in this work.

Han et al. (2012) on the other hand modified the surface of PSF support using bio-inspired polymer-polydopamine (PDA) through oxidant-induced dopamine polymerization. By varying the PDA coating time from zero to 5 h, they reported that 1-h PDA coating time was the most ideal condition to achieve a good combination of PRO water flux and reverse salt flux. The polyamide layer formed over the PDA-coated PSF substrate could attain water flux of 24 L/m^2.h compared to only 7.5 L/m^2.h shown by the polyamide layer formed over the unmodified substrate when both membranes were tested using 2 M NaCl as draw solution and DI water as feed solution under PRO mode. Furthermore, the reverse salt flux of the modified membrane was lower compared to the unmodified membrane, indicating the potential of PDA in overcoming the trade-off effect between water flux and reverse salt flux. The authors attributed the improved performance of the modified membranes to the enhanced hydrophilicity and smoother surface when PDA was coated on the PSF substrate.

Unlike the previous works that used different polymeric materials for substrate fabrication, Liu and Ng (2014) fabricated a two-layer PSF substrate using the double-blade casting technique. In this work, the polymer concentration of substrate's bottom layer that attached onto the PET mesh was varied in the range of 5–9 wt% while the top layer contained a fixed PSF concentration (10 wt%). Compared to the TFC membrane made of the single-blade cast substrate, the membrane made of the double-blade cast substrate showed increased water flux without compromising the reverse salt flux. With a 1 M NaCl draw solution and DI water feed solution, the best TFC membrane (containing 7 wt% PSF in bottom layer) achieved water flux (J_v) of 31.1 L/m^2.h and reverse salt flux (J_s) of 8.5 g/m^2.h in the FO orientation, and J_v of 60.3 L/m^2.h and J_s of 17.6 g/m^2.h in the PRO orientation. Moreover, the membrane made of double-blade cast substrate showed reduced apparent structural (*S*) parameter's value (i.e., 195 μm) and retained a relatively low J_s/J_v ratio. This could be attributed to the fact that the double-blade cast substrate exhibited a thinner and more permeable structure to reduce ICP, while retaining ideal surface pore structures for the formation of an intact, dense, and highly salt rejecting polyamide active layer.

Recently, Ong et al. (2017) fabricated a novel double-skinned polyamide TFC membranes by creating an additional thin layer of zwitterionic polyelectrolyte brush-poly(3-(N-2-methacryloxyethyl-N,N-dimethyl) ammonato propanesultone (PMAPS) on the bottom surface of microporous PES substrate. The developed membrane was then subjected to oily wastewater treatment under FO mode. The resultant double-skinned membrane exhibited high water flux of 13.7 L/m^2.h and reverse salt flux of 1.6 g/m^2.h using 2 M NaCl as the draw solution and emulsified oily solution as the feed. The double-skinned membrane outperformed the single-skinned membrane with much lower fouling propensity for emulsified oil-water separation. Figure 1.1 compares the water flux profile of two membranes over a period of 480 min in three-step filtration process using DI water and oil-water solution.

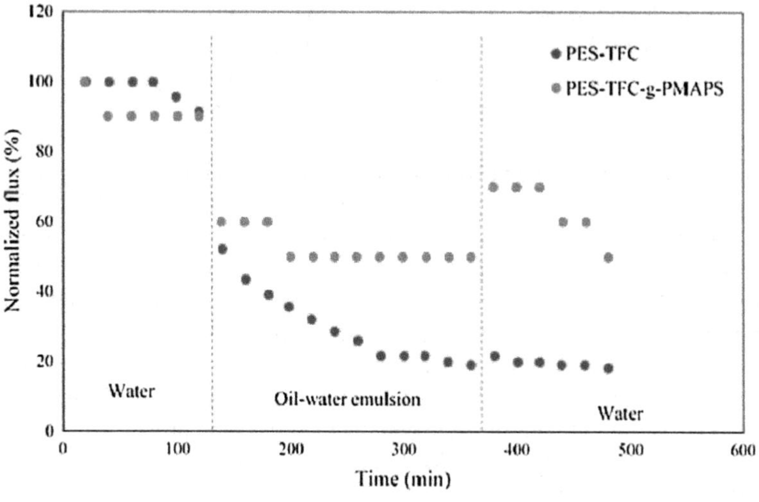

FIGURE 1.1 Comparison between the water flux recovery of single-skinned (PES-TFC) and double-skinned TFC membranes (PES-TFC-g-PMAPS) during three steps: (1) pure water flux for 120 min, (2) oil-water emulsion water flux for 260 min, and (3) pure water for 120 min after washing with DI water. (From Ong, C.S. et al., Scientific Reports. 7, 6904, 2017.)

The higher water flux recovery of the double-skinned membrane (PES-TFC-g-PMAPS) can be explained by the superior hydrophilic properties of PMAPS, which serves as an antifouling layer to prevent internal fouling and reduce the effects of ICP. When PMAPS is immersed into water, the PMAPS chain is hydrated and water molecules are thus trapped within the PMAPS and form a water layer. This water layer could prevent oil from adsorbing onto the substrate made of PES.

1.3.2 Polymer/Inorganic Nanomaterials Substrate

The use of inorganic nanomaterials as filler for substrate synthesis was first demonstrated by Pendergast et al. (2010) to study the effects of nanomaterials on the compaction behavior of RO membrane. In that study, zeolite or silica (SiO_2) nanoparticles were incorporated into the PSF substrate, which was then used in the IP process to prepare composite membranes. Besides exhibiting higher water flux, the prepared membranes experienced lower flux decline during the compaction when compared with the original TFC membrane. The existence of nanofillers was believed to have provided necessary mechanical support to mitigate the collapse of the porous structure and thickness reduction upon high-pressure operation. Electron microscope images meanwhile supported the hypothesis that the nanofillers-incorporated substrate (also known as the nanocomposite substrate) resisted physical compaction.

Compared with TFC NF or RO membranes, the effect of nanomaterials on the substrate properties of FO membranes is much more significant as both the top polyamide selective layer and bottom substrate are simultaneously contacted with aqueous solutions during the filtration process, but of different osmotic pressures. Developing a substrate with enhanced hydrophilicity and reduced S value (which is controlled by thickness, tortuosity, and porosity) is, therefore, important to mitigate ICP during the filtration process that occurs inside the substrate membrane.

It must be noted that substrate modification using appropriate loading of inorganic nanofillers does improve chemistry of the polyamide layer formed over the nanocomposite substrate to a certain extent, leading to improved water permeability without affecting excellent solute separation. Kim et al. (2012) and Son et al. (2015) have incorporated multiwalled carbon nanotubes (MWCNTs) into microporous substrates made of PSF and PES, respectively, aiming to develop an enhanced performance composite membrane for water purification and separation. In comparison to the control TFC membrane, the newly developed composite membrane displayed higher water flux without sacrificing salt rejection, owing to the improved surface hydrophilicity, average pore width, and porosity in the MWCNTs-modified substrate (Son et al., 2015). Higher negative surface charge of the polyamide layer was also reported in that work, but the authors did not perform in-depth analysis to explain how the MWCNTs-modified substrate with increased charge properties could render the coated polyamide layer with thickness of several hundreds of nanometers to possess a greater negative charge.

With respect to the nanofiller loadings used in the nanocomposite substrate synthesis, Emadzadeh et al. (2014) found that 0.6 wt% titanium dioxide (TiO_2) nanoparticles is the optimum loading to produce a composite membrane with a good balance of water permeability and salt rejection. When that amount of TiO_2 was introduced into the PSF substrate, the water permeability of the new composite membrane was significantly increased to 6.57 L/m^2.h, which was 100% higher than that of the control TFC membrane (3.27 L/m^2.h) when tested using 20 mM NaCl feed solution at 2.5 bar. This flux enhancement can be explained by the improved hydrophilicity of the substrate together with the increased overall porosity upon TiO_2 incorporation. Excessive use of TiO_2 nanoparticles (0.9 wt%) is not recommended, as it tends to cause significant nanoparticle agglomeration on the substrate surface, leading to increase in surface roughness. This, as a result, negatively affects the cross-linking degree of polyamide and further reduces the salt removal rate. The impact of adding zeolite NaY nanoparticles into the substrate matrix on the RO membrane performance can also be found elsewhere (Ma et al., 2013), where it was reported that 0.5 wt% was the ideal nanofiller loading for nanocomposite substrate in order to achieve a good combination of membrane water flux and solute rejection.

To further improve water flux and salt rejection of RO membrane, Ahmad et al. (2015) incorporated SiO_2 nanoparticles into an optimized cellulose acetate/polyethylene glycol-600 (CA/PEG) substrate. In addition to increase in substrate hydrophilicity, the incorporation of SiO_2 led to enhanced mechanical and thermal properties of substrate. More importantly, the TFC membrane made of CA/PEG/SiO_2 (16/4/5 wt%) substrate was reported to have higher water flux (2.46 L/m^2.h) and better salt rejection (90%) compared to the control membrane (0.35 L/m^2.h and 81.5%) when tested using 10,000 ppm NaCl solution at 6.5 bar.

An attempt was also made to incorporate GO into the PSF substrate for the preparation of a new type of TFC NF membrane (Lai et al., 2016b). The results revealed that the presence of a small quantity of GO (0.3 wt%) in the PSF substrate was able to improve the physicochemical properties of the PSF substrate and polyamide selective layer with respect to structural integrity (of the microporous substrate), surface hydrophilicity, roughness, and charge. When benchmarking with the control TFC membrane, the pure water flux of the best-performing GO-embedded TFC membrane was reported to improve by 50.9%. This membrane also showed high rejection towards multivalent salts, that is, 95.2% and 91.1% for Na_2SO_4 and $MgSO_4$, respectively. Furthermore, it was capable of rejecting close to 60% of NaCl, that is, 88.5% higher than that of the control TFC membrane. The promising outcome from the filtration experiments can be mainly attributed to the unique characteristics of GO nanosheets such as superior hydrophilicity and high negative charge. The charge value of GO (–30 to –58 mV at pH 5–8) is generally much higher than that of TiO_2 nanoparticles (30 to –15 mV at pH 5–7). Other nanofillers that have been successfully used for NF/RO membrane fabrication include zinc oxide (Pal et al., 2015), zeolite (Fathizadeh et al., 2011), and CNTs (Song et al., 2016). All these works have demonstrated the positive impacts of nanofillers on substrate properties and thus composite membrane performance, provided the loading of nanofillers added is at an appropriate range.

On the other hand, to enhance the performance of composite membrane during FO process, Wang and Xu (2015) studied the effect of montmorillonite (MMT) addition on the physicochemical properties and morphology of a PES/sulphonated polyethersulfone (SPES) blended substrate and the performance of a prepared TFC membrane. The TFC membrane prepared using MMT-incorporated PES/SPES substrate exhibited 28.39 L/m^2.h water flux and 3.53 g/m^2.h reverse solute flux with DI water as feed solution and 2 M NaCl as draw solution in FO mode. Compared to the membrane based on PES/SPES substrate, the FO water flux of the membrane with the MMT-incorporated substrate was improved by about 4 folds and the reverse salt leakage was reduced by about half. The authors attributed the promising results to the enhanced wettability of substrate and narrower pore size distribution of MMT-modified substrate.

Wang et al. (2013) also experienced water flux enhancement of FO membrane when surface-modified MWCNTs were embedded in the matrix of a PES-based substrate. Their results showed that the presence of an appropriate amount of MWCNTs could produce a substrate with higher porosity, which led to a significant decrease in ICP and greater osmotic water flux. Moreover, the tensile strengths of the modified substrates incorporated with loadings studied in this work were all better than that of the control substrate.

In the work conducted by Wang et al. (2015), reduced graphene oxide modified graphitic carbon nitride (rGO/CN) was used to alter the characteristics of a PES-based substrate. The addition of such nanofillers was found to have a positive impact on the membrane properties. With the presence of 0.5 wt% rGO/CN in the substrate of TFC membrane, the FO water flux of the control membrane (without rGO/CN) could be improved by about 20% when tested using DI water as feed solution and 2 M NaCl as draw solution. One of the main factors contributing to the improved performance is due to the formation of a substrate with a lower structural parameter (163 μm) compared to the control membrane (217 μm). This as a consequence resulted in reduced ICP effect.

Another potential nanofiller for the substrate of TFC FO membrane is TiO_2 nanoparticle. TiO_2 is one of the most widely used nanofillers for improving surface hydrophilicity of microporous UF membranes whether in the past or present. Emadzadeh et al. (2014) fabricated TiO_2-modified PSF substrates by incorporating different amounts of TiO_2 (ranging from zero to 1 wt%) into substrates made of 16.5 wt% PSF. The results showed that a substrate made of 0.60 wt% TiO_2 is the best condition to produce a composite membrane with a good balance of water flux and reverse salt flux. Compared to the control TFC membrane, the osmotic water flux of the TFC membrane made of the best TiO_2-loading substrate was reported to increase significantly from 4.2 to 8.1 L/m^2.h (FO mode) and from 6.9 to 13.8 L/m^2.h (PRO mode) when 0.6 M NaCl was used as feed solution and 2 M NaCl was used as draw solution. The increase in water flux is linked to the decreased S value as a result of the formation of finger-like macrovoids that connect the top and bottom layer of the substrate. Such desired structure is of low tortuosity and could play a positive role in decreasing ICP effect.

Unlike the previous works that introduced nanoparticles into the entire structure of microporous substrate, Kim et al. (2016) performed coating on the bottom surface (substrate) of commercial TFC membranes with TiO_2 nanoparticles using a sol-gel-derived spray method. The presence of the TiO_2 coating imparted hydrophilic properties and a negative charge to the substrate bottom surface, leading to higher water flux and lower reverse salt flux of the developed TFC membrane. More importantly, the antifouling resistance of the developed membrane was much better compared to the commercial membrane when tested with feed solution containing humic acid foulants, recording 32% less flux reduction. The hydrophilic nanoparticles-coated UF membranes have also been reported elsewhere as an approach to enhance water flux and anti-organic fouling properties (Moghimifar et al., 2015).

1.3.3 Nanofiber Substrate

Recent developments in nanofiber production technology have opened up the possibilities of applying nanofibers for various process improvements, including as a substrate for TFC membranes. Compared to the microporous substrate made via the phase inversion technique, the nanofiber substrate that is electrospun exhibits very unique characteristics such as superior porosity (up to 90%), low tortuousness, a very large surface area to volume ratio, and excellent mechanical properties.

Bui et al. (2011) introduced a novel flat-sheet TFC membrane supported by nanofiber substrate for engineered osmosis. The nanofibers were electrospun onto nonwoven polyester fabric followed by a polyamide layer coating onto the nanofiber substrate. Two polymeric materials (PSF and PES) were selected to prepare the nanofiber substrates and the results showed the PSF-based nanofibers demonstrated strong adhesion with the top polyamide layer while delamination occurred on the PES nanofibers. Bui et al. explained that the presence of Bisphenol A (BPA) moiety in the PSF structure was the main factor contributing to better adhesion between the substrate and polyamide derived from MPD and TMC through a specific chemical interaction as illustrated in Figure 1.2(a). The cross section and top surface of the TFC membrane made of PSF nanofiber substrate are shown in Figure 1.2(b) and 1.2(c),

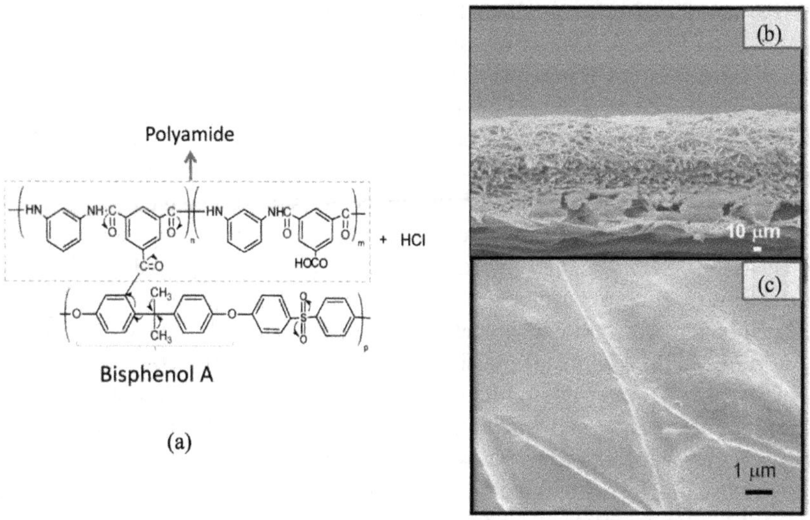

FIGURE 1.2 (a) Diagram of a possible chemical interaction between polyamide and the BPA group of PSF (arrows show the proposed reaction mechanism), and scanning electron microscopy (SEM) images of (b) cross section and (c) top surface of PSF nanofiber supported-TFC membrane. (From Bui, N.-N. et al., Journal of Membrane Science. 385–386, 10–19, 2011.)

respectively. With respect to performance, the best nanofiber supported-TFC membranes exhibited 2–5 times higher flux than a standard commercial membrane.

Wang et al. (2014) compared the performance of a TFC membrane made of multilayered nanofiber structure with the control membrane made of PAN microporous substrate. The newly fabricated membrane that consisted of a three-layered structure, including the nonwoven support, PAN electrospun mid-layer, and composite barrier layer based on cellulose nanofiber (CN) and polyamide matrix showed significantly higher flux than that of the membrane prepared by the conventional phase inversion substrate, possibly because of the water-channel structure induced at the interface between CN and polyamide, which increased the permeability of the barrier layer.

Yang et al. (2016) fabricated novel TFC NF membranes consisting of a polyamide layer via interfacial polymerization based on electrospun double-layer nanofibrous substrates, which have an ultrathin poly(acrylonitrile-co-acrylic acid) (PAN-AA) nanofibrous layer as top layer and a thicker PAN nanofiber layer as bottom support layer. Immersing the PAN/PAN-AA nanofibrous substrate into 1-ethyl-(3-3-dimethylaminopropyl) carbodiimide hydrochloride/N-hydroxysuccinimide (EDC/NHS) aqueous solution and PIP aqueous solution (0.2 wt%) sequentially for 4 h, the carboxyl groups on PAN-AA nanofibers were activated by the carbodiimide followed by a reaction with the amide groups. The newly developed composite membrane has an integrated structure with high rejection rate (98.0%) and permeate flux (40.4 $L/m^2.h$) for 2 g/L of $MgSO_4$ aqueous solution at 7 bar.

Nanofiber made of nylon 6,6 was also used as substrate for a TFC membrane owing to its intrinsic hydrophilicity and superior strength compared to other nanofiber materials. Huang and McCutcheon (2013) reported that in addition to exhibiting lower water contact angle (around 38°), the tensile strength of nylon 6,6 nanofiber (10 MPa) was found to be significantly higher than that of nanofibers made of PSF (1.8 MPa) and PAN (5.7 MPa), making it highly desirable as TFC membrane substrate. When tested in FO process, the best TFC membrane made of nylon 6,6 nanofiber outperformed the standard commercial membrane by exhibiting a 1.5–2-fold enhanced water flux and an equal or lower reverse salt flux. The authors believed that such performance and tolerance to low selectivity could make this membrane an excellent candidate for further exploration in the pressure-driven membrane process.

A hydrophobic/hydrophilic interpenetrating network composite nanofiber (HH-IPN-CNF) was designed by Tian et al. (2014) to enhance hydrophilicity and permeability of support layer of FO membrane. The composite nanofibers were fabricated using hydrophobic polyethylene terephthalate (PET) and hydrophilic polyvinyl alcohol (PVA). By increasing the PVA content in the HH-IPN-CNF substrate layer, the FO water flux of the membrane was increased accordingly. The best performing membrane (made of PET/PVA ratio of 1/4) could achieve water flux as high as 47.2 $L/m^2.h$ and a low reverse flux of 9.5 $g/m^2.h$ under PRO mode when DI water was used as feed solution and 0.5 M NaCl as draw solution. This water flux was about 6.4 times greater than that of an FO membrane derived from neat PET nanofiber substrate. The improved wetting performance of the substrate layer and the water-transferring function due to the existence of the HH-IPN-CNF structure formed between PVA and PET nanofibers are believed to be the main factors leading to lower ICP effect and enhanced water flux.

The potential for using nanomaterials to enhance the properties of nanofiber substrate was also demonstrated for FO and PRO applications. Zhang et al. (2017) synthesized a TFC membrane using electrospun PSF nanofiber incorporated with different TiO_2 loadings (0.25, 0.5, 0.75, and 1 wt%), and the results indicated that all the TFC membranes prepared from PSF/TiO_2 nanofiber substrates had much higher water flux compared to the TFC membrane without TiO_2. For example, the water fluxes of the TFC membrane prepared using the PSF substrate embedded with 0.25 wt% TiO_2 were recorded at 52 L/m^2.h in FO mode and 59 L/m^2.h in PRO mode in comparison to 42.5 L/m^2.h and 50.5 L/m^2.h, respectively, in the unmodified membrane when both membranes were evaluated using DI water as feed solution and 1 M NaCl as draw solution. Although higher TiO_2 loading (>0.5 wt%) could lead to greater water flux, the developed TFC membranes suffered from increasing reverse solute flux.

The use of nanomaterials-embedded nanofiber was also reported in the work of Tian et al. (2015) in which PEI nanofiber incorporated with functionalized multi-walled carbon nanotubes (f-CNTs) was used as substrate for TFC membrane. Three membrane substrates were fabricated in this work, namely PEI-0, CNT-PEI, and CNT-PEI-L. PEI-0 and CNT-PEI denoted the control substrate (without CNTs) and the CNT-filled substrate, respectively. CNT-PEI-L meanwhile referred to the CNT-reinforced substrate with larger pores, which was synthesized by varying dope composition and electrospinning conditions. The incorporation of f-CNTs in the nanofibers was found to increase the substrate porosity by 18% and reduced S value by 30%, and significantly improved the substrate tensile strength by 53%. When tested under FO and PRO mode, the two TFC membranes made of CNT-filled substrate exhibited at least 20% higher water flux compared to the control membrane made of PEI-0 substrate. This is mainly due to the reduced ICP effect. Further comparison showed that the TFC membrane made of CNT-PEI-L substrate was the best performing membrane by taking into consideration its lowest J_s/J_v (~0.1), which resulted from the larger pores and higher porosity of the substrate.

Song et al. (2013) investigated the effects of SiO_2 nanoparticles incorporation and PVA coating on the characteristics of PAN nanofiber substrate of TFC membrane for PRO application. Three different types of nanofiber substrates were fabricated. They were a substrate without PVA coating and SiO_2 nanoparticles (designated as NSM-0), a substrate with PVA coating but no SiO_2 (NSM-1), and a substrate with both PVA coating and SiO_2 (NSM-2). It was found that the presence of SiO_2 nanoparticles in the nanofiber matrix could improve the strain at break from 9 MPa in the NSM-1 to 17 MPa in the NSM-2. This mechanical strength improvement could be explained by the adsorption affinity between the SiO_2 nanoparticles that are strapped to the PAN polymer chains. Furthermore, NSM-1 and NSM-2 exhibited lower S values compared to the control NSM-0. The improved mechanical properties coupled with lower S values are very important for high power density PRO operation. The results showed that the newly developed PRO membrane could achieve a power density of 15.2 W/m^2 and maximum energy recovery of 0.86 kWh/m^3, using synthetic brackish water (80 mM NaCl) and seawater brine (1.06 M NaCl) as feed and draw solution, respectively.

As the substrate of a TFC membrane can be synthesized from various polymer materials and in different formats (e.g., flat sheet, hollow fiber, and nanofiber), the selection of a substrate for TFC membrane preparation is, therefore, dependent on the area of industrial application as well as the manufacturing and material cost.

1.4 DEVELOPMENT OF POLYAMIDE SELECTIVE LAYER OF COMPOSITE MEMBRANE

1.4.1 Monomer/Monomer Selective Layer

Over the years, a lot of work has been done to improve TFC membrane separation performances by changing the surface chemistry and morphology of the polyamide selective layer. Selection of appropriate monomers is one of the key factors in determining the resultant membrane chemical properties and performances. It is generally agreed that water flux and salt rejection of TFC membranes are strongly correlated to surface chemistry, morphology, and thickness of polyamide layer (Ghosh et al., 2008). Therefore, a fundamental understanding of the effects of different monomers on composite membrane properties is necessary in order to tailor the desired membrane structure and separation performance. The TFC membranes with ultrathin, highly crosslinked, and good hydrophilic active layers appear to offer greater water flux/salt rejection and antifouling properties. Table 1.1 summarizes the active monomers that have been used for TFC membrane preparation in recent literature and their impacts on the membrane performance with respect to water flux and solute rejection. This section does not intend to provide an exhaustive review of all the monomers used to date because comprehensive reviews on the TFC membranes have been previously published by Petersen (1993) and Lau et al. (2012).

Although the crosslinked aromatic polyamide composite membrane produced by the interfacial polymerization of MPD and TMC is one of the most successful commercial products (Lau et al., 2015), the research on how to further improve the properties of polyamide skin layer using new type of monomers has remained high-priority over the years. For instance, Zhao et al. (2014a) prepared novel TFC NF membranes through interfacial polymerization of dopamine and TMC on PES substrate. They reported that the rejections of developed membranes against dyes and divalent salts were increased by increasing dopamine concentration from 0.05 to 0.5 wt.% At high dopamine concentration, more dopamine could diffuse into the interface zone of the water/oil two phases, producing a dense activity layer with higher rejection. Nevertheless, the increased rejection was associated with decreasing water flux owing to the trade-off effect between water permeability and solute rejection.

Using pentaerythrotol (PE) with multi-hydroxyl groups in its molecule, Cheng et al. (2017) fabricated a new series of TFC NF membranes by interfacial polymerization of PE with TMC on the PES substrate. Through the optimization process, they reported that the best performing NF membrane could be synthesized by reacting 5 w/v% PE with 0.2 w/v% TMC for up to 20 min reaction time. Although the optimized membrane could achieve 98.1% Na_2SO_4 rejection and exhibit better chlorine resistance, its water flux (6.1 L/m^2.h at 5 bar) was significantly lower compared to

TABLE 1.1
Newly Reported Monomers for TFC Membrane Preparation and Their Impacts on Performance

Aqueous Phase Monomer(s)	Organic Phase Monomer(s)	Application	Membrane Performance	Ref.
Dopamine; m-phenylenediamine (MPD)	Trimesoyl chloride (TMC)	NF	Water permeability (2 bar) – TFC without dopamine: 160 L/m^2.h TFC with 0.05 wt% dopamine: 60 L/m^2.h Salt rejection (2 bar, 100 ppm solution) – TFC without dopamine: N/A TFC with 0.05 wt% dopamine: 96% Congo Red; 37% Na_2SO_4, and 8% $MgSO_4$	(Zhao et al., 2014)
N,N'-dimethyl-m-phenylenediamine (DMMPD); 4-methylphenylenediamine (MMPD); MPD	5-choroformyloxyisophaloyl chloride (CFIC); TMC	RO	Water permeability and salt rejection (15.5 bar, 2000 ppm NaCl feed) – TFC without MMPD-CFIC and CFIC-DMMPD (via 2-step IF process): 31 L/m^2.h; 95% TFC with MMPD-CFIC and CFIC-DMMPD (via 2-step IP process): 23 L/m^2.h; 97% Improved chlorine stability up to 8000 ppm using TFC with MMPD-CFIC and CFIC-DMMPD (via 2-step IP process)	(Liu et al., 2014)
MPD	2,4,6-pyridinetricarboxylic acid chloride (PTC); TMC	RO	Water permeability and salt rejection (200 psi, 1500 ppm NaCl) – TFC without PTC: 36 L/m^2.h; 73% TFC with PTC: 52.7 L/m^2.h; 93% (0.06:0.04 ratio of TMC: PTC)	(Jewrajka et al., 2013)

(Continued)

TABLE 1.1 (CONTINUED)
Newly Reported Monomers for TFC Membrane Preparation and Their Impacts on Performance

Aqueous Phase Monomer(s)	Organic Phase Monomer(s)	Application	Membrane Performance	Ref.
1,3–diamino-2-hydroxypropane (DAHP), MPD	TMC	RO	Water permeability and salt rejection (15 bar, 2000 ppm NaCl) – TFC without DAHP: 2.18 L/m^2.h.bar; 96–98% TFC with DAHP: 2.67 L/m^2.h.bar; 96–98% (at DAHP/MPD ratio of 12.8%)	(Perera et al., 2015)
MPD terminated PEG (MeO-PEG-MPD); melamine; MPD	TMC	RO	Water permeability and salt rejection (14 bar, 2000 ppm NaCl) – TFC without MeO-PEG-MPD: 30 L/m^2.h; 78% TFC with MeO-PEG-MPD: 38 L/m^2.h; 93% (at 1:1:0.5 w/w ratio of MPD: melamine: MeO-PEG-MPD)	(Bera et al., 2015)
MPD	Ethane diamine (EDA) and 2-[(2-aminoethyl) amino]-ethane sulfonic acid monosodium salt (SEA)	FO	Osmotic water flux and reverse salt flux (Draw solution: 0.5–2 M NaCl; Feed solution: DI water) – Control TFC: 4–8 L/m^2.h; 0.1–0.25 g/m^2.h TFC modified by EDA: 5–10 L/m^2.h; 0.15–0.4 g/m^2.h TFC modified by SEA: 8–17.5 L/m^2.h; 0.1–0.3 g/m^2.h	(Wang et al., 2016)

most of the results published in literature. They provided no suggestion for how to improve the water flux of the membrane without affecting salt rejection.

Zhao et al. (2014b) incorporated o-aminobenzoic acid-triethylamine (o-ABA-TEA) into the aqueous MPD (3 wt%) solution to react with TMC (0.21 wt%) in the organic solution during the interfacial polymerization on a PSF substrate and studied the effect of o-ABA-TEA (zero-0.25 wt%) concentration on the polyamide layer properties. o-ABA-TEA is considered hydrophilic since it contains a $-COO^-(HNEt_3)^+$ group as the hydrophilic portion and a $-NH_2$ group as the reactive portion to react with –COCl group from TMC. Through the same reaction mechanism between MPD and TMC, o-ABA-TEA can be chemically bonded onto the PA layer as illustrated in Figure 1.3. At optimal loading (1 wt%), the incorporation of o-ABA-TEA salt could reduce salt passage by providing electrostatic repulsion via its hydrophilic group. Nevertheless, when the concentration of o-ABA-TEA salt was high enough, it tended to interfere with the main reaction between MPD and TMC, negatively affecting desalination performance. The TFC membrane modified by 1 wt% o-ABA-TEA was also found to be able to reject negatively charged alginate (organic foulant in seawater) more effectively, demonstrating a better fouling resistance in the long run.

Wang et al. (2010) introduced triamine monomer, 3,5-diamino-N-(4-aminophenyl) benzamide (DABA), into the polyamide layer of a TFC RO membrane to enhance the crosslinking degree of the selective layer and to produce a smoother and thinner active layer with greater hydrophilicity. When tested with 2000 ppm NaCl feed solution at 20 bar, the water flux of control TFC membrane (without DABA) reportedly

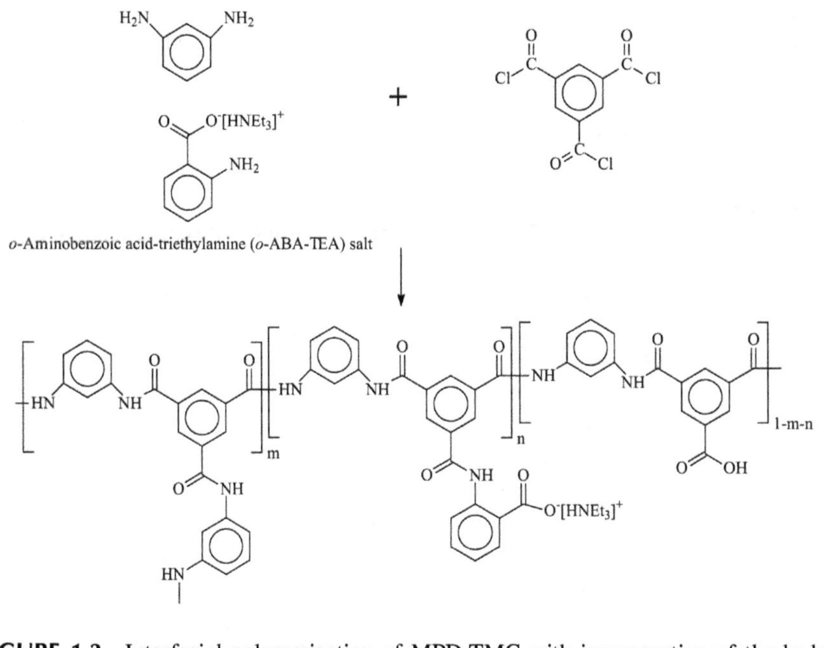

FIGURE 1.3 Interfacial polymerization of MPD-TMC with incorporation of the hydrophilic o-ABA-TEA. (From Zhao, L. et al., Journal of Membrane Science. 455, 44–54, 2014.)

increased from approximately 37.5 to above 55 $L/m^2.h$ upon addition of 0.25 w/v% DABA in MPD solution. The significant improvement in water flux was accompanied with a minimal decrease in NaCl rejection (~0.3%).

To simultaneously enhance the anti-organic and anti-biofouling properties of a TFC membrane, Bera et al. (2015) incorporated MPD-terminated PEG (MeO-PEG-MPD) into a mixture of MPD and melamine and interacted with TMC. They reported that the incorporated PEG could enhance the anti-organic fouling property whereas the triazine ring could lower the growth of bacteria on the TFC membrane surface. With respect to filtration performance, the TFC membrane prepared by interfacial polymerization between TMC (0.125 w/v%) and mixture of MPD, melamine, and MeO-PEG-MPD in MPD:melamine:MeO-PEG-MPD of 1:1:0.5 (w/w) exhibited reasonably high water flux of 38 $L/m^2.h$ and NaCl rejection of 93% when tested using feed solution of 2000 ppm NaCl at 14 bar.

To improve membrane chlorine resistance, Liu et al. (2014) developed a new type of TFC RO membrane by reacting 5-choroformyloxyisophaloyl chloride (CFIC) with 4-methylphenylenediamine (MMPD) and modified functional diamine N,N-dimethyl-m-phenylenediamine (DMMPD) via a two-step interfacial polymerization technique. The performance of the developed poly(amide-urethane@imide) membrane (MMPD-CFIC@CFIC–DMMPD) was then compared with the conventional polyamide (MPD-TMC) membranes. Although the water flux and salt rejection rate of the MMPD-CFIC@CFIC-DMMPD membrane were observed to be lower than that of the MPD-TMC membrane, the second round of modification using DMMPD on the polyamide surface worked perfectly as a chlorine-resistant protective layer owing to the increased thickness of the active layer and the decreased hydrophilicity of the surface.

On the other hand, Xiong et al. (2016) developed a novel TFC membrane for engineered osmosis via interfacial polymerization on the PAN substrate using an organic-inorganic hybrid compound –N-[3-(trimethoxysilyl) propyl] ethylenediamine (NPED) as the amine monomer and mixed it with commonly used MPD in aqueous phase. As shown in Figure 1.4, the water fluxes of the control membrane (TFC-0) when tested at FO and PRO mode were gradually increased with increasing NPED content from zero to 1.5 wt%. However, the addition of NPED tended to produce a less-perfect selective layer and form larger fractional free volume, causing the increase in reverse salt flux. The reverse salt flux was in agreement with the reduced salt rejection when the membranes were tested using RO setup.

Wang et al. (2016) developed novel TFC FO membranes via interfacial polymerization on the PES substrate by mixing MPD in aqueous phase with ethane diamine (EDA) or 2-[(2-aminoethyl) amino]-ethane sulfonic acid monosodium salt (SEA). The possible cross-linking polyamide layer of MPD-TMC-EDA and MPD-TMC-SEA is shown in Figure 1.5. The introduction of aliphatic diamine monomers to an FO membrane surface can remarkably enhance the membrane water flux owing to the increase in membrane hydrophilicity and the decrease in membrane fouling tendency, as shown in Figure 1.6, when these membranes were evaluated at different concentrations of draw solution using DI water as feed solution. It is clearly shown that the TFC membrane modified with SEA performed better than that of the control

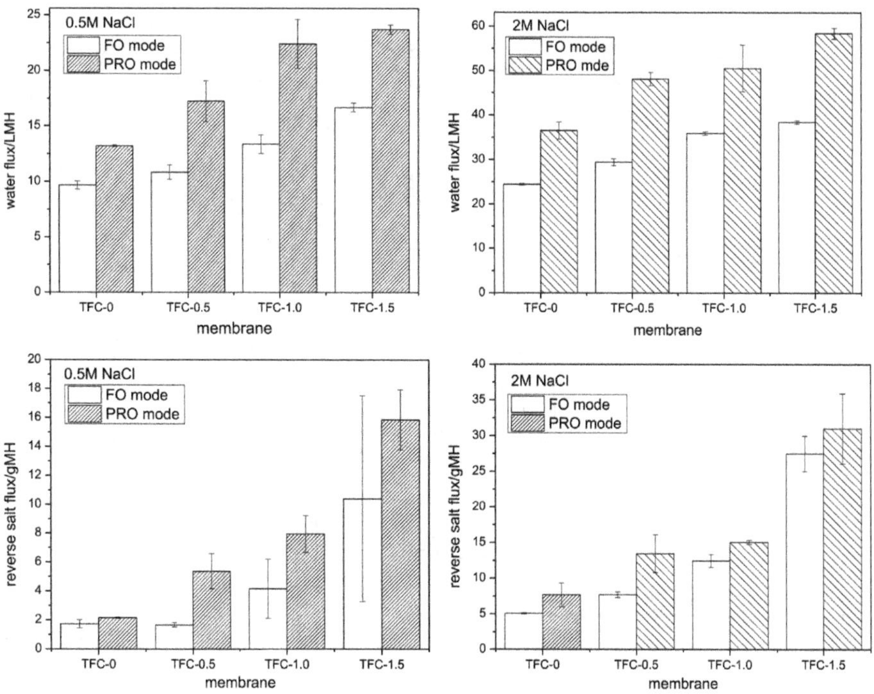

FIGURE 1.4 Performance of TFC membranes made of different NPED content (0, 0.5, 1.0, and 1.5 v/w%) in the selective layer, (a) water flux and (b) reverse salt flux (feed solution: DI water; draw solution: 0.5 and 2 M NaCl; flow rate: 0.3 L/min; temperature: 20°C). (From Xiong, S. et al., Journal of Membrane Science. 520, 400–414, 2016.)

and EDA-modified TFC membranes at the same draw concentration. This is likely due to the greater hydrophilic property of the MPD-TMC-SEA layer with the presence of ionic sulfonated groups.

Recently, Shen et al. (2017) incorporated novel tripodal amine (TAEA) (ranging from zero to 3 wt%) into MPD aqueous solution to develop a TFC FO membrane with improved surface properties. It was reported that TAEA not only played an active role as an amine monomer but also acted as a catalyst to accelerate the reaction rate of interfacial polymerization. In comparison with the control TFC membrane, both the water flux and reverse salt flux of the TAEA-modified TFC membranes were increased with increasing TAEA content from zero to 1 wt%, but decreased with further increase in TAEA content. At 3 wt% TAEA, the water flux of the TFC membrane was even lower than that of the control membrane. Further comparison showed that the TFC membrane modified by 1 wt% TAEA exhibited remarkably higher flux recovery rate (>95%) compared to only 60% recorded by the control membrane. This indicated the improved surface characteristics of the modified membrane such as higher hydrophilicity, smoother surface, and fewer adsorption sites that led to increased antifouling resistance.

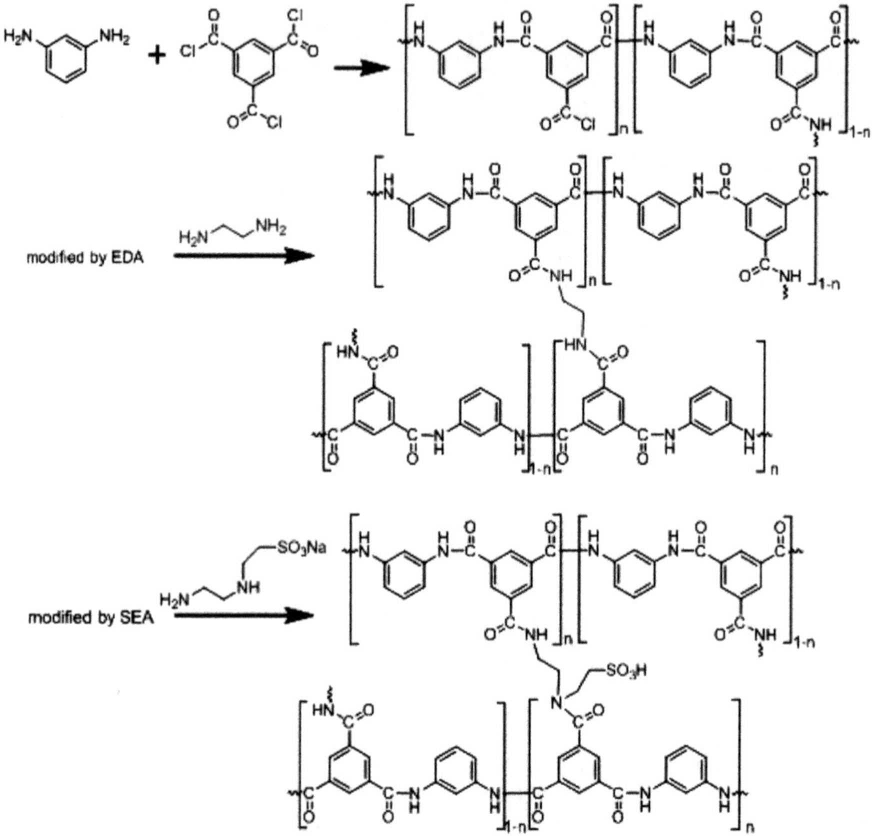

FIGURE 1.5 Reaction schematic of the cross-linking MPD-TMC PA layer with EDA or SEA as modifier. (From Wang, Y. et al., Journal of Membrane Science. 498, 30–38, 2016.)

1.4.2 Monomers/Inorganic Nanomaterial Selective Layer

Research interest in embedding inorganic nanomaterials within the thin polyamide layer of the TFC membrane has received a great deal of scientific attention since the publication of the first report by Hoek's research group (Jeong et al., 2007). Although the history of the new generation TFN membrane is still relatively short in comparison to the commercial TFC membrane, which has been known since the 1960's, the growing interest in TFN membrane development among scientists has been obvious over the past several years (Lau et al., 2015).

Table 1.2 highlights several important works about the synthesis and performance of TFN membranes incorporated with different types of inorganic nanomaterials for either pressure-driven process or osmotically-driven process. As can be seen, the roles of nanomaterials embedded within the polyamide layer of composite membrane are not only limited to water flux and salt rejection enhancement but also to improve membrane surface resistances with respect to antibiofouling, antibacterial activity, chlorine stability, etc.

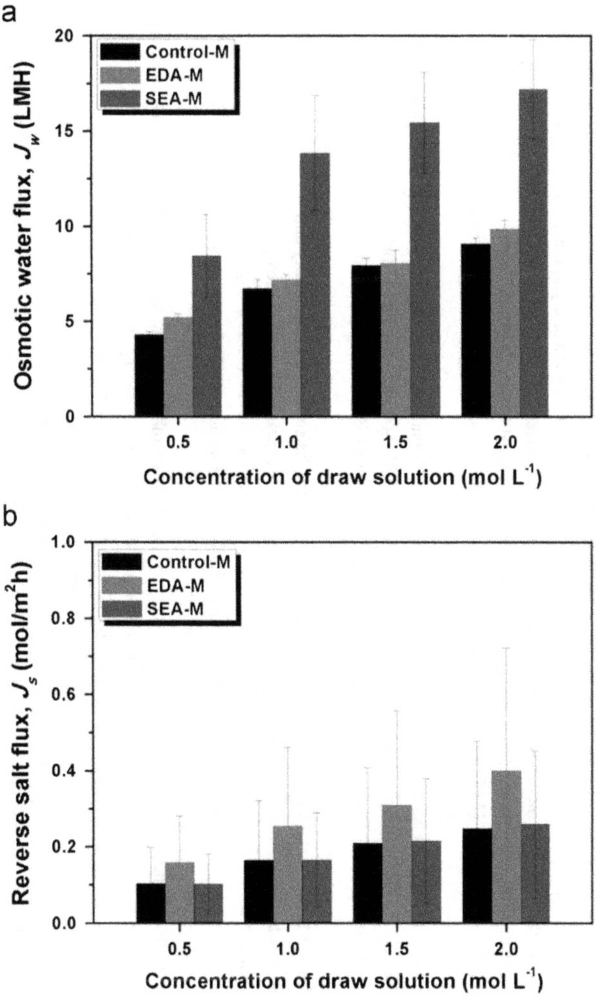

FIGURE 1.6 Comparison between the FO performance of Control-M (MPD-TMC), EDA-M (MPD-TMC-EDA), and SEA-M (MPD-TMC-SEA), (a) water flux and (b) reverse salt flux as the function of NaCl draw solution concentration. (From Wang, Y. et al., Journal of Membrane Science. 498, 30–38, 2016.)

To improve the surface charge of NF membrane, Pal et al. (2015) incorporated zinc oxide (ZnO) with different loading (0.05–0.2 w/v%) into the polyamide layer of membrane and studied the effect of the nanomaterials in either aqueous or organic phase on the properties of resultant membranes. They found that the water flux of the membrane tended to increase with increasing ZnO concentration regardless of the type of solvent the nanomaterials were introduced. However, the water flux of the membranes was relatively lower in the case where ZnO was dispersed in the organic phase. This was attributed to unstable dispersion of ZnO in the organic phase that increased the transport resistance of water through the membrane. With respect to salt rejection, the ZnO-incorporated TFN membranes exhibited the order

TABLE 1.2
Summary of the Recent Works Studying the Effect of Inorganic Nanofillers on the Performance of Composite Membranes

Inorganic Nanofillers	Interfacial Polymerization Conditions	Separation Properties	Ref.
ZnO	ZnO (0.05, 0.1, 0.2 wt.%) was dispersed in either TMC hexane solution or branched PEI aqueous solution. The nanocomposite film was then cured at 90°C for 30 min.	NF process: By incorporating 0.1 wt% ZnO into polyamide layer, the TFN membrane synthesized showed improved fluxes (23–40 $L/m^2.h$) with slightly lower NaCl rejection (50.1–57.2%) compared to the control membrane (22 $L/m^2.h$ and 58.4%) when tested at 10 bar using 1753 ppm NaCl feed solution.	(Pal et al., 2015)
GO	0.1–0.3 wt.% GO was introduced into MPD aqueous solution and the developed polyamide layer was cured at 60°C for 10 min.	NF process: By embedding appropriate GO loading (0.2 wt%) into polyamide layer, the developed TFN membrane could show not only flux improvement but also better antifouling properties towards BSA and humic acid.	(Bano et al., 2015)
SiO_2	SiO_2 (0.1 wt.%) was introduced into MPD/glycerol/NMP/SDS/TEA aqueous solution. The nanocomposite film was then cured at 80°C for 5 min.	RO process: Upon incorporation of silica, the surface hydrophilicity and roughness of the TFN membrane were increased. This led to greater water flux (50 $L/m^2.h$) compared to the control membrane (32 $L/m^2.h$) when tested at 44 bar using NaCl feed solution. The rejection of TFN membrane meanwhile remained at >90%.	(Peyki et al., 2015)
Zeolitic Imidazolate Framework (ZIF)-8	0.4 wt.% ZIF-8 was dispersed in TMC/hexane solution and the developed polyamide film was cured at 23°C for 1 min.	RO process: The ZIF-8 incorporated TFN membrane was found to have better hydrophilicity but lower degree of surface cross-linking. Its water flux and NaCl rejection when tested using 2000 ppm NaCl solution at 15.5 bar were ~52 $L/m^2.h$ and 98.5%, respectively. As a comparison, the control TFC membrane showed 20 $L/m^2.h$ water flux and 98.1% rejection.	(Duan et al., 2015)
MWCNTs modified with acid and diisobutyryl peroxide	0.001–0.1 wt.% modified MWCNTs were introduced in MPD aqueous solution and the developed polyamide film was cured at 60°C for 20 min.	RO process: Besides showing higher water flux with slight decrease in NaCl rejection, the MWCNTs-incorporated membrane also exhibited better antifouling properties and chlorine stability in comparison to the control TFC membrane.	(Zhao et al., 2014)

(Continued)

TABLE 1.2 (CONTINUED)
Summary of the Recent Works Studying the Effect of Inorganic Nanofillers on the Performance of Composite Membranes

Inorganic Nanofillers	Interfacial Polymerization Conditions	Separation Properties	Ref.
Amino-functionalized TNTs	0.01–0.1 wt% modified TNTs were added into TMC/cyclohexane solution and the formed polyamide layer was further post-treated at 90°C for 10 min.	RO process: When tested using 2000 ppm NaCl at 15 bar, the newly developed membranes could show improved fluxes (26–58 $L/m^2.h$) and comparable solute rejection (>80%) compared to the control membrane (19 $L/m^2.h$ and 95%). Furthermore, these membranes also displayed better antifouling property against 200 ppm BSA.	(Emadzadeh et al., 2015b)
GO	15, 38, 76 ppm GO was added into MPD aqueous solution and the polyamide layer was further dried at room temperature for 10 min.	RO process: In addition to flux improvement without compromising NaCl rejection, the GO-incorporated membranes were reported to have better anti-biofouling properties as well as greater chlorine resistance in comparison to the control TFC membranes.	(Chae et al., 2015b)
Polyoctahedral oligomeric silsequioxanes (POSS)	POSS was dispersed in MPD (The concentration of POSS is 5 mol% of MPD) aqueous solution. The curing temperature and time for the formation of polyamide layer is not given.	RO process: POSS-incorporated membrane was reported to have greater water flux (44.6 $L/m^2.h$) and better solute rejection (99.6%) compared to typical TFC membrane (33.7 $L/m^2.h$ and 99.0%) when tested at 32,000 ppm NaCl feed and 55 bar.	(Moon et al., 2014)
TiO_2/HNTs	TiO_2/HNTs (0.01, 0.05, 0.1 w/v%) were added to TMC organic solution and the developed polyamide layer was dried in an oven (90°C) for 10 min.	FO process: The TFN membrane incorporated with 0.1 w/v% TiO_2/HNTs was reported to have greater water flux (35 $L/m^2.h$ in FO mode and 55 $L/m^2.h$ in PRO mode) compared to the control membrane (13 $L/m^2.h$ in FO mode and 24 $L/m^2.h$ in PRO mode) when both membranes were tested using 2 M NaCl draw solution and 10 mM NaCl feed solution.	(Ghanbari et al., 2015b)

of $MgSO_4$ > NaCl > Na_2SO_4, revealing the positive surface charge that was induced by the ZnO incorporation.

A literature search revealed that most of the work on TFN membranes was focused on the RO application by removing NaCl from synthetic brackish water or seawater solution. Dong et al. (2015) embedded the zeolite-Y (NaY) nanoparticles into the polyamide layer via interfacial polymerization to form a novel nanocomposite layer on top of a microporous substrate. They reported that upon incorporation of the optimal 0.15 wt.% NaY nanoparticles, the water flux of the developed RO membrane increased to 74.76 L/m^2.h from 39.2 L/m^2.h as shown in the control TFC membrane when tested using 2000 ppm NaCl solution at 15.5 bar. The significant enhancement in the water flux of the TFN membrane did not compromise solute separation efficiency because the newly developed membrane exhibited similar NaCl rejection (98.8%) as the control TFC membrane.

Using GO as nanofillers in the polyamide layer, Yin and his co-workers (2016) also reported remarkable water flux improvement in the TFN membrane in comparison to the control membrane. The rejection of the TFN membrane against NaCl and Na_2SO_4, however, was slightly decreased from 95.7% to 93.8% and from 98.1% to 97.3%, respectively, with increasing GO loading from 0 to 0.02 wt.%. Their findings revealed that the TFN membrane with 0.015 wt.% of GO was the best performing membrane owing to its good balance between water permeability and salt rejection. The increased water permeability could be due to water channels created by the interlayer spacing of GO nanosheets as well as improvement in membrane surface hydrophilicity.

To improve the compatibility between inorganic nanofillers and the PA layer, Zhao et al. (2014c) grafted the surface of MWCNTs with abundant long chains of aliphatic acid and studied the effect of modified MWCNTs loading on the characteristics of TFN membranes. Owing to the formation of water channels and voids between modified MWCNTs and polyamide layer, more water molecules could pass through the membrane with minimum transport resistance. As a consequence, this led the TFN membranes to have higher water flux than the control TFC membrane. Moreover, the enhanced hydrophilicity and surface charge properties of the TFN membrane could reduce severe fouling on the membrane surface, thus positively influencing the water permeability of the membrane. Zhao et al. also reported that there were no significant changes in both water flux and rejection for the TFN membrane compared to the control TFC membrane when tested using chlorine concentration of 100 ppm hypochlorous acid in the feed solution, indicating that the antioxidant properties of the TFN membrane are stronger than those of the typical TFC membrane. They attributed the enhanced chlorine resistance to the protection of active site in MPD residual by the electron-rich MWCNTs from being attacked by the free chlorine.

Chae et al. (2015) on the other hand also demonstrated that the composite membrane incorporated with GO could lead to better resistance against chlorination. As shown in Figure 1.7, the salt rejection and water flux of the TFN membrane modified by the highest loading of GO were almost at the same level before and after the 24 h chlorination. It was reported that the existence of GO could protect the polyamide layer from chlorine attack and inhibit the replacement of amidic hydrogen with chlorine (Choi et al., 2013; Kim et al., 2013). In addition to the improved chlorination

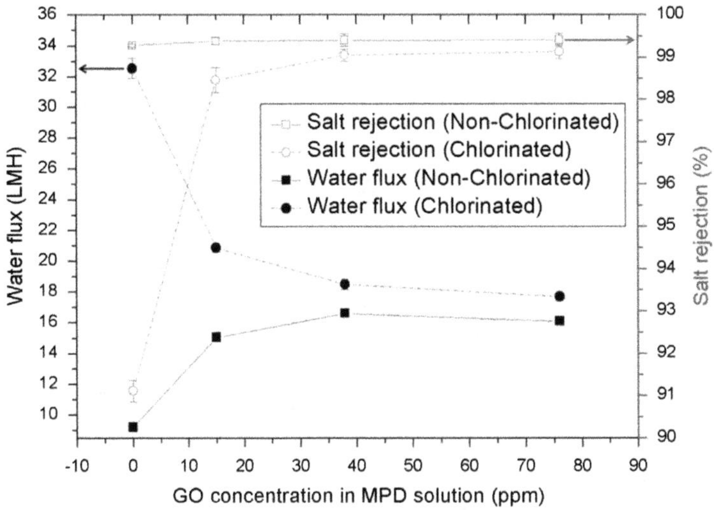

FIGURE 1.7 Salt rejections and water fluxes of the TFN membranes modified with 0, 15, 38, and 76 ppm GO before and after the 24-h chlorination using 2000-ppm NaOCl solution at pH 7. (From Chae, H.-R. et al., Journal of Membrane Science. 483, 128–135, 2015.)

resistance, Bano et al. (2015) reported that the introduction of GO into the polyamide layer of a membrane could enhance the antifouling properties towards bovine serum albumin (BSA) and humic acid, which may be ascribed to the enhanced surface hydrophilicity introduced by the hydrophilic groups of GO.

Emadzadeh et al. (2017) modified the polyamide layer of RO membrane with silane-modified nanoporous titanate nanoparticles (mNTs) (0.01, 0.05, and 0.1 w/v%) and found that the composite membrane incorporated with 0.01 w/v% mNTs always demonstrated a lower degree of flux decline compared to the control membrane when the membranes were tested with organic, inorganic, and multicomponent synthesized water. The best performing TFN membrane also displayed significantly higher water flux (40 L/m^2.h) compared to the control membrane (30 L/m^2.h) with almost similar salt rejection (>98%) when tested using 30,000 ppm NaCl solution at 55 bar.

Similar to the NF and RO process, the potential of using nanofillers for making TFN membranes for FO/PRO process was also demonstrated. Amini et al. (2013) synthesized a TFN membrane by incorporating functionalized MWCNTs (f-MWCNTs) into the polyamide layer. As shown in Figure 1.8, they found that the water fluxes of TFN membranes were always higher than that of the control TFC membrane when tested at two different membrane orientations. The range of water fluxes changed from 25–40 L/m^2.h in AL-FS to 37–95 L/m^2.h in AL-DS since the ICP is more severe in the AL-FS compared to the AL-DS model. The improved water flux in the TFN membranes could be attributed to the existence of f-MWCNTs that react as nanochannels in the top surface of the membranes. Both inner cores of nanotubes (internal nanochannels) and the interfacial gap between f-MWCNTs and polymer at the interface of the polyamide layer (external nanochannels) create further passage for solvent transfer. It must be noted that the TFN membranes in

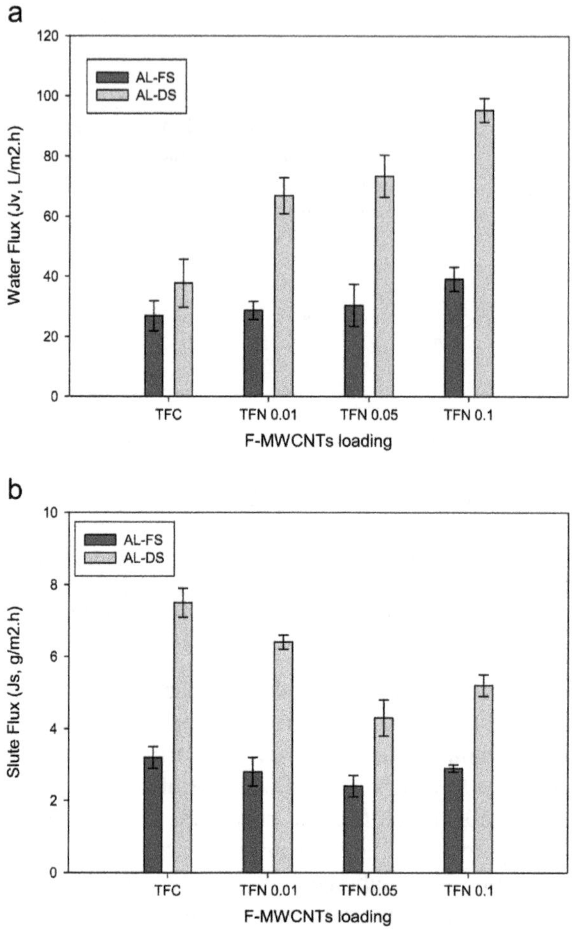

FIGURE 1.8 Performance comparison between TFC and TFN membranes under FO (AL-FS) and PRO (AL-DS) mode, (a) water flux and (b) reverse solute flux. (From Amini, M. et al., Journal of Membrane Science. 435, 233–241, 2013.)

general showed lower reverse solute flux compared to the TFC membrane, indicating the potential of nanofillers in overcoming the trade-off effect between water flux and reverse solute flux.

Ghanbari et al. (2015a) also experienced a similar water flux increment by increasing the loading of nanofiller - halloysite nanotubes (HNTs) into a polyamide layer of composite membrane. Nevertheless, they also found that the membrane reverse solute flux was increased with the presence of higher quantities of HNTs in the selective layer. The larger pores formed at higher concentration of HNT caused more dissolved ions to pass through the membranes. When tested using feed solution containing 200 ppm BSA under FO mode, they found that the TFN membrane incorporated with optimal HNTs loading (0.05 wt%) could achieve complete water recovery after subjecting it to 30 min cleaning process. The control TFC membrane meanwhile only obtained ~83% water flux recovery under the same conditions.

The better antifouling properties of the TFN membrane could be ascribed to the enhanced hydrophilicity of nanocomposite layer that weakens the interaction between foulant and polyamide dense layer.

Emadzadeh et al. (2015a) on the other hand incorporated NH_2-TNTs (0.01, 0.05, and 0.1 w/w%) into the polyamide layer of TFC FO membrane to enhance the membrane hydrophilicity. They reported that by adding only 0.05 wt% of NH_2-TNTs, the water flux of the TFC membrane was significantly improved from 11.58 L/m^2.h of TFC (control) to 20.79 L/m^2.h in FO mode and from 21.0 L/m^2.h of TFC (control) to 31.5 L/m^2.h in PRO mode. However, further increasing NH_2-TNTs loading to 0.1 wt% tended to form microvoids between the nanotube and the polymer matrix in the skin layer, causing increased reverse solute flux.

1.5 SURFACE MODIFICATION OF POLYAMIDE SELECTIVE LAYER

1.5.1 Poly(Ethylene Glycol) Grafts for Fouling Resistance Surface Modification

Covalent grafting of polyethylene glycol (PEG)-like polymers on the membrane active layer is an effective avenue to impart membranes with low fouling propensity. Specifically, two approaches were implemented (Romero-Vargas Castrillón et al., 2014; Shaffer et al., 2015): (1) *in situ* Jeffamine protocol, in which a block copolymer of PEG and polypropylene glycol (PPG) is grafted on the nascent polyamide active layer immediately after interfacial polymerization and (2) the amino-poly(ethylene glycol) diglycidyl ether (PEGDE) protocol, in which PEGDE, an epoxide PEG derivative, is grafted to an amino-rich polyamide active layer.

1.5.1.1 *In Situ* Jeffamine Fouling-Resistant Polyamide Membranes

The *in situ* modification exploits the presence of reactive acyl chloride groups on the nascent polyamide layer in order to covalently bond Jeffamine (*O,O*'-Bis(2-aminopropyl)PPG-*block*-PEG-*block*-PPG), an amine terminated block copolymer of ethylene glycol and propylene glycol, to the membrane active layer (Figure 1.9). This modification protocol was previously applied to modify RO membranes. To prepare *in situ* Jeffamine membranes, the nascent polyamide layer was contacted with 25.0 mL of aqueous Jeffamine (molecular weight 1,900 g/mol) immediately after the post-TMC draining step. This reaction was allowed to proceed for 2 min. With the exception of this step, the modification protocol was analogous to that of the control membranes.

1.5.1.2 Amino-PEGDE Surface Modification

The amino-PEGDE modification protocol relies on the existence of reactive primary amine groups on the polyamide surface, to which PEGDE, an epoxy-terminated polyethylene glycol derivative, is grafted. Following interfacial polymerization between TMC and MPD over the PSF support and subsequent draining of excess TMC, the nascent polyamide layer was contacted with ethylene diamine. The reaction between the EDA primary amine groups and the dangling acyl chloride groups resulted in a polyamide active layer with a high surface concentration of primary amine groups.

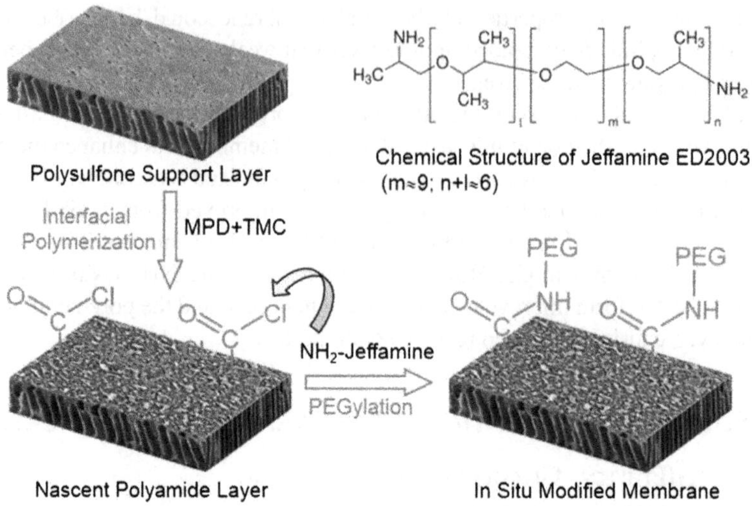

FIGURE 1.9 *In situ* Jeffamine membrane preparation scheme. (From Lu, X. et al., Environmental Science & Technology. 47(21), 12219–12228, 2013.)

The membrane was then subjected to a curing step in DI water (95°C) for 2 min. Next, PEGDE, pre-heated to 40°C, was contacted with the membrane active layer. The membrane was rinsed with DI water several times before being stored at 4°C in DI water. Figure 1.10 summarizes the amino-PEGDE membrane synthesis protocol.

The aforementioned approaches resulted in membranes with low fouling propensity, which were examined in the cross-flow filtration in FO. Flux decline was significantly reduced for *in situ* Jeffamine modified membranes (Figure 1.11(A)). The modified membranes showed a flux decline of 7.4%, nearly half of that observed in the control membranes (14.1%). Similarly, the flux decline exhibited by amino-PEGDE membranes was less than that of the control (Figure 1.11(B)). Once more, the flux decline in amino-PEGDE membranes was reduced by one-half relative to the controls (7.2% vs. 14.1% for controls), showing that this modification protocol resulted in membranes with improved fouling resistance. The flux recovery after cleaning was ~100% for both modified membranes and the controls, confirming that alginate fouling was reversible.

To mechanistically investigate the surface modification for fouling mitigation, AFM adhesion force measurements were applied to the measurement of intermolecular forces between carboxylate groups on the functionalized cantilever tip and the surface chemistry of control and modified membrane surfaces. Those measurements are relevant to the early stages of membrane fouling, and therefore provide information about the intrinsic fouling propensity of different surfaces. Figure 1.12 compares histograms of the maximum adhesion force recorded during the AFM measurements corresponding to *in situ* Jeffamine and amino-PEGDE membranes. The adhesion force measurements provided further evidence of the efficacy of the modification protocols in mitigating fouling phenomena. In all cases, adhesion forces between the functionalized particle at the cantilever tip and the modified membrane materials are weakened compared to the control polyamide membranes. The presence of

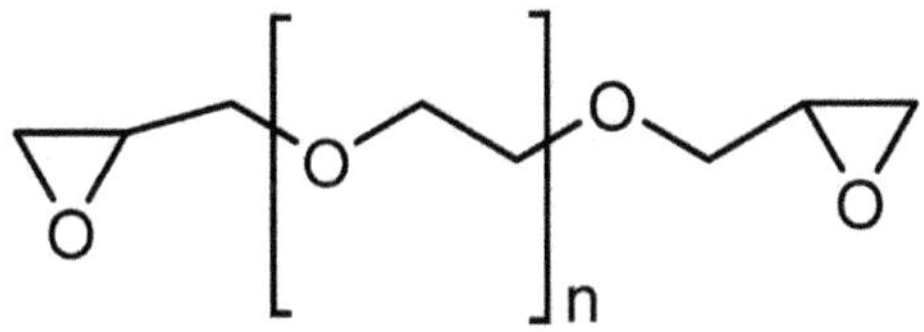

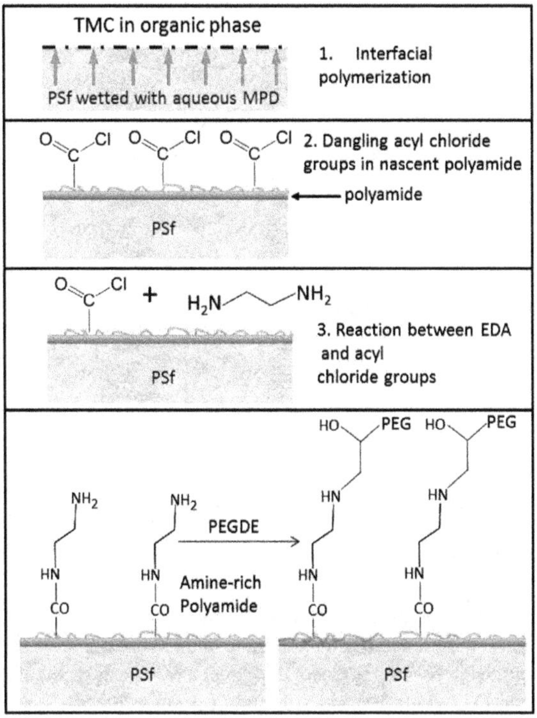

FIGURE 1.10 Schematic diagram of amino-PEGDE membrane preparation. (From Romero-Vargas Castrillón, S. et al., Journal of Membrane Science. 450, 331–339, 2014.)

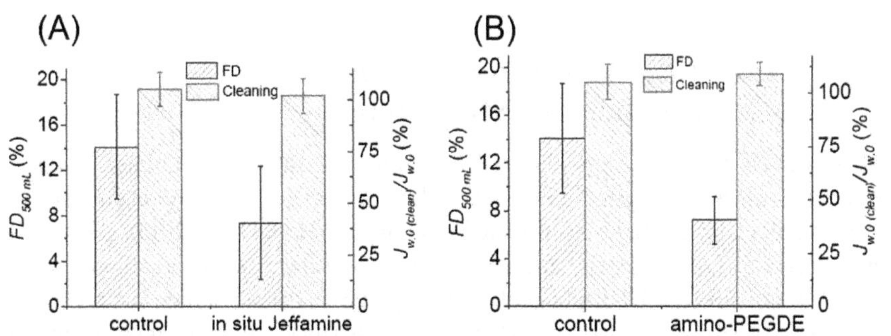

FIGURE 1.11 Flux profile comparison of (A) *in situ* Jeffamine modification and (B) amino-PEFGE modification. (From Romero-Vargas Castrillón, S. et al., Journal of Membrane Science. 450, 331–339, 2014; Shaffer, D.L. et al., Journal of Membrane Science. 490, 209–219, 2015.)

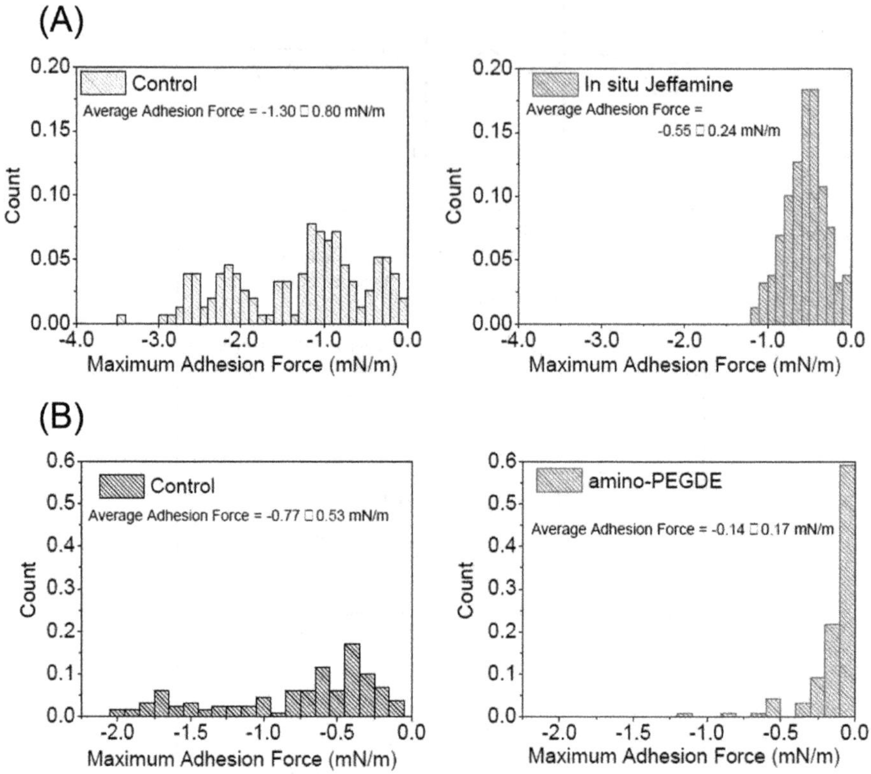

FIGURE 1.12 Distribution of adhesion forces of a carboxyl-modified latex particle on membranes. (A) control and Jeffmine modified membrane and (B) control and amino-PEGDE modified membrane. (From Romero-Vargas Castrillón, S. et al., Journal of Membrane Science. 450, 331–339, 2014; Shaffer, D.L. et al., Journal of Membrane Science. 490, 209–219, 2015.)

PEG-type grafts on the different modified membrane surfaces results in a barrier for foulant adsorption that significantly weakens the interactions between the functionalized particle and the underlying polyamide.

1.5.2 Surface Modification by Graphene Oxide or Graphene Oxide/Silver Nanocomposite

Post-fabrication modification is focused on the immobilization of nanomaterials at the membrane surface via physical interactions (Ben-Sasson et al., 2014a), chemical binding (Perreault et al., 2014), or layer-by-layer techniques (Hu and Mi, 2013). Because the nanomaterials are placed specifically at the membrane surface, post-fabrication functionalization is unlikely to affect significantly the properties of the polyamide layer (Perreault et al., 2014; Tiraferri et al., 2011). This technique is also material- and cost-efficient since fewer nanomaterials are required to tailor the membrane surface chemistry.

The growth of bacteria as biofilms can affect membrane performance by decreasing permeate water flux and salt rejection. Furthermore, biofouling development can lead to an increase in energy consumption. Ordinary procedures such as pre-treatment and chemical cleaning are being used to mitigate biofouling. However, no pre-treatment can completely eliminate biofouling, and it is well known that the polyamide layer of TFC membranes undergoes degradation in the presence of chemical oxidants such as chlorine. Therefore, there is a critical need to develop innovative strategies to control microbial proliferation at the membrane surface.

Several studies have proposed modifying the surface of TFC membranes with polymers (Ye et al., 2015), bio-active molecules (Saeki et al., 2013), or antimicrobial nanomaterials (Ben-Sasson et al., 2014b) in order to impart antimicrobial activity and biofouling resistance to the membrane. For instance, it has been shown that the TFC membranes functionalized with silver or copper nanoparticles present a diminished susceptibility to biofouling (Yin et al., 2013; Ben-Sasson et al., 2014a, 2014b). Alternatively, carbon-based nanomaterials such as CNTs and GO have also been linked to the polyamide layer to generate TFC membranes with enhanced antimicrobial properties.

The use of GO or GO/Ag nanocomposite for biofouling mitigation in FO process were conducted in which the GO- or GO/Ag-functionalized membranes were exposed to an artificial secondary wastewater feed, to which the biofilm-forming bacterium *Pseudomonas aeruginosa* was added, and tested in a bench-scale cross-flow FO unit (Perreault et al., 2016; Faria et al., 2017). Analysis of the structure and composition of the biofilm formed on the membrane, in conjunction with a characterization of the change in surface properties imparted by GO or GO/Ag nanocomposite, provided insights on the mechanisms involved in biofouling mitigation.

The binding of GO and GO/Ag nanocomposites to TFC membranes was developed through a reaction mediated by EDC and NHS. The polyamide layer of TFC membranes possesses native carboxyl groups that can react with ethylene diamine (ED) via EDC and NHS to yield an amine-terminated surface. Similarly, the carboxyl groups on the GO layer are activated when exposed to EDC and NHS in a buffered solution. During this activation, the carboxyl groups on GO are converted to intermediate esters that readily react with amine groups on ED-functionalized TFC membranes. GO and GO/Ag sheets are covalently linked to the polyamide layer through the formation of an amide bond between carboxyl groups of GO and the amine groups on ED-functionalized TFC membranes. A scheme in Figure 1.13 illustrates the reaction mechanism involved in the binding of GO/Ag sheets to the top polyamide surface of TFC membrane (Perreault et al., 2016; Faria et al., 2017).

Antimicrobial activity was first evaluated after exposing the membrane surface to *P. aeruginosa* cells for 3 h. In comparison to the pristine TFC, the TFC-GO membrane displayed no toxic effect towards *P. aeruginosa* (Figure 1.14). The TFC-GO/Ag membrane, on the other hand, exhibited a bacterial inactivation rate of around 80% against *P. aeruginosa*, relative to the non-modified TFC membranes. In other words, the number of viable cells on TFC-GO/Ag was significantly lower than that of the unmodified control, implying that functionalization with GO/Ag imparted a strong antimicrobial activity to the membrane surface.

The anti-biofouling properties of TCF and TFC-GO/Ag membranes were investigated by allowing *P. aeruginosa* cells to grow on the membrane surface for 24 h in a dynamic cross-flow biofouling test. One of the consequences of biofilm formation

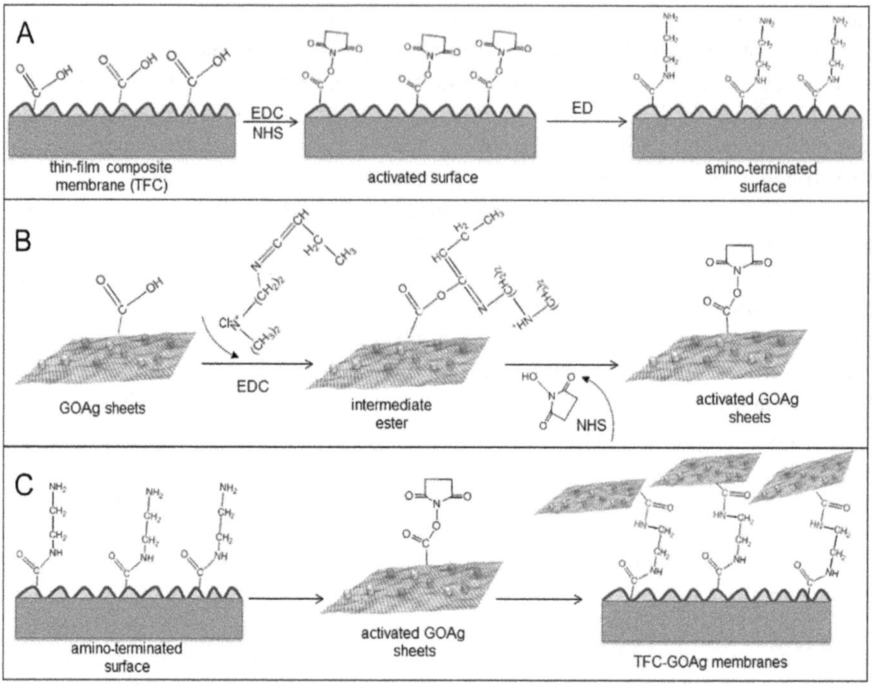

FIGURE 1.13 Scheme illustrating the three-sequential steps (A, B, and C) for binding GO/Ag sheets to the surface of thin-film composite membranes: (A) carboxylic groups on the polyamide layer are converted into primary amine groups; the native carboxylic groups are activated by EDC and NHS to generate a highly reactive ester that spontaneously reacts with ED to allow an amine-terminated surface; (B) carboxylic functional groups on GO/Ag sheets are activated in presence of EDC and NHS; and (C) amine-terminated TFC membrane contacts the activated GO/Ag sheets. (From Perreault, F. et al., Environmental Science & Technology. 50(11), 5840–5848, 2016; Faria, A.F. et al., Journal of Membrane Science. 525, 146–156, 2017.)

on TFC membranes is the decrease in permeate water flux. As shown in Figure 1.15, the development of biofilm on the pristine TFC membrane resulted in a flux decline of approximately 50%. However, when the TFC membrane was functionalized with GO/Ag nanocomposites, the flux decline was significantly reduced. The difference in the water flux behavior is attributable to differences in the structure and composition of the biofilms on the pristine TFC and TFC-GO/Ag membranes.

To obtain information about the biofilm properties, the biofouled membranes were characterized by confocal microscopy. Dead cells, represented by the color red, are more abundant on TFC-GO/Ag than on the TFC-GO or pristine TFC membranes. This pronounced increase in dead cells indicates that GO/Ag nanosheets played a key role in mitigating biofilm development on TFC membranes.

Table 1.3 summarizes the biofilm properties for TFC, TFC-GO, and TFC-GO/Ag membranes. For instance, the biofilm on the TFC-GO/Ag membrane was almost two times thinner than that on the pristine TFC membrane. Furthermore, the live cell biovolume on TFC-GO/Ag was decreased by almost 50% compared to the non-modified TFC membrane. The biofilm contents of protein and total carbon were also drastically

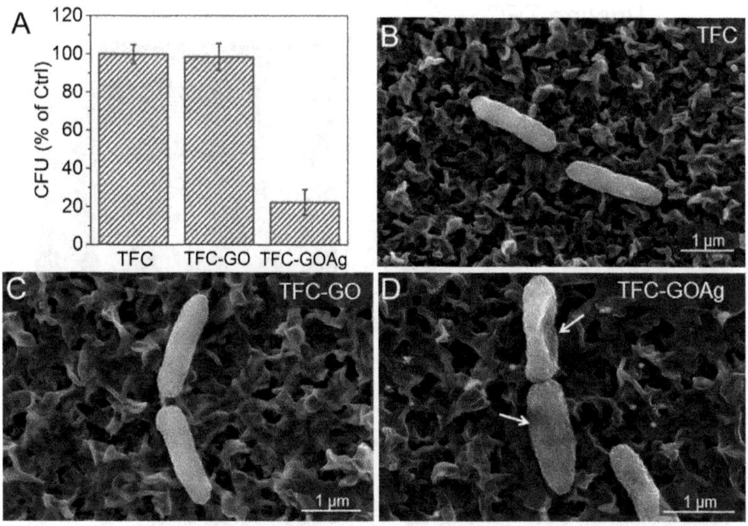

FIGURE 1.14 (A) Viable cells of *P. aeruginosa* after 3 h contact with the surface of pristine and graphene modified membranes. The viability of *P. aeruginosa* cells is expressed as the percentage of colony-forming units (CFU) relative to the pristine TFC control membrane. SEM images of bacteria cells attached to the polyamide active layer of (B) pristine TFC, (C) TFC-GO, and (D) TFC-GO/Ag membranes. Severe morphological damage for bacteria cells on TFC-GO/Ag is highlighted by white arrows on the image (D). (From Perreault, F. et al., Environmental Science & Technology. 50(11), 5840–5848, 2016; Faria, A.F. et al., Journal of Membrane Science. 525, 146–156, 2017.)

reduced after modification of the TFC membranes with GO/Ag nanocomposites. The total protein mass was diminished from 18.7 to 9.1 pg/μm^2 after binding GO/Ag sheets to the membrane surface.

Results suggested that bacterial growth on the TFC membrane surface was strongly inhibited by GO/Ag nanocomposites. The decrease in the number of live cells on the TFC-GO/Ag membrane led to a significant reduction in biofilm thickness, live cell biovolume, and EPS production. The results demonstrated that the development of anti-biofouling TFC membranes can benefit from the physicochemical and biological properties of GO/Ag nanocomposites. Recognizing that the antimicrobial activity of GO/Ag nanocomposites is partially dependent on the release of Ag^+ ions, their anti-biofouling properties can be improved by maximizing membrane coverage and/or by tuning the size, shape, and content of Ag nanoparticles in the GO/Ag nanocomposites.

1.5.3 Surface Pattern by Nanoimprint as a Novel Approach for Surface Modification

Nanoimprint, a simple and versatile nanofabrication technique, is able to manufacture structures from micro- to nano-scale on material surfaces, with a wide range of applications from photonics (Li et al., 2000), micro- and nano-fluidics (Lee et al., 2011), to chip-based sensors (Zankovych et al., 2001; Guo, 2007). Over the last

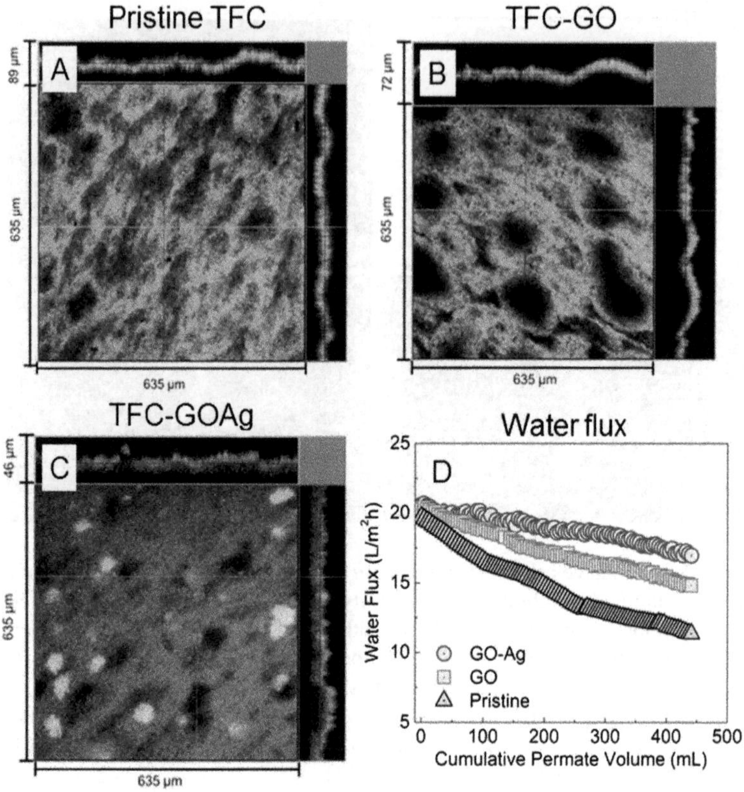

FIGURE 1.15 (A) Water flux decline caused by the formation of biofilm during biofouling experiments in a cross-flow cell (Feed: Synthetic wastewater with glucose as a carbon source, temperature: 25°C and cross-flow velocities: 9.56 cm/s); confocal laser scanning microscopy (CLSM) images of *P. aeruginosa* biofilm developed on the polyamide active layer of (B) pristine TFC, (C) TFC-GO, and (D) TFC-GOAg membranes. Live cells, dead cells, and exopolysaccharides were stained with Syto 9 (green), propidium iodide (red), and Con A (blue) dyes, respectively. (From Perreault, F. et al., Environmental Science & Technology. 50(11), 5840–5848, 2016; Faria, A.F. et al., Journal of Membrane Science. 525, 146–156, 2017.)

decade, nanoimprint fabrication allows production of structured surfaces with greater geometrical complexity (del Campo and Arzt, 2008): for example, a patterned polymer with elongated features in the vertical dimension, exhibiting several hierarchy levels, or in intricate tilted, suspended, or curved three-dimensional arrangements. These patterned features allow for biosensors with increased sensitivity, or surfaces with controlled adhesion, nonbiofouling coatings for undersea pipelines. Figure 1.16 shows the schematic of the nanoimprinting process.

Such nanofabrication techniques can be innovatively applied to water treatment membrane surfaces. Imparting membrane surfaces with highly ordered features can potentially advance membrane fouling mitigation (Ding et al., 2017). For instance, Maruf et al. (2013) patterned an UF membrane and reported a higher critical flux for colloidal particle fouling. Rickman and co-authors (2017) also used nanoimprint on

TABLE 1.3
Characteristics of *P. aeruginosa* Biofilm Grown on Pristine TFC, TFC-GO, and TFC-GO/Ag Membranes after 24 h. All Parameters Were Determined from Confocal Laser Scanning Microscopy (CLSM) Images

Membrane	Average Biofilm Thickness (μm)[a]	"Live" Cell Biovolume (μm³/μm²)	"Dead" Cell Biovolume (μm³/μm²)	EPS Biovolume (μm³/μm²)	Total Protein Mass (pg/μm²)[b]	TOC Biomass (pg/μm²)[b]
Pristine TFC	89 ± 5	21.2 ± 4.1	12.1 ± 2.3	20.9 ± 2.2	18.7 ± 2.5	1.57 ± 0.05
TFC-GO	72 ± 2	27.2 ± 5.1	12.1 ± 2.3	12.3 ± 3.6	12.1 ± 4.5	1.11 ± 0.03
TFC-GO/Ag	46 ± 3	12.5 ± 5.1	29.6 ± 1.1	8.3 ± 3.6	9.1 ± 6.2	0.82 ± 0.07

Sources: Perreault, F. et al., Environmental Science & Technology 50(11), 5840–5848, 2016; Faria, A.F. et al., Journal of Membrane Science 525, 146–156, 2017.

[a] Biofilm thickness and biovolume were averaged, with standard deviations calculated from ten random samples in duplicated experiments.

[b] TOC and protein biomasses were presented with standard deviations calculated from four measurements by two membrane coupons.

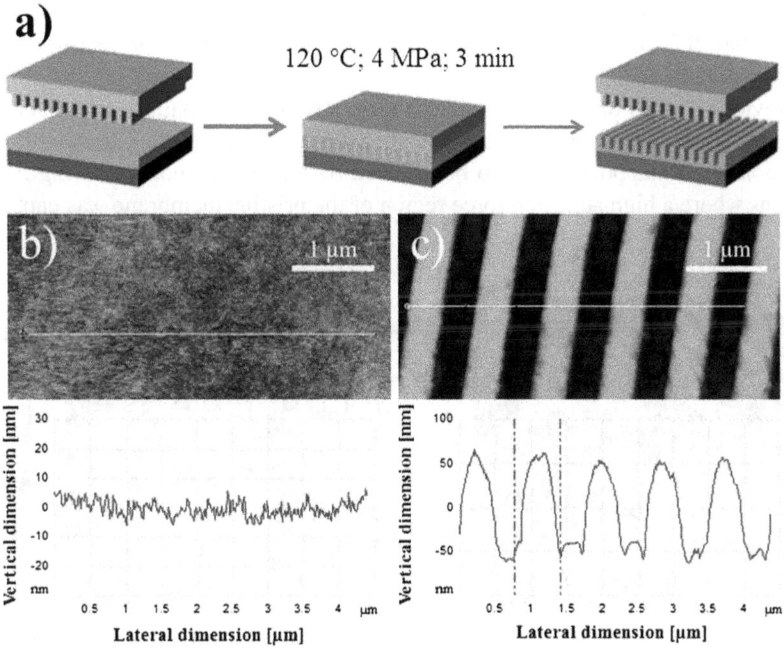

FIGURE 1.16 (a) Schematic of the NIL process for membrane patterning. The scale is not proportional. (b) and (c) are topographic AFM image of the pristine and patterned membranes, and the corresponding cross-sectional profiles of both membranes are shown below the AFM images. (From Maruf, S.H. et al., Journal of Membrane Science. 428, 598–607, 2013.)

NF membrane for BSA fouling mitigation. In addition, Maruf and co-workers (2014) used a two-step fabrication process with nanoimprinting to produce a PES support and formed a thin dense film via interfacial polymerization on the PES surface. The results showed that the resultant patterned TFC membrane exhibited water flux and salt rejection comparable to that of commercial TFC RO membranes, while scaling experiments revealed that gypsum distribution on the surface of the patterned membranes was more widely scattered in comparison to that on the non-patterned membranes.

Prior studies have shown that patterned membranes demonstrated strong fouling resistance to various colloidal fouling. It was suggested that presence of surface patterns could alter the flow profile and local streamlines in the vicinity of the patterns, exhibiting an oriented fouling region as a function of the pattern feature. Brownian dynamics simulation showed a reduced particle deposition on the apex of the surface pattern (Jung et al., 2015).

Xie and co-authors (2017) employed an atomic force microscopy (AFM) to measure the interaction between the patterned membrane surface and a model foulant, BSA protein. They mapped adhesion force in a 5 μm by 5 μm membrane area to reveal the spatial distribution of force interaction between the patterned membrane and BSA protein (Figure 1.17). Adhesion force mapping allows a more objective comparison between pristine and patterned membrane in regard to the BSA deposition. From a micro-scale perspective, the adhesion force of the patterned membrane was lower than that of the pristine membrane. In some regions (e.g., y-axis from 2.25 to 3.25 μm), the adhesion force of the pristine membrane was one order of magnitude higher than that of the patterned one, which favoured the BSA protein deposition onto the pristine membrane.

Previous studies on microbial adhesion on a sub-micron patterned surface also indicated a similar mechanism (Yang et al., 2015; Lu et al., 2016). The reliability of adhesion force mapping may also be affected by membrane manufacturing imperfections where a high adhesion force region of the pristine membrane was captured.

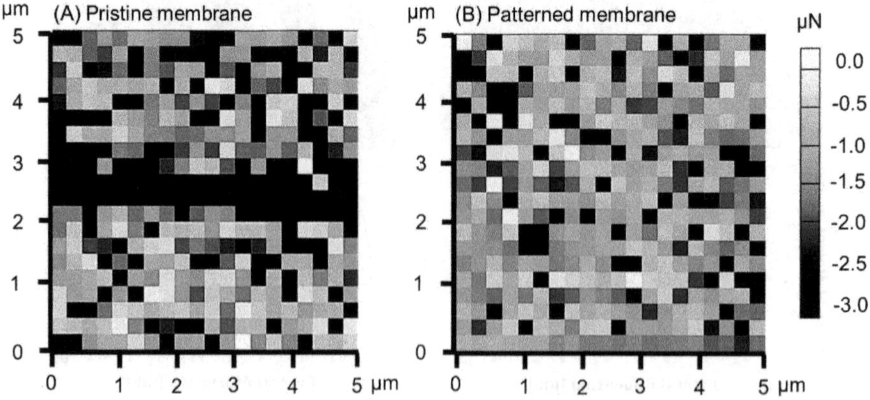

FIGURE 1.17 Adhesion force mapping for (A) pristine membrane and (B) patterned membrane. The adhesion forces measured within 64 cycles were then plotted as a distribution map. The colloidal AFM probe was functionalised by adsorption of BSA protein. (From Xie, M. et al., Water Research. 124, 238–243, 2017.)

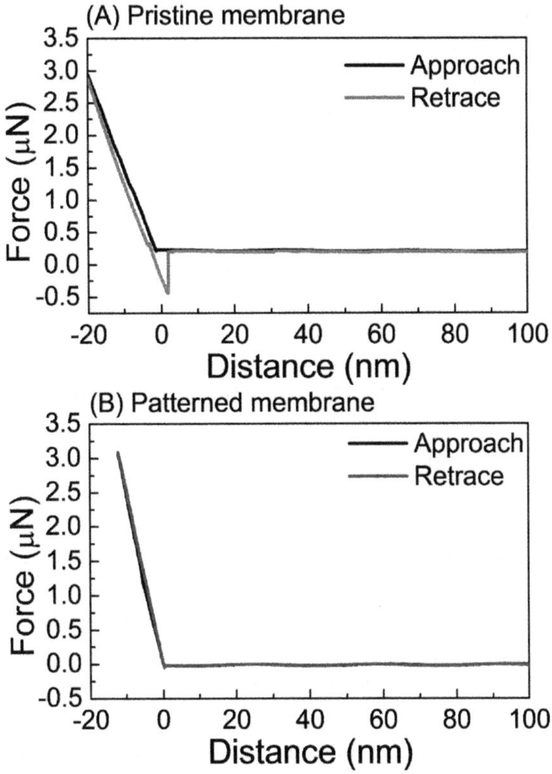

FIGURE 1.18 Representative force-distance curve for (A) pristine membrane and (B) patterned membrane. The colloidal atomic force spectroscopy (AFM) probe was functionalised by adsorption of BSA protein. Each curve is the average of 20 independent force measurements. (From Xie, M. et al., Water Research. 124, 238–243, 2017.)

In order to further prove the role of surface pattern in surface energy, manifested as adhesion force, representative force-distance curves for both pristine and patterned membranes during the adhesion force measurements were extracted (Figure 1.18), allowing for interpretation of the details in the interaction between the patterned membrane and BSA protein. Unlike the patterned membrane, a strong attractive depletion force was observed for the pristine membrane during the retrace phase. The adhesion force of the engineered nanostructure on the patterned membrane was significantly smaller than that of the pristine membrane. Together with adhesion force mapping, the force curve further highlighted that the membrane with surface pattern resulted in a much lower adhesion force, thereby substantially reducing the protein deposition during filtration.

1.6 CONCLUDING REMARKS

TFC membranes have experienced tremendous development since the concept of interfacial polymerization was first introduced in 1965. The membranes were developed to overcome the low water permeability and poor biological stability of cellulose-based RO membranes that were made via phase inversion technique for seawater desalination. In the mid-1980s, the TFC membranes with reduced separation efficiency for smaller and less charged ions such as sodium and chloride were commercialized for NF applications. Owing to the unique advantages of engineered osmosis that can operate at low or no hydraulic pressure, the application of TFC membranes has been expanded to FO and PRO processes since 2010. Although the existing TFC membranes could be utilized for various industrial processes without having major problems, concerted effort has been made continuously over the years to further improve the membrane performance, aiming to overcome the trade-off effect between water flux and salt rejection. With respect to worldwide research and development activities, the growth of TFC membrane research publications has increased significantly over the last 5 years. Most of the publications are concerned with the fabrication and modification of TFC membranes using advanced materials and focused on improving membrane resistances with respect to fouling, chlorine, chemical, and thermal performance, in addition to water permeability. This chapter provides a comprehensive review on the latest development of TFC membrane fabrication and modification in which both a microporous substrate and a top polyamide layer can be independently optimized to achieve desirable properties for the water separation process. Since a large number of advanced materials (either organic or inorganic) have been previously shown to have positive impacts on the TFC membrane properties, the selection and use of these materials should depend on the compounds such as the type of ions, organic foulants, microorganisms, etc., present in the feed solution. No TFC membrane is universally applicable for any type of water applications.

ABBREVIATIONS

AFM	Atomic force spectroscopy
BPA	Bisphenol A
BSA	Bovine serum albumin
CA	Cellulose acetate
CFIC	5-choroformyloxyisophaloyl chloride
CFU	Colony-forming units
CLSM	Confocal laser scanning microscopy
CN	Cellulose nanofiber
CTA	Cellulose triacetate
DABA	3,5-diamino-N-(4-aminophenyl) benzamide
DAHP	1,3-diamino-2-hydroxypropane
DMF	Dimethylformamide
DMMPD	N,N-dimethyl-m-phenylenediamine
ED	Ethylene diamine
EDA	Ethane diamine
EDC/NHS	1-ethyl-(3-3-dimethylaminopropyl) carbodiimide hydrochloride/N-hydroxysuccinimide
f-CNTs	Functionalized multi-walled carbon nanotubes

FO	Forward osmosis
GO	Graphene oxide
rGO/CN	Graphitic carbon nitride
HH-IPN-CNF	Hydrophobic/hydrophilic interpenetrating network composite nanofiber
HNTs	Halloysite nanotubes
ICP	Internal concentration polarization
MF	Microfiltration
MMPD	4-methylphenylenediamine
MMT	Montmorillonite
mNT	Silane-modified nanoporous titanate nanoparticle
MPD	M-phenylenediamine
MWCNTs	Multiwalled carbon nanotubes
NaY	Zeolite-Y
NF	Nanofiltration
NMP	*N*-methyl-2-pyrrolidone
NPED	N-[3-(trimethoxysilyl) propyl] ethylenediamine
o-ABA-TEA	o-aminobenzoic acid-triethylamine
PAN	Polyacrylonitrile
PAN-AA	Poly (acrylonitrile-co-acrylic acid)
PDA	Polydopamine
PE	Pentaerythrotol
PEG	Polyethylene glycol
PEGDE	Poly(ethylene glycol) diglycidyl ether
PES	Polyethersulfone
PET	Polyethylene terephthalate
PMAPS	Polyelectrolyte brush - poly(3-(N-2-methacryloxyethyl-N, N-dimethyl) ammonato propanesultone
PMDA/ODA	Poly(pyromellitic dianhydride-co-4,4′-oxydianiline)
POSS	Polyoctahedral oligomeric silsequioxanes
PP	Polypropylene
PPG	Polypropylene glycol
PRO	Pressure retarded osmosis
PSF	Polysulfone
PTC	2,4,6-pyridinetricarboxylic acid chloride
PTFE	Polytetrafluoroethylene
PVA	Polyvinyl alcohol
PVC	Polyvinyl chloride
PVDF	Polyvinylidene fluoride
PVP	Polyvinylpyrrolidone
PRO	Pressure retarded osmosis
RO	Reverse osmosis
sDCDPS	3,30-di-sodiumdisulfate-4,40-dichlorodiphenyl sulfone
SEA	2-[(2-aminoethyl) amino]-ethane sulfonic acid monosodium salt
SEM	Scanning electron microscopy
SPEEK	Sulfonated polyetheretherketone

SPES	Sulphonated polyethersulfone
SPPESK	Sulfonated poly (phthalazinone ether sulfone ketone)
sPPSU	Sulfonated polyphenylenesulfone
SPSF	Sulfonated polysulfone
TAEA	Tripodal amine
TFC	Thin film composite
TFN	Thin film nanocomposite
TMC	Trimesoyl chloride
UF	Ultrafiltration
ZIF	Zeolitic imidazolate framework
ZnO	Zinc oxide

REFERENCES

Ahmad, A., Waheed, S., Khan, S.M., e-Gul, S., Shafiq, M., Farooq, M., Sanaullah, K., Jamil, T., 2015. Effect of silica on the properties of cellulose acetate/polyethylene glycol membranes for reverse osmosis. Desalination 355, 1–10.

Amini, M., Jahanshahi, M., Rahimpour, A., 2013. Synthesis of novel thin film nanocomposite (TFN) forward osmosis membranes using functionalized multi-walled carbon nanotubes. Journal of Membrane Science 435, 233–241.

Ba, C., Economy, J., 2010. Preparation of PMDA/ODA polyimide membrane for use as substrate in a thermally stable composite reverse osmosis membrane. Journal of Membrane Science 363(1–2), 140–148.

Bano, S., Mahmood, A., Kim, S.-J., Lee, K.-H., 2015. Graphene oxide modified polyamide nanofiltration membrane with improved flux and antifouling properties. Journal of Materials Chemistry A 3(5), 2065–2071.

Ben-Sasson, M., Lu, X., Bar-Zeev, E., Zodrow, K.R., Nejati, S., Qi, G., Giannelis, E.P., Elimelech, M., 2014a. In situ formation of silver nanoparticles on thin-film composite reverse osmosis membranes for biofouling mitigation. Water Research 62, 260–270.

Ben-Sasson, M., Zodrow, K.R., Genggeng, Q., Kang, Y., Giannelis, E.P., Elimelech, M., 2014b. Surface functionalization of thin-film composite membranes with copper nanoparticles for antimicrobial surface properties. Environmental Science & Technology 48(1), 384–393.

Bera, A., Gol, R.M., Chatterjee, S., Jewrajka, S.K., 2015. PEGylation and incorporation of triazine ring into thin film composite reverse osmosis membranes for enhancement of anti-organic and anti-biofouling properties. Desalination 360, 108–117.

Bui, N.-N., Lind, M.L., Hoek, E.M.V., McCutcheon, J.R., 2011. Electrospun nanofiber supported thin film composite membranes for engineered osmosis. Journal of Membrane Science 385–386, 10–19.

Chae, H.-R., Lee, J., Lee, C.-H., Kim, I.-C., Park, P.-K., 2015. Graphene oxide-embedded thin-film composite reverse osmosis membrane with high flux, anti-biofouling, and chlorine resistance. Journal of Membrane Science 483, 128–135.

Cheng, J., Shi, W., Zhang, L., Zhang, R., 2017. A novel polyester composite nanofiltration membrane formed by interfacial polymerization of pentaerythritol (PE) and trimesoyl chloride (TMC). Applied Surface Science 416, 152–159.

Choi, W., Choi, J., Bang, J., Lee, J.-H., 2013. Layer-by-layer assembly of graphene oxide nanosheets on polyamide membranes for durable reverse-osmosis applications. ACS Applied Materials & Interfaces 5(23), 12510–12519.

Corvilain, M., Klaysom, C., Szymczyk, A., Vankelecom, I.F.J., 2017. Formation mechanism of sPEEK hydrophilized PES supports for forward osmosis. Desalination 419, 29–38.

Dai, Y., Jian, X., Zhang, S., Guiver, M.D., 2002. Thin film composite (TFC) membranes with improved thermal stability from sulfonated poly(phthalazinone ether sulfone ketone) (SPPESK). Journal of Membrane Science 207(2), 189–197.

Dalwani, M., Benes, N.E., Bargeman, G., Stamatialis, D., Wessling, M., 2011. Effect of pH on the performance of polyamide/polyacrylonitrile based thin film composite membranes. Journal of Membrane Science 372(1–2), 228–238.

del Campo, A., Arzt, E., 2008. Fabrication approaches for generating complex micro- and nanopatterns on polymeric durfaces. Chemical Reviews 108(3), 911–945.

Ding, Y., Maruf, S., Aghajani, M., Greenberg, A.R., 2017. Surface patterning of polymeric membranes and its effect on antifouling characteristics. Separation Science and Technology 52(2), 240–257.

Dong, H., Zhao, L., Zhang, L., Chen, H., Gao, C., Winston Ho, W.S., 2015. High-flux reverse osmosis membranes incorporated with NaY zeolite nanoparticles for brackish water desalination. Journal of Membrane Science 476, 373–383.

Duan, J., Pan, Y., Pacheco, F., Litwiller, E., Lai, Z., Pinnau, I., 2015. High-performance polyamide thin-film-nanocomposite reverse osmosis membranes containing hydrophobic zeolitic imidazolate framework-8. Journal of Membrane Science 476, 303–310.

Emadzadeh, D., Ghanbari, M., Lau, W.J., Rahbari-Sisakht, M., Rana, D., Matsuura, T., Kruczek, B., Ismail, A.F., 2017. Surface modification of thin film composite membrane by nanoporous titanate nanoparticles for improving combined organic and inorganic antifouling properties. Materials Science and Engineering: C 75, 463–470.

Emadzadeh, D., Lau, W.J., Matsuura, T., Ismail, A.F., Rahbari-Sisakht, M., 2014. Synthesis and characterization of thin film nanocomposite forward osmosis membrane with hydrophilic nanocomposite support to reduce internal concentration polarization. Journal of Membrane Science 449, 74–85.

Emadzadeh, D., Lau, W.J., Rahbari-Sisakht, M., Ilbeygi, H., Rana, D., Matsuura, T., Ismail, A.F., 2015a. Synthesis, modification and optimization of titanate nanotubes-polyamide thin film nanocomposite (TFN) membrane for forward osmosis (FO) application. Chemical Engineering Journal 281, 243–251.

Emadzadeh, D., Lau, W.J., Rahbari-Sisakht, M., Daneshfar, A., Ghanbari, M., Mayahi, A., Matsuura, T., Ismail, A.F., 2015b. A novel thin film nanocomposite reverse osmosis membrane with superior anti-organic fouling affinity for water desalination. Desalination 368, 106–113.

Faria, A.F., Liu, C., Xie, M., Perreault, F., Nghiem, L.D., Ma, J., Elimelech, M., 2017. Thin-film composite forward osmosis membranes functionalized with graphene oxide–silver nanocomposites for biofouling control. Journal of Membrane Science 525, 146–156.

Fathizadeh, M., Aroujalian, A., Raisi, A., 2011. Effect of added NaX nano-zeolite into polyamide as a top thin layer of membrane on water flux and salt rejection in a reverse osmosis process. Journal of Membrane Science 375(1–2), 88–95.

Ghanbari, M., Emadzadeh, D., Lau, W.J., Matsuura, T., Davoody, M., Ismail, A.F., 2015a. Super hydrophilic TiO_2/HNT nanocomposites as a new approach for fabrication of high performance thin film nanocomposite membranes for FO application. Desalination 371, 104–114.

Ghanbari, M., Emadzadeh, D., Lau, W.J., Matsuura, T., Ismail, A.F., 2015b. Synthesis and characterization of novel thin film nanocomposite reverse osmosis membranes with improved organic fouling properties for water desalination. RSC Advances 5(27), 21268–21276.

Ghosh, A.K., Hoek, E.M.V., 2009. Impacts of support membrane structure and chemistry on polyamide–polysulfone interfacial composite membranes. Journal of Membrane Science 336(1), 140–148.

Ghosh, A.K., Jeong, B.-H., Huang, X., Hoek, E.M.V., 2008. Impacts of reaction and curing conditions on polyamide composite reverse osmosis membrane properties. Journal of Membrane Science 311(1–2), 34–45.

Guo, L.J., 2007. Nanoimprint lithography: Methods and material requirements. Advanced Materials 19(4), 495–513.

Han, G., Chung, T.-S., Toriida, M., Tamai, S., 2012. Thin-film composite forward osmosis membranes with novel hydrophilic supports for desalination. Journal of Membrane Science 423–424, 543–555.

Han, G., Zhao, B., Fu, F., Chung, T.-S., Weber, M., Staudt, C., Maletzko, C., 2016. High performance thin-film composite membranes with mesh-reinforced hydrophilic sulfonated polyphenylenesulfone (sPPSU) substrates for osmotically driven processes. Journal of Membrane Science 502, 84–93.

Huang, L., McCutcheon, J.R., 2014. Hydrophilic nylon 6,6 nanofibers supported thin film composite membranes for engineered osmosis. Journal of Membrane Science 457, 162–169.

Jeong, B.-H., Hoek, E.M.V., Yan, Y., Subramani, A., Huang, X., Hurwitz, G., Ghosh, A.K., Jawor, A., 2007. Interfacial polymerization of thin film nanocomposites: A new concept for reverse osmosis membranes. Journal of Membrane Science 294(1–2), 1–7.

Jewrajka, S.K., Reddy, A.V.R., Rana, H.H., Mandal, S., Khullar, S., Haldar, S., Joshi, N., Ghosh, P.K., 2013. Use of 2,4,6-pyridinetricarboxylic acid chloride as a novel co-monomer for the preparation of thin film composite polyamide membrane with improved bacterial resistance. Journal of Membrane Science 439, 87–95.

Jung, S.Y., Won, Y.-J., Jang, J.H., Yoo, J.H., Ahn, K.H., Lee, C.-H., 2015. Particle deposition on the patterned membrane surface: Simulation and experiments. Desalination 370, 17–24.

Kim, E.-S., Kim, Y.J., Yu, Q., Deng, B., 2009. Preparation and characterization of polyamide thin-film composite (TFC) membranes on plasma-modified polyvinylidene fluoride (PVDF). Journal of Membrane Science 344(1–2), 71–81.

Kim, E.-S., Hwang, G., Gamal El-Din, M., Liu, Y., 2012. Development of nanosilver and multi-walled carbon nanotubes thin-film nanocomposite membrane for enhanced water treatment. Journal of Membrane Science 394–395, 37–48.

Kim, J., Suh, D., Kim, C., Baek, Y., Lee, B., Kim, H.J., Lee, J.-C., Yoon, J., 2016. A high-performance and fouling resistant thin-film composite membrane prepared via coating TiO_2 nanoparticles by sol-gel-derived spray method for PRO applications. Desalination 397, 157–164.

Kim, S.G., Hyeon, D.H., Chun, J.H., Chun, B.-H., Kim, S.H., 2013. Novel thin nanocomposite RO membranes for chlorine resistance. Desalination and Water Treatment 51(31–33), 6338–6345.

Kong, X., Zhou, M.-Y., Lin, C.-E., Wang, J., Zhao, B., Wei, X.-Z., Zhu, B.-K., 2016. Polyamide/PVC based composite hollow fiber nanofiltration membranes: Effect of substrate on properties and performance. Journal of Membrane Science 505, 231–240.

Korikov, A.P., Kosaraju, P.B., Sirkar, K.K., 2006. Interfacially polymerized hydrophilic microporous thin film composite membranes on porous polypropylene hollow fibers and flat films. Journal of Membrane Science 279(1–2), 588–600.

Lai, G.S., Lau, W.J., Gray, S.R., Matsuura, T., Gohari, R.J., Subramanian, M.N., Lai, S.O., Ong, C.S., Ismail, A.F., Emazadah, D., Ghanbari, M., 2016a. A practical approach to synthesize polyamide thin film nanocomposite (TFN) membranes with improved separation properties for water/wastewater treatment. Journal of Materials Chemistry A 4(11), 4134–4144.

Lai, G.S., Lau, W.J., Goh, P.S., Ismail, A.F., Yusof, N., Tan, Y.H., 2016b. Graphene oxide incorporated thin film nanocomposite nanofiltration membrane for enhanced salt removal performance. Desalination 387, 14–24.

Lau, W.J., Gray, S., Matsuura, T., Emadzadeh, D., Paul Chen, J., Ismail, A.F., 2015. A review on polyamide thin film nanocomposite (TFN) membranes: History, applications, challenges and approaches. Water Research 80, 306–324.

Lau, W.J., Ismail, A.F., Misdan, N., Kassim, M.A., 2012. A recent progress in thin film composite membrane: A review. Desalination 287, 190–199.
Lau, W.J., Ismail, A.F., 2017. Nanofiltration membranes: Synthesis, characterization, and applications. CRC Press, Taylor and Francis Group.
Lee, S.-W., Lee, K.-S., Ahn, J., Lee, J.-J., Kim, M.-G., Shin, Y.-B., 2011. Highly sensitive biosensing using arrays of plasmonic Au nanodisks realized by nanoimprint lithography. ACS Nano 5(2), 897–904.
Li, M., Wang, J., Zhuang, L., Chou, S.Y., 2000. Fabrication of circular optical structures with a 20 nm minimum feature size using nanoimprint lithography. Applied Physics Letters 76(6), 673–675.
Li, D., Yan, Y., Wang, H., 2016. Recent advances in polymer and polymer composite membranes for reverse and forward osmosis processes. Progress in Polymer Science 61, 104–155.
Liu, L.-F., Cai, Z.-B., Shen, J.-N., Wu, L.-X., Hoek, E.M.V., Gao, C.-J., 2014. Fabrication and characterization of a novel poly(amide-urethane@imide) TFC reverse osmosis membrane with chlorine-tolerant property. Journal of Membrane Science 469, 397–409.
Liu, X., Ng, H.Y., 2014. Double-blade casting technique for optimizing substrate membrane in thin-film composite forward osmosis membrane fabrication. Journal of Membrane Science 469, 112–126.
Lu, N., Zhang, W., Weng, Y., Chen, X., Cheng, Y., Zhou, P., 2016. Fabrication of PDMS surfaces with micro patterns and the effect of pattern sizes on bacteria adhesion. Food Control 68, 344–351.
Lu, X., Romero-Vargas Castrillón, S., Shaffer, D.L., Ma, J., Elimelech, M., 2013. In situ surface chemical modification of thin-film composite forward osmosis membranes for enhanced organic fouling resistance. Environmental Science & Technology 47(21), 12219–12228.
Ma, N., Wei, J., Qi, S., Zhao, Y., Gao, Y., Tang, C.Y., 2013. Nanocomposite substrates for controlling internal concentration polarization in forward osmosis membranes. Journal of Membrane Science 441, 54–62.
Maruf, S.H., Greenberg, A.R., Pellegrino, J., Ding, Y., 2014. Fabrication and characterization of a surface-patterned thin film composite membrane. Journal of Membrane Science 452, 11–19.
Maruf, S.H., Wang, L., Greenberg, A.R., Pellegrino, J., Ding, Y., 2013. Use of nanoimprinted surface patterns to mitigate colloidal deposition on ultrafiltration membranes. Journal of Membrane Science 428, 598–607.
Moghimifar, V., Livari, A.E., Raisi, A., Aroujalian, A., 2015. Enhancing the antifouling property of polyethersulfone ultrafiltration membranes using NaX zeolite and titanium oxide nanoparticles. RSC Advances 5(69), 55964–55976.
Moon, J.H., Katha, A.R., Pandian, S., Kolake, S.M., Han, S., 2014. Polyamide–POSS hybrid membranes for seawater desalination: Effect of POSS inclusion on membrane properties. Journal of Membrane Science 461, 89–95.
Ong, C.S., Al-anzi, B., Lau, W.J., Goh, P.S., Lai, G.S., Ismail, A.F., Ong, Y.S., 2017. Antifouling double-skinned forward osmosis membrane with zwitterionic brush for oily wastewater treatment. Scientific Reports 7, 6904.
Pal, A., Dey, T.K., Singhal, A., Bindal, R.C., Tewari, P.K., 2015. Nano-ZnO impregnated inorganic-polymer hybrid thinfilm nanocomposite nanofiltration membranes: An investigation of variation in structure, morphology and transport properties. RSC Advances 5(43), 34134–34151.
Park, S.J., Ko, T.-J., Yoon, J., Moon, M.-W., Oh, K.H., Han, J.H., 2018. Highly adhesive and high fatigue-resistant copper/PET flexible electronic substrates. Applied Surface Science 427, 1–9.

Pendergast, M.T.M., Nygaard, J.M., Ghosh, A.K., Hoek, E.M.V., 2010. Using nanocomposite materials technology to understand and control reverse osmosis membrane compaction. Desalination 261(3), 255–263.

Perera, D.H.N., Song, Q., Qiblawey, H., Sivaniah, E., 2015. Regulating the aqueous phase monomer balance for flux improvement in polyamide thin film composite membranes. Journal of Membrane Science 487, 74–82.

Perreault, F., Jaramillo, H., Xie, M., Ude, M., Nghiem, L.D., Elimelech, M., 2016. Biofouling mitigation in forward osmosis using graphene oxide functionalized thin-film composite membranes. Environmental Science & Technology 50(11), 5840–5848.

Perreault, F., Tousley, M.E., Elimelech, M., 2014. Thin-film composite polyamide membranes functionalized with biocidal graphene oxide nanosheets. Environmental Science & Technology Letters 1(1), 71–76.

Petersen, R.J., 1993. Composite reverse osmosis and nanofiltration membranes. Journal of Membrane Science 83(1), 81–150.

Peyki, A., Rahimpour, A., Jahanshahi, M., 2015. Preparation and characterization of thin film composite reverse osmosis membranes incorporated with hydrophilic SiO2 nanoparticles. Desalination 368, 152–158.

Rickman, M., Maruf, S., Kujundzic, E., Davis, R.H., Greenberg, A., Ding, Y., Pellegrino, J., 2017. Fractionation and flux decline studies of surface-patterned nanofiltration membranes using NaCl-glycerol-BSA solutions. Journal of Membrane Science 527, 102–110.

Romero-Vargas Castrillón, S., Lu, X., Shaffer, D.L., Elimelech, M., 2014. Amine enrichment and poly(ethylene glycol) (PEG) surface modification of thin-film composite forward osmosis membranes for organic fouling control. Journal of Membrane Science 450, 331–339.

Saeki, D., Nagao, S., Sawada, I., Ohmukai, Y., Maruyama, T., Matsuyama, H., 2013. Development of antibacterial polyamide reverse osmosis membrane modified with a covalently immobilized enzyme. Journal of Membrane Science 428, 403–409.

Shaffer, D.L., Jaramillo, H., Romero-Vargas Castrillón, S., Lu, X., Elimelech, M., 2015. Post-fabrication modification of forward osmosis membranes with a poly(ethylene glycol) block copolymer for improved organic fouling resistance. Journal of Membrane Science 490, 209–219.

Shen, L., Zuo, J., Wang, Y., 2017. Tris(2-aminoethyl)amine in-situ modified thin-film composite membranes for forward osmosis applications. Journal of Membrane Science 537, 186–201.

Son, M., Choi, H.-G., Liu, L., Celik, E., Park, H., Choi, H., 2015. Efficacy of carbon nanotube positioning in the polyethersulfone support layer on the performance of thin-film composite membrane for desalination. Chemical Engineering Journal 266, 376–384.

Song, X., Liu, Z., Sun, D.D., 2013. Energy recovery from concentrated seawater brine by thin-film nanofiber composite pressure retarded osmosis membranes with high power density. Energy & Environmental Science 6(4), 1199–1210.

Song, X., Wang, L., Mao, L., Wang, Z., 2016. Nanocomposite membrane with different carbon nanotubes location for nanofiltration and forward osmosis applications. ACS Sustainable Chemistry & Engineering 4(6), 2990–2997.

Sun, Y., Xue, L., Zhang, Y., Zhao, X., Huang, Y., Du, X., 2014. High flux polyamide thin film composite forward osmosis membranes prepared from porous substrates made of polysulfone and polyethersulfone blends. Desalination 336, 72–79.

Tang, H., He, J., Hao, L., Wang, F., Zhang, H., Guo, Y., 2017. Developing nanofiltration membrane based on microporous poly(tetrafluoroethylene) substrates by bi-stretching process. Journal of Membrane Science 524, 612–622.

Tian, M., Wang, Y.-N., Wang, R., 2015. Synthesis and characterization of novel high-performance thin film nanocomposite (TFN) FO membranes with nanofibrous substrate reinforced by functionalized carbon nanotubes. Desalination 370, 79–86.

Tian, M., Wang, Y.-N., Wang, R., Fane, A.G., 2017. Synthesis and characterization of thin film nanocomposite forward osmosis membranes supported by silica nanoparticle incorporated nanofibrous substrate. Desalination 401, 142–150.

Tian, E.L., Zhou, H., Ren, Y.W., Mirza, Z.A., Wang, X.Z., Xiong, S.W., 2014. Novel design of hydrophobic/hydrophilic interpenetrating network composite nanofibers for the support layer of forward osmosis membrane. Desalination 347, 207–214.

Tiraferri, A., Vecitis, C.D., Elimelech, M., 2011. Covalent binding of single-walled carbon nanotubes to polyamide membranes for antimicrobial surface properties. ACS Applied Materials & Interfaces 3(8), 2869–2877.

Wang, H., Li, L., Zhang, X., Zhang, S., 2010. Polyamide thin-film composite membranes prepared from a novel triamine 3,5-diamino-N-(4-aminophenyl)-benzamide monomer and m-phenylenediamine. Journal of Membrane Science 353(1–2), 78–84.

Wang, K.Y., Chung, T.-S., Amy, G., 2012. Developing thin-film-composite forward osmosis membranes on the PES/SPSf substrate through interfacial polymerization. AIChE Journal 58(3), 770–781.

Wang, X., Fang, D., Hsiao, B.S., Chu, B., 2014. Nanofiltration membranes based on thin-film nanofibrous composites. Journal of Membrane Science 469, 188–197.

Wang, Y., Li, X., Cheng, C., He, Y., Pan, J., Xu, T., 2016. Second interfacial polymerization on polyamide surface using aliphatic diamine with improved performance of TFC FO membranes. Journal of Membrane Science 498, 30–38.

Wang, Y., Ou, R., Ge, Q., Wang, H., Xu, T., 2013. Preparation of polyethersulfone/carbon nanotube substrate for high-performance forward osmosis membrane. Desalination 330, 70–78.

Wang, Y., Ou, R., Wang, H., Xu, T., 2015. Graphene oxide modified graphitic carbon nitride as a modifier for thin film composite forward osmosis membrane. Journal of Membrane Science 475, 281–289.

Wang, Y., Xu, T. 2015. Anchoring hydrophilic polymer in substrate: An easy approach for improving the performance of TFC FO membrane. Journal of Membrane Science 476, 330–339.

Widjojo, N., Chung, T.-S., Weber, M., Maletzko, C., Warzelhan, V., 2013. A sulfonated polyphenylenesulfone (sPPSU) as the supporting substrate in thin film composite (TFC) membranes with enhanced performance for forward osmosis (FO). Chemical Engineering Journal 220, 15–23.

Xie, M., Luo, W., Gray, S.R., 2017. Surface pattern by nanoimprint for membrane fouling mitigation: Design, performance and mechanisms. Water Research 124, 238–243.

Xiong, S., Zuo, J., Ma, Y.G., Liu, L., Wu, H., Wang, Y., 2016. Novel thin film composite forward osmosis membrane of enhanced water flux and anti-fouling property with N-[3-(trimethoxysilyl) propyl] ethylenediamine incorporated. Journal of Membrane Science 520, 400–414.

Yang, M., Ding, Y., Ge, X., Leng, Y., 2015. Control of bacterial adhesion and growth on honeycomb-like patterned surfaces. Colloids and Surfaces B: Biointerfaces 135, 549–555.

Yang, Y., Wang, X., Hsiao, B.S., 2016. Preparation of thin film nanofibrous composite NF membrane based on EDC/NHS modified PAN-AA nanofibrous substrate. IOP Conference Series: Materials Science and Engineering 137(1), 1–5.

Ye, G., Lee, J., Perreault, F., Elimelech, M., 2015. Controlled architecture of dual-functional block copolymer brushes on thin-film composite membranes for integrated "Defending" and "Attacking" strategies against biofouling. ACS Applied Materials & Interfaces 7(41), 23069–23079.

Yin, J., Yang, Y., Hu, Z., Deng, B., 2013. Attachment of silver nanoparticles (AgNPs) onto thin-film composite (TFC) membranes through covalent bonding to reduce membrane biofouling. Journal of Membrane Science 441, 73–82.

Yin, J., Zhu, G., Deng, B., 2016. Graphene oxide (GO) enhanced polyamide (PA) thin-film nanocomposite (TFN) membrane for water purification. Desalination 379, 93–101.

Zankovych, S., Hoffmann, T., Seekamp, J., Bruch, J.U., Torres, C.M.S., 2001. Nanoimprint lithography: Challenges and prospects. Nanotechnology 12(2), 91.

Zhang, C., Huang, M., Meng, L., Li, B., Cai, T., 2017. Electrospun polysulfone (PSF)/titanium dioxide (TiO_2) nanocomposite fibers as substrates to prepare thin film forward osmosis membranes. Journal of Chemical Technology & Biotechnology 92(8), 2090–2097.

Zhao, J., Su, Y., He, X., Zhao, X., Li, Y., Zhang, R., Jiang, Z., 2014a. Dopamine composite nanofiltration membranes prepared by self-polymerization and interfacial polymerization. Journal of Membrane Science 465, 41–48.

Zhao, L., W.S. Winston Ho., 2014b. Novel reverse osmosis membranes incorporated with a hydrophilic additive for seawater desalination. Journal of Membrane Science 455, 44–54.

2 Stimuli-Responsive Materials for Membrane Fabrication

Ida Amura, Salman Shahid, and Emma A.C. Emanuelsson

CONTENTS

We would like to dedicate this chapter to the late Dr. Darrell Patterson, an expert on stimuli-responsive membranes and separation processes, who was meant to write this chapter. His work and devotion to the membranes field gave the authors the encouragement and motivation to write this chapter.

2.1 INTRODUCTION

Stimuli-responsive membranes, based on "smart" polymers, have recently gained much academic interest as their separation performances can be adjusted according to changes in environmental conditions. Polymers undergo a reversible change, either physical or chemical in its properties, that is based on external stimuli and that induces a macroscopic response in the membrane. Using stimuli allows for a specific conformational transition on a microscopic level and then the stimuli-responsiveness occurs, thus amplifying these transitions into macroscopically measurable changes in membrane properties. The mutual interactions among pore structure and change in conformation, polarity, and reactivity of functional groups of the responsive polymers in the membrane bulk or its surface are the key factors that allow for the responsiveness of the membrane.

These membranes can respond to external stimuli such as pH (Dai et al., 2008; Tufani et al., 2017), temperature (Park et al., 2016; Zhang et al., 2016; Kamoun et al., 2017), ionic strength (Huang et al., 2009), light (Duong et al., 2017; Kaner et al., 2017), electric (Loh et al., 1990; Ahmed et al., 2016) and magnetic fields (Wen et al., 2017), and chemical cues (Miyata et al., 2002). The most common stimuli are pH and temperature that occur spontaneously in biological systems and chemical reactions, and the more novel stimuli responsive membranes are based on an electric or magnetic field. All these responsive systems have the potential to overcome traditional membrane issues in which the pore size is defined during the fabrication process and the permeability cannot be adjusted to exploit external environmental conditions. Furthermore, membranes based on electrical polymeric systems are showing potential in water treatment, fuel cells, catalysis, and as sensors due to the possibility of combining the electrical conductivity of metals with the strength and processability of synthetic polymers (Pile et al., 2002; Reece et al., 2005). These electrically tuneable responsive membranes can be potentially used to obtain systems with controllable transport properties and *in situ* self-cleaning mechanisms as demonstrated in recent studies (Sun et al., 2013; Lalia et al., 2015). Temperature, pH, and electrical-responsive polymers can also be introduced on pre-formed membrane supports as surface functionalities using different surface modification methods, and the resulting membrane systems are, therefore, called gating membranes (Luo et al., 2015; Liu et al., 2016).

Until now, both non-porous and porous responsive membranes have found applications in sensors, separation processes, drug delivery devices, and complex technical systems in general (Gugliuzza, 2013; Darvishmanesh et al., 2015).

In this chapter, we examine the many recent contributions to the fast growing field of stimuli-responsive membranes with specific attention to the polymer science and the material properties. The focus will be first on the progress and advances of pH and temperature-responsive systems since they have been the most researched in selective filtrations and biomaterials. Next, we will discuss electrical-responsive membranes as novel smart materials in order to provide the reader with the current state of knowledge. This chapter is meant to provide a general understanding of stimuli-responsive membranes in terms of mechanism of action and preparation methods, but more detailed reviews on the topic are available for the reader. Wandera et al. (2010) have covered in an excellent review all type of stimuli and fabrication methods, and Husson (2012) has given a detailed overview about design and fabrication of responsive membranes. Furthermore, Ahmed et al. (2016) have covered the recent developments in the application of conductive polymeric membranes to water treatment and desalination. To complement these reviews, we have presented the already established protocols for preparing pH, temperature, and electrical-responsive membranes and highlight some of features of these membranes.

The chapter is organized in three main sections: Section 2.2 is a description of materials, their basic features, and the responsive mechanism; Section 2.3 reviews fabrication methods and the latest advances are covered with specific examples. The conclusive Section 2.4 includes a discussion about future challenges for responsive membranes.

2.2 STIMULI-RESPONSIVE MATERIALS FOR MEMBRANES SYSTEMS: BASIC PRINCIPLES AND MECHANISM OF ACTION

Temperature and pH-responsive membranes are made with macromolecular systems containing acidic or basic ionizable groups called polyelectrolytes, or soluble polymers able to undergo a phase transition at a specific temperature. Electrical-responsive membranes can be fabricated from conducting polymers that are sensitive to electrical field and electrical voltage. The responsive behavior for the three stimuli is explained in Section 2.2.1 while Sections 2.2.2, 2.2.3, and 2.2.4 will cover properties of such responsive polymers.

2.2.1 Responsive Mechanism for Stimuli Responsive Membranes

Membrane stimuli-responsive behavior can be explained based on phase transition mechanism of the membrane materials in specific controlled environments (Minko, 2006).

2.2.1.1 Mechanism of pH Responsive Membranes

Membranes sensitive to pH are based on polyelectrolytes containing acidic and basic groups, which can either swell or shrink in response to a change in environmental pH caused by protonation and deprotonation of these ionizable side groups. Electrostatic interactions are one of the main factors that govern the structural change of polyelectrolytes chains, along with solvation forces and excluded volume effects (Azzaroni et al., 2007; Su et al., 2007). In particular, electrostatic repulsions and excluded volume effects support a stretched conformation in which the polymer chains repulse each other. In difference, the minimization of electrostatic charge leads to an entropically favorable collapsed conformation.

Polyelectrolytes have also been used in the development of surfaces with switchable characteristics in the form of polymer brushes, which are assemblies of macromolecular chains tethered or grafted by one of their extremities to a surface (Rühe et al., 2004; Azzaroni et al., 2005; Minko, 2006; Das et al., 2015). In terms of mechanism of action, these brushes exploit the strong segment-segment repulsions and electrostatic interactions to switch reversibly from a collapsed to a stretched state allowing permeability control.

2.2.1.2 Mechanism of Thermo-Responsive Membranes

Thermo-responsive membranes respond to variations in the external temperature by changing accordingly their permeability, selectivity, and absorption abilities (Tang et al., 2016).

Generally, thermo-responsive membranes are dependent on the lower critical solution temperature (LCST) transition and upper critical solution temperature (UCST) transition (Figure 2.1).

The LCST is defined as the lowest temperature of the binodal curve while the UCST is defined as the highest temperature of the binodal curve. Polymers that undergo LCST transition in water phase separate upon increasing the temperature above their LCST, and the polymer chains collapsed in a shrunken state. For systems with UCST, an increase in temperature above the temperature of the UCST transition causes the system to merge into a single homogeneous phase and polymer chains swell. It is this shrinking and swelling behavior of the thermosensitive polymer chains that influence solute permeability by altering the pore size. The most studied temperature responsive polymer is poly(N-isopropylacrylamide) (PNIPAM), which will be reviewed in Section 2.2.3.

2.2.1.3 Mechanism of Electrical-Responsive Membranes

Electrical-responsive membranes can be either prepared by incorporating polyelectrolytes hydrogels onto membrane supports or by using conducting polymers as bulk materials and with other supporting systems.

Electrical-responsive membranes that are made from polyelectrolyte hydrogels exploit their ionizable side groups: the side groups undergo reversible contraction

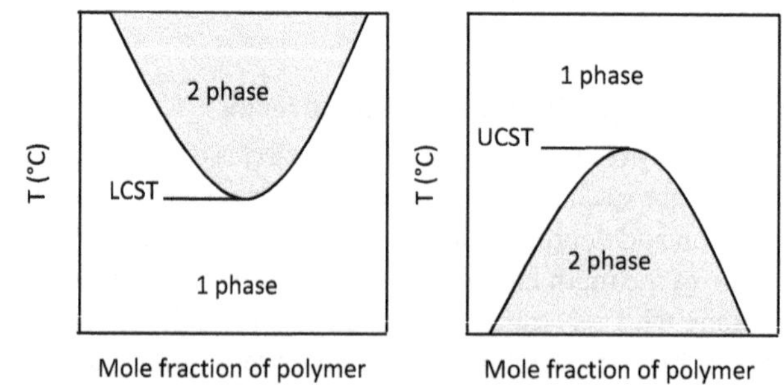

FIGURE 2.1 Phase diagrams for polymers exhibiting LCST behavior (left) and UCST behavior (right).

and expansion in response to an electric field (Haider et al., 2008). The transition between the expanded and coiled state allow the membrane pore to close and open, respectively, in response to the electrical stimulus. On the other hand, for conducting polymers the pore size actuation is triggered by changing the electrochemical state of the conducting polymer through doping (oxidation) and dedoping (reduction) (Loh et al., 1990). Doping of the conductive backbone with negative charges/anions enhance the conductivity, as dopants introduce a charge carrier into the polymer conjugated system by removing or adding electrons and relocalize them as polarons or bipolarons (part of a macromolecular chain containing two positive charges in a conjugated system). It is the polaron structure that is responsible for the transfer of the electric charge since, when an electrical potential is applied, the polarons and bipolarons start to move along the backbone, passing through the charge (Bredas et al., 1985; Ahmed et al., 2016).

2.2.1.4 Mechanism of Responsive Gating Systems based on pH, Temperature, and Electrical Stimuli

Gating membranes combine the advantages of the porous nature of substrates and the smart responsive gates for advanced performance and applications. The mechanism of action can be described as follows: first, by applying the external stimulus, hydrogen bonding and charge repulsion interactions between molecules occur, which cause swelling and de-swelling of polymer chains. Therefore, the membrane pores are selectively open or closed in response to the environmental change. This open/close tuneability can be explained for gating membranes as positively responsive gating and negatively responsive gating as shown in Figure 2.2 (Liu et al., 2016). For positively responsive gating membranes, permeability increases when the stimulus appears or increases. The responsive gates undergo swelling/shrinking transition with the application of the stimulus, and consequently the membrane pores open influencing the permeability. Negatively responsive gating results in reversed properties that decrease permeability when the stimulus increases.

For instance, thermo-responsive polymers with positively responsive gating swell at a temperature below the LCST due to hydrophilic interactions, thus the membrane pores are closed. Increasing the temperature above the LCST makes the pore open due to dominant hydrophobic interactions. On the contrary, UCST thermo-responsive

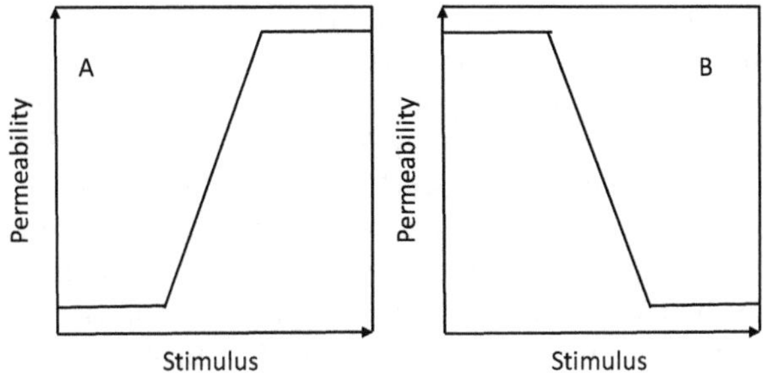

FIGURE 2.2 Positively (A) and negatively (B) responsive gating mechanism.

polymers such as interpenetrating networks of polyacrilamide and polyacrilic acid can be used as negatively responsive gates (Chu et al., 2005). When the temperature is lower than the UCST of the polymer, the pores open due to the formation of hydrogen bonding, while increasing the temperature above the UCST can cause the polymeric network to swell because of the breakage of the hydrogen bonding, leading to pore closing.

2.2.2 Properties of pH-Responsive Polymers

For both porous and non-porous membranes, pH-responsive functional groups such as carboxyl, pyridine, imidazole, and dibuthylamine groups are responsible for the conformational changes due to pH change. The carboxyl and the pyridine groups are most commonly used in the fabrication of pH-responsive membranes (Zhao et al., 2011). Pyridine groups swell at low pH, and they are often called acid-swellable groups or cationic groups (the same for imidazole and dibuthylamine groups) in contrast to alkali-swellable or anionic carboxyl groups. Carboxyl groups with formula –C(=O)OH or COOH possess a carbonyl and an hydroxyl moiety. At low pH, carboxyl groups are protonated while at high pH they dissociate into carboxylate anions. In the first case, hydrophilic interactions dominate leading to shrinkage in volume of the polymer, while in the second the carboxyl group dissociates into carboxylate ions because of dominant hydrophobic interactions that result in a high charge density, causing the polymer to swell. Examples of pH-sensitive polymers with anionic groups are poly(carboxylic acids) as poly(acrylic acid) (PAA) or poly(methacrilic acid) (PMA) (Table 2.1). PAA possess ionizable hydrophilic properties due to the transformation between the de-ionized form of the -COOH group and the ionized form of the $-COO^-$ group at pH value near a pKa of 4.7 (Hendri et al., 2001). PAA gates show negatively responsive behavior, leading to a volume shrinking of the polymeric chain at low pH, due to the formation of intermolecular hydrogen bonding between carboxylic groups (Luo et al., 2014).

Pyridine group is a basic heterocyclic compound with formula C_5H_4N; its structure is similar to a benzene ring but one of the carbons is replaced by a nitrogen atom. At low pH, the nitrogen of the pyridine group is protonated causing internal charge repulsion that leads to an expansion of the dimension of the polymer. Higher pH values reduce the charge repulsion between the less ionized protonated pyridine groups while the polymer-polymer interactions increase, leading to a decrease in the overall hydrodynamic diameter of the polymer. The most common example of a pyridine-containing polymer is poly(vinyl pyridine) (PVP) with a pKa in solution of 3.5 (Tantavichet et al., 2001). Typical examples of polyelectrolytes containing cationic functional groups are poly(N,N-dialkylaminoethyl methacrylates), poly(L-lisine) (PLL), poly(ethylenimine) (PEI), and chitosan (CS).

A wide range of polyelectrolytes can be used to impart pH responsiveness by exploiting protonation/deprotonation of their ionizable side chains. Some of them, such as chitosan and PMA, have a good biocompatibility and are preferred for biological applications (Jayakumar et al., 2011; Dupré de Baubigny et al., 2017) while other examples have also found application as ion exchange membranes for desalination and in separations (Wei et al., 2009; Zou et al., 2010).

TABLE 2.1
Common pH, Temperature, and Electrical-Responsive Polymers and Their Properties

Stimulus	Common Materials		
pH		pKa	General Properties
	Poly(acrylic acid) (PAA)	~4.7	Reversible conformational transition around pH 5 is driven by carboxylic groups (Swift et al., 2016).
	Poly(methacrylic acid)	~5.5	Ionizable hydrophilic polymer. At pH value lower than 5.5, COOH are not ionized and the polymer is in its collapsed state. Increasing the pH value, the charged COO^- groups repel each other, leading to polymer swelling (Zhang et al., 2000).
	Poly(2-(dimethylamino) ethyl methacrylate)	7.3–7.5	Water-soluble polymer sensitive to both temperature and pH changes (Fournier et al., 2007).
	Poly(vinylpyridine)	3.5–4.5	Based on basic and cheap monomers. pKa depends on measurement method (Zhao et al., 2011).
	Chitosan	6.5	Abundant natural polymer, common in the fabrication of drug delivery systems (Park et al., 2001; Sinha et al., 2004) and membranes in combination with other functional materials to form polymer blends or composites (Pandis et al., 2014).
Temperature		**LCST**	
	Poly(N-isopropylacrylamide) (PNIPAm)	32	LCST lies between body and room temperature. Non-biodegradable and limited in vivo biocompatibility. High glass transition temperature (140–150°C) limits its processability (Van Durme et al., 2004; Aguilar et al., 2014)
	Poly(N-vinylcaprolactam)	31	High biocompatibility and low toxicity Soluble in both polar and non-polar solvents (Cortez-Lemus et al., 2016)
Electrical Field/ Electrical voltage	Polyelectrolytes		Poly(acrylic acid) (PAA) with Poly(vinyl alcohol) (PVA) hydrogels is the most common combination (Wandera et al., 2010)

(*Continued*)

TABLE 2.1 (CONTINUED)
Common pH, Temperature, and Electrical-Responsive Polymers and Their Properties

Stimulus	Common Materials	Conductivity (S/cm)	General Properties
	Polyaniline (PANI)	10	Good stability and processability (Bhadra et al., 2009). Cheap monomer and easy synthesis (Negi et al., 2002).
	Polypyrrole (PPy)	100	Chemical stability and good processability (Bhadra et al., 2009). Good in vivo and in vitro biocompatibility (Balint et al., 2014).

2.2.3 Properties of Thermo-Responsive Polymers

Thermo-responsive polymers have garnered a lot of interest in the last three decades since temperature variations can be applied and controlled externally in a non-invasive manner. Thermo-responsive membranes are specifically attractive where the chemical modification of the feed solution is either not practicable or unwanted. As mentioned before, PNIPAM (Figure 2.3) is one of the most common thermo-responsive polymers and has been applied broadly to develop thermo-responsive membranes. This polymer is soluble in water at room temperature with the key advantage of reversible phase transition at the LCST around 32°C (Table 2.1) (Schild, 1992). When the environmental temperature is lower that the LCST, the amide groups of PNIPAM can bind to water molecules by hydrogen-bonding interactions and the polymer chains swell, reaching a hydrophilic state. However, above 32°C, the PNIPAM is in a shrunken, hydrophobic state, due to thermal dissociation of water molecules from the polymer chains that now associate preferentially with each other, thus precipitating (Chu et al., 2011). Membrane structure and barrier properties are altered by such phase transition.

Poly(N-vinylcaprolactam) (PVCL) is another thermo-responsive polymer (Figure 2.3 and Table 2.1) that has been reported to possess similar characteristics to PNIPAM, such as biocompatibility and a LCST around 31°C (Vihola et al., 2005). PVCL has been less researched, but like PNIPAM it does not degrade to low molecular weight

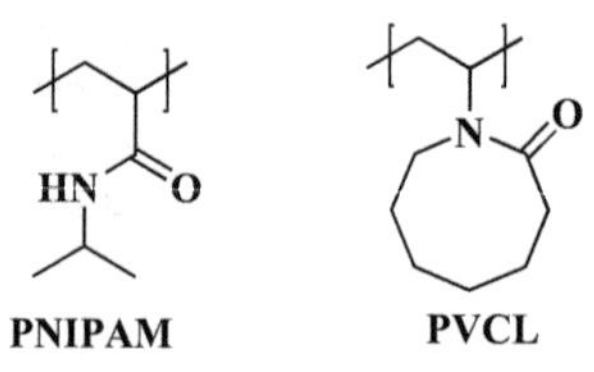

FIGURE 2.3 Examples of common temperature responsive polymers.

cytotoxic products, making both of them good candidates for biotechnological application.

Thermo-responsive polymers with UCST behavior in water are less common and generally based on polybetaines, which are zwitterionic polymers with a positive and a negative charge in every repeat unit (Kudaibergenov et al., 2006).

A recent example is reported in Section 2.3.2 where the latest progress in fabrication of thermo-responsive membranes as thermo-responsive hydrogels, copolymeric systems, and responsive gates are covered.

2.2.4 Properties of Electrical-Responsive Polymers

Generally, electrical-responsive polymers are intrinsically conducting polymers with a characteristic conjugated molecular structure (alternating double-bond system with planar conformation) of the polymer chain where the π-electrons are delocalized over the polymer backbone. The p-orbitals in the series of π-bonds overlap each other, and the electrons freely move between atoms (Ahmed et al., 2016). Representative conducting structures include polyaniline (PANI), polypyrrole (PPy), polyacetylene and polythiophene. Among these, PANI and PPy are widely investigated in membrane fabrication and tissue engineering for their interesting properties such as chemical stability and good processability (Balint et al., 2014). These macromolecular conjugated structures become conductive after doping. Polyaniline has a promising future because of its ease of synthesis, low cost monomer, tunable properties, and better stability compared to other conducting polymers (Bhadra et al., 2009). The main structure of PANI is an alternation of single and double bonds (Figure 2.4(A) and Table 2.1).

Its electrochemical stability depends on the pH conditions and the counterion of the Brönsted acid used as dopant (Negi et al., 2002).

Polypyrrole has a tight and rigid structure consisting of the repetition of a five-member ring with the formula C_4H_4NH (Figure 2.4(B) and Table 2.1). The polymer possesses a good stability in air and water, as well as good conductivity (Wang et al., 2001). However the polymer is difficult to further process into membranes because of its insolubility and infusible processing nature, therefore deposition of the conductive layer on a support is often employed for membrane preparation.

In addition, composite membranes have been developed from polyelectrolytes or soluble polymers and polymer-carbon nanotubes (CNT) composites. Carbon nanotubes

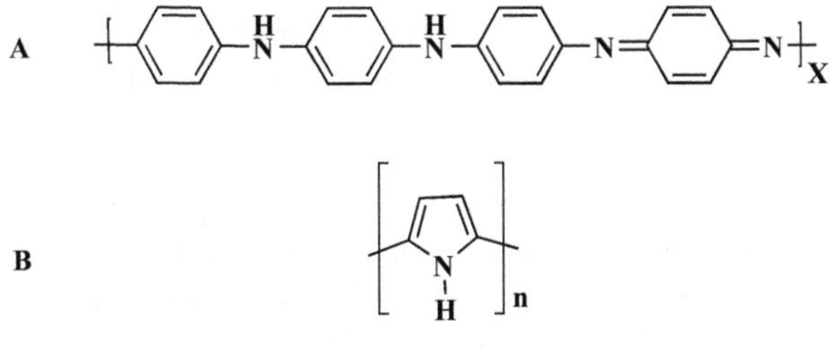

FIGURE 2.4 Schematic rapresentation of polyaniline (A) and polypyrrole (B) main chain structure.

are allotropes of carbon with a cylindrical structure and walls formed by graphene sheets with nanoscale diameters. CNTs membranes can be prepared as composites or self-supported systems; however, the latter tend to have irregular pores and limited control over the pore size. Moreover, potential leakage due to CNTs not being chemically bonded is reported as a drawback for these systems (Ahmed et al., 2016). Recent examples of the fabrication of composites carbon nanotube polymeric membranes will be reported in Section 2.3.4.

2.3 FABRICATION OF PH, TEMPERATURE, AND ELECTRICAL-RESPONSIVE MEMBRANES

Stimuli responsive membranes can be designed and fabricated with various styles and using three basic preparation approaches.

The responsive systems can be developed by (i) pre-synthesizing the polymers or copolymers and then processing these into membranes, (ii) involving *in-situ* polymerization to form membrane films or interpenetrating polymer networks, and (iii) chemical or physical modification of either the membrane surface or the membrane pores with stimuli responsive units, which has been widely employed for the preparation of smart gating membranes (Liu et al., 2016). Generally, modification imparts functionality that enhances membrane performance, maintaining the useful properties of the base membrane and introducing responsive sites on its surface.

2.3.1 Preparation Methods for Stimuli-Responsive Membranes

In order to prepare membranes from stimuli-responsive polymers and copolymers, these materials can be used either as pure single phase systems, as components of blends, or as additives during membrane formation. During the fabrication of stimuli responsive membrane systems, classic techniques such as solvent casting and phase inversion are generally employed (Ren et al., 2011) while surface modification can be performed to obtain membranes with responsive gates. A brief introduction to these techniques is given below.

2.3.1.1 Solvent Casting

The solvent casting process involves dissolving stimuli-responsive polymers or copolymers in an appropriate solvent and casting the obtained solution on a flat surface. The solvent is then allowed to evaporate, and membranes are dried and cross-linked by annealing them (Wandera et al., 2010; Husson, 2012). Solvent casting is normally used with robust polymers and generates relatively thick membranes.

2.3.1.2 Phase Inversion

The phase inversion technique is the most popular preparation method, not only for classic polymeric membranes but also for stimuli-responsive membrane systems. A solution of stimuli-responsive polymers and copolymers is cast on a flat surface and then immersed in a non-solvent coagulant bath to enable membrane formation. Phase inversion generally produces much thinner membranes with less diffusive resistance compared to solvent casting.

2.3.1.3 Membrane Surface Modification

In this method, membrane surfaces are modified following two distinctive surface-selective approaches: the "grafting to" approach and the "grafting from" approach (Minko, 2008; Wandera et al., 2010). The first approach introduces end-functionalized polymer chains on the membrane surface, whereas in the "grafting from" approach, the grafting reaction proceeds by polymerization from the surface and polymer chains grow from initiator sites by monomer addition from solution. The final result is the formation of a polymer brush layer on the membrane surface that, in the case of stimuli sensitive membranes, introduce sensitive functionalities. Comparing the above two approaches, the "grafting from" is a more promising method in the synthesis of polymer brushes with a high grafting density because of the limitation of immobilization of polymer "grafting to" approach (Zhao et al., 2000).

The "grafting from" modification is done using methods that differ from the mechanism used for radical generation. The initiator can be a photo-initiator or a redox species as in the photo-initiated polymerization and redox-initiated polymerization case, or the reaction can be induced by a radiation. Plasma-graft-filling and atom transfer radical polymerization represent two additional approaches.

The surface modification of a membrane by the "grafting to" method is carried out by physical adsorption or chemical grafting. Physical adsorption involves coating the membranes with a stimuli-responsive polymer; alternatively, stimuli-responsive additives can be incorporated into a thin film polymer composite coating that will be placed on the membranes. The modification of membranes by chemical grafting results in immobilization of responsive macromolecules onto the membrane surface by reaction with membrane functional groups, or photo-generated on it, and a reactive group of the additive polymer.

Surface modified membranes can also be designed as membranes with grafted stimuli responsive surface or membranes with porous substrates and stimuli-responsive gates (Husson, 2012; Liu et al., 2016). The first type of membranes is prepared by grafting or coating responsive materials onto the substrate membrane surface, using chemical or physical methods. The substrates provide mechanical strength, and the grafted chains possess freely mobile ends that respond quickly to environmental stimuli. The second type are prepared by grafting functional macromolecules either onto the external membrane surface or both the external surface and the inside surface of pores. The grafted chains in the membrane pores have freely mobile ends, and their length and density can be controlled to adjust the pore switching characteristic (Zhao et al., 2000).

2.3.1.4 In Situ Polymerization of Responsive Membranes: Preparation of Interpenetrating Polymer Networks IPNs

Interpenetrating polymer networks (IPNs) are a class of high performance multi-component polymer material that have been used as responsive membranes. IPNs comprise two or more miscible polymers whose networks are at least partially interlaced on a molecular scale but are not covalently bonded to each other and cannot be separated unless their chemical bonds are broken (Sperling, 1981). The stimuli-responsive IPN system is formed by polymerizing a stimuli-sensitive monomer within a physically entangled copolymer in the presence of an initiator and a cross-linker (Wandera et al., 2010; Husson, 2012). Chain interpenetration does not involve chemical bonding; therefore, the two networks may retain their own properties while combined together, and each network responds to stimuli in an independent manner. Membranes prepared by IPNs possess synergistic effects by sharing the properties of both the polymers. A disadvantage associated with the use of the IPN is that sometimes the polymers are interpenetrated to such an extent that it becomes less responsive.

Each of the above mentioned methods have their advantages and disadvantages. By choosing the appropriate fabrication approach based on the stimuli-responsive polymer, copolymer, and polymer-additive mixture, membranes with desired mechanical properties, pore structure (porosity, pore size and pore-size distribution), barrier structure (symmetric versus asymmetric), and layer thickness(es) can be prepared.

2.3.2 Formation of pH-Responsive Membranes

In terms of configuration, pH-sensitive membranes can be fabricated as flat sheet membrane and hollow-fiber membranes, and in both cases the pH sensitive polymers are either incorporated in the membrane matrix to prepare blended membranes or coated on the membrane surface (Zhao et al., 2011).

Often, blended membranes are prepared from mixtures of copolymers and polymers. Poly(vinylidene fluoride) (PVDF) is a recurrent material in membrane fabrication, with excellent thermal and good chemical stability (Prest Jr et al., 1975), and has gained popularity in the fabrication of blended pH-responsive membranes. Functional modification of PVDF is necessary in order to overcome its hydrophobic properties, which make it easily fouled. A large number of the pH-responsive membranes reported in literature in the last 10 years are based on a PVDF matrix blended with amphiphilic copolymers that possess functional and hydrophilic segments able to improve membrane's properties. A PVDF-g-PMMA copolymer blended with PVDF has been reported (Yang et al., 2017) in the preparation of a pH-responsive membrane. The key point of the synthesis method was blending the base polymer (PVDF) with its derivatives and tailoring the membrane casting conditions (Liu et al., 2013). The same copolymer blended with PVDF has been used to produce a series of novel pH-sensitive microfiltration membranes via ultraviolet irradiation-induced atom transfer radical polymerization (Hua et al., 2014). This "grafting from" approach provides the control growth of the polymer chains from the membrane surface.

Furthermore, microfiltration membranes with excellent pH-sensitivity and pH-reversible behavior based on PVDF blended with a functional ter-polymer have been reported (Ju et al., 2014). Poly(methyl methacrylate-2-hydroxyethyl methacrylate-acrylic acid) (PMMA-HEMA-AA) is a combination of hydroxyethyl methacrylate, a neutral monomer with hydroxyl groups, and poly(acrylic acid), the most commonly used pH responsive polymer, and it has been synthesized by free radical solution polymerization and blended with PVDF via the phase separation process. The resulting blended membrane showed changes in the permeation rate in response to the changes in the pH due to the switching between stretched and collapsed states in the PAA-modified membrane that altered the apparent size of the micropores.

Novel pH-sensitive membranes have also been prepared exploiting surface segregation method by blending pH-sensitive amphiphilic copolymers with PES (Su et al., 2015). Surface segregation is presented as a facile and simple *in-situ* method to add pH responsive segments on the membrane surface. The copolymer system, made of Pluronic F127 and PMMA segments, $PMMA_n$-F127-$PMMA_{n,}$ is synthesized by free radical polymerization and used as surface segregation additive during the membrane formation process.

2.3.3 Formation of Thermo-Responsive Membranes

Preparation protocols of temperature-responsive membranes involve the use of copolymers and hydrogels as common materials as well as the functionalization of existing membranes to develop responsive gating systems (Xie et al., 2007). Polymeric groups can be grafted on the external membrane surface or both the membrane surface and inside the pores; hydrogels and micro- or nanospheres, not only polymeric chains, can also be used as responsive groups (Chu et al., 2011; Liu et al., 2016).

Very recently, pore-filling *N*-isopropylacrylamide (NIPAM) polymer hydrogels have been grafted onto track-etched polycarbonate (PC) membranes by plasma-induced graft copolymerization. The PC-*g*-PNIPAM membrane shows thermo-responsive gating characteristics dependent on the pore size change, since the swelling and de-swelling of the cross-linked PNIPAAM hydrogels in the temperature range of its LCST are responsible for the switching on and off of the pores (Wang et al., 2010).

Membranes with nanoscale pores are fabricated by attaching a layer of poly(ethylene glycol methyl ether methacrylate)-b-polystyrene-b-poly(ethylene glycol methyl ether methacrylate) (PMENMA-b-PS-b-PMENMA) to a PVDF macroporous supporting layer. The water flux increase at a temperature higher than the LCST suggests that the pore size is temperature controllable. This polymer possesses oligo(ethylene glycol) side chains where the number of ethylene glycol units affects the LCST, providing a means of tuning the thermo-responsivity (Tang et al., 2016).

Another recent example reports that a membrane constructed by grafting a UCST polymer, such as poly(sulfobetaine), onto graphene oxide sheets, can exhibit changes in ion permeability in a non-aqueous electrolyte (Shen et al., 2014) . The attractive interactions between zwitterions that form hydrophobic aggregates on the membrane are disrupted above the UCST, causing the polymer chains to uncoil and expose the zwitterions to the electrolyte. The ionic permeability is potentially influenced and lowered by the obstruction of the dissolved polymer chains.

2.3.4 Membranes with both pH and Temperature Responsive Behavior

Numerous membranes that are both pH- and thermo-responsive have been reported in literature. These systems can be synthesized following the same methods already explained. A lot of work has been done to combine pH and temperature sensitivities into a single membrane by copolymerizing two monomers or forming IPNs. Very recently, interpenetrating membranes composed of poly(N-isopropylacrylamide) thermo-sensitive polymer and hyaluronan pH-sensitive polymer have been reported.

PNIPAM-HA hydrogels membranes display increased swelling/de-swelling based on HA content and both temperature and pH response (Kamoun et al., 2017).

Membranes based on hydrogels of interpenetrating polymeric network have been fabricated from different combination of temperature sensitive and pH-responsive polymers, such as PNIPAM/PMMA (Zhang et al., 2000), PNIPAM/PAAm (Hebeish et al., 2015), and β-cyclodextrin copolymerized with 2-methylacrylic acid and N,N'-methylene diacrylamide (Yang et al., 2016).

2.3.5 Formation of Electro-Responsive Membranes

Electro-responsive membranes for different purposes are prepared from polymer composites, generally conducting polymers, or PELs, and from polymer composites incorporating conducting fillers (e.g., carbon nanotube). In general these materials can be used as surface coatings or bulk materials during the fabrication process.

Recently, the fabrication of a nanoporous membrane based on PPy doped with dodecylbenzenesulfonate has been reported. The conducting polymer is electro-polymerized on an anodized aluminium oxide membrane with a regular pore size that is then electrically actuated by changing the electrochemical state of PPy. (Jeon et al., 2011).

However, conducting polymers are often not soluble in common solvents and are difficult to cast; therefore, the deposition of conductive layers on porous supports is an extensively employed technique. Examples include PVDF/PANI blends (Wang et al., 2002), sulfonated poly(phenylsulfone) SPPSU (Dyck et al., 2002), and a mixture of poly(carbonate)-poly(pyrrole) (Hacarlioglu et al., 2003).

Recent examples also report preparation of flexible conductive bacterial cellulose/polypyrrole membranes via *in situ* chemical synthesis of PPy nanoparticles on the surface of cellulose nanofibrils (Tang et al., 2015) and polypyrrole/polyacrylonitrile (PPy/PAN) conducting electroactive polymer membranes for potential electro-separation of synthetic azodyes from simulated wastewaters (Karimi et al., 2014). Electrically conductive polymeric membranes have been fabricated using carboxylated multiwalled nanotubes and poly(vinyl alcohol) (PVA) active layers. These composite membranes exhibit high electrical conductivity, high permeate flux, and a hydrophilic surface (de Lannoy et al., 2012).

PVA has also been employed to fabricate robust and permeable PVA-CNT-COOH thin films via sequential deposition and the cross-linking method. The thin films have been coated onto ultrafiltration supports and present potential for application in reducing fouling of UF treatment process (Dudchenko et al., 2014).

A CNT-PVDF mesh has been placed as a layer on a PES UF membrane in order to create a system able to capacitively-reduce negatively-charged organic matter fouling (Zhang et al., 2014).

2.4 CONCLUSION AND FUTURE DIRECTIONS

As seen from the above discussion, developing stimuli-responsive membranes with easy-to-tailor properties has extensively driven research in many fields. At present, pH and temperature responsive systems are widely investigated for the possibility of developing controlled drug delivery systems as well as chemical and bioseparations. These systems are based on polyelectrolytes and soluble polymers that undergo a reversible change in their properties that allows control of membrane performance. Electro-responsive membranes are fabricated from conducting polymers that possess a characteristic conjugated structure in which electrons can move freely. Changing the electrochemical states of these conducting polymers leads to changing their membrane pore size accordingly. The membranes from the above mentioned polymer systems can be prepared from the desired stimuli-responsive polymer materials exploiting classic fabrication methods such as solvent casting and phase inversion, or alternatively by *in-situ* polymerization to form IPNs. In addition, surface modification of a pre-formed membrane can be performed using a "grafting to" or a "grafting from" approach. In general, phase inversion produces thinner membranes with less diffusive resistance compared to solvent casting, and IPNs can be less responsive due to the strong interactions among the polymeric chains. Depending on the required pore topology, barrier structure, and mechanical properties, each of these methods can be chosen to fabricate the stimuli-responsive membranes.

Considering the current state of art, there are a number of opportunities in design and synthesis of stimuli-responsive membranes, but the main challenges that remain are large scale production of membrane responsive systems as well as the development of smart membranes adaptable to different environments. Future efforts should focus first on implementing robust stimuli-responsive systems with long-term stable performance in large scale applications by investigating easily up-scalable fabrication processes. In addition, the next generation stimuli-responsive membranes should be able to mimic natural systems with flexible and adaptable properties. For pH- and thermo-responsive membranes, their potential application in the biomedical field is often prevented by biocompatibility and cytotoxicity limitations of the systems, which need to be carefully tested. Similarly, the downside for electrical-responsive membranes is the limited processability of conducting polymers that makes them difficult to cast, and hence the preparation of reliable and robust systems with controllable properties is challenging.

Potential applications of stimuli-responsive membranes range from advanced separations to drug delivery and tissue engineering. In particular, electrical-responsive membranes can be designed as self-cleaning systems to prevent fouling by removing the unwanted layer and to extend the overall membrane performance. Another innovative application is the development of electro-catalytic systems, where electrical-responsive membrane can be coupled with chemical reactions to catalyze the reaction and separate the desired products or reactants. Furthermore, stimuli-responsive systems can be employed to regulate the transport and release of chemicals in biological systems where pH and temperature fluctuations normally occur. They can also be used as scaffolding to artificially grow cells and tissues exploiting the electrical stimulation triggered by conducting polymers.

Finally, membranes with switchable or responsive physiochemical properties have strong potential to improve many technological fields.

BIBLIOGRAPHY

Aguilar, M. R., San Román, J., 2014. Introduction to smart polymers and their applications. In M. R. Aguilar and J. San Román (Eds.), *Smart Polymers And Their Applications* (1–11). Woodhead Publishing.

Ahmed, F., Lalia, B. S., Kochkodan, V., Hilal, N., Hashaikeh, R., 2016. Electrically conductive polymeric membranes for fouling prevention and detection: A review. Desalination. 391, 1–15.

Azzaroni, O., Brown, A. A., Huck, W. T. S., 2007. Tunable wettability by clicking counterions into polyelectrolyte brushes. Advanced Materials. 19, 151–154.

Azzaroni, O., Moya, S., Farhan, T., Brown, A. A., Huck, W. T. S., 2005. Switching the properties of polyelectrolyte brushes via "hydrophobic collapse". Macromolecules. 38, 10192–10199.

Balint, R., Cassidy, N. J., Cartmell, S. H., 2014. Conductive polymers: Towards a smart biomaterial for tissue engineering. Acta Biomaterialia. 10, 2341–2353.

Bhadra, S., Khastgir, D., Singha, N. K., Lee, J. H., 2009. Progress in preparation, processing and applications of polyaniline. Progress in Polymer Science. 34, 783–810.

Bredas, J. L., Street, G. B., 1985. Polarons, bipolarons, and solitons in conducting polymers. Accounts of Chemical Research. 18, 309–315.

Chu, L.-Y., Li, Y., Zhu, J.-H., Chen, W.-M., 2005. Negatively thermoresponsive membranes with functional gates driven by zipper-type hydrogen-bonding interactions. Angewandte Chemie International Edition. 44, 2124–2127.

Chu, L., Xie, R., Ju, X., 2011. Stimuli-responsive membranes: Smart tools for controllable mass-transfer and separation processes. Chinese Journal of Chemical Engineering. 19, 891–903.

Cortez-Lemus, N. A., Licea-Claverie, A., 2016. Poly(n-vinylcaprolactam), a comprehensive review on a thermoresponsive polymer becoming popular. Progress in Polymer Science. 53, 1–51.

Dai, S., Ravi, P., Tam, K. C., 2008. Ph-responsive polymers: Synthesis, properties and applications. Soft Matter. 4, 435–449.

Darvishmanesh, S., Qian, X., Wickramasinghe, S. R., 2015. Responsive membranes for advanced separations. Current Opinion in Chemical Engineering. 8, 98–104.

Das, S., Banik, M., Chen, G., Sinha, S., Mukherjee, R., 2015. Polyelectrolyte brushes: Theory, modelling, synthesis and applications. Soft Matter. 11, 8550–8583.

de Lannoy, C. F., Jassby, D., Davis, D. D., Wiesner, M. R., 2012. A highly electrically conductive polymer–multiwalled carbon nanotube nanocomposite membrane. Journal of Membrane Science. 415, 718–724.

Dudchenko, A. V., Rolf, J., Russell, K., Duan, W., Jassby, D., 2014. Organic fouling inhibition on electrically conducting carbon nanotube-polyvinyl alcohol composite ultrafiltration membranes. Journal of Membrane Science. 468, 1–10.

Duong, P. H. H., Hong, P. Y., Musteata, V., Peinemann, K. V., Nunes, S. P., 2017. Thin film polyamide membranes with photoresponsive antibacterial activity. Chemistry Select. 2, 6612–6616.

Dupré de Baubigny, J., Trégouët, C., Salez, T., Pantoustier, N., Perrin, P., Reyssat, M., Monteux, C., 2017. One-step fabrication of ph-responsive membranes and microcapsules through interfacial h-bond polymer complexation. Scientific Reports. 7, 1265.

Dyck, A., Fritsch, D., Nunes, S. P., 2002. Proton-conductive membranes of sulfonated polyphenylsulfone. Journal of Applied Polymer Science. 86, 2820–2827.

Fournier, D., Hoogenboom, R., Thijs, H. M. L., Paulus, R. M., Schubert, U. S., 2007. Tunable ph- and temperature-sensitive copolymer libraries by reversible addition–fragmentation chain transfer copolymerizations of methacrylates. Macromolecules. 40, 915–920.

Gugliuzza, A., 2013. Intelligent membranes: Dream or reality? Membranes. 3, 151.

Hacarlioglu, P., Toppare, L., Yilmaz, L., 2003. Polycarbonate-polypyrrole mixed matrix gas separation membranes. Journal of Membrane Science. 225, 51–62.

Haider, S., Park, S.-Y., Lee, S.-H., 2008. Preparation, swelling and electro-mechano-chemical behaviors of a gelatin-chitosan blend membrane. Soft Matter. 4, 485–492.

Hebeish, A., Farag, S., Sharaf, S., Shaheen, T. I., 2015. Radically new cellulose nanocomposite hydrogels: Temperature and pH responsive characters. International Journal of Biological Macromolecules. 81, 356–361.

Hendri, J., Hiroki, A., Maekawa, Y., Yoshida, M., Katakai, R., 2001. Permeability control of metal ions using temperature- and pH-sensitive gel membranes. Radiation Physics and Chemistry. 60, 617–624.

Hua, H., Xiong, Y., Fu, C., Li, N., 2014. Ph-sensitive membranes prepared with poly(methyl methacrylate) grafted poly(vinylidene fluoride) via ultraviolet irradiation-induced atom transfer radical polymerization. RSC Advances. 4, 39273–39279.

Huang, R., Kostanski, L. K., Filipe, C. D. M., Ghosh, R., 2009. Environment-responsive hydrogel-based ultrafiltration membranes for protein bioseparation. Journal of Membrane Science. 336, 42–49.

Husson, S. M., 2012. Synthesis aspects in the design of responsive membranes. In D. Battacharyya, T. Schäfer, S. R. Wickramasinghe, S. Daunert (Eds.), *Responsive Membranes and Materials* (73–96). John Wiley & Sons, Ltd.

Jayakumar, R., Prabaharan, M., Sudheesh Kumar, P. T., Nair, S. V., Tamura, H., 2011. Biomaterials based on chitin and chitosan in wound dressing applications. Biotechnology Advances. 29, 322–337.

Jeon, G., Yang, S. Y., Byun, J., Kim, J. K., 2011. Electrically actuatable smart nanoporous membrane for pulsatile drug release. Nano Letters. 11, 1284–1288.

Ju, J., Wang, C., Wang, T., Wang, Q., 2014. Preparation and characterization of ph-sensitive and antifouling poly(vinylidene fluoride) microfiltration membranes blended with poly(methyl methacrylate-2-hydroxyethyl methacrylate-acrylic acid). Journal of Colloid and Interface Science. 434, 175–180.

Kamoun, E. A., Fahmy, A., Taha, T. H., El-Fakharany, E. M., Makram, M., Soliman, H. M. A., Shehata, H., 2017. Thermo-and ph-sensitive hydrogel membranes composed of poly(n-isopropylacrylamide)-hyaluronan for biomedical applications: Influence of hyaluronan incorporation on the membrane properties. International Journal of Biological Macromolecules. 106, 158–167.

Kaner, P., Hu, X. R., Thomas, S. W., Asatekin, A., 2017. Self-cleaning membranes from comb-shaped copolymers with photoresponsive side groups. ACS Applied Materials & Interfaces. 9, 13619–13631.

Karimi, M., Mohsen-Nia, M., Akbari, A., 2014. Electro-separation of synthetic azo dyes from a simulated wastewater using polypyrrole/polyacrylonitrile conductive membranes. Journal of Water Process Engineering. 4, 6–11.

Kudaibergenov, S., Jaeger, W., Laschewsky, A., 2006. Polymeric betaines: Synthesis, characterization, and application. *Supramolecular Polymers Polymeric Betains Oligomers* (157–224).

Lalia, B. S., Ahmed, F. E., Shah, T., Hilal, N., Hashaikeh, R., 2015. Electrically conductive membranes based on carbon nanostructures for self-cleaning of biofouling. Desalination. 360, 8–12.

Liu, B., Chen, C., Li, T., Crittenden, J., Chen, Y., 2013. High performance ultrafiltration membrane composed of pvdf blended with its derivative copolymer pvdf-g-pegma. Journal of Membrane Science. 445, 66–75.

Liu, Z., Wang, W., Xie, R., Ju, X.-J., Chu, L.-Y., 2016. Stimuli-responsive smart gating membranes. Chemical Society Reviews. 45, 460–475.

Loh, I.-H., Moody, R. A., Huang, J. C., 1990. Electrically conductive membranes: Synthesis and applications. Journal of Membrane Science. 50, 31–49.

Luo, F., Xie, R., Liu, Z., Ju, X.-J.,Wang, W., Lin, S., Chu, L.-Y., 2015. Smart gating membranes with in situ self-assembled responsive nanogels as functional gates. Scientific Reports. 5, 14708.

Luo, T., Lin, S., Xie, R., Ju, X.-J., Liu, Z., Wang, W., Mou, C.-L., Zhao, C., Chen, Q., Chu, L.-Y., 2014. Ph-responsive poly(ether sulfone) composite membranes blended with amphiphilic polystyrene-block-poly(acrylic acid) copolymers. Journal of Membrane Science. 450, 162–173.

Minko, S., 2006. Responsive polymer brushes. Journal of Macromolecular Science, Part C. 46, 397–420.

Minko, S., 2008. Grafting on solid surfaces: "Grafting to" and "grafting from" methods. In M. Stamm (Ed.), *Polymer Surfaces and Interfaces: Characterization, Modification And Applications* (215–234). Berlin, Heidelberg, Springer Berlin Heidelberg.

Miyata, T., Uragami, T., Nakamae, K., 2002. Biomolecule-sensitive hydrogels. Advanced Drug Delivery Reviews. 54, 79–98.

Negi, Y. S., Adhyapak, P. V., 2002. Development in polyaniline conducting polymers. Journal of Macromolecular Science, Part C. 42, 35–53.

Pandis, C., Madeira, S., Matos, J., Kyritsis, A., Mano, J. F., Ribelles, J. L. G., 2014. Chitosan–silica hybrid porous membranes. Materials Science and Engineering: C. 42, 553–561.

Park, S.-B., You, J.-O., Park, H.-Y., Haam, S. J., Kim, W.-S., 2001. A novel ph-sensitive membrane from chitosan-teos IPN; Preparation and its drug permeation characteristics. Biomaterials. 22, 323–330.

Park, Y., Gutierrez, M. P., Lee, L. P., 2016. Reversible self-actuated thermo-responsive pore membrane. Scientific Reports. 6, 10.

Pile, D. L., Hillier, A. C., 2002. Electrochemically modulated transport through a conducting polymer membrane. Journal of Membrane Science. 208, 119–131.

Prest Jr, W. M., Luca, D. J., 1975. The morphology and thermal response of high-temperature-crystallized poly(vinylidene fluoride). Journal of Applied Physics. 46, 4136–4143.

Reece, D. A., Ralph, S. F., Wallace, G. G., 2005. Metal transport studies on inherently conducting polymer membranes containing cyclodextrin dopants. Journal of Membrane Science. 249, 9–20.

Ren, J., Wang, R., 2011. Preparation of polymeric membranes. In L. K. Wang, J. P. Chen, Y.-T. Hung and N. K. Shammas (Eds.), *Membrane and Desalination Technologies* (47–100). Totowa, NJ: Humana Press.

Rühe, J., Ballauff, M., Biesalski, M., Dziezok, P., Gröhn, F., Johannsmann, D., Houbenov, N., Hugenberg, N., Konradi, R., Minko, S., Motornov, M., Netz, R. R., Schmidt, M., Seidel, C., Stamm, M., Stephan, T., Usov, D., Zhang, H., 2004. Polyelectrolyte brushes. In M. Schmidt (Ed.), *Polyelectrolytes with Defined Molecular Architecture I* (79–150). Berlin, Heidelberg: Springer Berlin Heidelberg.

Schild, H. G., 1992. Poly(n-isopropylacrylamide): Experiment, theory and application. Progress in Polymer Science. 17, 163–249.

Shen, J., Han, K., Martin, E. J., Wu, Y. Y., Kung, M. C., Hayner, C. M., Shull, K. R., Kung, H. H., 2014. Upper-critical solution temperature (ucst) polymer functionalized graphene oxide as thermally responsive ion permeable membrane for energy storage devices. Journal of Materials Chemistry A. 2, 18204–18207.

Sinha, V. R., Singla, A. K., Wadhawan, S., Kaushik, R., Kumria, R., Bansal, K., Dhawan, S., 2004. Chitosan microspheres as a potential carrier for drugs. International Journal of Pharmaceutics. 274, 1–33.

Sperling, L. H., 1981. An introduction to polymer networks and ipns. In *Interpenetrating Polymer Networks and Related Materials* (1–10). Boston, MA: Springer US.

Su, Y., Li, C., 2007. Tunable water flux of a weak polyelectrolyte ultrafiltration membrane. Journal of Membrane Science. 305, 271–278.

Su, Y. L., Liu, Y., Zhao, X. T., Li, Y. F., Jiang, Z. Y., 2015. Preparation of pH-responsive membranes with amphiphilic copolymers by surface segregation method. Chinese Journal of Chemical Engineering. 23, 1283–1290.

Sun, X., Wu, J., Chen, Z., Su, X., Hinds, B. J., 2013. Fouling characteristics and electrochemical recovery of carbon nanotube membranes. Advanced Functional Materials. 23, 1500–1506.

Swift, T., Swanson, L., Geoghegan, M., Rimmer, S., 2016. The pH-responsive behaviour of poly(acrylic acid) in aqueous solution is dependent on molar mass. Soft Matter. 12, 2542–2549.

Tang, L., Han, J., Jiang, Z., Chen, S.,Wang, H., 2015. Flexible conductive polypyrrole nanocomposite membranes based on bacterial cellulose with amphiphobicity. Carbohydrate Polymers. 117, 230–235.

Tang, Y., Ito, K., Hong, L., Ishizone, T., Yokoyama, H., 2016. Tunable thermoresponsive mesoporous block copolymer membranes. Macromolecules. 49, 7886–7896.

Tantavichet, N., Pritzker, M. D., Burns, C. M., 2001. Proton uptake by poly(2-vinylpyridine) coatings. Journal of Applied Polymer Science. 81, 1493–1497.

Tufani, A., Ince, G. O., 2017. Smart membranes with pH-responsive control of macromolecule permeability. Journal of Membrane Science. 537, 255–262.

Van Durme, K., Van Assche, G., Van Mele, B., 2004. Kinetics of demixing and remixing in poly(n-isopropylacrylamide)/water studied by modulated temperature dsc. Macromolecules. 37, 9596–9605.

Vihola, H., Laukkanen, A., Valtola, L., Tenhu, H., Hirvonen, J., 2005. Cytotoxicity of thermosensitive polymers poly(n-isopropylacrylamide), poly(n-vinylcaprolactam) and amphiphilically modified poly(n-vinylcaprolactam). Biomaterials. 26, 3055–3064.

Wandera, D., Wickramasinghe, S. R., Husson, S. M., 2010. Stimuli-responsive membranes. Journal of Membrane Science. 357, 6–35.

Wang, L.-X., Li, X.-G., Yang, Y.-L., 2001. Preparation, properties and applications of polypyrroles. Reactive and Functional Polymers. 47, 125–139.

Wang, P., Tan, K. L., Kang, E. T., Neoh, K. G., 2002. Preparation and characterization of semi-conductive poly(vinylidene fluoride)/polyaniline blends and membranes. Applied Surface Science. 193, 36–45.

Wang, W., Tian, X., Feng, Y., Cao, B., Yang, W., Zhang, L., 2010. Thermally on–off switching membranes prepared by pore-filling poly(n-isopropylacrylamide) hydrogels. Industrial & Engineering Chemistry Research. 49, 1684–1690.

Wei, Q., Li, J., Qian, B., Fang, B., Zhao, C., 2009. Preparation, characterization and application of functional polyethersulfone membranes blended with poly (acrylic acid) gels. Journal of Membrane Science. 337, 266–273.

Wen, H. Y., Gao, T., Fu, Z. Z., Liu, X., Xu, J. T., He, Y. S., Xu, N. X., Jiao, P., Fan, A., Huang, S. P., Xue, W. M., 2017. Enhancement of membrane stability on magnetic responsive hydrogel microcapsules for potential on-demand cell separation. Carbohydrate Polymers. 157, 1451–1460.

Xie, R., Li, Y., Chu, L.-Y., 2007. Preparation of thermo-responsive gating membranes with controllable response temperature. Journal of Membrane Science. 289, 76–85.

Yang, B., Yang, X., Liu, B., Chen, Z., Chen, C., Liang, S., Chu, L.-Y., Crittenden, J., 2017. Pvdf blended pvdf-g-pmaa ph-responsive membrane: Effect of additives and solvents on membrane properties and performance. Journal of Membrane Science. 541, 558–566.

Yang, K., Wan, S., Chen, B., Gao, W., Chen, J., Liu, M., He, B., Wu, H., 2016. Dual ph and temperature responsive hydrogels based on β-cyclodextrin derivatives for atorvastatin delivery. Carbohydrate Polymers. 136, 300–306.

Zhang, J., Peppas, N. A., 2000. Synthesis and characterization of ph- and temperature-sensitive poly(methacrylic acid)/poly(n-isopropylacrylamide) interpenetrating polymeric networks. Macromolecules. 33, 102–107.

Zhang, Q., Vecitis, C. D., 2014. Conductive cnt-pvdf membrane for capacitive organic fouling reduction. Journal of Membrane Science. 459, 143–156.

Zhang, Z., Xie, G. H., Xiao, K., Kong, X. Y., Li, P., Tian, Y., Wen, L. P., Jiang, L., 2016. Asymmetric multifunctional heterogeneous membranes for ph- and temperature-cooperative smart ion transport modulation. Advanced Materials. 28, 9613–9619.

Zhao, B., Brittain, W. J., 2000. Polymer brushes: Surface-immobilized macromolecules. Progress in Polymer Science. 25, 677–710.

Zhao, C., Nie, S., Tang, M., Sun, S., 2011. Polymeric ph-sensitive membranes—A review. Progress in Polymer Science. 36, 1499–1520.

Zou, W., Huang, Y., Luo, J., Liu, J., Zhao, C., 2010. Poly (methyl methacrylate–acrylic acid–vinyl pyrrolidone) terpolymer modified polyethersulfone hollow fiber membrane with ph sensitivity and protein antifouling property. Journal of Membrane Science. 358, 76–84.

3 Recent Progress in the Fabrication of Electrospun Nanofiber Membranes for Membrane Distillation

Yun Chul Woo, Minwei Yao, Leonard D. Tijing, Sung-Eun Lee, and Ho Kyong Shon

CONTENTS

3.1 INTRODUCTION

3.1.1 Background of Membrane Distillation (MD)

Water and energy have been recognized as the top two major challenges in the current era. Nowadays, many freshwater resources are becoming unrenewable due to climate change and massive human activities. Moreover, the water shortage issues are much more serious in countries with bad climate and poor economic conditions, so lack of clean water is a great threat to the hygiene of local residents (Tijing et al., 2014b).

Desalination is a viable option to obtain a supply of freshwater stably for coastal countries that are short of fresh water, and, currently, reverse osmosis (RO) is widely applied in desalination treatment plant due to its relatively high energy efficiency compared with thermal processes (Drioli and Macedonio, 2010). However, the capital and maintenance costs of RO plants are high and large amounts of electrical energy are required for generating high pressure in the process. In addition, RO has negative impacts on global environment via carbon dioxide as the generation of electrical energy is mostly fossil fuel based. RO brine disposal is another major issue that is of wide concerned to the world due to its impact on local ecological system. Therefore, the development of new generations of technology is imperative for replacement of or integration with RO technology (Alkhudhiri et al., 2012).

Recently, forward osmosis (FO), capacitive deionization (CDI), membrane distillation (MD), pressure retarded osmosis (PRO), and reverse electrodialysis (RED) have been developed to replace or augment conventional seawater desalination technology (Drioli et al., 2015). Among these processes, MD has high potential to replace RO in the desalination process in terms of reduction of the total volume of brine and total cost (Alkhudhiri et al., 2013; Boubakri et al., 2014; Geng et al., 2014a; Tian et al., 2014).

The driving force of MD is generated by vapor pressure formed by a difference in temperature between solutions on both sides of a hydrophobic membrane (Fan and Peng, 2012). Evaporation of liquid water occurs at the membrane interface where it comes in contact with the hot solution (Dong et al., 2014). Under the vapor pressure difference, the vapor then diffuses through the pores to the other side of interface where it condenses by cool permeate side (Tomaszewska et al., 1994).

However, MD still faces two main challenges, namely a high energy requirement for heating the feed solution and inadequate membranes designed specifically for MD. The first challenge could be addressed by utilizing renewable energy, such as solar heat, or by employing waste or low-grade heat, due to the fact that MD can be operated at low temperatures. The lack of appropriate membranes especially designed for MD is another challenge (Drioli et al., 2015; Geng et al., 2014b; Khayet, 2011).

In MD application, pure water vapor can penetrate via a high porous hydrophobic membrane from a hot feed side to cool permeate side (Li et al., 2014). Currently, microfiltration (MF) membranes are used for MD because they are hydrophobic membranes with suitable pore size distributions. However, the MF membranes have limitations in terms of water permeability and salt rejection for a long-term MD operation due to their low hydrophobicity (Su et al., 2017). Thus, there is a strong need to improve the design and structure of current MD membranes. In this chapter, we present a new MD membrane design via an electrospinning technique, and we summarize the recent updates of the potential of nanofiber membranes for MD application, including the details on various strategies for nanofiber membrane fabrication. Finally, we suggest future outlooks to improve membrane performances in MD, to reduce total costs of the fabrication, and to simplify membrane modification steps.

3.1.2 History of MD Development

The first MD patent was filed by B.R. Bodell in 1963, and five years later, he developed another patent using sweeping gas membrane distillation (SGMD) for desalination with a novel apparatus. Both theoretical and experimental studies were mentioned in his paper. Additionally, his paper first addressed vacuum membrane distillation (VMD) as an alternative configuration in which vacuum was applied on the permeate side (Khayet and Matsuura, 2011). In 1967, M.E. Findley tried various membrane materials, including aluminum foil, cellophane, glass fibers, paper plate, diatomaceous earth mat, nylon, paper hot cup, and gum wood, to coat the membranes for a hydrophobic surface. Hydrophobic materials such as Teflon and silicone were examined as well. By doing these tests, he concluded that a long-life membrane lasting at high temperature could be economical. In the same year, the concept of using waste heat and solar heat for MD was proposed in the 2nd European Symposium on Fresh Water from the Sea held in Athens (Khayet and Matsuura, 2011).

Soon after, MD lost the focus of the research field due to its much lower permeation performance than reverse osmosis (RO) process. The field's interest in MD recovered in the early 1980s as novel membranes with better characteristics became available (e.g., Gore-Tex membranes). Innovative types of MD membrane designs and their relative apparatuses, including composite membranes that comprised of both hydrophobic and hydrophilic membranes, had also been developed. Later, a research group improved the dual-layer fabrication technique, which involved coating a thin non-porous hydrophilic layer on the top of the hydrophobic membrane. Additionally, fluoro-substituted vinyl polymers such as polytetrafluoroethylene (PTFE) and polyvinylidene fluoride (PVDF) were proposed for the hydrophobic layers (Khayet and Matsuura, 2011).

Subsequently, the academic, not necessarily commercial, interest in MD grew rapidly, which was apparent through the increasing number of papers referenced in reviews of the literature concerning MD. For example, the 1997 MD review by Lawson and Lloyd referenced only 87 papers; in only ten years' time, however, the number increased to 168 in the MD review by El-Bourawi (2006). As of 2015, the estimated number of published paper regarding MD in international journals is more than 500 (Ahmed et al., 2015).

3.1.3 Design of MD Membrane

As an emerging technology, MD has not yet been widely applied in the global water industry due to lack of suitable membranes for long-term operation. Qualities such as strong resistance against wetting and fouling are lacking in the membranes that are currently available on the market (Francis et al., 2014). As mentioned above, membranes designed for microfiltration are utilized in MD, and they are mainly made of PVDF due to its high hydrophobicity, good solubility in common solvent, and high resistance against chemicals and heat (Dong et al., 2014; Hwang et al., 2011; Liao et al., 2013b). Non-solvent-induced phase separation (NIPS) and thermally induced phase separation (TIPS) are the two most common approaches applied to fabricate membranes. However, these membranes are still not good enough for MD processes due to their low flux performance and susceptibility to wetting (Nghiem and Cath 2011; Goh et al., 2013; Song and Jiang, 2013; Ge et al., 2014). Thus, there is a growing trend of new approaches to membrane fabrication for MD. Zhang et al. (2011) noted that the main challenge for those who manufacture the membranes used in MD is to design features including both a porous structure and a superhydrophobic surface for good filtration performance as well as high liquid entry pressure (LEP) for long-term operation (Francis et al., 2013; Lalia et al., 2013; Song and Jiang, 2013; Zhang et al., 2011). Electrospun membranes possess many appropriate advantages including high hydrophobicity, high porosity, adjustable pore size, and membrane thickness, which make them attractive candidates as MD membranes (Alkhudhiri et al., 2012; Feng et al., 2013; Francis et al., 2013). Compared with NIPS and TIPS, electrospinning is a relatively simple technique to fabricate membranes. By applying high electric fields to a polymer solution, millions of fibers are formed and then join together to become a nonwoven membrane sheet, collected on the rotating or static collector (Feng et al., 2013, Tijing et al., 2014a). Though electrospun membranes have many attractive properties for MD, they also have some drawbacks that limit their performance compared with the membranes fabricated by casting methods, including relatively large pore sizes, low mechanical properties, and LEP. Therefore, there is a need to improve these characteristics without sacrificing high porosity and hydrophobicity through some approaches to membrane modification.

3.1.4 Purpose of the Chapter

The main purpose of this chapter is to introduce the electrospinning technique and to explain the electrospinning method in detail regarding solution preparations, operating parameters, and examples of the MD electrospun nanofiber membrane. In addition, we have fully researched and summarized specific methods of electrospun nanofiber membrane fabrication and modification in order to obtain a membrane with high LEP or hydrophobicity and water vapor flux performance in MD. Finally, we suggest future steps to improve membrane performances in MD, to reduce total costs of the fabrication, and to simplify membrane modification steps.

3.2 CONFIGURATIONS OF MEMBRANE DISTILLATION (MD)

Membrane distillation (MD), a non-isothermal membrane separation technology, has been developed for more than 50 years (Alkhudhiri et al., 2012). However, MD still lacks adequate industrial implementations, as it requires further research (Tijing et al., 2014a). The separation of liquid from impurities are driven by the partial vapor pressure caused by the temperature difference between the feed side and the permeate side. Due to the uniqueness of this process, MD is fundamentally different from other membrane processes in which liquid molecules and not vapor pass through the membrane. Therefore, MD has some unique advantages (e.g., nearly 100% rejection of non-volatile impurities), and it is widely accepted as one of the next generation membrane separation processes (Peñate and García-Rodríguez, 2012).

Figure 3.1 describes the schematic diagram of the four different MD configurations: direct contact membrane distillation (DCMD), vacuum membrane distillation (VMD), sweeping gas membrane distillation (SGMD), and air gap membrane distillation (AGMD) (Wang and Chung, 2015). Several researchers have conducted massive studies and compared the four MD configurations in terms of energy efficiency, heat loss (HL), flux performance, and more (Essalhi and Khayet, 2012; 2014; Fan and Peng, 2012).

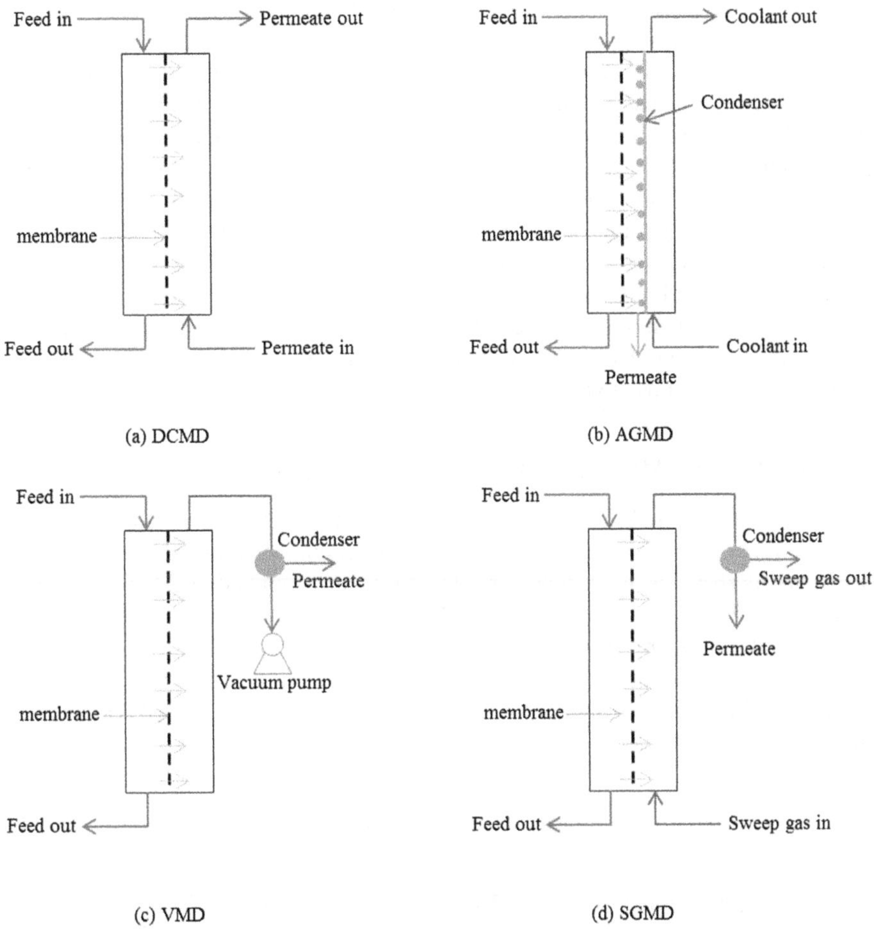

FIGURE 3.1 Different configurations of membrane distillation (MD) application: (a) Direct contact MD (DCMD), (b) Air gap MD (AGMD), (c) Vacuum MD (VMD), and (d) Sweep gas MD (SGMD).

SGMD was developed and patented at the earliest stage of MD research. However, until 2014, the number of papers published regarding applications of SGMD was the lowest among all types of the configurations (Tijing et al., 2014a). The need for an external condenser and additional cold sweeping gas equipment contributed to its unpopularity due to the increase in both cost and complexity of system design.

Similar to SGMD, VMD configuration requires an external condenser, which can result in higher design complexity and costs. Nevertheless, VMD has some benefits such as high thermal efficiency, which means the conductive heat loss (HL) through the membrane is minimalized as the applied vacuum on the permeate side leads to effective insulation (Fan and Peng, 2012). However, higher LEP of the applied membranes is required than other configurations, as the vacuum applied on the membrane greatly increases the pressure on the membrane.

The most studied configuration to date is DCMD due to its simple setup: both membrane and condensation plate can be incorporated into one single MD module. The mass transfer coefficient of DCMD is usually higher than other configurations as there is no air, gas, or other medium between the condenser and membrane; therefore, the vapor can condense directly after it passes through the membrane (Manawi et al., 2014). However, due to lack of insulation, DCMD has the highest HL among all the configurations, leading to higher requirement of membrane fabrication and module design to improve its thermal energy efficiency (Zhang et al., 2013a).

The fourth configuration is AGMD, which is the second most studied. Similar to DCMD, the condenser plate can be installed in the module along with the membrane. The only difference between AGMD and DCMD is an additional narrow air gap between the membrane and condensation plate. The vapor condenses on the condensation plate via natural convection in the air gap after diffusing through the membrane (Alkhudhiri et al., 2013).

3.3 ESSENTIAL MD MEMBRANE CHARACTERISTICS

To fabricate a suitable MD membrane, several significant factors should be considered. A proper MD membrane should be high hydrophobicity, liquid entry pressure (LEP) and porosity, as narrow as pore size distribution, and adoptable thickness to prevent membrane pore wetting and to enhance water vapor flux in MD (Table 3.1).

LEP is the pressure where the liquid water begins to thoroughly pass through the membrane pores as the penetrating water droplet can be observed. The values of LEP can be calculated by the following Laplace-Young equation (Alkhudhiri et al., 2012):

$$LEP = (-\,4\,B\,\sigma\cos\theta)/d_{max}, \tag{3.1}$$

where B is a pore geometric factor (B = 1 for cylindrical pores), σ is the surface tension of the solution, θ is the contact angle between the membrane surface and contact solution, and d_{max} is the maximum diameter of the pore in the membrane. High LEP is required for better wetting resistance and hence better permeation stability in long-term operation (Gryta et al., 2009).

TABLE 3.1
Significant Factors of the MD Membrane

Characteristics	Details	Remarks
LEP[a]	- High LEP[a] leads to better wetting resistance and better permeation stability in MD[b] - LEP[a] is related to a maximum pore size, contact angle, surface tension of the feed solution, and pore geometric factor	The Laplace-Young equation (Eq. 3.1) written in Section 2.3
PSD[c]	- A narrow PSD is preferred in terms of stable water vapor flux in MD[b] - A membrane with a broad PSD is likely to suffer pore wetting	- 0.1 μm ≤ PSD[c] ≤ 0.6 μm
Porosity	- A high porosity leads to high water vapor flux in MD[b] as high as possible	- Commercial PVDF[d] membranes have a porosity of 70%[e]
Thickness	- Membrane thickness determines the mass transfer resistance - Proper thickness in various operation scenarios must be determined	- Commercial PVDF[d] membranes haves a thickness of 107.4 μm[e]
Hydrophobicity	- Only vapor are is allowed to penetrate a high hydrophobic membrane in MD[b] - Membranes fabricated by electrospinning technique have improved hydrophobic properties	- PVDF[d], PP[f], and PTFE[g] have been usually used for membrane fabrication due to their low surface tension (high hydrophobicity) - Contact angle ≥ 90°

[a] LEP: Liquid entry pressure.
[b] MD: Membrane distillation.
[c] PSD: Pore size distribution.
[d] PVDF: Polyvinylidene fluoride.
[e] Woo et al., 2017b.
[f] PP: Polypropylene.
[g] PTFE: Polytetrafluoroethylene.

The mean and maximum pore sizes and pore size distribution (PSD) are significant factors in determining the water vapor flux performance of MD. To fabricate high water vapor permeable membrane, the range of pore sizes should be narrow for a uniform pore structure of the membrane (Lee et al., 2017a). Thus, PSD plays a primary role in the MD application although very few studies have investigated the effect of PSD on MD performance. High water vapor flux in MD and LEP of the membrane could be optimized by modifying pore size. To achieve both improved water flux and no pore wetting, Liao et al. suggested that a maximum pore size below 0.6 μm on a membrane is a suitable range for MD (Liao et al., 2013b).

Porosity is relevant to the void volume fraction in the membrane, which greatly affects the MD water vapor flux performance. To achieve more evaporation surface area and/or pore channels for molecular diffusion, porosity should be as high as possible. Therefore, high porosity leads to high water vapor permeate flux performance in MD application (Woo et al., 2016a, 2017a; Yao et al., 2017).

The membrane thickness has a significant effect on water flux performance, which then determines the mass transfer resistance (Cui et al., 2014). Generally, a low water permeate flux is caused by an increased mass transfer resistance due to the increased thickness of a membrane. Similarly, a thin membrane can have a high water permeate flux due to its low mass transfer resistance as well as short vapor transport paths. However, although a thin membrane has a lower mass transfer resistance, the thermal resistance is also reduced, which results in a low interface temperature difference. Thus, for MD membranes, a suitable thickness should be determined to find a balance between water flux performance and thermal resistance.

MD membranes should be fabricated with highly hydrophobic materials to prevent a penetration of liquids (Lin et al., 2014). Only vapor can penetrate highly hydrophobic MD membranes. The chemical and physical compositions of the materials and the geometrical structure determine the wettability of the membrane surface. In general, polypropylene (PP), PTFE, and PVDF are generally used to fabricate hydrophobic porous MD membranes in recent decades due to their good thermal and chemical resistance, which is required for MD application. Among these materials, PTFE MD membranes show the highest hydrophobic property on the membrane surface, which means that it has the lowest surface free energy (Shim et al., 2015). To enhance the hydrophobicity of the membrane surface, low surface free energy materials should be used. Another method to improve the hydrophobicity of the membrane is to apply surface modification techniques that increase the surface roughness with hierarchical nanostructure (Razmjou et al., 2012).

3.4 ELECTROSPINNING TECHNIQUE

Sharing the same equipment setup, electrospinning is considered a special state of electrospraying when certain conditions are met (Ahmed et al., 2015). Electrospinning involves applying a high voltage electric field to the polymer solution or melted polymer to form a solution jet. Fibers ranging from a few micrometers to nanometers can be obtained after the "whipping" process when the jet is elongated and the solvent evaporates, and the nanofibers are collected on the grounded collector and form nonwoven mats. The electrospun membranes can provide very large specific surface area, high porosity (>80%), and high degree of interconnection, which make them very suitable for MD process (Lalia et al., 2013). Moreover, the properties of an electrospun membrane, including its thickness, pore size, and porosity, can be controlled by changing the parameters in the electrospinning process (Liao et al., 2014a). Furthermore, additional required characteristics for the specific applications can easily be applied to the membrane through modifications during or after the fabrication process (Frenot and Chronakis, 2003). The electrospinning device consists of three parts (Figure 3.2): (i) the high voltage supply (0–30 kV), (ii) the syringe containing the polymeric solution on a pressure syringe pump with a metallic tip, and (iii) the rotating drum collector that collects the fibers (Baji et al., 2010). As the name

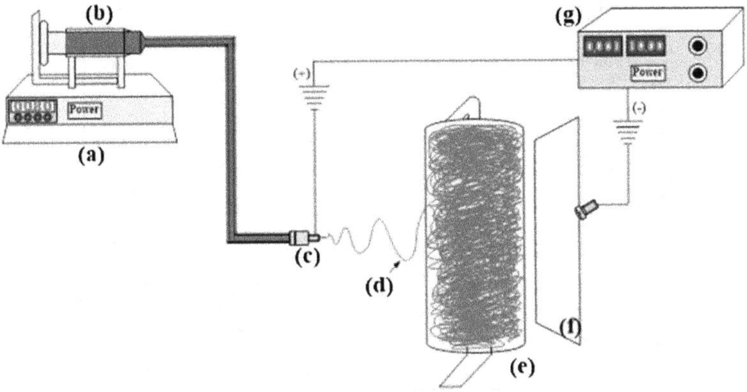

FIGURE 3.2 Schematic diagram of the electrospinning system: (a) syringe pump, (b) polymer solution, (c) tip, (d) nanofiber, (e) collector, (f) plate, and (g) high voltage power supply. (Adapted from Woo, Y.C. et al., Desalination. 403, 187–198, 2017.)

suggests, a syringe pressure pump pressurizes a syringe containing the polymeric solution, and the high voltage supply is applied to the metallic tip. When the applied electric field is high enough to overcome the surface tension of the solution in the metallic tip, a Taylor cone shape droplet of the solution will form on the metallic tip and fine nanofibers will be produced (Peining et al., 2012). The produced nanofibers are ejected from the tip and deposited on the rotating drum collector. The produced fibers can be greatly elongated by electrostatic repulsion and Coulombic forces and the solvent evaporates during the process.

3.4.1 Development of Electrospinning

Tracing the technique back to the 16th century, William Gilbert discovered the phenomenon of electrospraying when he placed an electrically charged piece of amber close to a droplet of water. Electrospinning, both the theory and the practice, were extensively studied from the beginning of the 20th century and was commercialized soon after when several important patents were registered in the years 1934–1944. It is interesting to note that the former Soviet Union had applied the electrospinning technique to fabricate a battlefield smoke filter for gas masks in military equipment beginning in the 1940s. In the 1960s, Sir Geoffrey Ingram Taylor developed the mathematical models of electrospinning; therefore, the shape of the extended fluid in front of the spinneret was named after him to honor his contributions (Frenot and Chronakis, 2003; Ahmed et al., 2015).

In the 1990s, some research groups found that many organic polymers could be electrospun into micro- or nanofibers. From then on, electrospinning garnered much attention as it was as a versatile technique for a broad range of applications. Since then, the number of papers regarding electrospinning increased greatly each year (Figure 3.3).

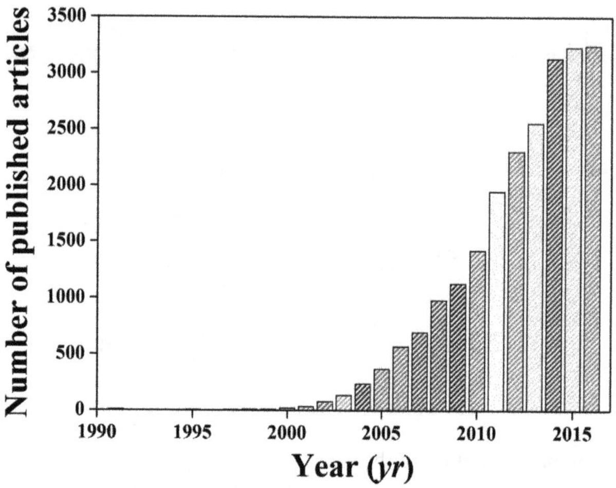

FIGURE 3.3 Since 1990, annual publication on the electrospinning as searched via Scopus (keywords: electrospinning and electrospun nanofiber membrane).

3.4.2 Mechanism of Electrospinning Technique

When the polymer solution is discharged under the pressure driven by the syringe pumps, the solution droplet at the tip of the spinneret will be under an electrostatic repulsion force, which is caused by the applied high-voltage-generated electrostatic field. A cone (conical shape distortion of the droplet) can be formed at the tip of the spinneret owing to the balance between solution surface tension and electrostatic repulsion force. When the electrostatic repulsion force is high enough to overcome the surface tension, a polymer jet can be ejected from the cone (at certain angles dependent on the solutions), which is called the Taylor cone (Frenot and Chronakis, 2003).

Under a high electrostatic field, the polymer jet is in a state of instability and experiences stretching and whipping (Tijing et al., 2014a). If the cohesion of polymer solution is high enough, there will be no breakup in the jet stream so that the electrospraying will not occur, and hence continuous solution fiber can be achieved. The solvent in the polymer jet will evaporate during the whipping process, and the evaporation rate is greatly affected by the solution properties and relative humidity (Eda et al., 2007). If the solvent evaporates adequately before the jet reaches the collector, the polymer string will eventually change from a liquid homogeneous solution into solid polymer nanofibers during the process of stretching and whipping, and nonwoven nanofiber membranes can be formed on the collector. These membranes are made of enormous amounts of nanofibers after several hours of the electrospinning operation.

3.4.3 Preparation of Polymeric Solution for Electrospinning

Feng et al. (2013) stated that several types of polymers worked well with electrospinning. The most commonly used polymers are polystyrene (PS), PVDF, polyacrylonitrile (PAN), polyethylene terephthalate (PET), and Nylon 6. In the application of MD process, high hydrophobicity is an essential requirement for membranes; therefore, a membrane made of a low surface energy polymer is more favorable. PVDF has relatively low surface energy and is easily dissolved in common organic solvents, while other hydrophobic polymer such as PTFE cannot be. Moreover, it is convenient to introduce a co-polymer into the PVDF polymer chains to further decrease the surface energy of the material, so the hydrophobicity of the membranes will be further increased. Recently, PVDF-co-HFP has been widely researched in the electrospinning process because PVDF-co-HFP membranes have better hydrophobicity and porosity than PVDF membranes, and the co-polymer remains highly solubile in common organic solvents (Ataollahi et al., 2012; Lalia et al., 2013). Solvent selection is a significant procedure for electrospinning process as well because it can affect the viscosity of the solution and thus the membrane morphology greatly. Generally, solvents can be classified into two types: strong and weak. A strong solvent, such as dimethylacetamide (DMAc), dimethylformamide (DMF), dimethyl sulfoxide (DMSO), and n-Methyl-2-pyrrolidone (NMP), has a higher boiling point and dielectric constant (Eda et al., 2007; Liu et al., 2011). Weak solvents, such as acetone and tetrahydrofuran (THF), have lower boiling points and dielectric constant. By controlling the ratio between the strong and weak solvents in the mixture, the researcher can precisely control the viscosity of the polymer and thus indirectly control the diameters of the electrospun fibers. Additionally, morphology of the electrospun membranes can be controlled in this way, because whether the morphology of membrane is beads dominant or fiber dominant is strongly dependent on the viscosity of the electrospinning solution (Tijing et al., 2014a).

3.4.4 Effect of Electrospinning Parameters

Parameters that can significantly affect the polymer morphology and characteristics have been extensively studied for understanding and improvement of electrospun membranes (Tijing et al., 2014a). In Liao et al.'s paper (2013b), the parameters that affected the properties of the nanofibers and membranes could be mainly classified into three groups: solution parameters, experimental parameters, and external environmental parameters. How each parameter affects morphology and characteristics of the electrospun membranes is summarized in Table 3.2.

3.4.4.1 Effect of Solution Parameters

Viscosity of the solution plays an important role in the effect on the morphology and characteristics of the electrospun membranes. Generally, a lower viscosity will result in an increase in nanofiber size and hence larger pore size (Liao et al., 2013b). There are, however, exceptions to the effect of viscosity on the electrospun membranes when changes of polymer concentration and molecular weight are involved. Increasing the concentration of the polymer can increase the viscosity of the solution, but it can increase the fiber size rather than decrease it (Frenot and Chronakis, 2003; Lalia et al., 2013). However, too high of a polymer concentration often makes the polymer solution block the needle and interrupts the process of continuous

TABLE 3.2
Summary of Electrospinning Parameters

Electrospinning Parameters	Details	Remarks
Solution parameters	- Viscosity of solutions plays an important role in the morphology and characteristics of the electrospun nano-fibrous membranes - Surface tension also affects the surface volume area, morphology, and characteristics - Solution conductivity has significant effect on the electrospinning process	- Low viscosity: increase in nanofiber size and pore size - High surface tension: decrease in nanofiber size and pore size - High solution conductivity: decrease in nanofiber size and pore size
Experimental parameters	- A proper voltage should be applied - An internal diameter of the electrospinning spinneret affects the thickness of a fabricated nanofiber - Tip-to-collector distance (TCD) has a direct influence on the electric field strength - Temperature of the solution plays an important role, which can affect the viscosity of the solution	- High voltage leads to a decrease in nanofiber size and pore size, but too low voltage is not strong enough to form polymer fibers - Decreased internal diameter of spinneret: thin thickness of the fabricated nanofiber
Environmental parameters	- Room humidity (RH) is one of the significant factors among all the parameters - Temperature of electrospinning chambers affects the membrane morphologies and characteristics	- Low RH: thick thickness of the nanofiber

electrospinning, while too low of a polymer concentration viscosity can increase the tendency to form beads and droplets (Liu et al., 2011; Pelipenko et al., 2013).

High surface tension can decrease the surface area per unit mass of fluid and affect the morphology and characteristics of membranes in the same way as viscosity. Increasing surface tension can result in a decrease in the membrane nanofiber diameter and membrane pore size. However, surface tension should be low enough to prevent the jet from collapsing to droplets before the solvent has evaporated (Frenot and Chronakis, 2003).

High solution conductivity leads to higher charges on the nanofibers in the electrospinning, which has a similar effect as power supply: the fiber diameter will be decreased due to additional stretching and elongation during the whipping process. Moreover, secondary jets can be formed as a result of an increase in the conductivity of polymer solution and thus the electrospun nanofiber size will be further decreased due to a decrease of solution volume in main jet (Pelipenko et al., 2013).

3.4.4.2 Effect of Experimental Parameters

Selecting the level of voltage to apply is important. Low voltage may be not strong enough to form the polymer solution jet, and increasing the voltage can reduce the diameter of solution fibers. However, voltage that is too high above the critical value may cause high frequency droplets due to strong pulling force on the solution jet, and beads will be formed on the membranes (Ahmed et al., 2015).

Decreasing the internal diameter of the spinneret can theoretically result in nanofibers with smaller diameters. However, if the internal diameter is too small, the polymer solution may easily block the spinneret, and the electrospinning jet cannot be formed (Tijing et al., 2014a).

Tip-to-collector distance (TCD) has a direct influence on the electric field strength and the flight time of the solution jet (Frenot and Chronakis, 2003). When the distance is too narrow, the solvent may not evaporate in a timely fashion, the fibers may fuse to each other, and the nanofiber structures can be lost. If the distance is too wide, the electric field will become too weak to form a continuous electrospinning jet, and beads and droplets may be the dominant morphology features on the membrane.

The temperature of the polymer solution plays an important role as well because it can affect the viscosity of solution. Generally, increasing the temperature of the solution can decrease the viscosity of the solution due to the increase in solubility of the solvent at a higher temperature (Tijing et al., 2014a).

3.4.4.3 Effect of Environmental Parameters

For an aqueous polymer solution, decreasing the room humidity (RH) will result in an increase in nanofiber diameter and pore size. When the RH is very low, the assumed rate of solvent evaporation will be high, thus increasing the local polymer concentration and decreasing the stretching of polymer chains (as there is less time for stretching). Therefore, large diameters of nanofiber can be obtained when the RH is low. Another possible explanation for the effect of RH on the aqueous polymer solution also relies on assumption: a decrease in fiber diameter can be caused by the formation of secondary jets due to an increase in conductivity at higher RH (Pelipenko et al., 2013).

The temperature of electrospinning chambers has effects on the membrane morphology and thus its characteristics. Increasing the temperature will decrease the local polymer viscosity and thus increase the elongation time, resulting in smaller nanofiber and pore size (Tijing et al., 2014a).

3.5 ELECTROSPINNING FOR MD MEMBRANE

Recently, flat-sheet and hollow fiber membranes fabricated by NIPS and TIPS have been studied. Both methods are still limited in their usefulness in MD; due to the low porosity and sponge-like structure of membranes fabricated via both methods, the membranes do not obtain high enough water permeate (Nejati et al., 2015). To solve this issue, the electrospinning technique has been used to fabricate membranes for use in MD.

Electrospun nanofiber membranes are suitable for the MD process due to their high hydrophobicity, high porosity, and proper pore sizes. Generally, the literature detailed improved permeation flux and salt rejection performance for nanofiber membranes compared to other conventional membranes (Lee et al., 2016b). However, issues of long-term stability and fouling formation have still not yet been clearly examined. Possible ways to address this issue is through the use of four different electrospun membrane designs and fabrication approaches: hydrophobic single-layer electrospinning, dual or triple-layer electrospinning, nano-materials embedded electrospinning, and surface modification after electrospinning.

3.5.1 Single-Layer Electrospun Nanofiber Membranes for MD

Similar to other fabrication techniques, the electrospun nanofiber membranes can also be designed to have different layer formations such as single-, dual-, and triple-layers (Figure 3.4 and Table 3.3). Initial reports focused on single-layer electrospun membranes. PVDF or PVDF-co-HFP (PH) is usually used for electrospun membranes due to its hydrophobicity and ease of fabrication (Yao et al., 2017). Lee et al., (2016a) used PVDF-co-HFP polymer to fabricate electrospun membranes for MD. Based on the results, a 20 wt% PVDF-co-HFP electrospun membrane showed flux ranging from 30 to 40 LMH for 50 hours in a DCMD test. During the DCMD operation, salt rejection sharply decreased after 30 hours of operation (Figure 3.5(a)). Generally, it is observed that single-layer neat PVDF or PVDF-co-HFP electrospun nanofiber membranes suffer severe pore wetting problems in MD process. To solve the problem, dual-layer or triple-layer electrospun membranes have been employed.

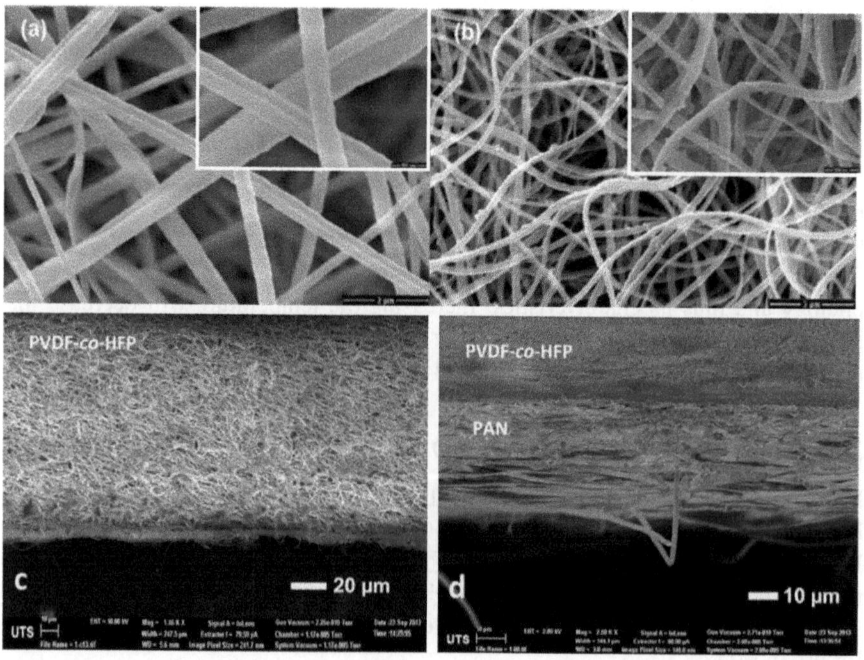

FIGURE 3.4 Surface SEM images of (a) 20 wt% PVDF-co-HFP and (b) 10 wt% PVDF-co-HFP with 10 wt% TiO_2 electrospun membranes. Cross-sectional SEM images of (c) PVDF-co-HFP single-layer and (d) PVDF-co-HFP/PAN dual-layer electrospun membranes. (From Tijing, L.D. et al., Chemical Engineering Journal. 256, 155–159, 2014; Lee, E.-J. et al., Journal of Membrane Science. 520, 145–154, 2016.)

TABLE 3.3
Reports in Literature on Electrospun Nanofiber Membrane for MD Application (Inlet Temperatures: Feed = 60°C, Permeate = 20°C)

MD Application	Layer Format	Active Layer	Pore Size (μm)	Porosity (%)	Contact Angle (°)	Flux (L/m^2 h)	Salt Rejection (%)	Reference
DCMD	Single-layer	PVDF	0.31	79	138	36	–	(Liao et al., 2013a)
DCMD	Single-layer	PVDF-co-HFP	0.73	~ 91	134	–	–	(Lalia et al., 2013)
AGMD	Single-layer	PVDF	–	–	–	4.2	–	(Feng et al., 2008)
DCMD	Dual-layer	PVDF/silica	0.36	64	150.0	16–27	99.99	(Liao et al., 2014a)
DCMD	Single-layer	PVDF-co-HFP with benzyltriethylammonium chloride (BTEAC)	0.42	–	~ 130	36	99.99	(Lee et al., 2016b)
DCMD	Single-layer	Matrimid 5218	2.15	–	130.0	~ 24	99.99	(Francis et al., 2013)
DCMD	Single-layer	PVDF	0.18	71.4	138	~ 10	–	(Liao et al., 2013b)
VMD	Single-layer	PTFE/PVA composite membrane	–	81.5	156.7	~ 10[a]	99.11	(Zhou et al., 2014)
DCMD	Single-layer	PVDF-co-HFP with nanocrystalline cellulose (NCC)	~ 0.2	70	~ 127	10.2–11.5	99	(Lalia et al., 2014)
DCMD	Single-layer	PVDF/SiO_2	0.23	79	152	25	99.99	(Su et al., 2017)
DCMD	Single-layer	PVDF with clay particles	0.64	81	154.2	~ 2	99.97	(Prince et al., 2012)
DCMD	Single-layer	PVDF-co-HFP	1.4	91	153	18	–	(Yao et al., 2017)
VMD	Single-layer	PVDF-SiO_2	0.22	55	121.5	15[b]	–	(Dong et al., 2015a)
DCMD	Single-layer	PVDF-co-HFP with carbon nanotubes			158.5	35	99.99	(Tijing et al., 2016)
AGMD	Single-layer	PVDF-co-HFP with graphene			162.x	22.6	100	(Woo et al., 2016b)

(Continued)

TABLE 3.3 (CONTINUED)
Reports in Literature on Electrospun Nanofiber Membrane for MD Application (Inlet Temperatures: Feed = 60°C, Permeate = 20°C)

MD Application	Layer Format	Active Layer	Pore Size (μm)	Porosity (%)	Contact Angle (°)	Flux (L/m² h)	Salt Rejection (%)	Reference
DCMD	Single-layer by dual-nozzle	PVDF-co-HFP with functionalized TiO_2 nanoparticles	0.76	89.8	153.4	40[c]	99.99	(Lee et al., 2017a)
DCMD	Single-layer	PVDF-co-HFP with fluorosilane coated TiO_2	0.75	91.6	149.0	37.8[c]	99.99	(Lee et al., 2016a)
DCMD	Single-layer	PTFE/PVA with PAN	0.79	81.1	143.6	14.59	99.8	(Huang et al., 2017)
DCMD	Single-layer	PVDF	0.68	85	142.8	~ 12	–	(Liao et al., 2014b)
	Dual-layer	PVDF/SiO_2	0.69	82	156.3	~ 19	99.99	
DCMD	Dual-layer	PVDF-co-HFP/PAN (hydrophobic/hydrophilic)	1.0	90	150	30	98.5	(Tijing et al., 2014b)
DCMD	Dual-layer	Polystyrene (PS) on PET support layer	0.76	69	150.2	~ 58	–	(Li et al., 2014)
DCMD	Dual-layer	PVDF with SiO_2 on nonwoven support (hydrophobic/hydrophilic dual-layer)	0.36	62	~ 150	21	99.99	(Liao et al., 2014a)

(Continued)

TABLE 3.3 (CONTINUED)
Reports in Literature on Electrospun Nanofiber Membrane for MD Application (Inlet Temperatures: Feed = 60°C, Permeate = 20°C)

MD Application	Layer Format	Active Layer	Pore Size (μm)	Porosity (%)	Contact Angle (°)	Flux (L/m^2 h)	Salt Rejection (%)	Reference
DCMD	Dual-layer	PVDF with surface modifying macromolecules (SMMs)/ PVDF phase-inversion support layer	0.06	–	148.4	10	99.98	(Prince et al., 2014b)
AGMD	Triple-layer (Casting with electrospinning)	PVDF	0.10	–	145.0	15.2	–	(Prince et al., 2013)
AGMD	Triple-layer	PVDF nanofiber (top)/ PVDF phase separation (middle)/ PET support (bottom)	0.1	–	145.02	~ 10	99.99	(Prince et al., 2013)

[a] Vacuum pressure is 30 kPa.
[b] Vacuum pressure is 9 kPa.
[c] Feed solution is 7.0 wt% NaCl.

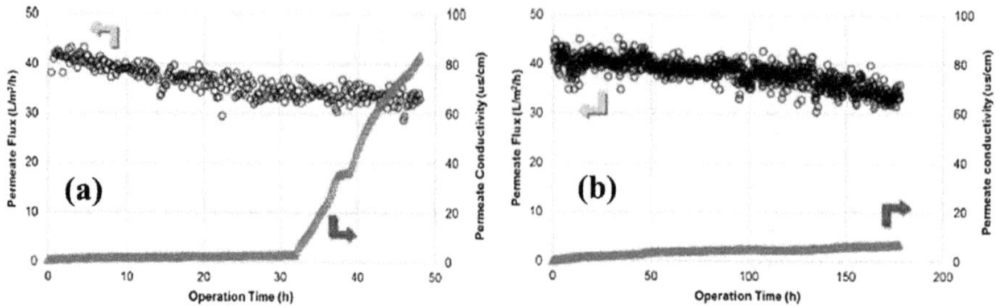

FIGURE 3.5 DCMD performance of (a) neat PVDF-co-HFP nanofiber membrane and (b) PVDF-co-HFP with 10 wt% TiO_2 nanofiber membranes. (Adapted from Lee, E.-J. et al., Journal of Membrane Science. 520, 145–154, 2016.)

3.5.2 Dual- or Triple-Layer Electrospun Nanofiber Membranes for MD

Numberous reports (Khayet et al., 2005; Prince et al., 2013; Tijing et al., 2014b; Woo et al., 2017b) noted that dual-layer membranes (hydrophobic/hydrophilic or hydrophobic/less hydrophobic) and triple-layer electrospun membranes had increased hydrophobicity of the active layer, thereby mitigating wetting issue (Figures 3.4(d) and 3.6(a)). In addition, a thinner hydrophobic active layer with a thicker hydrophilic layer (or a less hydrophobic layer) had decreased mass transfer resistance. Thus, the dual-layer and triple-layer electrospun membranes could achieve a higher water vapor flux performance with less wetting problems compared with single-layer electrospun membranes.

Figure 3.6(a) exhibits a novel designed hydrophobic/hydrophilic Janus-type dual-layer nanofiber membrane (Woo et al., 2017b). However, the hydrophobic and hydrophilic layers are not bound well in the dual-layer membrane. So, the heat-press post-treatment was used to create a good adhesion between the two layers. The heat-pressed dual-layer nanofiber membrane also exhibited increased mechanical properties, which is higher than the commercial PVDF membrane (GVHP, Millipore). Although the contact angle of the heat-pressed membrane was lower than the neat membrane (due to its smooth surface), it exhibited a thin thickness, uniform pore size, and narrow pore size distribution as well as increased LEP. Hence, the heat-pressed dual-layer membrane showed reduced mass transfer resistance and less heat transfer loss, leading to increased water vapor flux performance in MD application.

A triple-layer membrane is one of the other proper approaches to prevent pore wetting in MD (Prince et al., 2014a). Triple-layer membranes are composed of a PVDF with hydrophilic nanofiber bottom layer modified by a surface modifying macromolecule (SMM), a microporous PVDF intermediate phase-inverted layer, and a hydrophobic PVDF nanofiber top layer (Figure 3.6(b)). Based on the results, the triple-layer membrane had six times higher water vapor flux performance in DCMD than a neat electrospun nanofiber membrane. It is worth noting that the triple-layer membrane could be operated over 95 hours without wetting problems or decreased salt rejection performance. Therefore, the triple-layer membrane is regarded as a potential approach to prevent pore wetting problems while enhancing water vapor flux performance in MD.

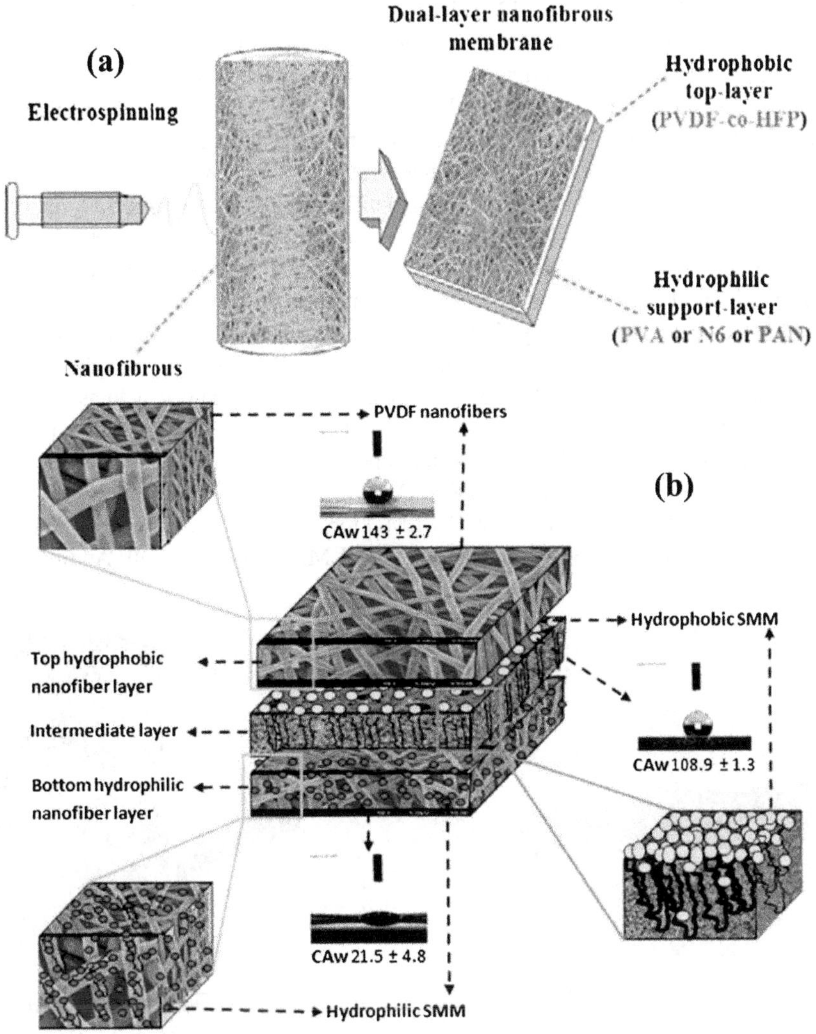

FIGURE 3.6 The configuration of (a) the dual-layer and (b) triple-layer membranes. (Adapted from Prince, J.A. et al., Scientific Reports. 4, 6949, 2014; Woo, Y.C. et al., Desalination. 403, 187–198, 2017.)

3.5.3 Nano-Materials Incorporated on/in Electrospun Nanofiber Membranes for MD

As mentioned above, single-layer PVDF electrospun membranes suffer severe wetting problems during MD operation, calling for modification of the membrane. To address this issue, several hydrophobic nanoparticles such as functionalized SiO_2 and TiO_2, clay particles, SMM, etc. have been individually incorporated in a single-layer electrospun membrane (Table 3.3). This resulted in greatly improved wetting resistance while maintaining the benefits of an electrospun membrane (such as high porosity) (Prince et al., 2012; Prince et al., 2014b; Lee et al., 2016a; Lee et al., 2017a). These studies indicated that hydrophobic nanoparticles incorporated with the electrospun membrane showed superhydrophobic properties (the contact angle was over 150°) on the membrane surface. Dumée et al. (2011) claimed that an increased hydrophobic property on the membrane led to not only prevention of wetting problems but also improved water vapor permeability in MD operation.

Li et al. (2015) developed a modification method using monodisperse SiO_2 nanoparticles with a size of 40 nm. The nanoparticles were first immersed in DMF and then mixed with PVDF/DMF solution in a weight ratio of 1:2, respectively. The contact angle and LEP of the PVDF/SiO_2 electrospun nanofiber membrane were 152.3 ± 2.0° and 1.65 ± 0.05 bar, respectively (Figure 3.7). Figure 3.7(A) and 3.7(B) show the TEM images of PVDF/SiO_2 electrospun nanofiber membranes. Figure 3.7(C) illustrates a set-up using a methyl orange PH indicator solution and hydrochloric acid as a volatile solution to evaluate the breathable performance of a PVDF/SiO_2 electrospun nanofiber membrane. Figure 3.7(E) exhibits the water vapor flux performance of the PVDF/SiO_2 electrospun nanofiber membrane in DCMD using 3.5 wt% NaCl as feed. It shows a high water vapor flux performance of 41.1 LMH for 24 hours, while the flux of the neat PVDF electrospun nanofiber membrane is 32.5 LMH (Figure 3.7(D)). Additionally,

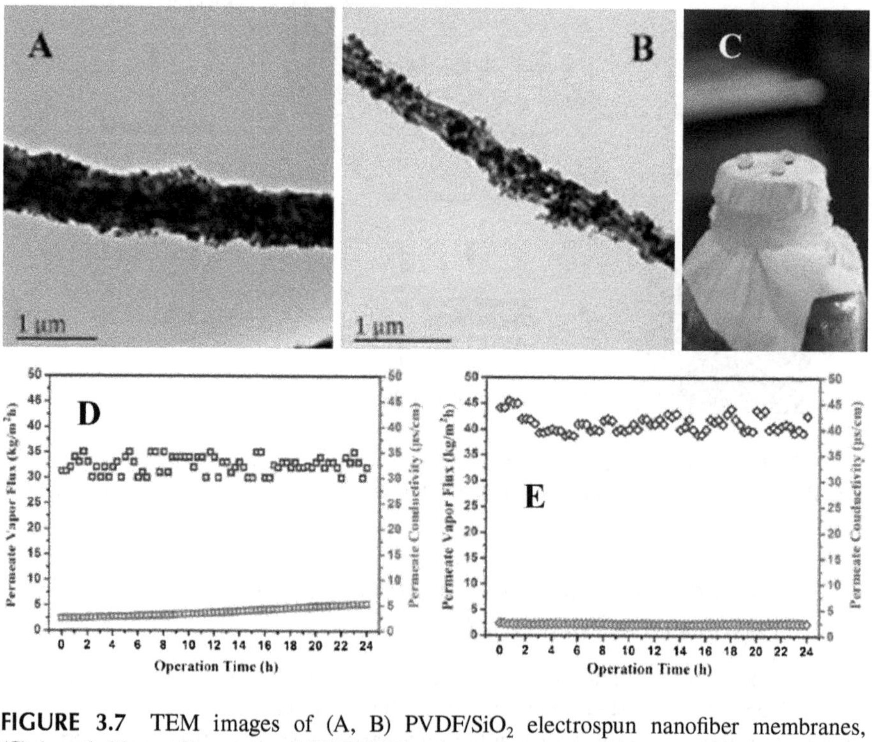

FIGURE 3.7 TEM images of (A, B) PVDF/SiO_2 electrospun nanofiber membranes, (C) breathable performance of PVDF/SiO_2 electrospun nanofiber membrane by using a methyl orange PH indicator solution and hydrochloric acid as a volatile solution and DCMD test of (D) the neat PVDF and (E) PVDF/SiO_2 electrospun nanofiber membranes. (Adapted from Li, X. et al., ACS Applied Material & Interfaces. 7(39), 21919–21930, 2015.)

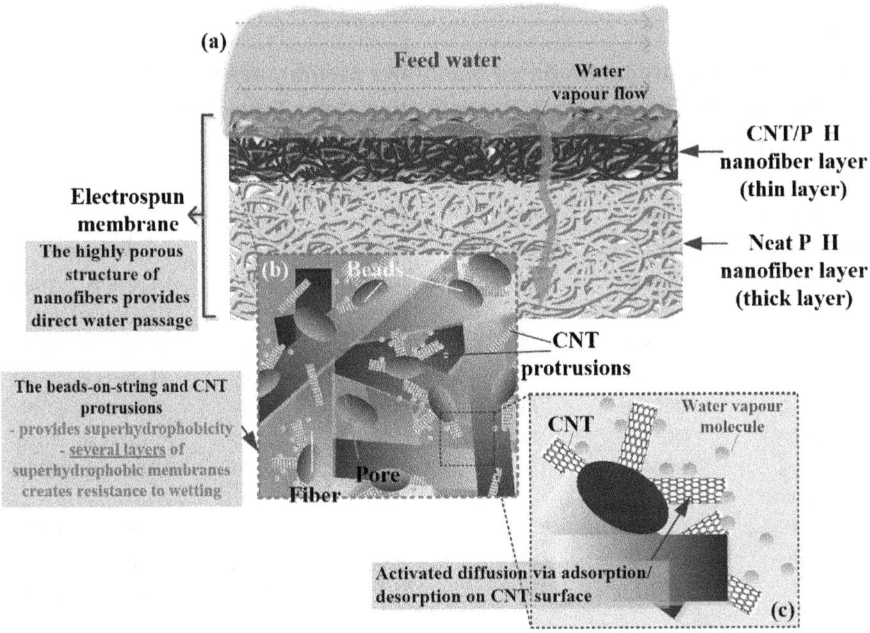

FIGURE 3.8 Schematic depiction of the mechanism of CNT/PH electrospun nanofiber membrane for MD. (Adapted from Tijing, L.D. et al., Journal of Membrane Science. 502, 158–170, 2016.)

the neat PVDF electrospun nanofiber membrane shows gradually increased permeate conductivity after 10 hours of DCMD operation, which indicates wetting issues.

To improve the hydrophobic property of the MD membrane, carbon based nanomaterials such as carbon nanotubes (CNTs) and graphene have been investigated (Tijing et al., 2016; Woo et al., 2016b). CNTs that incorporated PVDF-co-HFP electrospun nanofiber membrane had two unique features: a beads-on-string structure and the protrusion of CNTs on nanofiber (Figure 3.8). Both features led to improved hydrophobic properties, and the CNT nanofiber membrane eventually achieved a superhydrophobic property. The membrane therefore exhibited a high water vapor flux performance even though it suffered some scaling and fouling problems. On the other hand, the graphene incorporated nanofiber membrane showed a stable water vapor flux (~22.5 LMH in AGMD) and salt rejection performances for 60 hours without any fouling and wetting issues. Graphene nanoparticles led to enhanced surface hydrophobicity and roughness and hence an improved anti-wetting property. Also, the graphene nanoparticles provided diffusion path for water vapor (i.e., adsorption or desorption capacity) and enhanced mechanical properties and thermal stability of the mixed matrix membrane.

3.5.4 Surface Modification on Electrospun Membranes

In recent years, a number of research studies have been focused on membrane surface modification to enhance hydrophobicity and anti-fouling properties of MD

TABLE 3.4
Reports in Literature on Modified Nanofiber Membranes Using Various Modification Techniques for MD Application (Inlet Temperatures: Feed = 60°C, Permeate = 20°C)

Modification Technique	MD Application	Contact Angle (°)	Flux (L/m² h)	Salt Rejection (%)	Reference
Dip-coating	DCMD	150	~12	99.99	(Lee et al., 2016b)
Grafting	DCMD	158	25.2	99.99	(Dong et al., 2015b)
Dip-coating	DCMD	153	31.8	99.99	(Liao et al., 2013a)
Heat-press	DCMD	122	20–22	99.99	(Lalia et al., 2013)
Heat-press	DCMD	136	20.6	99.99	(Liao et al., 2013b)
Heat-press	DCMD	140	29	99.99	(Yao et al., 2016)
CF_4 plasma modification	AGMD	160.6	15.3	99.99	(Woo et al., 2017a)
Electrospraying	DCMD	157.8	31.5	99.99	(Lee et al., 2017b)
Annealing	DCMD	146	~35	99.99	(Yao et al., 2017)
Grafting	VMD	160.5	31.5[a]	99.99	(Dong et al., 2015a)

[a] Vacuum pressure is 9 kPa.

membranes. Membranes modified using various techniques for MD application and their reports in literature are outlined in Table 3.4.

Different superhydrophobic coatings were applied on various substrates, and results showed less biofouling formation on these coated substrates than the uncoated ones (Zhang et al., 2005; Privett et al., 2011). A previous study has reported the control of organic fouling by hydrophilization of the membrane surface.

Increasing the hydrophobicity of a membrane usually leads to higher LEP and consequently more resistance to pore wetting. Razmjou et al. (2012) fabricated a superhydrophobic PVDF membrane by incorporating TiO_2 nanoparticles via a low temperature hydrothermal process. DCMD tests were carried out including the effect of fouling formation using humic acid (HA) and calcium chloride as foulants. The deposition of TiO_2 formed hierarchical structure with multilevel roughness on the surface and increased the hydrophobicity up to 166°, consequently increasing the LEP to 195 kPa. Fouling tests revealed similar fouling behavior for both unmodified and modified membranes, but the modified membrane showed much higher flux recovery, indicating a better anti-fouling property.

Liao, Wang, and Fane (2013a) fabricated an electrospun nanofiber membrane using PVDF polymer. The electrospun nanofiber membrane showed flux of 36 LMH although it suffered wetting problems during DCMD test. As a result, they employed surface modification to improve the membrane's wetting resistance with enhanced hydrophobic property on the active layer of the electrospun nanofiber membrane. They also studied methods to prevent wetting issue by post-treatment such as heat-press (Liao et al., 2013b). Heat-press can enhance LEP, characteristics, tensile strength, and morphology of the membrane. Thus, the heat-press technique is strongly suggested for electrospun membranes to improve their characteristics and water vapor flux performance in MD operation (Yao et al., 2016).

In another study, Zhang et al. (2013b) fabricated a superhydrophobic PVDF flat-sheet membrane by casting and spraying a mixture of polydimethylsiloxane (PDMS) and hydrophobic SiO_2 nanoparticles onto the membrane surface. The modified membrane showed a contact angle of 156°, which was much higher than that of the original membrane (107°). The DCMD flux performance of the modified membrane, however, was lower than that of the original membrane but was balanced with high salt rejection efficiency. Fouling tests were carried out at very high concentration of 25 wt% NaCl solution. The flux profile was similar for both membranes in the first 40 hours; however, after 40 hours, there was a steep flux decline for the original membrane, indicating membrane wetting and fouling. The modified membrane showed more stable flux. By examining the membranes after the test, NaCl deposits were found on the original membrane surface, with some blocking the pores, while almost no deposits were found on the modified membrane surface. This study indicated the potential of surface modification by SiO_2 nanoparticles as a good method for fabricating anti-fouling MD membranes.

In recent years, MD has also been considered as a potential technology to treat challenging water sources such as textile wastewater, and dye and oil wastewaters. However, MD process suffers membrane wetting issues by the existence of low surface tension liquids in the feed, such as benzene, methanol, and other surfactants. Therefore, a new membrane should be developed to prevent wetting issues. One increasingly popular solution has been to give the membrane an omniphobic property. Omniphobic property means that all liquids can be repelled on the surface. Woo et al. (2017a) investigated CF_4 plasma modified electrospun nanofiber membranes. CF_4 plasma modification is one of the alternative methods to generate omniphobic property on the membrane surface due to the introduction of several low surface free energy chemical bonds such as CF_2-CF_2 and CF_3 (Figure 3.9(a)). CF_4 plasma treated nanofiber membrane exhibits a stable water vapor flux and high salt rejection (99.99%) against coal seam gas (CSG) RO brine containing sodium dodecyl sulfate (SDS) of 0.7 mM, while a neat nanofiber membrane suffers fast membrane pore wetting. CF_4 plasma treatment is expected to decrease surface tension of the membrane surface (Figure 3.9(b)), which contributes to both hydrophobicity and oleophobicity. Hence, CF_4 plasma modified nanofiber membranes showed anti-wetting property in MD applications.

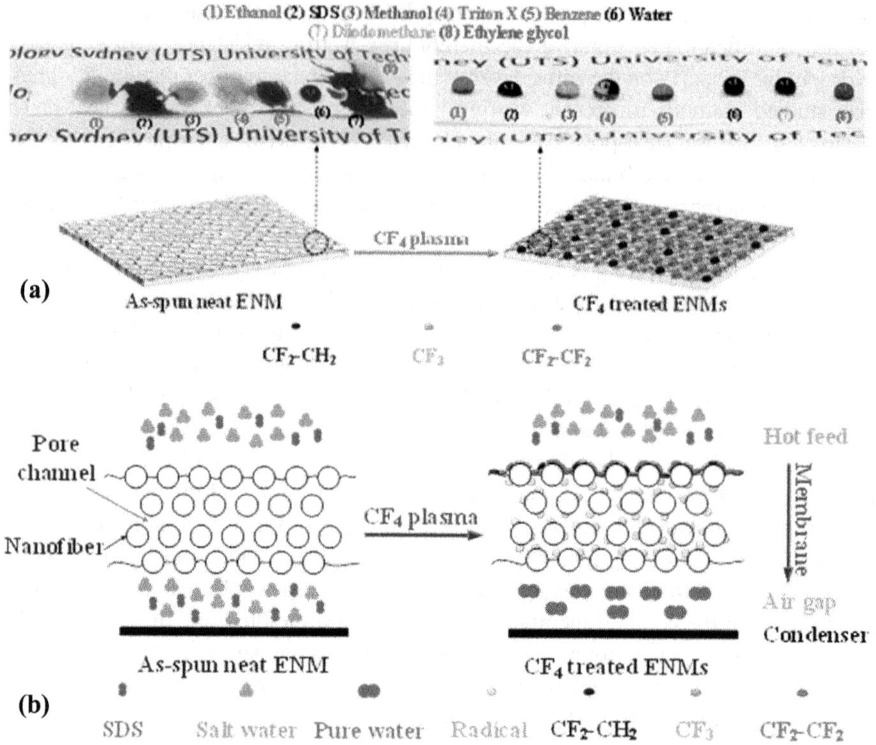

FIGURE 3.9 (a) Schematic illustrations and (b) comparison between the behavior of the neat and CF_4 plasma modified electrospun nanofiber membranes in AGMD process. (Adapted from Woo, Y.C. et al., Journal of Membrane Science. 529, 234–242, 2017.)

3.6 CONCLUSION AND FUTURE OUTLOOK

Recently, MD has emerged as one of the most promising alternative membrane based technologies to produce freshwater. It also has potentials for various other applications such as wastewater reclamation, desalination, and treatment of mining water and gas produced water. Although it is hard to completely replace RO technology in the current market, MD is attractive as a supplementary technology for desalination considering its high water recovery. However, an ideally-designed suitable membrane still needs to be developed for MD to improve the water permeability and overcome several issues such as membrane pore wetting, fouling, and scaling for commercial implementation. Currently, the electrospinning technique is being used for development of a suitable membrane for MD applications. So far, several studies have focused on three different concepts of the electrospun nanofiber membrane: dual- or triple-layer electrospun nanofiber membrane, nanomaterials embedded on/in the electrospun nanofiber membrane, and surface modified electrospun nanofiber membrane. However, several issues need to be addressed by further improving the membrane properties and performance before potential upscaling of electrospun nanofiber membranes for MD.

First, the dual- or triple-layer membrane can improve water permeability due to a low mass transfer resistance of the hydrophilic or less hydrophobic support layer compared with a common hydrophobic single layer membrane. However, it is a very complicated technique to successfully fabricate a membrane. So, a simple technique of dual or triple-layer membrane fabrication would be preferred to improve water vapor flux performance. A nanofiber membrane has internal interpenetrated pores between enormous nanofibers and increased porosity while maintaining high LEP. In addition, the porous morphologies on nanofiber membranes lead to a much higher surface area to volume ratio than conventionally (e.g., phase inversion) fabricated nanofiber membranes. Thus, the internal porous nanofiber membranes can have enhanced water vapor flux performance and anti-wetting properties in MD applications.

Second, several published reports used nanomaterials to enhance the hydrophobic property of the fabricated nanofiber membrane. Despite promises of high performance, nanomaterials are still expensive to synthesize. Thus, to solve the problem, other cost-effective nanoparticles will be used. One of the possible nanoparticles is silica aerogel, which can be integrated into membranes to improve their MD performances. Silica aerogel is a low-priced material and shows high hydrophobicity due to the existence of methyl and methoxy groups on its surface. It also exhibits a low thermal conductivity (below 0.05 W/mK), which is lower than PVDF (0.22 W/mK) and PP (0.25 W/mK). Thus, incorporation of silica aerogel may reduce the conductive heat losses, which leads to high energy efficiency of the operation.

Finally, a fabricated electrospun nanofiber membrane was modified to enhance hydrophobic property and even acquire omniphobic property on the membrane surface. Generally, surface modified electrospun nanofiber membranes had high and stable flux and salt rejection performances. Some of the modified electrospun nanofiber membranes also showed the formation of new CF_3 and CF_2-CF_2 bonds after modification through various measurements, which contributed to the acquirement of highly hydrophobic property. However, the modification technique is unlikely to be commercially applied to the membranes because of its high modification cost, energy usage, and limitations of the technique such as size of facilities and usage of hazardous plasma gas. For these reasons, another electrospinning technique, co-axial electrospinning, is being developed to achieve similar results in simple procedures. Via this technique, it is possible to fabricate a MD membrane while modifying it simultaneously. Compared to other techniques, it is much more efficient to fabricate a superhydrophobic and/or omniphobic MD membrane by forming three-dimensional structures. Therefore, the co-axial electrospinning technique has much more potential for commercialization of MD membrane fabrication.

REFERENCES

Ahmed, F.E., Lalia, B.S. and Hashaikeh, R. (2015) A review on electrospinning for membrane fabrication: Challenges and applications. Desalination 356(0), 15–30.

Alkhudhiri, A., Darwish, N. and Hilal, N. (2012) Membrane distillation: A comprehensive review. Desalination 287(0), 2–18.

Alkhudhiri, A., Darwish, N. and Hilal, N. (2013) Treatment of saline solutions using Air Gap Membrane Distillation: Experimental study. Desalination 323(0), 2–7.

Ataollahi, N., Ahmad, A., Hamzah, H., Rahman, M. and Mohamed, N. (2012) Preparation and characterization of PVDF-HFP/MG49 based polymer blend electrolyte. International Journal of Electrochemical Science 7(6693), e6703.

Baji, A., Mai, Y.-W., Wong, S.-C., Abtahi, M. and Chen, P. (2010) Electrospinning of polymer nanofibers: Effects on oriented morphology, structures and tensile properties. Composites Science and Technology 70(5), 703–718.

Boubakri, A., Hafiane, A. and Bouguecha, S.A.T. (2014) Application of response surface methodology for modeling and optimization of membrane distillation desalination process. Journal of Industrial and Engineering Chemistry 20(5), 3163–3169.

Cui, Z., Drioli, E. and Lee, Y.M. (2014) Recent progress in fluoropolymers for membranes. Progress in Polymer Science 39(1), 164–198.

Dong, Z.-Q., Ma, X.-H., Xu, Z.-L. and Gu, Z.-Y. (2015a) Superhydrophobic modification of PVDF–SiO2electrospun nanofiber membranes for vacuum membrane distillation. RSC Advances 5(83), 67962–67970.

Dong, Z.-Q., Ma, X.-H., Xu, Z.-L., You, W.-T. and Li, F.-B. (2014) Superhydrophobic PVDF–PTFE electrospun nanofibrous membranes for desalination by vacuum membrane distillation. Desalination 347(0), 175–183.

Dong, Z.Q., Wang, B.J., Ma, X.H., Wei, Y.M. and Xu, Z.L. (2015b) FAS grafted electrospun poly(vinyl alcohol) nanofiber membranes with robust superhydrophobicity for membrane distillation. ACS Applied Materials and Interfaces 7(40), 22652–22659.

Drioli, E., Ali, A. and Macedonio, F. (2015) Membrane distillation: Recent developments and perspectives. Desalination 356, 56–84.

Drioli, E. and Macedonio, F. (2010) Integrated membrane systems for desalination. In K.-V. Peinemann and S.P. Nunes (Eds.), *Membrane Technology* (93–146). Wiley-VCH Verlag GmbH & Co. KGaA.

Dumée, L., Germain, V., Sears, K., Schütz, J., Finn, N., Duke, M., Cerneaux, S., Cornu, D. and Gray, S. (2011) Enhanced durability and hydrophobicity of carbon nanotube bucky paper membranes in membrane distillation. Journal of Membrane Science 376(1–2), 241–246.

Eda, G., Liu, J. and Shivkumar, S. (2007) Solvent effects on jet evolution during electrospinning of semi-dilute polystyrene solutions. European Polymer Journal 43(4), 1154–1167.

Essalhi, M. and Khayet, M. (2012) Surface segregation of fluorinated modifying macromolecule for hydrophobic/hydrophilic membrane preparation and application in air gap and direct contact membrane distillation. Journal of Membrane Science 417–418, 163–173.

Essalhi, M. and Khayet, M. (2014) Application of a porous composite hydrophobic/hydrophilic membrane in desalination by air gap and liquid gap membrane distillation: A comparative study. Separation and Purification Technology 133, 176–186.

Fan, H. and Peng, Y. (2012) Application of PVDF membranes in desalination and comparison of the VMD and DCMD processes. Chemical Engineering Science 79, 94–102.

Feng, C., Khulbe, K.C., Matsuura, T., Gopal, R., Kaur, S., Ramakrishna, S. and Khayet, M. (2008) Production of drinking water from saline water by air-gap membrane distillation using polyvinylidene fluoride nanofiber membrane. Journal of Membrane Science 311(1–2), 1–6.

Feng, C., Khulbe, K.C., Matsuura, T., Tabe, S. and Ismail, A.F. (2013) Preparation and characterization of electro-spun nanofiber membranes and their possible applications in water treatment. Separation and Purification Technology 102(0), 118–135.

Francis, L., Ghaffour, N., Alsaadi, A.S., Nunes, S.P. and Amy, G.L. (2014) Performance evaluation of the DCMD desalination process under bench scale and large scale module operating conditions. Journal of Membrane Science 455(0), 103–112.

Francis, L., Maab, H., AlSaadi, A., Nunes, S., Ghaffour, N. and Amy, G.L. (2013) Fabrication of electrospun nanofibrous membranes for membrane distillation application. Desalination and Water Treatment 51(7–9), 1337–1343.

Frenot, A. and Chronakis, I.S. (2003) Polymer nanofibers assembled by electrospinning. Current Opinion in Colloid & Interface Science 8(1), 64–75.

Ge, J., Peng, Y., Li, Z., Chen, P. and Wang, S. (2014) Membrane fouling and wetting in a DCMD process for RO brine concentration. Desalination 344(0), 97–107.

Geng, H., He, Q., Wu, H., Li, P., Zhang, C. and Chang, H. (2014a) Experimental study of hollow fiber AGMD modules with energy recovery for high saline water desalination. Desalination 344(0), 55–63.

Geng, H., Wu, H., Li, P. and He, Q. (2014b) Study on a new air-gap membrane distillation module for desalination. Desalination 334(1), 29–38.

Goh, S., Zhang, J., Liu, Y. and Fane, A.G. (2013) Fouling and wetting in membrane distillation (MD) and MD-bioreactor (MDBR) for wastewater reclamation. Desalination 323(0), 39–47.

Gryta, M., Grzechulska-Damszel, J., Markowska, A. and Karakulski, K. (2009) The influence of polypropylene degradation on the membrane wettability during membrane distillation. Journal of Membrane Science 326(2), 493–502.

Huang, Y., Huang, Q.-L., Liu, H., Zhang, C.-X., You, Y.-W., Li, N.-N. and Xiao, C.-F. (2017) Preparation, characterization, and applications of electrospun ultrafine fibrous PTFE porous membranes. Journal of Membrane Science 523, 317–326.

Hwang, H.J., He, K., Gray, S., Zhang, J. and Moon, I.S. (2011) Direct contact membrane distillation (DCMD): Experimental study on the commercial PTFE membrane and modeling. Journal of Membrane Science 371(1–2), 90–98.

Khayet, M. (2011) Membranes and theoretical modeling of membrane distillation: A review. Advances in Colloid and Interface Sciend 164(1–2), 56–88.

Khayet, M. and Matsuura, T. (2011) *Membrane Distillation: Principles and Applications.* Elsevier.

Khayet, M., Matsuura, T. and Mengual, J. (2005) Porous hydrophobic/hydrophilic composite membranes: Estimation of the hydrophobic-layer thickness. Journal of Membrane Science 266(1–2), 68–79.

Lalia, B.S., Guillen-Burrieza, E., Arafat, H.A. and Hashaikeh, R. (2013) Fabrication and characterization of polyvinylidenefluoride-co-hexafluoropropylene (PVDF-HFP) electrospun membranes for direct contact membrane distillation. Journal of Membrane Science 428(0), 104–115.

Lalia, B.S., Guillen, E., Arafat, H.A. and Hashaikeh, R. (2014) Nanocrystalline cellulose reinforced PVDF-HFP membranes for membrane distillation application. Desalination 332(1), 134–141.

Lee, E.-J., An, A.K., Hadi, P., Lee, S., Woo, Y.C. and Shon, H.K. (2017a) Advanced multi-nozzle electrospun functionalized titanium dioxide/polyvinylidene fluoride-co-hexafluoropropylene (TiO2/PVDF-HFP) composite membranes for direct contact membrane distillation. Journal of Membrane Science 524, 712–720.

Lee, E.-J., An, A.K., He, T., Woo, Y.C. and Shon, H.K. (2016a) Electrospun nanofiber membranes incorporating fluorosilane-coated TiO2 nanocomposite for direct contact membrane distillation. Journal of Membrane Science 520, 145–154.

Lee, E.J., Deka, B.J., Guo, J., Woo, Y.C., Shon, H.K. and An, A.K. (2017b) Engineering the Re-Entrant Hierarchy and Surface Energy of PDMS-PVDF Membrane for Membrane Distillation Using a Facile and Benign Microsphere Coating. Environ Sci Technol.

Lee, J., Boo, C., Ryu, W.H., Taylor, A.D. and Elimelech, M. (2016b) Development of omniphobic desalination membranes using a charged electrospun nanofiber scaffold. ACS Applied Materials & Interfaces 8(17), 11154–11161.

Li, X., Wang, C., Yang, Y., Wang, X., Zhu, M. and Hsiao, B.S. (2014) Dual-biomimetic superhydrophobic electrospun polystyrene nanofibrous membranes for membrane distillation. ACS Applied Material & Interfaces 6(4), 2423–2430.

Li, X., Yu, X., Cheng, C., Deng, L., Wang, M. and Wang, X. (2015) Electrospun superhydrophobic organic/inorganic composite nanofibrous membranes for membrane distillation. ACS Applied Material & Interfaces 7(39), 21919–21930.

Liao, Y., Loh, C.H., Wang, R. and Fane, A.G. (2014a) Electrospun superhydrophobic membranes with unique structures for membrane distillation. ACS Applied Material & Interfaces 6(18), 16035–16048.

Liao, Y., Wang, R. and Fane, A.G. (2013a) Engineering superhydrophobic surface on poly(vinylidene fluoride) nanofiber membranes for direct contact membrane distillation. Journal of Membrane Science 440, 77–87.

Liao, Y., Wang, R. and Fane, A.G. (2014b) Fabrication of bioinspired composite nanofiber membranes with robust superhydrophobicity for direct contact membrane distillation. Environmental Science & Technology 48(11), 6335–6341.

Liao, Y., Wang, R., Tian, M., Qiu, C. and Fane, A.G. (2013b) Fabrication of polyvinylidene fluoride (PVDF) nanofiber membranes by electro-spinning for direct contact membrane distillation. Journal of Membrane Science 425–426(0), 30–39.

Lin, S., Nejati, S., Boo, C., Hu, Y., Osuji, C.O. and Elimelech, M. (2014) Omniphobic membrane for robust membrane distillation. Environmental Science & Technology Letters 1(11), 443–447.

Liu, F., Hashim, N.A., Liu, Y., Abed, M.R.M. and Li, K. (2011) Progress in the production and modification of PVDF membranes. Journal of Membrane Science 375(1–2), 1–27.

Manawi, Y.M., Khraisheh, M., Fard, A.K., Benyahia, F. and Adham, S. (2014) Effect of operational parameters on distillate flux in direct contact membrane distillation (DCMD): Comparison between experimental and model predicted performance. Desalination 336(0), 110–120.

Nejati, S., Boo, C., Osuji, C.O. and Elimelech, M. (2015) Engineering flat sheet microporous PVDF films for membrane distillation. Journal of Membrane Science 492, 355–363.

Nghiem, L.D. and Cath, T. (2011) A scaling mitigation approach during direct contact membrane distillation. Separation and Purification Technology 80(2), 315–322.

Peining, Z., Nair, A.S., Shengjie, P., Shengyuan, Y. and Ramakrishna, S. (2012) Facile fabrication of TiO2-graphene composite with enhanced photovoltaic and photocatalytic properties by electrospinning. ACS Applied Material Interfaces 4(2), 581–585.

Pelipenko, J., Kristl, J., Janković, B., Baumgartner, S. and Kocbek, P. (2013) The impact of relative humidity during electrospinning on the morphology and mechanical properties of nanofibers. International Journal of Pharmaceutics 456(1), 125–134.

Peñate, B. and García-Rodríguez, L. (2012) Current trends and future prospects in the design of seawater reverse osmosis desalination technology. Desalination 284, 1–8.

Prince, J.A., Anbharasi, V., Shanmugasundaram, T.S. and Singh, G. (2013) Preparation and characterization of novel triple layer hydrophilic–hydrophobic composite membrane for desalination using air gap membrane distillation. Separation and Purification Technology 118, 598–603.

Prince, J.A., Rana, D., Matsuura, T., Ayyanar, N., Shanmugasundaram, T.S. and Singh, G. (2014a) Nanofiber based triple layer hydro-philic/-phobic membrane—A solution for pore wetting in membrane distillation. Scientific Reports 4, 6949.

Prince, J.A., Rana, D., Singh, G., Matsuura, T., Jun Kai, T. and Shanmugasundaram, T.S. (2014b) Effect of hydrophobic surface modifying macromolecules on differently produced PVDF membranes for direct contact membrane distillation. Chemical Engineering Journal 242, 387–396.

Prince, J.A., Singh, G., Rana, D., Matsuura, T., Anbharasi, V. and Shanmugasundaram, T.S. (2012) Preparation and characterization of highly hydrophobic poly(vinylidene fluoride)—Clay nanocomposite nanofiber membranes (PVDF–clay NNMs) for desalination using direct contact membrane distillation. Journal of Membrane Science 397–398, 80–86.

Privett, B.J., Youn, J., Hong, S.A., Lee, J., Han, J., Shin, J.H. and Schoenfisch, M.H. (2011) Antibacterial fluorinated silica colloid superhydrophobic surfaces. Langmuir 27(15), 9597–9601.

Razmjou, A., Arifin, E., Dong, G., Mansouri, J. and Chen, V. (2012) Superhydrophobic modification of TiO2 nanocomposite PVDF membranes for applications in membrane distillation. Journal of Membrane Science 415–416(0), 850–863.

Shim, W.G., He, K., Gray, S. and Moon, I.S. (2015) Solar energy assisted direct contact membrane distillation (DCMD) process for seawater desalination. Separation and Purification Technology 143, 94–104.

Song, Z.W. and Jiang, L.Y. (2013) Optimization of morphology and perf ormance of PVDF hollow fiber for direct contact membrane distillation using experimental design. Chemical Engineering Science 101(0), 130–143.

Su, C., Chang, J., Tang, K., Gao, F., Li, Y. and Cao, H. (2017) Novel three-dimensional superhydrophobic and strength-enhanced electrospun membranes for long-term membrane distillation. Separation and Purification Technology 178, 279–287.

Tian, R., Gao, H., Yang, X.H., Yan, S.Y. and Li, S. (2014) A new enhancement technique on air gap membrane distillation. Desalination 332(1), 52–59.

Tijing, L.D., Choi, J.-S., Lee, S., Kim, S.-H. and Shon, H.K. (2014a) Recent progress of membrane distillation using electrospun nanofibrous membrane. Journal of Membrane Science 453(0), 435–462.

Tijing, L.D., Woo, Y.C., Johir, M.A.H., Choi, J.-S. and Shon, H.K. (2014b) A novel dual-layer bicomponent electrospun nanofibrous membrane for desalination by direct contact membrane distillation. Chemical Engineering Journal 256, 155–159.

Tijing, L.D., Woo, Y.C., Shim, W.-G., He, T., Choi, J.-S., Kim, S.-H. and Shon, H.K. (2016) Superhydrophobic nanofiber membrane containing carbon nanotubes for high-performance direct contact membrane distillation. Journal of Membrane Science 502, 158–170.

Tomaszewska, M., Gryta, M. and Morawski, A.W. (1994) A study of separation by the direct-contact membrane distillation process. Separations Technology 4(4), 244–248.

Wang, P. and Chung, T.-S. (2015) Recent advances in membrane distillation processes: Membrane development, configuration design and application exploring. Journal of Membrane Science 474, 39–56.

Woo, Y.C., Chen, Y., Tijing, L.D., Phuntsho, S., He, T., Choi, J.-S., Kim, S.-H. and Shon, H.K. (2017a) CF4 plasma-modified omniphobic electrospun nanofiber membrane for produced water brine treatment by membrane distillation. Journal of Membrane Science 529, 234–242.

Woo, Y.C., Kim, Y., Shim, W.-G., Tijing, L.D., Yao, M., Nghiem, L.D., Choi, J.-S., Kim, S.-H. and Shon, H.K. (2016a) Graphene/PVDF flat-sheet membrane for the treatment of RO brine from coal seam gas produced water by air gap membrane distillation. Journal of Membrane Science 513, 74–84.

Woo, Y.C., Tijing, L.D., Park, M.J., Yao, M., Choi, J.-S., Lee, S., Kim, S.-H., An, K.-J. and Shon, H.K. (2017b) Electrospun dual-layer nonwoven membrane for desalination by air gap membrane distillation. Desalination 403, 187–198.

Woo, Y.C., Tijing, L.D., Shim, W.-G., Choi, J.-S., Kim, S.-H., He, T., Drioli, E. and Shon, H.K. (2016b) Water desalination using graphene-enhanced electrospun nanofiber membrane via air gap membrane distillation. Journal of Membrane Science 520, 99–110.

Yao, M., Woo, Y.C., Tijing, L., Cesarini, C. and Shon, H.K. (2017) Improving Nanofiber Membrane Characteristics and Membrane Distillation Performance of Heat-Pressed Membranes via Annealing Post-Treatment. Applied Sciences 7(1), 78.

Yao, M., Woo, Y.C., Tijing, L.D., Shim, W.-G., Choi, J.-S., Kim, S.-H. and Shon, H.K. (2016) Effect of heat-press conditions on electrospun membranes for desalination by direct contact membrane distillation. Desalination 378, 80–91.

Zhang, H., Lamb, R. and Lewis, J. (2005) Engineering nanoscale roughness on hydrophobic surface—Preliminary assessment of fouling behaviour. Science and Technology of Advanced Materials 6(3-4), 236–239.

Zhang, J., Gray, S. and Li, J.-D. (2013a) Predicting the influence of operating conditions on DCMD flux and thermal efficiency for incompressible and compressible membrane systems. Desalination 323(0), 142–149.

Zhang, J., Li, J.-D. and Gray, S. (2011) Effect of applied pressure on performance of PTFE membrane in DCMD. Journal of Membrane Science 369(1–2), 514–525.

Zhang, J., Song, Z., Li, B., Wang, Q. and Wang, S. (2013b) Fabrication and characterization of superhydrophobic poly (vinylidene fluoride) membrane for direct contact membrane distillation. Desalination 324, 1–9.

Zhou, T., Yao, Y., Xiang, R. and Wu, Y. (2014) Formation and characterization of polytetrafluoroethylene nanofiber membranes for vacuum membrane distillation. Journal of Membrane Science 453, 402–408.

4 Preparation and Surface Modification of Porous Ceramic Membranes for Water Treatment

Raja Ben Amar, Samia Mahouche-Chergui, Abdallah Oun, Mouna Khemakhem, and Benjamin Carbonnier

CONTENTS

4.1 INTRODUCTION

Membrane separation is well-recognized as a mature technology well-suited to address issues of separation and treatment of complex mixtures and reject handling arising from the textile, chemical, food, and pharmaceutical industries, to name but a few. This large success can be rationalized by the intrinsic characteristics of membrane processes such as high stability and efficiency, low energy requirement, and ease of operation (Saufi et al., 2004; Tin et al., 2011).

Membrane materials can be classified as function of their chemical nature, organic *vs.* inorganic or pore size and related separation mechanism. Organic membranes are made of polymers including polyamide, polysulfone, polyethersulfone, and polydimethylsiloxane and are typically prepared by casting methods (air, immersion, and melt casting), track etching techniques, or controlled film stretching. Polymer-based microporous membranes are commercially available for use in ultrafiltration (UF) and microfiltration (MF) in health care for sample preparation prior to, for instance, chromatographic analysis, virus removal, and diagnosis purposes. They are also widely applied in the area of energy, as in lithium-ion batteries and fuel cells as well as in the petrochemical industry to treat produced water. Although polymeric membranes can be cost-effectively designed with well-controlled characteristics such as pore size and hydrophilic/hydrophobic balance, these constructed membranes suffer from limited chemical, thermal, and mechanical stability compared to their inorganic counterparts.

Porous inorganic membranes are mainly manufactured from metal oxides such as alumina (Al_2O_3), silica (SiO_2), zirconia (ZrO_2), and titania (TiO_2) (Chowdhury et al., 2003; Lenza et al., 2000; Van Gestel et al., 2006), while metals such as palladium, nickel, silver, zirconium, and their alloys are used for the preparation of dense membranes (Biswas et al., 2018). It is reported also that zeolite-based (alumino silicate) membranes (Ghouil et al., 2015) showed the characteristic of dense membranes. Inorganic membranes can resist harsh conditions making their use possible in wide pH (pH 1–14) and temperature (up to 500°C) ranges. These advantages also allow for regeneration of the membranes through washing with aggressive chemicals, organic solvents, or a hot water stream.

Ceramic membranes with a broad range of pores size can be elaborated through different preparation approaches, including slip casting, tape casting, pressing, extrusion, sol-gel process, dip coating, and chemical vapor deposition. Most inorganic membranes are made of a combination of a thin separation layer and a porous support and are mainly employed in pressure-driven separation processes (Guizard et al., 1996). Production of industrial ceramic membranes relies on the use of a limited choice of raw materials and generally involves a high temperature-sintering step to ensure strong bonding between membrane and support. The high sintering temperature used in the case of titania and alumina ceramic membranes may explain their relatively high price (Anderson et al., 1988). The presence of hydroxyl groups on the membrane surface governs the permeation properties of the membrane and allows for modification of surface properties such as surface roughness and hydrophilicity or hydrophobicity. Khemakhem et al. (2011a; 2011b), for instance, reported that chemical modification of a membrane surface with hydrophilic moieties may prevent membrane fouling.

To date, a number of methods have been proposed such as plasma-enhanced chemical vapor deposition (CVD) (Wu et al., 2002), plasma polymerization (Chen et al., 1999), and chemical grafting (Alami Younssi et al., 2003; Kujawa et al., 2013a; 2013b; 2014a). The research group of Ben Amar (Tunisia) described an elegant method for the chemical functionalization of ceramic materials through a grafting process of perfluoroalkylsilane (Khemakhem et al., 2013). The newly developed ceramic MF membranes that have 7 L/m^2.h bar as water permeability are very promising in the field of membrane distillation, exhibiting total salt rejection of 99% with permeate flux of 165 L/day.m^2 under ΔT of 85°C.

The use of aliphatic organosilanes such as trimethylsilyl chloride (Me_3SiCl) (Alami Younsi et al., 2003), triphenylsilyl chloride (Si(C6H5)3Cl), tert-butyldimethylsilyl chloride-($C_4H_9(CH_3)_2SiCl$) (Caro et al., 1998; Sah et al., 2004), dichlorodimethylsilane ($(CH_3)_2SiCl_2$), and dichloromethyloctylsilane ($C_8H_{17}CH_3SiCl_2$) (Van Gestel et al., 2003) was reported for hydrophobization purposes. It was notably shown that the degree of hydrophilic or hydrophobic modification depends on both the extent of grafting and the length of the grafted segment. Indeed, steric hindrance may limit the grafting reaction of bulky silanes at the internal surface of the membrane.

The aim of this chapter is to conclusively highlight the synergy of inorganic materials and surface functionalization for the design of microporous ceramic membranes with controlled permeability and selectivity. In this chapter, particular attention is paid to the chemical modification of membrane surface with the aim to control the hydrophilic/hydrophobic balance and control the pore size. We discuss the application of the so-chemically modified membranes for water treatment including membrane distillation, oil/water separations, and wastewater treatment.

4.2 DEVELOPMENT OF CERAMIC MEMBRANES DERIVED FROM LOW COST RAW MATERIALS

Ceramic membranes have been extensively applied in different industrial areas such as the food and beverage industry (Almandoz et al., 2015; Bogianchini et al., 2011; Cissé et al., 2011; Nandi et al., 2009; Zulewska et al., 2009), chemistry (Athanasekou et al., 2009; Padaki et al., 2015), water treatment and reuse (Van Der Bruggen et al., 2003), and biotechnology and pharmaceuticals (Díaz-Reinoso et al., 2009; Liu et al., 2017), as summarized in Table 4.1. In general, ceramic membranes have tubular and multi-tubular geometry. For instance, the commercial membranes

TABLE 4.1
Examples of Using Ceramic Membranes in Major Industrial Applications

Area	Process	Applications	References
Food and beverage industry	MF and UF	Concentration of milk, concentration of protein, clarification of fruit juice/wine, and microorganism separation from fermented juice.	Almandoz et al., 2015; Nandi et al., 2009; Zulewska et al., 2009; Bogianchini et al., 2011; Cissé et al., 2011
Biotechnology and pharmaceutical industry	MF and UF	Microorganism separation, cell debris filtration and plasma separation.	Liu et al., 2017; Díaz-Reinoso et al., 2009
Chemical and industrial applications	MF and UF	Oil-water separation, purification of used oil and removal of precipitated heavy metals and solids.	Padaki et al., 2015; Athanasekou et al., 2009
Recovery and recycling	UF and NF	Drinking water and wastewater treatment.	Van Der Bruggen et al., 2003

from Orelis (KERASEP membrane), made of Al_2O_3-TiO_2 or Al_2O_3-ZrO_2, have a multi-tubular geometry with 7 or 19 channels with each channel having a diameter of 6 mm (Wallberg et al., 2003). KLEANSEP ceramic membranes are commercially available in a wide range of molecular weight cut-off (MWCO), providing industrial solutions in the filtration of suspensions, emulsions, and solutions.

Crucial requirements for using ceramic membranes are high permeability, long lifetime, and selectivity. Undoubtedly, the high mechanical resistance and chemical stability of ceramic membranes toward harsh operating conditions under continuous flow are key features of their success. Indeed, ceramic materials enable the use of high pressures and temperatures, chemically aggressive cleaning methods, and back pulsing during processes (Ben Amar et al., 1990). MF and UF ceramic membranes have been used in a wide variety of filtration processes such as separation of micro-organisms, proteins, and colloidal solutes. Although nanofiltration (NF) membranes have been applied to water treatment, including wastewater containing heavy metals and synthetic dyes, the ceramic membranes are not widely applied to NF compared to organic membranes (Arkell et al., 2014). The application of ceramic membranes has also been extended to non-aqueous solution separation especially in petrochemical processing. Al_2O_3, SiO_2, and ZrO_2 UF membranes have been applied to the separation of asphaltene from crude oil in the temperature range of 155–180°C (Guizard, 1996). The treated oils were considered clean oils owing to the effective removal of heavy metals such as copper, lead, iron, and chromium.

A quick overview of the literature dedicated to the field of ceramic membranes reveals that various inorganic materials have been considered precursors for membrane design. Asymmetric membranes are typically manufactured from metal oxides among which pure or mixed Al_2O_3 (Gaber et al., 2013; Garmsiri et al., 2017; Ren et al., 2015), ZrO_2 (Bouzerara et al., 2015; Guo et al., 2016), TiO_2 (Cai et al., 2015; Gaber et al., 2013; Garmsiri et al., 2017; Guo et al., 2016; Palacio et al., 2009; Yacou et al., 2015), and SiO_2 (Lim et al., 2009; Yoshino et al., 2008) have been primarily used. Although metal oxide-based membranes have proved very efficient for a variety of separation processes associated with pollution treatment, they suffer from high preparation cost. The use of metal oxide powders and the high sintering temperature would make the commercial-quality membranes costly. This drawback slows down the development of this type of membranes.

Today, the trend is toward using low cost materials and, if possible, recycling by-products that arise from industrial activities with the aim to reduce cost and to develop green processes. Researchers and engineering firms are continuously working on the development of cost-effective processes, and one of the solutions relies on the use of low cost materials of natural origin.

Natural clays (Bouazizi et al., 2016; Garmsiri et al., 2017; Khemakhem et al., 2009; 2011a; Oun et al., 2017; Vinoth Kumar et al., 2015), zeolites (Kosinov et al., 2016; Lafleur et al., 2017), sand (Aloulou et al., 2017; Kouras et al., 2017), quartz (Kouras et al., 2017), fly ash (Jedidi et al., 2009a; 2011; Suresh et al., 2016), calcite (Kouras et al., 2017), apatite (Masmoudi et al., 2007; Sahoo et al., 2016), phosphate (Bouazizi et al., 2017; Palacio et al., 2009), and phosphate sub-products (Khemakhem et al., 2011c; 2015) have proven to be very promising. This is because natural raw powders may provide high chemical and thermal stability, enhanced

mechanical resistance, long-term durability, and ease of cleaning between each application run to prevent surface fouling phenomena (Garmsiri et al., 2017). Moreover, formulations derived from low cost materials may require low thermal treatment temperatures contributing significantly to cost reduction (Khemakhem et al., 2006a; 2009; 2011c; Sarkar et al., 2012; Tahri et al., 2013).

The research group of Ben Amar showed that the use of mud, a by-product of the mining industry in Tunisia, allows for a significant decrease of the sintering temperature to 700°C for the support and 650°C for the active layer (Khemakhem et al., 2015). This range of temperature is much lower than the one obtained with other raw materials (e.g., clay, apatite, sand, and fly ash), which have a sintering temperature higher than 1000°C (Khemakhem et al., 2015). Many researchers have discussed the potential use of natural Moroccan and Tunisian clays to produce membrane supports for MF and UF applications (Khemakhem et al., 2006b; 2007; 2011a; Saffaj et al., 2004). In addition, Masmoudi et al. (2007) studied the elaboration of UF membranes based on natural phosphates as well as synthetic apatite while Aloulou et al. (2017) reported on the design of MF membranes meant to be used for wastewater treatment using Tunisian natural sand. Other authors also demonstrated the possibility of developing ceramic membranes from abundant industrial sub-products, such as fly ash arising from coal fired power stations (Jedidi et al., 2009a; 2009b; 2011) and muds obtained from the washing steps of phosphates rocks (Khemakhem et al., 2015). The so-designed ceramic membranes were applied in different areas of water treatment.

4.3 SURFACE MODIFICATION OF CERAMIC MEMBRANE VIA STRUCTURATION

Tailoring the microstructure of ceramic membrane surface has been accomplished with a variety of oxide particles via depositing a thin layer on the membrane support. A broad range of membrane properties can be obtained through a judicious choice of both particle nature and size as well as thickness of the coating. Membrane structuration is a subject of much interest in the purification of oily wastewater because it yields small pores that favor efficient oil particles rejection. It is highly possible to provide anti-adhesive properties against oil to the membrane surface by increasing the membrane hydrophilicity.

Among the metal oxides that have been considered to date, TiO_2 nanoparticles emerged as promising candidates since their structural morphology and formed coating thickness can be easily adjusted by simply changing the experimental conditions. Chang et al. (2014) improved the hydrophilicity of 19-channel tubular ceramic MF membrane using a one-step nano-TiO_2 synthesis-deposition procedure. The TiO_2 nanoparticles were synthesized directly inside the membrane channels by filling them with a solution containing a 1:2 molar ratio of $Ti(SO_4)_2$/urea. The titane-containing membrane support was treated twice at 85°C for 3 h followed by a calcination at 950°C for 2 h. This produced an inner surface alumina membrane with a uniform TiO_2 nanoparticles array (30 nm TiO_2 spheres), causing a decrease of the contact angle from 33° to 8°. Such an increase in wettability of the composite ceramic membrane led to an improvement of the membrane flux in separation of a stable oil-in-water emulsions compared to the unmodified ceramic membrane for the optimized cross-flow MF conditions (trans-membrane pressure of 0.16 MPa, feed cross-flow velocity of 5 m/s and temperature of 40°C). Under these conditions, the maximum flux was reported to be about 330 L/m^{2}h.

A different route to prepare TiO_2 composite ceramic membrane was described by Suresh et al. (2017). It was based first on the co-precipitation of the TiO_2 nanoparticles using $TiCl_4$ in the presence of a solution of NH_4OH. Subsequently, the as-prepared titania nanoparticles suspension was placed together with the low-cost clay-based porous ceramic membrane support for hydrothermal treatment at 160°C for 12 h followed by a calcination step at 400°C for 3 h in a muffle furnace. The contact angle of a deposited water droplet on the surface of the as-designed TiO_2 ceramic membrane reached a value of about 15° revealing the significant surface hydrophilicity (see Figure 4.1). The average pore size as determined by N_2 adsorption-desorption decreased slightly from about 0.981 μm in the case of the unmodified membrane to 0.877 μm after TiO_2 coating. Remarkable improvement in both flux (648 L/m^2.h) and oil removal efficiency (>99%) was obtained for the elaborated TiO_2-ceramic membrane compared to the native ceramic support, which exhibited a permeate flux and oil removal efficiency of 313.2 L/m^2.h and 93–96%, respectively, when both membranes were tested

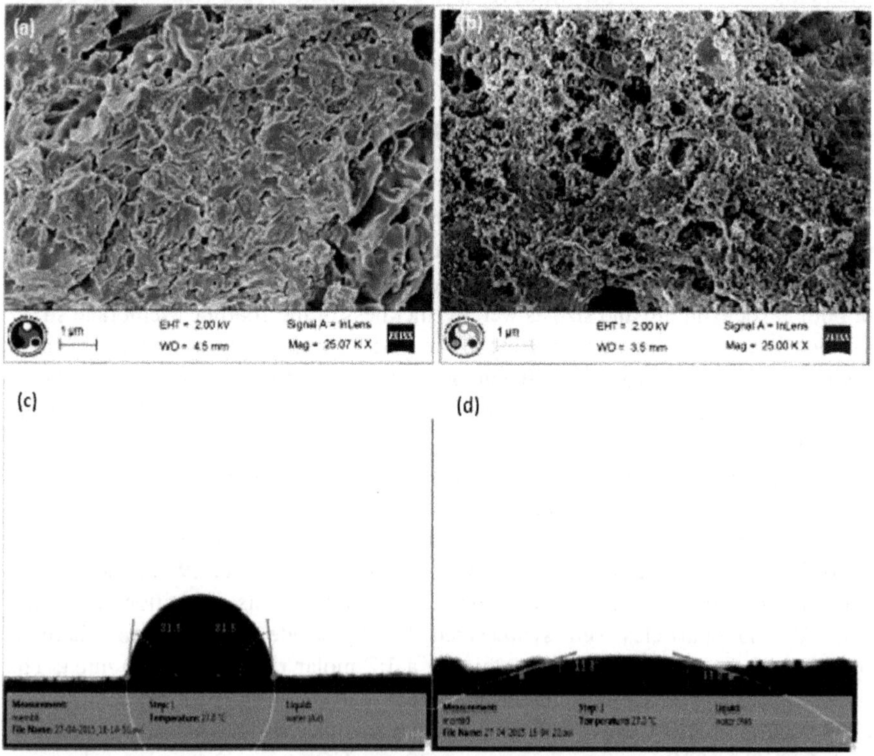

FIGURE 4.1 FESEM images showing (a and b) top and (c and d) cross-sectional views. (From Suresh, K., Pugazhenthi, G., Egyptian Journal of Petroleum. 26, 679–694, 2017.)

using feed oil concentration of 200 mg/L and at 2.07 bar. This effect was due to the hydrophilic nature of titania membrane surface, which prevents access of the oil droplets into the pores and significantly reduces membrane fouling.

4.3.1 Ceramic Membrane from Slip-Casting Method

TiO_2-modified ceramic membranes have also been largely used in the domain of the separation of dyes from wastewater because of their ability to improve permeate flow and antifouling capacity. Oun et al. (2017) developed novel asymmetric UF ceramic membranes by the deposition of a single separation layer of TiO_2 particles on the inner surface of extruded tubular porous clay-alumina membranes (ID/OD = 7 mm/10 mm) by the slip casting method. The separation UF layer was prepared via a one-step slip casting at room temperature of an aqueous suspension of TiO_2 particles (4 g), containing 12 wt% of polyvinyl alcohol (PVA) used as binder and 0.2 wt% of Dolapix CE64 used as dispersant, directly inside the tubular supports. SEM characterization indicated the presence of a dense and homogeneous 4.2 μm thin TiO_2 layer with a mean pore size of about 50 nm (see Figure 4.2). The UF TiO_2-coated tubular ceramic membranes showed high performance towards the removal of alizarin red dye (MW: 240.21 g/mol) at a transmembrane pressure of 5 bar and at pH of 9. In comparison to previous works on removal of dyes from wastewater, the membranes designed by Oun et al. (2016) exhibited greater performance in terms of water flux (117 L/m^2.h) and alizarin red dye rejection rate (99%) than the ceramic tubular membranes made of TiO_2-$ZnAl_2O_4$-clay (Saffaj et al., 2004), Kaolin-based membrane (Bouzerara et al., 2009), and TiO_2-ZrO_2-Al_2O_3 (Ma et al., 2017) that showed water flux of 9.42, 33 and 28.1 L/m^2.h with rejection rate of 90% (P = 10 bar), 98% (P = 3 bar), and 97% (P = 3 bar), respectively. This finding can be attributed to the contribution of the TiO_2 nanoparticles that enhance the filtration performances. The color retention is mainly due to the adsorption onto the membrane surface (Ganesh et al., 2016).

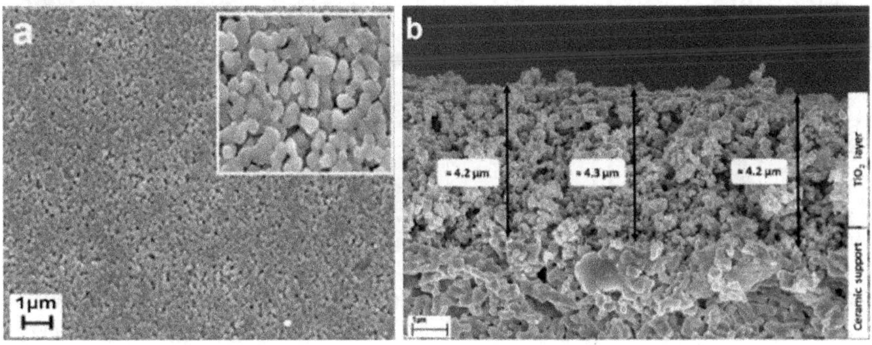

FIGURE 4.2 FESEM images showing (a) top-surface and (b) cross-sectional views of ceramic membrane support prepared from the composition clay/alumina/methocel 25/75/6 and coated with a titania layer. The inset in (a) shows a magnified image of the surface of the titania-based membrane. (From Oun, A. et al., Separation and Purification Technology. 188, 126–133, 2017.)

4.3.2 Ceramic Membrane from Atomic Layer Deposition

Metal oxide thin films with 3-D structures can also be deposited on ceramic membrane surface via the novel atomic layer deposition method (ALD), which is a gas phase deposition technique. This one-step pathway enables the generation of a uniform atomic-scale thin layer of oxides that strongly attaches to the membrane surface as well as to the pore walls. Moreover, both thickness and pore size of the atomic layer can easily be tuned by varying the number of deposition cycles and adjusting the exposure time of the membrane support and metal precursors to the oxidant source. This method was described for the first time when Li et al. (2012) deposited a thin film of Al_2O_3 on tubular ceramic membranes based on a layer of ZrO_2 nanoparticles-coated Al_2O_3 supports. The coating process was carried out at 250°C using trimethylaluminum and deionized water as precursors. SEM analysis, as shown in Figure 4.3, indicated that consecutive deposition steps induced a continuous decrease of the membrane pore size from about 50 to 6.8 nm, suggesting controllable conversion from MF to UF and NF and even dense membranes by increasing the number of ALD cycles. Furthermore, an increase of the number of ALD cycles from 0 to 800 led to a significant decrease in the membrane permeability from about 1698 to 118 L/m^2.h.bar. This result was attributed to the decrease of the pore size of the membranes after surface modification.

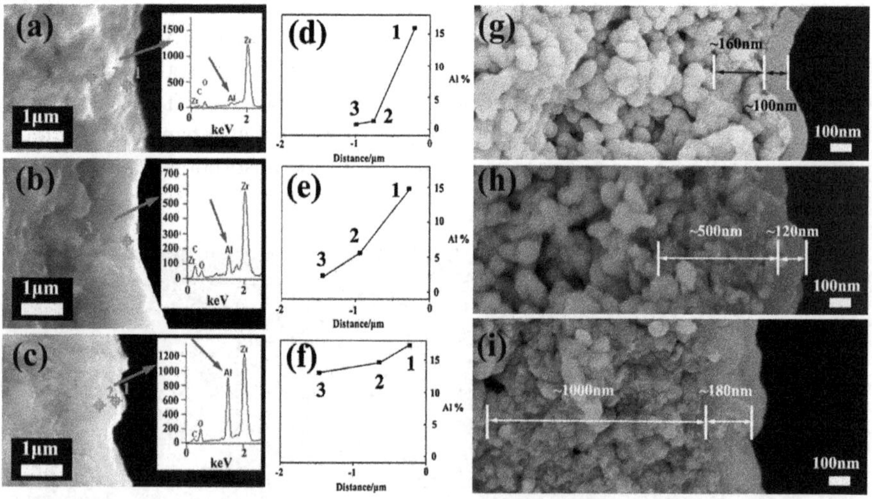

FIGURE 4.3 Cross sectional SEM images together with EDS spectra of ceramic membranes subjected to 600 atomic layer deposition cycles with varied exposure time: (a, d, and g) 0 s; (b, e, and h) 10 s; (c, f, and i) 40 s. The EDS spectra were recorded at three different points (Points 1–3 marked in the corresponding locations in SEM images). Only the EDS profile for Point 2 of each sample is shown in the inset of the corresponding SEM image. (d–f) Plots of the relative atomic percentages of Al recorded at different positions (Points 1–3 in the SEM images). (g–i) High magnification SEM images showing the surface layer and transition layer. (From Li, F. et al., Journal of Membrane Science. 397–398, 17–23, 2012.)

More recently, atmospheric pressure ALD has been explored for the fabrication of tight ceramic NF membranes (Shang et al., 2017). Compared to the work of Li et al. (2012), the deposition step could be successfully achieved without the need for an expensive vacuum system, and thus the atmospheric pressure ALD method is more cost-effective for the fabrication of modified membranes. The thickness of the TiO_2 coating was estimated to be 0.05 nm/cycle by AFM technique. It is worth noting that the increase of the number of cycles leads to the decrease of the pore size. This finding was fully evidenced by molecular weight cut-off (MWCO) characterization using polyethylene glycols (200 to 1000 g/mol). The performances were evaluated by measuring the water permeability before and after treatment. A slight improvement in water permeability was observed as the value increased from 11 to 16 L/m^2.h.bar.

4.4 SURFACE MODIFICATION OF CERAMIC MEMBRANES VIA CHEMICAL GRAFTING

In this section, we will discuss different synthetic strategies that are employed for the surface modification of ceramic membranes. Surface modification is considered the key step in tuning the morphological structure, the chemical composition, and thus the separation properties of ceramic membranes for water and wastewater treatment applications. Furthermore, membrane surface post-modification is shown to be an efficient route to solve one of the most critical issues in membrane technology, namely, the fouling tendency. To be able to prevent fouling problems while maintaining a correct flux is of high economic and ecological importance because it both increases the productivity and lifetime of membranes and reduces cleaning steps.

Significant research efforts have been devoted to developing surface modification strategies compatible with ceramic materials in order to enhance filtration performance. To this aim, many surface modifying agents such as Grignard reagents (Hosseinabadi et al., 2015), organophosphonic acids (Mustafa et al., 2014), metal/metal oxide particles (Chang et al., 2014; Li et al., 2012; Oun et al., 2017; Suresh et al., 2017), organosilanes (Gao et al., 2013; Khemakhem et al., 2014; Kujawa et al., 2013a; 2014b; Lee et al., 2016), and polyethylene glycols (Tanardi et al., 2016a) have been proposed. These materials can be divided into two categories: metal oxide particles and hydrophobic modifiers. The first category is frequently used for developing nanostructured membranes while the second is employed to prepare hydrophobic hybrid membranes. Table 4.2 summarizes the main characteristics of the different ceramic membranes made from natural and metal oxide materials after surface modification and their applications.

4.4.1 Membrane Hydrophobization

Although ceramic membranes have been successfully used in the industrial applications, fouling phenomena still remain highly challenging. To overcome this problem, the surface functionalization of ceramic membranes through efficient chemistry strategies and the use of different modifying agents have largely been considered. This has resulted in the production of tailor-made hybrid organo-modified ceramic membranes.

TABLE 4.2
Summary of the Mains Characteristic of Different Ceramic Membrane after Surface Modification and Their Applications

Modification Method	Modifying Agents	Membrane Materials	Water Permeability (J)	Contact Angle (CA)	Applications	References
Silanization	(3-aminopropyl)triethoxysilane, 3-(trihydroxysilyl)1-propanesulfonic acid and trimethoxy(propyl)silane	Alumina (Al_2O_3)	848–1037 L/m2h.bar	N /A	Humic acid filtration	Lee et al., 2016
	n-octyltrichlorosilane (C8Cl3) n-octyltriethoxysilane (C8OEt3) and trichloro(octadecyl)-silane (C18Cl3)	Titania (TiO_2) and alumina (Al_2O_3)	0.7–4.8L /m2h.bar	CA (TiO_2-C8) = 113° ± 2° CA (TiO_2-C8Cl3) = 128° ± 2° CA (TiO_2-C18Cl3) = 132° ± 2°	Water desalination	Kujawa et al., 2017
	1H,1H,2H,2H-perfluorooctyltriethoxysilane (C6), 1H,1H,2H,2H-perfluorotetradecyltri ethoxysilane (C12)	Titania (TiO_2)	J (TiO_2-C6) = 3.74 L/m2h.bar J (TiO_2-C12) = 3.86 L/m2h. bar	CA (TiO_2-C6) = 135° CA (TiO_2-C12) = 145°	Water desalination	Kujawa et al., 2014b
	Ethyltrimethoxysilane (C2), Hexyltrimethoxysilane (C6), Octyltrimethoxysilane (C8), and hexadecyltrimethoxy-silane (C12)	Alumina (Al_2O_3)	N/A	CA (unmodified membrane) = 9.1° CA (Al_2O_3-C6) = 126° ± 1° CA (Al_2O_3-C8) = 131.4° ± 1° CA (Al_2O_3-C16) = 140° ± 1°	Water-oil separation	Gao et al., 2013
	1H,1H,2H,2H-perfluorodecyltriethoxy-silane (C8)	Mud of hydrocyclone laundries of phosphates	J = 720 L/m2h.bar (unmodified membrane) J = 7 L/m2h.bar (functionalized membrane)	CA (unmodified membrane) = 25° CA (functionalized membrane) = 160°	Water-oil separation	Khemakhem et al., 2013
	1H,1H,2H,2H-perfluorodecyltriethoxy-silane (C8)	Tunisian clay	J = 155 L/m2 day (MF membrane) J = 110 L/m2 day (UF membrane)	CA (Clay-C8) = 180°	Water desalination	Khemakhem et al., 2011b

(Continued)

TABLE 4.2 (CONTINUED)
Summary of the Mains Characteristic of Different Ceramic Membrane after Surface Modification and Their Applications

Modification Method	Modifying Agents	Membrane Materials	Water Permeability (J)	Contact Angle (CA)	Applications	References
Phosphorylation	methyl phosphonic acid (MPA), phenyl phosphonic acid (PPA), and hexadecylphosphonicacid (HDPA)	Titania (TiO2)	J (TiO2-MPA) = 15 L/m2h.bar J (TiO2-MPA) = Low L/m2h.bar J (TiO2-MPA) = 0 L/m2h.bar	CA (TiO2-MPA) = 37° CA (TiO2-PPA) = 80° CA (TiO2-HDPA) = 124°	Drinking water production	Mustafa et al., 2014
	Methylmagnesium bromide (MGR), phenylmagnesium bromide (PGR) and Methylphosphonicacid (MPA)	Titania (TiO2)	J (unmodified UF membrane) = 170–200 L/m2h.bar J (TiO2-MGR) = 104 L/m2h.bar J (TiO2-MPA) = 118 L/m2h.bar J (unmodified NF membrane) = 10–30 L/m2h.bar J (TiO2-MGR) = 5–13 L/m2h.bar J (TiO2-PGR) = 5–14 L/m2h.bar J (TiO2-MPA) = 8–27 L/m2h.bar	UF membrane: N/A NF membranes: CA (unmodified NF membrane) = 10–20° CA (TiO2-MGR) = 50–60° CA (TiO2-PGR) = 50–60° CA (TiO2-MPA) = 35–45°	Wastewater treatment (paper mill effluents, olive oil wastewater)	Mustafa et al., 2016
Polymer grafting	Triglyceride	Mud of hydrocyclone laundries of phosphates	J (unmodified membrane) = 90 L/m2h.bar J (functionalized membrane) = 7 L/m2h.bar	CA (unmodified membrane) = 16° CA (functionalized membrane) = 121°	Sea water desalination	Derbel et al., 2017

Native ceramic membranes prepared using unique or mixed metal oxide(s) and clay are characterized by their high hydrophilic character due to the presence of abundant surface hydroxyl groups. This hydrophilicity may induce water adsorption and/or capillary condensation in the pores, reducing the number of pores. Numerous studies have been dedicated to the introduction of hydrophobic chemical moieties onto ceramic membrane surface through strong interfacial bonding involving the replacement of the -OH surface groups by organic moieties.

4.4.1.1 Grignard Reagents

Grignard reagents are a class of coupling agents proposed as a new class of surface modifiers of metal oxides (Hosseinabadi et al., 2015; Mustafa et al., 2016). These organometallic molecules have the generic formula of RMgX, where X is a halogen and R is an alkyl or aryl group. They were used for the first time by Van Heetvelde et al. (2013) for the chemical modification of titanium alkoxides as an alternative to organosilane coupling agents. Grignard modification yields non-hydrolysable bonding between the surface of the metal oxide and the functional group, since the Grignard organic groups are directly bonded through a Metal-C bond at room temperature, as schematically shown in Figure 4.4 (Mustafa et al., 2014). These bonds in the submonolayer coating exhibit higher stability towards hydrolysis than the M-O-Si bonds obtained with silanes. The benefit of such hybrid membranes relies on the combination of both organic and inorganic materials, which induces new interesting properties. Interestingly, Grignard modification method makes membrane amphiphilic due to the partial surface replacement of the membrane surface -OH groups by the Grignard hydrophobic groups.

Hosseinabadi et al. (2014) have successfully attached different organometallic Grignard molecules (C_1 to C_{12} groups) to 1 nm TiO_2 membranes in dry diethyl ether solvent, producing a series of hydrophobic titania ceramic membranes. After functionalization, removal of any remaining by-product, namely MgBr salt, requires multiple successive washings of the membranes in Et_2O, HCl (0.1 or 1 M), and H_2O. Examination of the modified membranes by ATR/FTIR spectroscopy revealed the presence of organic moieties on the membrane surface, leading to an increase in the contact angle from 19° (unmodified TiO_2 membrane) to 83° (TiO_2-C_8 membrane). This confirmed the partial hydrophobization of the ceramic membrane.

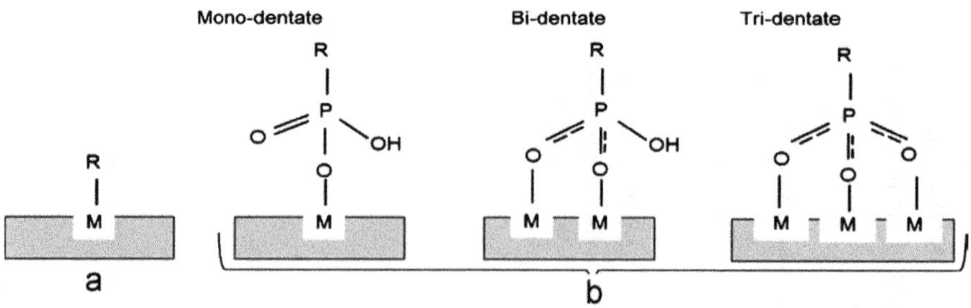

FIGURE 4.4 Schemes of the possible interactions of (a) Grignard reagents and (b) phosphonic acids with the surface of metal oxides. (From Mustafa, G. et al., Journal of Membrane Science. 470, 369–377, 2014.)

The as-prepared amphiphilic TiO_2 ceramic membranes were used to treat wastewater containing organic solvents, and they exhibited higher flux for tetrahydrofuran, toluene, n-hexane, and polyethylene glycol retentions in water than unmodified membranes.

This Grignard grafting method has been found to be very promising in developing-enhanced antifouling TiO_2 ceramic membranes applied in filtration of three challenging wastewater streams, namely pulp and paper mill effluent, olive oil wastewater, and oil/water emulsions (Mustafa et al., 2016). Mustafa et al. (2016) reported that the membranes grafted with methyl groups in particular showed an enhancement of filtration performances.

4.4.1.2 Organophosphonic Acids

Phosphorylation using organophosphonic acids has been applied to derivatize ceramic membranes with organic moieties that act as surface hydrophobization agents. Moreover, the phosphorylation strategy is known to be fast, cheap, and can be successfully performed in water in contrast with the Grignard surface modification, which requires anhydrous organic media. This method involves generally three reactive groups (P-OH, P-OR, and PO) resulting in the formation of strong mono-, bi-, or tridendate bonds with the membrane surface. However, implementation of the phosphorylation approach in surface modification of ceramic membranes remains limited because of the low commercial availability of organophosphonic acid derivatives.

Randon et al. (1995) applied this strategy for the functionalization of TiO_2 and ZrO_2 mesoporous ceramic membranes with phosphoric acid and alkyl (methyl, ethyl, and phenyl) phosphonic acids. The grafting reactions were performed in a 0.1 M toluene solution of phosphonic acid precursor at reflux during 4 h and were shown to proceed successfully by infrared analysis. The separation properties of the resulting hydrophobic membranes have been studied by UF process and demonstrated high efficiency in filtering bovine serum albumin (BSA).

Mustafa et al. (2014) have compared organophosphonic acids and Grignard reagents in the chemical modification of commercial TiO_2 membranes by using different functional groups. In the case of the Grignard method, alkyl and phenyl groups have been grafted onto the TiO_2 membranes under dry atmosphere using methyl and phenyl magnesium. The reaction was carried out in diethyl ether solvent for 48 h at room temperature. It should be noted that prior to the grafting reaction, the TiO_2 membranes were pre-treated in order to remove traces of adsorbed water. In the case of surface phosphorylation, methyl-, phenyl-, and hexadecyl phosphonic acid groups have been used. The authors showed an increase of the water contact angle and a decrease of the water permeability values after phosphorylation of the TiO_2 membranes with the different groups. These changes were attributed to the transformation of the surface characteristics of the membranes from hydrophilic to hydrophobic ones. Hydrophobization of the membranes induced a decrease in water flux. Interestingly, the fouling tendency of the functionalized tubular TiO_2 membranes, as determined using humic acid as model foulant, decreased obviously compared with the unmodified membrane. It has been shown that the effect of the surface functionalization on the membrane fouling depended strongly on both the grafting method and the chemical nature of the surface modifier. The most efficient antifouling effect was obtained with methyl Grignard reagent.

Caro et al. (1998) also compared silanization and phosphorylation of UF Al_2O_3 ceramic membranes by organic moieties. XPS analyses have shown that, in the case of silanization reaction, the surface modification was efficient when the Al_2O_3 membrane surface contained a Si-OH-rich thin layer. Phosphorylation was successful by reacting the phosphonic acid with the Al on the membrane surface. Investigation of the thermal stability of the phosphonic-modified membranes revealed that the silylated membranes were stable up to 230°C while the phosphonic-functionalized ones were stable up to 400°C. The effect of the surface modification on both single gases permeation and liquid mixtures pervaporation has been demonstrated.

4.4.1.3 Organosilanes

Organosilane agents have been shown to be useful for the functionalization of the surface of ceramic membranes with a high variety of both hydrophobic and hydrophilic moieties. These molecules with formula of either $X\text{-}CH_2CH_2CH_2Si(OR)_{3\text{-}n}$ or $X\text{-}CH_2CH_2CH_2SiCl_3$, where n = 0, 1, 2, OR is a hydrolysable alcoxy group, and X is an organo-functional group, bear two different reactive groups (OR or Cl and X) on their silicon atom and permit their coupling with a large variety of materials surface. The chemical grafting of an organosilane layer onto ceramic membranes is achieved through a silanization reaction in polar aprotic solvents between the surface hydroxyl groups arising from the ceramic membrane and the chloro or alkoxy groups coming from organosilane agents. The silanization reaction consists of two main steps involving (1) hydrolysis of the trialcoxy or chloro groups of silane that generates silanol (Si-OH) moieties and (2) reaction of the silanols with hydroxyl groups on the membrane surface that results in the chemical bonding of a silane layer onto the ceramic membrane. In order to avoid the polycondensation of organosilane molecules in the presence of traces of air or solvent moisture, the silanization reaction requires anhydrous solvent and an inert atmosphere.

The preparation of ceramic membranes meant to be used for desalination and oil-water separation requires a hydrophobic membrane surfaces. Generally, this modification leads to an increase of the membrane roughness and a decrease of membrane pore size. This can be explained by the condensation of silane moieties at the membrane surface, which may increase to a large extent the degree of densification.

Various examples of silanized ceramic membranes with hydrophobic surfaces have been reported in literature. Perfluoroalkyltriethoxysilanes and/or perfluoroalkyltrichlorosilanes are the most widely used hydrophobic agents to this aim. Khemakhem et al. (2011b) grafted triethoxy-1H,1H,2H,2H perfluorodecylsilane onto the MF (d_{pore} = 0.18 μm) and NF (d_{pore} = 15 nm) planar and tubular membranes made of Tunisian clay. The surface functionalization was performed under mild conditions (room temperature and 15 min of grafting time in silane solution) using 0.01 M silane solution. Water contact angle measurements and infrared characterization showed that all the membranes have been effectively grafted with super-hydrophobic moieties. After surface modification, the membranes exhibited non-wettable surfaces as ascertained by the contact angle value of approximately 180°. The chemical attachment of fluoroalkyl moieties onto the membrane surface has been demonstrated by the presence of the stretching vibrations of C-C, Si-C, and C-F at 1540 cm^{-1}, 1270 cm^{-1}, and 1240 cm^{-1}, respectively. The success of seawater desalination using air gap membrane distillation further suggested the great potential of these hydrophobic membranes.

The same research group, in pursuit of their investigations on developing hydrophobized ceramic membranes, has described the surface modification of ceramic microfiltration membrane made of zirconium/mud of hydrocyclone laundries of phosphates (Khemakhem et al., 2014). In the presence of triethoxy-1H,1H,2H,2H perfluorodecylsilane as a grafting agent, they reported a significant increase in the contact angle from 25° to values higher than 150°, as well as a sharp decrease in the pore size of the membrane as shown by SEM analysis. Similarly, Das et al. (2016) reported the hydrophobization of capillary clay-alumina-based ceramic membranes prepared using 55 wt% of clay with a pore size of 1.43 μm in diameter. Both the water contact angle (145°) and liquid entry pressure (1 bar) revealed the hydrophobic character of the clay-alumina functionalized membrane.

As an alternative to this type of organosilanes, the use of alkylsilanes with different length as hydrophobic surface modifiers has been proposed by Gao et al. (2013). The authors studied the influence of ceramic alumina membranes hydrophobicity on the filtration performance of water-in-oil emulsions. Ethyltrimethoxysilane, hexyltrimethoxysilane, octyltrimethoxysilane, and hexadecltrimethoxysilane were used as organic modifiers. The chemical composition as well as the morphological surface of the as-prepared membranes were studied using X-ray photoelectron spectroscopy, infrared characterization, thermogravimetric analysis, and SEM. Although no significant changes in surface morphology have been highlighted, a decrease in the pore size together with an increase of the surface hydrophobicity and change in the chemical composition were observed. Despite the small changes in the membrane surface morphology, the modified membranes exhibited greater hydrophobic character than the pristine membrane. Moreover, the membrane hydrophobicity increased with increasing the length of the silane alkyl chains.

To study the effect of the surface charge on antifouling properties of ceramic membranes, Lee at al. (2016) described a similar strategy. The authors used three different functional alkylsilanes, namely trimethoxy(popyl)silane, (3-aminopropyl) triethoxysilane, and 3-(trihydroxysilyl)-1-propanesulfonic acid that possessed neutral (-CH_3), positive (-NH_2), and negative (-SO_3) charge, respectively, to plane alumina ceramic membranes. The alumina membranes were first washed with ethanol and subsequently introduced separately in a 0.1 M silane solution prepared in anhydrous ethanol for 5 h at room temperature. SEM, mercury intrusion porosimetry, and thermogravimetric analysis have shown no change in the membrane surface morphology, and only a slight decrease in the average pore size of the alumina membranes was observed after surface silanization. The authors explained these negligible surface modifications by the ultrathin organosilane grafts (thickness of approximately 0.1 μm). When tested using humic acid feed solution, the negatively charged (SO_3-functionalized) membranes exhibited the lowest fouling (flux recovery rate of 80–85%), and a noticeable resistance to flux decline. As a comparison, the amine-grafted membrane and the unmodified membrane exhibited a higher degree of flux decline and a lower flux recovery rate (52%), which could be explained by their electrostatic interactions with the negatively charged humic acid (isoelectric point = 4.7).

4.4.1.4 Organic Polymers

The surface functionalization of ceramic membranes through the deposition or grafting of macromolecules holds huge potential in water treatment processes. Indeed, ceramic-polymer composite membranes are a very interesting class of membranes as they combine the high chemical, thermal, and mechanical stability of ceramic membranes with the ease of surface modification and the low cost of organic polymers. As an important result, this combination imparts high selectivity and thus provides efficient water purification.

Mittal et al. (2011) prepared membranes by dip-coating hydrophilic UF ceramic membranes using hydrophilic cellulose acetate (10% w/v in acetone). The effective pore size of the deposited polymer top-layer, calculated by a novel theoretical approach based on the mass of the polymer film and the hydraulic permeability of the composite membrane, was found to be in the UF range (28 nm). The authors indicated that the chemical resistance of the prepared composite membrane was highly dependent on pH, as a slight change was detected in the polymer film after acid treatment contrary to basic treatment, which induced approximately 19.6% change in porosity. It has been shown that the acetate cellulose-ceramic membranes are suitable for separation of oil molecules from low concentrated oily wastewater (<250 mg/L).

Taking advantage of the strong adhesion between organosilanes and ceramic membranes, organosilanes were used as coupling agents in the preparation of polymer-based ceramic membranes. More precisely, the reactive functionality of silane-grafted ceramic membranes enabled the chemical attachment of polymer chains onto the ceramic support. Two strategies consisting of two successive steps can be envisioned to graft polymer films on ceramic membranes, i.e., (1) modification of the ceramic membrane surface by organosilane moieties followed by the coupling of the polymer and (2) functionalization of the polymer chains by organosilane agents prior to their grafting onto ceramic membranes. The latter strategy has been used by Tanardi et al. (2016a) for the grafting of various hydrophilic silylated PEG agents on mesoporous γ-alumina membranes. PEGs were initially chemically modified by the reaction between amines and isocyanates groups. The silane terminated PEGs solution (3 mM in toluene) were then considered to accomplish the silylation reaction between the silane terminated PEGs and the mesoporous membrane for 24 h at 110°C, as schematically illustrated in Figure 4.5.

The effective grafting of PEG moieties to the alumina surface was proven by FTIR and ^{29}Si-NMR. The degree of self-condensation is about the same for most of the grafting agents. The grafting density decreased for PEG with long chain length.

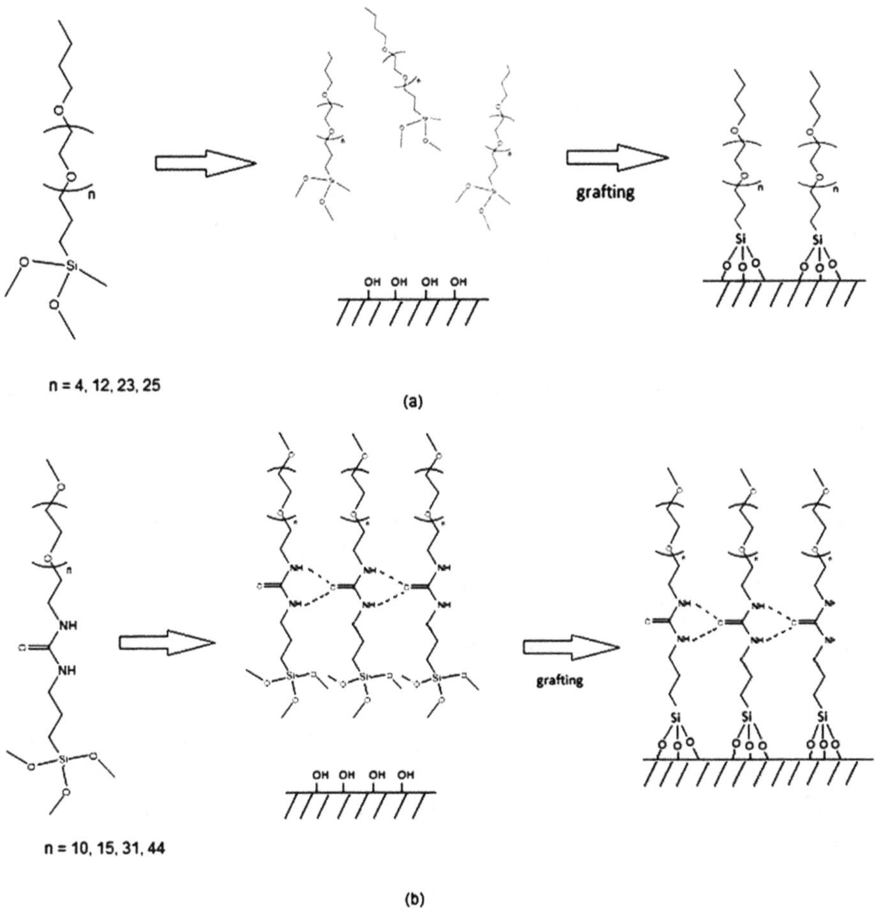

FIGURE 4.5 Schematic illustration of the grafting mechanism by (a) the non-self-assembling silylated PEG grafting agents versus (b) the self-assembling silylated ureido PEG grafting agents. (From Tanardi, C.R. et al., Microporous and Mesoporous Materials. 229, 106–116, 2016a.)

4.5 APPLICATION OF CHEMICALLY MODIFIED CERAMIC MEMBRANES

Functionalized ceramic membranes have gained increasing interest for separation and purification issues, especially in the context of water purification. The surface of ceramic membranes can be easily modified with a plethora of chemical modifiers that provide highly flexible and robust solutions for the separation of solutes from complex mixtures with enhanced selectivity, permeate water recycling and reuse, as well as improved antifouling properties. Two domains of application of chemically modified ceramic membranes are presented in the following subsections. They are (1) desalination using membrane distillation method and (2) water/oil separation.

4.5.1 Water Desalination

Membrane distillation (MD) is a very useful strategy for water desalination. The method consists in a thermally driven separation process of two solutions. The difference of temperature between the feed and the permeate generates a water vapor pressure gradient that acts as the driving force for the water transfer in the MD process. A three-step mechanism has been suggested for the mass transfer based on (1) evaporation of the feed solution and vapor formation at the membrane pores entrance, (2) transport of the vapor through the membrane pores, and (3) condensation of the vapor at the permeate side. In order to prevent the undesirable phenomenon of penetration of the aqueous solutions into the MD pores, non-wetted hydrophobic porous membranes are required.

Khemakhem et al. (2013) proposed an air-gap membrane distillation (AGMD) process using grafted MF tubular mud/zirconia membrane with 1H,1H,2H,2H-perfluorodecyltriethoxysilane (C_8) for water desalination as shown in Figure 4.6. The authors indicated that the modified membrane was able to reject salt from oily wastewater with a rejection rate higher than 99% and a membrane permeability of 7 L/m^2.h.bar. They also found that the change in the salt concentration in the feed solution did not alter the retention of salts in the AGMD process. Using the same experimental setup, Khemakhem et al. (2011b) found that the C_8-grafted MF and UF tubular clay membranes were suitable for seawater desalination. They reported that the membranes could achieve a permeate flux of 155 and 110 L/m^2.day, respectively, with a complete removal of salts.

Similarly, Kujawa et al. (2014b) prepared hydrophobic hybrid titania ceramic membranes (300 kD) by chemical grafting of $C_6F_{13}C_2H_4Si(OC_2H_5)_3$ and $C_{12}F_{25}C_2H_4Si(OC_2H_5)_3$ perfluorinated agents and used the resultant membranes for NaCl removal. The effects of operating conditions, namely feed temperature,

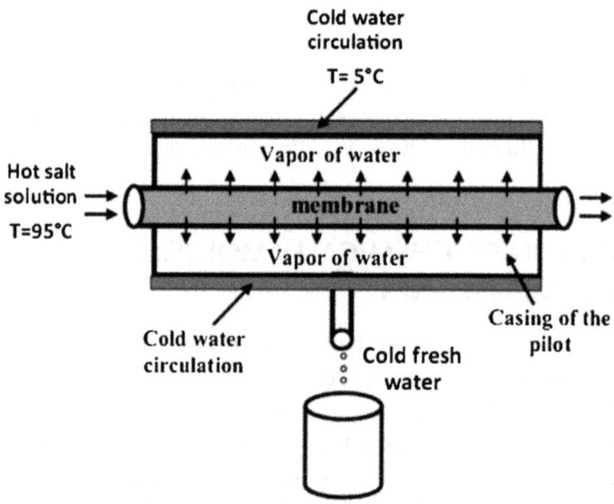

FIGURE 4.6 Schematic illustration of air-gap membrane distillation in counter current flow configuration. (From Khemakhem, M. et al., Colloids and Surfaces A: Physicochemical and Engineering Aspects. 436, 402–407, 2013.)

salt feed concentration, and MD mode (AGMD or direct contact MD (DCMD) on the membrane performance were investigated. They showed that a higher permeate flux could be obtained using the membrane grafted with a smaller alkyl group (C_6) regardless of MD mode, achieving >99% NaCl retention. Under the feed temperature of 90°C and salt feed concentration of 2 M, the membrane grafted with C6 exhibited water flux of 2300 g/m^2.h and 2100 g/m^2.h for AGMD and DCMD mode, respectively.

More recently, Kujawa et al. (2017) grafted three various non-fluorinated hydrophobic coupling agents to TiO_2 and Al_2O_3 ceramic membranes for the AGMD of desalination. They observed that the permeate fluxes were in the range of 0.7–4.8 kg/m^2.h for tests with pure water whereas the fluxes were in the range of 0.4–2.8 kg/m^2.h in contact with salt solution. Interestingly, all the hydrophobic prepared membranes showed a NaCl retention coefficient higher than 98%.

Derbel et al. (2017) used Tunisian olive oil to chemically modify ceramic UF membranes made of mud of hydrocyclone laundries from hydrophilic to hydrophobic. Increasing the contact angle values from 16° before modification to 121° after olive oil grafting revealed highly hydrophobic membranes. This membrane hydrophobization led to a significant decreasing of the water permeability from 90 L/m^2.h before modification to 7 L/m^2.h after modification. Upon grafting the olive oil on the UF membrane, 99% salt rejection was achieved using feed solution containing 30 g/L NaCl.

Ren et al. (2015) grafted 0.76 μm alumina planar membranes with fluoroalkyl silane using phase-inversion tape casting method for DCMD of water desalination. They tested the water desalination performance by exposing the hydrophobic separation layer to different concentrations of NaCl aqueous solution (2–6 wt%). They observed that the hydrophobic membranes could be operated up to 2 bar. When tested using a salt solution of 2 wt%, the permeate flux was 19.1 L/m^2.h under a feed temperature of 80°C. The hydrophobic membrane achieved >99.5% salt rejection. It is interesting to note that the MD flux exhibited a slight decrease when the concentration increased.

4.5.2 Oil/Water Separation

Separation of oil and water from stable oil/water emulsions is very challenging because of the sub-micrometer size of the oil droplets. Conventional oily wastewater treatment processes such as gravity separation, coagulation, thermal de-emulsification, free-floating oil, flocculation, and biological treatments show very poor removal of low oil concentrations (Zhu et al., 2014). Among the existing oil/water separation membranes, hydrophilic membranes exhibited the most effective fouling resistance behavior. As such, chemically-modified ceramic UF membranes are appealing candidates for oil/water separation.

Mittal et al. (2011) prepared cellulose acetate-grafted clay ceramic membranes using the dip-coating method for the treatment of oil-in-water emulsion with concentration of 50–200 mg/mL. The maximum rejection of 93% was obtained at a pressure of 138 kPa after 41 min of filtration using a feed solution with 200 mg/L oil concentration. The membrane water flux meanwhile was recorded at 6.66 L/m^2.h.bar.

Gao et al. (2013) grafted four non-fluorinated alkylsilanes (C_2, C_6, C_8, and C_{16}) to Al_2O_3 membranes and employed the resultant membranes for oily wastewater treatment at 0.1 MPa. The separation tests of kerosene from water revealed that the hydrophobic coated membranes showed improved permeate flux compared to the unmodified membranes (214.9 L/m^2.h) and that the highest oil flux could be obtained using the C_{16}-modified alumina membrane (668.7 L/m^2.h). The separation efficiency for the unmodified and the four grafted membranes were 75.4%, 77.8%, 85.2%, 84.2%, and 84.3%, respectively, which clearly indicates the increase of the retention of the modified membrane.

Hydrophilized ceramic membranes are also found to be useful for the separation of oil and water. Chen et al. (2016) designed superhydrophilic ceramic membranes for the treatment of oil-contaminated wastewater using in-situ surface membrane modification via the deposition of Stöber silica inorganic nanoparticles. The silica nanoparticles top separation layer has been obtained by immersion of ceramic membrane supports previously fabricated from a mixture of kaolin, quartz, alumina oxide, and corn starch raw materials in a homogeneous silica sol in aqueous media in the presence of ammonia as a catalyst. The resultant hierarchical-structured silica membrane has been reported to offer excellent separation efficiency for 658 ppm oil-in-water emulsion with oil droplet mean diameter of 1.54 μm, achieving >99.95% oil rejection under mild pressure-driven conditions (0.1 MPa).

4.6 CONCLUSION

For several decades, ceramic membrane materials have been synthesized using a plethora of raw materials and synthetic routes enabling the fine control of both membrane support properties and the surface chemistry of the active membrane layer with respect to pore size and size distribution, chemical nature, roughness, and hydrophilicity or hydrophobicity. The ceramic membrane support affords mechanical resistance for applications under pressure-driven conditions while the membrane selective layer provides specific separation mechanism and wettability properties. However, significant research is still needed related to the reduction of the production cost of ceramic membranes, reduction and prevention of membrane fouling and pores clogging during cake formation, enhancement of membrane life time through efficient cleaning, and regeneration processes. The wide range of potential applications of ceramic membranes in the water treatment area will continue to drive research in this field forward. Recent studies on surface modification of ceramic supports through either structuration with metal oxide nanoparticles or chemical grafting are very promising. The former approach allows for the design of asymmetric membranes with finely controlled pore size that depends on both the size of the metal oxide nanoparticles and the thickness of the membrane layer. A decrease in pore size down to some tens of nanometers gives access to NF process that is highly relevant for water and wastewater treatment. In the future, it is expected that metal oxide-coated ceramic membranes will be widely applied to the filtration process in combination with photocatalysis, preferably under visible light, for the treatment of effluents. We also summarized the recent grafting strategies involving Grignard reagents, organophosphonic acids, organosilanes, and polymers applied to the robust chemical modification of ceramic membrane surface. The covalent attachment of organic moieties on ceramic membrane surface ensures the robustness of the coating and enables a change in the intrinsic hydrophilic nature of ceramic membranes to make the membrane hydrophobic or in contrast, to enhance the membrane's hydrophilicity. Grafting time and density, together with the type (fluorinated versus aliphatic) of the grafting molecules as well as the grafting conditions (in water versus organic solvent), and hydrolytic and thermal stability of the bonds are key parameters to be considered. Considering membrane hydrophilization, the grafting of PEG derivatives proved highly efficient to minimize membrane fouling while hydrophobic modified membranes are very promising for MD of water desalination and the separation of oil and water.

REFERENCES

Alami Younssi, S., Iraqi, A., Rafiq, M., Persin, M., Larbot, A., Sarrazin, J., 2003. Alumina membranes grafting by organosilanes and its application to the separation of solvent mixtures by pervaporation. Separation Science Technology. 32, 175–179.

Almandoz, M.C., Pagliero, C.L., Ochoa, N.A., Marchese, J., 2015. Composite ceramic membranes from natural aluminosilicates for microfiltration applications. Ceramics International. 41, 5621–5633.

Aloulou, H., Bouhamed, H., Ben Amar, R., Khemakhem, S., 2017. New ceramic microfiltration membrane from Tunisian natural sand: Application for tangential wastewater treatment. Desalination and Water Treatment. 78, 41–48.

Anderson, M.A., Gieselmann, M.J., Xu, Q., 1988. Titania and alumina ceramic membranes. Journal of Membrane Science. 39, 243–258.

Arkell, A., Olsson, J., Wallberg, O., 2014. Process performance in lignin separation from softwood black liquor by membrane filtration. Chemical Engineering Research and Design. 92, 1792–1800.

Athanasekou, C.P., Papageorgiou, S.K., Kaselouri, V., Katsaros, F.K., Kakizis, N.K., Sapalidis, A.A., Kanellopoulos, N.K., 2009. Development of hybrid alginate/ceramic membranes for Cd^{2+} removal. Microporous and Mesoporous Materials. 120, 154–164.

Ben Amar, R., Gupta B.B., Jaffrin M.Y., 1990. Clarification of apple juice using crossflow microfiltration with mineral membranes: Membrane fouling control by backpulsing and pulsatil flow. Journal of Food Science. 55, 1620–1620.

Biswas, D.P., O'Brien-Simpson, N.M., Reynolds, E.C., O'Connor, A.J., Tran, P.A., 2018. Comparative study of novel in situ decorated porous chitosan-selenium scaffolds and porous chitosan-silver scaffolds towards antimicrobial wound dressing application. Journal of Colloid and Interface Science. 515, 78–91.

Bogianchini, M., Cerezo, A.B., Gomis, A., López, F., García-Parrilla, M.C., 2011. Stability, antioxidant activity and phenolic composition of commercial and reverse osmosis obtained dealcoholised wines. LWT - Food Science and Technology. 44, 1369–1375.

Bouazizi, A., Saja, S., Achiou, B., Ouammou, M., Calvo, J.I., Aaddane, A., Alami Younssi, S., 2016. Elaboration and characterization of a new flat ceramic MF membrane made from natural Moroccan bentonite. Application to treatment of industrial wastewater. Applied Clay Science. 132–133, 33–40.

Bouazizi, A., Breida, M., Karima, A., Achiou, B., Ouammou, M., Calvo, J.I., Aaddane, A., Khiat, K., Alami Younssi, S., 2017. Development of a new TiO_2 ultrafiltration membrane on flat ceramic support made from natural bentonite and micronized phosphate and applied for dye removal. Ceramics International. 43, 1479–1487.

Bouzerara, F, Harabi, A., Condom, S., 2009. Porous ceramic membranes prepared from kaolin. Desalination and Water Treatment. 12, 415–419.

Bouzerara, F., Boulanacer, S., Harabi, A., 2015. Shaping of microfiltration (MF) ZrO_2 membranes using a centrifugal casting method. Ceramics International. 41, 5159–5163.

Cai, Y., Wang, Y., Chen, X., Qiu, M., Fan, Y., 2015. Modified colloidal sol–gel process for fabrication of titania nanofiltration membranes with organic additives. Journal of Membrane Science. 476, 432–441.

Caro, J., Noack, M., Kolsch, P., 1998. Chemically modified ceramic membranes. Microporous and Mesoporous Materials. 22, 321–332.

Chang, Q., Wang, Y., Cerneaux, S., Zhou, J-E., Zhang, X., Wang, X., Dong, Y., 2014. Preparation of microfiltration membrane supports using coarse alumina grains coated by nano TiO_2 as raw materials. Journal of the European Ceramic Society. 34, 4355–4361.

Chen, T., Duan, M., Fang, S., 2016. Fabrication of novel superhydrophilic and underwater superoleophobic hierarchically structured ceramic membrane and its separation performance of oily wastewater. Ceramics International. 42, 8604–8612.

Chen, W., Fadeev, A.Y., Hsieh, M.C., Oner, D., Youngblood, J., McCarthy, T.J., 1999. Ultrahydrophobic and ultralyophobic surfaces: Some comments and examples. Langmuir. 15, 3395–3399.

Chowdhury, S., Schmuhl, R., Keizer, K., Ten Elshof, J., Blank, D., 2003. Pore size and surface chemistry effects on the transport of hydrophobic and hydrophilic solvents through mesoporous ɤ-alumina and silica MCM-48. Journal of Membrane Science. 225, 177–186.

Cissé, M., Vaillant, F., Pallet D., Dornier, M., 2011. Selecting ultrafiltration and nanofiltration membranes to concentrate anthocyanins from roselle extract (Hibiscus sabdariffa L.). Food Research International. 44, 2607–2614.

Das, R., Sondhi, K., Majumdar, S., Sarkar, S., 2016. Development of hydrophobic clay-alumina based capillary membrane for desalination of brine by membrane distillation. 4, 243–251.

Derbel, I., Khemakhem, M., Cerneaux, S., Cretin, M., Ben Amar, R., 2017. Grafting of low cost ultrafiltration ceramic membrane by Tunisian olive oil molecules and application to air gap membrane distillation. Desalination and Water Treatment. 82, 20–25.

Díaz-Reinoso, B., Moure, A., Domínguez H., Parajó, J.C., 2009. Ultra- and nanofiltration of aqueous extracts from distilled fermented grape pomace. Journal of Food Engineering. 91, 587–593.

Gaber, A.A.A., Ibrahim, D.M., Fawzia Fahm Abd-Almohsen, F.F., El-Zanati E.M., 2013. Synthesis of alumina, titania, and alumina-titania hydrophobic membranes via sol–gel polymeric route. Journal of Analytical Science and Technology. 4, 1–20.

Gao, N., Ke, W., Fan, Y., Xu, N., 2013. Evaluation of the oleophilicity of different alkoxysilane modified ceramic membranes through wetting dynamic measurements. Applied Surface Science. 283, 863–870.

Garmsiri, E., Rasouli, Y., Abbasi, M., Izadpanah, A.A., 2017. Chemical cleaning of mullite ceramic microfiltration membranes which are fouled during oily wastewater treatment. Journal of Water Process Engineering. 19, 81–95.

Ghouil, B., Harabi, A., Bouzerara, F., Boudaira, B., Figoli, A., 2015. Development and characterization of tubular composite ceramic membranes using natural alumino-silicates for microfiltration applications. Materials Characterization. 103, 18–27.

Guizard, C., 1996. Sol-Gel Chemistry and its Application to Porous Membrane Processing. In Burggraaf A.J., Cot, L. (Eds.), Fundamentals of Inorganic Membrane Science and Technology, 689. The Netherlands: Elsevier Science.

Guo, H., Zhao, S., Wu, X., Qi, H., 2016. Fabrication and characterization of TiO_2/ZrO_2 ceramic membranes for nanofiltration. Microporous and Mesoporous Materials. 260, 125–131.

Hosseinabadi, S.R., Wyns, K., Meynen, V., Carleer, R., Adriaensens, P., Buekenhoudt, A., Vander Bruggen, B., 2014. Organic solvent nanofiltration with Grignard functionalised ceramic nanofiltration membranes. Journal of Membrane Science. 454, 496–504.

Hosseinabadi, S.R., Wyns, K., Buekenhoudt, A., Van der Bruggen, B., Ormerod, D., 2015. Performance of Grignard functionalized ceramic nanofiltration membranes. Separation and Purification Technology. 147, 320–328.

Jedidi, I., Saïdi, S., Khemakhem, S., Larbot, A., Elloumi-Ammar, N., Fourati, A., Charfi, A., Ben Salah, A., Ben Amar, R., 2009a. Elaboration of new ceramic microfiltration membranes from mineral coal fly ash applied to waste water treatment. Journal of Hazardous Materials. 172, 152–158.

Jedidi, I., Khemakhem, S., Larbot, A., Ben Amar, R., 2009b. Elaboration and characterisation of fly ash based mineral supports for microfiltration and ultrafiltration membrane. Ceramics International. 35, 2747–2753.

Jedidi, I., Khemakhem, S., Saïdi, S., Larbot, A., Elloumi-Ammar, N., Fourati, A., Charfi, A., Ben Salah, A., Ben Amar, R., 2011. Preparation of a new ceramic microfiltration membrane from mineral coal fly ash: Application to the treatment of the textile dying effluents. Powder Technology. 208, 427–432.

Khemakhem, M., Khemakhem, S., Ben Amar, R., 2013. Emulsion separation using hydrophobic grafted ceramic membranes by Air Gap Membrane Distillation process. Colloids and Surfaces A: Physicochemical and Engineering Aspects. 436, 402–407.

Khemakhem, M., Khemakhem, S., Ben Amar, R., 2014. Surface modification of microfiltration ceramic membrane by fluoroalkylsilane. Desalination and Water Treatment. 52, 1786–1791.

Khemakhem, M., Khemakhem, S., Ayedi, S., Cretin, M., Ben Amar, R., 2015. Development of an asymmetric ultrafiltration membrane based on phosphates industry sub-products. Ceramics International. 41, 10343–10348.

Khemakhem, S., Ben Amar, R., Ben Hassen, R., Larbot, A., Ben Salah, A., Cot, L., 2006a. Fabrication of mineral supports of membranes for microfiltration/ultrafiltration from Tunisian clay. Annales de Chimie Science des Matériaux. 31, 169–181.

Khemakhem, S., Larbot, A., Ben Amar, R., 2006b. Study of performances of ceramic microfiltration membrane from Tunisian clay applied to cuttlefish effluents treatment. Desalination. 200, 307–309.

Khemakhem, S., Ben Amar, R., Larbot, A., 2007. Synthesis and characterization of a new inorganic ultrafiltration membrane composed entirely of Tunisian natural illite clay. Desalination. 206, 210–214.

Khemakhem, S., Larbot, A., Ben Amar, R., 2009. New ceramic microfiltration membranes from Tunisian natural materials: Application for the cuttlefish effluents treatment. Ceramics International. 35, 55–61.

Khemakhem, S., Ben Amar, R., 2011a. Grafting of fluoroalkylsilanes on microfiltration Tunisian clay membrane. Ceramics International. 37, 3323–3328.

Khemakhem, S., Ben Amar, R., 2011b. Modification of Tunisian clay membrane surfaceby silane grafting: Application for desalination with air process. Colloids and Surfaces A: Physicochemical and Engineering Aspects. 387, 79–85.

Khemakhem, M., Khemakhem, S., Ayedi, S., Ben Amar, R., 2011c. Study of ceramic ultrafiltration membrane support based on phosphate industry subproduct: Application for the cuttlefish conditioning effluents treatment. Ceramics International. 37, 3617–3625.

Kosinov, N., Gascon, J., Kapteijn, F., Hensen, E.J.M., 2016. Recent developments in zeolite membranes for gas separation. Journal of Membrane Science. 499, 65–79.

Kouras, N., Harabia, A., Bouzerara, F., Foughali, L., Policicchio, A., Stelitano, S., Galiano, F., Figoli, A., 2017. Macro-porous ceramic supports for membranes prepared from quartz sand and calcite mixtures. Journal of the European Ceramic Society. 37, 3159–3165.

Kujawa, J., Kujawski, W., Koter, S., Jarzynka, K., Rozicka, A., Bajda, K., Cerneaux, S., Persin, M., Larbot, A., 2013a. Membrane distillation properties of TiO_2 ceramic membranes modified by perfluoroalkylsilanes. Desalination and Water Treatment. 51, 1352–1361.

Kujawa, J., Kujawski, W., Koter, S., Rozicka, A., Cerneaux, S., Persin, M., Larbot, A., 2013b. Efficiency of grafting of Al_2O_3, TiO_2 and ZrO_2 powders by perfluoroalkylsilanes. Colloids and Surfaces A: Physicochemical and Engineering Aspects. 420, 64–73.

Kujawa, J., Cerneaux, S., Kujawski, W., 2014a. Investigation of the stability of metal oxide powders and ceramic membranes grafted by perfluoroalkylsilanes. Colloids and Surfaces A: Physicochemical and Engineering Aspects. 443, 109–117.

Kujawa, J., Cerneaux, S., Koter, S., Kujawski, W., 2014b. Highly Efficient Hydrophobic Titania Ceramic Membranes for Water Desalination. ACS Applied Materials and Interfaces. 6, 14223–14230.

Kujawa, J., Cerneaux, S., Kujawski, W., Knozowska, K., 2017. Hydrophobic ceramic membranes for water desalination. Applied Sciences. 7, 402–412.

Lafleur, M., Bougie, F., Guilhaume, N., Larachi, F., Fongarland, P., Iliuta, M.C., 2017. Development of a water-selective zeolite composite membrane by a new pore-plugging technique. Microporous and Mesoporous Materials. 237, 49–59.

Lee, J., Ha, J.H., Song, I.H., 2016. Improving the antifouling properties of ceramic membranes via chemical grafting of organosilanes. Separation Science and Technology. 51, 2420–2428.

Lenza, R.F.S., Vasconcelos, W.L., 2000. Synthesis and properties of microporous sol-gel silica membranes. Journal of Non-Crystalline Solids. 273, 164–169.

Li, F., Yang, Y., Fana, Y., Xinga, W., Wanga, Y., 2012. Modification of ceramic membranes for pore structure tailoring: The atomic layer deposition route. Journal of Membrane Science. 397–398, 17–23.

Lim, G.L., Jeong, H.G., Hwang, I.S., Kim, D.H., Park, N., Cho, J., 2009. Fabrication of a silica ceramic membrane using the aerosol flame deposition method for pretreatment focusing on particle control during desalination. Desalination. 238, 53–59.

Liu, B., Qu, F., Liang, H., Gan, Z., Yu, H., Li, G., Van der Bruggen, B., 2017. Algae-laden water treatment using ultrafiltration: Individual and combined fouling effects of cells, debris, extracellular and intracellular organic matter. Journal of Membrane Science. 528, 178–186.

Ma, X., Chen, P., Zhou, M., Zhong, Z., Zhang, F., Xing, W. 2017. Tight ultrafiltration ceramic membrane for separation of dyes and mixed salts (both NaCl/Na2SO4) in textile wastewater treatment. Industrial & Engineering Chemistry Research. 56, 7070–7079.

Masmoudi, S., Larbot, A., El Feki, H., Ben Amar, R., 2007. Elaboration and characterisation of apatite based mineral supports for microfiltration and ultrafiltration membranes. Ceramics International. 33, 337–344.

Mittal, P., Jana, S., Mohanty, K., 2011. Synthesis of low-cost hydrophilic ceramic–polymeric composite membrane for treatment of oily wastewater. Desalination. 282, 54–62.

Mustafa, G., Wyns, K., Vandezande, P., Buekenhoudt, A., Meynen, V., 2014. Novel grafting method efficiently decreases irreversible fouling of ceramic nanofiltration membranes. Journal of Membrane Science. 470, 369–377.

Mustafa, G., Wyns, K., Buekenhoudt, A., Meynen, V., 2016. Antifouling grafting of ceramic membranes validated in a variety of challenging wastewaters. Water Research. 104, 242–253.

Nandi, B.K., Das, B., Uppaluri, R., Purkait, M.K., 2009. Microfiltration of mosambi juice using low cost ceramic membrane. Journal of Food Engineering. 95, 597–605.

Oun, A., Tahri, N., Mahouche-Chergui, S., Carbonnier, B., Majumdar, S., Sarkar, S., Sahoo, G.C., Ben Amar, R., 2017. Tubular ultrafiltration ceramic membrane based on titania nanoparticles immobilized on macroporous clay-alumina support: Elaboration, characterization and application to dye removal. Separation and Purification Technology. 188, 126–133.

Padaki, M., Surya Murali, R., Abdullah, M.S., Misdan, N., Moslehyani, A., Kassim, M.A., Hilal, N., Ismail, A.F., 2015. Membrane technology enhancement in oil-water separation. A review. Desalination. 357, 197–207.

Palacio, L., Bouzerdib, Y., Ouammoub, M., Albizaneb, A., Bennazhab, J., Hernándeza, A., Calvo, J.I., 2009. Ceramic membranes from Moroccan natural clay and phosphate for industrial water treatment. Desalination. 245, 501–507.

Randon, J., Blanc, P., Paterson, R., 1995. Modification of ceramic membrane surfaces using phosphoric acid and alkyl phosphonic acids and its effects on ultrafiltration of BSA protein. Journal of Membrane Science. 98, 119–129.

Ren, C., Fang, H., Gu, J., Winnubst, L., Chen, C., 2015. Preparation and characterization of hydrophobic alumina planar membranes for water desalination. Journal of the European Ceramic Society. 35, 723–730.

Saffaj, N., Loukili, H., Alami Younssi, S., Albizane, A., Bouhria, M., Persin, M., Larbot A., 2004. Filtration of solution containing heavy metals and dyes by means of ultrafiltration membranes deposited on support made of Moroccan clay. Desalination. 168, 301–306.

Sah, A., Castricum, H.L., Bliek, A., Blank, D.H.A., ten Elshof, J.E., 2004. Hydrophobic modification of y-alumina membranes with organochlorosilanes. Journal of Membrane Science. 243, 125–132.

Sahoo, G.C., Halder, R., Jedidi, I., Oun, A., Nasri, H., Roychoudhurry, P., 2016. Preparation and characterization of microfiltration apatite membrane over low cost clay-alumina support for decolorization of dye solution. Desalination and Water Treatment. 57, 27700–27709.

Sarkar, S., Bandyopadhyay, S., Larbot, A., Cerneaux, S., 2012. New clay–alumina porous capillary supports for filtration application. Journal of Membrane Science. 392–393, 130–136.

Saufi, S.M., Ismail, A.F., 2004. Fabrication of carbon membranes for gas separation - A review. Carbon. 42, 241–259.

Shang, R., Goulas, A., Tang, C.Y., de Frias Serra, X., Rietveld, L.C., Heijman, S.G.J., 2017. Atmospheric pressure atomic layer deposition for tight ceramic nanofiltration membranes: Synthesis and application in water purification. Journal of Membrane Science. 528, 163–170.

Suresh, K., Pugazhenthi, G., Uppaluri, R., 2016. Fly ash based ceramic microfiltration membranes for oil-water emulsion treatment: Parametric optimization using response surface methodology. Journal of Water Process Engineering. 13, 27–43.

Suresh, K., Pugazhenthi, G., 2017. Cross flow microfiltration of oil-water emulsions using clay based ceramic membrane support and TiO_2 composite membrane. Egyptian Journal of Petroleum. 26, 679–694.

Tahri, N., Jedidi, I., Cerneaux, S., Cretin, M., Ben Amar, R., 2013. Development of an asymmetric carbon microfiltration membrane: Application to the treatment of industrial textile wastewater. Separation and Purification Technology. 118, 179–187.

Tanardi, C.R., Catana, R., Barboiu, M., Ayral, A., Vankelecom, I.F.J., Nijmeijer, A., Winnubst L., 2016a. Polyethyleneglycol grafting of y-alumina membranes for solvent resistant nanofiltration. Microporous and Mesoporous Materials. 229, 106–116.

Tin, P.S., Lin, H.Y., Ong, R.C., Chung, T.S., 2011. Carbon molecular sieve membranes for biofuel separation. Carbon. 49, 369–375.

Van Der Bruggen, B., Vandecasteele, C., Van Gestel, T., Doyen, W., Leysen, R., 2003. A review of pressure-driven membrane processes in wastewater treatment and drinking water production. Environmental Progress. 22, 46–56.

Van Gestel, T., Van der Bruggen, B., Buekenhoudt, A., Dotremont, C., Luyten, J., Vandecasteele, C., Maes, G., 2003. Surface modification of γ -Al_2O_3/TiO_2 multilayer membranes for applications in non-polar organic solvents. Journal of Membrane Science. 224, 3–10.

Van Gestel, T., Kruidhof, H., Blank, D.H.A., Bouwmeester, H.J.M., 2006. ZrO_2 and TiO_2 membranes for nanofiltration and pervaporation Part1. Preparation and characterization of a corrosion resistant ZrO_2 nanofiltration membrane with a MWCO<300. Journal of Membrane Science. 284, 128–136.

Van Heetvelde, P., Beyers, E., Wyns, K., Adriaensens, P., Maes, B.U.W., Mullens, S., Buekenhoudt, A., Meynen, V., 2013. A new method to graft titania using Grignard reagents. Chemical Communications. 49, 6998–7000.

Vinoth Kumar, R., Kumar Ghoshal, A., Pugazhenthi, G., 2015. Elaboration of novel tubular ceramic membrane from inexpensive raw materials by extrusion method and its performance in microfiltration of synthetic oily wastewater treatment. Journal of Membrane Science. 490, 92–102.

Wallberg, O., Jarwon, A.S., Wimmerstedt, R., 2003. Ultrafiltration of kraft black liquor with a ceramic membrane. Desalination. 156, 145–153.

Wu, Y., Sugimura, H., Inoue, Y., Takai, O., 2002. Thin films with nanotextures for transparent and ultra water-repellent coatings produced from trimethylmethoxysilane by microwave plasma CVD. Chemical Vapor Deposition. 8, 47–50.

Yacou, C., Smart, S., Diniz da Costa, J.C., 2015. Mesoporous TiO_2 based membranes for water desalination and brine processing. Separation and Purification Technology. 147, 166–171.

Yoshino, Y., Suzuki, T., Taguchi, H., Nomura, M., Nakao, S.-I., Itoh, N., 2008. Development of an all-ceramic module with silica membrane tubes for high temperature hydrogen separation. Separation Science and Technology. 43, 3432–3447.

Zhu, Y., Wang, D., Jiang, L., Jin, J., 2014. Recent progress in developing advanced membranes for emulsified oil/water separation. NPG Asia Materials. 6, 1–11.

Zulewska, J., Newbold,M., Barbano, D.M., 2009. Efficiency of serum protein removal from skim milk with ceramic and polymeric membranes at 50 1C. Journal of Dairy Science. 92, 1361–1377.

5 Advanced Materials for Polymeric Ultrafiltration Membranes Fabrication and Modification

Mimi Suliza Muhamad, Noor Aina Mohd Nazri, Woei-Jye Lau, and Ahmad Fauzi Ismail

CONTENTS

5.1 INTRODUCTION

The successful application of ultrafiltration (UF) membranes for different industrial separation processes, particularly water and wastewater treatment, greatly depends on the membrane surface properties. In general, materials with superior hydrophilicity and excellent antifouling properties are preferable for use during membrane synthesis in order to produce a high-performance UF membrane that has a significantly lower surface fouling tendency and requires minimum cleaning.

Although polysulfone (PSF), polyethersulfone (PES), and polyvinylidene fluoride (PVDF) are the most commonly used materials for commercial UF membrane fabrication, their (semi-)hydrophobic characteristics are the main factors that cause rapid fouling when these membranes are used to treat water or wastewater that contain different types of foulants. In order to improve the membrane antifouling properties without compromising water flux and solute separation efficiency, the use of advanced materials (either polymeric materials or inorganic nanoparticles) has been widely researched to alter the surface properties of polymeric UF membranes via various modification techniques such as blending, coating, grafting, etc.

The effort toward reducing and minimizing fouling problems of UF membranes has intensified over the past decade as evidenced by the large number of publications in this research topic. Statistics from the *Scopus* indicate that of the 576 total papers related to UF membranes and antifouling properties, >95% of the works were published between 2006 and 2017. In this regard, UF membranes with superior antifouling properties are being developed and employed in order to minimize flux decline, reduce cleansing frequency, and achieve better membrane stability, with the goal of reducing operating cost and prolonging membrane lifespan.

This chapter aims to provide a comprehensive review on the role of advanced materials in improving surface properties of UF membranes, aiming to achieve greater antifouling resistance and water flux without sacrificing solute rejection rate and structural integrity. In order to facilitate better understanding of the content, this chapter is organized into two main sections: (1) organic-based materials for membrane making and modification and (2) inorganic materials for polymeric-based membrane modification.

5.2 POLYMER/POLYMER UF MEMBRANES

In this section, the use of various types of polymeric-based materials for making UF membranes will be comprehensively reviewed in order to understand the role of these advanced materials in producing membranes with enhanced surface chemistries for water separation. These polymeric-based materials could be used either as a main membrane-forming material or as a secondary additive, depending on their functions and characteristics.

5.2.1 Amphiphilic Copolymers

Amphiphilic copolymers are composed of both hydrophilic and hydrophobic segments in their organic structure and, depending on the characteristics of the segments, the properties of the copolymers vary in terms of their chemical nature, molecular interaction, and affinities in aqueous solvent (Yang et al., 2010). Generally, amphiphilic copolymers are produced from the polymerization of one or more monomers without jeopardizing the original properties of the parent polymer. Similar to the other commercial polymers, the amphiphilic copolymers can be used as either the main membrane-forming material or secondary additive.

The most unique properties of the amphiphilic copolymer lie in its self-organizing behavior at polymer-water interfaces. Because of this mechanism, the surface properties of the membrane such as hydrophilicity, charge, and biorecognition could be easily altered (Park et al., 2006). Figure 5.1 depicts the formation of a hydrophilic

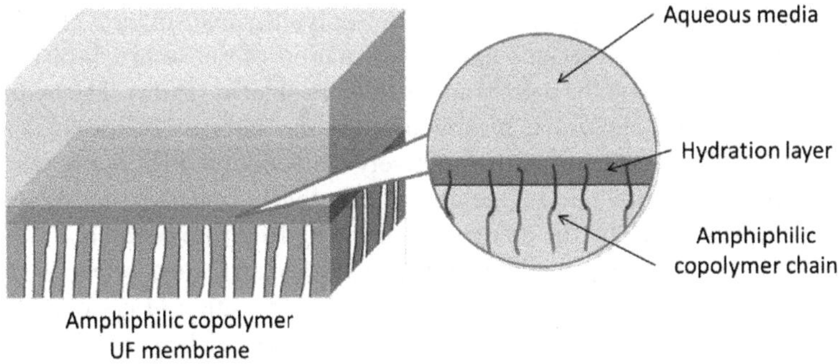

FIGURE 5.1 Reconstruction behavior of amphiphilic copolymer in aqueous environment. (From Liu, F. et al., Journal of Membrane Science. 345, 331–339, 2009.)

layer on the membrane surface by the migration of the hydrophilic block to aqueous environment. Besides improving membrane surface hydrophilicity, previous work has reported that the amphiphilic copolymer could act as a pore-forming agent and improve membrane water permeability (Loh and Wang, 2011).

Adding an amphiphilic copolymer to a membrane matrix involves migration of the hydrophilic segments to the aqueous environment due to thermodynamic incompatibility of the hydrophilic chain with the hydrophobic membrane matrix and anchoring of the hydrophobic segments on the membrane matrix (Su et al., 2009; Loh et al., 2011). Figure 5.2 depicts the migration of

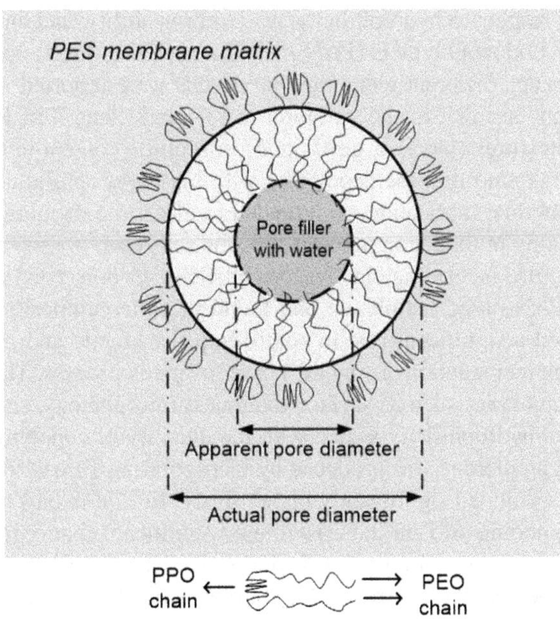

FIGURE 5.2 Formation of hydration layer on membrane surface and internal pores during surface segregation. (From Loh, C.H. et al., Journal of Membrane Science. 380, 114–123, 2011.)

poly(ethylene oxide)-poly(propylene oxide)-poly(ethylene oxide) (PEO-PPO-PEO) to polymer-water interface and the formation of dense brush of the PEO hydrophilic side chain on the membrane surface (Loh et al., 2011). The hydrophobic segments of the copolymer, meanwhile, aid the stability of the hydrophilic segments by anchoring them on the relatively hydrophobic PES membrane. The anchoring of the hydrophobic segments will eventually contribute to the stability of the hydrophilic layer by the elongation of the hydrophilic segments.

In addition to enhancing membrane hydrophilicity and water permeability, the use of amphiphilic copolymers for UF membrane fabrication offers other feature such as tunable membrane morphology. However, it must also be noted that the addition of amphiphilic copolymers as additives might cause poor compatibility between its hydrophobic segment and the membrane matrix, leading to the production of undesirable membrane properties. Table 5.1 presents some of the amphiphilic copolymers that have been synthesized and used as either membrane-forming material or additive in UF membrane fabrication and their impacts on the membrane properties.

Chen et al. (2011) investigated the filtration and antifouling performances of UF membranes made of polyacrylonitrile-*block*-polyethylene glycol (PAN-*b*-PEG) copolymer. Both water permeability and protein resistance were reported to increase when PEG content in the copolymer increased. The enhanced membrane performance is attributed to the enrichment of PEG segments on the membrane surface and its pores as a result of self-migration of the hydrophilic segments. The developed membrane also exhibited low hydraulic-resistance increase rate, indicating its low susceptibility to fouling. Using poly(ethylene glycol)-*graft*-polyacrylonitrile (PEG-*g*-PAN), Su et al. (2009) also reported an enhancement in the performances of UF membrane with respect to hydrophilicity, water permeability, and antifouling properties. Although both copolymers (PAN-*b*-PEG and PEG-*g*-PAN) contain the same polymeric segments, different membrane properties were reported in these works. The main reasons are due to the variation in the chain length of PEG segments, molecular architectures (block or graft), and hydrophilic content in the copolymer. Cho et al. (2011) found that the copolymer with high hydrophilic content was not suitable for membrane fabrication, as it tended to affect the membrane mechanical properties, resulting in reduction in water flux and rejection.

Nazri et al. (2015) incorporated polyacrylonitrile-*graft*-poly(vinyl alcohol) (PAN-*g*-PVA) into a PAN-based membrane and found that the composition of the graft copolymer played a significant role in controlling the kinetic and thermodynamic factors of the polymer solution during the phase inversion process. The properties of the resultant membranes such as surface roughness, morphology, surface chemical composition, and hydrophilicity varied with the quantity of copolymer added. The best performing membrane was produced by incorporating 10 wt% of PAN-*g*-PVA. This membrane exhibited significantly higher flux (140 L/m^2.h.bar) than that of the control PAN membrane (41 L/m^2.h.bar) without a significant change in bovine serum albumin (BSA) rejection (almost 100% removal). The modified membrane also showed a lower percentage of irreversible fouling, indicating the improved surface chemistry was resistant to permanent deposition or adsorption of organic foulants.

Incorporation of a self-assembly amphiphilic copolymer into PES-based membranes has also been reported in literature. Loh et al. (2011) blended the PES

TABLE 5.1
Summary of the Use of Amphiphilic Copolymer for UF Membrane Fabrication and Modification and Its Effects on Membrane Properties

Membrane	Copolymer Synthesis Method	Dope Formulation (wt%)	[c]Properties of the Best Performing Membrane	Effects on Membrane Properties
[a]Main polymer: PVDF-*g*-POEM (Shen et al., 2017)	Atom transfer radical polymerization (ATRP)	Polymer content: 20.0 wt%	PWF: 568 L/m^2.h.bar MWCO: 40 kDa BSA rejection: 86.5% Flux recovery: 91.1% Contact angle: 85.0°	PVDF-g-POEM membrane made of lower PVDF molecular weight (~275 kDa) resulted in larger pore size, higher pure water flux and larger MWCO.
[a]Main polymer: PSF-*b*-PEG (Chen et al., 2017)	N/A	Polymer content: 15.0 wt%	PWF: 455 L/m^2.h.bar MWCO: 70 kDa BSA rejection: 90% Flux recovery: 86%	The content of PEG in the copolymer should be kept low in order to maintain the membrane mechanical strength and thermal stability.
[a]Main polymer: PES-*g*-PEO (Hou et al., 2015)	N/A	Polymer content: 18.0 wt%	PWF: 339 L/m^2.h.bar BSA retention: 98% BSA adsorption: Nil Contact angle: 59°	Higher amount of PEO in PES-g-PEO significantly increased membrane hydrophilicity, water flux, and antifouling properties.
[a]Main polymer: PSF-*r*-PEO (Cho et al., 2011)	Polycondensation reaction via nucleophilic substitution	Polymer content: 15.0 wt%	PWF: 135 L/m^2.h.bar MWCO: 20.0 kDa Oil rejection: 94.7% Contact angle: 60°	Presence of high PEO content in copolymer could lead to membrane flux reduction and poor mechanical property.
[a]Main polymer: PAN-*b*-PEG (Chen et al., 2011)	Combined redox polymerization and reversible addition fragmentation	Polymer content: 10.0 wt% (*The content of PEG (5000, 10200, and 23500 Da) in copolymer was varied in the range of 0–11.1 mol%*).	PWF: 319 L/m^2.h.bar BSA rejection: >98% Protein adsorption: 24.3 μg/cm^2 Contact angle: 45°	Copolymer composition played a significant role in controlling membrane structure, properties and performance. Effect of PEG MW in copolymer is almost negligible.

(*Continued*)

TABLE 5.1 (CONTINUED)
Summary of the Use of Amphiphilic Copolymer for UF Membrane Fabrication and Modification and Its Effects on Membrane Properties

Membrane	Copolymer Synthesis Method	Dope Formulation (wt%)	[c]Properties of the Best Performing Membrane	Effects on Membrane Properties
[b]Main polymer: PAN Copolymer additive: PAN-*g*-PVA (Nazri et al., 2015)	Ce(IV)-initiated free radical polymerization	Polymer content: 12.0 wt% (PAN:PAN-*g*-PVA mass ratio: 95:05)	PWF: 137 L/m^2.h.bar Pore size: 15 nm Surface BSA rejection: 100% Contact angle: 53°	Increasing content of PAN-*g*-PVA in dope solution could make membrane surface with higher PVA content, leading to higher water flux and better fouling resistance.
[a]Main polymer: PVDF Copolymer additive: PVDF-*g*-PEGMA (Liu et al., 2013)	Atom transfer radical polymerization (ATRP)	Total polymer content: 18.0 wt% PVDF-g-PEGMA in total polymer: 5 wt%	[c]PWF: 1080 L/m^2.h (at 0.7 bar) Sodium alginate retention: 98.0% Flux recovery ratio (Sodium alginate): 56.0% Contact angle: 67.0°	High amount of PVDF-*g*-PEGMA could result in rougher membrane surface and lower antifouling properties.
[a]Main polymer: PSF Copolymer additive: PSF-*g*-PNMG (Shi et al., 2013b)	Atom transfer radical polymerization (ATRP)	Total polymer content: 13.0 wt% PSF/PSF-*g*-PNMG weight ratio: 7:3	PWF: ~470 L/m^2.h.bar Flux recovery (BSA): 90% Dynamic contact angle: 86°–29° (in 15 s)	Addition of higher content of PSF-*g*-PNMG increased membrane pure water flux, hydrophilicity and antifouling properties.
[b]Main polymer: PLA Copolymer additive: PLA-PEG-PLA (Shen et al., 2013)	Ring-opening polymerization	Total polymer content: 20.0 wt% PSF/PSF-*g*-PNMG weight ratio: 97:3	PWF: 115 L/m^2.h.bar BSA rejection: 92% Contact angle: 53.0°	PLA-PEG-PLA with longer PEG segments significantly improved hydrophilicity and antifouling properties.

(Continued)

TABLE 5.1 (CONTINUED)
Summary of the Use of Amphiphilic Copolymer for UF Membrane Fabrication and Modification and Its Effects on Membrane Properties

Membrane	Copolymer Synthesis Method	Dope Formulation (wt%)	[c]Properties of the Best Performing Membrane	Effects on Membrane Properties
[a]Main polymer: PES Copolymer additive: PMAA-F127-PMAA (Su et al., 2015)	Free radical polymerization	Total polymer content: 25.6 wt% PMAA-F127-PMAA in total polymer: 3.76 wt%	PWF: 239 L/m^2.h.bar (at pH 2) and 117 L/m^2.h.bar (at pH 10) BSA rejection: 85–98% Contact angle: 39.4°	PMAA endowed the membranes with pH-responsive property by its conformational transition. Higher degree of PMAA surface coverage resulted in stronger pH-responsive property.
[a]Main polymer: PSF Copolymer additive: PSF-*b*-PEG (Chen et al., 2016)	Atom transfer radical polymerization (ATRP)	Total polymer content: 16.0 wt% Copolymer content: 40.0 wt% (based on total polymer content)	PWF: 400 L/m^2.h.bar BSA rejection: 96% Flux recovery: 85% Contact angle: 69°	The incorporation of PSF-*b*-PEG resulted in thicker membrane skin layer and further enhanced its hydrophilicity and antifouling resistance.
[a]Main polymer: PES Copolymer additive: MF-*g*-PEGn (Liu et al., 2015)	Etherification	Total polymer content: 26.4 wt% Copolymer content: 0.36 wt% (based on MF-g-PEG6000 % in the total amount of polymers)	Contact angle: 69.7° PWF: 165 L/m^2.h.bar BSA rejection: 100% Flux recovery: 91.6% Contact angle: 69.7°	Increasing the average molecular weight of PEG in the copolymer could enhance the membrane porosity, leading to higher and pure water flux. The modified membrane was found to have good antifouling resistance against BSA.

[a] Flat sheet membrane.
[b] Hollow fiber membrane.
[c] PWF: Pure water flux; MWCO: Molecular weight cut-off; BSA: Bovine serum albumin.

membrane with tri-block copolymers of different molecular architectures—poly(ethylene oxide)-poly(propylene oxide)-poly(ethylene oxide) (PEO-PPO-PEO) and PPO-PEO-PPO. Their results indicated that the PPO-PEO-PPO-modified membranes experienced lower hydrophilicity and permeability compared to the PEO-PPO-PEO-modified membranes. This is due to the steric hindrance effect on the hydration as the PEO is located at the middle of the molecule. Although Loh et al. (2011) reported that the membrane water flux was improved by incorporating a higher amount of the tri-block copolymer during dope preparation, Wang et al. (2005) found that the tri-block copolymer concentration had little impact on the membrane permeability. The contradictory results between these two studies could be attributed to the different dope formulation and membrane fabrication conditions.

Many other works also evaluated the potential of different types of copolymers on the performance enhancement of PES-based membrane. For instance, Ma et al. (2007) incorporated polystyrene-*block*-poly(ethyl glycol) (PS-*b*-PEG) into PES membranes and reported that the membranes with enhanced water flux (up to 140.9 L/m^2.h.bar), low protein adsorption (0.5 μg/cm^2), and promising flux recovery rate (88.2%) were able to produce when 4.5 wt% of the copolymer was added into the dope solution containing 13.5 wt% PES. In addition to the enhanced antifouling resistance, incorporation of poly(butyl methacrylate)-*b*-poly(methacrylic acid)-*b*-poly(hexafluorobutyl methacrylate) (PBMA-*b*-PMAA-*b*-PHFBM) into the PES membrane was reported to alter membrane performance according to the feed pH (Zhao et al., 2011). Figure 5.3 compares the pH-responsive behavior of the modified membranes incorporated with PBMA-*b*-PMAA and PBMA-*b*-PMAA-*b*-PHFBM. As can be seen, the water flux of the membrane incorporated with PBMA-*b*-PMAA-*b*-PHFB demonstrated higher flux than the PBMA-*b*-PMAA-modified membranes in the pH ranging from 3.5 to 8.2. This is because the PBMA-*b*-PMAA-modified membranes exhibited lower effective pore size due to the repelling interaction between fluorine-containing segments that may restrict the extension of hydrophilic PMAA in the membrane pores. This novel membrane showed a unique pH-dependent flux and fouling release property due to the characteristics of hydrophilic PMAA and fluorine containing segments.

Modification of PVDF-based membranes using amphiphilic copolymers has also been reported for water treatment process in several works. Liu et al. (2013) modified the PVDF membrane by incorporating poly(vinylidene fluoride)-*graft*-poly(ethylene glycol) methyl ether methacrylate (PVDF-*g*-PEGMA) during dope solution preparation and reported that the modified PVDF membrane showed improved surface hydrophilicity (contact angle of 67°) and enhanced pure water flux (1080 L/m^2.h at 0.7 bar) in comparison to the control PVDF membrane (78° and 116 L/m^2.h). Despite the improvement of the surface hydrophilicity, the water flux of the copolymer-modified PVDF membranes demonstrated lower flux recovery (56.0%) during filtration of sodium alginate (SA) solution, likely due to formation of a rougher surface. Rougher membrane surface is generally more prone to foulant adsorption and/or deposition.

Moghareh Abed et al. (2013) used poly(vinylidene fluoride)-*graft*-poly(oxyethylene methacrylate) (PVDF-*g*-POEM) to improve the water flux antifouling property of PVDF-based membrane. Compared to the control PVDF membrane that showed a contact angle of 90.1° and water flux of 0.3 L/m^2.h.bar, the modified PVDF

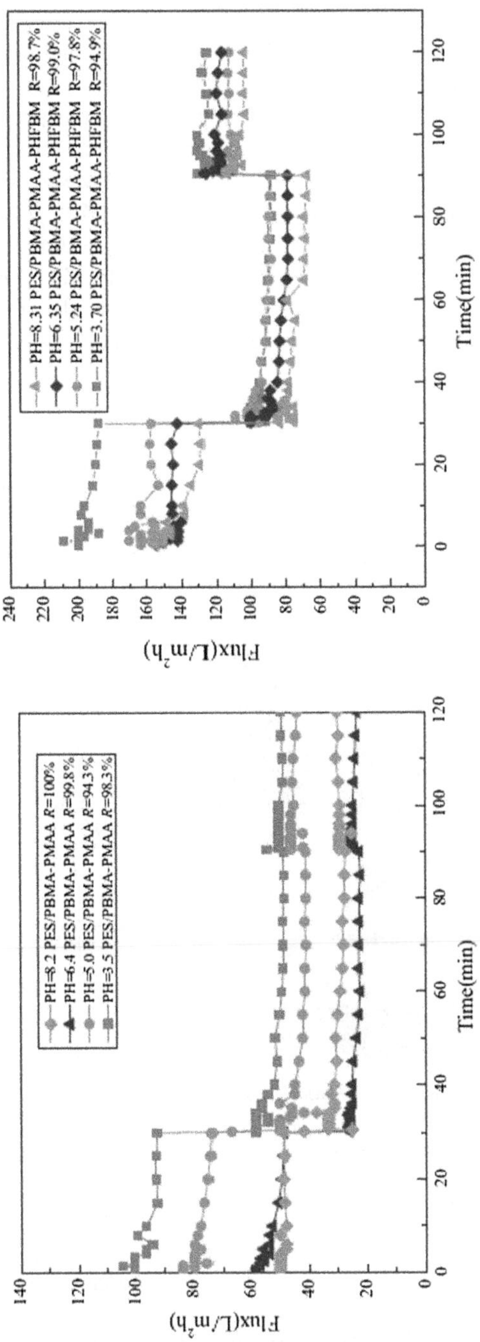

FIGURE 5.3 Time-dependent flux of PES/PBMA-*b*-PMAA and PES/PBMA-*b*-PMAA-*b*-PHFBM membranes during UF process at different pH values. (From Zhao, X. et al., Journal of Membrane Science. 382, 222–230, 2011.)

membrane was reported to exhibit better results. It recorded a 63.0° contact angle and close to 130.0 L/m^2.h.bar water flux. The promising results are attributed to the presence of a hydrophilic POEM side chain in the copolymer. The experimental data also indicated the high flux recovery rate (FRR) of modified membrane (up to 95%) after subjecting to 30-min filtration of 1.0 g/L BSA solution.

Bera et al. (2015) proposed a convenient approach for the preparation of pH- or temperature- or simultaneous pH- and temperature-responsive PVDF-based membranes by incorporating poly(methyl methacrylate) (PMMA)-co-X-copolymer [where X = poly(dimethylaminoethyl methacrylate) (PDMA), poly(acrylic acid) (PAA), poly(N-isopropyl acrylamide) (PNIPAM), poly(N,N-dimethyl acrylamide) (PDMAA), polyvinylpyrrolidone (PVP), PDMA-co-PNIPAM, PAA-co-PNIPAM, or PDMA-co-PDMAA], as shown in Figure 5.4. The positive effect of PMMA for the preparation of blend membranes is attributed to the compatibility of PVDF with the

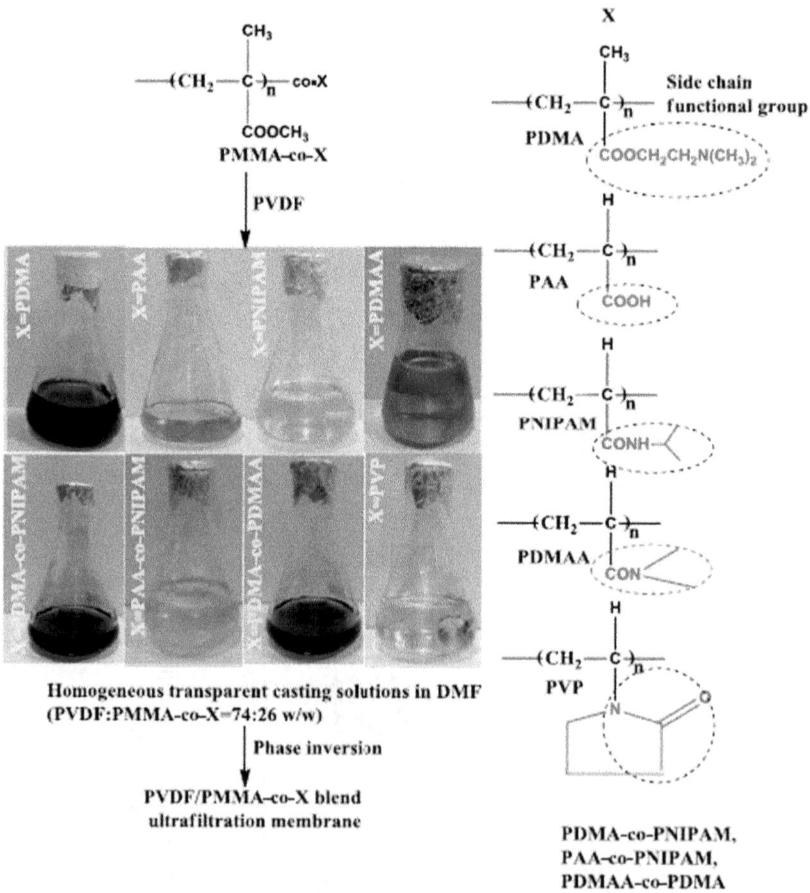

FIGURE 5.4 Preparation of PVDF/PMMA-co-X blend UF membranes using homogeneous and transparent casting solutions. Dotted circles show free functional groups of copolymers. (From Bera, A. et al., Journal of Membrane Science. 481, 137–147, 2015.)

PMMA-part of the copolymers that lead to enhanced interfacial adhesion between PVDF and copolymers. Besides exhibiting pH- and/or temperature-responsive behaviors, all the modified PVDF membranes also showed higher water flux and excellent antifouling resistance in comparison to the control PVDF membrane owing to the enhanced water wetting behavior resulted from the hydrogen bonding with the ester, hydroxyl, amide, pyrrolidone, or carboxylic acid groups of the copolymers and increased membrane porosity. Further comparison indicated that of all the membranes prepared, the membranes containing PMMA-co-PDMA or PMMA-co-PAA gave the highest flux whereas the membrane of PMMA-co-PDMAA gave the lowest flux. Su et al. (2015), on the other hand, studied the role of PMAA in endowing PES membranes with pH-responsive behavior. The extended and contracted conformation of the PMAA by the ionization and protonation of the PMAA segment in response to the pH values of feed solutions significantly improved the pH-responsive behavior of the modified membranes.

Most of the works blended the dope solution with a copolymer concentration of less than 20 wt% for UF membrane fabrication, as excessive use of a copolymer might affect membrane structural integrity and mechanical properties due to poor compatibility between the main polymer and the copolymer. Chen et al. (2016) attempted to prepare a PSF-based membrane that contained a large quantity of a copolymer in the casting solution (up to 70.0 wt% in the total polymer concentration) in order to understand the role and effect of high concentration PSF-*g*-PEG copolymer on the membrane characteristics and performance. The water flux of the PSF-based membrane was improved gradually by increasing the content of PSF-*g*-PEG from zero to 40 wt%. The water flux of the membrane blended with 40% PSF-*g*-PEG (400 $L/m^2.h.bar$) was 10 times greater than that of the unmodified PSF membrane. A further increase in PSF-*g*-PEG content to 70 wt%, however, reduced the membrane water flux to 54.0 $L/m^2.h.bar$. The formation of a thicker skin layer as a result of the use of the high PSF-*g*-PEG concentration causes a delayed demixing process. At low copolymer concentration (<40 wt%), the formation of microdomain connecting pores is favorable and beneficial to water flux. Unlike previously mentioned works that reported the improved antifouling property of a membrane upon modification with a copolymer, the FRR of the modified membrane blended with 20–50 wt% copolymer remained stable at 85% when tested using 0.5 g/L BSA solution. The findings suggested that a further increase in copolymer content would have little effect on the protein adsorption on the membrane surface.

Other types of amphiphilic copolymers such as polysulfone-*g*-poly(N-methyl-D-glucamine) (PSF-*g*-PNMG) (Shi et al., 2013), polylactic acid-polyethylene glycol-polylactic acid (PLA-PEG-PLA) (Shen et al., 2013), and poly(phthalazinone ether sulfone ketone)-*g*-polyethylene glycol (PPESK-g-PEG) (Zhu et al., 2007) have also been reportedly used as an additive to modify the membranes made of PPPESK and PLA. In general, these copolymers play a vital role in enhancing membrane properties and performance particularly with respect to surface hydrophilicity, water permeability, and antifouling resistance. However, it must be pointed out that in order to produce desirable membrane characteristics for particular types of feed properties, factors such as the composition and molecular structure of the copolymer, its concentration in the dope solution, and membrane synthesis conditions need to be taken into consideration.

5.2.2 Hydrophilic/Non-Hydrophilic Polymers

The blending technique, using a secondary polymer in a polymeric solution, is one of the simplest strategies to produce new types of membranes that combine positive features of the main and secondary polymers. The blending method is highly attainable with minimal preparation and allows concurrent modification on the membrane during phase inversion process. It is generally agreed that the inherent property of the secondary polymer and its quantity in the dope solution are the main factors that determine the ultimate structural morphology and surface characteristics of the UF membranes for effective separation.

Hydrophilic polymers, i.e., PVP and PEG, have been reported in literature more often than any other polymeric additives for improving the hydrophilicity of membranes and thus their water flux and antifouling properties. These two hydrophilic polymers exhibit very good compatibility with the membranes made of PES, PSF, PVDF, and PAN and can dissolve easily in most of the organic solvents without causing agglomeration. Until now, many types of polymeric materials have been evaluated as potential additives to alter the properties of UF membranes, aiming to improve not only water flux and solute rejection but also antifouling resistance against various types of foulants.

Behboudi et al. (2017) blended polyvinyl chloride (PVC) membranes with polycarbonate (PC) in an effort to promote greater fouling resistance of the PVC-based membrane. Of all the blended membranes prepared, the membrane blended with 50 wt% PC (based on the total polymer weight) demonstrated the highest antifouling resistance (89.5% FRR) after 4-h filtration of 1.0 g/L BSA solution. In addition to the excellent separation rate (99% BSA rejection), this best performing membrane (531 L/m^2.h.bar) also exhibited significantly higher water flux than that of the neat PVC membrane (202 L/m^2.h.bar) owing to the enhancement in hydrophilicity coupled with an increased number of surface pores as shown in Figure 5.5.

Fan et al. (2014), on the other hand, synthesized high-antifouling PVC UF membrane by adding a different concentration of polyvinyl formal (PVF). The incorporation of hydrophilic PVF to the PVC-based membrane could promote pore formation as well as enhance the membrane's hydrophilicity and antifouling properties. Additionally, the good compatibility between PVC and PVF improved the stability of PVF on the membrane surface. The results revealed that PVF played a significant role as a pore forming agent in which the mean pore size and porosity of the PVC-based membranes increased from 14.9 to 36.3 nm and 85.6 to 94.2%, respectively, with increasing PVF concentration from 2 to 8 wt%. The resultant PVC membranes blended with 8 wt% PVF also demonstrated superior antifouling properties in filtrating different feed properties, recording 95%, 99%, and 98% FRR for BSA (1 mg/mL), humic acid (1 mg/mL) and yeast solution (1 mg/mL), respectively, in addition to excellent water permeability (323 L/m^2.h.bar).

Alsalhy (2012) blended PVC membrane with polystyrene (PSR) and found that the rejection and mechanical strength of the modified membrane were able to be

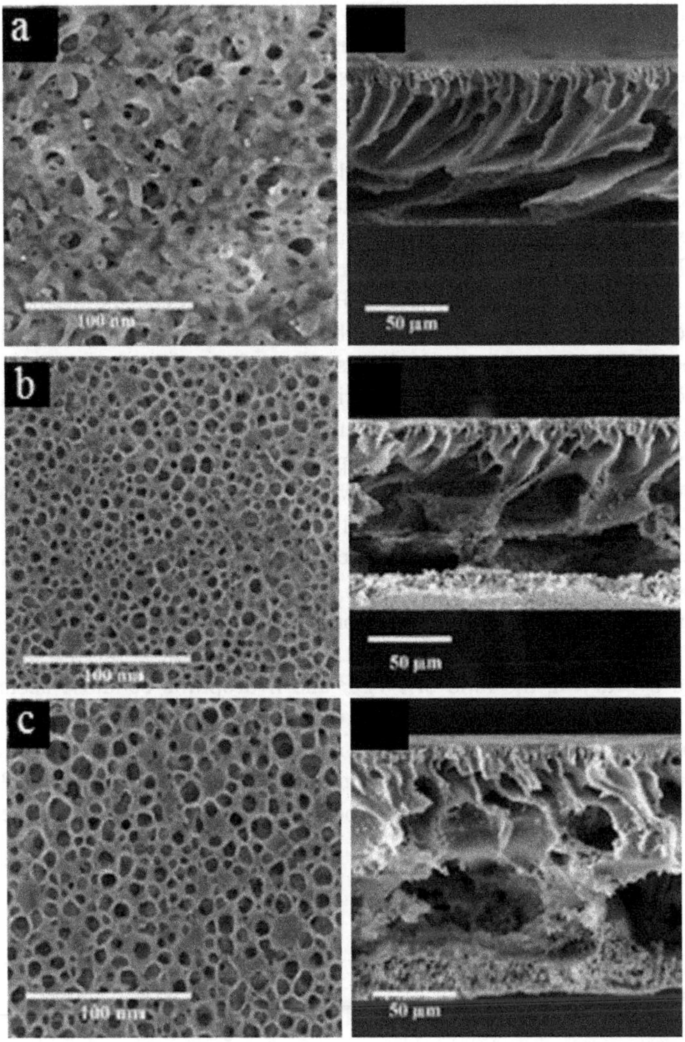

FIGURE 5.5 FESEM images of membrane surface (left) and cross section (right), (a) Neat PVC, (b) PVC blended with 50% PC, and (c) PVC blended with 70% PC. (From Behboudi, A. et al., Journal of Membrane Science. 534, 18–24, 2017.)

improved, provided the quantity of secondary polymer added did not exceed 3 wt% in the dope solution containing 14 wt% PVC. Hydrophilic polyvinyl alcohol (PVA) has also been utilized as additive by Zhang et al. (2014a) to improve the antifouling properties of a PVDF/PES blend membrane. At the optimum PVA concentration (0.3% PVA content in 10 wt% PVDF/PES), the modified PVDF/PES membrane exhibited improved water flux (1376 $L/m^2.h$ at 0.3 bar) and lower fouling rate (0.09×10^{12} m^{-1}/h) compared to the unmodified PVC/PES membrane (1306 $L/m^2.h$ and 0.21×10^{12} m^{-1}/h). The introduction of a high amount of PVA to the PVC/PES was not recommended as it negatively affected the membrane properties such as flux and antifouling properties.

In order to improve the compatibility of the PVA with the main membrane-making polymer, Yuan and Ren (2017) used acetalized PVA (PVAd) to modify a PVDF-based membrane. The PVAd was synthesized from acetalization of PVA by acetaldehyde under acidic conditions. In comparison to the pristine PVDF membrane, a more hydrophilic membrane with higher porosity, larger pore size, and thinner skin layer were produced upon PVAd incorporation. The results indicated the membrane water flux was increased from 40 to 550 L/m^2.h.bar by increasing the PVAd content from 0 to 30 wt% (value was determined based on the total polymer weight). The authors found that the ideal quantity of PVAd added to the dope solution was at 20 wt%, as excessive use of PVAd tended to cause a dramatic decline in BSA rejection, even though such membrane offered the highest water flux. The membrane modified by 20 wt% PVAd was found to exhibit better antifouling properties (FRR up to 87%) than that of the pristine membrane (73%) when tested using 300 mg/L BSA solution. This was attributed to the formation of a hydration layer on the membrane surface that minimized pore plugging and protein adhesion on the membrane surface. Figure 5.6 shows the water flux and BSA rejection of the PVDF/PVAd membranes made at different PVAd content.

As the membranes made of PVDF are relatively hydrophobic in comparison to the membranes made of PSF and PES, much research has been conducted to enhance the hydrophilicity of PVDF-based membranes using secondary hydrophilic polymers. For instance, Hossein Razzaghi et al. (2014) altered the properties of the PVDF membrane by varying the ratio of PVDF and hydrophilic cellulose acetate (CA), aiming to obtain a membrane with improved permeability and antifouling resistance. The best performing membrane (total polymer weight in the dope solution: 20 wt%) was produced by blending two polymers at the weight ratio of 8 (PVDF):2 (CA). In addition to the improved water permeability, the best CA-modified PVDF

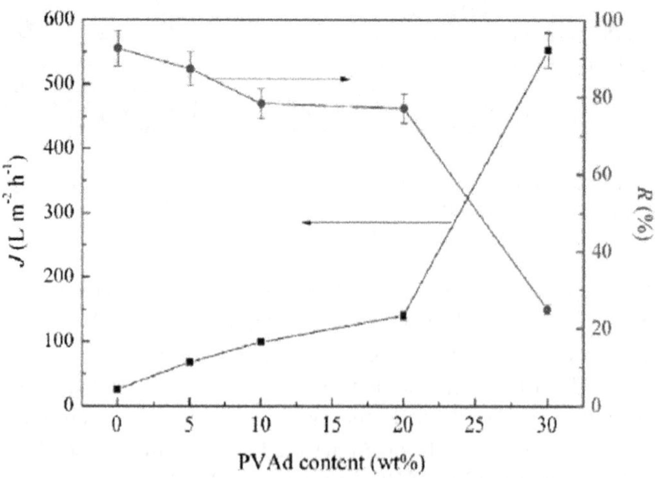

FIGURE 5.6 Effect of PVAd content on the permeation performance of PVDF/PVAd membrane. (From Yuan, H., Ren, J., Chemical Engineering Research and Design. 121, 348–359, 2017.)

membrane also showed the lowest total fouling compared to other membranes made at different PVDF:CA ratios.

Poly(vinyl butyral) (PVB) is a promising material for UF membrane preparation due to its hydrophilicity and low manufacturing cost. Yan and Wang (2011) incorporated PVB into a PVDF-based membrane and reported that the membrane water flux was improved from 20 to 430 L/m^2.h (tested at 2 bar) by varying weight ratio of PVB:PVDF from 0:10 to 7:3 while maintaining the total polymer weight in the dope solution at 23 wt%. Nevertheless, the PVDF/PVB blend membranes showed lower rejection of BSA in comparison to the membrane made of either PVDF or PVB as a result of formation of more finger-like structures at the top membrane layer. With respect to hydrophilicity, it was reported that the membrane hydrophilicty could only be improved upon addition of a large amount of PVB in the dope solution. Separately, Yuan et al. (2009) utilized perfluorosulfonic acid (PFSA) to modify a PVDF-based membrane, aiming to achieve greater water flux. However, the water fluxes of the developed membranes (the highest flux obtained was 99 L/m^2.h at 2 bar) were significantly lower compared to other works and no explanation was provided.

Table 5.2 lists some of the previous works that have evaluated the roles of secondary polymers in improving the characteristics of membranes made of PSF, PES, or PAN. As can be seen, the addition of polymeric additives plays an important role in enhancing the membrane properties and performance for effective UF process. The improvement in membrane antifouling resistance is particularly important for industrial process as it can extend membrane lifespan and minimize cleaning frequency, leading to reduced operating and maintenance costs. However, the major issues related to the blending approach are the poor compatibility that might occur between two polymers (mainly in the case where a high concentration of secondary polymer is used) and the leaching problem (mainly when the superior hydrophilic polymers have a relatively small molecular weight). Therefore, optimizing the dope formulation and membrane fabrication conditions should be undertaken in order to produce desirable membrane properties that are suitable for UF process.

5.2.3 Polymer-Based Surface Coating Materials

Surface modification on the existing membranes can be achieved by introducing hydrophilic moieties on the membrane surface through covalent bond interaction (e.g., grafting polymerization) or physical interaction (e.g., coating). The advantage of surface modification is that the bulk structure of the membrane remains intact and only the top skin layer is altered to enhance hydrophilicity degree and antifouling resistance. Compared to the chemical modification approach, the physical coating is usually associated with several drawbacks. These include an easily removable coating layer, instability of the coating layer, complicated coating process, and low efficiency (Ma et al., 2015).

Ma et al. (2015) coated the surface PVDF membranes with poly(2-methacryloyloxyethyl phosphorylcholine)-co-(n-butylmethacrylate) (poly(MPC-co-BMA)) using different coating methods, the flow-through method and the immersion method, in an effort to enhance the membrane antifouling properties. MPC is a zwitterionic monomer that can promote electrostatic interaction with surrounding water to impart

TABLE 5.2
Summary of the Polymeric Additives Used for UF Membrane Modification and Its Effects on Membrane Properties

Membrane	[c]Dope Formulation	Best Performing Membrane Properties	Effects on Membrane Performance
[a]Main polymer: PES Additive: PAN (Amirilargani et al., 2010)	Total polymer content: 16 wt% Copolymer content: 40 wt% in total polymer Solvent: DMF	Contact angle: 69° PWF: 400 L/m^2.h.bar BSA rejection: 96% Flux recovery: 85%	The incorporation of PSF-*b*-PEG resulted in thicker membrane skin layer and further enhanced its hydrophilicity and antifouling resistance.
[b]Main polymer: PAN Polymer additive: PVA (Nazri et al., 2015)	Total polymer and additive composition: 12 wt% Ratio of PAN:PVA: 85:15 Solvent: DMSO	Contact angle: 60° PWF: 251 L/m^2.h.bar BSA rejection: 100% FRR: 72.5% Surface roughness: 53.46 nm	Further addition of PVA beyond the optimum ratio (PAN:PVA 85:15) could negatively affect the membrane properties and performance.
[a]Main polymer: PSF Polymer additive: Sulfobetaine polyimides (PI) (Gao et al., 2017)	Total polymer and additive composition: 21.6 wt% Additive content: 0.6 wt% Solvent: NMP	Contact angle: 67° PWF: 135 L/m^2.h.bar BSA rejection: 94% Porosity: 80% FRR: 93%	Addition of sulfobetaine PI significantly enhanced the membrane pure water flux, BSA rejection, porosity, and flux recovery. Hydration of sulfobetaine moieties on the membrane surface significantly enhanced the membrane antifouling properties.
[a]Main polymer: PES Polymer additive: Sulfonated polysulfone (SPSf) (Li et al., 2016)	Total polymer and additive composition: 18 wt% Additive content in total polymer: 16 wt% Solvent: DMAc	Contact angle: 70.2° PWF: 1467 L/m^2.h.bar BSA rejection: 91.4% Porosity: 78.2% Surface pore size: 22.3 nm FRR: 89%	Addition of SPSf favorably enhanced the membrane hydrophilicity and consequently resulted in higher pure water flux (1.8 times higher than PES control membrane) and antifouling properties (FRR for control PES was 58%).

(Continued)

TABLE 5.2 (CONTINUED)
Summary of the Polymeric Additives Used for UF Membrane Modification and Its Effects on Membrane Properties

Membrane	[c]Dope Formulation	Best performing Membrane Properties	Effects on Membrane Performance
[a]Main polymer: PES Polymer additive: aromatic polyamide (Shockravi et al., 2017)	Total polymer and additive composition: 18.3 wt% Additive content in total polymer: 2 wt% Solvent: DMAc	Contact angle: 68.1° PWF: 80.4 L/m^2.h.bar BSA rejection: 98% FRR: 64% Porosity: 86% Pore radius: 7.21 nm	Higher content of aromatic polyamide could reduce water flux due to agglomeration of additive on the pore and formation of smaller finger-like structure. Further addition of aromatic polyamide beyond optimum ratio (2 wt%) could negatively affect the membrane properties and performance.
[a]Main polymer: PSF Polymer additive: N-succinyl chitosan (NSCS) (Kumar et al., 2013)	Total polymer and additive composition: 9.8 wt% Additive content in total polymer: 20.0 wt% Solvent: NMP	Contact angle: 60.9° PWF: 117.5 L/m^2.h.bar BSA rejection: 93.5% FRR: 70%	Presence of NSCS enhanced the membrane hydrophilicity and permeation flux (3 times higher compared to pristine PSF membrane) and consequently improved antifouling properties.

[a] Flat sheet membrane.
[b] Hollow fiber membrane.
[c] DMF: Dimethylformamide; DMSO: Dimethly sulfoxide; NMP: N-methyl-2-pyrrolidone; DMAc: Dimethylacetamide.

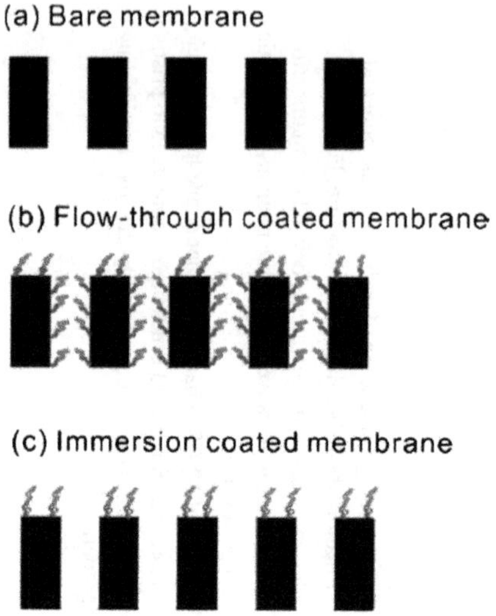

FIGURE 5.7 Schematic representation of (a) bare membrane, (b) flow-through coated membrane, and (c) immersion coated membranes. (From Ma, W. et al., Applied Surface Science. 357, 1388–1395, 2015.)

better fouling resistance; BMA enhances stability of the copolymer on the membrane surface because of its affinity with PVDF. Figure 5.7 shows the schematic representation of the bare membrane and the membranes coated with different methods. As can be seen, the flow-through coating method was effective to introduce the coating material not only onto the membrane surface but also to its pore channels. Because of this, the PVDF membranes coated via the flow-through method displayed better antifouling properties than that of the membrane prepared via the immersion method. Nevertheless, the modification of the pore channels affected the membrane permeability because the deposition of poly(MPC-co-BMA) on the pore wall was linked to reduced pore size. With respect to coating layer stability, the experimental results indicated that the coated membrane surface only experienced marginal loss of poly(MPC-co-BMA) after subjecting to 1-h ultrasonic cleaning using 1.0 wt% sodium dodecyl sulfate (SDS) aqueous solution. This further suggested that the fouling resistance of the poly(MPC-co-BMA)-coated membrane could be maintained under harsh conditions.

Over the past several years, polydopamine (PDA) has become a favorable coating material for membrane surface modification owing to its self-polymerization, high anchoring capability, and special recognition. Li et al. (2014a) reported that a tight and stable PD coating layer could be formed on the membrane surface via a tris(hydroxymethyl)aminomethane hydrochloride treatment. To impart better antifouling resistance, functional molecules such as $-NH_2$ or –SH were immobilized on top of the PD coating layer surface. It is found that a PD coating, followed by grafting of PEG-NH_2 onto the PD layer to create a PD-g-PEG modified membrane, was able to improve the stability and antifouling properties of the PES membrane. In the work of Li et al. (2014a), the PES membranes coated with PD and PD-g-PEG were synthesized and evaluated with respect to their characteristics for UF process.

Despite the coated membranes experiencing flux reduction upon surface modification (due to reduced pore size), the experimental results indicated that the surface coating and surface grafting methods were able to improve the fouling resistant property of the PES membrane. In comparison to the unmodified PES membranes, the modified membranes had less adsorbed BSA under the same condition owing to the improved surface hydrophilicity. The PD-g-PEG modified membrane was reported to have the lowest adsorptive fouling potential (~28 μg/cm^2) than the PD-coated membrane (~35 μg/cm^2) and unmodified membrane (~55 μg/cm^2) when tested with 40 g/L BSA solution for 24 h. This is likely due to the formation of a hydrophilic PEG "brush" that prevented BSA molecules from adhering to the membrane surface.

Li et al. (2015a) grafted the surface of the PD-coated PES membranes with either amine-terminated polyethylene glycol (mPEG-NH_2) or glycine-functionalized PVA and compared the efficiency of the two different grafting materials on the membrane performance. Figure 5.8 depicts the schematic diagram of the surface modification for the PES membranes. Although the water flux of the membrane grafted with glycine-functionalized PVA was reduced by increasing the glycine-functionalized PVA content and grafting time (as a result of reduced pore size), the membrane antifouling resistance (FRR: 90%) against oil molecules was increased tremendously compared to the membrane without grafting (FRR: <50%). Furthermore, the grafted membrane with glycine-functionalized PVA also possessed better stability compared to the mPEG-NH_2 modified membrane when exposed to sodium hydroxide (pH 13), hydrochloric acid (pH 2) and sodium hypochlorite solutions (chloride concentration: 400–499 mg/L) for up to 6 h. It was mainly due to presence of glycine-functionalized PVA on the modified membrane surface that acted as protecting layer for the PD layer. The findings showed that the glycine-functionalized PVA is more favorable for UF membrane modification as it offered good stability in both acidic and alkaline conditions as well as exhibited better antifouling properties.

Recently, Hou et al. (2017) developed integrated antifouling and antimicrobial carboxylated cardopoly(aryl ether ketone) (PEK-COOH) UF membranes by grafting PEO-NH_2 of different molecular weights (120, 350, and 550 g/mol) onto the membrane surface. The presence of a carboxyl group (–COOH) in the functionalized PEK-COOH is believed to be very helpful for membrane surface modification via

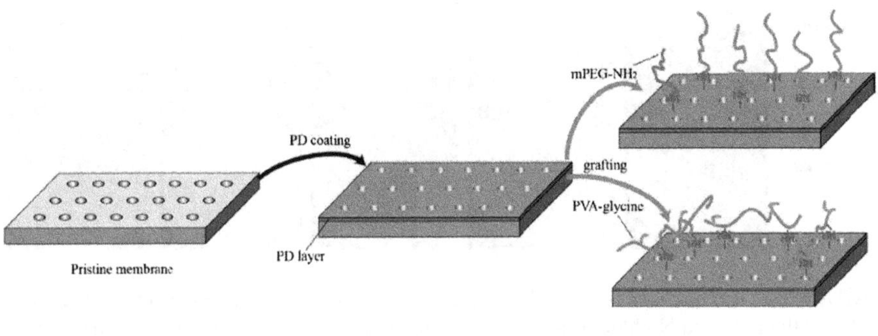

FIGURE 5.8 Schematic diagram for the surface modification of the PES membranes based on PD coating followed by grafting. (From Li, F. et al., Applied Surface Science. 345, 301–309, 2015a.)

grafting method. The results showed that all the grafted membranes demonstrated greater water flux (124–141 L/m^2.h.bar) than that of unmodified membrane (96 L/m^2.h.bar), with no significant change in BSA rejection. Using a confocal fluorescence microscope, no protein adsorption was observed on the membrane surface modified by PEO-NH_2 of 350 and 550 g/mol. The excellent antifouling property should be attributed to the increased hydrophilicity and reduced electronegativity of membrane surface as a result of the presence of PEO segments that inhibits the protein adsorption. More importantly, the functionalized membranes were found to offer complete resistance against bacterial adhesion of *Escherichia coli (E. coli)* and *Staphylococcus aureus* (*S. aureus*) as shown in Figure 5.9.

An attempt was also made by Xueli et al. (2013) to modify surface properties of a PSF membrane using acrylamide vinyl monomers containing natural capsaicin (N-(3-tert-butyl-2-hydroxy-5-methylbenzy)), with the goal of increasing both antifouling and antibacterial resistance. The authors found that the modified membranes prepared with the presence of benzophenone (BP) could offer better antifouling properties (FRR: 50%) compared to the unmodified membrane (FRR: 20%) when tested with 1 g/L BSA solution for 30 min. Wang et al. (2014b), on the other hand, performed a surface modification on the commercial PSF membrane via UV-assisted graft polymerization of acrylic acid (AA) and N-(5-methyl-3-tert-butyl-2-hydroxy benzyl) acrylamide (MBHBA). The effects of the MBHBA:AA ratio and irradiation during the grafting process were studied in order to determine the best conditions for production of membranes with excellent antifouling and antibacterial properties. As reported, an MBHBA:AA ratio of 1:10 and 12-min irradiation time were the optimized conditions to produce a membrane with excellent fouling resistance against protein (BSA), achieving complete flux recovery after being subjected to 1 g/L BSA filtration for 30 min. 100% antibacterial efficiency was observed for all the modified membranes at 12-min irradiation time when subjected to 0.1 mL *E. coli* suspension with a concentration of 6×10^5 cfu/mL for 24 h. They also found that the antibacterial activity was not influenced by the presence of AA in the membrane surface.

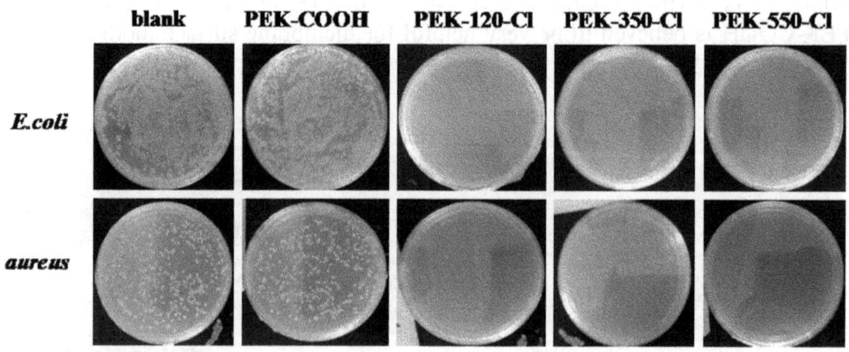

FIGURE 5.9 Comparison between the biocidal test results of the PEK-COOH membrane (control) and the PEO and N-chloramine functional group modified membranes (Note: The number indicates the molecular weight of PEO-NH_2 used during grafting process). (From Hou, S. et al., Journal of Colloid and Interface Science. 500, 333–340, 2017.)

Although the improved surface hydrophilicity upon modification could enhance membrane antifouling properties, the water flux of the membrane was compromised as a result of pore reduction.

Overall, it can be seen that in most of the studies, the membrane water fluxes were negatively affected upon surface modification (either grafting or physical coating), although other properties such as hydrophilicity, antifouling, and antibacterial resistances were improved. The reduction of membrane surface pore size and/or pore blockage is unavoidable through surface modification, but the extent of flux reduction should be minimized through optimization of synthesis conditions.

5.3 POLYMER/INORGANIC UF MEMBRANES

As discussed in the previous section, polymeric materials such as PSF, PES, and PVDF are very commonly used to produce UF membranes with a wide range of pore size. These polymeric membranes, however, are prone to fouling problem and flux reduction owing their low hydrophilic property. To overcome the issue, polymeric membranes incorporated with hydrophilic inorganic nanomaterials are attempted to improve membrane antifouling properties without sacrificing water flux and solute separation efficiency.

Advancement in nanotechnology in recent years has led to the development of a wide range of inorganic nanomaterials that can potentially be incorporated into polymeric membrane matrix. The forming of an organic and inorganic network can create a nanocomposite membrane with synergistic effects as the incorporation of inorganic additives is able to not only produce the desired membrane structure but also improve its thermal, mechanical, and hydrophilic properties, leading to enhanced performance with respect to permeability, selectivity, and antifouling properties.

Inorganic nanomaterials can be categorized into two different types: dense and porous materials. Some of the examples of dense-structured nanomaterials that have been previously used in polymeric membrane synthesis are titanium dioxide (TiO_2), zinc oxide (ZnO), manganese oxide (MnO), silver nitrate ($AgNO_3$), zirconium dioxide (ZrO_2), and iron oxide (Fe_2O_3). Carbon nanotubes (CNTs), halloysite nanotubes (HNTs), graphene oxide (GO), titania nanotubes (TNTs), zeolite, mesoporous silicon dioxide (SiO_2), and TiO_2 are some of the examples of porous nanomaterials. These nanomaterials can be incorporated in the polymeric membranes via either the blending method (for bulk modification) or the grafting/coating technique (for surface modification) as shown in Figure 5.10. Upon incorporation of nanomaterials, the modified polymeric membranes are more effective in repelling foulants due to an improved hydrophilic property that attracts more water molecules than foulants. The improved surface property will ultimately reduce the adherence of foulants on the membrane, leading to an enhancement in water permeability. In the following subsections, the effects of bulk modification and surface modification on the membrane properties upon incorporation of nanomaterials will be discussed in detail. Focus will be placed on how nanomaterial incorporation could improve membrane structures and their surface chemistry that leads to greater water permeability and antifouling propensity.

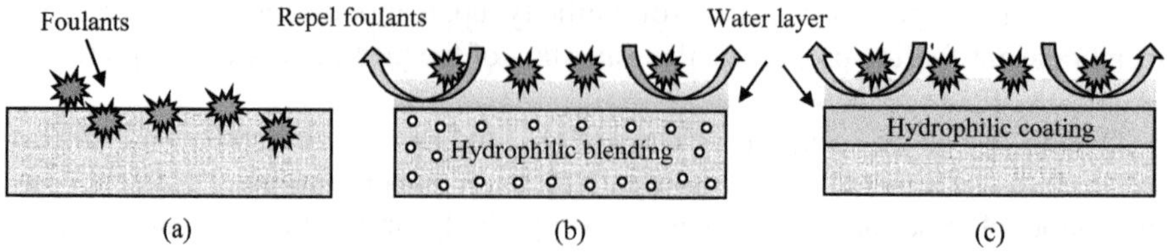

FIGURE 5.10 Mechanism of fouling occurrences, (a) control membrane, (b) membrane with bulk modification, and (c) membrane with surface modification.

5.3.1 Polymer/Inorganic Blending

The introduction of nanomaterials in polymeric membranes by blending is the most commonly used method to produce polymer/inorganic UF membranes owing to the facile preparation process. Undoubtedly, the presence of nanomaterials can increase the hydrophilicity of a membrane, resulting in higher water flux. Nevertheless, in many cases the membrane solute rejection efficiency could be negatively affected owing to the trade-off effect between flux and rejection properties. The negative impact however could be minimized or avoided if the selection of inorganic nanomaterials and its loadings used are properly conducted.

It is agreed by many that TiO_2 nanoparticle is one of the most widely used nanomaterials for improving surface hydrophilicity of polymeric membranes and thus its antifouling properties. TiO_2 is a transition metal oxide that polymorphs naturally and can be categorized into four main types: anatase (tetragonal), brookite (orthorhombic), rutile (tetragonal), and TiO_2 (B) (monoclinic) (Mital and Manoj, 2011). It has been reported in literature perhaps more often than any other nanomaterial as it is desired for its powerful photocatalyst property that has high stability, low cost and is non-toxic to both humans and the environment (Bet-moushoul et al., 2016). Although porous TiO_2 exhibits higher surface area and greater photocatalytic activity than nonporous TiO_2, its synthesis process via chemical reaction is rather complicated (Liu et al., 2016).

5.3.1.1 Nonporous TiO_2

Asgarkhani et al. (2013) investigated the effect of nonporous TiO_2 loading on the properties of cellulose acetate butyrate (CAB) membranes. Increasing the TiO_2 content to 4 wt% was reported to produce a membrane of greater water flux, achieving pure water flux of 190 L/m^2.h at operating pressure of 5 bar. This is mainly due to the formation of macrovoids and/or improved surface hydrophilicity. Further investigation using a BSA feed solution (1000 ppm) revealed that the CAB/TiO_2 membranes showed improved antifouling properties, recording a much lower degree of flux decline compared to the neat CAB membrane. At 4 wt% TiO_2 content, the CAB/TiO_2 membrane could achieve 98% BSA rejection with FRR recorded at 70%.

The use of nonporous TiO_2 for polymeric membrane fabrication was also reported in the work of Méricq et al. (2015). The resultant PVDF/TiO_2 membranes were aimed to limit the fouling tendency and minimize cleaning frequency during operation. The improved membrane water flux upon addition of nanoparticles was well correlated with the contact angle and SEM results that showed improved surface hydrophilicity and development of more finger-like structures. At optimum TiO_2 content (2.4 wt%), the resultant membrane exhibited pure water flux as high as 130 L/m².h.bar. Excessive use of TiO_2 should be avoided as it tended to form a severe agglomeration that might block membrane pores and affect water flux. As TiO_2 could act as a photocatalyst, the PVDF/TiO_2 membrane demonstrated better flux recovery compared to the neat PVDF membrane when UV irradiation was introduced during filtration process.

Li et al. (2014b) reported the blending of TiO_2 nanoparticles mediated by triblock copolymer (Pluronic F127) in the PES membrane. They reported that the water flux of the pristine PES membrane (51 L/m^2.h.bar) increased four-fold (236 L/m^2.h.bar) upon incorporation of appropriate amount of TiO_2. The remarkable improvement in water flux did not compromise the separation performance as the modified PES membrane could still achieve 96% BSA removal rate. With respect to antifouling resistance, the PES/TiO_2 membrane displayed FRR of 79% compared to 52% shown by the pristine membrane. The good interaction between the nanoparticles and Pluronic F127 was claimed to be the main factor restraining the leaching of nanoparticles from the membrane and leading to a stable and sustainable antifouling property.

5.3.1.2 Nanoporous TiO_2/Titania Nanotubes

Nanoporous TiO_2 nanoparticles were also considered as potential filler in improving the properties of polymeric membranes. Low et al. (2015), for instance, incorporated self-synthesized nanoporous TiO_2 into a PES membrane and studied the effect of nanoporous TiO_2 on the membrane antifouling properties. Although adding a higher loading of nanoporous TiO_2 could make the resultant membrane less susceptible to fouling, it at the same time compromised the rejection rate. Thus, the authors recommended lower nanoparticle loading (<0.1 wt%) in order to achieve a good balance of separation and antifouling properties. Bidsorkhi et al. (2016) also agreed that the loading of nanoporous TiO_2 added into the polymeric membranes should be properly controlled to minimize the negative impacts caused by nanoparticle aggregates.

Padaki et al. (2015) studied the role of TNTs on the characteristics of PSF membranes and reported that the increase of TNTs loading from zero to 0.5 wt% tended to improve remarkably the membrane pure water flux by 400%, achieving a value of 160 L/m^2.h.bar. At 0.5 wt% TNTs, the modified PSF membrane showed remarkable antifouling properties with 100% FRR and 3% R_{ir} (irreversible resistance) compared to 73% FRR and 23% R_{ir} in the pristine PSF membrane. The improved antifouling properties are mainly due to the presence of abundant hydroxyl groups on the membrane surface upon addition of TNTs.

5.3.1.3 Modification of TiO_2/Nanoporous TiO_2

Research has been carried out to modify the surface of the TiO_2 nanoparticle for better distribution in the membrane matrix. Poor dispersion of a nanomaterial in the polymeric dope solution is likely to cause severe particle agglomeration that results in membrane surface defects and/or reduced mechanical property. Teli et al. (2013) modified TiO_2 nanoparticles using polyaniline (PANI) via *in-situ* polymerization prior to blending with PSF membrane. The PSF membrane incorporated with PANI-modified TiO_2 was found to improve not only membrane surface hydrophilicity (higher water permeability) but also enhance BSA rejection and antifouling property. Results showed that the PSF membrane blended with 1 wt% PANI-modified TiO_2 could achieve 79% FRR compared to 60% shown by the neat PSF membrane when both types of membranes were tested using 0.7 g/L BSA solution for 180 min at 2 bar.

Zhang et al. (2013), on the other hand, demonstrated that the agglomeration between TiO_2 nanoparticles in the membrane matrix could be greatly reduced by grafting the nanoparticles with 2-hydroxyethyl methacrylate (HEMA). By incorporating the PSF membrane with 2 wt% HEMA-modified TiO_2, the resultant membrane achieved a good balance of water flux (150 L/m^2.h.bar) and BSA rejection (>90%). As a comparison, the neat PSF membrane only showed 40 L/m^2.h.bar water flux and 88% BSA rejection when tested under the same conditions. With respect to antifouling ability, the modified PSF showed 83% FRR compared to 72% FRR shown in the neat membrane. The enhanced antifouling ability of the modified PSF membrane was attributed to the even distribution of the nanomaterial coupled with its improved compatibility with polymeric matrix after HEMA grafting.

Wu et al. (2017a) used PDA to modify TiO_2 nanoparticles to enhance the immobilization of TiO_2 in the membrane matrix. It was found that upon addition of optimum content of PDA-modified TiO_2 (0.8 wt%), the developed membrane showed 1.5 times higher pure water flux than that of the control PSF membrane while maintaining high level of BSA rejection (~90%). This is because of the synergistic effects of hydrophilic TiO_2 and PDA that contribute to the increase hydrophilicity in the modified membrane, thus limiting the fouling action of BSA on membrane surface.

Alsohaimi et al. (2017) found that the PSF membrane modified by sulfonic acid-functionalized TNTs (TNTs-SO_3H) could achieve better filtration performance in terms of water permeability and fouling resistance than the pristine PSF membrane. Upon addition of optimum TNTs-SO_3H loading (3 wt%), the water flux as high as 233 L/m^2.h.bar was attained owing to the increase in porosity, hydrophilicity, and water uptake. The modified membrane also demonstrated improved antifouling properties and achieved FRR of 63% and R_{ir} of 14% after 240-min BSA filtration. As a comparison, the control membrane showed FRR and R_{ir} of 47% and 33%, respectively.

5.3.1.4 Other Nanomaterials

In addition to the commonly used TiO_2 nanoparticles, other types of inorganic nanomaterials that possess similar effects in improving membrane water flux and antifouling property are SiO_2, ZnO, and CNTs. Li et al. (2015b) investigated the effect of hollow mesoporous silica sphere (HMSS) loading on the properties of PES membrane. Unlike nonporous SiO_2, HMSS had a porous and hollow channel diameter structure in which the inner cavity facilitates water transport and enhances membrane water permeability without affecting the membranes selectivity. In addition, HMSS is high in surface area and possesses tunable pore dimensions. It was reported that the loading of HMSS for membrane synthesis should be controlled at <2 wt% as high HMSS loading is likely to cause the viscosity of the casting solution to increase and further affect the dispersibility of nanoparticles in membranes. At optimum HMSS loading of 1.5 wt%, the resultant membranes exhibited almost twice pure water flux (195.7 L/m^2.h.bar) and better antifouling properties (FRR of 76.7%) compared to the membrane made of typical mesoporous SiO_2 that exhibited water flux of 98.7 L/m^2.h.bar and FRR of 72.3%. As both types of SiO_2 are hydrophilic in nature, the authors attributed the improvement in water flux of the newly developed membranes to the unique hollow structure of HMSS that provides a shortcut for water to pass through with minimal transport resistance.

Separately, Huang et al. (2012) examined the effects of mesoporous SiO_2 on the characteristics of PES membranes with respect to hydrophilicity, water permeability, antifouling properties, and thermal stability. Results showed that the BSA adsorption of the developed membrane decreased remarkably from 45.8 mg/cm^2 to 21.4 mg/cm^2 with increasing mesoporous SiO_2 loading from zero to 2 wt%. The reduced BSA adsorption led to the improved FRR of the membrane: the membrane incorporated with 2 wt% mesoporous SiO_2 displayed 76.2% FRR compared to 45.2% in the neat PES membrane. Furthermore, it was also found that mesoporous SiO_2 improves membrane thermal properties. The addition of nanofillers strengthens the interaction and chemical bonding between the polymer matrix and silica networks and contributes to the enhancement of membrane thermal stability (Muhamad et al., 2015).

Similar to TiO_2 nanoparticle, which has a surface that can be chemically modified, Zhi et al. (2014) evaluated the effect of grafted SiO_2 on the PVDF-based membrane properties. SiO_2 was first modified by grafting poly(hydroxyethyl methacrylate)-block-poly(methyl methacrylate) (PHEMA-b-PMMA) brushes onto the nanoparticle surface prior to use. Upon incorporation of PHEMA-b-PMMA-modified SiO_2, the resultant membrane displayed improved water permeability as a result of increased membrane porosity and hydrophilicity. The improved water permeability did not compromise separation efficiency as the resultant membrane could still achieve 95% BSA removal rate. More importantly, its water flux and FRR were better than that of the neat PVDF membrane. Water flux in particular increased 2.5 times when compared with the neat PVDF membrane. The increase in FRR indicated lower BSA adsorption and can be explained by the fact that the hydroxyl groups of PHEMA can bind water molecules to form a hydration layer and prevent protein adsorption. Wang et al. (2016b), on the other hand, evaluated the properties of a PVDF membrane incorporated with polydimethylsiloxane (PDMS)-modified SiO_2 and reported that the modified membrane exhibited promising fouling repulsion and anti-adsorption ability, recording more than 90% FRR during fouling test of BSA.

Guo et al. (2017) functionalized SBA-15 mesoporous silica by doping it with Zr and Ti prior to use in PES membrane fabrication. The best antifouling property was observed in the membrane incorporated with SBA-15 doped with Ti. This membrane showed the lowest BSA adsorption of 55 μg/cm^2, compared to 65 μg/cm^2 shown by the membrane in which Zr was used to dope SBA-15. Although both membranes exhibited a high degree of hydrophilicity, the PES-Ti membrane with slightly smaller pore size (4.5 nm) than the PES-Zr membrane (4.8 nm) was better in terms of BSA adsorption.

Grafting SiO_2 nanoparticles using polyvinylpyrrolidone (PVP) was also attempted and reported in the work of Song and Kim (2013). The PSF membrane incorporated with PVP-g-SiO_2 showed improvement in filtration performance. Even though the membrane water flux was 2.3 times higher than that of the PSF membrane incorporated with unmodified SiO_2, its excellent separation efficiency was not compromised as the membrane rejection against PEG (40 kDa) remained high at 98%. Furthermore, Song and Kim observed that the use of PVP-g-SiO_2 could increase the interfacial adhesion between the organic and inorganic network, which played a key role in improving its dispersibility throughout the membrane structure as shown in Figure 5.11. Similar results (increased water flux and better particle dispersibility) were also reported in the work of Muhamad et al. (2015) in which the anionic surfactant-sodium dodecyl sulphate (SDS) was used to modify SiO_2 followed by its incorporation into the PES membrane.

In another study by Yu et al. (2013), SiO_2-Ag composite was synthesized via reduction reaction prior to use in UF membrane snythesis. This composite material tended to have synergistic effects in improving both organic antifouling and antibacterial properties of PES membrane. Although the developed membrane could demonstrate excellent results by completely preventing the growth of *E. coli* and *S. aureus*, its separation rates against PEG20,000 and PVA30,000–70,000 were compromised owing to the increase of membrane surface pore size upon addition of SiO_2-Ag composite.

Similarly, Wang et al. (2017) modified a PVDF membrane by embedding N-halamine functionalized silica nanospheres (HFSNs) at different quantities into the polymer matrix, aiming to enhance antifouling and antibacterial properties of the membranes. Optimal loading was reported to be 0.6 wt% of HFSNs, as the membrane made of this loading achieved the highest pure water flux (600 $L/m^2.h.bar$) without compromising BSA rejection (>96%). More importantly, the newly developed membrane could achieve high FRR and improved antibacterial properties against *E. coli* and *S. aureus*.

In addition to SiO_2 and modified SiO_2 nanoparticles, Shen et al. (2012) utilised ZnO to improve the properties of PES membranes. By introducing 0.4 wt% ZnO in the membrane matrix, water flux as high as 120 $L/m^2.h$ could be achieved at

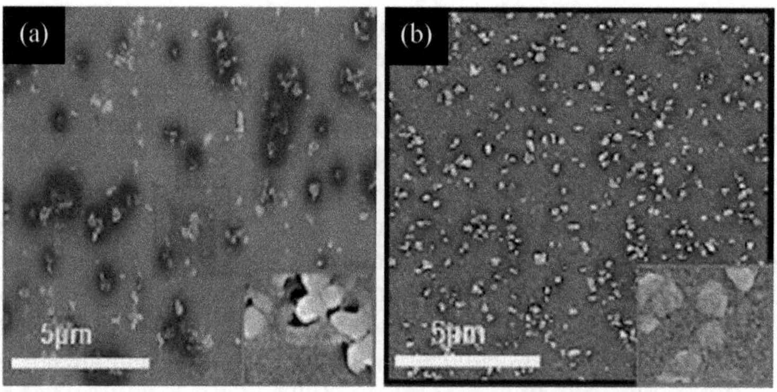

FIGURE 5.11 SEM images of surface of PSF membrane incorporated with (a) PSF/SiO_2 membrane and (b) PSF/PVP-g-SiO_2 membrane. (From Song, H.J., Kim, C.K., Journal of Membrane Science. 444, 318–326, 2013.)

operating pressure of 1 bar. The flux was tremendously improved after taking into account the flux of the pristine PES membrane, i.e., 30 L/m^2.h.bar. Besides being able to maintain the BSA rejection at >90%, the ZnO-incorporated membrane also exhibited a much lower flux decline ratio, suggesting improved surface properties against foulant adsorption or deposition.

ZnO is also reported to be useful in improving the hydrophilicity of membranes made of PVDF. A study by Liang et al. (2012) found that the water permeability of a neat PVDF membrane was improved by almost two-fold upon incorporation of optimum ZnO loading (6.7 wt%). Another study showed that the ZnO-modified PVDF membrane could have the additional feature of self-cleaning capability under mercury lamp irradiation (Hong and He, 2014). By exposing the used membranes to 30-min low-pressure mercury lamp irradiation, Hong and He found that 94.8% of initial water flux of ZnO-modified PVDF membrane could be retrieved. This recovery rate was obviously higher than that of the neat PVDF membrane that showed only 63.3% flux recovery. The improvement of the membrane self-cleaning property is mainly due to the photo-catalysis mechanism of the ZnO embedded.

In order to improve the dispersibility of ZnO in the membrane matrix, Jo et al. (2016) grafted the ZnO with PVP prior to the blending process. They proved that the PVP-grafted ZnO could disperse much better on the membrane surface compared to the unmodified ZnO as shown in Figure 5.12. The improved nanomaterial dispersibility has caused the membrane to have better filtration performance, achieving 50 L/m^2.h (3 bar) water flux, >90% PEG rejection, and 40% flux reduction during filtration of polyoxyethylene glycol alkylether ($C_{16}E_8$) solution of 1×10^{-6} mol/L. The control membrane (incorporated with unmodified ZnO), meanwhile, displayed 35 L/m^2.h (3 bar) water flux, <80% PEG rejection, and 80% flux reduction.

The incorporation of single-walled carbon nanotubes (SWCNTs) or multi-walled carbon nanotubes (MWCNTs) in polymeric membranes has also been shown to

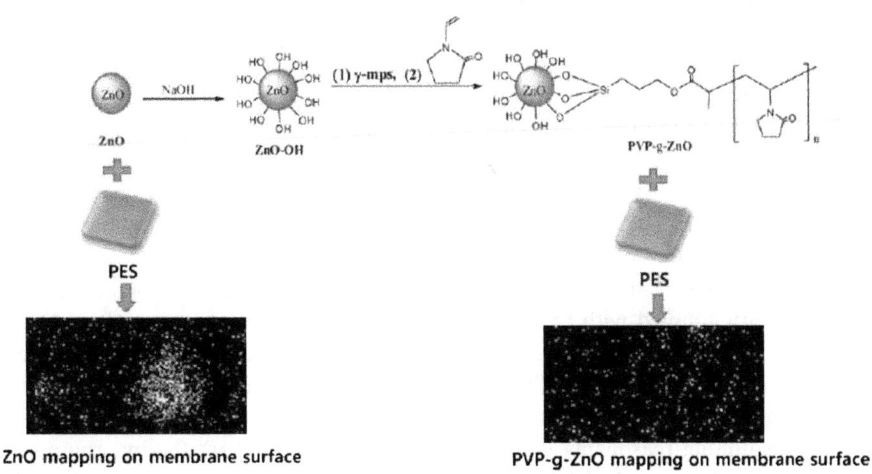

FIGURE 5.12 Comparison between the dispersibility of ZnO and ZnO-g-PVP on the PES membrane surface. (From Jo, Y.J. et al., Industrial & Engineering Chemistry Research. 55, 7801–7809, 2016.)

improve the membrane overall performance. Zhao et al. (2012a) investigated the blending of hyperbranched poly(amine-ester) functionalized MWCNTs with a PVDF polymer. The functionalized MWCNTs showed good dispersibility at the individual nanotube levels in the PVDF matrix. The functional groups of HPAE groups further contributed to the membrane hydrophilicity that reduced protein adsorption from 70 mg/cm^2 to, at the lowest, 20 mg/cm^2 upon addition of 2 wt% functionalized MWCNTs. Filtration experiments revealed that besides improving membrane water flux and FRR, functionalized MWCNTs maintained the good rejection of the membrane against BSA. Similarly, Qiu et al. (2009) reported that the water flux of the PSF membrane could be further enhanced by incorporating 0.19 wt% MWCNTs functionalized by isocyanate and isophthaloyl chloride groups into the polymeric dope solution. With respect to BSA adsorption, the static adsorption test demonstrated that the membrane incorporated with functionalized MWCNTs had lower BSA adsorption (<7 mg/L) than that of the control membrane (12 mg/L) at pH 8.2. At a higher pH environment, the modified membrane showed lower BSA adsorption rate owing to the increase in its surface charge (negative) that repulsed more BSA molecules.

Daraei et al. (2013a), on the other hand, evaluated the performance of the PES membrane incorporated with polycitric acid (PCA)-functionalized MWCNTs. This newly developed membrane in general showed good reusability with only slight decrease in pure water flux after three cycles of whey proteins filtration. Nevertheless, it must be pointed out that the pure water flux (22 L/m^2.h at 4 bar) of the MWCNTs-incorporated membrane was significantly lower compared to many works reported in literature. The authors did not provide any explanation for the poor water permeability.

In another study, Wang et al. (2014a) introduced 0.6 wt% oxidized MWCNTs in the poly(vinyl butyral) (PVB) matrix and found that the pure water flux of the membrane was improved 1.7 times compared to the control membrane. Additionally, the modified PVB membrane displayed lower BSA adsorption (21.6 μg/cm^2) and higher FRR (98.3%) owing to its improved surface hydrophilicity. High FRR indicated that the fouled membrane could be easily regenerated with simple washing process.

Zhang et al. (2014a) made an attempt to improve the properties of PVDF membranes by introducing both perfluorosulfonicacid (PFSA) and oxidized MWCNTs into the polymer matrix. The PVDF membranes made of 0.75 wt% oxidized MWCNTs and 3 wt% PFSA showed outstanding performances, attaining pure water flux as high as 229.4 L/m^2.h.bar with 93% FRR after 3 cycles of fouling and cleaning. The excellent results are mainly caused by the increased membrane surface hydrophilicity coupled with reduced surface roughness. The presence of oxidized MWCNT is likely to form zonal structures on the membrane surface that reduces protein adsorption rate. MWCNT can also be modified by coating its surface with PDA before using it to enhance membrane performance (Sianipar et al., 2016). Besides exhibiting reasonably good water flux (81.3 L/m^2.h.bar), the incorporation of 0.1 wt% PDA-coated MWCNTs into PSF membranes also showed relatively high FRR (83%) and minimum R_{ir} (17.4%).

Another type of nanoparticle of interest particularly for its antibacterial and biofouling resistances is the Ag nanoparticle. Li et al. (2013b) investigated the immobilization of Ag nanoparticles in the PVDF membrane matrix with the aims of alleviating both organic and biofouling problem. The results revealed that the membrane incorporated with 0.5 wt% Ag could attain the highest BSA rejection of 93%. Further increase in the Ag loading from 0.5 to 1.5 wt% tended to increase the membrane water flux to the maximum value of 109 L/m^2.h.bar with 88% BSA rejection. More importantly, the Ag-incorporated membrane (1.5 wt%) showed better antifouling properties, i.e., 1.3 times higher than that of the neat PVDF membrane in terms of FRR. The addition of Ag nanoparticles could increase the membrane surface hydrophilicity, creating a strong attraction towards water molecules and preventing protein adsorption.

Similarly, Zhang et al. (2014b) and Ananth et al. (2014) found that the performance of a PES membrane could be improved upon incorporation of Ag. Zhang et al. (2014b) reported that the pure water flux as high as 350 L/m^2.h (at 2 bar) could be achieved by using the modified PES membrane incorporated with 1 wt% Ag without lowering BSA rejection (98%). Compared with the neat PES membrane, the Ag-incorporated membrane exhibited better resistance against BSA adsorption by showing a lower degree of flux decline. Further investigation indicated that the leaching of Ag from the membrane was found to be in the safe range and had little impact on the water flux. This could be attributed to the stable immobilized Ag in the membrane matrix.

Over the past decade, many works have reported the potential of using GO as an inorganic additive during synthesis of polymeric membranes intended for the applications of water separation. GO is popular mainly because of the presence of abundant oxygen-containing functional groups that could contribute greatly to membrane hydrophilicity and filtration performances. In addition, the high surface area and good chemical stability of GO might be considered as additional positive features for it to be used as an additive in polymeric membranes (Zhao et al., 2013a).

A study conducted by Wang et al. (2012) demonstrated that GO was not only capable of improving membrane water flux and antifouling resistance but also its mechanical properties. Compared to the properties of neat PVDF membrane, the water flux, FRR and tensile strength of 0.2 wt% GO-incorporated membrane was improved by 196%, 123%, and 228%, respectively. More importantly, the modified membrane also exhibited a good rejection rate against BSA (>90%).

Zhao et al. (2013a), on the other hand, varied the loading of GO in order to study the effect of inorganic nanomaterials on the PVDF membrane properties. As shown in Figure 5.13, the BSA adsorption rate of PVDF membrane decreased significantly from 165.1 mg/m^2 to about 35 mg/m^2 by increasing the GO loading from zero to 2 wt%. A decrease in BSA adsorption rate is strongly linked to improved membrane antifouling resistance owing to the existence of more –OH groups on the GO-incorporated membrane surface that form a hydrated layer to prevent BSA adsorption. With respect to water permeability, the PVDF membrane incorporated with 2 wt% GO displayed higher pure water flux of 26.5 L/m^2.h than that of 14.8 L/m^2.h by the pure PVDF membrane at operating pressure of 1 bar.

The use of functionalized graphene oxide (f-GO) was also attempted to synthesize a new type of UF membrane (Xu et al., 2014). 3-aminopropyltriethoxysilane

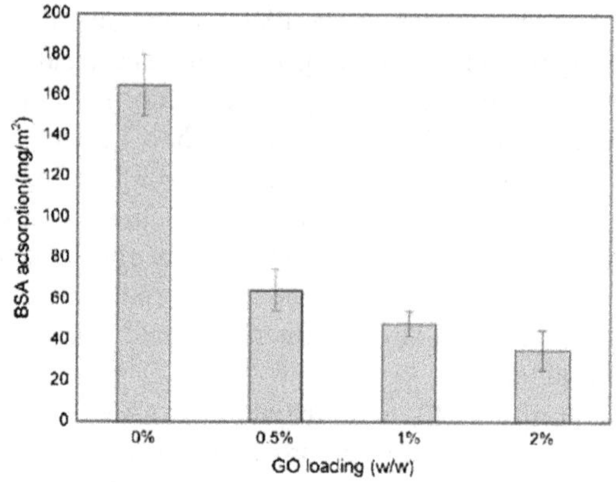

FIGURE 5.13 Effect of GO loading on BSA adsorption of PVDF membrane. (From Zhao, C. et al., Journal of Environmental Chemical Engineering. 1, 349–354, 2013a.)

(APTES) was used to functionalize GO in order to improve the interfacial interaction between inorganic nanomaterials and polymer matrix. Owing to the improved dispersion quality of f-GO in the PVDF membrane, greater membrane performances with respect to water flux, BSA rejection, antifouling resistance, and mechanical properties were observed. Upon addition of 1 wt% f-GO, the resultant membrane showed water flux of 361 L/m^2.h.bar, BSA rejection of 58% and FRR of 82%. As a comparison, the neat PVDF membrane displayed 240 L/m^2.h.bar, 42% and 62%, respectively. The results suggested the better antifouling performance of PVDF membrane upon f-GO incorporation following the reduction of interaction between foulants and membrane surface. Recently, Zambare et al. (2017) and Ayyaru and Ahn (2017) fabricated modified UF membranes using ethylenediamine functionalized GO and sulfonated GO, respectively, and both also reported improved membrane performance in terms of water flux and FRR without negatively affecting BSA separation efficiency. The improved antifouling properties can be attributed to high forces of hydrogen bonding and electrostatic repulsion of GO against the BSA, preventing it from depositing on the membrane surface.

Besides being functionalized, GO can also be synthesized with other nanoparticles to form a nanohybrid with better dispersion and compatibility with polymeric membrane. This was demonstrated by Wu et al. (2014) through doping of a SiO_2-GO nanohybrid in the PSF membrane. The unique property of the SiO_2-GO nanohybrid particles increased the membrane water permeability, BSA rejection, as well as antifouling properties. With the presence of only 0.3 wt% SiO_2-GO, the resultant membrane could attain water flux as high as 375 L/m^2.h (2 bar) with BSA rejection recorded at > 98%. Meanwhile, the control PSF and the PSF with only GO possessed lower water fluxes of 200 L/m^2.h and 230 L/m^2.h, respectively. The membrane incorporated with SiO_2-GO also showed higher FRR (72%) than that of the control PSF (62%) and the PSF with GO (64%) when all membranes were evaluated under same conditions. Other nanohybrid particles that have been recently utilized to improve UF membrane performance are Ag-GO (Wu et al., 2017b) and TiO_2-N-doped GO (Xu et al., 2017).

Fe_3O_4 is one of the potential nanoparticles that can be used together with GO to form a nanohybrid particle. Xu et al. (2016) studied the effect of the Fe_3O_4-GO nanocomposite on the PVDF-based membrane that was developed via the combination of magnetic field induced casting and a phase inversion technique as shown in Figure 5.14(a). Due to the magnetic attraction effect, the Fe_3O_4-GO nanohybrid particle could migrate toward the membrane top surface (shown in Figure 5.14(b)) and thereby render the surface highly hydrophilic with robust resistance to fouling. Results showed that the water flux of the Fe_3O_4-GO-incorporated PVDF membrane fabricated with magnetic field induced casting was 595 L/m^2.h.bar, or 206% higher than that of the neat PVDF membrane (195 L/m^2.h.bar) and 33% higher than the membrane made of 1 wt% Fe_3O_4-GO without magnetic field induced casting (447 L/m^2.h.bar). Furthermore, the newly developed hybrid membranes showed high rejection towards BSA (>92%) and high FRR (up to 86.4%) as well as with a lower adhesion force between foulants and membrane surface compared to other membranes.

Daraei et al. (2013b) also studied the effect of a magnetic field on the PES membranes blended with magnetic nanoparticles. By employing a magnetic field, the PES membrane incorporated with 0.1 wt% PANI-coated Fe_3O_4 possessed pure water flux of 52 L/m^2.h (at 4 bar) and a contact angle of 51.1°. The water flux was higher compared to the neat PES membrane (36 L/m^2.h) and the PES membrane made of PANI-coated Fe_3O_4 without magnetic effect (45 L/m^2.h). Besides the improved surface hydrophilicity, the authors explained that the changes in the membrane morphology in terms of skin layer thickness and surface pore size/porosity could partly contribute to higher flux of the membrane incorporated with PANI-coated Fe_3O_4. Further investigation using a highly concentrated whey solution found that the membrane fabricated under magnetic field tended to minimize adhesion of hydrophobic whey proteins on membrane surface due to the improved membrane surface hydrophilicity. Its FRR and reversible fouling rate were recorded at 80% and 58%, respectively. These were significantly higher compared to 52% and 34% showed by the neat PES membrane, respectively. Hwa et al. (2015) also found that the PEG-coated cobalt doped Fe_2O_3 nanoparticles had potential to improve the PES

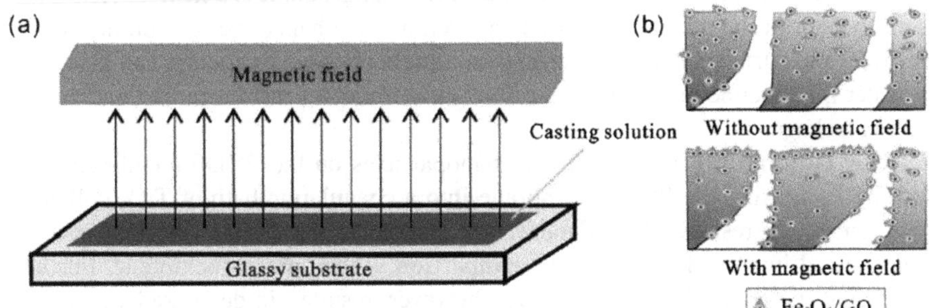

FIGURE 5.14 (a) Employing magnetic field on the surface of casting film composed of Fe_3O_4/GO nanoparticles and (b) the distribution of Fe_3O_4/GO nanoparticles within membrane matrix with and without magnetic field effect. (From Xu, Z. et al., ACS Applied Materials & Interfaces. 8, 18418–18429, 2016.)

membrane antifouling properties against BSA, but all the membranes were fabricated without the influence of magnetic field.

Other nanomaterials that have been previously utilized for UF membrane fabrication over the past several years are summarized in Table 5.3. These include mesoporous carbon nanoparticles (MCNs), salicylate-alumoxane (SA), hydrous manganese oxide (HMO), HNTs, boehmite (γ-AlOOH), titanium silicon oxide ($TiSiO_4$), zeoliticimidazolate, etc. As shown, there are many types of nanoparticles available for membrane fabrication and they are different in terms of size, structure, and characteristics. Generally, it has been demonstrated that the incorporation of nanoparticles into the polymeric membranes has huge potential to overcome the trade-off effect between water flux and solute rejection, provided the amount added is at the appropriate range and does not cause severe agglomeration within the membrane structure and/or top surface. It must also be noted that most of the studies only investigated a single type of nanomaterial on the membrane properties and comparison with other types of nanomaterials for the same membrane composition is rarely reported. Thus, it is worthwhile to research this further.

5.3.2 Membrane Surface Modification by Inorganic-Based Materials

The surface modification of polymeric UF membranes using inorganic-based materials is another way to achieve good compatibility between polymeric and inorganic materials. In comparison to the blending technique, the surface modification process normally involves complex reaction, high chemical usage, and longer time. Furthermore, there is a high risk for the inorganic nanomaterials to detach from the top membrane surface under high cross flow velocity during the operation. Because of these drawbacks, its study in research publications is much lower compared to the study of the blending method. Nevertheless, the recent advancements in nanomaterial synthesis and modification have made the top membrane surface alteration more feasible and practical than before.

Surface coating is a technique that establishes a thin layer of colloidal solution on the top layer of the membrane surface via either physical adsorption, interpenetration, or macroscopic entanglement of the functional group. It is a flexible technique used to smooth the membrane surface as well as to enhance the membrane hydrophilicity and surface charge. Nevertheless, a higher dosage of coating can also lead to water flux decline due to the increased water transport resistance and membrane pores blockage.

Teow et al. (2012) introduced TiO_2 nanoparticles on the PVDF membrane surface by immersing the PVDF membrane into a coagulation bath of TiO_2 colloidal suspension. The resultant PVDF membrane made of 0.01 g/L TiO_2 colloidal suspension showed improved water flux in comparison to the control membrane. Further increasing the concentration to 0.1 g/L, however, resulted in decreased water flux due to membrane pore blockage caused by the severe particle agglomeration.

Many agree that the interaction between the coating layer and membrane surface can be relatively weak and might cause the coating layer to peel off easily during long-term operation. Thus, in order to improve the stability of nanoparticles on the membrane surface, Shao et al. (2014) introduced PDA as the "bio-glue" layer on the

TABLE 5.3
Comparison of Selected UF Membranes Prepared from Different Nanomaterials between 2013 and 2017

Nanomaterial[a]	Properties of Nanomaterials/Precursors	Optimized Dope Composition (wt%)[b]	Format	Effects on Membrane Properties
MCNs (Orooji et al., 2017)	Self-synthesized MCNs using silica powder as precursor	PES/NMP/PVP/MCN 16/79.8/4/0.2	Flat sheet	The use of MCN of different quantities could maintain membrane BSA rejection at >99%, but the membrane incorporated with 0.2 wt% MCN showed a good balance between pure water flux (258 L/m^2.h.bar) and lowest BSA adsorption (7.8 μg/cm^2) due to increased hydrophilicity and improved antifouling resistance.
SA (Mokhtari et al., 2017)	Self-synthesized SA using aluminum nitrate 9-hydrate ($Al(NO_3)_3 \cdot 9H_2O$) as precursor	PSF/DMAc/ TritonX100/PEG/SA NA/NA/2/1/1	Flat sheet	PSF membrane made of 1 wt% SA nanoparticles had the highest FRR (87%) and the lowest R_{ir} (13%) compared to other PSF membranes incorporated with different loading, due to better distribution of nanoparticles and better membrane structural integrity.
HNTs-CS@Ag (Chen et al., 2013)	Halloysite clay from Henan Province, China. Particle size = 5 nm	PES/DMAc/PVP/ HNTs-CS@Ag 18/71/8/3	Flat sheet	The PES membrane incorporated with 3 wt% nanomaterials displayed enhanced water flux (375.6 L/m^2.h.bar) and higher FRR (97.6%) compared to the control PES membrane. However, the higher the loading of nanomaterials added, the lower the solute rejection owing to the formation of large surface pores. This newly developed membrane also showed good antibacterial activity against *E. coli* and *S. aureus*.
γ-AlOOH (Boehmite) (Vatanpour et al., 2012)	Self-synthesized Boehmite from aluminum nitrate and sodium hydroxide obtained from Merck (Germany). Particle size = 30 nm	PES/DMAc/PVP/ γ-AlOOH 21/77.5/1/0.5	Flat sheet	It was reported that the loading of nanomaterials added into PES membrane should be controlled at <0.5 wt%, as high nanomaterial loading tended to affect water flux due to nanoparticle agglomeration. Although the modified membrane displayed improved FRR owing to the presence of hydroxyl groups on the γ-AlOOH surface, the membrane's pure water fluxes were quite low (5.24–4.14 L/m^2 h at 5 bar).

(Continued)

TABLE 5.3 (CONTINUED)
Comparison of Selected UF Membranes Prepared from Different Nanomaterials between 2013 and 2017

Nanomaterial[a]	Properties of Nanomaterials/Precursors	Optimized Dope Composition (wt%)[b]	Format	Effects on Membrane Properties
$TiSiO_4$ (Dasgupta et al., 2014)	$TiSiO_4$ from Central Drug House Pvt. Ltd., India. Mean particle size <50 nm	CA/NMP/ $TiSiO_4$ 18/62/20	Flat sheet	The addition of 20 wt% $TiSiO_4$ in the CA membrane displayed maximum pure water flux of 67.2 L/m^2.h.bar and moderate BSA rejection of 90%. Besides enlarging the membrane pore size and improving structure interconnectivity, $TiSiO_4$ also increased CA membrane resistance against BSA fouling.
ZrO_2 (Pang et al., 2014)	Zirconyl chloride from Sinopharm Reagent Co. Ltd. Particle size = 5–10 nm	PES/DMF/PVP/ ZrO_2 16/80.5/2/1.5	Flat sheet	The water flux of the PES membrane was improved significantly from 8.2 to 179 L/m^2.h.bar upon addition of 1.5 wt% ZrO_2. Its BSA rejection, however, decreased due to larger pore size and porosity. With respect to antifouling ability, the smoother surface of the PES/ ZrO_2 membrane could minimize the degree of foulants being adsorbed on membrane surface.
Ag loaded zeolite (Shi et al., 2013a)	NaY from Anhui MingmeiMinChem Co., Ltd., China. Particle size = 0.5–1 µm Silver nitrate ($AgNO_3$) from Aladdin Chemistry Co., Ltd. China.	PVDF/H_2O/DMAc/ TEP/PVP/Ag loaeded zeolites 15/3/3/54.6/23.4/1	Hollow fiber	The presence of zeolite in the membrane showed better fouling resistance against BSA due to the improved negative surface charge that inhibited the adhesion of BSA to membrane surface. Moreover, the water flux of the modified PVDF membrane was also improved compared to the control PVDF membrane. The Ag-loaded zeolites membrane also showed excellent antibacterial property by achieving almost complete elimination of *E. coli* cells.

(Continued)

TABLE 5.3 (CONTINUED)
Comparison of Selected UF Membranes Prepared from Different Nanomaterials between 2013 and 2017

Nanomaterial[a]	Properties of Nanomaterials/Precursors	Optimized Dope Composition (wt%)[b]	Format	Effects on Membrane Properties
ZIF-L (Low et al., 2014)	Zinc nitratehexahydrateand 2-methy-limidazole from Sigma-Aldrich.	PES/NMP/PVP/ZIF-L 16/79.5/4/0.5	Flat sheet	The introduction of 0.5 wt% ZIF-L in the PES membrane could overcome the trade-off between water flux and solute rejection as the modified membrane showed improved water flux (378 L/m^2.h.bar) without significantly affecting molecular weight cut-off. Furthermore, the combined effects of increased membrane hydrophilicity and surface charge, along with reduced surface roughness, played an important role in reducing BSA adsorption on the membrane surface.
HMO (Jamshidi Gohari et al., 2014)	Self-synthesized HMO from manganese (II) sulfate monohydrate purchased from Merck. Particle size = 12 nm	PES/NMP/PVP/HMO 12.24/68.18/1.22/18.36	Flat sheet	The membrane water flux was improved remarkably from 39.2 L/m^2.h.bar in the control PES to ~500 L/m^2.h.bar in the modified PES membrane following the incorporation of a high amount of HMO at HMO:PES weight ratio of 1.5. The presence of the –OH groups of HMO has contributed greatly to the increased hydrophilicity and antifouling property of the PES/HMO membrane.

[a] AlOOH: aluminum oxide hydroxide/boehmite, HNTs-CS: halloysite nanotube-chitosan, HMO: hydrous manganese dioxide, $Mg(OH)_2$: magnesium hydroxide, MCNs: mesoporous carbon nanoparticles, SA: salicylate-alumoxane, $TiSiO_4$: titanium silicon oxide, ZIF-L: zeoliticimidazolate leaf-shape, ZrO_2: zirconium dioxide.
[b] CA: cellulose acetate, DMAc: dimethylacetamide, DMF: dimethylformamide, NMP: N-Methyl-2-pyrrolidone, PEG: polyethyleneglycol, PES: polyethersulfone, PSF: Polysulfone; PVDF: polyvinylidene fluoride, PVP: polyvinylpyrrolidone, TEP: triethyl phosphate, N/A: not available.

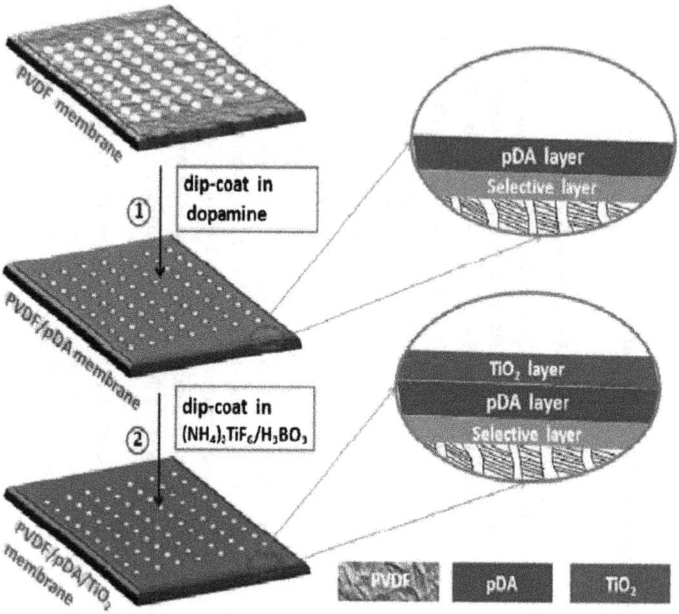

FIGURE 5.15 Schematic diagram of the binding process of TiO_2 layer via coating of pDA layer onto PVDF membrane. (From Shao, L. et al., Journal of Membrane Science. 461, 10–21, 2014.)

PVDF membrane surface prior to TiO_2 coating, as illustrated in Figure 5.15. The self-polymerization of PDA has assisted the homogeneous and tight bonds of the TiO_2 nanoparticles onto the PVDF membrane surface owing to the coordination bond (C-O···Ti) formed between TiO_2 and PDA by the deprotonation of catechol groups. The bonds improved the stability and performances of the membrane as evidenced by the simultaneous enhancement of water flux (228 L/m^2.h.bar) and BSA rejection (98%). The modified membranes also demonstrated improved antifouling properties as shown by the lower degree of flux decline, increased FRR, and decreased BSA adsorption under dynamic conditions. These promising results are mainly attributed to the tremendous increase of membrane hydrophilicity (contact angle decreased from 80.3° to <25°).

Qin et al. (2015) modified the surface of PVDF membranes by chemically binding TiO_2 nanoparticles and crosslinking with a PVA layer. They observed that the TiO_2 nanoparticles could be dispersed uniformly on the membrane surface using this modification approach. Although increasing TiO_2 concentration on the membrane surface tended to decrease water contact angle, the membrane water flux did not improve accordingly. This is mainly due to the formation of TiO_2 layer that increased water transport resistance. Results also showed that the cross-linked PVA might have partially blocked the membrane pores, leading to an increase in BSA rejection. At optimum loading of 0.08 wt% TiO_2, the PVA cross-linked PVDF membrane could achieve water flux of 420 L/m^2.h.bar, BSA rejection of 88%, and FRR of 80.5%.

The use of Ag nanoparticle to modify membrane surface properties was also attempted due to its antibacterial effect. Park et al. (2013) modified the PVDF membrane surface by covalent immobilization of Ag nanoparticles, which was mediated by thiol-end functionalized amphiphilic block copolymeric linker as shown in Figure 5.16. The Ag-modified PVDF membrane displayed excellent binding stability as no Ag was detected from the membrane via a leaching test. The formation of the sulfur-metal stable covalent bond between the Ag and the membrane was the main factor contributing to the excellent binding stability. The experimental results indicated that there were not many differences between the Ag-modified membrane and the neat membrane, as both membranes showed very similar water flux (approximately 1770 L/m^2.h.bar) and globulin rejection (100%). Nevertheless, the Ag-modified membrane was superior compared to the neat PVDF membrane in terms of antifouling and antibacterial performance.

The surface modification of a membrane using Ag nanoparticles was also studied by Sawada et al. (2012), in which the PES membrane was initially grafted with acrylamide (Am) prior to deposition of Ag nanoparticles. As a comparison, the normalized flux of the Ag-Am modified membrane was much more stable than that of neat PES membrane which showed sharp decrease of flux within a short period of time. The improved normalized flux was attributed to the increased membrane hydrophilicity following Am grafting. The BSA rejection of the Ag-Am modified membrane was also reported to increase owing to the formation of additional grafted layer. Li et al. (2013a) also reported a decrease in pore size and porosity after the PVDF membrane was grafted with PAA followed by Ag deposition, but the modified membrane showed improved antifouling performances compared to the neat PVDF membrane.

Another potential hydrophilic nanoparticle that was used to improve membrane surface properties was SiO_2. Liang et al. (2013) modified the PVDF membrane through a simple dip-coating technique that bound SiO_2 nanoparticles onto the membrane surface. To provide sufficient carboxyl groups as anchor sites for the nanoparticles

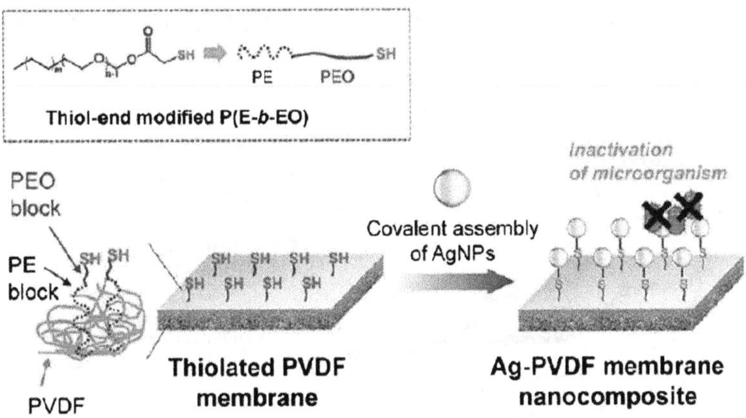

FIGURE 5.16 The covalent assembly of Ag nanoparticles via thiol-end functionalized block copolymeric linker. (From Park, S.Y. et al., ACS Applied Materials & Interfaces. 5, 10705–10714, 2013.)

binding, poly(methacrylic acid) (PMAA) was first grafted on the PVDF membrane surface through plasma induced graft copolymerization. The PMAA-grafted membrane was then dipped into the SiO_2 suspension in the presence and absence of N-(3-dimethylaminopropyl)-N-ethylcarbodiimide hydrochloride/N-hydroxysuccinimide (EDC/NHS) cross-linker. The functionalized PVDF/PMAA-grafted membrane in the presence and absence of EDC/NHS had a superhydropihilic property with water contact angle as low as 20° and 17°, respectively. The neat PVDF membrane (without any cross-linker) meanwhile displayed 76°. No significant changes were observed in permeability and selectivity between the functionalized and the neat membrane. However, the antifouling performance of the functionalized membrane was much better, with more than 80% water flux recovery as shown in Figure 5.17. The formation of a tightly bound hydration layer on the functionalized membrane surface played an important role in repelling protein molecules and reducing BSA adsorption onto the membrane surface.

Using the simple dip-coating method, Pan et al. (2017) altered the surface properties of a PVDF membrane by immobilizing Ag/SiO_2 nanoparticles. Increasing the Ag/SiO_2 suspension concentration from zero to 0.1 wt% decreased the contact angle of the membrane (Figure 5.18), suggesting improved hydrophilicity property that is likely to result in higher water flux and better antifouling properties. Lower BSA rejection rate, however, was observed in the Ag/SiO_2-modified membrane owing to the cracks formation and/or increased surface pore size. This caused the BSA rejection to decrease from 85% in the neat PVDF membrane to less than 78% for the PVDF-Ag/SiO_2 membranes.

Another way to modify membrane surface is through UV photografting as reported by Garcia-Ivars et al. (2014). The PES membrane was first immersed in a grafting solution of PEG/Al_2O_3 nanoparticles followed by UV irradiation exposure. The results showed that the modified membrane had improved hydrophilicity, permeability, solute rejection, and antifouling properties. The UV-grafted PES membrane with 2.0 wt% PEG and 0.5 wt% Al_2O_3 displayed the best antifouling performances among all the membranes fabricated. Its final water flux and normalized

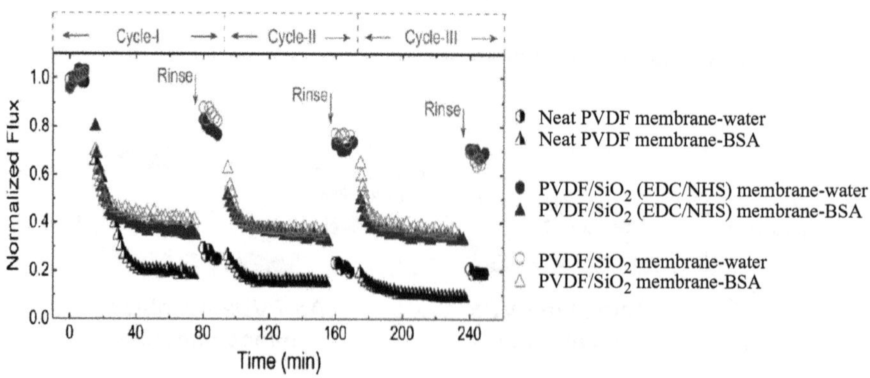

FIGURE 5.17 Pure water and BSA solution flux of the neat and functionalized PVDF membranes in three cycle filtration. (From Liang, S. et al., ACS Applied Materials & Interfaces. 5, 6694–6703, 2013.)

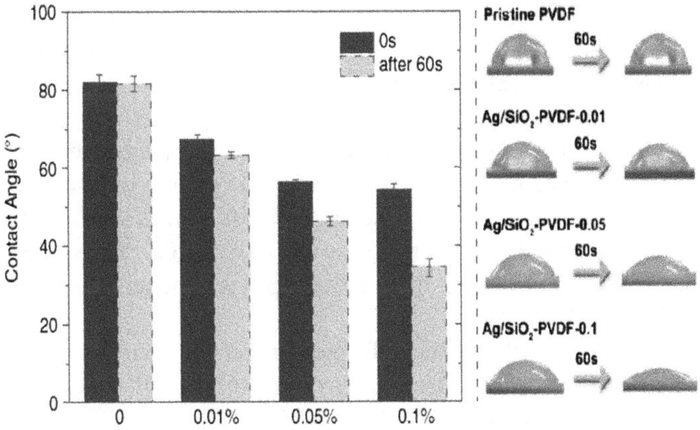

FIGURE 5.18 The contact angles of pristine PVDF and PVDF/Ag-SiO_2 membranes. (From Pan, Y. et al., Journal of Chemical Technology and Biotechnology. 92, 562–572, 2017.)

flux ratio were recorded at 388 L/m^2.h and 0.81, respectively, when being tested with 5000 ppm PEG (35 kDa) feed solution for 120 min at 2 bar. The neat PES membrane meanwhile showed 320 L/m^2.h and 0.70, respectively, under the same conditions.

Dong et al. (2015) performed surface modification on PSF membranes via *in-situ* embedment technique where the membrane was immersed in a water bath dispersed with various types of nanoparticles, i.e., Al_2O_3, SiO_2, SWCNTs, and zeolite, at a concentration of 500 ppm. Analysis revealed that both the Al_2O_3 and zeolite nanoparticles were uniformly dispersed on the membrane surface with a high coverage rate of 38% and 49%, respectively. The improved surface hydrophilicity upon the nanoparticles' embedment led to the decrease in water transport resistance in which the Al_2O_3-modified and zeolite-modified PSF membranes showed lower membrane resistance (0.91×10^{12} m^{-1} and 0.94×10^{12} m^{-1}) compared to the neat PSF membrane (1.3×10^{12} m^{-1}). Other nanoparticles that have been used as coating material for microporous membranes are functionalized SBA-15 SiO_2 particles (Díez et al., 2017) and copper nanoparticles (Gul et al., 2017). Both studies also demonstrated the positive role of nanoparticles in improving antifouling resistance of PES-based membranes.

5.4 MODIFICATION OF COMMERCIAL POLYMERS/NEWLY DEVELOPED POLYMERS

For decades, many works have been conducted to enhance the flux and fouling resistance of membranes by modifying their properties either through bulk modification (blending) or surface modification approach. There are cases in which the intrinsic characteristics of the commercial polymers (their molecular structure) are altered in order to further improve the degree of hydrophilicity. Some polymers are only recently developed for the primary purpose of combating the adherence of protein to membranes without compromising water flux and solute separation.

Modification of commercial polymers can be carried out by introducing functional groups such as sulfonic ($-SO_3H$), hydroxyl ($-OH$), carboxyl ($-COOH$), amino ($-NH_2$), or fluorobenzene to the existing organic structure of the polymers. Sulfonation is a chemical reaction that takes place when a sulfonic acid group ($-SO_3H$) is introduced into the structure of a molecule or ion to replace a hydrogen atom. For example, the sulfonation of PES can be performed by adding SO_3H groups to the (aromatic) backbone of PES using sulfonating agents such as chlorosulfonicacid ($ClSO_3H$), sulfuric acid (H_2SO_4) and trimethyl silylchlorosulfate ($(CH_3)_3SiSO_3Cl$). The successful sulfonation of PES (SPES) will make it more advantageous to be used for UF membrane fabrication (Zhao et al., 2013b).

Xiang et al. (2014) modified the PSF membrane via a surface chemical reaction involving various types of functional groups. Although the water fluxes for all the modified membranes were almost the same (approximately 228 $L/m^2.h.bar$), the adsorbed amounts of BSA by these membranes were quite different. It was reported that the BSA adsorbed amount of the unmodified membrane was significantly decreased from 18.3 $\mu g/cm^2$ to 9.0, 6.2, and 6.0 $\mu g/cm^2$ after the introduction of $-OH$, $-COOH$, and $-SO_3H$ groups, respectively. These revealed the improvement of the membrane antifouling property. Zhao et al. (2012b), on the other hand, grafted perfluoroalkyl groups onto an aminated PAN membrane surface via acylation reaction. The fluorinated-PAN membranes displayed superior antifouling ability with almost 99% BSA rejection and FRR along with the lowest total flux decline rate of 16%. The improved filtration properties can be explained by the fact that the fluorinated PAN membranes have relatively low surface energy that could weaken the adhesion force toward foulants and further minimize fouling susceptibility.

Poly(dimethylsiloxane) (PDMS) is widely used in many biomedical devices and implants because it is non-toxic, and its ease of fabrication, high oxygen permeability and mechanical properties, as well as biological and chemical inertness. Since the surface of PDMS can be modified to increase its hydrophilicity, its application can be potentially extended to membrane fouling minimization (Wu et al., 2012). Tu et al. (2013) grafted the surface of PDMS with a quaternized poly(dimethylaminoethyl methacrylate) (QPDMAEMA) and found that the contact angle of the control membrane could be remarkably reduced from 117.2° to <50° in the QPDMAEMA-modified membrane. Following this great improvement in surface hydrophilicity, the modified membrane also displayed 3.3 times lower BSA adsorption than the control membrane. This indirectly indicated the better antifouling property of the modified membrane.

Another commercial polymer that requires modification prior to use is poly(lactic acid) (PLA). PLA is a green material derived from plant resources, but its hydrophobicity makes it not suitable for UF membrane fabrication as it tends to foul easily (Moriya et al., 2012). An attempt was made by Shen et al. (2013) to enhance the antifouling property of PLA membrane by introducing length to the PEG segment in the PLA-PEG-PLA triblock copolymer. Although the increased PEG segment had significantly reduced the contact angle from 80° in the pure PLA membrane to the lowest of 53°, the pure PLA membrane had the highest pure water flux of 300 $L/m^2.h.bar$ in comparison to that of 115 $L/m^2.h.bar$ by the modified PLA/PEG membrane. The decreased water flux in the modified membrane is mainly governed by decreasing pore size upon addition of the triblock polymer. The effect of increased hydrophilicity in the modified PLA/PEG membrane was however confirmed in the BSA filtration process as the membrane displayed higher relative permeability (>0.8) compared to the pure PLA membrane (<0.5).

The development of new polymers for the purpose of mitigating fouling is now focused on aromatic polyamides (also known as aramid polymers). However, it must be pointed out that aromatic polyamides such as polyisophthalamides and polyterephthalamides cannot dissolve in common organic solvents. Thus, in order to make them more processable and soluble, flexible linkages such as polyalkylene or PEO are introduced within the polymer backbone (Carretero et al., 2013). Using this approach, Molina et al. (2014) investigated the performance of UF membrane made of aramid-g-PEO. Results revealed that the aramid-g-PEO membrane could attain much higher pure water flux (229 L/m^2.h.bar) than the commercial PSF (125 L/m^2.h.bar) and the polyamide membrane (92.3 L/m^2.h.bar). In addition, the incorporation of PEO side chains has greatly enhanced the hydrophilicity and fouling resistance of the resultant membranes with FRR recorded at >90%.

Poly(p-phenyleneterephthamide) (PPTA) is one of the newly developed polymers that has shown potential for synthesis of UF membrane with improved surface properties. As PPTA has a high melting point and insoluble properties in typical organic solvents, inorganic acids (e.g., sulfuric acid) are used to dissolve the polymer for dope preparation (Wang et al., 2015). According to Wang et al. (2016a), the pure water flux of the resultant PPTA membranes remained unchanged with BSA rejection >90%, even though they were subjected to high operating temperature (90°C) for up to 7 h. This confirmed the excellent thermal stability of the membranes in dealing with hot solution. Furthermore, the PPTA membrane could achieve almost complete flux recovery after three alternate cycles of BSA filtration and backwash, suggesting its superior antifouling performance.

Besides PPTA, the use of poly(m-phenyleneisophthalamide) (PMIA) to overcome selectivity-permeability trade-off of UF membranes was also reported in the work of Lin et al. (2016). The developed membranes demonstrated the water permeability and BSA separation factor in the range of 2.5–8 × 10^{-9} m/s.Pa and 400–10, respectively. In brief, the membrane permeability at a given separation factor increased two-fold, whereas the separation factor at a given permeability was improved almost 10 times. These promising results were attributed to the high porosity, narrow pore radius distribution, and excellent membrane hydrophilicity.

5.5 CONCLUSIONS

This chapter provides a comprehensive review of the materials that could be used to modify the morphology and surface chemistry of UF membranes for better performance in water separation process. There is a wide range of materials that can be considered for the UF membrane fabrication and modification. These materials are generally categorized into polymeric-based materials (e.g., amphiphilic copolymers and hydrophilic polymers) and inorganic nanomaterials (e.g., TiO_2, SiO_2, CNT, GO, zeolite, etc.). The introduction of such materials into polymeric membranes can be performed either by direct blending (during dope preparation) or top surface modification/coating (on the UF membrane surface). There are pros and cons of each method but in general, the membranes modified by secondary materials must be able to overcome the major drawbacks of pristine membranes. These include low water permeability, poor antifouling and antibacterial effect, and low mechanical strength. Developing zero-fouling membranes might seem impossible right now, but the degree of membrane fouling can be greatly reduced with the use of advanced materials. The selection and use of advanced materials for UF membrane fabrication is of utmost importance and should depend on the compounds to be removed from the feed solution. It is because there is no UF membrane that is universally applicable for any types of water applications.

ABBREVIATIONS

AA	Acrylic acid
$AgNO_3$	Silver nitrate
AlOOH	Aluminum oxide hydroxide/boehmite
APTES	3-aminopropyltriethoxysilane
BP	Benzophenone
BSA	Bovine serum albumin
CA	Cellulose acetate
CAB	Cellulose acetate butyrate
CNTs	Carbon nanotubes
DMAc	Dimethylacetamide
DMF	Dimethylformamide
DMSO	Dimethly sulfoxide
EDC/NHS	N-(3-dimethylaminopropyl)-N-ethylcarbodiimide hydrochloride/N-hydroxysuccinimide
Fe_2O_3	Iron oxide
f-GO	Functionalized graphene oxide
FRR	Flux recovery rate
GO	Graphene oxide
HMO	Hydrous manganese oxide
HMSS	Hollow mesoporous silica sphere
HNTs	Halloysite nanotubes
HNTs-CS	Halloysite nanotube-chitosan
MBHBA	N-(5-methyl-3-tert-butyl-2-hydroxy benzyl) acrylamide
MCNs	Mesoporous carbon nanoparticles
$Mg(OH)_2$	Magnesium hydroxide
MnO	Manganese oxide
$mPEG-NH_2$	Amine-terminated polyethylene glycol
MWCO	Molecular weight cut-off
MWCNTs	Multi-walled carbon nanotubes
NMP	N-methyl-2-pyrrolidone
PANI	Polyaniline
PAN-*b*-PEG	Polyacrylonitrile-*block*-polyethylene glycol
PAN-*g*-PVA	Polyacrylonitrile-*graft*-poly(vinyl alcohol)

PAA	Poly(acrylic acid)
PBMA-*b*-PMAA-*b*-PHFBM	Poly(butyl methacrylate)-*b*-poly(methacrylic acid)-*b*-poly(hexafluorobutyl methacrylate)
PC	Polycarbonate
PCA	Polycitric acid
PDMA	Poly(dimethylaminoethyl methacrylate)
PDMAA	Poly(N,N-dimethyl acrylamide)
PDA	Polydopamine
PDMS	Poly(dimethylsiloxane)
PEG-*g*-PAN	Poly(ethylene glycol)-*graft*-polyacrylonitrile
PEO-PPO-PEO	Poly(ethylene oxide)-poly(propylene oxide)-poly(ethylene oxide)
PEK-COOH	Carboxylated cardopoly(aryl ether ketone)
PEG	Polyethyleneglycol
PES	Polyethersulfone
PEO	Poly(ethylene oxide)
PFSA	Perfluorosulfonic acid
PHEMA-b-PMMA	Poly(hydroxyethyl methacrylate)-block-poly(methyl methacrylate)
PLA-PEG-PLA	Polylactic acid-polyethylene glycol-polylactic acid
PLA	Poly (lactic acid)
PMMA	Poly(methyl methacrylate)
PNIPAM	Poly(N-isopropyl acrylamide)
PMAA	Poly(methacrylic acid)
PMIA	Poly(m-phenyleneisophthalamide)
PPTA	Poly(p-phenyleneterephthamide)
PPO	Poly(propylene oxide)
PPESK-g-PEG	Poly(phthalazinone ether sulfone ketone)-*g*-polyethylene glycol
PS-*b*-PEG	Polystyrene-*block*-poly (ethyl glycol)
PSF-*g*-PNMG	Polysulfone-*g*-poly(N-methyl-D-glucamine)
PSF	Polysulfone
PSR	Polystyrene
PVA	Polyvinyl alcohol
PVAd	Acetalized PVA
PVB	Poly(vinyl butyral)
PVC	Polyvinyl chloride
PVDF	Polyvinylidene fluoride
PVDF-*g*-PEGMA	Poly(vinylidene fluoride)-*graft*-poly(ethylene glycol) methyl ether methacrylate
PVDF-*g*-POEM	Poly(vinylidene fluoride)-*graft*-poly(oxyethylene methacrylate)
PVF	Polyvinyl formal
PVP	Polyvinylpyrrolidone
PWF	Pure water flux

QPDMAEMA	Quaternized poly(dimethylaminoethyl methacrylate)
SA	Salicylate-alumoxane
SDS	Sodium dodecyl sulphate
SiO_2	Silicon dioxide
SWCNTs	Single-walled carbon nanotubes
TEP	Triethyl phosphate
TiO_2	Titanium dioxide
$TiSiO_4$	Titanium silicon oxide
TNTs	Titania nanotubes
UF	Ultrafiltration
ZIF-L	Zeoliticimidazolate leaf-shape
ZnO	Zinc oxide
ZrO_2	Zirconium dioxide

REFERENCES

Alsalhy, Q.F., 2012. Hollow fiber ultrafiltration membranes prepared from blends of poly (vinyl chloride) and polystyrene. Desalination 294, 44–52.

Alsohaimi, I.H., Kumar, M., Algamdi, M.S., Khan, M.A., Nolan, K., Lawler, J., 2017. Antifouling hybrid ultrafiltration membranes with high selectivity fabricated from polysulfone and sulfonic acid functionalized TiO_2 nanotubes. Chemical Engineering Journal 316, 573–583.

Amirilargani, M., Sadrzadeh, M., Mohammadi, T., 2010. Synthesis and characterization of polyethersulfone membranes. Journal of Polymer Research 17, 363–377.

Ananth, A., Arthanareeswaran, G., Ismail, A.F., Mok, Y.S., Matsuura, T., 2014. Effect of bio-mediated route synthesized silver nanoparticles for modification of polyethersulfone membranes. Colloids Surfaces A Physicochemical and Engineering Aspects 451, 151–160.

Asgarkhani, M.A.H., Mousavi, S.M., Saljoughi, E., 2013. Cellulose acetate butyrate membrane containing TiO_2 nanoparticle: Preparation, characterization and permeation study. Korean Journal of Chemical Engineering 30, 1819–1824.

Ayyaru, S., Ahn, Y.-H., 2017. Application of sulfonic acid group functionalized graphene oxide to improve hydrophilicity, permeability, and antifouling of PVDF nanocomposite ultrafiltration membranes. Journal of Membrane Science 525, 210–219.

Behboudi, A., Jafarzadeh, Y., Yegani, R., 2017. Polyvinyl chloride/polycarbonate blend ultrafiltration membranes for water treatment. Journal of Membrane Science 534, 18–24.

Bera, A., Kumar, C.U., Parui, P., Jewrajka, S.K., 2015. Stimuli responsive and low fouling ultrafiltration membranes from blends of polyvinylidene fluoride and designed library of amphiphilic poly(methyl methacrylate) containing copolymers. Journal of Membrane Science 481, 137–147.

Bet-moushoul, E., Mansourpanah, Y., Farhadi, K., Tabatabaei, M., 2016. TiO_2 nanocomposite based polymeric membranes: A review on performance improvement for various applications in chemical engineering processes. Chemical Engineering Journal 283, 29–46.

Bidsorkhi, H.C., Riazi, H., Emadzadeh, D., Ghanbari, M., Matsuura, T., Lau, W.J., Ismail, A.F., 2016. Preparation and characterization of a novel highly hydrophilic and antifouling polysulfone/nanoporous TiO_2 nanocomposite membrane. Nanotechnology 27, 415706.

Carretero, P., Molina, S., Sandı, R., Rodrı, J., Lozano, A.E., Abajo, J. De, 2013. Hydrophilic Polyisophthalamides Containing poly (ethylene oxide) side chains: Synthesis, characterization, and physical properties. Journal of Polymer Science Part A: Polymer Chemistry 51, 963–976.

Chen, W., Wei, M., Wang, Y., 2017. Advanced ultrafiltration membranes by leveraging microphase separation in macrophase separation of amphiphilic polysulfone block copolymers. Journal of Membrane Science 525, 342–348.

Chen, X., Su, Y., Shen, F., Wan, Y., 2011. Antifouling ultrafiltration membranes made from PAN-b-PEG copolymers: Effect of copolymer composition and PEG chain length. Journal of Membrane Science 384, 44–51.

Chen, Y., Wei, M., Wang, Y., 2016. Upgrading polysulfone ultrafiltration membranes by blending with amphiphilic block copolymers: Beyond surface segregation. Journal of Membrane Science 505, 53–60.

Chen, Y., Zhang, Y., Zhang, H., Liu, J., Song, C., 2013. Biofouling control of halloysite nanotubes-decorated polyethersulfone ultrafiltration membrane modified with chitosan-silver nanoparticles. Chemical Engineering Journal 228, 12–20.

Cho, Y.H., Kim, H.W., Nam, S.Y., Park, H.B., 2011. Fouling-tolerant polysulfone-poly(ethylene oxide) random copolymer ultrafiltration membranes. Journal of Membrane Science 379, 296–306.

Daraei, P., Madaeni, S.S., Ghaemi, N., Khadivi, M.A., Astinchap, B., Moradian, R., 2013a. Enhancing antifouling capability of PES membrane via mixing with various types of polymer modified multi-walled carbon nanotube. Journal of Membrane Science 444, 184–191.

Daraei, P., Madaeni, S.S., Ghaemi, N., Khadivi, M.A., Astinchap, B., Moradian, R., 2013b. Fouling resistant mixed matrix polyethersulfone membranes blended with magnetic nanoparticles: Study of magnetic field induced casting. Separation and Purification Technology 109, 111–121.

Dasgupta, J., Chakraborty, S., Sikder, J., Kumar, R., Pal, D., Curcio, S., Drioli, E., 2014. The effects of thermally stable titanium silicon oxide nanoparticles on structure and performance of cellulose acetate ultrafiltration membranes. Separation and Purification Technology 133, 55–68.

Díez, B., Roldán, N., Martín, A., Sotto, A., Perdigón-melón, J.A., 2017. Fouling and biofouling resistance of metal-doped mesostructured silica/polyethersulfone ultrafiltration membranes. Journal of Membrane Science 526, 252–263.

Dong, L.-X., Yang, H.-W., Liu, S.-T., Wang, X.-M., Xie, Y.F., 2015. Fabrication and antibiofouling properties of alumina and zeolite nanoparticle embedded ultrafiltration membranes. Desalination 365, 70–78.

Fan, X., Su, Y., Zhao, X., Li, Y., Zhang, R., Zhao, J., Jiang, Z., Zhu, J., Ma, Y., Liu, Y., 2014. Fabrication of polyvinyl chloride ultrafiltration membranes with stable antifouling property by exploring the pore formation and surface modification capabilities of polyvinyl formal. Journal of Membrane Science 464, 100–109.

Gao, H., Sun, X., Gao, C., 2017. Antifouling polysulfone ultrafiltration membranes with sulfobetaine polyimides as novel additive for the enhancement of both water flux and protein rejection. Journal of Membrane Science 542, 81–90.

Garcia-Ivars, J., Iborra-Clar, M.I., Alcaina-Miranda, M.I., Mendoza-Roca, J.A., Pastor-Alcaniz, L., 2014. Development of fouling-resistant polyethersulfone ultrafiltration membranes via surface UV photografting with polyethylene glycol/aluminum oxide nanoparticles. Separation and Purification Technology 135, 88–99.

Gul, S., Ali, S., Akhtar, K., Ali, M., Khan, M.I., Imtiaz, M., Asiri, A.M., Bahadar, S., 2017. Antibacterial PES-CA-Ag_2O nanocomposite supported Cu nanoparticles membrane toward ultrafiltration, BSA rejection and reduction of nitrophenol. Journal of Molecular Liquids 230, 616–624.

Guo, J., Sotto, A., Martín, A., Kim, J., 2017. Preparation and characterization of polyethersulfone mixed matrix membranes embedded with Ti- or Zr-incorporated SBA-15 materials. Journal of Industrial and Engineering Chemistry 45, 257–265.

Hong, J., He, Y., 2014. Polyvinylidene fluoride ultrafiltration membrane blended with nano-ZnO particle for photo-catalysis self-cleaning. Desalination 332, 67–75.

Hossein Razzaghi, M., Safekordi, A., Tavakolmoghadam, M., Rekabdar, F., Hemmati, M., 2014. Morphological and separation performance study of PVDF/CA blend membranes. Journal of Membrane Science 470, 547–557.

Hou, S., Xing, J., Dong, X., Zheng, J., Li, S., 2017. Integrated antimicrobial and antifouling ultrafiltration membrane by surface grafting PEO and N-chloramine functional groups. Journal of Colloid and Interface Science 500, 333–340.

Hou, S., Zheng, J., Zhang, S., Li, S., 2015. Novel amphiphilic PEO-grafted cardo poly(aryl ether sulfone) copolymer: Synthesis, characterization and antifouling performance. Polymer (United Kingdom). 77, 48–54.

Huang, J., Zhang, K., Wang, K., Xie, Z., Ladewig, B., Wang, H., 2012. Fabrication of polyethersulfone-mesoporous silica nanocomposite ultrafiltration membranes with antifouling properties. Journal of Membrane Science 423–424, 362–370.

Hwa, K., Ting, E., Irfan, M., Idris, A., Mohd, N., 2015. Enhanced Cu (II) rejection and fouling reduction through fabrication of PEG-PES nanocomposite ultrafiltration membrane with PEG-coated cobalt doped iron oxide nanoparticle. Journal of the Taiwan Institute of Chemical Engineers 47, 50–58.

Jamshidi Gohari, R., Halakoo, E., Nazri, N.A., Lau, W.J., Matsuura, T., Ismail, A.F., 2014. Improving performance and antifouling capability of PES UF membranes via blending with highly hydrophilic hydrous manganese dioxide nanoparticles. Desalination 335, 87–95.

Jo, Y.J., Choi, E.Y., Choi, N.W., Kim, C.K., 2016. Antibacterial and Hydrophilic Characteristics of Poly(ether sulfone) Composite Membranes Containing Zinc Oxide Nanoparticles Grafted with Hydrophilic Polymers. Industrial & Engineering Chemistry Research 55, 7801–7809.

Kumar, R., Isloor, A.M., Ismail, A.F., Matsuura, T., 2013. Performance improvement of polysulfone ultrafiltration membrane using N-succinyl chitosan as additive. Desalination 318, 1–8.

Li, F., Meng, J., Ye, J., Yang, B., Tian, Q., Deng, C., 2014a. Surface modification of PES ultrafiltration membrane by polydopamine coating and poly(ethylene glycol) grafting: Morphology, stability, and anti-fouling. Desalination 344, 422–430.

Li, F., Ye, J., Yang, L., Deng, C., Tian, Q., Yang, B., 2015a. Surface modification of ultrafiltration membranes by grafting glycine-functionalized PVA based on polydopamine coatings. Applied Surface Science 345, 301–309.

Li, J.H., Shao, X.S., Zhou, Q., Li, M.Z., Zhang, Q.Q., 2013a. The double effects of silver nanoparticles on the PVDF membrane: Surface hydrophilicity and antifouling performance. Applied Surface Science 265, 663–670.

Li, Q., Pan, S., Li, X., Liu, C., Li, J., Sun, X., Shen, J., Han, W., Wang, L., 2015b. Hollow mesoporous silica spheres/polyethersulfone composite ultrafiltration membrane with enhanced antifouling property. Colloids Surfaces A Physicochemical and Engineering Aspects 487, 180–189.

Li, S., Cui, Z., Zhang, L., He, B., Li, J., 2016. The effect of sulfonated polysulfone on the compatibility and structure of polyethersulfone-based blend membranes. Journal of Membrane Science 513, 1–11.

Li, X., Fang, X., Pang, R., Li, J., Sun, X., Shen, J., Han, W., Wang, L., 2014b. Self-assembly of TiO_2 nanoparticles around the pores of PES ultrafiltration membrane for mitigating organic fouling. Journal of Membrane Science 467, 226–235.

Li, X., Pang, R., Li, J., Sun, X., Shen, J., Han, W., Wang, L., 2013b. In situ formation of Ag nanoparticles in PVDF ultrafiltration membrane to mitigate organic and bacterial fouling. Desalination 324, 48–56.

Liang, S., Kang, Y., Tiraferri, A., Giannelis, E.P., Huang, X., Elimelech, M., 2013. Highly hydrophilic polyvinylidene fluoride (PVDF) ultrafiltration membranes via postfabrication grafting of surface-tailored silica nanoparticles. ACS Applied Materials & Interfaces 5, 6694–6703.

Liang, S., Xiao, K., Mo, Y., Huang, X., 2012. A novel ZnO nanoparticle blended polyvinylidene fluoride membrane for anti-irreversible fouling. J Journal of Membrane Science 394–395, 184–192.

Lin, C., Wang, J., Zhou, M., Zhu, B., Zhu, L., Gao, C., 2016. Poly (m-phenylene isophthalamide) (PMIA): A potential polymer for breaking through the selectivity-permeability trade-off for ultrafiltration membranes. Journal of Membrane Science 518, 72–78.

Liu, B., Chen, C., Li, T., Crittenden, J., Chen, Y., 2013. High performance ultrafiltration membrane composed of PVDF blended with its derivative copolymer PVDF-g-PEGMA. Journal of Membrane Science 445, 66–75.

Liu, F., Xu, Y.Y., Zhu, B.K., Zhang, F., Zhu, L.P., 2009. Preparation of hydrophilic and fouling resistant poly(vinylidene fluoride) hollow fiber membranes. Journal of Membrane Science 345, 331–339.

Liu, X., Duan, W., Chen, Y., Jiao, S., Zhao, Y., Kang, Y., Li, L., Fang, Z., Xu, W., 2016. Porous TiO_2 assembled from monodispersed nanoparticles. Nanoscale Research Letters 11, 1–8.

Liu, Y., Su, Y., Zhao, X., Li, Y., Zhang, R., Jiang, Z., 2015. Improved antifouling properties of polyethersulfone membrane by blending the amphiphilic surface modifier with cross-linked hydrophobic segments. Journal of Membrane Science 486, 195–206.

Loh, C.H., Wang, R., 2011. Effects of additives and coagulant temperature on fabrication of high performance pvdf/pluronic f127 blend hollow fiber membranes via nonsolvent induced phase separation. Chinese Journal of Chemical Engineering 20, 71–79.

Loh, C.H., Wang, R., Shi, L., Fane, A.G., 2011. Fabrication of high performance polyethersulfone UF hollow fiber membranes using amphiphilic Pluronic block copolymers as pore-forming additives. Journal of Membrane Science 380, 114–123.

Low, Z.X., Razmjou, A., Wang, K., Gray, S., Duke, M., Wang, H., 2014. Effect of addition of two-dimensional ZIF-L nanoflakes on the properties of polyethersulfone ultrafiltration membrane. Journal of Membrane Science 460, 9–17.

Low, Z.X., Wang, Z., Leong, S., Razmjou, A., Dumée, L.F., Zhang, X., Wang, H., 2015. Enhancement of the Antifouling Properties and Filtration Performance of Poly(ethersulfone) Ultrafiltration Membranes by Incorporation of Nanoporous Titania Nanoparticles. Industrial & Engineering Chemistry Research 54, 11188–11198.

Ma, W., Rajabzadeh, S., Matsuyama, H., 2015. Preparation of antifouling poly(vinylidene fluoride) membranes via different coating methods using a zwitterionic copolymer. Applied Surface Science 357, 1388–1395.

Ma, X., Su, Y., Sun, Q., Wang, Y., Jiang, Z., 2007. Preparation of protein-adsorption-resistant polyethersulfone ultrafiltration membranes through surface segregation of amphiphilic comb copolymer. Journal of Membrane Science 292, 116–124.

Méricq, J., Mendret, J., Brosillon, S., Faur, C., 2015. High performance PVDF-TiO_2 membranes for water treatment. Chemical Engineering Science 123, 283–291.

Mital, G.S., Manoj, T., 2011. A review of TiO_2 nanoparticles. Chinese Science Bulletin 56, 1639–1657.

Moghareh Abed, M.R., Kumbharkar, S.C., Groth, A.M., Li, K., 2013. Economical production of PVDF-g-POEM for use as a blend in preparation of PVDF based hydrophilic hollow fibre membranes. Separation and Purification Technology 106, 47–55.

Mokhtari, S., Rahimpour, A., Shamsabadi, A.A., Habibzadeh, S., Soroush, M., 2017. Enhancing performance and surface antifouling properties of polysulfone ultrafiltration membranes with salicylate-alumoxane nanoparticles. Applied Surface Science 393, 93–102.

Molina, S., Carretero, P., Teli, S.B., de la Campa, J.G., Lozano, Á.E., de Abajo, J., 2014. Hydrophilic porous asymmetric ultrafiltration membranes of aramid-g-PEO copolymers. Journal of Membrane Science 454, 233–242.

Moriya, A., Shen, P., Ohmukai, Y., Maruyama, T., Matsuyama, H., 2012. Reduction of fouling on poly(lactic acid) hollow fiber membranes by blending with poly (lactic acid)-polyethylene glycol–poly (lactic acid) triblock copolymers. Journal of Membrane Science 415–416, 712–717.

Muhamad, M.S., Salim, M.R., Lau, W.-J., 2015. Preparation and characterization of PES/SiO_2 composite ultrafiltration membrane for advanced water treatment. Korean Journal of Chemical Engineering 32, 2319–2329.

Nazri, N.A.M., Lau, W.J., Ismail, A.F., 2015. Improving water permeability and anti-fouling property of polyacrylonitrile-based hollow fiber ultrafiltration membranes by surface modification with polyacrylonitrile-g-poly(vinyl alcohol) graft copolymer. Korean Journal of Chemical Engineering 32, 1853–1863.

Orooji, Y., Faghih, M., Razmjou, A., Hou, J., Moazzam, P., Emani, N., Aghababaie, M., Nourisfa, F., Chen, V., Jin, W., 2017. Nanostructured mesoporous carbon polyethersulfone composite ultrafiltration membrane with significantly low protein adsorption and bacterial adhesion. Carbon 111, 689–704.

Padaki, M., Emadzadeh, D., Masturra, T., Ismail, A.F., 2015. Antifouling properties of novel PSf and TNT composite membrane and study of effect of the flow direction on membrane washing. Desalination 362, 141–150.

Pan, Y., Yu, Z., Shi, H., Chen, Q., Zeng, G., Di, H., Ren, X., He, Y., 2017. A novel antifouling and antibacterial surface-functionalized PVDF ultrafiltration membrane via binding Ag/SiO_2 nanocomposites. Journal of Chemical Technology and Biotechnology 92, 562–572.

Pang, R., Li, X., Li, J., Lu, Z., Sun, X., Wang, L., 2014. Preparation and characterization of ZrO_2/PES hybrid ultrafiltration membrane with uniform ZrO2 nanoparticles. Desalination 332, 60–66.

Park, J.Y., Acar, M.H., Akthakul, A., Kuhlman, W., Mayes, A.M., 2006. Polysulfone-graft-poly(ethylene glycol) graft copolymers for surface modification of polysulfone membranes. Biomaterials 27, 856–865.

Park, S.Y., Chung, J.W., Chae, Y.K., Kwak, S.Y., 2013. Amphiphilic thiol functional linker mediated sustainable anti-biofouling ultrafiltration nanocomposite comprising a silver nanoparticles and poly(vinylidene fluoride) membrane. ACS Applied Materials & Interfaces 5, 10705–10714.

Qin, A., Li, X., Zhao, X., Liu, D., He, C., 2015. Engineering a highly hydrophilic PVDF membrane via binding TiO_2 nanoparticles and a PVA layer onto a membrane surface. ACS Applied Materials & Interfaces 7, 8427–8436.

Qiu, S., Wu, L., Pan, X., Zhang, L., Chen, H., Gao, C., 2009. Preparation and properties of functionalized carbon nanotube/PSF blend ultrafiltration membranes. Journal of Materials Science 342, 165–172.

Sawada, I., Fachrul, R., Ito, T., Ohmukai, Y., Maruyama, T., Matsuyama, H., 2012. Development of a hydrophilic polymer membrane containing silver nanoparticles with both organic antifouling and antibacterial properties. Journal of Membrane Science 387–388, 1–6.

Shao, L., Wang, Z.X., Zhang, Y.L., Jiang, Z.X., Liu, Y.Y., 2014. A facile strategy to enhance PVDF ultrafiltration membrane performance via self-polymerized polydopamine followed by hydrolysis of ammonium fluotitanate. Journal of Membrane Science 461, 10–21.

Shen, J., Zhang, Q., Yin, Q., Cui, Z., Li, W., Xing, W., 2017. Fabrication and characterization of amphiphilic PVDF copolymer ultrafiltration membrane with high anti-fouling property. Journal of Membrane Science 521, 95–103.

Shen, L., Bian, X., Lu, X., Shi, L., Liu, Z., Chen, L., Hou, Z., Fan, K., 2012. Preparation and characterization of ZnO/polyethersulfone (PES) hybrid membranes. Desalination 293, 21–29.

Shen, P., Moriya, A., Rajabzadeh, S., Maruyama, T., Matsuyama, H., 2013. Improvement of the antifouling properties of poly (lactic acid) hollow fiber membranes with poly (lactic acid) – polyethylene glycol – poly (lactic acid) copolymers. Desalination 325, 37–39.

Shi, H., Liu, F., Xue, L., 2013a. Fabrication and characterization of antibacterial PVDF hollow fibre membrane by doping Ag-loaded zeolites. Journal of Membrane Science 437, 205–215.

Shi, Q., Meng, J.Q., Xu, R.S., Du, X.L., Zhang, Y.F., 2013b. Synthesis of hydrophilic polysulfone membranes having antifouling and boron adsorption properties via blending with an amphiphilic graft glycopolymer. Journal of Membrane Science 444, 50–59.

Shockravi, A., Vatanpour, V., Najjar, Z., Bahadori, S., Javadi, A., 2017. A new high performance polyamide as an effective additive for modification of antifouling properties and morphology of asymmetric PES blend ultrafiltration membranes. Microporous and Mesoporous Materials 246, 24–36.

Sianipar, M., Hyun, S., Min, C., Tijing, L.D., Shon, H.K., 2016. Potential and performance of a polydopamine-coated multiwalled carbon nanotube/polysulfone nanocomposite membrane for ultrafiltration application. Journal of Industrial & Engineering Chemistry Research 34, 364–373.

Song, H.J., Kim, C.K., 2013. Fabrication and properties of ultrafiltration membranes composed of polysulfone and poly (1-vinylpyrrolidone) grafted silica nanoparticles. Journal of Membrane Science 444, 318–326.

Su, Y., Liu, Y., Zhao, X., Li, Y., Jiang, Z., 2015. Preparation of pH-responsive membranes with amphiphilic copolymers by surface segregation method. Chinese Journal of Chemical Engineering 23, 1283–1290.

Su, Y.L., Cheng, W., Li, C., Jiang, Z., 2009. Preparation of antifouling ultrafiltration membranes with poly(ethylene glycol)-graft-polyacrylonitrile copolymers. Journal of Membrane Science 329, 246–252.

Teli, S.B., Molina, S., Sotto, A., Garc, E., Abajo, J. De, 2013. Fouling Resistant Polysulfone – PANI/ TiO_2 Ultrafiltration Nanocomposite Membranes. Journal of Industrial & Engineering Chemistry Research 52, 9470–9479.

Teow, Y.H., Ahmad, A.L., Lim, J.K., Ooi, B.S., 2012. Preparation and characterization of PVDF/TiO2 mixed matrix membrane via in situ colloidal precipitation method. Desalination 295, 61–69.

Tu, Q., Wang, J., Liu, R., He, J., Zhang, Y., Shen, S., Xu, J., Liu, J., Yuan, M., Wang, J., 2013. Antifouling properties of poly (dimethylsiloxane) surfaces modified with quaternized poly (dimethylaminoethyl methacrylate). Colloids Surfaces B Biointerfaces 102, 361–370.

Vatanpour, V., Siavash, S., Rajabi, L., Zinadini, S., Ashraf, A., 2012. Boehmite nanoparticles as a new nanofiller for preparation of antifouling mixed matrix membranes. Journal of Membrane Science 401–402, 132–143.

Wang, C., Xiao, C., Chen, M., Huang, Q., Liu, H., Li, N., 2016a. Unique performance of poly(p-phenylene terephthamide) hollow fiber membranes. Journal of Materials Science 51, 1522–1531.

Wang, C., Xiao, C., Huang, Q., Pan, J., 2015. A study on structure and properties of poly(p-phenylene terephthamide) hybrid porous membranes. Journal of Membrane Science 474, 132–139.

Wang, H., Wang, Z.-M., Yan, X., Chen, J., Lang, W.-Z., Guo, Y.-J., 2017. Novel organic-inorganic hybrid polyvinylidene fluoride ultrafiltration membranes with antifouling and antibacterial properties by embedding N-halamine functionalized silica nanospheres. Journal of Industrial & Engineering Chemistry Research 52, 295–304.

Wang, H., Zhao, X., He, C., 2016b. Enhanced antifouling performance of hybrid PVDF ultrafiltration membrane with the dual-mode SiO_2-g-PDMS nanoparticles. Separation and Purification Technology 166, 1–8.

Wang, J., Lang, W.Z., Xu, H.P., Zhang, X., Guo, Y.J., 2014a. Improved poly(vinyl butyral) hollow fiber membranes by embedding multi-walled carbon nanotube for the ultrafiltrations of bovine serum albumin and humic acid. Chinese Journal of Chemical Engineering 260, 90–98.

Wang, J., Sun, H., Gao, X., Gao, C., 2014b. Enhancing antibiofouling performance of Polysulfone (PSf) membrane by photo-grafting of capsaicin derivative and acrylic acid. Applied Surface Science 317, 210–219.

Wang, Y.Q., Wang, T., Su, Y.L., Peng, F.B., Wu, H., Jiang, Z.Y., 2005. Remarkable reduction of irreversible fouling and improvement of the permeation properties of poly(ether sulfone) ultrafiltration membranes by blending with pluronic F127. Langmuir 21, 11856–11862.

Wang, Z., Yu, H., Xia, J., Zhang, F., Li, F., Xia, Y., Li, Y., 2012. Novel GO-blended PVDF ultrafiltration membranes. Desalination 299, 50–54.

Wu, H., Liu, Y., Mao, L., Jiang, C., Ang, J., Lu, X., 2017a. Doping polysulfone ultrafiltration membrane with TiO_2-PDA nanohybrid for simultaneous self-cleaning and self-protection. Journal of Membrane Science 532, 20–29.

Wu, H., Tang, B., Wu, P., 2014. Development of novel SiO_2-GO nanohybrid/polysulfone membrane with enhanced performance. Journal of Membrane Science 451, 94–102.

Wu, Q., Chen, G., Sun, W., Xu, Z., Kong, Y., Zheng, X., Xu, S.-J., 2017b. Bio-inspired GO-Ag/PVDF/F127 membrane with improved anti-fouling for natural organic matter (NOM) resistance. Chemical Engineering Journal 313, 450–460.

Wu, Z., Tong, W., Jiang, W., Liu, X., Wang, Y., Chen, H., 2012. Poly (N-vinylpyrrolidone)-modified poly (dimethylsiloxane) elastomers as anti-biofouling materials. Colloids Surfaces B Biointerfaces 96, 37–43.

Xiang, T., Xie, Y., Wang, R., Wu, M., Sun, S., Zhao, C., 2014. Facile chemical modification of polysulfone membrane with improved hydrophilicity and blood compatibility. Materials Letters 137, 192–195.

Xu, H., Ding, M., Liu, S., Li, Y., Shen, Z., Wang, K., 2017. Preparation and characterization of novel polysulphone hybrid ultrafiltration membranes blended with N-doped GO/TiO_2 nanocomposites. Polymer 117, 198–207.

Xu, Z., Wu, T., Shi, J., Wang, W., Teng, K., Qian, X., Shan, M., Deng, H., Tian, X., Li, C., Li, F., 2016. Manipulating Migration Behavior of Magnetic Graphene Oxide via Magnetic Field Induced Casting and Phase Separation toward High-Performance Hybrid Ultrafiltration Membranes. ACS Applied Materials & Interfaces 8, 18418–18429.

Xu, Z., Zhang, J., Shan, M., Li, Y., Li, B., Niu, J., Zhou, B., Qian, X., 2014. Organosilane-functionalized graphene oxide for enhanced antifouling and mechanical properties of polyvinylidene fluoride ultrafiltration membranes. Journal of Membrane Science 458, 1–13.

Xueli, G., Haizeng, W., Jian, W., Xing, H., Congjie, G., 2013. Surface-modified PSf UF membrane by UV-assisted graft polymerization of capsaicin derivative moiety for fouling and bacterial resistance. Journal of Membrane Science 445, 146–155.

Yan, L., Wang, J., 2011. Development of a new polymer membrane - PVB/PVDF blended membrane. Desalination 281, 455–461.

Yang, Y.F., Li, Y., Li, Q.L., Wan, L.S., Xu, Z.K., 2010. Surface hydrophilization of microporous polypropylene membrane by grafting zwitterionic polymer for anti-biofouling. Journal of Membrane Science 362, 255–264.

Yu, H., Zhang, Y., Zhang, J., Zhang, H., Liu, J., 2013. Preparation and antibacterial property of SiO_2–Ag/PES hybrid ultrafiltration membranes. Desalination and Water Treatment 51, 3584–3590.

Yuan, G.L., Xu, Z.L., Wei, Y.M., 2009. Characterization of PVDF-PFSA hollow fiber UF blend membrane with low-molecular weight cut-off. Separation and Purification Technology 69, 141–148.

Yuan, H., Ren, J., 2017. Preparation of poly(vinylidene fluoride) (PVDF)/acetalyzed poly(vinyl alcohol) ultrafiltration membrane with the enhanced hydrophilicity and the anti-fouling property. Chemical Engineering Research and Design 121, 348–359.

Zambare, R.S., Dhopte, K.B., Patwardhan, A. V, Nemade, P.R., 2017. Polyamine functionalized graphene oxide polysulfone mixed matrix membranes with improved hydrophilicity and anti-fouling properties. Desalination 403, 24–35.

Zhang, G., Lu, S., Zhang, L., Meng, Q., Shen, C., Zhang, J., 2013. Novel polysulfone hybrid ultrafiltration membrane prepared with TiO_2-g-HEMA and its antifouling characteristics. Journal of Membrane Science 436, 163–173.

Zhang, J., Wang, Q., Wang, Z., Zhu, C., Wu, Z., 2014a. Modification of poly(vinylidene fluoride)/polyethersulfone blend membrane with polyvinyl alcohol for improving antifouling ability. Journal of Membrane Science 466, 293–301.

Zhang, M., Field, R.W., Zhang, K., 2014b. Biogenic silver nanocomposite polyethersulfone UF membranes with antifouling properties. Journal of Membrane Science 471, 274–284.

Zhang, X., Lang, W.Z., Xu, H.P., Yan, X., Guo, Y.J., Chu, L.F., 2014c. Improved performances of PVDF/PFSA/O-MWNTs hollow fiber membranes and the synergism effects of two additives. Journal of Membrane Science 469, 458–470.

Zhao, C., Xu, X., Chen, J., Yang, F., 2013a. Effect of graphene oxide concentration on the morphologies and antifouling properties of PVDF ultrafiltration membranes. Journal of Environmental Chemical Engineering 1, 349–354.

Zhao, W., Mou, Q., Zhang, X., Shi, J., Sun, S,, Zhao, C., 2013b. Preparation and characterization of sulfonated polyethersulfone membranes by a facile approach. European Polymer Journal 49, 738–751.

Zhao, X., Ma, J., Wang, Z., Wen, G., Jiang, J., Shi, F., Sheng, L., 2012a. Hyperbranched-polymer functionalized multi-walled carbon nanotubes for poly (vinylidene fluoride) membranes: From dispersion to blended fouling-control membrane. Desalination 303, 29–38.

Zhao, X., Su, Y., Chen, W., Peng, J., Jiang, Z., 2012b. Grafting perfluoroalkyl groups onto polyacrylonitrile membrane surface for improved fouling release property. Journal of Membrane Science 415–416, 824–834.

Zhao, X., Su, Y., Chen, W., Peng, J., Jiang, Z., 2011. PH-responsive and fouling-release properties of PES ultrafiltration membranes modified by multi-functional block-like copolymers. Journal of Membrane Science 382, 222–230.

Zhi, S., Xu, J., Deng, R., Wan, L., Xu, Z., 2014. Poly (vinylidene fluoride) ultrafiltration membranes containing hybrid silica nanoparticles: Preparation, characterization and performance. Polymer 55, 1333–1340.

Zhu, L.P., Xu, L., Zhu, B.K., Feng, Y.X., Xu, Y.Y., 2007. Preparation and characterization of improved fouling-resistant PPESK ultrafiltration membranes with amphiphilic PPESK-graft-PEG copolymers as additives. Journal of Membrane Science 294, 196–206.

Section II

Membranes for Gas Separation Process

6 Carbon Membranes for Gas Separation

Miki Yoshimune

CONTENTS

6.1 INTRODUCTION

This chapter reviews carbon molecular sieve (CMS) membranes as a promising material for use in gas separation processes. CMS membranes have micropores with diameters of approximately 0.3–0.5 nm and are characterized by high selectivities in the separation of gases such as H_2/CH_4, He/N_2, O_2/N_2, CO_2/CH_4, CO_2/N_2, and C_3H_6/C_3H_8. The high selectivities of CMS membranes originate from the selective permeation of smaller gas molecules through the micropores (sieving effect). Because the separation performance of CMS membranes exceeds that of conventional polymeric membranes, they have attracted consistently high levels of research and are the subject of a number of excellent reviews and books (Ismail and David, 2001; Saufi and Ismail, 2004; Kita et al. 2006; Ismail and Li, 2008; Williams et al., 2008; Salleh et al., 2011).

The preparation process of CMS membranes mainly consists of three steps: (1) precursor polymer selection, (2) preparation of a polymer membrane, and (3) pyrolysis (also represented as carbonization); this process will be described in detail later. The configuration of CMS membranes can be grouped into two categories: (1) unsupported or freestanding carbon membranes (flat film, capillary tube, or hollow fiber) and (2) composite or supported carbon membranes (flat or tubular). Some representative examples of membranes from these two categories are shown in Figure 6.1. Pyrolysis (or carbonization) is the process whereby a precursor membrane is heated to the pyrolysis temperature in a controlled atmosphere, such as vacuum or inert gas, at a specific heating rate and then held at the pyrolysis temperature for a sufficiently long thermal soak time (Saufi and Ismail, 2004). Gaseous decomposition products are evolved during the pyrolysis of the polymeric precursor, resulting in considerable weight loss and the formation of micropores. The pore structure of CMS membranes as represented by Xu et al. (2011) is shown in Figure 6.2; an idealized "slit-like" pore structure can be described by a bimodal pore distribution with micropores (7–20 Å) connected by ultramicropores (<7 Å). This combination of micropores and ultramicropores provides both high flux and high separation efficiency in CMS membranes via a molecular sieving function. Therefore, the nature of the pyrolysis process and

FIGURE 6.1 Configurations of carbon membranes: (a) flat film, (b) tubular supported membrane, and (c) hollow fiber membrane.

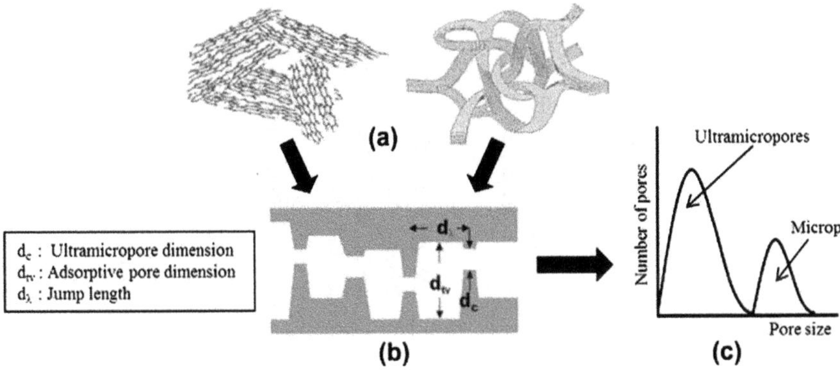

FIGURE 6.2 (a) Structure of pyrolytic carbon, (b) idealized pore structure of a CMS membrane, and (c) idealized bimodal pore size distribution of a CMS membrane. (From Xu, L. et al., Journal of Membrane Science. 380, 138–147, 2011.)

of the precursor selection are the most important factors affecting the preparation of CMS membranes with high performance for gas separation. Other treatments such as pre- or post-oxidation can also have considerable effects on the performance of the final membrane.

6.2 PRECURSOR POLYMER SELECTION

CMS membranes are typically prepared by pyrolysis of polymeric precursors, which have been extensively studied over the past three decades. Koresh and Soffer (1983) prepared CMS membranes by carbonizing polymeric hollow fiber precursors. Since this pioneering work, many other studies have investigated the use of continuous polymer matrixes as precursors for carbon membranes. Polymers such as cellulose derivatives, polyacrylonitrile, polyimides, phenolic resins, polyfurfuryl alcohol (PFA), polyvinylidene chloride (PDVC), and polyphenylene oxide (PPO) have been used as precursors to produce CMS membranes. Tables 6.1 and 6.2 list the reported carbon membranes according to their configurations, precursors, and pyrolysis conditions.

6.2.1 Cellulose Derivatives

Koresh and Soffer (1983) prepared CMS membranes by carbonizing a polymeric hollow fiber, the composition of which was not disclosed in their original paper. The precursor is likely to have been a hollow cellulose fiber, because Carbon Membranes Ltd. commercialized cellulose-derived CMS membranes developed from the research of Koresh and Soffer (1983); however, the company has now ceased trade. Lagorsse et al. (2004) characterized the CMS membranes produced by Carbon Membranes Ltd. in detail. Scanning electron microscopy (SEM) images of the CMS membranes are presented in Figure 6.3, which show that their structure is radially symmetric, dense, and crack-free. Lagorsse and Hägg (2012) reported hollow fiber carbon membranes prepared from different deacetylated cellulose acetate

TABLE 6.1
Preparation Conditions for Unsupported Microporous Carbon Membranes

Precursor	Morphology	Pyrolysis Condition (Final Temperature; Heating Rate; Soak Times; Atmosphere)	Other Notes	Reference
Cellulose derived	HF	800, 950°C	post-oxidation in some cases	Koresh and Soffer, 1983
Cellulose derived	HF	550°C; 4°C/min; 2 h; CO_2		He and Hägg, 2012
Polyacrylonitrile	HF	500°C; - ; 10–180 min; Inert	preoxidation in air at 250°C for 30 min	David and Ismail, 2003
Polyimide				
6FAD/BPDA-DAM	HF	500, 550°C; 13.3→3.85→0.25°C/min; 2 h; Vacuum		Jones and Koros, 1994a; Jones and Koros, 1994b; Jones and Koros, 1995a; Jones and Koros, 1995b
6FAD/BPDA-DAM	HF	500, 550, 800°C; 13.3→3.85→0.25°C/min; 2 h; Ar, He, CO_2, and Vacuum		Geiszler and Koros, 2000
6FAD/BPDA-DAM	FF	535, 550, 800°C; 13.3→3.85→0.25°C/min; 2 h; Vacuum		Ghosal and Koros, 2000
6FAD/BPDA-DAM, Matrimid	HF	550, 800°C; 13.3→3.85→0.25°C/min; 2 h; Vacuum		Vu et al., 2002
Kpton film	FF	600–1000°C; 1.33, 4.5, 10, 13.3°C/min; 2 h; Vacuum, Ar		Suda and Haraya, 1995; Suda and Haraya, 1997

(Continued)

TABLE 6.1 (CONTINUED)
Preparation Conditions for Unsupported Microporous Carbon Membranes

Precursor	Morphology	Pyrolysis Condition (Final Temperature; Heating Rate; Soak Times; Atmosphere)	Other Notes	Reference
Kpton film	FF	1000°C; 10°C/min; 2 h; Vacuum	post-treatment: exposing water vapor at 400°C for 10 min	Fuertes et al., 1999
Kapton film	FF	900–1100°C; - ; - ; Ar		Hatori et al., 2004
Kapton film	FF	550–1000°C; 5→0.5°C/min; - ; N_2, Vacuum		Lua and Su, 2006
Kapton film	FF	600, 800°C; 5→0.5°C/min; 2 h; Vacuum, Ar, He, N_2		Su and Lua, 2007
Matrimid	FF	550, 800°C; 4°C/min; - ; Vacuum		Steel and Koros, 2003
P84	HF	600, 700, 800, 900°C; 1°C/min; 1 h; N_2	preoxidation in air at 300°C for 1 h	Barsema et al., 2002
P84	FF	550, 650, 800°C; 13.3→2→0.2°C/min; 2 h; Vacuum		Tin et al., 2004a; Tin et al., 2004c
P84	HF	900°C; 5°C/min; 5, 30 60min; Ar		Favvas et al., 2015
Matrimid	FF	800°C; 13.3→3.8→2.5→0.2°C/min; 2 h; Vacuum	pretreatment: crosslinking, nonsolvent	Tin et al., 2004a; Tin et al., 2004b
Matrimid	FF, HF	500-800°C; 13.3→4→0.25°C/min; 2 h; Vacuum		Xu et al., 2011
Br-Matrimid	FF	550, 800°C; 10→0.2°C/min; 2 h; Vacuum		Xiao et al., 2005a
Matrimid	FF	300-525°C; 50→5→1°C/min; 5, 30 min ; N_2		Barsema et al., 2004

(Continued)

TABLE 6.1 (CONTINUED)
Preparation Conditions for Unsupported Microporous Carbon Membranes

Precursor	Morphology	Pyrolysis Condition (Final Temperature; Heating Rate; Soak Times; Atmosphere)	Other Notes	Reference
BPDA-aromatic diamine	HF	600–900°C; - ; 3.6 min; N_2		Kusuki et al., 1997; Tanihara et al., 1999
BPDA-DDBT/DABA	HF	500–700°C; 5°C/min; 0 h; N_2	preoxidation in air at 400°C for 30 min	Okamoto et al., 1999
6FDA/BPDA-DDBT	HF	500–700°C; 5°C/min; 0, 0.25, 1 h; N_2, Ar, CO_2, Vacuum		Yoshino et al., 2003
BTDA-ODA	FF	550, 700, 800°C; 3°C/min; 0, 0.5, 1 h; Ar		Kim et al., 2005a
BTDA-ODA	FF	550, 700°C; 3°C/min; 1 h; Ar	PI/PVP blend	Kim et al., 2005b
BTDA-ODA/m-PDA, BTDA-ODA/2,2-DAT, BTDA-ODA/m-TMPD	FF	300→600, 300→800°C; 3→5°C/min; 2 h; Ar	PI/PVP blend	Park et al., 2004
BTDA-ODA/m-PDA, BTDA-ODA/DBA	FF	300→700°C; 3°C /min; 1 h; Ar		Kim et al., 2004
BTDA-DAI, ODPA-DAI, BPDA-DAI, 6FDA-DAI	FF	550, 800°C; 13→3.8→2.5→0.2°C/min; 2 h; Vacuum		Xiao et al., 2005b
6FDA-Durene/p-intA	FF	800°C; 0.2°C/min; 2 h; Vacuum	crosslinked at 400°C	Xiao et al., 2007
6FDA-mPDA/DABA	FF	550, 675, 800°C; 10→3→0.25°C/min; 2 h; Ar		Qiu et al., 2014

(Continued)

TABLE 6.1 (CONTINUED)
Preparation Conditions for Unsupported Microporous Carbon Membranes

Precursor	Morphology	Pyrolysis Condition (Final Temperature; Heating Rate; Soak Times; Atmosphere)	Other Notes	Reference
6FDA/DETDA, 6FDA:BPDA/DETDA, 6FDA/DETDA:DABA, 6FDA/1,5-ND:ODA	FF	550°C; 13.3→3.85→0.25°C /min; 2 h; Ar		Fu et al., 2015
6FDA/PMDA-TMMDA, 6FDA-TMMDA	FF	550, 650, 800°C; 13→3.8→2.5→0.2°C/min; 2 h; Vacuum		Shao et al., 2004
6FDA-duren	FF	325–800°C; 3, 1°C/min; 1 h; Vacuum		Shao et al., 2005
6FDA-DABZ	FF	500–800°C; 5–8°C/min; 1–5 h; N_2		Kita et al., 2007
NTDA-based sulfonated polyimides	FF	450°C; 5°C/min; 1.5 h; N_2		Chen and Yang, 1994
Phenolic resin	FF, Asymmetric	800–950°C; - ; 2–3 h; N_2		Shusen et al., 1996
Cellophane	FF	400–850°C; 0.5°C/min; 0–8 h; N_2,Ar, CO_2		Campo et al., 2010
PPO and its derivatives	HF, symmetric	550–750°C; 10°C/min; 2 h; Vacuum	preoxidation in air at 280°C for 45 min	Yoshimune et al., 2005; Yoshimune et al., 2007
PPESK	FF	500–950°C; 1°C/min; 1 h; Ar	preoxidation in air at 460°C for 30 min	Zhang et al., 2006b; Zhang et al., 2006c
PAEK/Azide	FF	450, 550, 650°C; 2→1→0.2°C/min; 2 h; Vacuum		Chng et al., 2009

HF: follow fiber FF: flat film

TABLE 6.2
Preparation Conditions for Microporous Carbon Membranes Supported on Porous Substrates

Precursor	Morphology/Pore Diam. of Substrate	Pyrolysis Condition (Final Temperature; Heating Rate; Soak Times; Atmosphere)	Other Notes	Reference
Polyimides				
PMDA-ODA	F / 1 μm, Asymmetric	500–700°C; 0.5°C/min; 1 h; Vacuum		Fuertes et al., 1999
Matrimid	F / 1 μm, Asymmetric	475–700°C; 0.5°C/min; 1 h; Vacuum		Fuertes et al., 1999
Matrimid	T / 2–3 nm	550–700°C; 1–10°C/min; - ; N_2		Briceño et al., 2012
Polyetherimide	F / 1 μm	800°C; 0.5°C/min; 1 h; Vacuum	single coating	Fuertes and Centeno, 1998
Polyetherimide	T	600°C; 1°C/min; 4 h; Ar		Sedigh et al., 1999
Polyetherimide	F / 1 μm	600°C; 5°C/min; 2 h; Vacuum		Tseng et al., 2012
BPDA–ODA	T / 0.14 μm	500–900°C; 5°C/min; 0 h; N_2	3 time c/i cycles	Hayashi, 1995; Xiao et al., 2005
BPDA–ODA	T / 0.14 μm	600–900°C; 5°C/min; 0 h; N_2	2–3 time c/i cycles, post-oxidation	Hayashi et al., 1997a; Kusakabe et al., 1998
BPDA-ODA	T / 0.14 μm	700–800°C; 5°C/min; 0 h; Ar	3 time c/i cycles, post-treated by carbon CVD at 650°C for 1 h	Hayashi et al., 1997b
BPDA-pPDA	F	550–700°C; 0.5°C/min; - ; Vacuum	spin-coating, 1–3 time c/p cycles	Fueres and Centeno, 1999
Phenolic resin	T / 1 μm	600°C; 5°C/min; 1 h; N_2	4–5 time c/p cycles	Kita et al., 1997
Phenolic resin	T / 0.14 μm	250–800°C; 5°C/min; 1.5 h; N_2	single c/p cycle, multiple c/p cycles	Zhou et al., 2001; Zhou et al., 2003

(Continued)

TABLE 6.2 (CONTINUED)
Preparation Conditions for Microporous Carbon Membranes Supported on Porous Substrates

Precursor	Morphology/Pore Diam. of Substrate	Pyrolysis Condition (Final Temperature; Heating Rate; Soak Times; Atmosphere)	Other Notes	Reference
Phenolic resin	T / 5 nm	700°C; 0.5°C/min; - ; Vacuum	1 or 3 time coats, pre or post-oxidation and CVD in some cases	Fuertes and Menendez, 2002
Phenolic resin	T / 5 nm	700°C; 0.5°C/min; - ; Vacuum	post-oxidation at 300, 350, 400, 475°C for 30 min in some cases	Fuertes, 2000
Phenolic resin	T / 5 nm	700–1000°C; 0.5, 1, 5, 7, 10°C/min; 1–8 h; Vacuum, N_2		Centeno et al., 2004
Phenolic resin	T	800°C; 0.5°C/min; 1 h; inert gas	1 to 3 time coats, post-treat: exposed CO_2	Zhang et al., 2006
Phenolic resin	T / 0.2 µm	550°C; 1°C/min; 2 h; N_2		Teixeira et al., 2011
PFA	F / 5–10 µm	300→500°C; 1.5°C/min; 6 h; N_2	5 time c/p cycles	Chen and Yang, 1994
PFA	T / 4 nm	100→200→350→450→600°C; 1°C /min; 4 h; Ar	dip coating, multiple c/p cycles	Sedigh et al., 1998
PFA	T / 3.5 nm, 5 nm	450–600°C; 1°C/min; 1 h; -	2 time c/p cycles, vapor deposition polymerization	Wang et al., 2000
PFA	F / 0.2 µm	600°C; 10°C/min; 2 h; -	5 time coats, brush coat	Acharya et al., 1997
PFA	F / 0.2 µm	600°C; 10°C/min; 1–2 h; -	Spray coating	Acharya and Foley, 1999
PFA	T / 0.2 µm	200–600°C; 5°C/min; 2 h; He	Ultrasonic spray 3–6 time coats	Shiflett and Foley, 1999
PFA	T / 0.2 µm, 2 µm	150–600°C; 5°C/min; 0–2 h; He	Ultrasonic spray 3–6 time coats	Shiflett and Foley, 2000

(Continued)

TABLE 6.2 (CONTINUED)
Preparation Conditions for Microporous Carbon Membranes Supported on Porous Substrates

Precursor	Morphology/Pore Diam. of Substrate	Pyrolysis Condition (Final Temperature; Heating Rate; Soak Times; Atmosphere)	Other Notes	Reference
PFA	T / 0.2 μm	450°C; 5°C/min; 2 h; He	Ultrasonic spray 3–4 time coats	Shiflett and Foley, 2001
PFA	F / 0.2 μm	400–600°C; 5°C/min; 2 h; Ar	Spray and spin-coating 4 time coats	Anderson et al., 2008
PVDC	F / 0.7 μm	1000°C; 1°C/min; 3 h; N_2	4–5 time c/p cycles	Rao and Sircar, 1993
PVDC	T / <1 μm	600°C; - ; - ; N_2	single c/p cycle including postoxidation	Sircar et al., 1999a
PVDC-PVC	F / 1 μm	500–1000°C; 1°C/min; - ; Vacuum	spin-coating, pre-oxidation in some cases	Sircar et al., 1999b
PPO	T / 0.1 μm	500–800°C; 5°C/min; 1 h; Ar	2 time c/p cycles including preoxidation	Centeno and Fuertes, 2000
PPO/PVP	F / 0.14 μm	400–700°C; 5°C/min; 1 h; Vacuum		Itta et al., 2011
Resorcinol/ Formaldehyde	F / 0.1 μm	800°C; 5°C/min; 8 h; N_2	pre-coated with colloidal silica, 3–4 time coats	Nishiyama et al., 2006
F: flat	T: tubular			

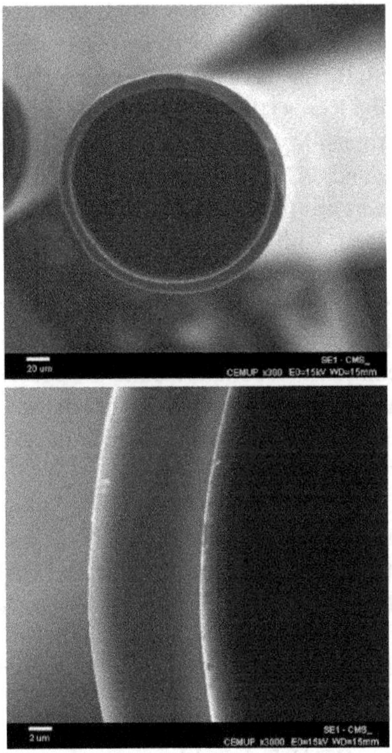

FIGURE 6.3 SEM images of a CMS membrane produced by Carbon Membranes Ltd. (From Lagorsse, S. et al., Journal of Membrane Science. 241, 275–287, 2004.)

precursors. The precursors were carbonized using CO_2 as the purge gas, a heating rate of 4°C/min, final temperature of 550°C, and final soak time of 2 h. Single gas permeation tests of H_2, CO_2, O_2, N_2, and CH_4 were conducted, and the influences of temperature and feed pressure on the gas separation performance of the membranes were also studied.

6.2.2 Polyimides

6.2.2.1 6FDA/BPDA-DAM Polyimide

Polyimides are the precursor of choice for many researchers, probably because of their high glass-transition temperatures (T_g), ease of processing, and good separation performance as polymeric membranes (Williams et al., 2008). Jones and Koros (1994a) prepared CMS membranes by carbonizing commercially available asymmetric hollow fiber membranes of a copolyimide prepared from 4,4'-(hexafluoroisopropylidene) diphthalic anhydride (6FDA) and 3,3',4,4'-biphenyltetracarboxylic acid dianhydride (BPDA) with 2,4,6-trimethyl-1,3-phenylenediamine (mTMPD or DAM). Carbonization of the 6FDA/BPDA-DAM polyimide was carried out at 500 or 550°C. An ideal separation factor (selectivity) for O_2/N_2 in the range of 11.0–14.0 and ideal CO_2/N_2 separation factor of about 55 were achieved. Exposure of the resulting membranes to volatile organic compounds at ambient temperature resulted in decreases in both permeability and selectivity (Jones and Koros (1994b)), but the membranes could be partially regenerated by exposing them to propylene at a pressure of 1.03 MPa. Jones and Koros (1995a) also studied the effects of humidity on the O_2/N_2 selectivity and permeability of their membranes by using feeds with relative humidities between 23% and 85%. Some performance degradation occurred at all humidity levels, but this was minimized by rendering the surfaces of the membranes hydrophobic by coating them with thin layers of Teflon AF1600 or AF2400 (Jones and Koros, 1995b). In a subsequent study, Geizler and Koros (1996) examined the effects of various pyrolysis atmospheres on the separation performance of asymmetric CMS hollow fiber membranes. Vacuum pyrolysis provided more selective but less productive CMS membranes in comparison to the pyrolysis performed under an inert atmosphere. Ghosal and Koros (2000) prepared dense CMS films from 6FDA/BPDA-DAM polyimides and studied their changes in intrinsic permeability and selectivity during the pyrolysis process.

Vu et al. (2002) carbonized two types of polyimide hollow fiber membranes prepared from 6FDA/BPDA-DAM and Matrimid 5218 (a polyimide produced from 3,3',4,4'-benzophenone tetracarboxylic dianhydride (BTDA) and 5(6)-amino-1-(4'-aminophenyl)-1,3-trimethylindane (DAPI)). They investigated the separation of CO_2/CH_4 at pressures of up to 6.89 MPa by the two types of membranes to confirm that their mechanical properties, permeability, and selectivity were stable at high pressures.

6.2.2.2 Kapton (PMDA-ODA) Polyimide

Kapton is a polyimide obtained by curing the polyamic acid prepared by condensation of pyromellitic dianhydride (PMDA) with oxydianiline (ODA). Because both Kapton and its precursor polyamic acid are available commercially, they are frequently used as starting materials for CMS membranes. Suda and Haraya (1995; 1997a; 1997b) prepared flat carbon membranes by carbonizing Kapton films. When the pyrolysis process was suitably controlled, the Kapton CMS membrane showed high performance, with H_2/N_2 and O_2/N_2 selectivities of 4700 and 36, respectively, at 35°C. Fuertes et al. (1999) prepared asymmetric flat CMS membranes from PMDA-ODA polyamic acid membranes by spin-coating the polymer on a porous carbon disk with subsequent phase inversion.

Hatori et al. (2004) obtained a Kapton CMS membrane by pyrolysis at 1000°C; this membrane showed an ideal H_2/CO separation factor of 5900 and it lowered the CO content of H_2 for use in fuel cells from 1% to 2 ppm. Lua and Su (2006; 2007) investigated the effects of the pyrolysis temperature and atmosphere on the final pore structure and gas permeation properties of Kapton CMS membranes.

6.2.2.3 Matrimid and P84 Polyimides

Matrimid (a polyimide prepared from BTDA and DAPI) and P84 (a polyimide prepared from BTDA and 80% methylphenylenediamine (TDI) + 20% methylenediamine (MDI)) are commercially available polyimides that are sometimes used as precursors of CMS membranes (Fuertes et al., 1999; Steel and Koros, 2003; Tin et al., 2004a; 2004b; 2004c; Xiao et al., 2005a; Xu et al., 2011; Briceño et al., 2012;

Favvas et al., 2015). These polyimides can be conveniently cast into any form because they are soluble in various solvents. Fuertes et al. (1999) prepared asymmetric flat CMS membranes from P84 and from Matrimid. Steel and Koros (2003) produced dense freestanding CMS membranes from Matrimid and examined the effects of the pyrolysis temperature on the ultramicropore distribution, which was related to the performance of the membrane in separating O_2/N_2, CO_2/CH_4, and C_3H_6/C_3H_8 mixtures. Vu et al. (2002) prepared CMS hollow fiber membranes as described above. Tin et al. (2004a; 2004b; 2004c) prepared flat CMS membranes from P84 films. The membrane obtained by pyrolysis at 800°C showed a CO_2 permeability of 500 Barrer (1 Barrer = 10^{-10} cm^3 (STP) cm cm^{-2} s^{-1} cmHg^{-1} = 3.35×10^{-16} mol m m^{-2} s^{-1} Pa^{-1}) and a CO_2/CH_4 selectivity of 89, which was the highest efficiency among a group of CMS membranes derived from four commercially available polyimides (P84, Matrimid, Kapton, and Ultem (a polyetherimide)) (Tin et al., 2004c). Xu et al. (2011) prepared Matrimid-derived dense flat films and hollow fiber CMS membranes for ethylene/ethane separation. The CMS membranes showed high separation performance for several gas pairs, such as O_2/N_2 and CO_2/CH_4, and especially high selectivity of 12 for C_2H_4/C_2H_6. Briceño et al. (2012) pyrolyzed Matrimid coated on tubular ceramic supports at 550, 650 and 700°C to fabricate CMS membranes. After the preparation conditions were optimized, the obtained membrane exhibited ideal selectivities of 2.37, 4.70, and 10.62 for H_2/CO_2, H_2/CO, and H_2/CH_4, respectively. Recently, Favvas et al. (2015) produced hollow fiber CMS membranes by pyrolysis of P84 hollow fibers under Ar atmosphere at 900°C. The produced CMS membranes showed high selectivities for He/CH_4, H_2/CH_4, He/N_2, and H_2/N_2 of 2925, 5500, 350, and 600, respectively.

6.2.2.4 Polyetherimides

Most polyimides used as precursors of CMS membranes are either very expensive commercial materials or available only on a laboratory scale. In this context, one polyimide-based material that can be used economically is the commercially available polyetherimide Ultem 1000. Fuertes et al. (1998) used Ultem 1000 to prepare a CMS membrane supported on a macroporous carbon substrate. Sedigh et al. (1999) produced tubular CMS membranes by pyrolyzing polyetherimide coated on the inside of a mesoporous tubular support. Salleh and Ismail (2012) fabricated hollow fiber CMS membranes derived from a polymer blend of polyetherimide and polyvinylpyrrolidone (PVP) by carbonization under N_2 atmosphere. The highest CO_2/CH_4 and CO_2/N_2 selectivities of 55.3 and 41.5, respectively, were obtained for the membrane derived from polyetherimide blended with 6 wt% PVP. Tseng et al. (2012) used polyetherimide as a polymeric precursor to prepare CMS membranes on a porous Al_2O_3 ceramic disk. The microstructure of the CMS selective layer was influenced by the pore structure and surface roughness of the ceramic support, which in turn affected the gas separation performance of the CMS membranes pyrolyzed at 600°C under vacuum for 2 h.

6.2.2.5 BPDA-Aromatic Diamine Polyimide

Kusuki and co-workers (Kusuki et al., 1997; Tanihara et al., 1999) prepared asymmetric CMS membranes from asymmetric hollow fiber membranes composed of a polyimide derived from BPDA and an unspecified aromatic diamine produced by UBE Industries. They developed a pyrolysis method for the continuous preparation of hollow fiber CMS membranes. For a feed gas mixture of 50% H_2 in CH_4 at 80°C, a CMS membrane carbonized at 700°C showed a H_2 permeability of 1000 GPU (1 GPU = 10^{-6} cm^3 (STP) cm^{-2} s^{-1} $cmHg^{-1}$ = 3.35 × 10^{-10} mol m^{-2} s^{-1} Pa^{-1}) and H_2/CH_4 selectivity of 132, whereas a membrane carbonized at 850°C displayed a H_2 permeability of 180 GPU and H_2/CH_4 selectivity of 631. The same group (Tanihara et al., 1999) examined the influence of trace levels of toluene vapor (7500 ppm) and concluded that trace toluene had little effect on the permeation properties of their CMS membranes.

Okamoto et al. (1999) prepared asymmetric CMS hollow fiber membranes by pyrolysis of asymmetric hollow fiber membranes of a copolyimide of BPDA with dimethyl-3,7-diaminodiphenyl-thiophene-5-,5'-dioxide (DDBT) and 3,5-diaminobenzoic acid (DABA) and then evaluated their olefin/paraffin separation properties. At 100°C, the membrane permeabilities for C_3H_6 and C_4H_6 were 50 and 80 GPU, respectively, and the selectivities for C_3H_6/C_3H_8 and C_4H_6/C_4H_{10} were 13 and 50, respectively. Another asymmetric hollow fiber CMS membrane was prepared from 6FDA/BPDA-DDBT copolyimide (Yoshino et al., 2003). The CMS membranes pyrolyzed at 540°C for 1 h displayed the best performance in terms of C_3H_6 permeability (26 GPU) and selectivity (22) for a 50:50 C_3H_6/C_3H_8 mixture at 100°C.

6.2.2.6 Laboratory-Synthesized Polyimides

Many laboratory-synthesized polyimides have also been used as precursors for carbonized membranes. Hayashi et al. (1995, 1996, 1997a, 1998, 1997b) synthesized BPDA-ODA polyimide and then used it to produce CMS membranes supported on porous alumina tubes. The permeability to CO_2 and CO_2/CH_4 selectivity of the CMS membrane pyrolyzed at 800°C were 300 and 100 GPU, respectively, at 30°C (Hayashi et al., 1995). The same researchers also reported the possibility of separating olefins from paraffins using a CMS membrane derived from BPDA-ODA carbonized at 700°C (Hayashi et al., 1996). The CMS membrane exhibited permeabilities of approximately 30 GPU for C_2H_4 and 6 GPU for C_3H_6 at 100°C. The selectivities of the membrane were 4–5 for C_2H_4/C_2H_6 and 25–29 for C_3H_6/C_3H_8 systems.

Fuertes and Centeno (1999) used a polyamic acid prepared from BPDA and *p*-phenylenediamine (pPDA) to prepare precursor membranes, which were converted into polyimide membranes that were subsequently pyrolyzed to give flat supported CMS membranes.

Lee and co-workers (Kim et al., 2004; 2005a; 2005b; Park et al., 2004;) performed a series of studies on CMS membranes derived from BTDA-aromatic diamine polyimides. They synthesized BTDA-ODA polyimide and used it as a precursor for flat CMS membranes (Kim et al., 2005a). These membranes exhibited attractive separation potential compared with that of CMS membranes derived from PMDA-ODA.

6.2.2.7 Other Rigid Polyimides

Kita et al. (1997) synthesized a polypyrrolone from 6FDA and 3,3'-diaminobenzidine (DABZ). The backbone chains of this polymer have a ladder structure that provided enhanced permeability and maintained the permselectivity of gases through inhibition of both chain packing and intermolecular motion. The CMS membranes carbonized at 700°C under N_2 showed better membrane performance than that of carbonized polyimides.

Xiao et al. (2005b) synthesized four polyimides from BTDA, 6FDA, 3,3'4,4'-oxydiphthalic dianhydride (ODPA), and 5,7-diamino-1,1,4,6-tetramethyl-indan (DAI): BTDA-DAI, ODPA-DAI, BPDA-DAI, and 6FDA-DAI. They investigated the effects of the chemical structure and physical properties of these rigid polyimides on the performance of the derived carbon membranes. At low pyrolysis temperatures, polyimides with a high fractional free volume (FFV) and low thermal stability gave carbon membranes with large pores and high gas permeabilities. Xiao et al. (2007) also synthesized crosslinked copolyimides from 6FDA, 2,3,5,6-tetramethyl-1,4-phenylenediamine (durene), and 4,4'-diaminodiphenyl-acetylene (p-intA) and then carbonized the resulting films at 800°C under vacuum. Thermally induced crosslinking occurred through the acetylene groups of the p-intA units, which were believed to form naphthalene structures through a Diels-Alder-type reaction. Carbon membranes derived from copolyimides with large numbers of internal acetylene units showed much better gas separation performance than did those derived from polyimides without internal acetylene units.

Qiu et al. (2014) synthesized uncrosslinked 6FDA-mPDA/DABA (3:2) copolyimides (mPDA is *m*-phenylenediamine). The gas transport properties of the uncrosslinked membrane, thermally cross-linked membrane, and CMS membranes were evaluated. The CMS membranes pyrolyzed at 550 and 800°C showed very high CO_2 permeabilities of 14750 and 2610 Barrer, respectively, with high CO_2/CH_4 selectivities of 52 and 118, respectively. Fu et al. (2015) synthesized four novel polyimide precursors using FDA, BPDA, DABA, diethyltoluenediamine (DETDA), and 1,5-diaminonaphthalene (1,5-ND) and then prepared dense films referred to as 6FDA/DETDA, 6FDA:BPDA (1:1)/DETDA, 6FDA/DETDA:DABA (3:2), and 6FDA/1,5-ND:ODA (1:1). The separation performance of the CMS membranes obtained by pyrolysis of the dense films at 550°C were examined. The CMS membrane derived from 6FDA/DETDA:DABA (3:2) showed the highest permeability of 10,689 Barrer for CO_2 and 2384 Barrer for O_2, respectively, with comparable selectivity before and after aging for one month. The separation performance of these CMS membranes was superior to that of other CMS membranes because of their higher FFV, as shown in Figure 6.4.

6.2.3 Phenolic Resins

Phenolic resins, which are very popular and inexpensive polymers, have also been used as precursors to prepare CMS membranes. Shusen et al. (1996) produced free-standing asymmetric flat carbon membranes by carbonizing thermosetting phenol–formaldehyde resin films at 800–950°C under N_2 atmosphere. The resulting films were oxidized on one face to give membranes with molecular sieve-like flow properties. Kita et al. (1997) prepared tubular CMS membranes by subjecting a phenolic resin coated on the surface of a porous tube of α-alumina to pyrolysis at 600°C under N_2 atmosphere. This coating-carbonization cycle was repeated four or five times. The resulting membranes showed excellent separation performance for alkenes/alkanes and CO_2/N_2.

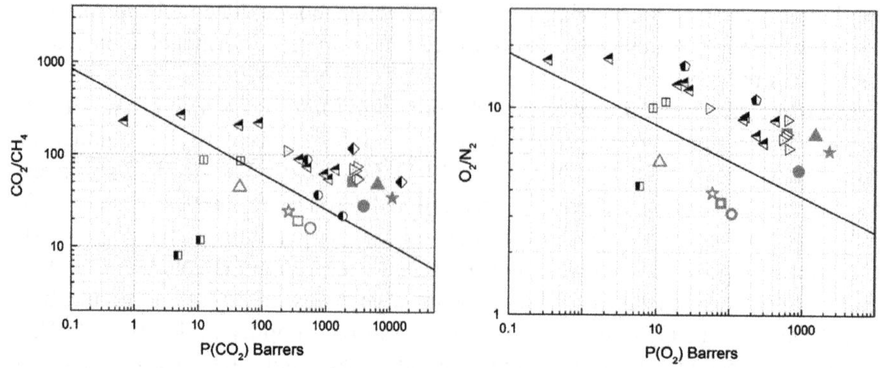

FIGURE 6.4 Pure gas separation performance of polymer films and CMS films (aged for one month) derived from: (a) 6FDA/DETDA (squares); (b) 6FDA:BPDA (1:1)/DETDA (circles); (c) 6FDA/1,5-ND:ODA (1:1) (triangles); and (d) 6FDA/DETDA:DABA (3:2) (stars); open symbols indicate polymer films and solid symbols indicate CMS films. (From Fu, S. et al., Journal of Membrane Science. 487, 60–73, 2015.)

Zhou et al. (2001; 2003) produced highly permeable CMS membranes from a sulfonated phenolic resin and investigated the effects of the pyrolysis temperature, dip-coating conditions, and number of coating/pyrolysis cycles on the gas permeation properties of the membranes. Membranes obtained under optimal preparation conditions exhibited an O_2 permeability of 30 GPU and ideal O_2/N_2 separation factor of 12 at 35°C.

Fuertes and colleagues (Fuertes and Menendez, 2002; Fuertes, 2000; Centeno et al., 2004) published a series of studies on CMS membranes made of phenolic resin. They produced CMS membranes consisting of a microporous carbon layer with a thickness of 2 μm obtained by pyrolysis of a film of a novolac-type phenolic resin supported on a macroporous carbon disk substrate with a pore size of 1 μm and 30% porosity. The carbon membrane prepared by carbonization at 700°C showed high selectivities for the separation of permanent gases, such as the O_2/N_2 system (selectivity of 10 at 25°C).

The same group (Centeno et al., 2004) also investigated how processing variables (heat-treatment temperature, heating rate, soaking time, and atmosphere) affected the pyrolysis of phenolic resin-based carbon membranes. The membranes they obtained at temperatures of around 700°C behaved as selective-adsorption and surface-diffusion membranes that proved highly effective for the recovery of hydrocarbons from hydrocarbon/N_2 mixtures. An increase in the carbonization temperature to 800°C caused a marked decrease in the gas permeability of the CMS membrane, resulting in a membrane that displayed good capability for the separation of O_2/N_2, CO_2/CH_4, and olefin/paraffin mixtures. Heat treatment of phenolic resin films at temperatures of around 900–1000°C produced CMS membranes that showed high permselectivities for mixtures of gases with molecular sizes smaller than 0.4 nm.

Zhang et al. (2006a) produced a CMS mixed-matrix carbon membrane. A coating solution was obtained by dispersing CMSs with a median particle size of 0.37 μm, prepared by high-temperature pyrolysis of walnut hulls, in an ethanolic solution of a phenol–formaldehyde novolac resin (PFNR). A green porous tubular support made of the same resin was dipped in the mixed-matrix coating solution and then carbonized at 800°C. After carbonization, the tube was activated in CO_2 for 20–60 min at the same temperature as that used for carbonization.

Teixeira et al. (2011) prepared composite CMS membranes derived from a resol phenolic resin loaded with boehmite nanoparticles on α-Al_2O_3 supports. After pyrolysis at 550°C, boehmite-derived Al_2O_3 nanowires with thickness of 1–2 nm and length of 10–30 nm were well dispersed in the carbon matrix. The CMS membrane exhibited high C_3H_6 permeability and C_3H_6/C_3H_8 ideal selectivity.

6.2.4 Polyfurfuryl Alcohol

PFA has been used extensively as a precursor for CMS membranes. Because PFA is a liquid at room temperature, all membranes derived from PFA are composite membranes supported by porous substrates. Chen and Yang (1994) carbonized PFA coated on a macroporous graphite disk support to obtain a PFA-CMS membrane. Sedigh et al. (1998) also used PFA as a precursor to prepare supported CMS films. The separation properties of the membranes were tested using single gases (H_2, CO_2, CO, CH_4, and Ar), binary mixtures of CO_2 and CH_4, and a quaternary mixture of CO_2, CO, H_2, and CH_4. Separation factors for CO_2/CH_4 in the range 34–37 were obtained for the binary and quaternary mixtures.

Wang et al. (2000) used the vapor deposition polymerization (VDP) technique to coat furfuryl alcohol (FA) on γ-Al_2O_3/α-Al_2O_3 and glass/α-Al_2O_3 support tubes. The support tubes were pretreated with an acid catalyst and exposed to FA vapor at 90°C. The tubes were then heated to 200°C to crosslink the PFA polymer before being carbonized at 600°C. Compared with those of PFA-CMS membranes prepared by dip-coating techniques, the membranes prepared by VDP had similar CO_2/CH_4 selectivities but lower CO_2 permeabilities.

Foley and co-workers (Acharya et al., 1997; Acharya and Foley, 1999; Shiflett and Foley, 1999; 2000; 2001) published a series of studies on PFA-CMS membranes formed on sintered stainless-steel supports. Solutions of PFA in acetone (50–60 wt%) were coated by hand brushing (Acharya et al., 1997) or spray coating (Acharya and Foley, 1999) on porous stainless-steel disks with a0.2-μm pore size. The same group (Shiflett and Foley, 1999) developed an ultrasonic spray-coating method using a 25 wt% solution of PFA in acetone to make the PFA precursor. The PFA-CMS membranes formed on sintered stainless-steel tubes with a pore size of 0.2 μm exhibited high capabilities for the separation of O_2/N_2, He/N_2, and H_2/N_2. Shiflett and Foley (2003) also explored a protocol involving high-temperature pyrolysis of the initial layers followed by lower-temperature pyrolysis of subsequent layers to give membranes with a higher permeation flux. In addition, they examined the modification of PFA-CMS membranes with additives such as titanium dioxide, small-pore high-silica zeolite, and polyethylene glycol. An automated ultrasonic spray system has been used to prepare uniform PFA films to improve the reproducibility of PFA-CMS membrane formation (Shiflett and Foley, 2001).

Anderson et al. (2008) used a similar method to make PFA-CMS membranes supported on porous stainless-steel disks. Positron annihilation lifetime spectrometry and wide-angle X-ray diffraction studies showed that the size of the micropores decreased and the membrane porosity increased with increasing pyrolysis temperature. The determined performances of the resulting membranes supported these findings, in that marked increases in permeability, related to increased porosity, were observed with rising pyrolysis temperature.

6.2.5 Vinylidene Chloride Copolymers

Rao and Sircar (1993; 1996) obtained a carbon membrane by pyrolysis of PDVC–acrylate terpolymer latex coated on a porous graphite support. The resulting membrane separated H_2/hydrocarbon mixtures by selective adsorption and surface diffusion of the larger component (the hydrocarbon). Thus, the membrane was called a selective surface flow (SSF) membrane. The mean pore diameter of the membrane was in the range of 0.5–0.6 nm, which is larger than those of CMS membranes (Rao and Sircar, 1996). The researchers extended their method to produce tubular membranes and also demonstrated the possibility of an SSF membrane/pressure-swing adsorption hybrid process to produce pure H_2 (Sircar et al., 1999a; 1999b).

Centeno and Fuertes (2000) formed carbon membranes by pyrolyzing poly(vinylidene chloride-co-vinyl chloride) (PVDC-PVC) films supported on porous carbon disks. These membranes showed molecular sieving properties; for example, a high permselectivity of 14 for the O_2/N_2 pair. Pre-oxidation in air at 200°C for 6 h improved the permselectivity of the membranes, but resulted in a decrease in gas permeability.

6.2.6 Polymers with Modified Fractional Free Volumes (FFV)

Xiao et al. (2005a) examined the effects of bromination of a Matrimid precursor before it was carbonized to produce carbon membranes on their performance. The lower thermal stability and higher FFV of brominated Matrimid resulted in higher gas permeabilities for carbon membranes pyrolyzed at lower pyrolysis temperatures, whereas their selectivities remained similar to those of membranes obtained by pyrolysis of the original Matrimid precursor under the same conditions.

Park et al. (2004) synthesized polyimides from BTDA with a 9:1 ratio of ODA to mPDA, 2,4-diaminotoluene (2,4-DAT), or mTMPD; these co-monomers possess no methyl substituent (mPDA), one methyl substituent (2,4-DAT), or three methyl substituents (mTMPD). The resulting films were carbonized to produce flat dense CMS membranes. The introduction of methyl substituents to the rigid polyimide backbone increased the FFV of the polyimides, and gas permeabilities typically increased with the FFV. CMS membranes prepared by pyrolysis of each of these polyimides in an inert atmosphere at 600 and 800°C showed similar gas permeation behavior.

6.2.7 Other Polymer Precursors

Yoshimune et al. (2005) produced novel CMS membranes from PPO and its derivatives. They synthesized PPO derivatives with functional groups such as SO_3H, CO_2H, Br, $SiMe_3$, and PPh_2 in one-step reactions, and cast these derivatives into hollow fiber configurations with symmetric dense structures. The gas permeabilities and selectivities for He, H_2, CO_2, O_2, and N_2 of the pyrolyzed membranes were as high as those observed for polyimide-based CMS membranes. The highest performance was exhibited by the $SiMe_3$-substituted PPO CMS membrane pyrolyzed at 650°C, for which the O_2 permeability was 125 Barrers and the O_2/N_2 selectivity was 10 at 25°C. The same group also conducted further detailed investigations of $SiMe_3$-substituted PPO CMS membranes (Yoshimune et al., 2007) and supported CMS membranes derived from dip-coated PPO on a porous ceramic tube with a pore size of 0.1 μm (Lee et al., 2006). In addition, flexible carbon hollow fiber membranes with excellent gas separation performance were produced using sulfonated PPO as a precursor polymer (Yoshimune and Haraya, 2010). Itta et al. (2011) fabricated CMS membranes from PPO with the thermally labile polymer PVP coated on alumina disks using a spin-coating technique. The effect of PPO content of the PPO/PVP blends on membrane performance was evaluated. A high PPO content increased the permeance at 700°C.

Campo et al. (2010) prepared CMS membranes from commercial cellophane films in a single heating step. Cellophane is a thin sheet of regenerated cellulose, which has advantages such as low cost, ready availability, and biodegradability. The effects of pyrolysis temperature, soaking time, and pyrolysis atmosphere on membrane performance were investigated extensively. The maximum permeability was reached for the membrane pyrolyzed at 550°C and its separation performance was above the upper bound.

Zhang et al. (2006b; 2006c) prepared CMS membranes using poly(phthalazinone ether sulfone ketone) (PPESK) as a novel polymeric precursor. The maximum permselectivities for H_2/N_2, CO_2/N_2, and O_2/N_2 gas pairs of these membranes were 278.5, 213.8, and 27.5, respectively. Oxidative stabilization of PPESK before carbonization was beneficial to provide carbon membranes with high gas separation performance. This is because oxidative stabilization shifted the pore size distribution to a smaller pore size and increased the maximum pore volume of the carbon matrix. These researchers also prepared sulfonated PPESK-derived CMS membranes and investigated the effects of sulfonation degree on gas permeation properties (Zhang et al., 2006c).

Chung et al. (2009) have explored carbon membranes derived from an interpenetrating network of poly(aryl ether ketone) and 2,6-bis-(4-azidobenzylidene)-4-methylcyclohexanone (PAEK/azide). Their concept involved a polymer precursor consisting of a thermally stable part and thermally labile part at the molecular level to create CMS membranes with relatively large pores that were suitable for C_3H_6/C_3H_8 separation. The membrane obtained from PAEK/azide (80:20) pyrolyzed at 550°C exhibited the best C_3H_6/C_3H_8 separation performance with a C_3H_6 permeability of 48 Barrer and ideal C_3H_6/C_3H_8 selectivity of 44.

Nishiyama et al. (2006) produced microporous carbon membranes on alumina supports by pyrolysis of cationic tertiary amine/anionic polymer composites. The precursor solutions contained a thermosetting resorcinol/formaldehyde polymer and cationic tertiary amine. Three types of cationic tertiary amines with different chain lengths were used: tetramethylammonium bromide (TMAB), tetrapropylammonium bromide (TPAB), and cetyl(trimethyl)ammonium bromide (CTAB). Study of the permeation of pure gases of various molecular sizes through the membranes revealed that the pore sizes of the carbon membranes prepared using TMAB, TPAB, and CTAB were 0.4, 0.5, and >0.55 nm, respectively.

Recently, ultrathin graphene oxide (GO) membranes were successfully prepared by Li et al. (2013) for selective H_2 separation. GO membranes with thicknesses of ~1.8, 9, and 18 nm were prepared by vacuum filtration of GO flakes on anodic aluminum oxide (AAO) supports. These ultrathin GO membranes showed superior separation performance far above that achieved previously for polymeric membranes, with selectivities as high as 3400 and 900 for H_2/CO_2 and H_2/N_2 mixtures, respectively.

6.3 PYROLYSIS PROCESSES

In an inert or vacuum atmosphere, the heat treatment of polymers can be separated into three processes: (1) annealing at 100–400°C, (2) intermediate heating at 400–500°C, and (3) pyrolysis to form carbon at 500–1000°C (Barsema et al., 2004). The pyrolysis process is governed by several parameters, including the heating rate, final pyrolysis temperature, thermal soak time, and pyrolysis atmosphere. The specific conditions used previously by researchers are listed in Tables 6.1 and 6.2.

6.3.1 Pyrolysis at Intermediate Temperature

Barsema et al. (2004) subjected films of Matrimid polyimide to various heat treatments between 300 and 525°C to investigate the intermediate structures that evolved at temperatures between the annealing and carbonization temperatures. The T_g of this polymer was 323°C. The permeability of the polymer to noncondensable gases (N_2 and O_2) was lowered by heat treatment below T_g of the polymer. High permeability was observed for the polymer treated at 350°C. Above 350°C, increasing formation of charge-transfer complexes and the resulting densification of the polymer structure led to a gradual decrease in permeability. A marked increase in permeability occurred when the films were exposed to temperatures above 475°C because of the onset of thermal decomposition.

Shao et al. (2005) synthesized a 6FDA/durene polyimide (T_g = 425°C) that was pyrolyzed by heat treatment at temperatures from 50 to 800°C under vacuum. The gas permeability of the resulting membrane increased with treatment temperature. The maximum O_2, N_2, and CH_4 permeabilities were obtained at 475°C. Carbon membranes pyrolyzed at various heating rates (1 and 3°C/min) showed different transport performance at low pyrolysis temperatures. However, at a high pyrolysis temperature (800°C), the gas-transport performance of carbon membranes obtained from pyrolysis using different protocols showed similar. Shao et al. (2004) observed that the casting solvent affected the morphologies and gas-transport properties of membranes made from a novel copolyimide of 6FDA with PMDA and tetramethylmethylenedianiline (TMMDA), along with those of its derived carbon membranes. The differences between CMS membranes derived from precursors with different morphologies decreased as the pyrolysis temperature was increased. At low pyrolysis temperatures, the structure and separation performance of the CMS membranes were strongly affected by the decomposition temperature of the precursor. Conversely, at higher pyrolysis temperatures, the factor that dominated the structure and performance of the CMS membranes was the pyrolysis temperature because of the complete degradation of the polymeric precursor.

As mentioned above, intermediate temperatures were used for the pyrolysis of precursors incorporating porogens, such as polymers containing sulfonic acid groups (Zhou et al., 2001; 2003), with the aim of retaining high permeability by preventing the loss of porosity that typically occurs during high-temperature pyrolysis.

6.3.2 Pyrolysis Temperature

The pyrolysis temperature is generally chosen to be above the decomposition point of the polymer but below the graphitization temperature (500–1000°C). Koresh and Soffer (1983) studied the effects of the carbonization temperature on membrane separation performance by preparing membranes at 800 and 950°C. They found that membranes pyrolyzed at 950°C exhibited lower permeabilities but higher permselectivities than those pyrolyzed at 800°C. In their study on 6FDA/BPDA-DAM polyimide-derived hollow fiber CMS membranes, Geiszler and Koros (1996) found that increasing the final pyrolysis temperature from 500 to 800°C decreased membrane permeability but raised its permselectivity. Suda and Haraya (1997a) reported that Kapton CMS membranes showed a decrease in gas permeabilities but an increase in permselectivities as the final pyrolysis temperature was elevated in the range of 600–1000°C. Similar trends in the relationships between permeabilities, permselectivities, and pyrolysis temperature have been reported for P-84 polyimide CMS membranes (Tin et al., 2004a), Matrimid polyimide CMS membranes (Xiao et al., 2005a), and BTDA-ODA polyimide CMS membranes (Park et al., 2004; Kim et al., 2005). As a general rule, an increase of pyrolysis temperature decreases the permeability of CMS membranes but increases their selectivity (Williams and Koros, 2008). High pyrolysis temperatures increase the crystallinity and density and decrease the average interplanar spacing of CMS membranes (Suda and Haraya, 1995; Xiao et al., 2005a; Lua and Su, 2006; Anderson et al., 2008).

Several researchers have conducted detailed studies on the effects of the pyrolysis temperature on membrane performance and have reported exact results. Increasing the pyrolysis temperature above 500°C results in an increase of the gas permeability of carbonized membranes by one or two orders of magnitude, with maximum permeability obtained at around 650–750°C; upon further increasing the pyrolysis temperature, the resulting carbon membranes become less permeable. Hayashi et al. (1995) reported that BPDA-ODA polyimide CMS membranes pyrolyzed at 550–700°C showed maximum permeability, whereas those pyrolyzed at 800°C exhibited the highest He/N_2 selectivity. Kusuki et al. (1998) found that a BPDA-aromatic diamine polyimide hollow fiber CMS membrane pyrolyzed at 650°C displayed maximum H_2 permeability, whereas one pyrolyzed at 850°C exhibited the highest H_2/CH_4 selectivity. In their study on 6FDA/BPDA-DDBT copolyimide hollow fiber CMS membranes, Yoshino et al. (2003) observed maximum permeability for the membranes pyrolyzed at 550°C, whereas the peak permselectivity occurred at 650°C. Yoshimune et al. (2005; 2007) found that CMS membranes based on PPO and PPO derivatives pyrolyzed at 650°C exhibited maximum permeability, whereas maximum permselectivity was observed at a different pyrolysis temperature.

These results suggest that pores appear at about 500°C and enlarge as their number increases in the temperature range of 550–700°C; heating to a higher temperature causes the pores to shrink or disappear. This behavior certainly depends on the physical properties of the polymers, so a suitable carbonization temperature needs to be chosen for a selected polymer precursor to attain optimal performance in gas separation.

6.3.3 Thermal Soak Time

The thermal soak time can have various effects on the performance of the final membrane. Varying the thermal soak time, particularly at the final pyrolysis temperature, can be used to fine-tune the permeation properties of a CMS membrane in an effective manner. Kim et al. (2005) showed that lengthening the thermal soak time for BPDA-ODA polyimide-based CMS membranes increased their selectivity but decreased their permeability. Yoshino et al. (2003) reported that thermal soaking increased the C_3H_6/C_3H_8 selectivity of 6FDA/BPDA-DDBT-based CMS hollow fiber membranes.

6.3.4 Heating Rate

The heating rate determines the rate of evolution of volatile components from a polymeric membrane during pyrolysis, and consequently affects the nature of the pores formed in the resulting carbon membranes. Widely different heating rates have been used, ranging from 0.2 to 13.3°C/min. Lower heating rates favor the formation of small pores and increase the crystallinity of the resulting carbon, thereby giving carbon membranes with higher selectivity (Suda and Haraya, 1997a). Higher heating rates can lead to the formation of pinholes, microscopic cracks, blisters, and distortions, which in extreme cases may render the membranes useless for gas separation (Saufi and Ismail, 2004). From a practical standpoint, an optimal heating rate that is not too low should be chosen, because low heating rates increase both the cost and time involved to produce carbon membranes.

6.3.5 Atmosphere

Pyrolysis is generally conducted under vacuum or inert gas to prevent undesired burn-off and chemical damage to the final carbonized membranes. Geizler and Koros (1996) examined the pyrolysis of 6FAD/BPDA-DAM polyimide-based hollow fiber CMS membranes under vacuum and inert gas. They concluded that vacuum pyrolysis produced membranes that were more selective but less productive than those obtained by pyrolysis in an inert atmosphere of He, Ar, or CO_2. Su and Lua (2007) examined the effects of the carbonization atmosphere (Ar, He, N_2, and vacuum) on the membrane structure and transport properties of Kapton-derived CMS membranes pyrolyzed at 600 and 800°C. They found that carbonization under vacuum at 600 and 800°C gave membranes with low gas permeabilities but maximum ideal selectivities for O_2/N_2 of 9.37 and 17.76, respectively, whereas carbonization at these temperatures under Ar gave membranes with maximum selectivities for CO_2/CH_4 of

93.35 and 476.74, respectively. Kiyono et al. (2010) investigated the effects of pyrolysis atmosphere on the gas separation performance of CMS membranes using 6FDA/BPDA-DAM as a precursor. The polymer was pyrolyzed in a gas mixture containing a specific amount of O_2 (4, 8, 30, or 50 ppm) and the separation performance of the resulting membranes was evaluated. A strong relationship was observed between the total amount of oxygen and the transport properties. Selectivity increases and permeability decreases as the amount of oxygen in the inert gas increases from 4 ppm to 30 ppm. However, both selectivity and permeability decreased when using a mixed gas containing 50 ppm of oxygen.

6.4 PRE- AND POST-TREATMENT

In some cases, pretreatment processes have been used to condition polymer precursors before pyrolysis. The most common pretreatment is preoxidation, which allows the polymeric precursor to retain its form during pyrolysis through the formation of crosslinks in the polymer that increase its thermal stability. The main purpose of post-treatment is to adjust the pore-size distribution or repair flaws in carbon membranes, thereby enhancing their performance. In general, oxidation processes are used to modify pore size and CVD treatment to repair flaws.

6.4.1 Pre-Treatment

Kusuki et al. (1997), Tanihara et al. (1999), and Okamoto et al. (1999) treated asymmetric hollow fiber membranes composed of a polyimide derived from BPDA and an aromatic diamine by oxidation in air at 400°C for 30 min before carbonization. They found that this preoxidation treatment process effectively prevented softening of the precursors during pyrolysis, which would otherwise have resulted in carbon membranes with poor performance. Barsema et al. (2002) pretreated P84 polyimide hollow fiber membranes by oxidation in air at 300°C for 1 h before carbonization. David and Ismail (2003) showed that polyacrylonitrile hollow fiber membranes were thermally stabilized by heating at 250°C in air or oxygen for 30 min. Yoshimune et al. (2005; 2007) pretreated precursors of PPO and its derivatives by oxidation in air at 280°C for 45 min to prevent melting during pyrolysis. PPESK precursors have also been preoxidized in air at 400°C for 30 min for the same reason (Zhang et al., 2006b; 2006c; 2011).

Besides oxidation, chemical modification has also been used as a pretreatment approach. Tin et al. (2004a; 2004b) examined the effects of crosslinking modification and non-solvent pretreatment of Matrimid and P81 precursors on the properties of the final CMS membranes.

6.4.2 Post-Treatment

Koresh and Soffer (1983) used an activation process in an oxidizing gas to enlarge the pore size and thus improve the performance of cellulose-derived carbon membranes. Hayashi et al. (1997a) and Kusakabe et al. (1998) examined the oxidation of carbonized BPDA-ODA polyimide-based CMS membranes by treatment in O_2/N_2

mixtures at 300°C. This treatment broadened the pore-size distribution of the membranes, resulting in markedly improved gas permeation accompanied by retention of the O_2/N_2 permselectivity. Hayashi et al. (1997b) investigated the post-treatment of BPDA-ODA polyimide CMS membranes by CVD through pyrolysis of propylene at 650°C. The treated membranes showed increased permselectivities of 14 for O_2/N_2 and 73 for CO_2/N_2 at 35°C.

Post-oxidation is also an effective method to increase the permeability of CMS membranes to larger gas molecules, such as hydrocarbons. For example, treatment of Kapton-based CMS membranes with water vapor at 400°C increased their permeabilities by several orders of magnitude and resulted in notably high selectivity of more than 100 for C_3H_6/C_3H_8 at 35°C (Suda and Haraya, 1997b). Fuertes and Menendez (2002) produced supported CMS membranes by pyrolyzing a novolac-type phenolic resin, which was deposited on the inner face of a tubular ceramic ultrafiltration membrane, at 700°C under vacuum. In some cases, pre- or post-treatment by aerial oxidation at 75–350°C was examined. The separation performance for olefin/paraffin hydrocarbon mixtures was increased by preoxidation, post-oxidation, or CVD treatment of the carbonized membrane. Fuertes (2000) also used air oxidation of supported tubular carbon membranes at 300–475°C to transform their gas permeation properties to those of selective-adsorption and surface-diffusion carbon membranes.

Hirota et al. (2013) demonstrated the pore size control of carbon membranes prepared from FA by a post-synthesis activation using various gases such as H_2, CO_2, O_2, and steam. After activation in H_2 and steam, the pore sizes of the CMS membranes increased from 0.3 to 0.45 nm, resulting in increased H_2 permeance.

6.5 CONCLUDING REMARKS

Recent technological developments in membrane science have facilitated the fabrication of a huge variety of carbon membranes with promising gas separation performance, as discussed in this chapter. Data for the O_2/N_2 separation performance of carbon membranes taken from literature sources are plotted in Figure 6.5, which also shows the upper limits for polymer membranes (Robeson, 1991). Results for O_2/N_2 selectivity and O_2 permeability are commonly used as benchmarks for the performance of membranes as a function of the synthesis variables. It can be clearly seen that most CMS membranes have O_2/N_2 selectivities that are roughly three times higher than the upper limit for polymer membranes when selectivities are compared at the same O_2 permeability. CMS membranes have considerable potential for commercial applications in not only O_2/N_2 separation, but also H_2 purification and the separation of CO_2/CH_4, CO_2/N_2, or C_3H_6/C_3H_8, which are processes in which conventional polymer membranes do not perform adequately.

Scale up of carbon membranes is one of the largest challenges to surmount for their commercialization. It will be necessary to prepare high-quality carbon membranes with large surface areas in a reliable and cost-effective manner and to integrate these membranes into process modules with high-temperature sealing. Some researchers have reported the scaled-up production of CMS membranes or modules (Karvan et al., 2013; He and Hägg, 2013; Parsley et al., 2014). Karvan et al. (2013)

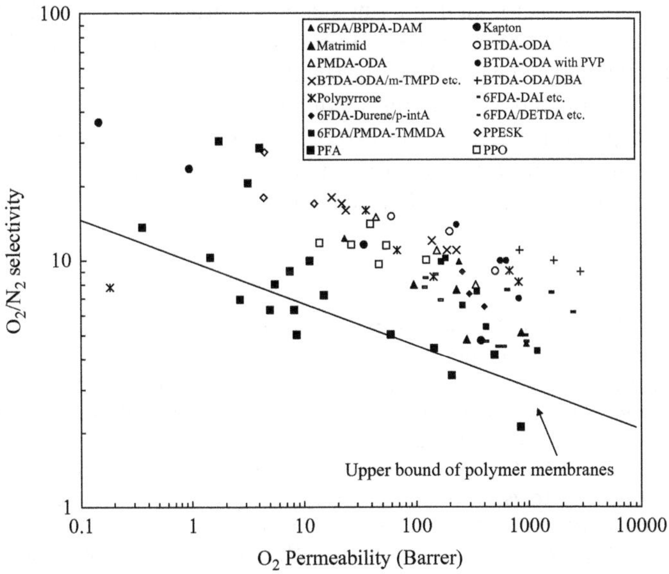

FIGURE 6.5 Comparison of the oxygen permeabilities and permselectivities of various CMS membranes with the upper limit for polymer membranes. (From Robeson, L.M., Journal of Membrane Science. 62, 165–185, 1991.)

devised a simple concept to scale up the manufacture of high-performance CMS hollow fiber pyrolyzed Matrimid precursors to produce pilot-scale membrane modules. Karvan and Hägg (2013) prepared a customized membrane module using 91 hollow fiber carbon membranes derived from cellulose acetate. This module had an effective length of 150 mm and effective membrane area of 86 cm^2. Parsley et al. (2014) reported the fabrication of a multitubular supported CMS module with 86 single tubes and an effective membrane area of 0.76 m^2, as depicted in Figure 6.6. The field test of this module was conducted above 250°C, and the membrane performance remained unchanged over several hundred cumulative hours. The long-term stability

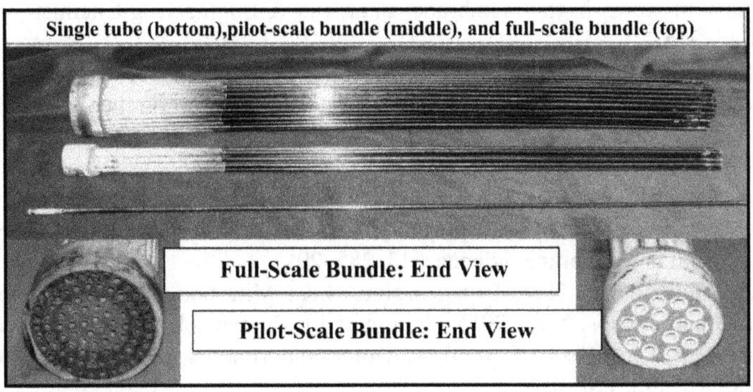

FIGURE 6.6 Photograph of single tube and pilot-scale and full-scale bundles of CMS membranes. (From Parsley, D. et al., Journal of Membrane Science. 450, 81–92, 2014.)

of CMS membranes is also an important matter to consider. Xu et al. (2014) studied the physical aging of CMS membranes prepared from Matrimid, 6FDA-DAM, and 6FDA/BPDA-DAM and revealed that the stabilized membranes demonstrated excellent stability under practical operating conditions.

The challenges at the forefront of carbon membrane development are considerable, but they may be overcome by using new polymer precursors, new production methods, and new module fabrication techniques.

REFERENCES

Acharya, M., Foley, H.C., 1999. Spray-coating of nanoporous carbon membranes for air separation. Journal of Membrane Science. 161, 1–5.

Acharya, M., Raich, B.A., Foley, H.C., Harold, M.P., Lerou, J.J., 1997. Metal-supported carbogenic molecular sieve membranes: Synthesis and applications. Industrial & Engineering Chemistry Research. 36, 2924–2930.

Anderson, C.J., Pas, S.J., Arora, G., Kentish, S.E., Hill, A.J., Sandler, S.I., Stevens, G.W., 2008. Effect of pyrolysis temperature and operating temperature on the performance of nanoporous carbon membranes. Journal of Membrane Science. 322, 19–27.

Barsema, J.N., Klijnsta, S.D., van der Vegt, N.F.A., Koops, G.H., Wessling, M., 2004. Intermediate polymer to carbon gas separation membranes based for Matrimid PI. Journal of Membrane Science. 238, 93–102.

Barsema, J.N., van der Vegt, N.F.A., Koops, G.H., Wessling, M., 2002. Carbon molecular sieve membranes prepared from porous fiber precursor. Journal of Membrane Science. 205, 239–246.

Briceño, K., Montané, D., Garcia-Valls, R., Iulianelli, A., Basile, A., 2012. Fabrication variables affecting the structure and properties of supported carbon molecular sieve membranes for hydrogen separation. Journal of Membrane Science. 415–416, 288–297.

Campo, M.C., Magalhães, F.D., Mendes, A., 2010. Carbon molecular sieve membranes from cellophane paper. Journal of Membrane Science. 350, 180–188.

Centeno, T.A., Fuertes, A.B., 2000. Carbon molecular sieve gas separation membranes based on poly(vinylidene chloride-co-vinyl chloride). Carbon. 38, 1067–1073.

Centeno, T.A., Vilas, J.L., Fuertes, A.B., 2004. Effects of phenolic resin pyrolysis conditions on carbon membrane performance for gas separation. Journal of Membrane Science. 228, 45–54.

Chen, Y.D., Yang, R.T., 1994. Preparation of carbon molecular sieve membrane and diffusion of binary mixtures in the membrane. Industrial & Engineering Chemistry Research. 33, 3146–3153.

Chung, M.L., Xiao, Y., Chung, T.S., Toriida, M., Tamai, S., 2009. Enhanced propylene/propane separation by carbonaceous membrane derived from poly (aryl ether ketone)/2,6-bis(4-azidobenzylidene)-4-methyl-cyclohexanone interpenetrating network. Carbon. 47, 1857–1866.

David, L.I.B., Ismail, A.F., 2003. Influence of the thermostabilization process and soak time during pyrolysis process on the polyacrylonitrile carbon membranes for O_2/N_2 separation. Journal of Membrane Science. 213, 285–291.

Favvas, E.P., Heliopoulos, N.S., Papageorgiou, S.K., Mitropoulos, A.C., Kapantaidakis, G.C., Kanellopoulos, N.K., 2015. Helium and hydrogen selective carbon hollow fiber membranes: The effect of pyrolysis isothermal time. Separation and Purification Technology. 142, 176–181.

Fuertes, A.B., 2000. Adsorption-selective carbon membrane for gas separation. Journal of Membrane Science. 177, 9–16.

Fuertes, A.B., Centeno, T.A., 1998. Carbon molecular sieve membranes from polyetherimide. Microporous and Mesoporous Materials. 26, 23–26.

Fuertes, A.B., Centeno, T.A., 1999. Preparation of supported carbon molecular sieve membrane. Carbon. 37, 679–684.

Fuertes, A.B., Menendez, I., 2002. Separation of hydrocarbon gas mixtures using phenolic resin-based carbon membranes. Separation and Purification Technology. 28, 29–41.

Fuertes, A.B., Nevskaia, D.M., Centeno, T.A., 1999. Carbon composite membranes from matrimid and kapton polyimide. Microporous and Mesoporous Materials. 33, 115–125.

Fu, S., Sanders, E.S., Kulkarni, S.S., Koros, W.J., 2015. Carbon molecular sieve membrane structure–property relationships for four novel 6FDA based polyimide precursors. Journal of Membrane Science. 487, 60–73.

Geiszler, V.C., Koros, W.J., 1996. Effect of polyimide pyrolysis conditions on carbon molecular sieve membrane properties. Industrial & Engineering Chemistry Research. 35, 2999–3003.

Ghosal, A.S., Koros, W.J., 2000. Air separation properties of flat sheet homogeneous pyrolytic carbon membranes. Journal of Membrane Science. 174, 177–188.

Hatori, H., Takagi, H., Yamada, Y., 2004. Gas separation properties of molecular sieving carbon membranes with nanopore channels. Carbon. 42, 1169–0073.

Hayashi, J., Mizuta, H., Yamamoto, M., Kusakabe, K., Morooka, S., 1996. Separation of ethane/ethylene and propane/propylene system with a carbonized BPDA-pp'ODA polyimide. Industrial & Engineering Chemistry Research. 35, 4176–4181.

Hayashi, J., Mizuta, H., Yamamoto, M., Kusakabe, K., Morooka, S., 1997. Pore size control of carbonized BPDA-pp'ODA polyimide membrane by chemical vapor deposition of carbon. Journal of Membrane Science. 124, 243–251.

Hayashi, J., Yamamoto, M., Kusakabe, K., Morooka, S., 1995. Simultaneous improvement of permeance and permselectivity of 3,3',4,4'-biphenyltetracarboxylic dianhydride-4,4'-oxydianiline polyimide membrane by carbonization. Industrial & Engineering Chemistry Research. 34, 4364–4370.

Hayashi, J., Yamamoto, M., Kusakabe, K., Morooka, S., 1997. Effect of oxidation on gas permeation of carbon molecular sieving membranes based on BPDA-pp'ODA polyimide. Industrial & Engineering Chemistry Research. 36, 2134–2140.

He, X., Hägg, M.-B., 2012. Structural, kinetic and performance characterization of hollow fiber carbon membranes. Journal of Membrane Science. 390–391, 23–31.

He, X., Hägg, M.-B., 2013. Hollow fiber carbon membranes: From material to application. Chemical Engineering Journal. 215–216, 440–448.

Hirota, Y., Ishikado, A., Uchida, Y., Egashira, Y., Nishiyama, N., 2013. Pore size control of microporous carbon membranes by post-synthesis activation and their use in a membrane reactor for dehydrogenation of methylcyclohexane. Journal of Membrane Science. 440, 134–139.

Ismail, A.F., David, L.I.B., 2001. A review on the latest development of carbon membranes for gas separation. Journal of Membrane Science. 193, 1–18.

Ismail, A.F., Li, K., 2008. From polymeric precursors to hollow fiber carbon and ceramic membranes. *Inorganic Membranes: Synthesis, Characterization and Applications.* Mallada, R., Menendez, M. (Eds.), 81–120, Elsevier, Inc., Oxford.

Itta, A.K., Tseng, H.-H., Wey, M.-Y., 2011. Fabrication and characterization of PPO/PVP blend carbon molecular sieve membranes for H_2/N_2 and H_2/CH_4 separation. Journal of Membrane Science. 372, 387–395.

Jones, C.W., Koros, W.J., 1994a. Carbon molecular sieve gas separation membranes-I. Preparation and characterization based on polyimide precursors. Carbon. 32, 1419–1425.

Jones, C.W., Koros, W.J., 1994b. Carbon molecular sieve gas separation membranes-II. Regeneration following organic exposure. Carbon. 32, 1427–1432.

Jones, C.W., Koros, W.J., 1995a. Characterization of ultramicroporous carbon membranes with humidified feeds. Industrial & Engineering Chemistry Research. 34, 158–163.

Jones, C.W., Koros, W.J., 1995b. Carbon composite membranes: A solution to adverse humidity effects. Industrial & Engineering Chemistry Research. 34, 164–167.

Karvan, O., Johnson, J.R., Williams, P.J., Koros, W.J., 2013. A pilot-scale system for carbon molecular sieve hollow fiber membrane manufacturing. Chemical Engineering Technology. 36, 53–61.

Kim, Y.K., Lee, J.M., Park, H.B., Lee, Y.M., 2004. The gas separation properties of carbon molecular sieve membranes derived from polyimides having carboxylic acid groups. Journal of Membrane Science. 235, 139–146.

Kim, Y.K., Park, H.B., Lee, Y.M., 2005a. Preparation and characterization of carbon molecular sieve membranes derived from BTDA-ODA polyimide and their separation properties. Journal of Membrane Science. 255, 265–273.

Kim, Y.K., Park, H.B., Lee, Y.M., 2005b. Gas separation properties of carbon molecular sieve membranes derived from polyimide/polyvinylpyrroldone blends: Effect of the molecular weight of polyvinylpyrrolidone. Journal of Membrane Science. 251, 159–167.

Kita, H., 2006. Gas and vapor separation membranes based on carbon membranes. *Material Science of Membranes for Gas and Vapor Separation*. Yampolskii, Y., Pinnau, I., Freeman, B.D. (Eds.), 337–354, John Wiley & Sons, Inc., Chichester.

Kita, H., Maeda, H., Tanaka, K., Okamoto, K., 1997. Carbon molecular sieve membrane prepared from phenolic resin. Chemistry Letters. 26, 179–180.

Kita, H., Yoshino, M., Tanaka, K., Okamoto, K., 1997. Gas permselectivity of carbonized polypyrrolone membrane. Chemical Communications. 1051–1052.

Kiyono, M., Williams, P.J., Koros, W.J., 2010. Effect of pyrolysis atmosphere on separation performance of carbon molecular sieve membranes. Journal of Membrane Science. 359, 2–10.

Koresh, J.E., Soffer, A., 1983. Molecular sieve carbon permselective membrane. Part 1. Presentation of a new device for gas mixture separation. Separation Science Technology. 18, 723–734.

Kusakabe, K., Yamamoto, M., Morooka, S., 1998. Gas permeation and micropore structure of carbon molecular sieving membranes modified by oxidation. Journal of Membrane Science. 149, 59–67.

Kusuki, Y., Shimazaki, H., Tanihara, N., Nakanishi, S., Yoshinaga, T., 1997. Gas permeation properties and characterization of asymmetric carbon membranes prepared by pyrolyzing asymmetric polyimide hollow fiber membrane. Journal of Membrane Science. 134, 245–253.

Lagorsse, S., Magalhaes, F.D., Mendez, A., 2004. Carbon molecular sieve membranes sorption, kinetic and structural characterization. Journal of Membrane Science. 241, 275–287.

Lee, H.J., Yoshimune, M., Suda, H., Haraya, K., 2006. Gas permeation properties of poly(2,6-dimethyl-1,4-phenylene oxide) (PPO) derived carbon membranes prepared on a tubular ceramic support. Journal of Membrane Science. 279, 372–379.

Li, H., Song, Z., Zhang, X., Huang, Y., Li, S., Mao, Y., Ploehn, H.J., Bao, Y., Yu, M., 2013. Ultrathin, molecular-sieving graphene oxide membranes for selective hydrogen separation. Science. 342, 95–98.

Lua, A.C., Su, J., 2006. Effects of carbonization on pore evolution and gas permeation properties of carbon membranes from Kapton polyimide. Carbon. 44, 2964–2972.

Nishiyama, N., Dong, Y.-R., Zheng, T., Egashira, Y., Ueyama, K., 2006. Tertiary amine-mediated synthesis of microporous carbon membranes. Journal of Membrane Science. 280, 603–609.

Okamoto, K., Kawamura, S., Yoshino, M., Kita, H., Hirayama, Y., Tanihara, N., 1999. Olefin/paraffin separation through carbonized membranes derived from an asymmetric polyimide hollow fiber membrane. Industrial & Engineering Chemistry Research. 38, 4424–4432.

Park, H.B., Kim. T.K., Lee, J.M., Lee, S.Y., Lee, Y.M., 2004. Relationship between chemical structure of aromatic polyimides and gas permeation properties of their carbon molecular sieve membranes. Journal of Membrane Science. 229, 117–127.

Parsley, D., Ciora Jr., R.J., Flowers, D.L., Laukaitaus, J., Chen, A., Liu, P.K.T., Yu, J., Sahimi, M., Bonsu, A., Tsotsis, T.T., 2014. Field evaluation of carbon molecular sieve membranes for the separation and purification of hydrogen from coal-and biomass-derived syngas. Journal of Membrane Science. 450, 81–92.

Qiu, W., Zhang, K., Li, F.S., Zhang, K., Koros, W.J., 2014. Gas separation performance of carbon molecular sieve membranes based on 6FDA-mPDA/DABA (3:2) polyimide. ChemSusChem. 7, 1186–1194.

Rao, M.B., Sircar, S., 1993. Nanoporous carbon membranes for separation of gas mixtures by selective surface flow. Journal of Membrane Science. 85, 253–264.

Rao, M.B., Sircar, S., 1996. Performance and pore characterization of nanoporous carbon membrane for gas separation. Journal of Membrane Science. 110, 109–118.

Robeson, L.M., 1991. Correlation of separation factor versus permeability for polymeric membranes. Journal of Membrane Science. 62, 165–185.

Salleh, W.N.W., Ismail, A.F., 2012. Fabrication and characterization of PEI/PVP-based carbon hollow fiber membranes for CO_2/CH_4 and CO_2/N_2 separation. AIChE Journal. 58, 3167–3175.

Salleh, W.N.W., Ismail, A.F., Matsuura, T., Abdullah, M.S., 2011. Precursor selection and process conditions in the preparation of carbon membrane for gas separation: A review. Separation and Purification Reviews. 40, 261–311.

Saufi, S.M., Ismail, A.F., 2004. Fabrication of carbon membranes for gas separation—A review. Carbon. 42, 241–259.

Sedigh, M.G., Onstot, W.J., Xu, L., Peng, W.L., Tsotsis, T.T., Sahimi, M., 1998. Experiments and simulation of transport and separation of gas mixtures in carbon molecular sieves membranes. Journal of Physical Chemistry A. 102, 8580–8589.

Sedigh, M.G., Xu, L., Tsotsis, T.T., Sahimi, M., 1999. Transport and morphological characteristics of polyetherimide-based carbon molecular sieve membranes. Industrial & Engineering Chemistry Research. 38, 3367–3380.

Shiflett, M.B., Foley, H.C., 1999. Ultrasonic deposition of high-selectivity nanoporous carbon membranes. Science. 285, 1902–1905.

Shiflett, M.B., Foley, H.C., 2000. On the preparation of supported nanoporous carbon membranes. Journal of Membrane Science. 179, 275–282.

Shiflett, M.B., Foley, H.C., 2001. Reproducible production of nanoporous carbon membranes. Carbon. 39, 1421–1446.

Shao, L., Chung, T.S., Pramoda, K.P., 2005. The evolution of physiochemical and transport properties of 6FDA-durene toward carbon membranes; from polymer, intermediate to carbon. Microporous and Mesoporous Materials. 84, 59–68.

Shao, L., Chung, T.S., Wensley, G., Goh, S.H., Pramoda, K.P., 2004. Casting solvent effects on morphologies, gas transport properties of a novel 6FDA/PMDA-TMMDA copolyimide membrane and its derived carbon membranes. Journal of Membrane Science. 244, 77–87.

Shusen, W., Meiyun, Z., Zhizhong, W., 1996. Asymmetric molecular sieve carbon membranes. Journal of Membrane Science. 109, 267–270.

Sircar, S., Rao, M.B., Thaeron, C.M.A., 1999a. Selective surface flow membrane for gas separation. Separation Science and Technology. 34(10), 2081–2093.

Sircar, S., Waldron, W.E., Rao, M.B., Anand, M., 1999b. Hydrogen production by hybrid SMR-PSA-SSF membrane system. Separation and Purification Technology. 17, 11–20.

Steel, K.M., Koros, W.J., 2003. Investigation of porosity of carbon materials and related effects on gas separation properties. Carbon. 41, 253–266.

Su, J., Lua, A.C., 2007. Effects of carbonization atmosphere on the structural characteristics and transport properties of carbon membranes prepared from Kapton polyimide. Journal of Membrane Science. 305, 263–270.

Suda, H., Haraya, K., 1995. Molecular sieving effect of carbonized kapton polyimide membrane. Journal of the Chemical Society, Chemical Communications. 1179–1180.

Suda, H, Haraya, K., 1997a. Gas permeation through micropores of carbon molecular sieve membranes derived from kapton polyimide. Journal of Physical Chemistry B. 101, 3988–3994.

Suda, H., Haraya, K., 1997b. Alkene/alkene permselectivities of a carbon molecular sieve membrane. Chemical Communications. 93–94.

Tanihara, N., Shimazaki, H., Hirayama, Y., Nakanishi, S., Yoshinaga, T., Kusuki, Y., 1999. Gas permeation properties of asymmetric carbon hollow fiber membranes prepared from asymmetric hollow fiber. Journal of Membrane Science. 160, 179–186.

Teixeira, M., Campo, M.C., Pacheco Tanaka, D.A., Llosa Tanco, M.A., Magen, C., Mendes, A., 2011. Composite phenolic resin-based carbon molecular sieve membranes for gas separation. Carbon. 49, 4348–4358.

Tin, P.S., Chung, T.S., Hill, A.J., 2004a. Advanced fabrication of carbon molecular sieve membranes by nonsolvent pretreatment of precursor polymers. Industrial & Engineering Chemistry Research. 43, 6476–6483.

Tin, P.S., Chung, T.S., Kawi, S., Guiver, M.D., 2004b. Novel approaches to fabricate carbon molecular sieve membranes based on chemical modified and solvent treated polyimides. Microporous and Mesoporous Materials. 73, 151–160.

Tin, P.S., Chung, T.S., Liu, Y., Wang, R., 2004c. Separation of CO_2/CH_4 through carbon molecular sieve membranes derived from P84 polyimide. Carbon. 42, 3123–3131.

Tseng, H.-H., Shih, K., Shiu, P.-T., Wey, M.-Y., 2012. Influence of support structure on the permeation behavior of polyetherimide-derived carbon molecular sieve composite membrane. Journal of Membrane Science. 405–406, 250–260.

Vu, D.Q., Koros, W.J., Miller, S.J., 2002. High pressure CO_2/CH_4 separation using carbon molecular sieve hollow fiber membranes. Industrial & Engineering Chemistry Research. 41, 367–380.

Wang, H., Zhang, L., Gavalas, G.R., 2000. Preparation of supported carbon membranes from furfuryl alcohol by vapor deposition polymerization. Journal of Membrane Science. 177, 25–31.

Williams, P.J., Koros, W.J., 2008. Gas separation by carbon membranes. *Advanced Membrane Technology and Applications*. Li, N.N., Fane, A.G., Ho, W.S.W., Matsuura, T. (Eds.), 599–632, John Wiley & Sons, Inc., New Jersey.

Xiao, Y., Chung, T.-S., Chung, M.L., Tamai, S., Yamaguchi, A., 2005a. Structure and properties relationships for aromatic polyimides and their derived carbon membranes: Experimental and simulation approaches. Journal of Physical Chemistry B. 109, 18741–18748.

Xiao, Y., Chung, T.-S., Guana, H.M., Guiver, M.D., 2007. Synthesis, cross-linking and carbonization of co-polyimides containing internal acetylene units for gas separation. Journal of Membrane Science. 302, 254–264.

Xiao, Y., Dai, Y., Chung, T.S., Guiver, M.D., 2005b. Effects of brominating matrimid polyimide on the physical and gas transport properties of derived carbon membranes. Macromolecules. 38, 10042–10049.

Xu, L., Rungta, M., Hessler, J., Qiu, W., Brayden, M., Martinez, M., Barbay, G., Koros, W.J., 2014. Physical aging in carbon molecular sieve membranes. Carbon. 80, 155–166.

Xu, L., Rungta, M., Koros, W.J., 2011. Matrimid® derived carbon molecular sieve hollow fiber membranes for ethylene/ethane separation. Journal of Membrane Science. 380, 138–147.

Yoshimune, M., Fujiwara, I., Haraya, K., 2007. Carbon molecular sieve membranes derived from trimethylsilyl substituted poly(phenylene oxide) for gas separation. Carbon. 45, 553–560.

Yoshimune, M., Fujiwara, I., Suda, H., Haraya, K., 2005. Novel carbon molecular sieve membranes derived from poly(phenylene oxide) and its derivatives for gas separation. Chemistry Letters. 34, 958–959.

Yoshimune, M., Haraya, K., 2010. Flexible carbon hollow fiber membranes derived from sulfonated poly(phenylene oxide). Separation and Purification Technology. 75, 193–197.

Yoshino, M., Nakamura, S., Kita, H., Okamoto, K., Tanihara, N., Kusuki, Y., 2003. Olefin/paraffin separation performance of carbonized membranes derived from an asymmetric hollow fiber membrane of 6FDA/BPDA-DDBT copolyimide. Journal of Membrane Science. 215, 169–183.

Zhang, X., Hu, H., Zhu, Y., Zhu, S., 2006a. Effect of carbon molecular sieve on phenol formaldehyde novolac resin based carbon membranes. Separation and Purification Technology. 52, 261–265.

Zhang, B., Wang, T., Liu, S., Zhang, S., Qiu, J., 2006b. Structure and morphology of microporous carbon membrane materials derived from poly(phthalazinone ether sulfone ketone). Microporous and Mesoporous Materials. 96, 79–83.

Zhang, B., Wang, T., Zhang, S., Qiu, J., Jian, X., 2006c. Preparation and characterization of carbon membranes made from poly(phthalazinone ether sulfone ketone). Carbon. 44, 2764–2769.

Zhang, B., Wu, Y., Wang, T., Qiu, J., Zhang, S., 2011. Microporous carbon membranes from sulfonated poly(phthalazinone ether sulfone ketone): Preparation, characterization, and gas permeation. Journal of Applied Polymer Science. 122, 1190–1197.

Zhou, W., Yoshino, M., Kita, H., Okamoto, K., 2001. Carbon molecular sieve membranes derived from phenolic resin with a pendant sulfonic acid group. Industrial & Engineering Chemistry Research. 40, 4801–4807.

Zhou, W., Yoshino, M., Kita, H., Okamoto, K., 2003. Preparation and gas permeation properties of carbon molecular sieve membranes based on sulfonated phenolic resin. Journal of Membrane Science. 217, 55–67.

7 Zeolite Membrane for Gas Separation

Motomu Sakai, Kei Yoshihara,
Masahiro Seshimo, and Masahiko Matsukata

CONTENTS

7.1 INTRODUCTION

Gas separations and water treatment are two of the most attractive targets for membrane separation technologies. Many industrial fields need gas separation, and many reports have appeared about gas separation using zeolite membranes such as CO_2 recovery, H_2 purification, natural gas upgrading, and air separation. Gas mixture has currently been separated by using cryogenic distillation, absorption, and/or adsorption. However, the cooling step in cryogenic distillation and the regeneration steps in absorption and adsorption require large energy consumption. To lower the energy consumption for gas separation, many kinds of membrane materials, such as polymeric, molecular sieving carbon, amorphous silica, organosilica, zeolite, metal organic framework, and mixed matrix membranes, have been developed over the past decades. Inorganic membranes including zeolite membranes generally have strong advantages for operations at high temperature and pressure because of their good thermal and mechanical stability.

Zeolites are composed of tetrahedral SiO_4 units. Part of Si^{4+} can be replaced by Al^{3+}, and the skeleton shows anionic property. Aluminosilicate zeolites have an anionic framework and thus use cation exchange property to compensate electronic balance. Aluminophosphate ($AlPO_4$-n) and silicoalminophosphate (SAPO-n) zeolites are also used for gas separation. $AlPO_4$-n is made by alternately ordered tetrahedral AlO_4 and PO_4 units. $AlPO_4$-n framework does not have cation exchange property unlike aluminosilicate because the frameworks hold electric neutrality based on AlO_4^- and PO_4^+ units. When P and Al in $AlPO_4$-n are replaced by Si, the resulting zeolite is called SAPO-n. Counter cations are occluded in their framework of SAPO-n like aluminosilicate zeolites as well. In this chapter, we collectively refer to aluminosilicate, aluminophosphate, and silicoaluminophosphate as zeolite. In addition, zeolites have ordered micropore channels in their crystalline structure and exhibit unique adsorption and molecular sieving properties based on their micropore systems. Such distinct characters contribute to their unique permselectivities of zeolite membranes. Figure 7.1 shows the types of zeolites widely used for gas separation.

Permeability, selectivity, and life are three important factors for permselective membranes, and thus most of membrane developments focus on improving them. Zeolite membranes have structures in which zeolite crystals are accumulated unlike other amorphous inorganic membranes such as silica and carbon membranes. Hence, it is easily assumed that there are two kinds of pathways across a membrane, intracrystalline and intercrystalline pathways (Nomura et al., 2001). Figure 7.2 shows the schematic diagram of intra- and intercrystalline pathways, the pathways through zeolite micropores in crystals, and defects among crystals, respectively. Permeation through a few intercrystalline pathways easily spoils separation performance. Thus, reducing the intercrystalline pathways would help zeolite membranes to improve permselectivity. Additionally, we should focus on improving permeability through the intracrystalline pathways to increase permeation properties.

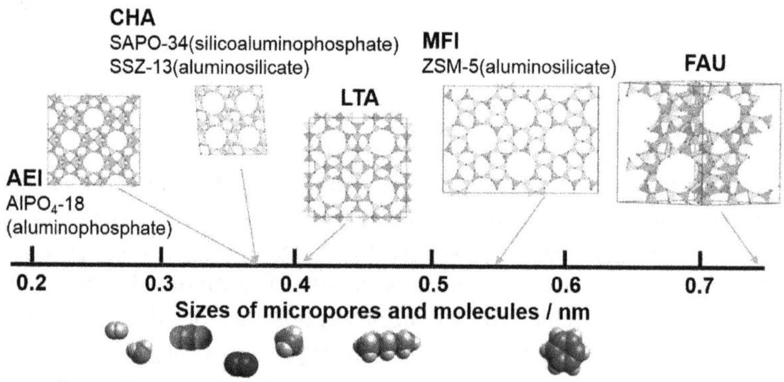

FIGURE 7.1 Typical zeolite types used for separation membrane (Original).

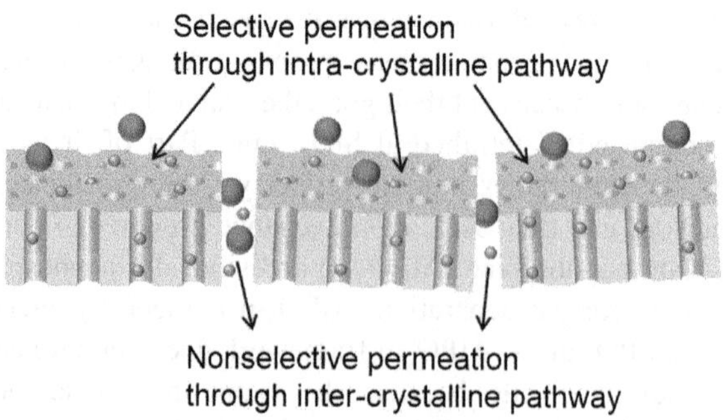

FIGURE 7.2 Schematic diagram of intra- and intercrystalline pathways (Original).

The principle of gas separation using zeolite membranes is mainly based on the molecular sieving effect. Zeolite membranes are able to separate molecules smaller than the micropore of the zeolite and larger molecules that cannot enter the micropore. In this case, the selectivity of zeolite membrane without defects would ideally reach infinity. Zeolite membranes can also separate molecules by the difference of diffusivities in the micropore on the basis of their molecular size difference even though both types can enter the micropore. Small pore zeolites that have 8-membered ring openings are often used for gas separations because these pore sizes (~0.36–0.42 nm) are suitable for molecular sieving in such applications.

In addition, differences in the affinities between gas molecules and the zeolite membrane can be utilized for separation. Molecules that have a strong affinity for zeolite preferentially penetrate through the membrane. In this case, zeolite membranes can exhibit selectivity for larger molecules, which is different from separation by molecular sieving effects (Sawamura et al., 2009; Sandström et al., 2010).

CO_2 and H_2 separation is a major target of gas separation by using zeolite membranes. In addition, air, noble gas, and hydrocarbon separations have been attempted. In this chapter, we introduce the state of the art of zeolite membrane research for gas separation.

7.2 CARBON DIOXIDE SEPARATION

CO_2 recovery from gas mixtures with N_2, O_2, H_2, H_2O, and CH_4, etc., is one of the hottest topics for membrane separation in the past decade. CO_2 recovery from N_2 and gaseous light hydrocarbons for the treatment of flue gas from heat power plants and steel manufacturing has been extensively investigated due to increasing interest in global warming in recent years, in addition to the applications such as CO_2/H_2 mixture in water gas shift reaction and CO_2/CH_4 mixture in natural gas upgrading. In other words, the demands of CO_2 separation are rapidly expanded from only for the purification of products to CO_2 capture.

Carbon capture and storage (CCS) is a technology designed for CO_2 emission reduction. Cost reduction in CO_2 recovery step is urgently needed to proceed social implementation of CCS because it is estimated that the cost for CO_2 recovery occupies more than 50% of the total cost of CCS, above \$60 per t-CO_2. CO_2 recovery from gas mixture is carried out by physical absorption, chemical absorption, adsorption, and membrane separation. The costs for CO_2 recovery is expected to decrease to \$25 and \$15 per t-CO_2 by the improvement of absorption and membrane separation processes, respectively. Regeneration steps in absorption and adsorption require large energy consumption. Hence, membrane separation draws attention as an ace in the hole.

Table 7.1 lists typical applications in which CO_2 separation using membrane separation is expected to be introduced. CO_2 separation is required in natural gas purification, treatment of flue gas from power plant, and syngas purification before and after water gas shift reaction. The CO_2 concentration in natural gas depends on the gas field, and at most reaches 70%. Although the CO_2 separation using polymeric membranes from methane has been commercialized in natural-gas purification processes, plasticization of polymeric membranes occurs in high CO_2 concentration atmosphere exceeding 10%. For this reason, development of chemically stable

TABLE 7.1
Typical Applications of CO_2 Separation (Original)

Application	Gas Mixture	Temperature (K)	Pressure (MPa)	CO_2 Conc. in Feed (%)
Natural gas purification	$CO_2/CH_4/(H_2O)$	<373	<7	~70
Flue gas from power plant	CO_2/N_2	<473	<1	12–15
Syngas from natural gas	$H_2/CO/CO_2$	200–400	2–4	~3
Syngas from coal				~10
WGS syngas from natural gas	H_2/CO_2	323–423	2–4	~30
WGS syngas from coal				~40

inorganic membranes contributes to expand the applicability of membrane separation and to reduce the consumption energy for CO_2 removal at CO_2 rich gas fields. In addition, the separation performance of membrane is a very important factor to determine the yields of products in the cases of purification like CO_2/CH_4 separation in natural gas upgrading. Improvement of selectivity by developing inorganic membrane goes directly to improvement of product yield even under low CO_2 concentration conditions in which polymeric membranes are able to be used currently. On the other hand, an important property for CO_2 capture such as CO_2/N_2 separation in the treatment of flue gas is not so much selectivity as permeability. Necessary CO_2 purity to reduce CO_2 emission is limited to ~95%. Extremely high separation performance is not required for membranes. Since a huge amount of gas should be treated in both natural gas purification and CO_2 capture processes, improvement of permeability is essential.

SAPO-34, a silicoaluminophosphate CHA-type zeolite membrane for gas separation, was first reported in 1997 by Enze and his coworkers (Lixiong et al., 1997). They reported the single gas permeances of H_2, CO_2, N_2, and n-C_4H_{10} at 323 K. The permeances decreased with increasing molecular size and finally n-C_4H_{10} permeance was difficult to observe. Noble and co-workers have been developing SAPO-34 membrane for gas separation for the last dozen years (Poshusta et al., 2000; Carreon et al., 2007, 2008). They prepared an SAPO-34 membrane on an alumina tubular support (Poshusta et al., 1998). The membrane exhibited molecular sieving properties. The permeances through the membrane were in the order of $CO_2 > N_2 > CH_4 > n$-C_4H_{10}, and the membrane showed CO_2 selectivity of 30 from CO_2/CH_4 equimolar mixture at 300 K. They investigated permeation and separation performances of SAPO-34 membrane under highly pressurized conditions up to 3.1 MPa (Li et al., 2005a). The CO_2 permeance through the SAPO-34 membrane was 2.4×10^{-7} mol m^{-2} s^{-1} Pa^{-1} with the separation factor of 95 at 295 K with a low pressure drop of 0.14 MPa. Even under the high pressure drop condition of 3 MPa, the membrane kept the high CO_2 permeance of 1.0×10^{-7} mol m^{-2} s^{-1} Pa^{-1} and the separation factor of 60. In addition, the effect of cation species in the SAPO-34 membrane on permselectivity for CO_2/CH_4 was also studied (Hong et al., 2007). Ion exchange from H^+ to Li^+, Na^+, K^+, and NH_4^+ in the SAPO-34 membrane increased the separation factor up to 60% for CO_2/CH_4 mixture. The permeance, in particular CH_4, was lowered by the ion exchange, apparently due to steric hindrance occluded in the micropores of SAPO-34.

SSZ-13, an aluminosilicate CHA-type zeolite, is also a promising material for CO_2 separation (Kosinov et al., 2015; Zheng et al., 2015). Falconer and co-workers prepared a SSZ-13 membrane inside of a porous stainless tube and investigated its permeation properties (Kalipcilar et al., 2002). The SSZ-13 membrane showed the CO_2/CH_4 and H_2/CH_4 ideal selectivities of 11 and 9.0, respectively. Hensen and co-workers prepared a SSZ-13 membrane on an α-alumina hollow fiber support (Kosinov et al., 2014a). The membrane exhibited the CO_2 separation factors of 42 and 12 for CO_2/CH_4 and CO_2/N_2 mixture, respectively, with the CO_2 permeance of 3.0×10^{-7} mol m^{-2} s^{-1} Pa^{-1}.

Figure 7.3 shows relative permeances of small gas molecules through an $AlPO_4$-18 membrane, small pore aluminophosphate, at 313 K (Yoshihara et al., 2017). The permeances are divided by that of CO_2 for normalization. The permeances decrease in order of molecular size increases, other than CO_2. The order of permeances is $CO_2 > H_2 > N_2 > CH_4 \approx C_2 > C_3 \approx n\text{-}C_4 \approx i\text{-}C_4$. Various small pore zeolites in addition to $AlPO_4$-18 exhibit this order of permeances (Kalipcilar et al., 2002; Kosinov et al., 2014b; Zheng et al., 2015). This result suggests the permeances through small pore zeolite are mainly dominated by molecular size, in other words diffusivities in micropore. The CO_2 permeance is generally larger than that of H_2. As in the case of CO_2, when a type of molecule has a strong affinity with the membrane material, its permeance would overwhelm that assumed from a molecular size. By contrast, the CO_2 permeance can be smaller than that of H_2 under conditions in which the interaction between CO_2 and the membrane is weak.

MFI-type zeolite having ~0.55 nm micropores is also studied for CO_2 selective membrane material (Lindmark and Hedlund, 2010a, 2010b; Sandström et al., 2011; Sjoberg et al., 2015). Hedlund's group reported the preparation of uniformly oriented MFI membrane and its permeation property (Zhou et al., 2014). This sort of MFI membrane exhibits high CO_2/H_2 separation performance at relatively low temperatures around 273 K. The separation factor reached 109 at 238 K with a high CO_2

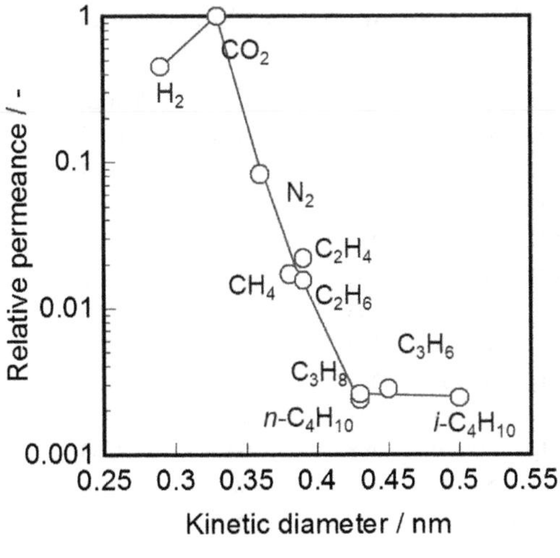

FIGURE 7.3 Relative permeances of small gas molecules through AEI-type zeolite membrane (Original).

permeance of 5.1×10^{-6} mol m^{-2} s^{-1} Pa^{-1}. Competitive adsorption contributed to such CO_2 selective permeation in CO_2/H_2 mixtures.

Some post-treatment and modification methods were developed to improve CO_2 permselectivity of zeolite membranes (Lindmark and Hedlund, 2010b; Venna and Carreon, 2011; Karimi et al., 2015). Kita and co-workers prepared a SSZ-13 membrane modified by ionic liquid which had strong interaction with CO_2 (Bo et al., 2017). The separation factor to CO_2 for a CO_2/CH_4 mixture through SSZ-13 membrane increased from 13 to 81 by such modification. It might be due to the strongly adsorbed CO_2 inhibited CH_4 permeation.

The effects of impurities such as water vapor and light hydrocarbons on CO_2 permselectivity are also investigated. In many reports, CO_2 permeance was lowered by water because water strongly adsorbs on micropore of zeolite and inhibits CO_2 permeation. Falconer and co-workers further studied the effect of impurities in a feed stream on CO_2 permeance through SAPO-34 membrane (Li et al., 2005b). The permselectivity for CO_2/CH_4 mixtures with some impurities such as H_2O, C_2, C_3, and C_4 were evaluated. The CO_2 permeance decreased by 12% after 12 days of exposure to 170 ppm of water vapor. Adding 1% of hydrocarbons depressed both permeance and selectivity, indicating that such impurities adsorbed on the membrane and inhibited CO_2 permeation. Hensen and co-workers also studied the effect of humidity on CO_2/CH_4 permselectivity through SSZ-13 membrane (Kosinov et al., 2014b). The permeances of both CO_2 and CH_4 decreased in the presence of 2.2 kPa of water vapor. At higher temperatures, the permeances increased close to the values under dry conditions because the water coverage on a membrane decreased.

Tables 7.2 through 7.4 summarize the typical results of CO_2 separation with various zeolite membranes. Zeolites having an 8-membered ring opening tended to show higher

TABLE 7.2
Typical Results of CO_2/CH_4 Separation by Using Zeolite Membranes (Original)

Membrane	Partial Pressure (kPa)	Temp. (K)	CO_2 Selectivity (–)	CO_2 Permeance (10^{-7} mol m^{-2} s^{-1} Pa^{-1})	Ref.
SAPO-34	CO_2/CH_4 = 112/112	295	290	4.1	Li et al., 2010
SAPO-34	CO_2/CH_4 = 100/100	298	160	12	Chen et al., 2017
Ba-SAPO-34	CO_2/CH_4 = 50/50	303	103	3.8	Chew et al., 2011
Li-SAPO-34	CO_2/CH_4 = 111/111	295	87	0.80	Hong et al., 2007
CrAPSO-34	CO_2/CH_4 = 111/111	302	156	7.9	Bing et al., 2016
SSZ-13	CO_2/CH_4 = 100/100	303	300	2.0	Zheng et al., 2015
$AlPO_4$-18	CO_2/CH_4 = 152/152	298	240	5.9	Wang et al., 2015
DDR	CO_2/CH_4 = 100/100	298	500	0.35	Wang et al., 2017
T-type	CO_2/CH_4 = 50/50	308	400	0.46	Cui et al., 2003
Silicalite-1	CO_2/CH_4 = 50/50	200	15	0.75	van den Broeke et al., 1999
Na-Y	CO_2/CH_4 = 50/50	303	20	1.0	Kusakabe et al., 1997

TABLE 7.3
Typical Results of CO_2/H_2 Separation by Using Zeolite Membranes (Original)

Membrane	Partial Pressure (kPa)	Temp. (K)	CO_2 or H_2 Selectivity (–)	CO_2 or H_2 Permeance (10^{-7} mol m^{-2} s^{-1} Pa^{-1})	Ref.
LTA	CO_2/H_2 = 50/50	373	12.5 (H_2)	1.4 (H_2)	Huang et al., 2012
SAPO-34	CO_2/H_2 = 688/912	253	140 (CO_2)	0.26 (CO_2)	Hong et al., 2008
$AlPO_4$-5/ $AlPO_4$-34	CO_2/H_2 = 50/50	308	9.7 (H_2)	2.0 (H_2)	Guan et al., 2003
MFI	CO_2/H_2 = 450/450	238	109 (CO_2)	51 (CO_2)	Zhou et al., 2013
H-ZSM-5	CO_2/H_2 = 450/450	235	210 (CO_2)	62 (CO_2)	Korelskiy et al., 2015
B-ZSM-5	CO_2/H_2 = 111/111	773	60 (H_2)	1.3 (H_2)	Hong et al., 2005

TABLE 7.4
Typical Results of CO_2/N_2 Separation by Using Zeolite Membranes (Original)

Membrane	Partial Pressure (kPa)	Temp. (K)	CO_2 Selectivity (–)	CO_2 Permeance (10^{-7} mol m^{-2} s^{-1} Pa^{-1})	Ref.
SAPO-34	CO_2/N_2 = 120/120	295	32	12	Li and Fan, 2010
T	CO_2/N_2 = 50/50	308	104	0.38	Cui et al., 2003
Silicalite-1	CO_2/N_2 = 50/50	293	69	7.1	Guo et al., 2006
ZSM-5	CO_2/N_2 = 50/50	298	54.3	0.36	Shin et al., 2005
ETS-10	CO_2/N_2 = 55/55	298	10	0.28	Tiscornia et al., 2010
K-Y	CO_2/N_2 = 50/50	308	67	13	Kusakabe et al., 1999
Na-Y	CO_2/N_2 = 50/50	303	100	0.5	Kusakabe et al., 1997
Na-X	CO_2/N_2 = 50/50	296	8.4	0.5	Weh. et al., 2002

permeability and selectivity compared with those having larger pores. Small pore zeolites are suitable materials for CO_2 selective membrane. In particular, CHA-type zeolites, SAPO-34, and SSZ-13 of small pore zeolites exhibited superior permselectivity.

Figure 7.4 shows Robeson plots of CO_2/CH_4 separation. Permselectivities of polymeric and zeolite membranes are compared in Figure 7.4(a). As observed, zeolite membranes clearly overwhelm the upper bound of polymeric membranes. The results of zeolite membranes were classified by age in Figure 7.4(b). One can see the change of research trends from this figure. The main objectives of zeolite membrane development shifted from improvement of permselectivity to durability. In the early stage of development, permselectivity improved year by year. In contrast, the upper bound of membrane performance hardly changed in recent years although the number of reports does not decrease. Recent reports have focused on membrane durability. The permselectivities of zeolite membranes for CO_2/CH_4 separation has become

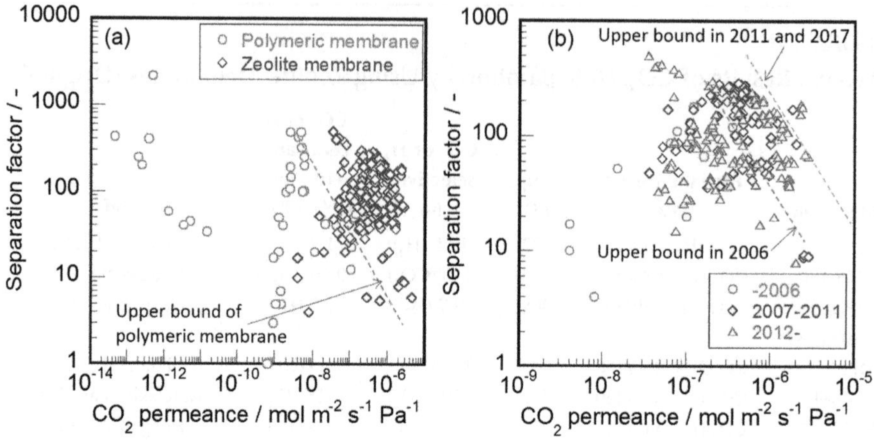

FIGURE 7.4 Robeson plot for CO_2/CH_4 separation. (a) Polymeric and zeolite membranes, (b) zeolite membranes classified by age (Original).

nearly sufficient in laboratory scale. In fact, Mitsubishi Chemical Corp. has started the demonstration test for natural gas upgrading by CHA-type zeolite membrane at Kurosaki, a northern part of Fukuoka prefecture in Japan, since 2016 (Nikkei Asian Review, 2016).

7.3 NITROGEN SEPARATION FOR NATURAL GAS UPGRADING

N_2 separation from CH_4 for natural gas upgrading comes under the same spotlight as CO_2/CH_4 separation. Most of the research about N_2/CH_4 separation are carried out in tandem with CO_2/CH_4 separation. In general, the permeance and selectivity of N_2 from N_2/CH_4 mixture are less than those of CO_2 from CO_2/CH_4 mixture. N_2 does not have as strong an interaction between membrane materials as CO_2, and its diffusivity in micropores is smaller than that of CO_2 due to its molecular size. Thus the N_2 permselectivity tends to be low. Noble's group reported N_2/CH_4 separation results through a SSZ-13 membrane (Wu et al., 2015). The membrane that had a

TABLE 7.5
Typical Results of N_2/CH_4 Separation by Using Zeolite Membranes (Original)

Membrane	Partial Pressure (kPa)	Temp. (K)	N_2 Selectivity (–)	N_2 Permeance (10^{-7} mol m^{-2} s^{-1} Pa^{-1})	Ref.
SAPO-34	N_2/CH_4 = 111/111	298	8.6	7.2	Zong et al., 2017a
$AlPO_4$-18	N_2/CH_4 = 111/111	298	4.6	10	Zong et al., 2017b
SSZ-13	N_2/CH_4 = 135/135	293	13	0.18	Wu et al., 2015
ETS-4	N_2/CH_4 = 50/50	308	5.1	0.50	Guan et al., 2001b

CO_2 separation factor of 280 for CO_2/CH_4 mixture at 293 K exhibited only 13 of the separation factors to N_2 for N_2/CH_4 mixture at the same temperature. Table 7.5 gives the typical results of N_2/CH_4 separation by using various zeolite membranes.

7.4 HYDROGEN SEPARATION

Membrane separation for H_2 purification has also been focused on in recent years. H_2 is an important feedstock in petroleum and petrochemical industries and a novel energy carrier. Research about H_2 membrane separation is mainly classified into H_2 generation process, H_2 carrier system, and dehydrogenation reaction. H_2 is often generated by steam reforming and water gas shift reaction (WGSR), as shown in following equations, and thus H_2 is recovered from mixtures of CO, CO_2, H_2O, and hydrocarbons:

Steam reforming,

$$CnHm + nH_2O \rightleftarrows nCO + (m/2 + n)H_2 \tag{7.1}$$

and water gas shift reaction,

$$CO + H_2O \rightleftarrows CO_2 + H_2 \tag{7.2}$$

There have been reports on CO_2 selective membranes for CO_2/H_2 mixture separation by using affinity, as described in the above section. In contrast, some reports about H_2 selective zeolite membrane for CO_2/H_2 mixture separation on the basis of molecular sieving effect. Silylated ZSM-5 membranes prepared by catalytic cracking of methyldietoxysilane (MDES) were proposed for H_2 separation from H_2/CO_2 and H_2/CH_4 mixtures (Hong et al., 2005). They prepared boron-substituted ZSM-5 membrane on which MDES reacted in the micropores of B-ZSM-5 and reduced its effective pore diameter. The silylation at 623 K for 10 h increased the H_2 separation factor from 1.4 and 1.6 to 37 and 33 for H_2/CO_2 and H_2/CH_4 mixtures, respectively. MFI membranes silylated by the catalytic cracking of MDES were also reported by Xu and his coworkers (Hong et al., 2013). In this case, the separation factor of H_2/CO_2 at 773 K increased from 3.4 to 45.6 by this treatment. Huang and Caro (2011) reported the preparation and permeation property of an $AlPO_4$ membrane that had LTA topology. Their membrane exhibited separation performance by a molecular sieving effect. The separation factors for H_2/CO_2, H_2/N_2, H_2/CH_4, and H_2/C_3H_8 mixtures were 10.9, 8.6, 8.3, and 142, respectively, with the H_2 permeance ~1.9×10^{-7} mol m^{-2} s^{-1} Pa^{-1}.

Organic hydride is one of the materials proposed as a H_2 carrier in energy use. Toluene/cyclohexane system has often been chosen for H_2 transfer due to the ease of operation because both saturated and unsaturated hydrocarbons keep liquid phase. Then, H_2/toluene separation by various inorganic membranes has been widely studied. Kita and co-workers prepared an MFI membrane on the outer surface of tubular mullite support (Kumakiri et al., 2016). Their membrane showed the H_2 permeance of 1.6×10^{-7} mol m^{-2} s^{-1} Pa^{-1} with the separation factor (H_2/toluene) of 4.1 from the mixture of H_2/toluene = 98/2. As a result, the hydrogen purity in the permeate reached 99.5%.

Extractor-type membrane reactors for steam reforming and WGSR with H_2 selective membranes have also been studied in both experimental and simulation environments (Yu et al., 2007; Bernardo et al., 2010). The levels of conversion of both reactions are limited by thermodynamic equilibrium. In accordance with Le Chatelier's principle, equilibrium values shift by removing H_2 from a reaction system, allowing conversion, and yields in the membrane reactors exceed those in the conventional reactors. Membrane reactors for WGSR using a H_2 selective MFI membrane modified with deposited silica were reported (Tang et al., 2010; Kim et al., 2012). At 823 K, the packed bed membrane reactor achieved 81.7% of the level of CO conversion, which was higher than the equilibrium conversion, 65%. Xu and co-workers studied a H_2-selective zeolite membrane reactor for WGSR as well (Zhang et al., 2012). They demonstrated that their membrane reactor exhibited a high CO conversion of 95.4%, which was higher than the equilibrium conversion of 93% at 573 K. Table 7.6 shows the typical results of H_2/CH_4 separation by various zeolite membranes.

Dehydrogenation of paraffin for olefin production is also strongly limited by thermodynamic equilibrium. Membrane reactor for olefin production has been studied in recent years. In particular, propane dehydrogenation with membrane separation is currently focused on, and one can expect expansion of its use for various reactants in the future, such as ethane, butane, cyclohexane, and ethylbenzene. In any reaction, superior thermal and mechanical stability is required for H_2 selective membranes under severe conditions for dehydrogenation (>773 K, <2 MPa). Nair and co-workers demonstrated the availability of propane dehydrogenation membrane reactor by using H_2 selective SAPO-34 membrane (Kim et al., 2016). Propane conversion reached 70% exceeding equilibrium conversion with a high propylene selectivity of 85% by their membrane reactor operated at 873 K. In addition, they also studied the membrane reactor system for the propylene dehydrogenation by simulation (Choi et al., 2017). They reported that the replacement of a conventional packed bed reactor with a packed bed membrane reactor increases the space-time yield of propylene

TABLE 7.6
Typical Results of H_2/CH_4 Separation by Using Zeolite Membranes (Original)

Membrane	Partial Pressure (kPa)	Temp. (K)	H_2 Selectivity (–)	H_2 Permeance (10^{-7} mol m^{-2} s^{-1} Pa^{-1})	Ref.
LTA	H_2/CH_4 = 50/50	293	4.5	5.9	Li et al., 2016
SAPO-34	H_2/CH_4 = 102/87	293	29	0.21	Hong et al., 2008
Li-SAPO-34	H_2/CH_4 = 111/111	295	16	0.33	Hong et al., 2007
SSZ-13	H_2/CH_4 = 300/300	293	22	2.0	Kosinov et al., 2014a
B-ZSM-5	H_2/CH_4 = 111/111	523	1.6	1.4	Hong et al., 2005
FAU	H_2/CH_4 = 50/50	323	9.9	1.9	Zhou et al., 2015

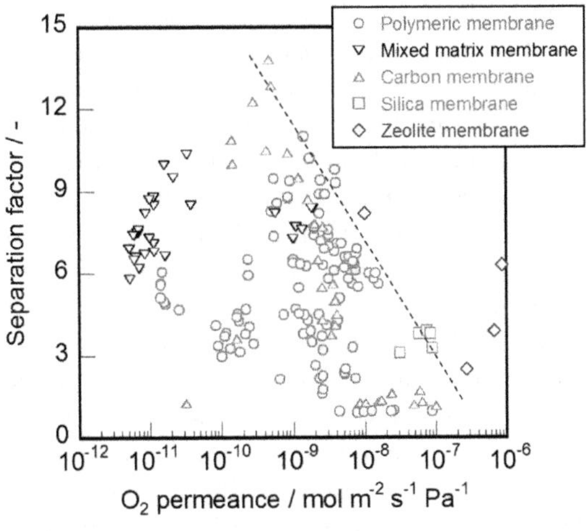

FIGURE 7.5 Robeson plots for air separation (Original).

production up to 45%. Kapteijn and co-workers studied *i*-butane dehydrogenation by using a H_2 permselective DD3R zeolite membrane reactor (van der Bergh et al., 2011) and reported that *i*-butene yield reached 41%, while the equilibrium yield was 28%.

7.5 AIR SEPARATION

Air separation is a previously well-known target for membrane separation. Air separation by polymeric membranes has been studied for many decades. Oxygen rich air, pure O_2, and N_2 are widely used in industrial and medical fields. Zeolite has been studied as adsorbate for air separation, and some sorts of zeolites like Li-X exhibited superior adsorption selectivity for O_2/N_2.

Some zeolites have been studied as membrane materials for air separation (Guan et al., 2001a, 2001b, 2003; Wang et al., 2002; Kusakabe and Sotowa, 2004; Ye et al., 2015). Amoros and co-workers studied LTA/carbon membrane for air separation and reported that their membrane separated air with an O_2 permeance of 2.7×10^{-7} mol m^{-2} s^{-1} Pa^{-1} with the separation factor of 2.7 at 293 K (Dominguez et al., 2008). Air separation at cryogenic temperature was studied with MFI membrane, and the O_2 separation factor of 3.9 with its permeance of 6.7×10^{-7} mol m^{-2} s^{-1} Pa^{-1} was reported at 70 K (Ye et al., 2014).

Figure 7.5 shows the Robeson plots for air separation. In general, although the selectivities of zeolite membranes for air are not very high, the permeances are magnitudes larger than those of polymeric membranes. As a result, zeolite membrane would be relatively suitable for production of oxygen-enriched air.

7.6 NOBLE GAS SEPARATION

Zeolite membranes for noble gas separation have recently been reported. Helium is an important and scarce resource. There is a large and increasing market of

He for industrial and medical use all over the world. The main He resource is natural gas fields that contain small concentrations of He. In addition, a small number of natural gas fields can produce the gas mixture with economically feasible concentrations of He higher than 0.4%. At present, He recovery from natural gas fields has been carried out via a combined process of cryogenic distillation, adsorption, and membrane separation. Although polymeric membranes are used in these processes, a highly permeable membrane is required to increase efficiency increase. Development of separation technology for He from N_2 and CH_4 would contribute to not only price decline but also improving recoverable reserves. Noble and co-workers reported the permselectivity of SAPO-34 membrane for He/CH_4 mixture (Funke et al., 2014). The He permeance was about 4.5×10^{-7} mol m^{-2} s^{-1} Pa^{-1} with the separation factor >20 at 293 K. Their membrane shows He selective permeation by a molecular sieving effect. In addition, the separation factor increased with decreasing He molar fraction in feed gas mixture, and reached 30 at 0.3 of the He molar fraction. Hedlund and co-workers investigated separation performance of MFI membrane for N_2/He mixture under very low temperature conditions (Ye et al., 2016). Although the He permeance in the single gas system was greater than that of N_2, it drastically decreased in binary separation experiments and dropped below the N_2 permeance. Strong adsorption of N_2 inhibited the permeation of He in the binary system. Their membrane shows the N_2 permeance of 3.9×10^{-6} mol m^{-2} s^{-1} Pa^{-1} with the separation factor of 75.7 at 124 K.

Xe/Kr separation with a zeolite membrane has been reported from several groups. Both ^{136}Xe and ^{85}Kr are released as off-gas with other species such as ^{129}I, ^3H_2O, NO, NO_2, and CO_2 from used nuclear fuel recycling process. While ^{136}Xe is a stable isotope, ^{85}Kr has to be captured because of its long decay half-lives. Thus Xe/Kr separation is required to reduce the volume of radiation waste. Carreon and co-workers prepared a SAPO-34 membrane on tubular alumina support and reported that their SAPO-34 membrane exhibited Kr selectivity with the Kr permeance of 1.0×10^{-7} mol m^{-2} s^{-1} Pa^{-1} and a mixture of Kr/Xe = 9/1 (Feng et al., 2016). The separation factor was reported to be 35. In addition, their membrane showed the Kr permeance of 1.2×10^{-7} mol m^{-2} s^{-1} Pa^{-1} with 45 of the separation factor under Kr lean conditions, Kr/Xe = 9/91. Nair and co-workers also reported a Kr selective SAPO-34 membrane showing the Kr permeance of 3.8×10^{-9} mol m^{-2} s^{-1} Pa^{-1} with 30 of separation factor at 255 K for the mixture of Kr/Xe = 1/9 (Kwon et al., 2017). They reported that the membrane showed Kr selective permeation whereas the amount of Xe adsorbed on SAPO-34 was larger than that of Kr, strongly indicating that Kr has a larger diffusivity in the micropore of SAPO-34 in comparison with Xe. This is a good example to show that the difference in diffusivity contributes to preferential permeation.

7.7 HYDROCARBON SEPARATION

Light hydrocarbon (~C_4) gas mixtures are also important targets for separation with inorganic membranes. In recent years, ethylene/ethane and propylene/propane separations with inorganic membranes have extensively been studied to save separation energies consumed in distillation processes. Most inorganic membranes such as carbon, silica, and MOF separate these mixtures on the basis of molecular sieving effect. Ethylene and propylene are slightly smaller than ethane and propane, respectively, and thus ethylene and propylene preferentially penetrate through such molecular sieving membranes. In addition, some sorts of zeolite membranes separate saturated and unsaturated hydrocarbons by using affinity difference. Cations occluded in the micropores of zeolite membrane have strong interaction with unsaturated hydrocarbons, and thus the membranes often exhibit olefin selective permeation (Kita et al., 2001; Matsukata et al., 2015a). In particular, it is known that Ag(I) cation strongly interacts with π-orbital electrons of unsaturated hydrocarbon. Ag(I)-containing zeolite membranes show excellent separation properties for propylene/propane mixtures (Matsukata et al., 2015b). In such separation systems based on affinity difference, the permeation of molecules that have relatively weak interactions with a membrane is often inhibited by preferentially adsorbed molecules. Figure 7.6 gives the permeances of propylene and propane through Ag-FAU membrane in unary and binary systems (Sakai et al., 2017a). These results clearly indicate that the propane permeance through Ag-FAU membrane drastically decreased with the coexistence of propylene. Figure 7.7 presents the typical results of saturated/unsaturated hydrocarbon separations through Ag-FAU membrane (Sakai et al., 2017b). Ag-FAU membrane exhibited ethylene, propylene, and benzene permselectivity from ethylene/ethane, propylene/propane, and benzene/cyclohexane mixtures, respectively, suggesting that such separation using the difference in the affinity between molecules and membrane is suitable for a wide variety of hydrocarbon mixtures.

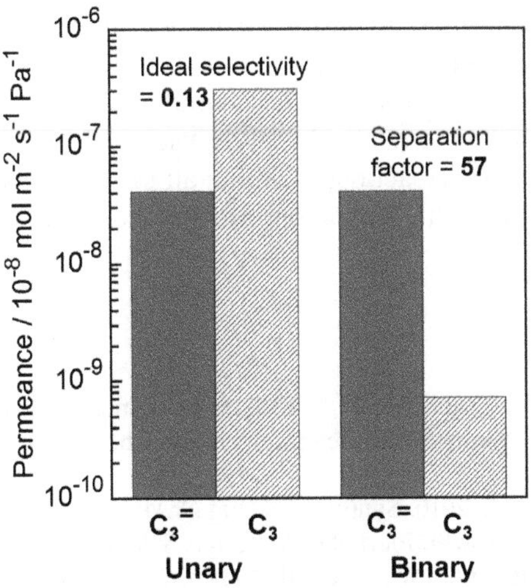

FIGURE 7.6 Permeances of propylene and propane through Ag-FAU membrane in unary and binary systems (Original).

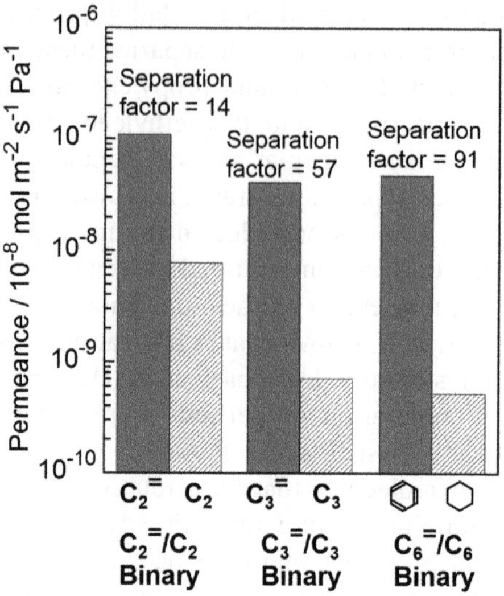

FIGURE 7.7 Typical results of separation measurements for saturated/unsaturated hydrocarbon mixtures by Ag-FAU membrane. Partial pressures of saturated and unsaturated hydrocarbons were fixed at 50 kPa each. Measurements were carried out at 313 (C_2), 353 (C_3), and 473 K (C_6), respectively (Original).

7.8 SUMMARY

Zeolite membranes including SAPO-34, SSZ-13, and MFI exhibit nearly sufficient permselectivity for CO_2 separation in laboratory scale especially for natural gas upgrading. Scale up and durability test are the next important issues for practical applications. For hydrogen separation, membrane development will continue for some time in lab scale. In addition, the study of membrane reactor systems will continue as well. The selectivity of zeolite membranes to hydrogen would be difficult to overwhelm those of carbon and Pd membranes for the future, so that zeolite membranes will be applied to relatively severe uses such as membrane reactors for WGSR and dehydrogenation reactions. Additionally, some novel applications such as noble gas separation and olefin separations are also expected. One can look forward to future development.

REFERENCES

Bernardo, P., Barbieri, G., Drioli, E., 2010. Evaluation of membrane reactor with hydrogen-selective membrane in methane steam reforming. Chemical Engineering Science. 65, 1159–1166.

Bing, L., Liu, X., Zhang, B., 2016. Synthesis of thin CrAPSO-34 membranes by microwave-assisted secondary growth. Journal of Materials Science. 51, 1476–1483.

Bo, L., Rongfei, Z., Na, B., Qing, W., Shenglai, Z., Bin, W., Kita, H., 2017. Room-temperature ionic liquids modified zeolite SSZ-13 membranes for CO_2/CH_4 separation. Journal of Membrane Science. 524, 12–19.

Carreon, M.A., Li, S., Falconer, J.L., Noble, R.D., 2007. SAPO-34 seeds and membranes prepared using multiple structure directing agents. Advanced Materials. 20, 729–732.

Carreon, M.A., Li, S., Falconer, J.L., Noble, R.D., 2008. Alumina-supported SAPO-34 membranes for CO_2/CH_4. Journal of American Chemical Society. 130, 5412–5413.

Chen, Y., Zhang, Y., Zhang, C., Jiang, J., Gu, X., 2017. Fabrication of high-flux SAPO-34 membrane on α-Al_2O_3 four-channel hollow fibers for CO_2 capture from CH_4. Journal of CO_2 Utilization. 18, 30–40.

Chew, T.L., Ahmad, A.L., Bhatia, S., 2011. Ba-SAPO-34 membrane synthesized from microwave heating and its performance for CO_2/CH_4 gas separation. Chemical Engineering Journal. 171, 1053–1059.

Choi, S.W., Sholl, D.S., Nair, S., 2017. Modeling and process simulation of hollow fiber membrane reactor systems for propane dehydrogenation. AIChE Journal. 63, 4519–4531.

Cui, Y., Kita, H., Okamoto, K., 2003. Preparation and gas separation properties of zeolite T membrane. Chemical Communications. 0, 2154–2155.

Dominguez, S.D., Murcia, A.B., Morallon, E., Solano, A.L., Amoros, D.C., 2008. Zeolite LTA/carbon membranes for air separation. Microporous and Mesoporous Materials. 115, 51–60.

Feng, X., Zong, Z., Elsaidi, S.K., Jasinski, J.B., Krishna, R., Thallapally, P.K., Carreon, M.A., 2016. Kr/Xe Separation over a Chabazite Zeolite Membrane. Journal of American Chemical Society. 138, 9791–9794.

Funke, H.H., Chen, M.Z., Prakash, A.N., Falconer, J.L., Noble, R.D., 2014. Separating molecules by size in SAPO-34 membranes. Journal of Membrane Science. 456, 185–191.

Guan, G., Kusakabe, K., Morooka, S., 2001a. Separation of N_2 from O_2 and other gases using FAU-type zeolite membranes. Journal of Chemical Engineering of Japan. 34, 990–997.

Guan, G., Kusakabe, K., Morooka, S., 2001b. Synthesis and permeation properties of ion-exchanged ETS-4 tubular membranes. Microporous and Mesoporous Materials. 50, 109–120.

Guan, G., Tanaka, T., Kusakabe, K., Sotowa, K., Morooka, S., 2003. Characterization of $AlPO_4$-type molecular sieving membranes formed on a porous α-alumina tube. Journal of Membrane Science. 214, 191–198.

Guo, H., Zhu, G., Li, H., Zou, X., Yin, X., Yang, W., Qiu, S., Xu, R., 2006. Hierarchical growth of large-scale ordered zeolite silicalite-1 membranes with high permeability and selectivity for recycling CO_2. Angewandte Chemie International Edition. 45, 7053–7056.

Hong, M., Falconer, J.L., Noble, R.D., 2005. Modification of zeolite membranes for H_2 separation by catalytic cracking of methyldiethoxysilane. Industrial & Engineering Chemistry Research. 44, 4035–4041.

Hong, M., Li, S., Falconer, J.L., Noble, R.D., 2008. Hydrogen purification using a SAPO-34 membrane. Journal of Membrane Science. 307, 277–283.

Hong, M., Li, S., Funke, H.H., Falconer, J.L., Noble, R.D., 2007. Ion-exchanged SAPO-34 membranes for light gas separations. Microporous and Mesoporous Materials. 106, 140–146.

Huang, A., Caro, J., 2011. Highly oriented, neutral and cation-free $AlPO_4$ LTA: From a seed crystal monolayer to a molecular sieve membrane. Chemical Communications. 47, 4201–4203.

Huang, A., Wang, N., Caro, J., 2012. Synthesis of multi-layer zeolite LTA membranes with enhanced gas separation performance by using 3-aminopropyltriethoxysilane as interlayer. Microporous and Mesoporous Materials. 164, 294–301.

Kalipcilar, H., Bowen, T.C., Noble, R.D., Falconer, J.L., 2002. Synthesis and separation performance of SSZ-13 zeolite membranes on tubular supports. Chemistry of Materials. 14, 3458–3464.

Karimi, S., Korelskiy, D., Yu, L., Mouzon, J., Khodadadi, A.A., Mortazavi, Y., Esmaeili, M., Hedlund, J., 2015. A simple method for blocking defects in zeolite membranes. Journal of Membrane Science. 489, 270–274.

Kim, S.J., Liu, Y., Moore, J.S., Dixit, R.S., Pendergast Jr, J.G., Sholl, D., Jones, C.W., Nair, S., 2016. Thin hydrogen-selective SAPO-34 zeolite membranes for enhanced conversion and selectivity in propane dehydrogenation membrane reactors. Chemistry of Materials. 28, 4397–4402.

Kim, S.J., Xu, Z., Reddy, G.K., Smirniotis, P., Dong, J., 2012. Effect of pressure on high-temperature water gas shift reaction in microporous zeolite membrane reactor. Industrial & Engineering Chemistry Research. 51, 1364–1375.

Kita, H., Fuchida, K., Horita, K., Asamura, H., 2001. Preparation of faujasite membranes and their permeation properties. Separation and Purification Technology. 25, 261–268.

Kosinov, N., Auffret, C., Borghuis, G.J., Sripathi, V.G.P., Hensen, E.J.M., 2015. Influence of the Si/Al ratio on the separation properties of SSZ-13 zeolite membranes. Journal of Membrane Science. 484, 140–145.

Kosinov, N., Auffret, C., Gucuyener, C., Szyja, B.M., Gascon, J., Kapteijn, F., Hensen, E.J.M., 2014b. High flux high-silica SSZ-13 membrane for CO_2 separation. Journal of Materials Chemistry A. 2, 13083–13092.

Kosinov, N., Auffret, Sripathi, V.G.P., Gucuyener, C., Gascon, J., Kapteijn, F., Hensen, E.J.M. 2014a. Influence of support morphology on the detemplation and permeation of ZSM-5 and SSZ-13 zeolite membranes. Microporous and Mesoporous Materials. 197, 268–277.

Korelskiy, D., Ye, P., Fouladvand, S., Karimi, S., Sjoberg, E., Hedlund, J., 2015. Efficient ceramic zeolite membranes for CO_2/H_2 separation. Journal of Materials Chemistry A. 3, 12500–12506.

Kumakiri, I., Qiu, L., Tanaka, K., Kita, H., Saito, T., Nishida, R., 2016. Application of MFI zeolite membrane prepared with fluoride ions to hydrogen/toluene separation. Journal of Chemical Engineering of Japan. 49, 753–755.

Kusakabe, K., Kuroda, T., Murata, A., Morooka, S., 1997. Formation of a Y-type zeolite membrane on a porous α-alumina tube for gas separation. Industrial & Engineering Chemistry Research. 36, 649–655.

Kusakabe, K., Kuroda, T., Uchino, K., Hasegawa, Y., Morooka, S., 1999. Gas permeation properties of ion-exchanged faujasite-type zeolite membranes. AIChE Journal. 45, 1220–1226.

Kusakabe, K., Sotowa, K., 2004. Development of zeolite membrane for air separation. Chemical Engineering. 49, 291–295.

Kwon, Y.H., Kiang, C., Benjamin, E., Crawford, P., Nair, S., Bhave, R., 2017. Krypton-Xenon separation properties of SAPO-34 zeolite materials and membranes. AIChE Journal. 63, 762–769.

Li, S., Alvarado, G., Falconer, J.L., Noble, R.D., 2005b. Effects of impurities on CO_2/CH_4 separations through SAPO-34 membranes. Journal of Membrane Science. 251, 59–66.

Li, S., Carreon, M.A., Zhang, Y., Funke, H.H., Noble, R.D., Falcone, J.L., 2010. Scale-up of SAPO-34 membranes for CO_2/CH_4 separation. Journal of Membrane Science. 352, 7–13.

Li, S., Falconer, J.L., Noble, R.D., 2004. SAPO-34 membranes for CO_2/CH_4 separation. Journal of Membrane Science. 241, 121–135.

Li, S., Falconer, J.L., Noble, R.D., 2006. Improved SAPO-34 membranes for CO_2/CH_4 separations. Advanced Materials. 18, 2601–2603.

Li, S., Falconer, J.L., Noble, R.D., 2008. SAPO-34 membranes for CO_2/CH_4 separations: Effect of Si/Al ratio. Microporous and Mesoporous Materials. 110, 310–317.

Li. S., Fan, C.Q., 2010. High-flux SAPO-34 membrane for CO_2/N_2 separation. Industrial & Engineering Chemistry Research. 49, 4399–4404.

Li, S., Martinek, J.G., Falconer, J.L., Noble, R.D., Gardner, T.Q., 2005a. High-pressure CO_2/CH_4 separation using SAPO-34 membranes. Industrial & Engineering Chemistry Research. 44, 3220–3228.

Li, X., Li, K., Tao, S., Ma, H., Xu, R., Wang, B., Wang, P., Tian, Z., 2016. Ionothermal synthesis of LTA-type aluminophosphate molecular sieve membranes with gas separation performance. Microporous and Mesoporous Materials. 228, 45–53.

Lindmark, J., Hedlund, J., 2010a. Carbon dioxide removal from synthesis gas using MFI membranes. Journal of Membrane Science. 360, 284–291.

Lindmark, J., Hedlund, J., 2010b. Modification of MFI membranes with amine groups for enhanced CO_2 selectivity. Journal of Materials Chemistry. 20, 2219–2225.

Lixiong, Z., Mengdong, J., Enze, M., 1997. Synthesis of SAPO-34/ceramic composite membranes. Studies in Surface Science and Catalysis. 105, 2211–2216.

Matsukata, M. Sasaki, Y., Sakai, M., Tomono, T., 2015b. Propylene/propane separation properties through FAU-type zeolite membrane. Proceeding of international symposium on zeolite and microporous crystals, P1-105, Sapporo, Japan.

Matsukata, M., Seshimo, M., Sakai, M., Kimura, N., Adachi, M., Waku, T., 2015a. Olefin separation method and zeolite membrane complex. Patent no: WO 2015141686.

Nikkei Asian Review, 2016. Mitsubishi Chemical strains out profit with zeolites. https://asia.nikkei.com/magazine/20160512-WEALTHIER-UNHEALTHIER/Tech-Science/Mitsubishi-Chemical-strains-out-profit-with-zeolites.

Nomura, M., Yamaguchi, T., Nakao, S., 2001. Transport phenomena through intercrystalline and intracrystalline pathways of silicalite zeolite membranes. Journal of Membrane Science. 187, 203–212.

Poshusta, J.C., Tuan, V.A., Falconer, J.L., Noble, R.D., 1998. Synthesis and permeation properties of SAPO-34 tubular membranes. Industrial & Engineering Chemistry Research. 37, 3924–3929.

Poshusta, J.C., Tuan, V.A., Pape, A., Noble, R.D., Falconer, J.L., 2000. Separation of light gas mixtures using SAPO-34 membranes. AIChE Journal. 46, 779–789.

Sakai, M., Yasuda, N., Sasaki, Y., Matsukata, M., 2017a. Permeances of propylene and propane through Ag-FAU membrane in unary and binary systems.

Sakai, M., Sasaki, Y., Tomono, T., Matsukata, M., 2017b. Typical results of separation measurements for saturated/unsaturated hydrocarbon mixtures by Ag-FAU membrane.

Sandström, L., Lindmark, J., Hedlund, J., 2010. Separation of Methanol and Ethanol from Synthesis Gas Using MFI Membranes. Journal of Membrane Science. 360, 265–275.

Sandström, J., Sjoberg, E., Hedlund, J., 2011. Very high flux MFI membrane for CO_2 separation. Journal of Membrane Science. 380, 232–240.

Sawamura, K., Izumi, T., Kawasaki, K., Daikohara, S., Ohsuna, T., Takada, M., Sekine, Y., Kikuchi, E., Matsukata, M., 2009. Reverse-selective microporous membrane for gas separation. Chemistry An Asian Journal. 4, 1070–1077.

Shin, D.W., Hyun, S.H., Cho, C.H., Han, M.H., 2005. Synthesis and CO_2/N_2 gas permeation characteristics of ZSM-5 zeolite membranes. Microporous and Mesoporous Materials. 85, 313–323.

Sjoberg, E., Barnes, S., Korelskiy, D., Hedlund, J., 2015. MFI membranes for separation of carbondioxide from synthesis gas at high pressures. Journal of Membrane Science. 486, 132–137.

Tang, Z., Kim, S.J., Reddy, G.K., Dong, J., Smirniotis, P., 2010. Modified zeolite membrane reactor for high temperature water gas shift reaction. Journal of Membrane Science. 354, 114–122.

Tiscornia, I., Kumakiri, I., Bredesen, R., Tellez, C., Coronas, J., 2010. Microporous titanosilicate ETS-10 membrane for high pressure CO_2 separation. Separation and Purification Technology. 73, 8–12.

van den Broeke, L.J.P., Bakker, W.J.W., Kapteijn, F., Moulijn, J.A., 1999. Binary permeation through a silicalite-1 membrane. AIChE Journal. 45, 976–985.

van der Bergh, J., Gucuyener, C., Gascon, J., Kapteijn, F., 2011. Isobutane dehydrogenation in a DD3R zeolite membrane reactor. Chemical Engineering Journal. 166, 368–377.

Venna, S.R., Carreon, M.A., 2011. Amino-functionalized SAPO-34 membranes for CO_2/CH_4 and CO_2/N_2 separation. Langmuir. 27, 2888–2894.

Wang, B., Hu, N., Wang, H., Zheng, Y., Zhou, R., 2015. Improved AlPO-18 membranes for light gas separation. Journal of Materials Chemistry A. 3, 12205–12212.

Wang, H., Huang, L., Holmberg, B.A., Yan, Y., 2002. Nanostructured zeolite 4A molecular sieving air separation membranes. Chemical Communications. 0, 1708–1709.

Wang, L., Zhang, C., Gao, X., Peng, L., Jiang, J., Gu, X., 2017. Preparation of defect-free DDR zeolite membranes by eliminating template with ozone at low temperature. Journal of Membrane Science. 539, 152–160.

Wu, T., Diaz, M.C., Zheng, Y., Zhou, R., Funke, H.H., Falconer, J.L., Noble, R.D., 2015. Influence of propane on CO_2/CH_4 and N_2/CH_4 separations in CHA zeolite membranes. Journal of Membrane Science. 473, 201–209.

Ye, P., Grahn, M., Korelskiy, D., Hedlund J., 2016. Efficient separation of N_2 and He at low temperature using MFI membranes. AIChE Journal. 62, 2833–2842.

Ye. P., Korelskiy, D., Grahn, M., Hedlund, J., 2015. Cryogenic air separation at low pressure using MFI membranes. Journal of Membrane Science. 487, 135–140.

Ye, P., Sjoberg, E., Hedlund, J., 2014. Air separation at cryogenic temperature using MFI membranes. Microporous and Mesoporous Materials. 192, 14–17.

Yoshihara, K., Sakai, M., Matsukata, M., 2017. Relative permeances of small gas molecules through AlPO-18 membrane.

Yu, W., Ohmori, T., Yamamoto, T., Endo, A., Nakaiwa, M., Itoh, N., 2007. Optimal design and operation of methane steam reforming in a porous ceramic membrane reactor for hydrogen production. Chemical Engineering Science. 62, 5627–5631.

Zhang, Y., Wu, Z., Hong, Z., Gu, X., Xu, N., 2012. Hydrogen-selective zeolite membrane reactor for low temperature water gas shift reaction. Chemical Engineering Journal. 197, 314–321.

Zheng, Y., Hu, N., Wang, H., Bu, N., Zhang, F., Zhou, R., 2015. Preparation of steam-stable high-silica CHA (SSZ-13) membranes for CO_2/CH_4 and C_2H_4/C_2H_6 separation. Journal of Membrane Science. 475, 303–310.

Zhou, C., Yuan, C., Zhu, Y., Caro, J., Huang, A., 2015. Facile synthesis of zeolite FAU molecular sieve membranes on bio-adhesive polydopamine modified Al_2O_3 tubes. Journal of Membrane Science. 494, 174–181.

Zhou, M., Korelskiy, D., Ye, P., Grahn, M., Hedlund, J., 2014. A uniformly oriented MFI membrane for improved CO_2 separation. Angewandte Chemie International Edition. 53, 3492–3495.

Zong, Z., Carreon, M.A., 2017a. Thin SAPO-34 membranes synthesized in stainless steel autoclaves for N_2/CH_4 separation. Journal of Membrane Science. 524, 117–123.

Zong, Z., Elsaidi, S.K., Thallapally, P.K., Carreon, M.A., 2017b. Highly permeable AlPO-18 membranes for N_2/CH_4 separation. Industrial & Engineering Chemistry Research. 56, 4113–4118.

8 Silica, Template Silica and Metal Oxide Silica Membranes for High Temperature Gas Separation

David K. Wang and João C. Diniz da Costa

CONTENTS

8.1 INTRODUCTION

Silica derived membranes were borne out of a desire to produce high quality inorganic membranes for gas separation. The initial precursor for silica membranes was glass-based materials. As early as 1992, Ma's group (Shelekhin et al., 1992) reported permselectivities as high as 11,674 for He/CH_4 separation. Although this separation result was outstanding, gas fluxes were extremely low thus making glass membranes undesirable for gas separation processes. This led to the development of the first generation of silica membranes, which took place mainly in the 1990s.

The first generation of silica membranes was accompanied by fundamental studies on porous substrates and interlayers, silica sol-gel synthesis, and thin film coating methods in the 1980s. Several groups that greatly contributed to the first-generation development were Burggraaf at the University of Twente in the Netherlands, Cot at the European Institute of Membranes in France, Gavalas at the California Institute of Technology, Brinker at the University of New Mexico in the United States, and Nakao at the University of Tokyo in Japan. Major achievements appeared at the end of the first generation. Verweij's group (de Vos and Verweij, 1998a; 1998b) at the University of Twente developed high quality silica membranes in clean rooms delivering He fluxes in the order of 1×10^{-6} mol m^{-2} s^{-1} Pa^{-1} and He/N_2 permselectivities of ~100. This became the benchmark standard for silica membranes. A second achievement was the use of the chemical vapor deposition (CVD) method to prepare silica membranes. Nakao's group (Nomura et al., 2005) developed a counter diffusion CVD method that led to silica membranes delivering H_2/N_2 permselectivities in excess of 1000, though lower H_2 fluxes in the order of 1×10^{-7} mol m^{-2} s^{-1} Pa^{-1}. During this first generation, high quality silica membranes were characterised by temperature-dependent flux of gases, thus complying with the model developed by Barrer (1990) for activated transport.

The second generation of silica membranes started in the late 1990s, mainly due to the advent of nanotechnology that employed a large number of templates to produce nanomaterials with tailored pore sizes. This concept was translated into the preparation of silica membranes, as pore size control is very important to separate gases. Initially, the silica membranes were calcined in air so the carbon template was burned off and left a cavity similar to the template dimensions in the silica film. Brinker's group (Raman and Brinker, 1995) and Morooka's group (Kusakabe et al., 1999) from Kyushu University in Japan demonstrated this concept by using a ligand templated silica precursor. The performance of the membranes showed high gas fluxes, though permselectivity was low. The problem here is that the cavity left by the template was too large, thus affecting gas separation. Subsequently, two major works showed the validity of using templates in silica membranes, and both carbonised the templates embedded in the silica thin film. The first work by Verweij's group (de Vos et al., 1999), using a ligand templated silica precursor, and the second work by Diniz da Costa's group (Duke et al., 2004b), using a non-ligand anionic surfactant with a silica precursor, have both resulted in good gas permeance in addition to good gas permselectivities. A major advantage of the work from these two groups is the improvement of the hydrostability of the silica membranes, and thus adding new functionalities not previously available in the pure silica membranes.

The third generation of silica membranes started around 2005 and is associated with the development of metal oxide membranes led by Asaeda's and Tsuru's groups from Hiroshima University in Japan and Diniz da Costa's group at the University of Queensland in Australia. By embedding metal oxide particles into silica thin films, permselectivities catapulted to values in excess of 1000, similar to pure silica membranes prepared by the CVD method. Nickel oxide (Kanezashi et al., 2005) and cobalt oxide (Battersby et al., 2008) were initially the metal oxide of choice in silica films since 2005, though this approach has diversified into other metal oxides and metals. Lately, this third generation is evolving towards binary metal oxides where different types of functionalities can be conferred. Two major developments in membrane preparation were reported in this third generation (Ballinger et al., 2014; Ballinger et al., 2015; Darmawan et al., 2015a; Darmawan et al., 2015b). Moreover, recent reports by Wang and co-workers (2013, 2014) showed for the first time that rapid thermal processing (RTP) could be used to prepare silica membranes in a few hours, compared to previous conventional thermal processing (CTP) that took over one week to make silica membranes. Also recently, Liu and co-workers (2015c) showed the preparation of interlayer-free silica membranes, thus reporting silica films coated directly on porous substrates. These recent developments greatly reduced the time and manufacturing costs of silica membranes.

The fourth and last generation is relatively young, beginning around 2010, and relates to silica membranes derived from organosilanes. A number of works from Nijmeijer's group at the University of Twente and Tsuru's group that focus on morphology, structure, and pore size tailoring using different types of organo-silica precursors have been reported. Organo-silica membranes show interesting properties for separating oleofins. For instance, Kanezashi and co-workers (2012) separated butane from butene, which is a very difficult gas separation process due to the similar thermodynamic properties and molecular sizes of these gases.

It is the aim of this chapter to cover all the important developments of silica-based membranes from the first to the third generation. In fact, these generations of silica membranes are still evolving with new findings and applications. The fourth generation related to organo-silica membranes is not covered in this chapter. This chapter also addresses important recent developments in the preparation methods related to interlayer-free and RTP silica-based membranes.

8.2 SOL-GEL SYNTHESIS AND CHARACTERIZATION

8.2.1 Synthesis Methods

The sol-gel chemistry is a wet chemistry method that is well-recognized to be highly reproducible and can be easily controlled by processing conditions such as type and concentrations of silica precursors, catalyst, solvent and co-solvents, water concentration, reaction temperature, and time. Due to these reasons, the synthesis of silica-based membranes by sol-gel process has been extensively reported in order to tailor the morphology and pore size of the silica microstructure.

Tetraethyl orthosilicate (TEOS) has been the most utilized silica monomer in sol-gel and vapor deposition methods to synthesize high quality silica membranes with the molecular sieving capacity. The separation performance relies on the careful control of the sol-gel reaction conditions. Principally, the silica sol-gel process is governed by a continuous progress of hydrolysis reaction and condensation reaction according to the following:

Hydrolysis reaction:

$$\equiv \mathrm{Si-OR} + \mathrm{H_2O} \xrightleftharpoons[\text{Esterification}]{\text{Hydrolysis}} \equiv \mathrm{Si-OH} + \mathrm{R-OH} \tag{8.1}$$

Polymerisation/Condensation reaction:

$$\equiv \mathrm{Si-OR} + \mathrm{HO-Si} \equiv \xrightleftharpoons[\text{Alcoholysis}]{\text{Alcohol Condensation}} \equiv \mathrm{Si-O-Si} \equiv + \mathrm{R-OH} \tag{8.2}$$

$$\equiv \mathrm{Si-OH} + \mathrm{HO-Si} \equiv \xrightleftharpoons[\text{Siloxane Hydrolysis}]{\text{Water Condensation}} \equiv \mathrm{Si-O-Si} \equiv + \mathrm{H_2O} \tag{8.3}$$

where OR is a hydrolysable alkoxy group bonded to the silicon atom.

First, hydrolysis reaction involving the functional alkoxysilane precursors occurs in Eq. 8.1. In this reaction, the alkoxy groups of silane are hydrolyzed by water to form hydroxyl groups via silanol bonds (Si–OH) and produce the alcohol as byproducts. This is followed by condensation reactions, where the silanol bonds further react with either an alkoxy group (alcohol condensation; Eq. 8.2) or another silanol group (water condensation; Eq. 8.3) to produce siloxane bonds (Si–O–Si) along with the by-products of alcohol (ROH) or water, respectively. Since alkoxysilanes are insoluble in water, a mutual co-solvent such as alcohol is required to form a homogeneous solution (Brinker and Scherer, 1990). Therefore, sol-gel mixtures generally contain an alkoxysilane (monomer), water (hydrolyzing agent), ethanol (co-solvent), and a catalyst involving an organic acid or base.

Another important consideration of the sol-gel system is the pH of the solution in relation to the isoelectric point (IEP) of the silica sol species. Silica polymeric particles have a point of zero charge around pH 2, which is the isoelectric point where the electrical mobility of the particles is zero (Iler, 1979). At this boundary, the silica particles are not ionized and therefore are not positive or negative. When pH of the solution is changed from acidic (below IEP 2) to basic (above IEP 2), silica chain growth proceeds from an acid-catalyzed mechanism to a base-catalyzed mechanism, which can have a significant effect on the final gel morphology (Brinker and Scherer, 1990). To achieve efficient gas separation, porous silica membranes must have pore sizes less than 1 nm to enable molecular sieving capability and thus the polymeric sol-gel route is the most widely applied technique. Therefore, silica membranes are generally prepared by an acid catalyzed sol-gel process, which promotes a high degree of hydrolysis and thus follows the polymeric sol route. Under the acidic condition, the sol consists of long, weakly-branched silica polymer chains that undergo limited condensation to form a loosely interconnected silica gel network (Brinker, 1988; Brinker and Scherer, 1990).

To achieve smooth silica membrane film with excellent pore size control and molecular sieving capacity, the molar ratio of water to silicon (H_2O:Si) is widely reported. The guiding principle of this sol-gel protocol is when the H_2O:Si ratio is kept relatively low around 5, hydrolysis dominates throughout the process and condensation is minimized. As a consequence, silica membranes prepared using a low H_2O:Si ratio are highly microporous and characterised by a high degree of silanol groups (Si–OH) which has been reported to generate superior separation selectivity of gases of similar size (Diniz da Costa et al., 2002). As this ratio is increased from 1 to 7, as shown in Figure 8.1, pore size systematically increases, and the system shifts from a highly microporous to a mesoporous region (Wang, 2016). Hence, it is very important to control the amount of water in the silica sol-gel system in order to achieve the desired pore size distribution for the targeted gas separation.

Furthermore, the silica sol-gel process is significantly influenced by the concentration of the solvent in the system. As Eq. 8.2 shows, the alcohol condensation reaction generates ethanol by-product, which is commonly the choice of the solvent for silica sol-gel chemistry. There are several reasons for this. First, TEOS is very soluble in ethanol as the functional groups of TEOS (ethoxides) are highly compatible with ethanol. Second, as more ethanol by-product is generated, the system will continue to remain in homogeneity without phase separation. In this respect, water by-product arising from Eq. 8.3 is miscible in ethanol. Therefore, the whole sol-gel system is homogeneous and a higher reaction efficiency can be expected. Since ethanol solvent plays such an important role in the sol-gel system, it is generally kept constant throughout the sol-gel process and subsequently used as a diluent to control the final sol viscosity prior to the membrane coating, hence film thickness.

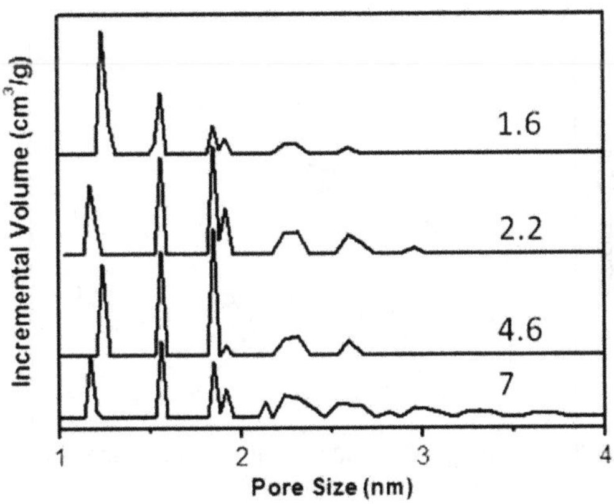

FIGURE 8.1 Pore size distribution profiles of silica gels obtained by different H_2O:Si ratios. (From Wang, S., 2016. *High performance ES40-derived silica membranes for desalination* (Doctoral Thesis). The University of Queensland, School of Chemical Engineering.)

8.2.2 Characterization Techniques

The structure of sol-gel derived silica materials and membranes is highly dependent on the various parameters (ratios of water and ethanol, temperature) and conditions (acid or base catalyzed, humidity), as well as thermal processing protocol. It is therefore necessary to discuss the structure and physicochemical properties of the silica gels that are intended for membranes for gas separation applications. Since it is not possible to directly examine the silica membrane layer attributed to the separation without the influence of the support, the pragmatic approach to gain an understanding into the structural evolution has been devoted to silica xerogels materials. These materials are produced using the same sol-gel process and thermal calcination but they are dried on petri dishes instead of on ceramic supports before calcination. Therefore, even though it is widely acknowledged that the properties of the xerogels and the thin films are non-equivalent, they are generally studied to allow for a qualitative and semi-quantitative comparison. In this section, the distinction of the two from the outset will be highlighted to avoid any ambiguity.

8.2.2.1 Solid-State ^{29}Si NMR

Solid-state ^{29}Si NMR has been employed extensively to evaluate the chemical structure and concentration of condensed (or uncondensed) silica species in the sol-gel materials. According to the Q notation for describing silica species, Q_n represents $(SiO)_nSi(OH)_{4-n}$ (Brinker and Scherer 1990). For instance, Q_1 and Q_4 represent one and four siloxane groups in the silica gel structure; in other words, there are three uncondensed species (or silanol groups) for the former and none for the latter. Figure 8.2 shows the NMR spectra of silica xerogels prepared by different H_2O/Si ratios and their deconvoluted peaks as shown by the dash line (Wang et al., 2017). Such information is useful to understand the degree of condensation as a function of sol-gel processing parameter, which can be used to infer material properties such as water adsorption, hydrophilicity, and hydrostability of the silica membranes.

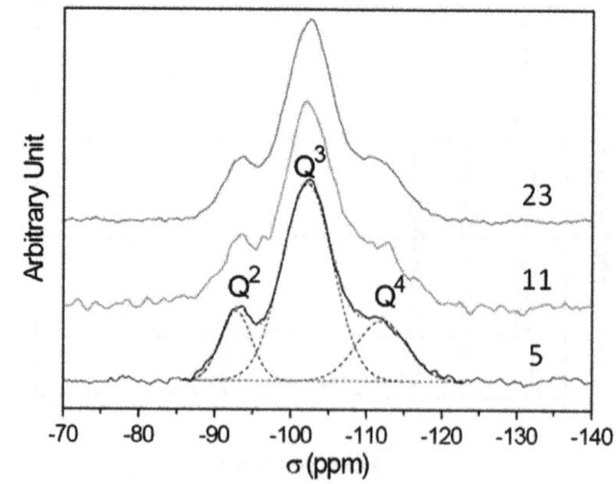

FIGURE 8.2 ^{29}Si NMR spectra of silica xerogels prepared by different H_2O/Si ratios (5, 11, and 23). The deconvoluted peaks for 5 H_2O/Si ratio are demonstrated as an example. (From Wang, S. et al., Desalination. 402, 25–32, 2017.)

Some considerations should be given in performing solid-state ^{29}Si NMR on amorphous silica xerogels with metal oxides incorporated. Due to the strong dipolar and quadrupolar interactions of the metal oxides, poor signal-noise ratio and spectral resolution of the cobalt oxide silica matrices are expected with an increase in the cobalt concentration. To overcome this effect, measurement conditions must be optimized by using a sample-customized SP-hpdec technique, which is a single pulse, high-power proton decoupling to enhance the spectral signal of the silicon peaks for each sample. Moreover, solid-state ^{29}Si NMR is often accompanied by FTIR as a complementary technique to comprehensively characterize the silica xerogels.

8.2.2.2 Attenuated Total Reflection Fourier-Transformed Infrared (FTIR)

FTIR is widely performed for qualitatively examining the evolution of the silica frameworks through their functional groups for a sol-gel reaction system (Lee et al., 1997; Riegel et al., 1998; Ambati and Rankin, 2011; Olguin et al., 2014) and xerogel characterization (Tripp and Hair, 1995, Liu et al., 2014; Masuda et al., 2014). In addition to the chemical information that could be gained from other functional groups, the degree of silica condensation can be qualitatively and semi-quantitatively determined from FTIR analyses. Also, it is necessary to understand how the reactions influence the porous structure formation, arising from the formation of silanol (Si–OH) groups via hydrolysis and siloxane (Si–O–Si) bridges via the condensation reaction. By carefully applying several constraints in the deconvolution process to minimize errors during the peak fitting analysis, one can semi-quantitatively determine the degree of condensation of the silica materials, as listed in Table 8.1. More simply, as shown in Figure 8.3, peaks VI and VII correspond to the products of the hydrolysis reaction for Si–OH and Si–O⁻ bonds, respectively. While for the broad band in the range 1250 to 1000 cm^{-1}, five fitted peaks centered at ~1205, 1146, 1105, 1065, and 1035 cm^{-1} relate to the products of the condensation reaction Si–O–Si bond. The ratio of these two

TABLE 8.1
Assignments of the FTIR Deconvoluted Peaks of the Silica Xerogel in Figure 8.3

Deconvoluted Peaks	Wavenumber (cm^{-1})	Vibration Mode	Chemicals
I	~1205	LO_3 mode of ν_{as}(Si-O-Si)	6-ring siloxane $(SiO)_6$
II	~1146	LO_4 mode of ν_{as}(Si-O-Si)	4-ring siloxane $(SiO)_4$
III	~1105	TO_4 mode of ν_{as}(Si-O-Si)	4-ring siloxane $(SiO)_4$
IV	~1065	TO_3 mode of ν_{as}(Si-O-Si)	6-ring siloxane $(SiO)_6$
V	~1035	ν_{as}(Si-O-Si)	chain silicate
VI	~962	ν(Si-OH)	silanol
VII	~934	ν(Si-O⁻)	silica open rings

Sources: Innocenzi, P., Journal of Non-Crystaline Solids. 316(2–3), 309-319, 2003; Fidalgo, A., Ilharco, L.M., Chemistry-a European Journal. 10(2), 392–398, 2004; Wang, S. et al., Scientific Reports. 5, 14560, 2015.

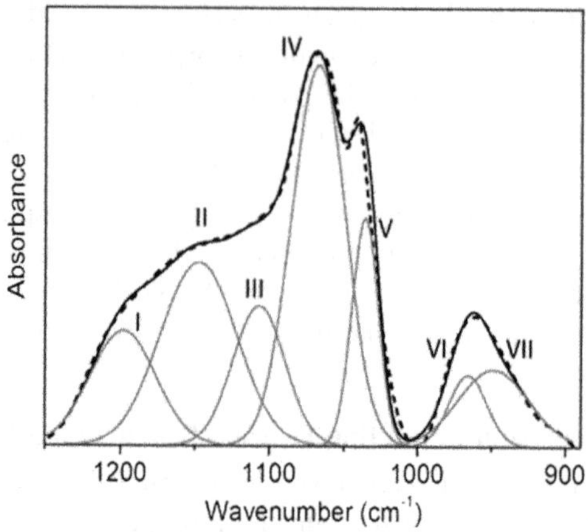

FIGURE 8.3 FTIR spectrum (dotted line) and deconvoluted peaks of silica xerogel. The solid lines are summation (black) of the fitted peaks (grey) with an R^2 fitting value ≥0.995.

distinct FTIR regions by comparing the areas of uncondensed silicon species (peaks VI and VII) to that of the condensed silica (mainly IV) can be used as an indicator of the degree of hydrolysis and condensation reactions. These studies strongly suggest that FTIR is a useful characterization tool for assessing the silica sol-gel process.

8.2.2.3 Nitrogen Physisorption

Nitrogen physisorption of silica xerogel samples offer the opportunity to study the porosity and textural properties of the materials prior to membrane coating as the fundamental science underpins the structural evolution of the final membrane matrices. As a preliminary characterization tool, it is a simple and fast way of inferring the membrane pore sizes, pore volume, and pore size distribution of the silica matrices. For gas separation, it is crucial that the silica microstructure is microporous with pore sizes less than 2 nm. As shown in Figure 8.4(a), microporous amorphous silica

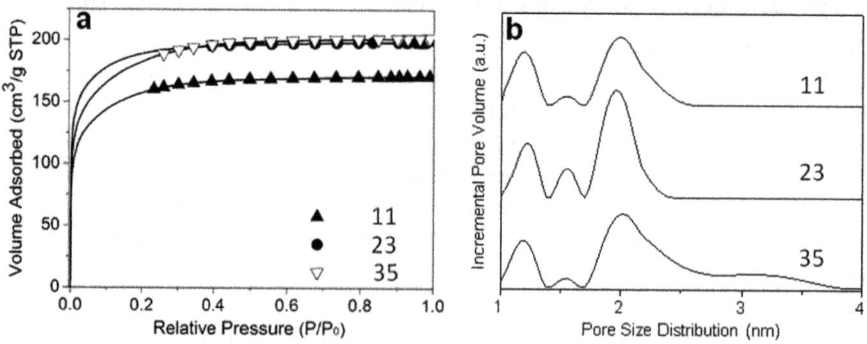

FIGURE 8.4 (a) N_2 adsorption (solid line) and desorption (symbols) isotherms and (b) pore size distribution of the silica xerogels prepared using different H_2O/Si ratios (11, 23, 35). (From Wang, S. et al., RSC Advances. 5(8), 6092–6099, 2015.)

typically displays a type I nitrogen adsorption isotherm with no hysteresis loop, and the pore size distribution in Figure 8.4(b) shows that there is a trimodal pore size distribution for silica xerogels synthesized with different H_2O/Si ratios (Wang et al. 2015a). A similar finding was also reported by Duke et al. (2008), who found that amorphous silica structures have a ternary pore size distribution (~3, 8, and 12 Å) via positron annihilation spectroscopy. Although these pore sizes are much larger than the average size of gases of interest for separation such as He (2.7 Å), H_2 (2.89 Å), CO_2 (3.4 Å), and N_2 (3.65 Å), there is a good correlation that silica xerogels with average pore sizes of 10 Å correspond to membranes with high molecular sieving transport (Igi et al., 2008; Boffa et al., 2009; Uhlmann et al., 2009; Kanezashi and Tsuru, 2011; Yacou et al., 2012). Therefore, N_2 physisorption characterization of silica xerogels is widely applied to demonstrate the synthesis parameters on the material porosity. The results should serve as a good correlation study for comparing with the gas separation selectivity of silica membranes.

8.2.2.4 X-Ray Diffraction (XRD)

X-ray diffraction (XRD) technique is routinely used for silica-based matrices with metal oxide incorporated to identify the presence and phase of metal oxides if the concentration of metal is significantly detectable. Although it is not suitable for examining the pure amorphous silica xerogels, XRD is a powerful tool to probe the effect of the sol-gel parameters and gas processing conditions on the metal species. For example, as demonstrated in Figure 8.5, crystalline cobalt oxide phase can be confirmed as Co_3O_4 by XRD within amorphous silica xerogels, which was evidenced at 20 mol% Co concentration (Martens et al., 2015). In other works, XRD showed that cobalt was impregnated into the silica matrix, which conferred a superior stability

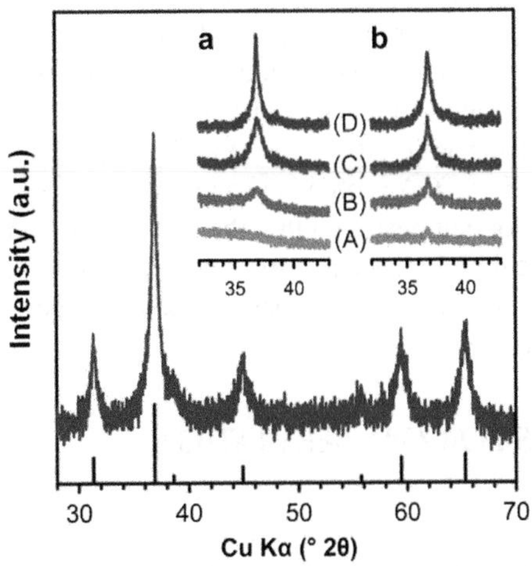

FIGURE 8.5 XRD pattern of cobalt oxide silica xerogels and JCPDF # 76-1802 stick pattern (Co_3O_4). Insets are focussed scans of the (311) reflection of the (a) sol-gel derived and (b) xerogel derived samples where (A) 15, (B) 20, (C) 30, and (D) 40 mol% Co incorporation. (From Martens, D.L. et al., Scientific Reports. 5, 7970, 2015.)

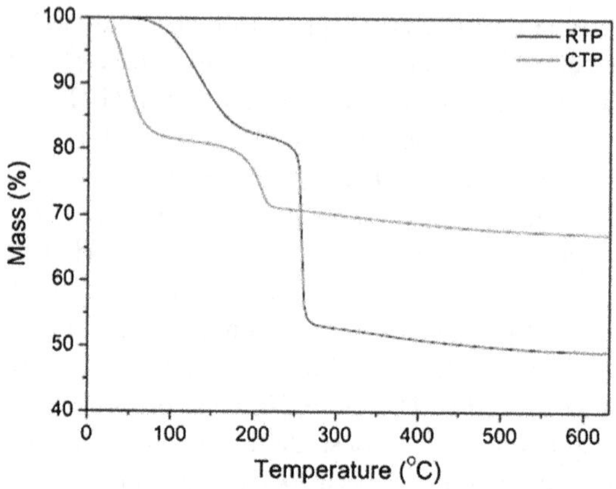

FIGURE 8.6 TGA mass loss profiles of silica xerogels prepared using rapid thermal processing (RTP; 100°C/min) or conventional thermal process (CTP; 1°C/min) techniques. (From Wang, S. et al., Journal of Membrane Science. 516, 94–103, 2016.)

under hydrothermal conditions and in reductive/oxidative environment (Igi et al., 2008; Tsuru et al., 2011), and, likewise, nickel doped silica membranes were found to improve the stability of the matrix against water and steam (Kanezashi et al., 2005).

8.2.2.5 Thermo-Gravimetric Analysis

For pure silica gels, thermo-gravimetric analysis (TGA) has been widely carried out to elucidate the degree of silica condensation as a function of calcination temperature to control the amount of hydroxyl groups within the silica matrices (James, 1988; Kim et al., 2009; Yamamoto et al., 2015). From the temperature-dependent mass loss of the materials, the profile indicates the extent of solvent (water and ethanol) evaporation at a low temperature range (<120°C) as well as the material densification (200–600°C) via thermal stress shrinkage and thermal-induced condensation. Wang et al. (2016) reported the thermal effect of calcination on amorphous silica matrix densification using RTP (100°C/min) and conventional thermal processing (1°C/min) techniques. The TGA profile is shown in Figure 8.6 for the silica gel prepared at pH 6. This work demonstrated that the material has undergone a significant degree of both chemical and physical restructuring prior to 300°C via slow or fast calcination, such that the rate of heating has a strong effect on structural stability of the membrane matrices.

8.3 MEMBRANES FOR GAS SEPARATION

8.3.1 General Principles

Membranes operate like filters by separating substances like gases, as schematically depicted in Figure 8.7. However, unlike typical filters that operate in a batch process with one input and one output stream, membranes operate in continuous processes by having one input and two output streams. The input is the feed stream, and the outputs are called permeate and retentate streams. The permeate stream is related

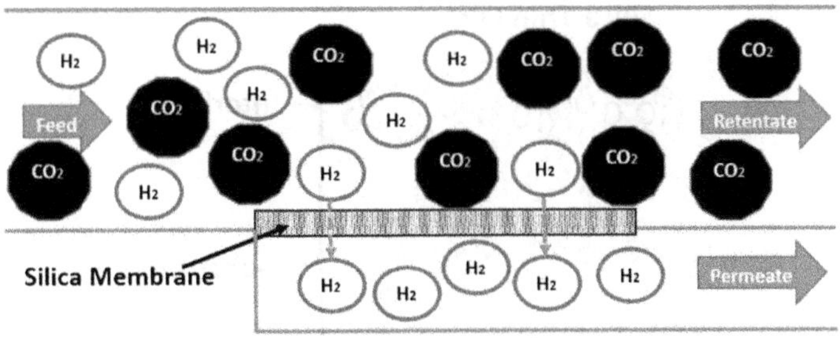

FIGURE 8.7 Schematic of a silica membrane for gas separation.

to the gases that preferentially diffuse from the feed side to the permeate side of the membrane, while the retentate stream contains non-permeable gases. A driving force is required for a gas to permeate or to diffuse through the membrane. In the case of gases, the driving force is the difference in pressure or partial pressure from the feed side to the permeate side of the membrane. If the pressures are the same on both sides of the membrane, gas can still diffuse through the membrane if its partial pressures (or concentration) are different. In the case of similar partial pressures, then there is no driving force for gas permeation or diffusion.

In terms of silica membrane operation, two engineering parameters are important, namely: permeance (as production) and separation (as quality or product purity). Permeance is a mass transfer coefficient and for inorganic membranes it is measured as the mass (mol) of a gas that diffuses through a membrane area (m^2) per time (s) per driving force, which is a pressure gradient (Pa). The unit for permeance is generally given as mol m^{-2} s^{-1} Pa^{-1}. Separation is a measure or factor that displays the capability of the membrane to separate different gases. This separation factor is also referred to as permselectivity, which is the permeance ratio of two different gases. In the case of separating a gas from several gases, the term gas purity is preferred. These two engineering parameters are often conflicting as there is a trade off when selecting silica membranes.

8.3.2 Preparation Method

Conventionally silica membranes are prepared in asymmetric configuration, as schematically depicted in Figure 8.8. In this configuration, a thin silica film is deposited on a porous ceramic substrate containing smooth interlayers, which has several advantages. First, silica thin films typically of ~250 nm thickness are mechanically weak, and thus mechanical strength is conferred by coating silica films on robust substrates. Second, the flux of gases through the membrane is inversely proportional to the thickness of the silica film; hence the smooth interlayer provides an ideal surface for preparing thin silica films. Third, the asymmetric configuration reduces the resistance of gas flow, as the pore sizes for the interlayers and substrate are much larger than those of molecular sieving dimensions of the silica films.

Alpha alumina (α-Al_2O_3) substrates have been extensively used in the preparation of silica membranes. These substrates can be easily purchased from suppliers

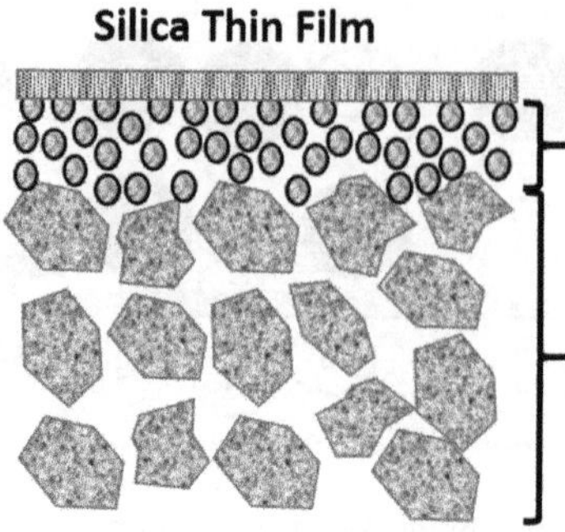

FIGURE 8.8 Schematic of conventional silica membranes in asymmetric configuration.

around the world and are reasonably cheap. Other materials used to prepare substrates include titania (Wang et al., 2007) and stainless steel (Brunetti et al., 2007; Brands et al., 2010). The substrates are generally prepared with large α-Al_2O_3 particles and the surface of these supports tends to be rough and inhomogeneous for coating thin silica films. These surfaces tend to propagate thin film defects, as pin holes and uneven film stresses cause microcracks (Bonekamp, 1996), thus rendering silica membranes ineffective for gas separation. To address this problem, substrates are traditionally coated with interlayers of small colloidal particles such as gamma alumina (γ-Al_2O_3) to improve surface smoothness for the coating of silica thin films.

There are several techniques available for coating silica thin films on top of substrates with interlayers. The most common and flexible technique is the sol-gel dip coating method, which can be used for small and large surfaces and for flat and tubular geometries. The majority of the high quality silica membranes published in literature have been prepared by the sol-gel dip coating method. The sol-gel spin coating method has also been used, though it is limited to small and flat geometries. CVD has been used for the preparation of high quality silica membranes, though this method is more complex and generally only applied for small surface areas.

8.3.3 Transport Phenomena

High quality silica membranes are porous, and the size of the pores is tailored to separate gas molecules such as H_2 from CO_2. Hence, silica membranes are often referred to as molecular sieving silica membranes. The transport of gases through porous materials complies with different transport mechanisms depending on the size of the pores. If silica membranes have gas selectivities of ~1, such membranes are not capable of separating gases. This could be attributed to silica thin film defects such as microcracks. These defects are characterised by large pores (i.e., they are mesoporous or macroporous), which create a parallel viscous flow of gases through a membrane described by the Poiseuille equation. In the case that silica membranes deliver gas selectivities of ~2–4, then the transport mechanism is described by Knudsen diffusion. This is a special case, as gas selectivities are a function of the square root of their molecular weight. However, the low gas selectivities of membranes complying with viscous flow and Knudsen diffusion show no major likelihood for industrial application.

High quality silica membranes are characterised by a porous matrix in the region of ultra micropores (<0.5 nm). These silica membranes deliver gas selectivities ranging from 30 all the way to very high values in excess of 1000. Let us consider two ideal gases, A and B, with different kinetic diameters (d_k). These gases need energy to enter and diffuse through the very small pores of the silica membrane with defined pore diameters (d_p). True molecular sieving transport mechanism occurs when the smaller gas molecule, A, can enter and diffuse through the pore, while the larger gas molecule, B, is completely excluded based on its kinetic diameter, so the relationship $d_kA < d_p < d_k B$ applies. In this case, the permeation of gas A is proportional to $e^{(-1/T)}$, therefore increasing with temperature.

This mechanism is also referred to as activated transport. According to the Lennard Jones potential, the gas A interaction with the silica is very high when d_kA approaches the same size as d_p. Hence, gas A will need energy to diffuse through the silica membranes, and this energy will increase significantly as the pore size of the silica membrane decreases. The activated transport model has been adapted from the Barrer's model of transport through micropores (Barrer, 1990) as follows:

$$J_x = -D_o K_o e^{\left(\frac{-E_{act}}{RT}\right)} \frac{dp}{dx} \tag{8.4}$$

where J_x is the flux (mol m^{-2} s^{-1}) through the membrane, D_o is a temperature independent proportionality constant associated with energy of diffusion (E_d), K_o is a temperature independent proportionality constant associated with the isosteric heat of adsorption (Q_{st}), E_{act} is an apparent activation energy (kJ mol^{-1}), R the gas constant (J mol^{-1} K^{-1}), and T the absolute temperature (K).

Fundamentally, E_{act} indicates that the diffusion of a gas molecule through a silica micropore depends upon the energy available to the molecule to diffuse (E_d) less the energy that holds the gas molecule to the pore of the silica surface (Q_{st}), as $E_{act} = E_d - Q_{st}$. In the case of gas separation at high temperature, sorption complies with Henry's law (i.e., linear sorption capacity), so the effect of sorption is small though affecting the diffusion of gases with stronger sorption such as CO_2, and to a lesser extent N_2 and H_2 gases, while He is a non-adsorbing gas at these high temperature conditions. At the same time, the d_k values increase from He (2.6 Å) to H_2 (2.89 Å), CO_2 (3.3 Å), and N_2 (3.64 Å) thus affecting the energy required by these gases to diffuse through silica micropores. For high quality silica membranes, the permeances of He and H_2 generally increase with temperature, also called positive E_{act}, while the permeances of CO_2 and N_2 decrease, or negative E_{act}.

The major assumption in this model for silica membrane transport is that microporous diffusion is assumed to be the rate-limiting step, thus the silica thin film controls the flux of gases and therefore has the highest transport resistance, while the intermediate layer and substrate have almost negligible resistances. Further, silica membranes are amorphous materials containing pores with different sizes. Duke et al. (2008) used Positron Annihilation Spectroscopy (PALS) and reported that silica thin films had a tri-modal pore size distribution, with discrete sizes of 2–3 Å, 7–8 Å, and 12–14 Å. For high quality silica membranes with high gas selectivities, gas transport is dominated by the smaller pores of 2–3 Å, which are also known as constrictions or bottle necks. These pores exclude the permeation of CO_2 and N_2 gases. However, due to the pore size distribution of silica thin films, there are a very few number of pathways which are continuously interlinked at the intermediate (7–8 Å) and larger (12–14 Å) pore regions, thus allowing for a very small permeation of the CO_2 and N_2 gases.

8.4 SILICA MEMBRANES FOR GAS SEPARATION

8.4.1 First Generation—Pure Silica Membranes

The first generation of silica membranes is where fundamental knowledge on silica sol-gel, thin film coating, and substrate technology development was consolidated. Armed with knowledge in these areas, several research groups around the world were able to develop silica membranes ranging from modest gas separation of ~10 to very high values of >1000. One fundamental aspect of this development was understanding how to control the pore size of silica thin films to molecular sieving dimensions using the sol-gel method. This was achieved by inhibiting the condensation reactions to a degree, so the silica thin film contained a higher concentration of silanol groups than the concentration of siloxane groups.

A major body of research in the first generation resulted in the production of silica membranes derived from sol-gel processes using TEOS as a silica precursor and a silica to water ratio of ~5:1. As water is under-supplied in this reaction, there is not enough water to drive the reaction to full condensation (Eqs. 8.2 and 8.3). Therefore, as TEOS hydrolyzes (Eq. 8.1), there is a major drive for the formation of silanols and a minor drive for the generation of siloxane groups; the reaction is completed when the water is fully consumed in the reaction. Upon drying and sintering the silica membranes to high temperatures (up to 600°C), condensation reactions still occur (Abidi et al., 1998) through silanol groups as they are still prevalent (Diniz da Costa et al., 2002). The control of the pore size was explained based on the Fractal theory as proposed by Mandelbrot (Mandelbrot, 1977). The sol-gel containing a high concentration of silanols is known as weakly branched system, as demonstrated by Brinker et al. (1988) using ^{29}Si NMR technique. Silanol (Si–OH) species contain terminal hydroxyl (OH) groups, which are not linked to other groups. Siloxane bridges (Si–O–Si) are rigid structures, as silicon is always connected to oxygen and to silicon and so forth. Therefore, silanols are able to interpenetrate each other and the siloxane bridges during gelation and sintering, thus forming smaller pores.

The first generation was characterised by two membrane preparation methods: dip coating and CVD. The major representative results from these preparation methods are plotted in Figure 8.9. The first plot (Figure 8.9(a)) shows a typical Robeson plot of selectivity versus permeance where two clusters of membrane performance are observed: (i) low selectivity and high permeance and (ii) high selectivity and low permeance. These clusters are associated with the traditional trade-off performance between selectivity and permeance. Although dip coated silica membranes showed good overall performance, the CVD silica membranes delivered the best results.

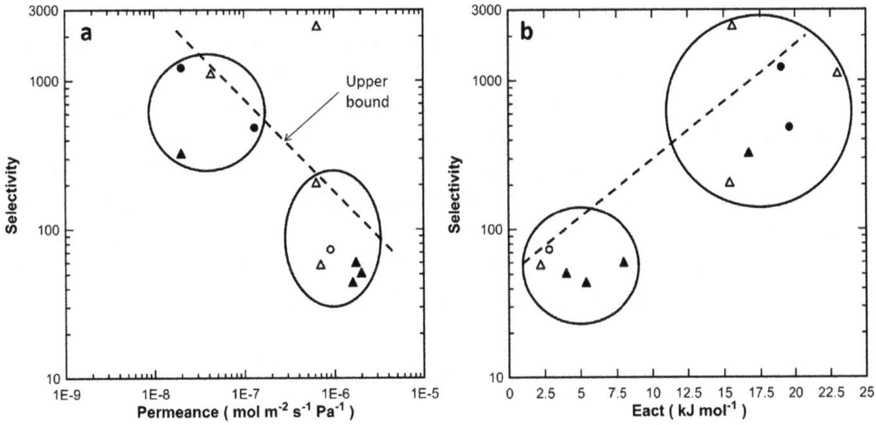

FIGURE 8.9 First generation silica membranes (a) selectivity of He or H_2 over N_2 versus permeance and (b) selectivity versus the apparent energy of activation. Dip coated silica membranes full symbols He (●) and H_2(▲) (From de Lange, R.S.A. et al., Journal of Membrane Science. 104(1–2), 81–100, 1995; Nair, B.N. et al., Langmuir. 16(10), 4558–4562, 2000; Yoshioka, T. et al., AIChE Journal. 47(9), 2052–2063, 2001; Diniz da Costa, J.C. et al., Journal of Membrane Science. 198(1), 9–21, 2002.) and CVD silica membranes hollow symbols He (○) and H_2 (△). (From Gopalakrishnan, S. et al., Journal of Membrane Science. 297(1–2), 5–9, 2007; Gopalakrishnan, S. and J.C. Diniz da Costa, Journal of Membrane Science. 323(1), 144–147, 2008; Amanipour, M. et al., Journal of Membrane Science. 423–424, 530535, 2012; Han, H.H. et al. Journal of Membrane Science. 431, 7278. 58, 2013.)

Of particular note, it is the counter-diffusion CVD method developed by Nakao's group (Gopalakrishnan et al., 2007) with the CVD silica membrane that showed the results well above the upper bound, delivering H_2/N_2 selectivity of 2300 while achieving a high H_2 permeance of 6.4×10^{-7} mol m^{-2} s^{-1} Pa^{-1}.

The second plot (Figure 8.9(b)) displays the selectivity versus the apparent energy of activation. The latter is based on the activated transport model as described in Section 8.3. Again two clusters of performance are observed and related to (iii) low selectivity and E_{act} and (iv) high selectivity and E_{act}. These results clearly indicate that the silica membranes that deliver the highest selectivity also have the highest E_{act} values. The clusters in Figure 8.9(b) are the same as the cluster in Figure 8.9(a). Therefore, as the pore size reduces and likewise the gas selectivity increases, the gas molecules need much more energy to diffuse through the membrane. These are known as the potential functions in micropores as presented by Everett and Powl (1976) based on the Lennard-Jones potential of a gas molecule as a function of the distance to the pore surface. In other words, as the kinetic diameter of the molecule approaches the same size of the pore, more energy is required for diffusion (de Lange et al., 1995).

8.4.2 Second Generation—Templated Silica Membranes

The second generation of silica membranes is closely associated with the advent of nanotechnology by developing micelles to form mesoporous ordered materials or by using carbon derived templates to control the pore size of silica membranes. The problem here is that the pores of templated silica membranes were too large for molecular sieving applications. However, a very interesting outcome from this second generation addressed a major weakness in the first generation associated with hydrostability. On one hand, controlling the pore size of silica membranes through a high concentration of silanol groups is beneficial in obtaining high selectivities. On the other hand, silanol groups are hydrophilic and adsorb water molecules. This results in the breakage of siloxane groups (hydrolysis) as per Eq. 8.5 (Burneau and Gallas, 1998), causing the rehydration of the silica surface where physisorbed water molecules react with nearby siloxane groups (Iler, 1979).

$$\equiv Si-O-Si\equiv +H_2O \rightarrow 2\equiv Si-OH \tag{8.5}$$

The rehydration of the silica surfaces caused silica membranes to lose their gas separation capabilities due to silica structural re-arrangement. The initial approach to this problem was developed by Verweij's group (de Vos et al., 1999) by using TMOS (tetramethoxysilane) instead of TEOS. TMOS contains a methyl group in the silica precursor. The TMOS derived silica membranes were calcined in an inert atmosphere, resulting in the carbonisation of the methyl group in the silica membrane. As carbon was kept in the silica matrix, the carbon imparted hydrophobic properties to the membranes. These membranes were called methylated silica membranes and showed extremely good stability in the case of liquid separation lasting for 18 months (Campaniello et al., 2004).

A second work of interest was the development of carbonised template silica membranes by Duke and co-workers (2004b) for hydrogen separation from a methanol reforming process, which also contained water vapor. The membranes were prepared from a sol-gel method using TEOS and a C6 surfactant (hexyl tri-ethyl ammonium bromide), followed by calcination in vacuum. This work showed that H_2/CO selectivity increased for the carbonised template silica membrane while it decreased for the pure silica membrane. In a subsequent work, Duke et al. (2006) reported that the carbonised silica matrix was actually hydrophobic though hydrostable. Based on these results, a mechanism was proposed as the carbon moieties embedded in the silica matrix were barriers for mobile silanol structure, which are susceptible to hydrolytic attack during water treatment. In this case, the carbon barrier maintained the pore size integrity of the membrane matrix, contrary to pure silica membranes as mobile silanol groups closed the small pores while enlarging other pores, thus the mechanism caused a deterioration of the silica matrix and loss of gas selectivity.

An interesting outcome of the second generation was the development of hierarchical membrane structures. Ayral's group at the European Institute of Membranes in France used surfactants and polystyrene spheres to prepare silica membranes with macro- and mesopores (Yacou et al. 2008; Yacou et al., 2009). Although these pores are too large for gas separation, hierarchical structures were reported to have good applications by incorporating catalysts within the structure of the silica membranes. These membranes, therefore, worked well as membrane reactor contactors for the oxidation reactions of propane (Yacou et al., 2010).

8.4.3 Third Generation—Metal Oxide Silica Membranes

The initial development of the third generation of silica derived membranes was based on incorporating metal oxides during silica sol-gel preparation using nickel nitrate (Kanezashi et al. 2005; Kanezashi and Asaeda, 2006) or cobalt nitrate (Battersby et al., 2008). Initially, it was found that these metal oxide silica membranes had superior hydrothermal stability as compared to the pure silica membranes. This was attributed to the metal oxide particle domains embedded in the silica matrix (Igi et al., 2008; Uhlmann et al., 2009), which restricted the mobility of unrestrained silanol groups. This superior property allowed the silica membranes to be coupled with reactors in a single unit, known as membrane reactors. This was demonstrated by Battersby and co-workers (2008), who used cobalt oxide silica membranes with catalysts for the water gas shift reaction, where CO conversion improved as H_2 permeated preferentially through the membrane. In the case of an equilibrium limited reaction such as the water gas shift and dehydrogenation reactions, higher conversions can be achieved by removing H_2 in the reactor chamber as soon as it is produced (Battersby et al., 2006). Liu et al. (2015a) exposed cobalt oxide silica membrane to harsh hydrothermal steam conditions of steam of 25 mol% up to 500°C. It was found that after steam exposure, the membranes containing a high concentration of tetrahedral cobalt coordination silica were unstable whilst those with a high content of Co_3O_4 silica exhibited superior hydrothermal stability.

Another major outcome of the third generation of silica membranes was very high gas selectivities (~1000), which were previously only afforded by CVD silica membranes. However, these highly selective membranes can also be prepared by thin film dip coating methods. Uhlmann et al. (2011) studied the pore sizes of cobalt oxide silica membrane by probing with different gas molecules, including H_2S, which has a propensity to react with metal oxides at high temperatures. Their work showed that H_2S did actually permeate through the membrane, but sulfur could not be detected by XPS analyses. Hence, it was postulated that the silica amorphous structures around the cobalt oxide particles had pore sizes below the kinetic diameter (D_k = 3.6 Å) of H_2S, as this gas could not access and react with the cobalt oxide particles in the silica membrane.

Metal oxides in silica membranes provided additional flexibility in pore size tailoring. Miller et al. (2013) showed that pore sizes could be tailored by reducing the tricobalt tetroxide to cobalt oxide and cobalt hydroxide, the result of which changed permeance and selectivity values. Remarkably, the membrane reverted to its original permeance and selectivity values upon oxidation, and the redox effect was reversible. This was lately attributed to the crackling effect of metal oxides which tend to break down into small particles under reduction (see Figure 8.10), thus adding an extra percolation pathway for gas permeation in the silica membrane (Ji et al., 2015). In an interesting experimental work, Ballinger et al. (2014) incorporated a binary mixture of palladium oxide and cobalt oxide into a silica membrane by a sol-gel method and under reduction found that only palladium reduced preferentially while cobalt oxide was maintained. Upon testing the membranes, the loss of oxygen from the palladium oxide slightly opened a molecular gap thus increasing gas permeances and reducing gas selectivities. Cobalt oxide has

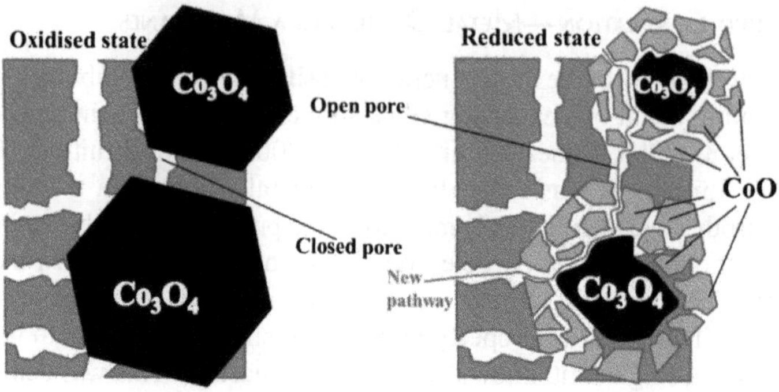

FIGURE 8.10 Schematic of the network connectivity under oxidised and reduced state. (From Ji, G. et al., Separation and Purification Technology. 154, 338–344, 2015.)

been the metal oxide of choice in a large number of publications. Other metal oxides that have been reported in silica derived membranes include niobia (Boffa et al., 2008a; Boffa et al., 2008b; Qi et al., 2012), palladium (Kanezashi et al., 2013), zirconia (Tsuru et al., 1998; Yoshida et al., 2001), titania (Gu and Oyama, 2009), and alumina (Fotou et al., 1995; Lee et al., 1999). Lately, new binary mixed metal oxide silica membranes such

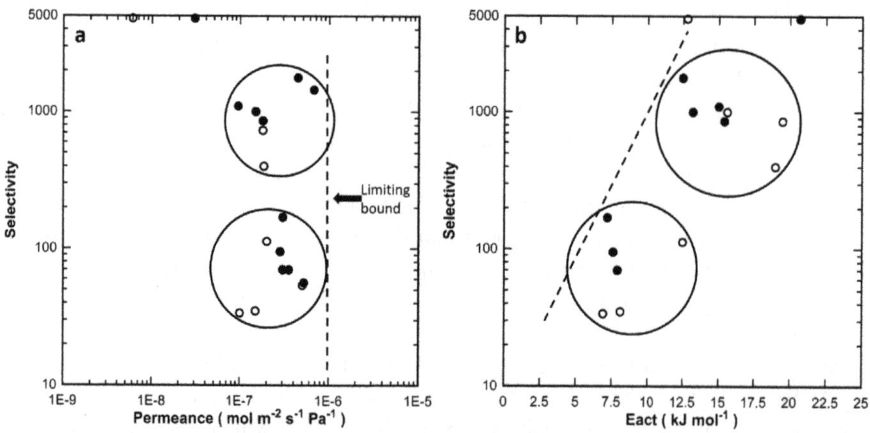

FIGURE 8.11 Third generation silica membranes (a) selectivity of He (●) or H_2(○) over N_2 versus permeance and (b) selectivity versus the apparent energy of activation for the permeance of He (●) or H_2(○). (From Kanezashi, M. et al., Separation Science and Technology. 40(1–3), 225–238, 2005; Igi, R. et al., Journal of the American Ceramic Society. 91(9), 2975–2981, 2008; Battersby, S. et al., Journal of Membrane Science. 329(1–2), 91–98, 2009; Uhlmann, D. et al., Journal of Membrane Science. 326(2), 316–321, 2009; Smart, S. et al., International Journal of Hydrogen Energy. 37(17), 12700–12707, 2012; Yacou, C. et al., Energy & Environmental Science. 5(2), 5820–5832, 2012; Wang, D.K. et al., International Journal of Hydrogen Energy. 38(18), 7394–7399, 2013; Ballinger, B. et al., Journal of Membrane Science. 451, 185–191, 2014; Wang, D.K. et al., Journal of Membrane Science. 456, 192–201, 2014; Ballinger, B. et al., Journal of Membrane Science. 489, 220–226, 2015; Darmawan, A. et al., Journal of Membrane Science. 474, 32–38, 2015.)

as cobalt iron oxides (Darmawan et al., 2015a; Darmawan et al., 2015b; Darmawan et al., 2016b), palladium cobalt oxide (Ballinger et al., 2014), and lanthanum cobalt oxide (Ballinger et al., 2015) have also been reported.

The performance of metal oxide membranes in terms of selectivity versus permeance is plotted in Figure 8.11(a). It is interesting to observe that there is a limiting bound for the metal oxide silica membranes instead of the upper bound as conventionally displayed in the Robeson's plots. The limiting bound here is the permeance of He, which is near 1×10^{-6} mol m^{-2} s^{-1} Pa^{-1}. The cluster with higher He/N_2 and H_2/N_2 selectivities is associated with cobalt silica and nickel silica membranes prepared by the dip coating method. The lower cluster is related mainly to binary metal oxide membranes and/or single metal oxide membranes prepared by the RTP method. These clusters follow the same trend in Figure 8.11(b), where the membranes with high energy of activation for He or H_2 permeance also have the highest selectivities. This suggests that as the pore size becomes smaller, the gas selectivity increases and more energy is required for a molecule to permeate through the membranes. The highest selectivity of ~5000 is associated with low permeance membranes.

8.5 INTERLAYER-FREE MEMBRANES

A very recent game-changing approach reported in the literature is the preparation of interlayer-free membranes, a departure from the conventional method of preparing silica derived membranes which required very smooth surfaces (e.g, interlayers) on macroporous substrates to avoid silica thin film cracking or pin holing, thus rendering silica membranes ineffective for separation processes. In the last five years, Diniz da Costa's group in Australia has pioneered the interlayer-free approach by preparing thin film silica-based membranes with copolymer templating (Elma et al., 2015a), cobalt oxide (Elma et al., 2015b), nickel oxide (Darmawan et al., 2016a), and hybrid organo-silica (Yang et al., 2017) directly on macroporous α–alumina substrates for desalination applications. This recent achievement of interlayer-free silica derived membranes has been attributed to changing silica precursors and silica sol-gel synthesis. The advantage of interlayer-free membranes is the reduction of the overall membrane thickness.

8.5.1 SOL-GEL SYNTHESIS

As discussed in Section 8.2.1, silica-based membranes are typically prepared by acid catalyzed sol-gel method with pH below 2, i.e., the IEP of silica species occurs. At these conditions, amorphous silica structures with molecular sieving domains are formed, and the sol-gel synthesis favours the production of silanol species. The problem here is that silica membranes derived by the acid catalyzed sol-gel method generally experience cracking if the substrate surface (with or without interlayers) is rough.

The concept of "interlayer-free" silica membranes resulted in a rethinking of conventional acid catalyzed sol-gel synthesis, as a stronger silica structure was required to be designed physico-chemically. From a sol-gel perspective, the silica sols have to be opposed to the strong capillary stress of the macroporous substrate during the fast evaporation of thin gel coating. This was achieved by designing a two-step acid-base sol-gel process, where the first step was carried out under an acidic condition for 1 h

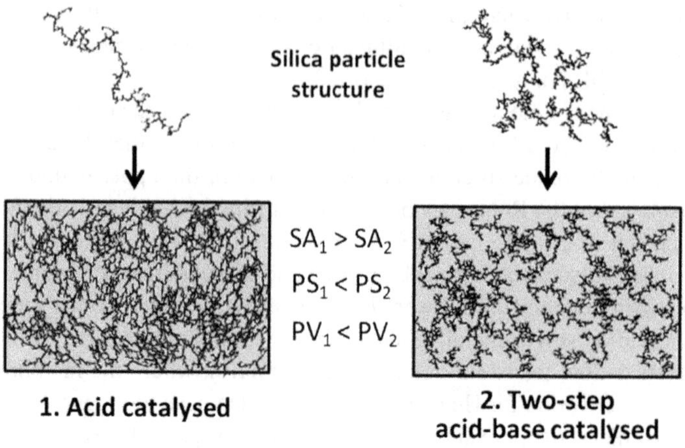

FIGURE 8.12 Schematic of silica matrices obtained from a single-step acid catalysed and a two-step acid-base catalysed sol-gel process, where SA, PS, and PV are surface area, pore size and pore volume respectively.

followed by a base catalyzed step by slowly adding ammonia ethanolic solution into the acidic sol to adjust the pH of the solution.

Acid catalyzed hydrolysis promotes a high production of silanol species from the TEOS precursor, and hence the final silica materials are ultra-microporous in texture as shown in Figure 8.12 (Wang et al., 2016). When pH of the sol is adjusted by the addition of the ammonia hydroxide in the second step, the sol pH increases rapidly to >4, which is much higher than the IEP (pH 1–3) of the silica species (Brinker and Scherer, 1990). At this instance, the silanol species are expected to be all deprotonated participating in the polycondensation reaction and generating a large concentration of highly condensed silica species with numerous siloxane bridges. Under a basic condition, the pathway of hydrolysis and condensation reactions preferentially takes on a monomer-cluster growth process which tends to produce mesoporous silica materials as demonstrated in Figure 8.12 (Wang et al., 2016). This has been shown to create a more viscous sol and thus mitigated sol infiltration effect (Elma et al., 2013; Wang et al., 2016). However, by operating the sol-gel in the basic regime, it should be cautioned that the polycondensation reaction will increase exponentially above pH 7 resulting in a cluster-cluster growth mechanism (Brinker and Scherer, 1990). This trend is also supported based on the N_2 physisorption results as shown in Figure 8.13.

As depicted in Figure 8.13(a) and 8.13(b), by controlling the pH of the sol in the second step, pore size distribution of the silica materials increases in both pore size and pore volume as well as broadening out into the mesoporous region with increasing sol pH. This is typically known as Ostwald ripening (Brinker et al., 1982; Brinker and Scherer, 1990) when sol-gel reactions occur in the basic regime causing the silica particles to grow in clusters. Particles grow even more rapidly between pH 8 and 9 resulting in a fast onset of gelation (Elma et al., 2013), which is not desirable for membrane dip coating.

From this collection of work in the optimisation of the two-step acid-base sol-gel process for the interlayer-free membrane concept, the pH range of the silica sol between 4 and 6 was found to produce the optimal condition for dip coating directly onto macroporous

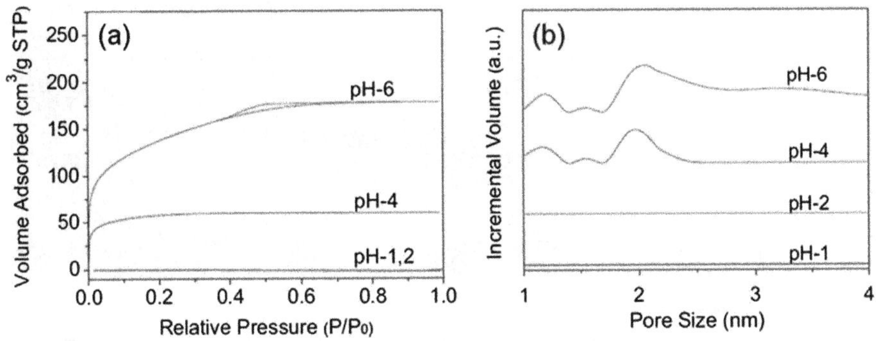

FIGURE 8.13 (a) Nitrogen physi-sorption and (b) DFT pore size distribution of the silica materials as a function of pH. (From Wang, S. et al., Journal of Membrane Science. 516, 94–103, 2016.)

substrates. This strategy was further developed to produce the desired final silica properties and structure of the membrane thin film for both the interlayer-free membranes and RTP membrane concepts. The requirements of these coating methods are not only that the silica thin film needs to be structurally strong on the macroporous substrates without peeling or pin-hole defects but that it also has to withstand the thermal shock and stress during the RTP technique, which will be discussed in the following section.

Liu et al. (2015c) demonstrated that interlayer-free silica thin film membranes can be prepared using a one-pot synthesis technique, called the seeding sol-gel technique. In their work, interlayer-free membranes were fabricated by incorporating a homogeneous colloidal silica into the polymeric cobalt silica sol before membrane dip coating. The silica colloid was prepared using the standard Stöber process in a basic sol-gel condition using ammonia, which often produced mesoporous silica structures (Costa et al., 2003; Topuz and Çiftçioğlu, 2010). This approach promoted heterogeneous nucleation sites for the silica source (TEOS) by creating structure-directing and anchoring properties inside the seeded gelling suspension.

As shown in Figure 8.14(a), dynamic light scattering shows the silica colloids and the seeded cobalt silica samples (SCoSi) have a single particle size distribution centered around 70 nm, apart from the highest concentrated SCoSi sample with 0.8 mole ratio (against TEOS moles). This is also confirmed in the TEM image of uncalcined and calcined colloid silica sample (Figure 8.14(b) and 8.14(c), respectively). Similarly, the morphologies of seeded samples of 0.4SCoSi and 0.8SCoSi (Figure 8.14(e) and 8.14(f)) were also observed to be similar to that of the 0.0SCoSi sample (Figure 8.14(d)), and no discrete silica particles could be clearly distinguished. This is because the silica sol penetrates into the colloidal silica seeds to form crosslinks (condensation reaction) at the surface and in the mesoporous structure of the seed. Upon gelling and calcination, this process leads to the formation of a 3D-continuous CoSi amorphous network with microporous texture. There is strong evidence that the seeds used in the work of Liu et al. (2015c) have heterogeneous nucleation sites for the silica source (TEOS) that could create structure-directing pathways, anchoring points inside the seeded gelling sol. The combined colloidal/polymeric sol can then be used to directly dip coat macroporous alumina substrates without further priming with interlayers.

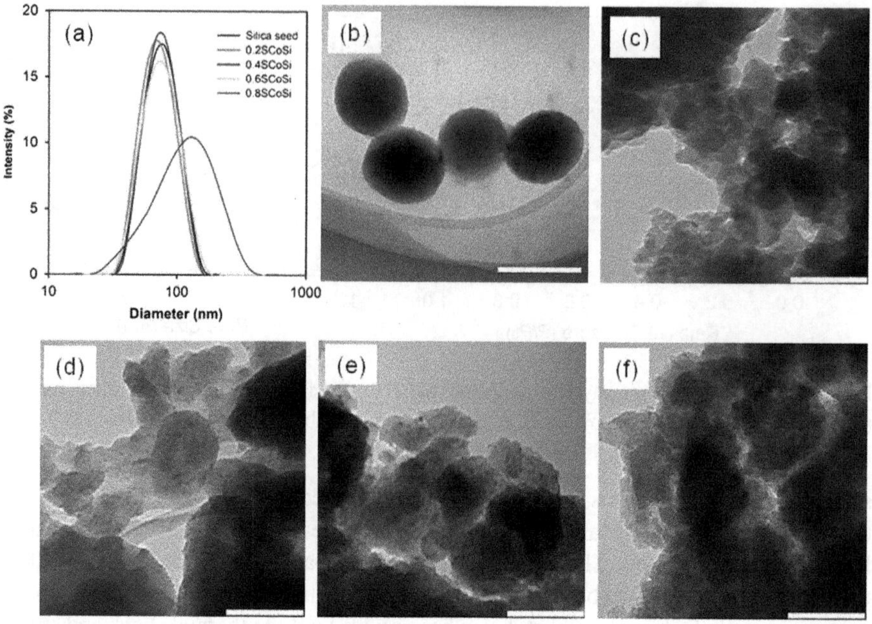

FIGURE 8.14 (a) Particle size distribution measured by DLS and TEM images of (b) uncalcined silica seed and (c) calcined silica seed and (d)–(f) calcined 0.0SCoSi, 0.4SCoSi and 0.8SCoSi (where seed concentration ranged between 0 to 0.8 mole ratio with respect to TEOS). Scale bar is 100 nm. (From Liu, L. et al., Journal of Membrane Science. 492, 1–8, 2015.)

8.5.2 Membrane Preparation

The advantage of interlayer-free microporous silica membranes is the reduction of the overall membrane preparation time and membrane thickness. Interlayer-free silica membranes, like any other silica thin film supported on membrane tubes, need to be high quality, defect-free, and continuous to enable a good separation performance. To achieve the target structure, deposition of silica sol on macroporous support is typically carried out by dip coating followed by a thermal calcination step, which is identical to the acid-catalysed sol-gel silica membrane preparation. As discussed in the previous section, this process of dip coating and calcination is repeated over several cycles to minimise any membrane defects. A recent series of studies on interlayer-free silica derived membranes has shown that the cycle of dip coating can be reduced from six and four to two layers by simply controlling the silica sol-gel synthesis as shown by Table 8.2.

Figures 8.15(a) and (b) show scanning electron microscope images of an interlayer-free silica membrane and the underlying alumina substrate (Elma et al., 2013). Due to the lack of an interlayer, the membrane shows a lack of clear boundary between the top layer and the macroporous substrate layer. This is due to the fact that the top layer sol is expected to infiltrate into the macropores of the substrate to a certain extent where the intermingling between the two layers is unavoidable. The thickness of the

TABLE 8.2
List of Interlayer-Free Silica-Based Membranes Reported in the Recent Literature

Membrane Type	Sol-Gel Process[a]	Calcination Temp. (°C)/ Type[b]	Number of Layers	Application	Year	Reference
Seeded cobalt oxide silica	1-step, one-pot A	630/CTP	6	Gas separation	2015	(Liu et al., 2015c)
Silica	2-step A-B	600/CTP	4	Desalination	2013	(Elma et al., 2013)
Cobalt oxide silica	2-step A-B	600/CTP	4	Desalination	2015	(Elma et al., 2015b)
Carbonised P123[c] silica	2-step A-B	450/CTP	4	Desalination	2015	(Elma et al., 2015a)
Nickel oxide silica	1-step H_2O_2	500/CTP	5	Desalination	2016	(Darmawan et al., 2016a)
Carbonised P123-TEVS[d] silica	2-step A-B	450/CTP	3	Desalination	2017	(Yang et al., 2017)
Silica	2-step A-B	630/RTP	2	Desalination	2016	(Wang et al., 2016)
Silica	2-step A-B	630/RTP	2	Desalination	2017	(Wang et al., 2017)

[a] One-step acid catalysis (1-step A) and two-step acid-base catalysis (2-step A-B).
[b] Conventional thermal processing (CTP; 1–2°C min^{-1}), rapid thermal processing (RTP; 100°C s-1).
[c] Pluronic triblock copolymer – poly(ethylene oxide)-poly(propylene oxide)-poly(ethylene oxide).
[d] TEVS – triethoxyvinylsilane.

silica-alumina intermix layer was estimated to be ~470 nm as shown by the inset of Figure 8.15(a). Also, the surface morphology of the membrane clearly shows that the roughness of the macroporous alumina substrate is translated onto the top, which essentially followed the morphology of the substrate (Figure 8.15(b)). Despite this, the top layer appears to be homogeneously covered by silica thin film without any major cracks or defects. Similarly, in subsequent works of interlayer-free carbonised P123 silica membranes (Figure 8.15(c) and (d)) and cobalt oxide silica membranes (Figure 8.15(e) and (f)), some degree of infiltration effect is observed with these membranes.

For the application of high temperature gas separation, interlayer-free cobalt oxide silica membranes seeded with silica colloids were reported for the first time by Liu et al. (2015c), as shown in Figure 8.16(a) and (b). SEM images of the cross-section and surface morphology of the unseeded membranes (Figure 8.16(c) and (e)) and seeded membranes (Figure 8.16(d) and (f)) show a marked difference. The unseeded membrane resulted in a deep infiltration of cobalt oxide silica gel into the macroporous substrate, and a distinct intermix film layer cannot be seen. Also the surface

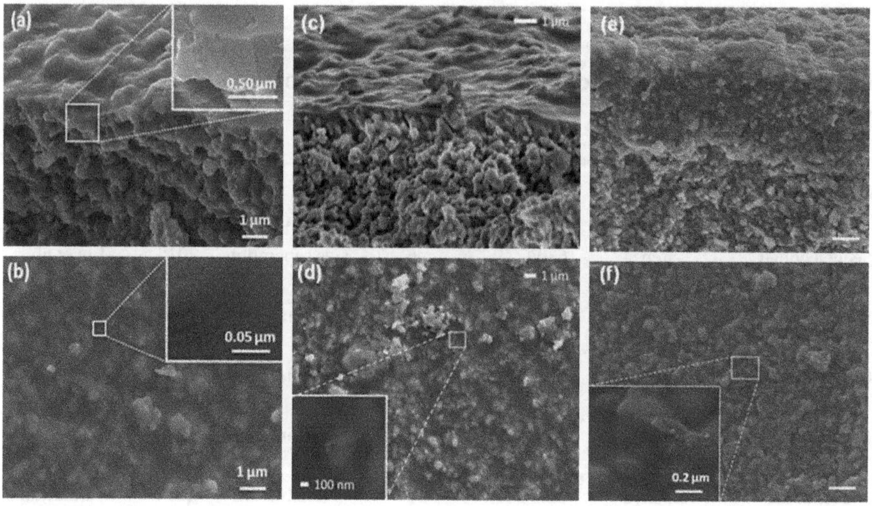

FIGURE 8.15 SEM images of the interlayer-free membranes in cross-section (top images) and top surface (bottom images) for pure silica (a) and (b), carbonised P123 silica (c) and (d), cobalt oxide silica (e) and (f), with inset images showing a close up of the respective morphology. (From Elma, M. et al., Journal of Membrane Science. 475, 376–383, 2015; Elma, M. et al., Desalination. 365, 308–315, 2015; Elma, M. et al., Membranes. 3(3), 136–150, 2013.)

can be seen to be very rough with very low surface coverage. This further supports that without an interlayer, the unseeded membrane prepared using the conventional cobalt silica sol infiltrates deep into the macroporous substrate and the possibility of forming major defects is very high, even after several cycles of dip coating and calcination processes.

In contrast, the seeded membranes show a clearer intermix film layer on the substrate and the infiltration effect is minimised somewhat, which offered additional stability from a structural standpoint. This work demonstrates for the first time that the preparation of ultra-microporous silica derived thin films achieved via a sol-gel seeding approach can be directly coated onto a macroporous substrate without any interlayers for gas separation applications. These relatively large silica colloids provided an anchoring point and effectively minimised the sol from infiltrating deeper into the macropores of the substrate, leading to the formation of a thin intermix film layer after several cycles of film priming. The resulting calcined intermix film was stabilized by the interconnected domains of pre-formed colloidal seeds and the microporous silica network with the cobalt oxide species.

8.5.3 Membrane Performance

Interlayer-free silica membranes are expected to reduce the resistance to molecular transport whilst keeping a good molecular selectivity as conferred by their molecular sieving properties. As shown in Table 8.2, the only work that shows that interlayer-free cobalt

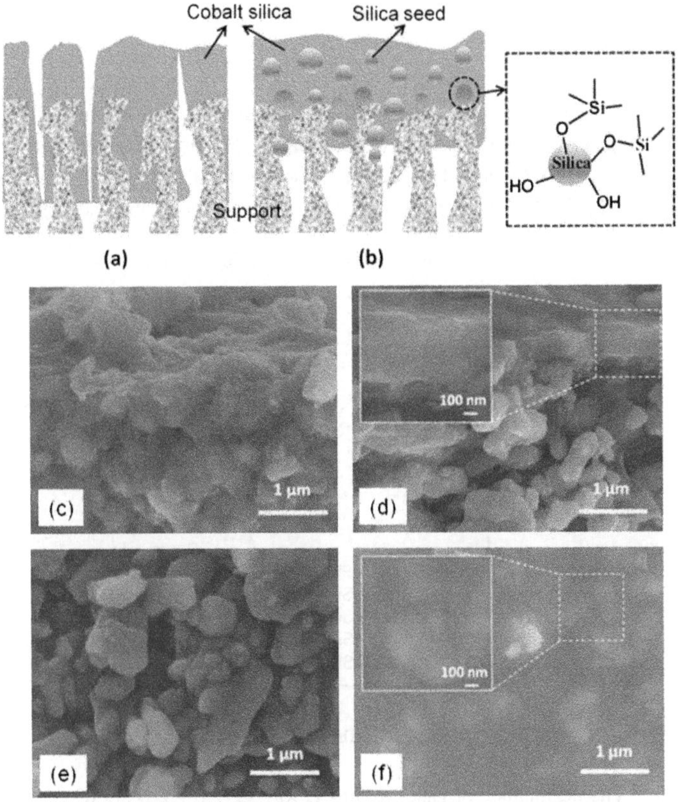

FIGURE 8.16 Schematic interlayer-free membrane structure of (a) unseeded membranes and (b) colloid seeded cobalt oxide silica membranes, and SEM images of (c) unseeded membranes and (d) seeded membranes in cross-sections and their corresponding surfaces in (e) and (f), respectively. (From Liu, L. et al., Journal of Membrane Science. 492, 1–8, 2015.)

oxide silica membranes can be produced via silica seeding sol-gel process was reported by Liu et al. (2015c) in 2015. In general, acid-catalyzed silica membranes with interlayers show excellent gas selectivities but are relatively low in transport flux, which is anticipated due to the total transport resistance of such asymmetric membrane configuration.

As discussed in the previous section, typical performance values of He and H_2 permeance in the region of 10^{-7} mol m^{-2} s^{-1} Pa^{-1} for conventional metal oxide silica membranes are reported, which is consistent with the majority of the published results (Smart et al., 2012; Miller et al., 2013; Darmawan et al., 2015a), and the He/N_2 and H_2/CO_2 permselectivities of 170 and 30 at 450°C are also relatively comparable. Although much higher values of H_2 permeance of 10^{-6} mol m^{-2} s^{-1} Pa^{-1} and the He/N_2 and H_2/CO_2 permselectivities (450°C) of >200 have also been reported for pure silica (de Vos and Verweij, 1998a, 1998b; de Vos et al., 1999; Tsuru et al., 2011), cobalt oxide silica (Igi et al., 2008; Tsuru et al., 2011; Yacou et al., 2012), and nickel oxide silica (Kanezashi and Asaeda, 2006) membranes. However, the best results are based on the evolution of conventional cobalt oxide membranes with interlayers for over a decade of research and development.

In comparison, as seen in Figures 8.17 and 8.18, the best interlayer-free seeded membrane with 0.4 mole ratio of the colloidal seeds calculated against TEOS mole ratio (0.4SCoSi) delivered a He and H_2 permeance of approximately 2.6 × 10^{-7} mol m^{-2} s^{-1} Pa^{-1} and a He/CO_2 and He/N_2 permselectivity of 80 and 97, respectively at 500°C (Liu et al., 2015c). On the other hand, the control, non-seeded membrane

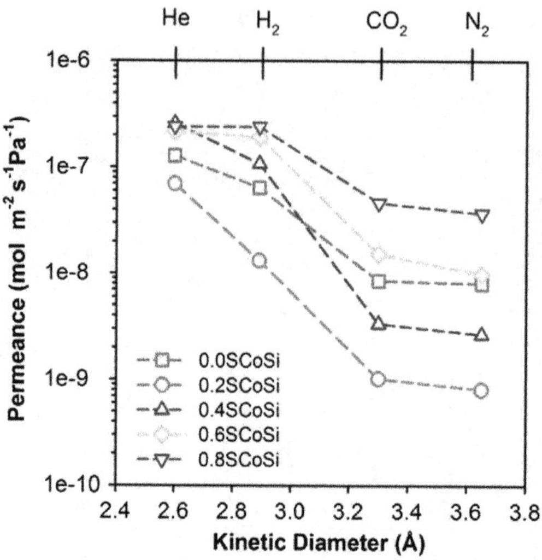

FIGURE 8.17 The permeance of gases at 500°C for He (2.6 Å), H_2 (2.89 Å), CO_2 (3.3 Å), and N_2 (3.65 Å) against their molecular kinetic diameters. (From Liu, L. et al., Journal of Membrane Science. 492, 1–8, 2015.)

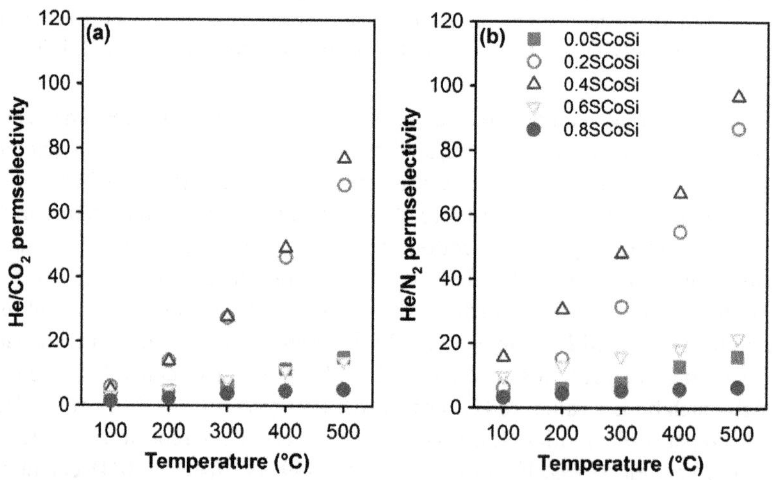

FIGURE 8.18 (a) He/CO_2 and (b) He/N_2 permselectivities of 0.0–0.8SCoSi membranes as a function of temperature from 100 to 500°C. (From Liu, L. et al., Journal of Membrane Science. 492, 1–8, 2015.)

showed only a He/N_2 permselectivity of 16 at 500°C and surprisingly with 50–70% lower permeance values than those of 0.4SCoSi membrane. Despite similar gas separation performances found between conventional silica membranes with interlayers and interlayer-free silica membranes, this seeding technique allowed the silica colloid to effectively block the macropores of the substrate and simultaneously assisted in the formation of a near defect-free selective membrane layer intermix by silica and alumina.

8.6 RAPID THERMAL PROCESSING

The majority of microporous, sol-gel derived, silica-based membranes have been prepared in the conventional way using a slow thermal processing technique which takes over 10–14 days to complete. Traditional belief for calcination of inorganic membranes imparts that a very slow heating rate reduces thermal stresses and therefore cracks on the silica thin film, which have employed heating rates of 1.0 or 0.5°C min^{-1} (Boffa et al. 2008a; Luiten et al., 2010; Darmawan et al., 2015a; Liu et al., 2015b) for calcination temperatures up to 650°C. In addition, a dwell time of 2–8 h followed by a slow cooling rates profile was used. As silica membranes typically need 3 to 6 layers of silica coating, the number of dipping cycles and slow calcination to achieve a near "defect-free" top layer is very time-consuming, and thus limits its large scale application. In view of this, it is very important to develop a processing technology to ensure that the ease, speed, and reproducibility are optimized.

RTP technique of inorganic membranes particularly with silica-based membranes is an emerging technique that vastly improves time and efficiency of inorganic membrane production. As first introduced by the semiconductor industry for thin film consolidation in the 1950s to principally "reduce thermal budget, process uniformity, high throughput, ease of process development, and low manufacturing cost" (Fiory, 2005), the concept of RTP is applied during the thermal treatment of inorganic membranes by using a simple rapid firing and short dwell time protocol, which dramatically decreases the time of thermal processing from several days to a few hours.

8.6.1 Sol-Gel Synthesis

The guiding principles of sol-gel chemistry for the RTP protocol are very similar to those of the conventional protocol. The chemistry is controlled by the hydrolysis and condensation mechanisms that generate silanol and siloxane groups, respectively, as described by Eqs 8.1–8.3. Similarly, both reactions can be accelerated by adding acidic or basic catalysts to direct the process of silicon network connectivity, which in turn governs the network pore size at the molecular scale. However, due to the fast heating rate and short firing time of the RTP protocol, the sol-gel process requires modification to accommodate the rapid thermal profile during thermal consolidation of the silica membranes. To date, the literature in this field details that the major difference lies in the use of silica oligomer precursor, known as ethyl silica 40 (ES40), instead of the commonly favored TEOS precursor. ES40 is an industrially derived silica oligomer from the pre-hydrolysis and partially-condensed TEOS with a general chemical structure as shown in Figure 8.19.

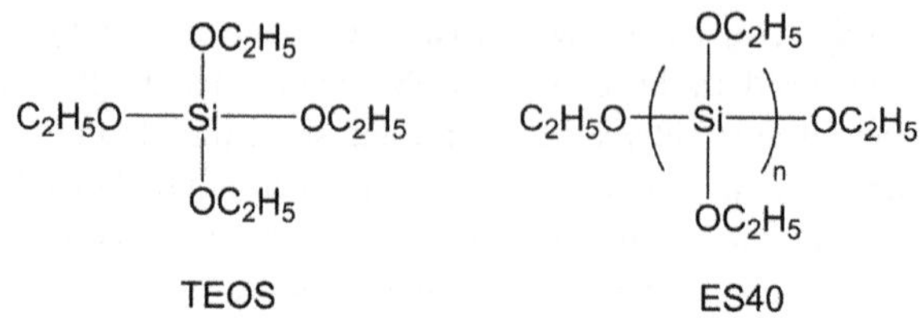

FIGURE 8.19 Chemical structures of TEOS and ES40 where n represents 2 to 9 repeating units.

As seen from its chemical structure, the ES40 oligomeric molecule with an average of five silicon atoms has a higher number of ethoxy functional groups per molecule, which gives rise to predominately Q^1 and Q^2 species. In contrast, the TEOS precursor consists of only the unhydrolyzed Q^0 silicon species, which are less acidic (reactive) than their ES40 counterparts. Due to this structural difference, the miscibility of the same given sol-gel process involving water, ethanol, and ES40 is quite different than that of the TEOS system, and it should be well understood before sol-gel reaction is carried out. In a study (Wang et al., 2015b) as displayed in Figure 8.20, ternary mixture of ES40 sol exhibited a reduced miscible area, which is shown in the region under the red curve. This important finding is crucial to the ES40 sol-gel process as it governs the miscibility of the system and the reaction pathway of silica polymerization. Also, the degree of phase separation of the ternary mixture is further increased in the case of ES40 system because of the propensity to form larger silica particles during the sol-gel process (Miller et al., 2013; Wang et al., 2013). Therefore, the ES40 sol-gel process should be closely monitored.

In addition, for the same given sol-gel process, condensation reaction preferentially occurs via the Q^1 and the more acidic Q^2 species leading to a terminus-pendant

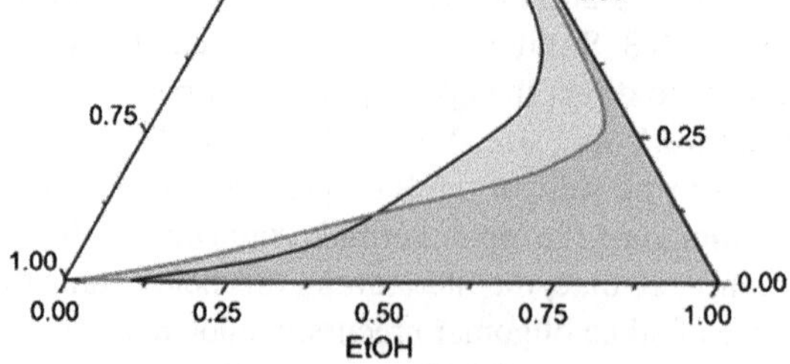

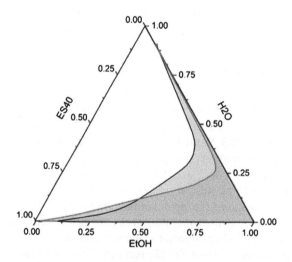

FIGURE 8.20 Ternary phase diagram of ES40-ethanol-water (gray line) and TEOS-ethanol-water system (black line) at 25°C. (From Wang, S. et al., Scientific Reports. 5, 14560, 2015.)

branching polymerization instead of a linear chain growth (as is the case with TEOS). Such polymerization behaviour would typically lead to a cluster-cluster growth of silica polymer with more siloxane bridges (Si–O–Si), whereby a thicker and more stable film is achieved (Mrowiec-Białoń et al., 2004; Miller et al., 2013; Wang et al., 2014). These properties are much desired for gas separation applications involving wet gases. As the stability of silica membranes in the presence of water or steam is an ongoing area of active research (Gavalas et al., 1989; Gallaher and Liu, 1994; Imai et al., 1997; Cassiers et al., 2002; Zhang et al., 2005), it has been frequently described that this behavior is caused by the adsorption of water molecules onto the hydrophilic silanol groups (Si–OH) located at the surface of silica, which promotes rehydration and recondensation of neighbouring siloxane bonds. Such behavior eventually leads to a reconstruction of the matrix to more a stable and denser structure (Iler, 1979), and hence leads to a loss of both gas flux and selectivity.

Furthermore, hydrothermal stability of microporous silica materials derived from ES40 sol-gel process has been investigated under a continuous 20-h exposure of high steam content of 75 mol% at 550°C (Wang et al., 2015a). It was found that the effects of sol-gel ratios of water (hydrolysing agent), acid (catalyst), and ethanol (solvent) on the microstructure of the silica matrices and their hydrothermal stability can be finely tuned by changing these ratios. Figure 8.21(a) clearly shows that the total pore volume of the ES40-derived silica matrices increases with increasing H_2O and HNO_3 ratios, yet increasing EtOH produced an opposite trend. In contrast, the structural integrity of these silica matrices as seen in Figure 8.21(b) exhibits that samples prepared with relatively high concentrations of H_2O or HNO_3 and low EtOH ratio led to a more hydrothermally stable structure. It was deduced from the results of small-angle X-ray scattering, FTIR, and solid-state ^{29}Si NMR that the silica microstructure and its hydrothermal stability are attributed to the formation of a more robust, open microstructure condensed by clusters of silica particles and a lower proportion of silanol groups in the matrix.

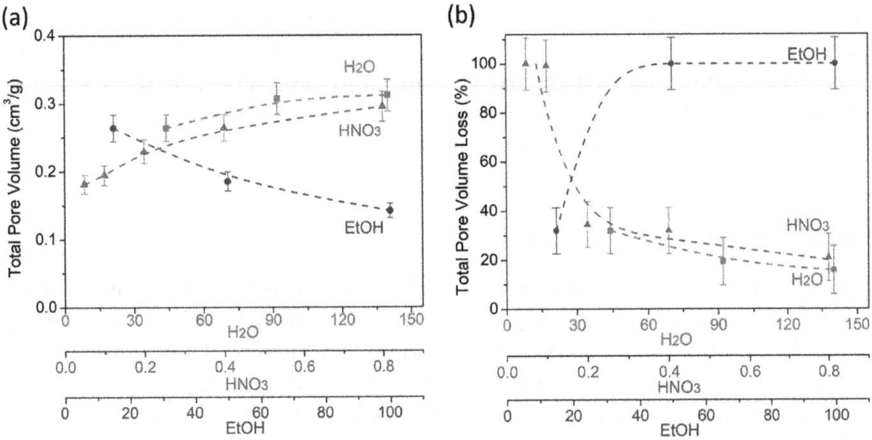

FIGURE 8.21 (a) Total pore volume and (b) total pore volume loss of silica xerogels prepared from different ratio of H_2O (medium gray ■), HNO_3 (dark gray ▲) and EtOH (black ●) to Si, respectively. (From Wang, S. et al., RSC Advances. 5(8), 6092–6099, 2015.)

8.6.2 Membrane Preparation

The route of silica membrane preparation via the RTP technique follows the same initial steps of dip coating of inorganic support using the ES40 sol. The dip coating process is a simple and versatile process for preparing flat and tubular membranes on small or large surfaces, which is covered in the previous section. This is followed by a short drying time of approximately 30 min in an oven to remove excess ethanol prior to RTP calcination to avoid the ethanol reaching its auto-ignition point during calcination.

RTP calcination is characterised by temperature ramping rates of >100°C s^{-1} as the silica membranes are subjected to a pre-heated furnace equilibrated at 600–630°C for a duration of 1–60 min. The first RTP silica membrane was made from a hybrid process in which the membranes were calcined in a pre-heated environment, albeit cooling down occurred in the furnace at 1°C min^{-1} (Wang et al., 2013). By changing the sol-gel processing condition, Wang et al. (2014) explored the addition of an excess amount of HNO_3, which is an important reactant in producing a microporous, stable matrix. This simple but necessary change to the sol-gel process paved the way to the first RTP silica membranes which were directly introduced into a furnace at 600°C, with a dwell time of 1 h, and then directly withdrawn from the furnace to cool down at room temperature conditions. The successful outcome is the preparation of RTP silica membranes within short a period of time (a few hours) instead of several days (6–12 days) as needed by the conventional thermal processing (CTP) silica membranes.

SEM micrographs of the prepared membranes, shown in Figure 8.22, exhibit that the silica film derived from CTP and TEOS has a homogeneous and smooth thin layer (Smart et al., 2012). In contrast, the hybrid RTP ES40 derived silica film showed the absence of cracks, and the presence of small micron sized blemishes seemed to be thin and isolated islets (Wang et al., 2013). In comparison, the SEM images of the RTP ES40 silica membrane closely resemble the morphology of the hybrid system, but the number of the blemishes increases significantly even though their size decreases to 20–100 nm (Wang et al., 2014). These craters and pits appear to be very shallow and isolated and were found to have very little impact on the apparent membrane selectivity, which agrees well with another RTP pure silica membranes study (Van Gestel et al., 2014).

8.6.3 Membrane Performance

Microporous silica membranes have attracted a great deal of attention due to their unique molecular sieving property for gas separation (de Vos and Verweij, 1998a; 1998b; Peters et al., 2005; Battersby et al., 2009). Contrary to the well-established applications of ES40 in mesoporous materials (Mrowiec-Białoń et al., 2004; Gaydhankar et al., 2005; Gaydhankar et al., 2007; Mrowiec-Białoń and Jarzębski, 2008), the synthesis of microporous silica membranes using ES40 for molecular sieving application is still in its infancy. Miller et al. (2013) showed a redox effect of CTP derived cobalt oxide ES40 silica membranes that exhibit reversible gas molecular sieving for high temperature gas separation. They attributed the robust structure

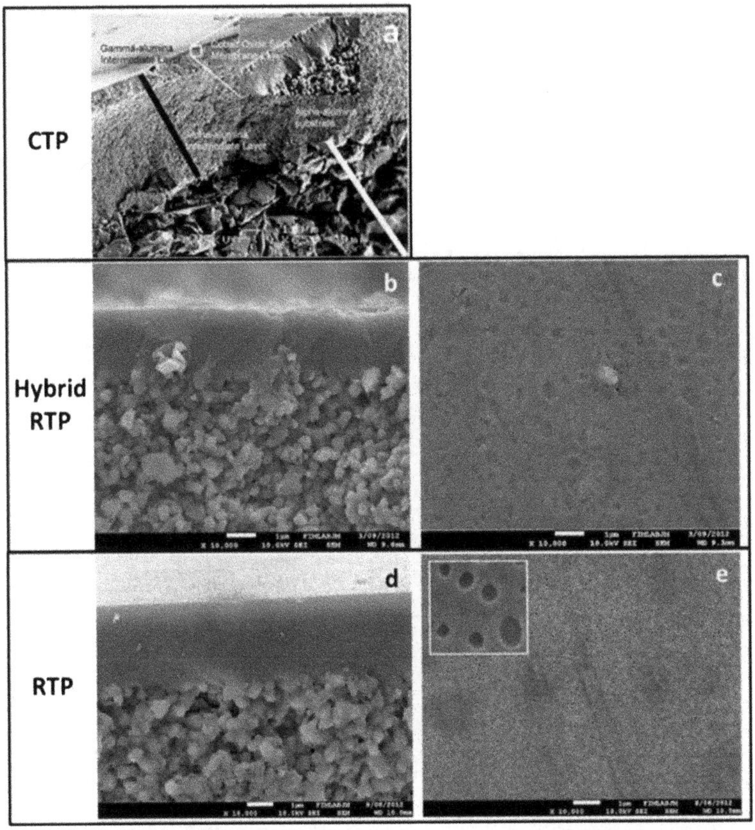

FIGURE 8.22 SEM micrographs of an (a) CTP cobalt silica (TEOS) membranes where the black and white bars are guides to identify the substrate, the various interlayers and the membrane layer itself (insert) (From Smart, S. et al., International Journal of Hydrogen Energy. 37(17), 12700–12707, 2012.); (b) cross-section and (c) top surface of a hybrid RTP cobalt silica (ES40) membrane (From Wang, D.K. et al., International Journal of Hydrogen Energy. 38(18), 7394–7399, 2013.); (d) cross-section and (e) top surface of an RTP cobalt silica (ES40) membrane (From Wang, D.K. et al., Journal of Membrane Science. 456, 192–201, 2014.).

to the formation of more open silica network in the case of ES40. Wang et al. (2013; 2014) later showed the development of RTP derived cobalt oxide ES40 silica membranes in which the gas permeation results of He and H_2 from CO_2 and N_2 gases were demonstrated to be superior by an order of magnitude to the TEOS counterpart. Issues anticipated from rapid calcination process, such as cracking and peeling, were not observed by employing ES40 as a precursor.

As shown in Figure 8.23, the gas transport of the RTP cobalt oxide ES40 silica membranes (CoES40 and CoES40_A) complies with an activated transport process with respect to temperature (Wang et al., 2014). The transport mechanism is characterised by a positive activation energy for smaller gases (He and H_2) and a negative activation energy for the larger gas molecules (CO_2 and N_2), which demonstrates that the separation is governed by molecular sieving mechanism. On the other hand,

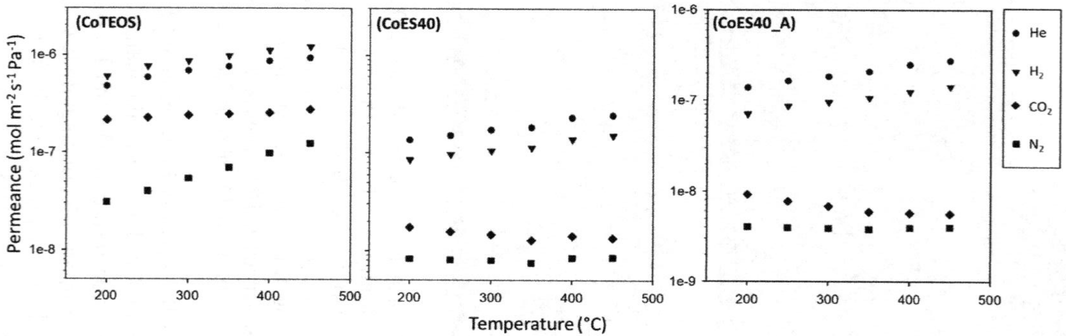

FIGURE 8.23 Gas permeances of He, H_2, CO_2 and N_2 as a function of temperature for RTP cobalt oxide silica membranes derived from TEOS, and ES40 without and with Acid addition (Si:HNO_3 mole ratio is 4:0.15). (From Wang, D.K. et al., Journal of Membrane Science. 456, 192–201, 2014.)

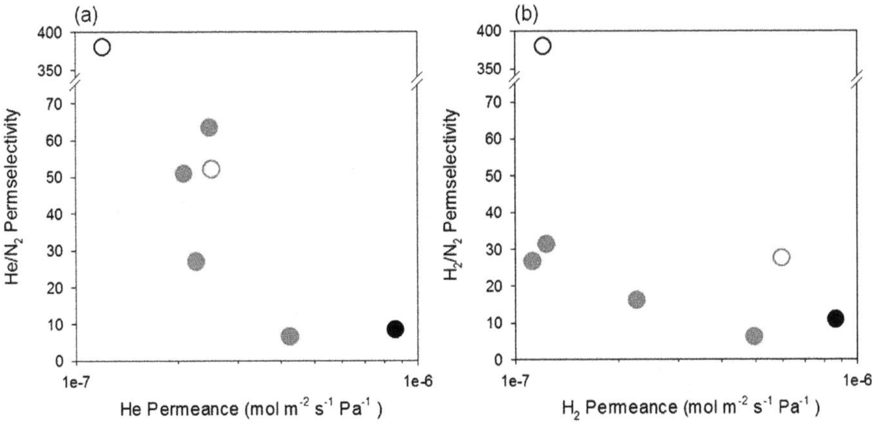

FIGURE 8.24 Gas permeances of (a) He and (b) H_2 versus He/N_2 and H_2/N_2 permselectivities at 400°C for RTP (filled circles) (From Wang, D.K. et al., Journal of Membrane Science. 456, 192–201, 2014.) and CTP (unfilled circles) (From Yacou, C. et al., Energy & Environmental Science. 5(2), 5820–5832, 2012; Miller, C.R. et al., Scientific Reports. 3, 1648, 2013.) derived cobalt oxide silica from TEOS (black circles) and ES40 (gray circles) membranes.

RTP cobalt oxide TEOS silica membrane (CoTEOS) showed positive energy of activation for all of the four gases with a much lower separation performance.

As further demonstrated in Figure 8.24, the state-of-the-art CTP derived cobalt oxide silica (TEOS) membranes (unfilled black circle) show He and H_2 permeance in the region of 10^{-7} mol m^{-2} s^{-1} Pa^{-1} and He/N_2 and H_2/N_2 permselectivities of ~380 at 400°C, which is consistent with the majority of the results published for CTP metal oxide silica membranes (Smart et al., 2012; Yacou et al., 2012). In comparison, the CTP cobalt oxide silica (ES40) membranes show an exceptional permeance with 200% (He) to 500% (H_2) increase in gas permeances yet permselectivities are an order of magnitude lower (Miller et al., 2013) due to the slightly enlarged pore size from the thermal consolidation of larger silica clusters. In contrast, the RTP derived CoES40 membranes with 95% of production time saved delivered very similar gas permeation results as the CTP counterpart. However, when nitric acid was incorporated into the ES40 sol gel, the RTP membranes (filled red circles) produced the best permselectivity performance and comparable gas permeances. Although the permselectivity results were not as high as the CoTEOS-CTP membrane, the processing time and reproducibility are significantly improved by using the ES40 substitute. Importantly, the scalability of RTP process is proven to be robust for high quality tubular silica membranes.

8.7 FUTURE OUTLOOK

A large body of work and knowledge on silica membranes for gas separation has been amassed over the last 30 years. Many of the initial challenges with pore size control, improving gas fluxes, and selectivities have been met. However, several challenges remain for the industrial uptake of this technology. Currently there is limited literature related to scale-up and long-term testing. One of the few works reported by

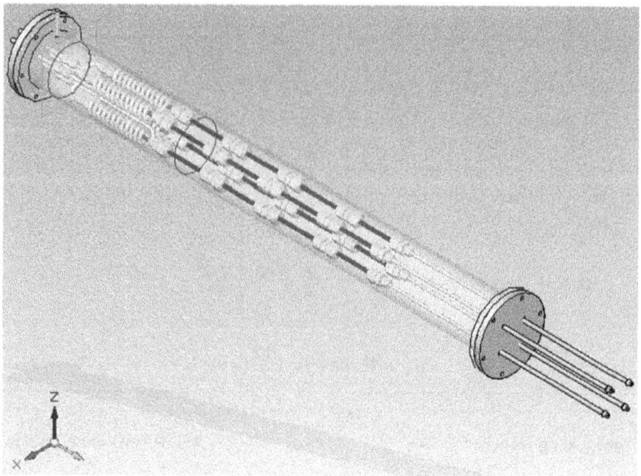

FIGURE 8.25 Mechanical design of a multi membrane module similar to the scale-up module reported by Yacou et al. (From Yacou, C. et al., Energy & Environmental Science. 5(2), 5820–5832, 2012.)

Yacou and co-workers (2012) showed cobalt silica membranes were stable for 2000 h operation for dry gas processing at a temperature up to 500°C. This work included scaling-up membranes to a multi-tube membrane module as depicted in Figure 8.25. This work also solved a major issue reported by Duke et al. (2004a) in the early 2000s related to the shearing of ceramic tubes. Yacou and co-workers (2012) also developed special graphite seals (see Figure 8.26) that could be accommodated as a ferrule in a Swagelok compression fitting. The challenge here is to design large membrane modules for gas separation that takes place at high temperatures and pressures. There are serious engineering problems to be dealt with, such as accommodating materials with different coefficients of thermal expansion. For instance, pressure vessels are generally steel built to sustain large pressures while membranes are ceramics. Future work should be directed towards close collaboration between mechanical engineers to work on design, materials engineers to work on sealing and robust ceramic tubes, and chemical engineers to address gas separation processes.

Scaling-up brings a new set of challenges in terms of module design. Computational fluid dynamics (CFD) is therefore an important tool to study the transport of gases in large membrane modules. CFD modeling for silica membranes was initially developed by Abdel-jawad et al. (2007), where both sides of the membrane were conjugated into the same simulation by linking mass transfer from the feed side to the permeate side. This allowed for the Navier Stokes equation to be solved, which was otherwise not possible as the membrane itself represents a discontinuity. However, this initial CFD modeling and validation work was based on small modules operating under ideal conditions, where the driving force was constant along the membrane axis. Ji and co-workers (2012; 2013; 2014) developed CFD models for large membrane modules that were validated using the experimental data from the multi-tube scale-up cobalt silica membrane module reported by Yacou et al. (2012). Very interesting results were found for a binary mixture containing He and Ar, as He preferentially permeated.

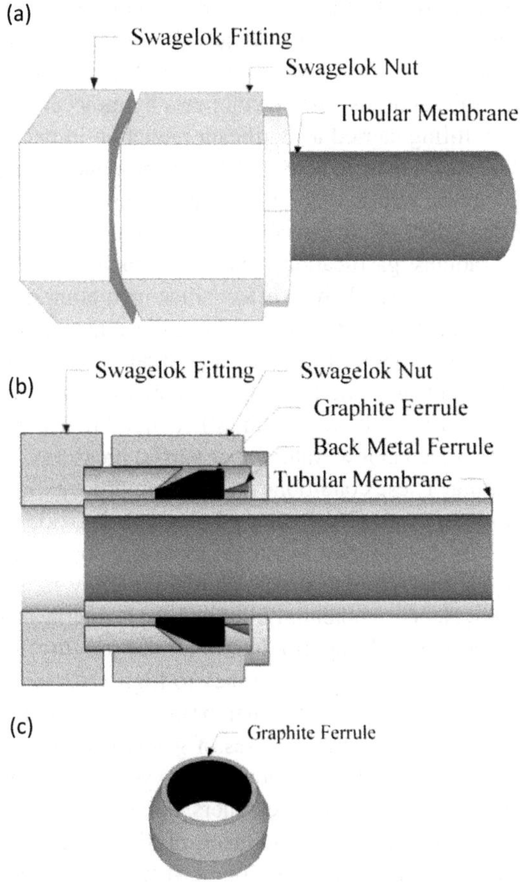

FIGURE 8.26 Membrane sealing design (a) assembled membrane, (b) cut-away of assembled membrane showing graphite ferrule and back metal ferrule in place, and (c) graphite ferrule. (From Yacou, C. et al., Energy & Environmental Science. 5(2), 5820–5832, 2012.)

As a result, the concentration of Ar increases and He decreases along the module axis, thus affecting the driving force for the permeation of He. As the concentration of Ar increases, there is back diffusion of Ar molecules from the outlet (retentate) to the inlet (feed) due to differential chemical potential. In a subsequent work, Ji et al. (2013) investigated how process conditions such as feed flow rate, pressure, and temperature, could affect permeation, gas purity in the permeate stream, and gas recovery. These initial works show the importance of CFD modeling to assist in designing large membrane modules and process engineering. However, more research is required as industrial gases generally contain water vapor and condensable vapors, which brings a different set of modeling conditions for CFD simulation.

A large body of work in silica membranes deals with permeating pure gases only. In terms of experimental work, this provides a simple approach to study the fundamental transport phenomena of gases though membranes. Several works on silica membranes also address simple gas mixtures by mixing gases prior to the membrane, or in as membrane reactors. Industrial gas separation is more challenging and condensable vapors are quite common. Yoshioka et al. (2003) studied the effect of condensable gases in microporous silica membranes by molecular dynamics simulation where micropore filling caused a significant reduction in permeation. Recently, Deonarine et al. (2017) used toluene as a model tar compound and reported for a feed gas containing 0.24 mol% toluene. The permeance decreased by an average value of 17%. Therefore, the effect of condensable gases can be significant. As there is an increasing interest in biomass gasification, which is a carbon neutral form of energy for H_2 production, biomass tars are likely to affect silica membranes. Therefore, further research is warranted on the effect of condensable gases and membrane regeneration.

Silica membranes work very well for dry gas separation, but hydrothermal stability still remains a problem. Although significant improvements have been achieved by carbonised template silica and metal oxide silica membranes, the majority of this work has in fact been carried at low water vapor partial pressures, low temperatures, and/or short term testing. These conditions are far from those expected for industrial gas separation. Uhlmann et al. (2010) showed that cobalt oxide silica membranes lost their gas selectivity capabilities by steam exposure to 450°C. Many of these problems have been attributed to the gamma alumina interlayers, which are unstable when exposed to water. Hence, a potential solution here is interlayer-free membranes, or interlayers with materials with superior hydrothermal stability. Recently, Liu et al. (2015b) exposed cobalt oxide silica membranes to high water vapor concentration (50 mol%) up to 500°C. Subsequently, the membranes were tested for single gas and gas mixture separation, showing minor losses of gas selectivity. However, proof of concept in long term testing of silica membranes has yet to be reported and remains a challenging solution that has eluded researchers so far.

Finally, cost reduction in the manufacturing of silica membranes is still necessary. The materials used in the preparation of silica membranes are reasonably inexpensive. However, the number of layers and film coating steps, in addition to slow thermal processes, make the production cost expensive. Recently reports on the preparation of interlayer-free membranes (Elma et al., 2015a; Elma et al. 2015b; Liu et al., 2015c; Darmawan et al., 2016a; Yang et al., 2017), rapid thermal treatment (Wang et al., 2013; Wang et al., 2014), and combined interlayer-free with rapid thermal treatment (Wang et al., 2016; Wang et al., 2017) are showing for the first time that cost reduction and fast production of silica membranes is possible. Nevertheless, many of these silica derived membranes have been developed for pervaporation process, and high gas temperature separation requires a better control of pore size. Hence, there is a need to develop cheaper silica derived membranes to be competitive for industrial applications that deal with high temperature gas separation.

ACKNOWLEDGMENTS

D.K. Wang gratefully thanks the support given by the Australian Research Council (ARC) Discovery Early Career Researcher Award (DE150101687), and J.C. Diniz da Costa gratefully thanks the support given by the ARC Future Fellowship Program (FT130100405).

REFERENCES

Abdel-jawad, M.M., Gopalakrishnan, S., Duke, M.C., Macrossan, M.N., Schneider, P.S., Diniz da Costa, J.C., 2007. Flowfields on feed and permeate sides of tubular molecular sieving silica (MSS) membranes. Journal of Membrane Science. 299(1–2), 229–235.

Abidi, N., Deroide, B., Zanchetta, J.V., De Menorval L.C., D'Espinose, J.B., 1998. ^{29}Si and ^{129}Xe NMR of Mn^{2+} doped silica xerogels. Journal of Non-Crystalline Solids. 231(1–2), 49–57.

Amanipour, M. et al., 2012. Effect of CVD parameters on hydrogen permeation properties in a nano-composite SiO2-A12O3 membrane. Journal of Membrane Science. 423–424, 530535.

Ambati, J., Rankin, S.E., 2011. Reaction-induced phase separation of bis(triethoxysilyl)ethane upon sot-gel polymerization in acidic conditions. Journal of Colloid and Interface Science. 362(2), 345–353.

Ballinger, B., Motuzas, J., Smart, S., Diniz da Costa, J.C., 2014. Palladium cobalt binary doping of molecular sieving silica membranes. Journal of Membrane Science. 451, 185–191.

Ballinger, B., Motuzas, J., Smart, S., Diniz da Costa, J.C., 2015. Gas permeation redox effect on binary lanthanum cobalt silica membranes with enhanced silicate formation. Journal of Membrane Science. 489, 220–226.

Barrer, R.M., 1990. Porous crystal membranes. Journal of the Chemical Society, Faraday Transactions. 86(7), 1123–1130.

Battersby, S., Duke, M.C., Liu, S., Rudolph V., Diniz da Costa, J.C., 2008. Metal doped silica membrane reactor: Operational effects of reaction and permeation for the water gas shift reaction. Journal of Membrane Science. 316(1–2), 46–52.

Battersby, S., Tasaki, T., Smart, S., Ladewig, B., Liu, S., Duke, M.C., Rudolph, V., Diniz da Costa, J.C., 2009. Performance of cobalt silica membranes in gas mixture separation. Journal of Membrane Science. 329(1–2), 91–98.

Battersby, S., Teixeira, P.W., Beltramini, J., Duke, M.C., Rudolph V., Diniz da Costa, J.C., 2006. An analysis of the Peclet and Damkohler numbers for dehydrogenation reactions using molecular sieve silica (MSS) membrane reactors. Catalysis Today. 116(1), 12–17.

Boffa, V., Blank, D.H.A., ten Elshof, J.E., 2008a. Hydrothermal stability of microporous silica and niobia-silica membranes. Journal of Membrane Science. 319(1–2), 256–263.

Boffa, V., ten Elshof, J.E., Garcia R., Blank, D.H.A., 2009. Microporous niobia-silica membranes: Influence of sol composition and structure on gas transport properties. Microporous and Mesoporous Materials. 118(1–3), 202–209.

Boffa, V., ten Elshof, J.E., Petukhov A.V., Blank, D.H., 2008b. Microporous niobia-silica membrane with very low CO2 permeability. ChemSusChem. 1(5), 437–443.

Bonekamp, B.C. (1996). Preparation of asymmetric ceramic membrane supports by dip-coating. In Burggraaf, A.J., Cot, L. (Eds.), *Fundamentals of Inorganic Membrane Science and Technology* (141–226). Elsevier.

Brands, K., Uhlmann, D., Smart, S., Bram M., Diniz da Costa, J.C., 2010. Long-term flue gas exposure effects of silica membranes on porous steel substrate. Journal of Membrane Science. 359(1–2), 110–114.

Brinker, C.J., 1988. Hydrolysis and condensation of silicates - Effects on structure. Journal of Non-Crystalline Solids. 100(1–3), 31–50.

Brinker, C.J., Keefer, K.D., Schaefer, D.W., Ashley, C.S., 1982. Sol-gel transition in simple silicates. Journal of Non-Crystalline Solids. 48(1), 47–64.

Brinker, C.J., Scherer, G.W., 1990. *Sol-Gel Science: The Physics and Chemistry of Sol-Gel Processing*. Boston: Academic Press.

Brunetti, A., Barbieri, G., Drioli, E., Lee, K.H., Sea, B., Lee, D.W., 2007. WGS reaction in a membrane reactor using a porous stainless steel supported silica membrane. Chemical Engineering and Processing: Process Intensification. 46(2), 119–126.

Burneau, A.E., Gallas, J.-P., 1998. The surface properties of silicas / edited by André P. Legrand. A. P. Legrand. Chichester ; New York, Chichester ; New York : John Wiley: 147–234.

Campaniello, J., Engelen, C.W.R., Haije, W.G., Pex, P.P.A.C., Vente, J.F., 2004. Long-term pervaporation performance of microporous methylated silica membranes. Chemical Communications. 10(7), 834–835.

Cassiers, K., Linssen, T., Mathieu, M., Benjelloun, M., Schrijnemakers, K., Van Der Voort, P., Cool P., Vansant, E.F., 2002. A detailed study of thermal, hydrothermal, and mechanical stabilities of a wide range of surfactant assembled mesoporous silicas. Chemistry of Materials. 14(5), 2317–2324.

Costa, C.A., Leite, C.A., Galembeck, F., 2003. Size dependence of Stöber silica nanoparticle microchemistry. Journal of Physical Chemistry B. 107(20), 4747–4755.

Darmawan, A., Karlina, L., Astuti, Y., Sriatun, Motuzas, J., Wang, D.K., and Diniz da Costa, J.C., 2016a. Structural evolution of nickel oxide silica sol-gel for the preparation of interlayer-free membranes. Journal of Non-Crystalline Solids. 447, 9–15.

Darmawan, A., Motuzas, J., Smart, S., Julbe A., Diniz da Costa, J.C., 2015a. Binary iron cobalt oxide silica membrane for gas separation. Journal of Membrane Science. 474, 32–38.

Darmawan, A., Motuzas, J., Smart, S., Julbe, A., Diniz Da Costa, J.C., 2015b. Temperature dependent transition point of purity versus flux for gas separation in Fe/Co-silica membranes. Separation and Purification Technology. 151, 284-291.

Darmawan, A., Motuzas, J., Smart, S., Julbe, A., Diniz da Costa, J.C., 2016b. Gas permeation redox effect of binary iron oxide/cobalt oxide silica membranes. Separation and Purification Technology. 171, 248–255.

de Lange, R.S.A., Keizer, K., Burggraaf, A.J., 1995. Analysis and theory of gas transport in microporous sol-gel derived ceramic membranes. Journal of Membrane Science. 104(1–2), 81–100.

de Vos, R.M., Maier, W.F., Verweij, H., 1999. Hydrophobic silica membranes for gas separation. Journal of Membrane Science. 158(1–2), 277–288.

de Vos, R. M., Verweij, H., 1998a. High-selectivity, high-flux silica membranes for gas separation. Science. 279(5357), 1710–1711.

de Vos, R.M., Verweij, H., 1998b. Improved performance of silica membranes for gas separation. Journal of Membrane Science. 143(1–2), 37–51.

Deonarine, B., Ji, G., Smart, S., Diniz da Costa, J.C., Reed G., Millan, M., 2017. Ultramicroporous membrane separation using toluene to simulate tar-containing gases. Fuel Processing Technology. 161, 259–264.

Diniz da Costa, J.C., Lu, G.Q., Rudolph, V., Lin, Y.S., 2002. Novel molecular sieve silica (MSS) membranes: Characterisation and permeation of single-step and two-step sol-gel membranes. Journal of Membrane Science. 198(1), 9–21.

Duke, M., Rudolph, V., Lu G.Q., Diniz Da Costa, J.C., 2004a. Scale-up of molecular sieve silica membranes for reformate purification. AIChE Journal. 50(10), 2630–2634.

Duke, M.C., Diniz da Costa, J.C, Do, D.D., Gray P.G., Lu, G.Q., 2006. Hydrothermally robust molecular sieve silica for wet gas separation. Advanced Functional Materials. 16(9), 1215–1220.

Duke, M.C., Diniz da Costa, J.C., Lu, G.Q., Petch, M., Gray, P., 2004b. Carbonised template molecular sieve silica membranes in fuel processing systems: Permeation, hydrostability and regeneration. Journal of Membrane Science. 241(2), 325–333.

Duke, M.C., Pas, S.J., Hill, A.J., Lin, Y.S., Diniz da Costa, J.C., 2008. Exposing the molecular sieving architecture of amorphous silica using positron annihilation spectroscopy. Advanced Funcional Materials. 18(23), 3818–3826.

Elma, M., Wang, D.K., Yacou, C., Diniz da Costa, J.C., 2015a. Interlayer-free P123 carbonised template silica membranes for desalination with reduced salt concentration polarisation. Journal of Membrane Science. 475, 376–383.

Elma, M., Wang, D.K., Yacou, C., Motuzas, J., Diniz da Costa, J.C., 2015b. High performance interlayer-free mesoporous cobalt oxide silica membranes for desalination applications. Desalination. 365, 308–315.

Elma, M., Yacou, C., Diniz da Costa J.C., Wang, D.K., 2013. Performance and long term stability of mesoporous silica membranes for desalination. Membranes. 3(3), 136–150.

Everett, D.H., Powl, J.C., 1976. Adsorption in slit-like and cylindrical micropores in the Henry's law region. A model for the microporosity of carbons. Journal of the Chemical Society, Faraday Transactions 1: Physical Chemistry in Condensed Phases. 72, 619–636.

Fidalgo, A., Ilharco, L.M., 2004. Chemical tailoring of porous silica xerogels: Local structure by vibrational spectroscopy. Chemistry-a European Journal. 10(2), 392-398.

Fiory, A.T. (2005). Rapid thermal processing for silicon nanoelectronics applications. JOM. 57(6), 21–26.

Fotou, G.P., Lin, Y.S., Pratsinis, S.E., 1995. Hydrothermal stability of pure and modified microporous silica membranes. Journal of Materials Science. 30(11), 2803–2808.

Gallaher, G.R., Liu, P.K.T., 1994. Characterization of Ceramic Membranes .1. Thermal and Hydrothermal Stabilities of Commercial 40 Angstrom Membranes. Journal of Membrane Science. 92(1), 29–44.

Gavalas, G.R., Megiris, C.E., Nam, S.W., 1989. Deposition of H-2-permselective Sio2-films. Chemical Engineering Science. 44(9), 1829–1835.

Gaydhankar, T.R., Samuel, V., Jha, R.K., Kumar, R., Joshi, P.N., 2007. Room temperature synthesis of Si-MCM-41 using polymeric version of ethyl silicate as a source of silica. Materials Research Bulletin. 42(8), 1473–1484.

Gaydhankar, T.R., Taralkar, U.S., Jha, R.K., Joshi, P.N., Kumar, R., 2005. Textural/structural, stability and morphological properties of mesostructured silicas (MCM-41 and MCM-48) prepared using different silica sources. Catalysis Communications. 6(5), 361–366.

Gopalakrishnan, S. and J.C. Diniz da Costa, 2008. Hydrogen gas mixture separation by CVD silica membrane. Journal of Membrane Science. 323(1), 144–147.

Gopalakrishnan, S., Yoshino, Y., Nomura, M., Nair B.N., Nakao, S.I., 2007. A hybrid processing method for high performance hydrogen-selective silica membranes. Journal of Membrane Science. 297(1–2), 5–9.

Gu, Y., Oyama, S.T., 2009. Permeation properties and hydrothermal stability of silica-titania membranes supported on porous alumina substrates. Journal of Membrane Science. 345(1–2), 267–275.

Han, H.H. et al., 2013. Gas permeation properties and preparation of porous ceramic membrane by CVD method using siloxane compounds. Journal of Membrane Science. 431, 7278. 58.

Igi, R., Yoshioka, T., Ikuhara, Y.H., Iwamoto, Y., Tsuru, T., 2008. Characterization of co-doped silica for improved hydrothermal stability and application to hydrogen separation membranes at high temperatures. Journal of the American Ceramic Society. 91(9), 2975–2981.

Iler, R.K. (1979). *The Chemistry of Silica: Solubility, Polymerization, Colloid and Surface Properties, and Biochemistry.* New York: Wiley.

Imai, H., Morimoto, H., Tominaga, A., Hirashima, H., 1997. Structural changes in sol-gel derived SiO2 and TiO3 films by exposure to water vapor. Journal of Sol-Gel Science and Technology. 10(1), 45–54.

Innocenzi, P., 2003. Infrared spectroscopy of sol-gel derived silica-based films: A spectra-microstructure overview. Journal of Non-Crystaline Solids. 316(2–3), 309–319.

James, P.F., 1988. The gel to glass transition: Chemical and microstructural evolution. Journal of Non-Crystalline Solids. 100(1–3), 93–114.

Ji, G., Smart, S., Bhatia, S.K., Diniz da Costa, J.C., 2015. Improved pore connectivity by the reduction of cobalt oxide silica membranes. Separation and Purification Technology. 154, 338–344.

Ji, G., Wang, G., Hooman, K., Bhatia, S., Diniz da Costa, J.C., 2012. Computational fluid dynamics applied to high temperature hydrogen separation membranes. Frontiers of Chemical Science and Engineering. 6(1), 3–12.

Ji, G., Wang, G., Hooman, K., Bhatia, S., Diniz da Costa, J.C., 2013. Simulation of binary gas separation through multi-tube molecular sieving membranes at high temperatures. Chemical Engineering Journal. 218, 394–404.

Ji, G., Wang, G., Hooman, K., Bhatia, S., Diniz da Costa, J.C., 2014. The fluid dynamic effect on the driving force for a cobalt oxide silica membrane module at high temperatures. Chemical Engineering Science. 111, 142–152.

Kanezashi, M., Asaeda, M., 2006. Hydrogen permeation characteristics and stability of Ni-doped silica membranes in steam at high temperature. Journal of Membrane Science. 271(1–2), 86–93.

Kanezashi, M., Fuchigami, D., Yoshioka, T., Tsuru, T., 2013. Control of Pd dispersion in sol-gel-derived amorphous silica membranes for hydrogen separation at high temperatures. Journal of Membrane Science. 439, 78–86.

Kanezashi, M., Fujita, T., Asaeda, M., 2005. Nickel-doped silica membranes for separation of helium from organic gas mixtures. Separation Science and Technology. 40(1–3), 225–238.

Kanezashi, M., Shazwani, W.N., Yoshioka, T., Tsuru, T., 2012. Separation of propylene/propane binary mixtures by bis(triethoxysilyl) methane (BTESM)-derived silica membranes fabricated at different calcination temperatures. Journal of Membrane Science. 415–416, 478–485.

Kanezashi, M., Tsuru, T., 2011. Gas permeation properties of helium, hydrogen, and polar molecules through microporous silica membranes at high temperatures. Correlation with silica network structure. Membrane Science and Technology. 14, 117–136.

Kim, J.M., Chang, S.M., Kong, S.M., Kim, K.S., Kim, J., Kim, W.S., 2009. Control of hydroxyl group content in silica particle synthesized by the sol-precipitation process. Ceramics International. 35(3), 1015–1019.

Kusakabe, K., Sakamoto, S., Saie, T., Morooka, S., 1999. Pore structure of silica membranes formed by a sol-gel technique using tetraethoxysilane and alkyltriethoxysilanes. Separation and Purification Technology. 16(2), 139–146.

Lee, J.H., Choi, S.C., Bae, D.S., Han, K.S., 1999. Synthesis and microstructure of silica-doped alumina composite membrane by sol-gel process. Journal of Materials Science Letters. 18(17), 1367–1369.

Lee, K., Look, J.L., Harris, M.T., McCormick, A.V., 1997. Assessing extreme models of the Stober synthesis using transients under a range of initial composition. Journal of Colloid and Interface Science. 194(1), 78–88.

Liu, L., Wang, D.K., Kappen, P., Martens, D.L., Smart, S., Diniz Da Costa, J.C., 2015a. Hydrothermal stability investigation of micro- and mesoporous silica containing long-range ordered cobalt oxide clusters by XAS. Physical Chemistry Chemical Physics. 17(29), 19500–19506.

Liu, L., Wang, D.K., Martens, D.L., Smart, S., Diniz da Costa, J.C., 2015b. Influence of sol-gel conditioning on the cobalt phase and the hydrothermal stability of cobalt oxide silica membranes. Journal of Membrane Science. 475, 425–432.

Liu, L., Wang, D.K., Martens, D.L., Smart, S., Diniz da Costa, J.C., 2015c. Interlayer-free microporous cobalt oxide silica membranes via silica seeding sol-gel technique. Journal of Membrane Science. 492, 1–8.

Liu, L., Wang, D.K., Martens, D.L., Smart, S., Strounina, E., Diniz da Costa, J.C., 2014. Physicochemical characterisation and hydrothermal stability investigation of cobalt-incorporated silica xerogels. RSC Advances. 4(36), 18862–18870.

Luiten, M.W.J., Benes, N.E., Huiskes, C., Kruidhof, H., Nijmeijer, A., 2010. Robust method for micro-porous silica membrane fabrication. Journal of Membrane Science. 348(1–2), 1–5.

Mandelbrot, B.B., 1977. *Fractals: Form, Chance, and Dimension*. San Francisco, CA: W. H. Freeman.

Martens, D.L., Wang, D.K., Motuzas, J., Smart, S., Diniz Da Costa, J.C., 2015. Modulation of microporous/mesoporous structures in self-templated cobalt-silica. Scientific Reports. 5, 7970.

Masuda, Y., Kugimiya, S., Kawachi, Y., Kato, K., 2014. Interparticle mesoporous silica as an effective support for enzyme immobilisation. RSC Advances. 4(7), 3573–3580.

Miller, C.R., Wang, D.K., Smart, S., Diniz da Costa, J.C., 2013. Reversible redox effect on gas permeation of cobalt doped ethoxy polysiloxane (ES40) membranes. Scientific Reports. 3, 1648.

Mrowiec-Białoń, J., Jarzębski, A.B., 2008. Fabrication and properties of silica monoliths with ultra large mesopores. Microporous and Mesoporous Materials. 109(1–3), 429–435.

Mrowiec-Białoń, J., Jarzębski, A.B., Pajak, L., Olejniczak Z., Gibas, M., 2004. Preparation and surface properties of low-density gels synthesized using prepolymerized silica precursors. Langmuir. 20(24), 10389–10393.

Nair, B.N. et al., 2000. Synthesis of gas and vapor molecular sieving silica membranes and analysis of pore size and connectivity. Langmuir. 16(10), 4558–4562.

Nomura, M., Ono, K., Gopalakrishnan, S., Sugawara, T., Nakao, S.I., 2005. Preparation of a stable silica membrane by a counter diffusion chemical vapor deposition method. Journal of Membrane Science. 251(1–2), 151–158.

Olguin, G., Yacou, C., Smart, S., Diniz da Costa, J.C., 2014. Influence of surfactant alkyl length in functionalizing sol-gel derived microporous cobalt oxide silica. RSC Advances. 4(76), 40181–40187.

Peters, T.A., Fontalvo, J., Vorstman, M.A.G., Benes, N.E., van Dam, R.A., Vroon, Z.A.E.P., van Soest-Vercammen, E.L.J., Keurentjes, J.T.F., 2005. Hollow fibre microporous silica membranes for gas separation and pervaporation: Synthesis, performance and stability. Journal of Membrane Science. 248(1–2), 73–80.

Qi, H., Chen, H., Li, L., Zhu, G., Xu, N., 2012. Effect of Nb content on hydrothermal stability of a novel ethylene-bridged silsesquioxane molecular sieving membrane for H_2/CO_2 separation. Journal of Membrane Science. 421–422, 190–200.

Raman, N.K., Brinker, C.J., 1995. Organic "template" approach to molecular sieving silica membranes. Journal of Membrane Science. 105(3), 273–279.

Riegel, B., Blittersdorf, S., Kiefer, W., Hofacker, S., Muller, M., Schottner, G., 1998. Kinetic investigations of hydrolysis and condensation of the glycidoxypropyltrimethoxysilane/aminopropyltriethoxy-silane system by means of FT-Raman spectroscopy I. Journal of Non-Crystalline Solids. 226(1–2), 76–84.

Shelekhin, A.B., Dixon A.G., Ma, Y.H., 1992. Adsorption, permeation, and diffusion of gases in microporous membranes. II. Permeation of gases in microporous glass membranes. Journal of Membrane Science. 75(3), 233–244.

Smart, S., Vente, J.F., Diniz da Costa, J.C., 2012. High temperature H_2/CO_2 separation using cobalt oxide silica membranes. International Journal of Hydrogen Energy. 37(17), 12700–12707.

Topuz, B., Çiftçioğlu, M., 2010. Preparation of particulate/polymeric sol–gel derived microporous silica membranes and determination of their gas permeation properties. Journal of Membrane Science. 350(1–2), 42–52.

Tripp, C.P., Hair, M.L., 1995. Reaction of methylsilanols with hydrated silica surfaces - The hydrolysis of trichloromethylsilanes, dichloromethylsilanes, and monochloromethylsilanes and the effects of curing. Langmuir. 11(1), 149–155.

Tsuru, T., Igi, R., Kanezashi, M., Yoshioka, T., Fujisaki, S., Iwamoto, Y., 2011. Permeation properties of hydrogen and water vapor through porous silica membranes at high temperatures. AIChE J. 57(3), 618–629.

Tsuru, T., Wada, S.I., Izumi S., Asaeda, M., 1998. Silica-zirconia membranes for nanofiltration. Journal of Membrane Science. 149(1), 127–135.

Uhlmann, D., Liu, S., Ladewig, B.P., Diniz da Costa, J.C., 2009. Cobalt-doped silica membranes for gas separation. Journal of Membrane Science. 326(2), 316–321.

Uhlmann, D., Smart, S., Diniz Da Costa, J.C., 2010. High temperature steam investigation of cobalt oxide silica membranes for gas separation. Separation and Purification Technology. 76(2), 171–178.

Uhlmann, D., Smart, S., Diniz da Costa, J.C., 2011. H_2S stability and separation performance of cobalt oxide silica membranes. Journal of Membrane Science. 380(1–2), 48–54.

Van Gestel, T., Hauler, F., Bram, M., Meulenberg, W.A., Buchkremer, H.P., 2014. Synthesis and characterization of hydrogen-selective sol-gel SiO 2 membranes supported on ceramic and stainless steel supports. Separation and Purification Technology. 121, 20–29.

Wang, D.K., Diniz da Costa, J.C., Smart, S., 2014. Development of rapid thermal processing of tubular cobalt oxide silica membranes for gas separations. Journal of Membrane Science. 456, 192–201.

Wang, D.K., Motuzas, J., Diniz da Costa, J.C., Smart, S., 2013. Rapid thermal processing of tubular cobalt oxide silica membranes. International Journal of Hydrogen Energy. 38(18), 7394–7399.

Wang, S., 2016. *High performance ES40-derived silica membranes for desalination* (Doctoral Thesis). The University of Queensland, School of Chemical Engineering.

Wang, S., Wang, D.K., Jack, K.S., Smart, S., Diniz da Costa, J.C., 2015a. Improved hydrothermal stability of silica materials prepared from ethyl silicate 40. RSC Advances. 5(8), 6092–6099.

Wang, S., Wang, D.K., Motuzas, J., Smart, S., Diniz da Costa, J.C., 2016. Rapid thermal treatment of interlayer-free ethyl silicate 40 derived membranes for desalination. Journal of Membrane Science. 516, 94–103.

Wang, S., Wang, D.K., Smart, S., Diniz Da Costa, J.C., 2015b. Ternary Phase-Separation Investigation of Sol-Gel Derived Silica from Ethyl Silicate 40. Scientific Reports. 5, 14560.

Wang, S., Wang, D.K., Smart, S., Diniz da Costa, J.C., 2017. Improved stability of ethyl silicate interlayer-free membranes by the rapid thermal processing (RTP) for desalination. Desalination. 402, 25–32.

Wang, Y.H., Liu, X.Q., Meng, G.Y., 2007. Preparation of asymmetric pure titania ceramic membranes with dual functions. Materials Science and Engineering A. 445–446, 611–619.

Yacou, C., Ayral, A., Giroir-Fendler, A., Baylet, A., Julbe, A., 2010. Catalytic membrane materials with a hierarchical porosity and their performance in total oxidation of propene. Catalysis Today. 156(3-4), 216–222.

Yacou, C., Ayral, A., Giroir-Fendler, A., Fontaine, M.L., Julbe, A., 2009. Hierarchical porous silica membranes with dispersed Pt nanoparticles. Microporous and Mesoporous Materials. 126(3), 222–227.

Yacou, C., Fontaine, M.L., Ayral, A., Lacroix-Desmazes, P., Albouy, P.A., Julbe, A., 2008. One pot synthesis of hierarchical porous silica membrane material with dispersed Pt nanoparticles using a microwave-assisted sol-gel route. Journal of Materials Chemistry. 18(36), 4274–4279.

Yacou, C., Smart, S., Diniz da Costa, J.C., 2012. Long term performance cobalt oxide silica membrane module for high temperature H_2 separation. Energy & Environmental Science. 5(2), 5820–5832.

Yamamoto, K., Ohshita, J., Mizumo, T., Kanezashi, M., Tsuru, T., 2015. Preparation of hydroxyl group containing bridged organosilica membranes for water desalination. Separation and Purification Technology. 156, 396–402.

Yang, H., Elma, M., Wang, D.K., Motuzas, J., Diniz da Costa, J.C., 2017. Interlayer-free hybrid carbon-silica membranes for processing brackish to brine salt solutions by pervaporation. Journal of Membrane Science. 523, 197–204.

Yoshida, K., Hirano, Y., Fujii, H., Tsuru, T., Asaeda, M., 2001. Hydrothermal stability and performance of silica-zirconia membranes for hydrogen separation in hydrothermal conditions. Journal of Chemical Engineering of Japan. 34(4), 523–530.

Yoshioka, T. et al., 2001. Experimental studies of gas permeation through microporous silica membranes. AIChE Journal. 47(9), 2052–2063.

Yoshioka, T., Tsuru, T., Asaeda, M., 2003. Condensable vapor permeation through miroporous silica membranes studied with molecular dynamics simulation. Separation and Purification Technology. 32(1–3), 231–237.

Zhang, F., Yan, Y., Yang, H., Meng, Y., Yu, C., Tu, B. Zhao, D., 2005. Understanding Effect of Wall Structure on the Hydrothermal Stability of Mesostructured Silica SBA-15. The Journal of Physical Chemistry B. 109(18), 8723–8732.

9 High Performance CO_2 Separation Thin Film Composite Membranes

Liang Liu and Sandra Kentish

CONTENTS

9.1 INTRODUCTION

Separating carbon dioxide (CO_2) from other gases is of great interest in many applications, such as in natural gas sweetening and biogas purification (CO_2/CH_4) (Zhang et al., 2013), syngas purification (CO_2/H_2) (Ockwig and Nenoff, 2007) and CO_2 capture from the flue gas of fossil fuel driven power plants (CO_2/N_2) (Brunetti et al., 2010). Membrane technology is promising compared with other commercial techniques (e.g., solvent absorption or cryogenic separation), as it can provide a smaller and more compact footprint, does not require chemicals to be on site and provides operational simplicity.

The enormous emissions of CO_2, mainly induced by the combustion of fossil fuels, is a major cause of global warming. CO_2 capture and storage (CCS) technologies have the potential to solve this problem and in this context, separating CO_2 from N_2 in the flue gas from power plants plays a key role. The post-combustion flue gas from a coal fired power station contains 10–15 vol% of CO_2, 78–80 vol% of N_2, impurities such as oxygen, sulphur oxides (SO_x) and nitrogen oxides (NO_x) and is saturated with water. The flue gases from a natural gas fired power station contain less CO_2, typically 4 vol%, but also significantly lower levels of NO_x and no SO_x. In both cases, the feed pressure is low, typically 1.5 Bar absolute pressure, while the temperature can range from ambient to 200°C.

Both coal and natural gas can also be converted to syngas, which is a mixture of carbon monoxide and hydrogen, through steam reforming or gasification. This mixture can be further converted to a mixture of carbon dioxide and hydrogen through the water gas shift reaction. This mixture can be separated using membrane technology to create a pure hydrogen stream, which can be combusted for electricity generation, used in chemical production or in fuel cell vehicles. The mixture composition depends upon whether the syngas is generated using pure oxygen or air. In the former case, the mixture is around 60% CO_2 and 40% H_2, while it is typically 60% N_2, with 25% CO_2 and 15% H_2 in the latter. Impurities include H_2S and NH_3 and again, the mixture is saturated with water. Feed pressures can be 20 to 60 Bar and temperatures can be up to 200°C. The membrane can be either H_2-selective or CO_2-selective (Scholes et al., 2010).

Raw natural gas varies widely in composition, depending on reservoir sources. It is generally composed of methane (30–90%), other light hydrocarbons (e.g., ethane and propane) and heavier hydrocarbons. Moreover, it also contains water, carbon dioxide (0–43%), hydrogen sulphide, helium and nitrogen (Scholes et al., 2012). Natural gas processing consists of two key processes: dehydration and gas sweetening. In the latter process, CO_2 and H_2S are removed from CH_4. Commercial natural gas sweetening membranes have been deployed since the 1980s (Schell et al., 1989) with cellulose acetate based membranes making up to 80% of the market. In the last ten years, these have begun to be challenged by newer membranes, such as those based on polyimides (Medal™, Air Liquide), polyether ether ketone (ALaS PEEK-Sep™, Air Liquide), perfluoropolymers (Baker et al., 2003) and polyethylene oxide (PolyActive™) (Brinkmann et al., 2015).

There are numerous reports of CO_2 separation membrane technologies, including polymeric and inorganic membrane systems and membrane contactors (membrane gas absorption). This chapter will focus on polymeric membrane based systems due to their low cost and greater technological maturity. Most of the data in the literature is collected from membranes that are more than 50 μm in thickness, and the membrane performance is generally compared with the well-known Robeson's upper bound (Robeson, 2008). Polymeric membranes with high gas permeability generally have low gas selectivity and vice versa. However, in order to obtain high flux (permeance), a membrane with a thin selective layer (<1 μm) is needed for industrial applications. The properties of this thin film can be very different from the thick bulk films tested in the laboratory. Fresh thin film composite membranes will lose 25% of their permeance within a few days (Baker and Low, 2014). This degradation is due to rearrangement of polymer chains, reducing the free volume that contributes to gas permeance. Second, fabrication of defect-free thin film membranes is challenging, in particular for some new membrane materials, which could limit their industrial deployment.

This chapter will mainly focus on the recent advances of polymer-based thin film composite membranes. Typical fabrication methods are introduced, and the concepts of solution-diffusion, facilitated transport and mixed matrix structures are discussed. The challenges of the deployment of membranes in industrial applications including physical aging, plasticization and impurity effects are also highlighted, followed by some conclusions and a brief perspective on future research directions.

9.2 TRANSPORT MECHANISMS OF POLYMERIC CO_2 SEPARATION MEMBRANES

For most polymeric membranes, the flux (N_i) of a gas (i) is governed by the solution-diffusion mechanism. This mechanism is based on Fick's Law (Eq. 9.1):

$$N_i = D_i \times \frac{\Delta c_i}{\Delta x} \tag{9.1}$$

where D_i is the diffusion coefficient (m²/s), Δc_i is the concentration gradient of the penetrant across the membrane and Δx is the membrane thickness. For rubbery polymers at low solute concentrations, there is a linear relationship between partial pressure of the gas (p_i) and concentration, known as Henry's Law. In this case, the concentration gradient across the membrane can be replaced directly by a partial pressure gradient (Eq. 9.2):

$$N_i = D_i S_i \times \frac{\Delta p_i}{\Delta x} \tag{9.2}$$

The gas permeability (P_i, barrer, 1 barrer = 10^{-10} cm³ (STP) cm^{-1} s^{-1} cmHg^{-1}) is then a simple product of the solubility coefficient (S_i, cm³ (STP) cm^{-3} cmHg^{-1}) and the diffusion coefficient (D_i, cm^{-2} s^{-1}) and is given by Eq. 9.3:

$$P_i = S_i \times D_i \tag{9.3}$$

For glassy polymers and rubbery polymers at higher penetrant concentrations, Eqs. 9.2 and 9.3 do not apply universally. However, if the penetrant concentration on the permeate side is zero (as is often the case in laboratory experiments) then:

$$N_i = D_i S_{ih} \times \frac{p_{ih}}{\Delta x} \tag{9.4}$$

where S_{ih} and p_{ih} are the solubility and partial pressure on the upstream or high pressure side of the membrane respectively.

For thin film composite or asymmetric membranes, the permeance (Q_i) is widely used, which is the permeability of the membrane divided by the thickness (l) of the selective layer (i.e., P_i/l). GPU (gas permeation unit) is widely used as the unit for gas permeance of polymeric membranes and 1 GPU = 10^{-6} cm³ (STP) cm^{-2} s^{-1} cmHg^{-1}.

The ideal selectivity is the permeability/permeance ratio of two pure gases, shown in Eq. 9.5.

$$\alpha_{ij} = \frac{p_i}{p_j} \tag{9.5}$$

The separation factor (β_{ij}) is commonly used for mixture separation tests and it is defined in Eq. 9.6:

$$\beta_{ij} = \frac{y_i / x_i}{y_j / x_j} \tag{9.6}$$

where y_i and y_j are the molar fraction of gas *i* and *j* in the permeate side, and x_i and x_j are the molar fraction of gas *i* and *j* in the feed side. It is worth noting that the separation factor is not necessarily identical to the ideal selectivity.

Facilitated transport membranes refer to membranes that have reactive carriers, which can reversibly react with some species and thus enhance the transport of these species. In particular, the acidic nature of CO_2 means that it can react with amine groups or carboxylate groups in a polymeric matrix, as shown in Figure 9.1. In this case, CO_2 passes through the membrane in two different forms, i.e., as CO_2 molecules through a solution-diffusion pathway and as CO_2 carrier complexes through a facilitated transport pathway. On the other hand, other gases like N_2, CH_4 and H_2 have no reaction with these groups and will only permeate through the polymer matrix. An enhanced CO_2 permeability and selectivity can be expected. Ideally, effective facilitated transport requires a balance of fast reaction kinetics and high CO_2 loading capacity. It is worthwhile nothing that these carriers normally cannot be operated unless the membrane also contains liquid water. Water is needed to accommodate the ionic species that form upon reaction with CO_2.

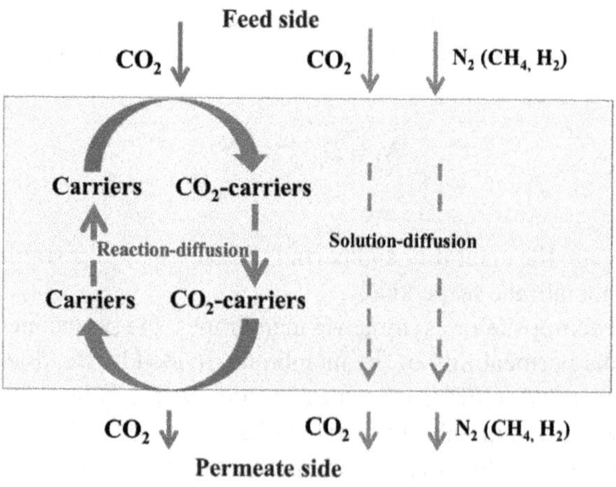

FIGURE 9.1 Schematic of CO_2-selective facilitated transport membranes. (From Wang, S. et al., Energy & Environmental Science. 9, 1863–1890, 2016.)

9.3 THIN FILM COMPOSITE MEMBRANE FABRICATION

In order to achieve high gas permeance, the membrane thickness should be as low as possible. Nevertheless, a stand-alone thin film membrane leads to poor mechanical strength, which is impractical for industrial use. Traditional cellulose acetate and polyimide membranes overcome this issue through the use of an asymmetric structure where a thin selective skin layer is first formed by exposing a polymer solution to air, but the remaining bulk is then immersed in an anti-solvent to form a spongy underlayer that is porous and adds little resistance to gas flow. This process is referred to as phase inversion. However, it is difficult to form an ultrathin selective layer using this approach.

More recently, thin film composite membranes have been developed (Figure 9.2). These consist of a porous support with negligible transport resistance and a thin film selective layer that is formed from a second polymer. This approach allows very expensive polymers to be used in the selective layer, as the porous support can use a cheaper material. To further improve membrane performance, a gutter layer is often applied on the porous support prior to coating the selective layer to reduce the surface roughness and to prevent the penetration of the selective layer polymer into the porous support. A protective layer can also be coated on top of the selective layer to seal any defects in the selective layer and to protect it from being damaged.

At an industrial scale, the gutter, selective and protective layers are usually applied consecutively to flat sheet membranes using a doctor blade, which forms a wet thin layer on the top of the support. The thickness of the layer can be controlled by the polymer concentration and the gap between the support and the casting blade. In the laboratory, spin-coating can also be used (Figure 9.3(b)). In this case, a uniform thin film is achieved with the assistance of centrifugal force. The membrane thickness can be controlled by the amount of coating solution or the spinning speed. The spin-coating process may need to be repeated a few times in order to eliminate defects.

Thin film composite hollow fibers are classically formed with the selective layer on the outside of the fiber, either by co-extrusion or by dip-coating. In the former case, the porous sublayer of the fiber and the selective skin layer are formed simultaneously by extruding two different polymer dopes and one bore fluid through a triple-orifice spinneret (Li et al., 2004). In the latter case, as shown in Figure 9.3(a), the porous support is dipped into a polymer solution for a period of time and then is

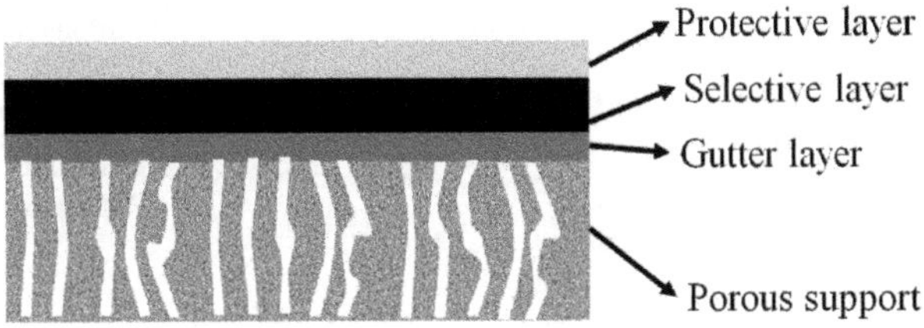

FIGURE 9.2 Typical structure of a thin film composite membrane.

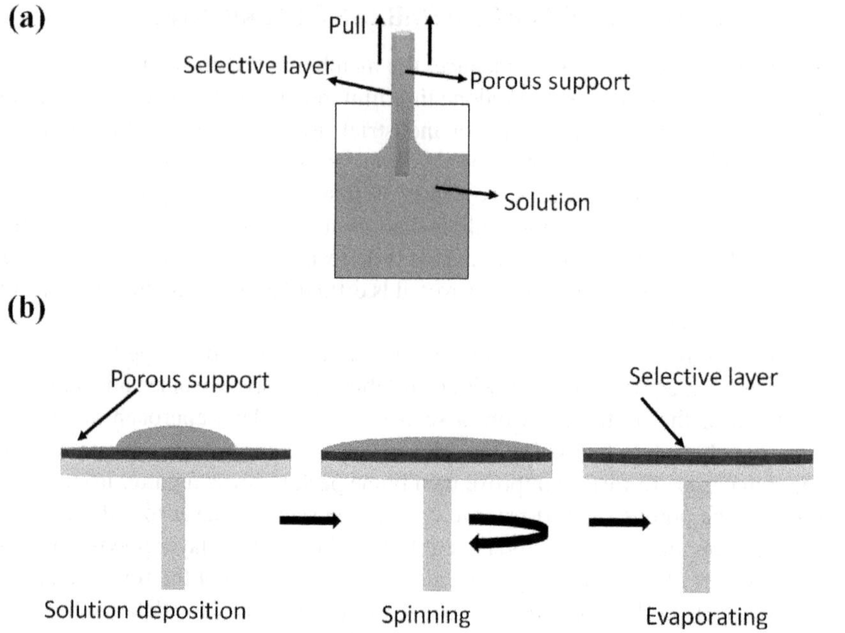

FIGURE 9.3 Schematic of: (a) dip-coating and (b) spin-coating.

pulled out at a controlled speed (Li et al., 2013). Although the process is straightforward, many parameters, such as membrane solution, holding time and withdrawal speed, play important roles in the final thin film formation.

Interfacial polymerization occurs at an interface between an aqueous solution containing one monomer and an organic solution containing a second monomer, as shown in Figure 9.4. Monomers with terminated amine groups can react with trimesoyl chloride (TMC), and a thin film layer can be formed. The process can be controlled by monomer concentration, reaction time and temperature. While interfacial

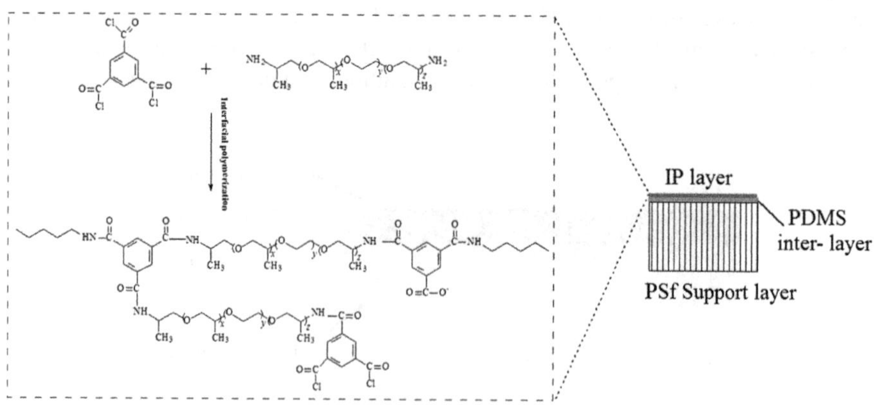

FIGURE 9.4 Schematic of the interfacial polymerization process. (From Salih, A.A.M. et al., Journal of Membrane Science. 472, 110–118, 2014.)

polymerization was classically only applied to flat sheet membranes, processes have recently been developed to extend this approach to hollow fibers. The interfacial layer can be applied to the lumen side of the fibers by simply pumping the aqueous phase through the fibers, followed by flushing with either air or an organic liquid, and finally the organic phase (Jo et al., 2017; Ren and McCutcheon, 2017; Veríssimo et al., 2005; Wang et al., 2010; Zhang et al., 2013). Wessling et al. (Gherasim et al., 2016; Wessling et al., 2015) have recently disclosed an approach where the porous support can be extruded simultaneously to the application of the interfacial skin layer, leading to better physical integration of the two layers.

A new approach to selective layer formation has recently been developed by (Fu et al., 2016). In this case, atom transfer radical polymerization (ATRP) is used to directly form an ultrathin polymer layer from macrocross-linkers contained within an aqueous solution that is applied to a gutter layer. The process, referred to as Continuous Assembly of Polymers (CAP) results in a surface confined thin film of ~100 nm.

9.4 MEMBRANE CHEMISTRY

According to the solution-diffusion mechanism, permeability can be manipulated by changing gas solubility and diffusivity. The solubility depends on the condensability of the penetrants and the interaction between the gas molecules and the membrane matrix. As CO_2 is more condensable than N_2 and CH_4, it has intrinsically greater solubility. Introduction of polar groups such as ethers can enhance this solubility. CO_2 has a smaller kinetic diameter than methane, and so has a greater diffusivity in CO_2/CH_4 separations. The difference in kinetic diameter between CO_2 and N_2 is disputed, with some authors arguing that CO_2 is smaller (Breck, 1973; Robeson et al., 2009) and others indicating that they are equal in size (Robeson et al., 2014). Increasing the fractional free volume (FFV) of the polymer matrix can enhance the diffusivity of all penetrants. Merkel et al. (2010) have indicated that for economic separation of CO_2 from N_2 in post combustion capture from a coal-fired power station, a permeance of at least 1000 GPU and a selectivity between 20 and 30 is optimal. Counter-intuitively, a very high selectivity in this case leads to reduced performance.

Poly(ethylene oxide) (PEO) based materials, thermally rearranged (TR) polymers, polymer of intrinsic microporosity (PIM) and facilitated transport membrane materials have attracted the most research attention for CO_2 separation in the last decade (see Table 9.1). Poly(ethylene oxide) (PEO) based polymers have demonstrated good CO_2 separation performance, but they have a strong tendency for crystallization and weak mechanical strength. Hence, several strategies including copolymerization, physical blending and crosslinking have been used to produce high performance CO_2 selective membranes.

The GKSS group has developed a block copolymer composed of PEO and PBT (Poly(ethylene oxide)-poly(butylene terephthalate)), commercialized under the PolyActive trademark, with some variants including modified PEG as additives to improve membrane performance (Yave et al., 2010a, 2010b). The properties of these materials can be controlled by the fraction of the PEO phase, its molecular weight or repeat unit length. The authors also observed a different organization behavior

TABLE 9.1
Performance of Thin Film Composite Polymeric Membranes for CO_2 Separation

Membrane Material	Fabrication Method	Operational Conditions	CO_2 Permeance (GPU)	Selectivity		Reference
				CO_2/N_2	CO_2/CH_4	
PEO-PBT	Dip-coating	Pure gas, 30°C, 100 kPa	1340	52	16	(Yave et al., 2010b)
PEO-PBT	Dip-coating	Pure gas	1830	60	14	(Yave et al., 2011)
PEO-PBT/PEG-DBE	Dip-coating	CO_2/N_2 (28/72 by volume)	~700	40		(Yave et al., 2010a)
PEO with methacrylate pendant groups	Continuous Assembly of Polymers	Pure Gas, 35°C, 340 kPa	1260	43		(Fu et al., 2016)
Pebax™ 2533	Spin-coating	Pure gas, 35°C, 340 kPa	370	26		(Fu et al., 2013)
Pebax™ 2533+high molecular weight amorphous PEO	Spin-coating	Pure gas, 35°C, 340 kPa	2290	38		(Fu et al., 2013)
Pebax™+soft polymer nanoparticles	Spin-coating	Pure gas, 35°C, 340 kPa	610–2210	26–19		(Fu et al., 2014)
Pebax™ 2533+ PEG-b-PDMS grafted star polymers	Spin-coating	Pure gas, 35°C, 340 kPa	4760	19		(Halim et al., 2014)
Pebax™ 2533+PDMS-b-PEG	Spin-coating	Pure gas, 35°C, 350 kPa	1650–1830	20–30		(Scofield et al., 2015)
Pebax™ 2533+PEG-b-PPFPA	Spin-coating	Pure gas, 35°C, 350 kPa	3330–1650	22–25		(Scofield et al., 2016a)
Pebax™ 1657+PEG-b-PPFPA	Spin-coating	Pure gas, 35°C, 350 kPa	1030–3250	41–20		(Scofield et al., 2016b)
TR-PBO450	Dry-jet wet spinning	Pure gas, room temperature, 100–500 kPa	1940	13	14	(Kim et al., 2012)
TR-PBO500	Dry-jet wet spinning	Pure gas, room temperature, 100–500 kPa	2330	20	22	(Kim et al., 2012)
PIM-250-1.0d	Casting	Pure gas, 35°C, 350 kPa	2220[a]	21	14.8	(Li et al., 2012)
PIM-300-1.0d	Casting	Pure gas, 35°C, 350 kPa	3080[a]	31	34	(Li et al., 2012)
PIM-300-2.0d	Casting	Pure gas, 35°C, 350 kPa	4000[a]	42	54.8	(Li et al., 2012)

(Continued)

TABLE 9.1 (CONTINUED)
Performance of Thin Film Composite Polymeric Membranes for CO_2 Separation

Membrane Material	Fabrication Method	Operational Conditions	CO_2 Permeance (GPU)	Selectivity		Reference
				CO_2/N_2	CO_2/CH_4	
PVAm	Casting	CO_2/N_2 (10/90 by volume), 35°C, humidified gas, 110 kPa	1830	500		(Kim et al., 2013)
PVAm+PIP	Casting	CO_2/N_2 (20/80 by volume), 25°C, humidified gas, 110 kPa	6500	277		(Qiao et al., 2012)
MEDA+TMC	Interfacial polymerization	CO_2/N_2 (15/85 by volume), 35°C, humidified gas, 110 kPa	1040	87		(Yuan et al., 2012)
P(DADMACA-*co*-PVAm)	Casting	CO_2/N_2 (15/85 by volume),CO_2/CH_4 (15/85 by volume), humidified gas, 110 kPa	1840	160	87	(Li et al., 2015b)
PVAm+PVA	Casting	CO_2/N_2 (10/90 by volume), humidified gas, 200–1500 kPa	210–50	174–194		(Deng et al., 2009)
PVAm+PDA/PDMS	Casting	CO_2/N_2 (15/85 by volume), 25°C, humidified gas, 110 kPa	1890	83		(Li et al., 2015a)
PVI with Zinc Complex	Casting	CO_2/N_2 (15/85 by volume), humidified gas, 110 kPa	1120	83		(Yao et al., 2012)
PEIE with Hydrotalcite	Casting	CO_2/N_2 (15/85 by volume), 25°C, humidified gas, 110 kPa	5690	268		(Liao et al., 2014)
PVAm with Hydrotalcite	Casting	CO_2/N_2 (15/85 by volume), 25°C, humidified gas, 110 kPa	3190	296		(Liao et al., 2015)
PVA with Zn-cyclen	Dip-coating	CO_2/N_2 (10/90 by volume), 25°C, humidified gas, 170 kPa	260	107		(Saeed and Deng, 2015)

[a] Gas permeability in Barrer.

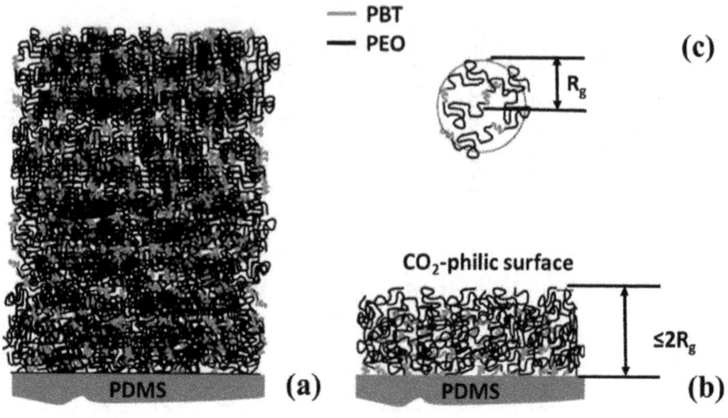

FIGURE 9.5 Schematic of: (a) the block copolymer organization in thick films (semicrystalline polymer and high T_g); (b) a polymer chain showing the radius of gyration, R_g; and (c) block copolymer organization within a super ultrathin film under the influence of the PDMS substrate (thin film mostly amorphous with high fractional free volume and low T_g of the PEO segment). (From Yave, W. et al., Energy & Environmental Science. 4, 4656–4661, 2011.)

within the confined nanospace of a thin film PEO-PBT membrane compared with thicker films which was attributed to the influence of the PDMS gutter layer as shown in Figure 9.5. The presence of the PDMS layer directed the re-organization of the multi-block copolymer due to the compatibility between PDMS and PET segments and thus hindered the crystallization of the PEO segment. This resulted in a CO_2-philic surface giving a highly permeable membrane (Yave et al., 2011).

Pebax™ is a similar block copolymer of PEO with a polyamide. While having high CO_2 selectivity, it suffers from low permeability due to crystallinity. Fu et al. (2013, 2014) blended soft polymeric particles with Pebax™ 2533 to achieve high gas separation performance, as shown in Figure 9.6. A modified PDMS gutter layer was firstly coated on the porous support. High molecular weight amorphous PEO copolymers were synthesized from the condensation polymerization of the PEO monomer with terephthaloyl chloride (TCL). The resulting polymeric nanoparticles were then added into the pristine Pebax™ solution for spin-coating. By optimizing additive structure and concentration, the blended membrane showed CO_2 permeance an order of magnitude greater than in the pure Pebax™, with little loss in selectivity (Table 9.1). This opens a new strategy to improve membrane separation performance as soft polymeric particles with different functionalities can be synthesized through complex polymer chemistry (Halim et al., 2014).

Scofield et al. (2016a, 2016b) investigated the influence of fluorinated additives on the performance of Pebax™ 2533 and Pebax™ 1657. A series of diblock polymers of poly(ethylene glycol)-block poly(pentafluoropropyl actylate) (PEG-b-PPFPA) were synthesized via the Reversible Addition Fragmentation Chain Transfer (RAFT) process with varying lengths of the fluorinated component. These additives significantly enhance CO_2 permeance with only a or slight compromise CO_2/N_2 selectivity.

Thermally rearranged polymers (TR polymers) first proposed by Park, Lee and coworkers have high free volume and narrow cavity size distribution based on their

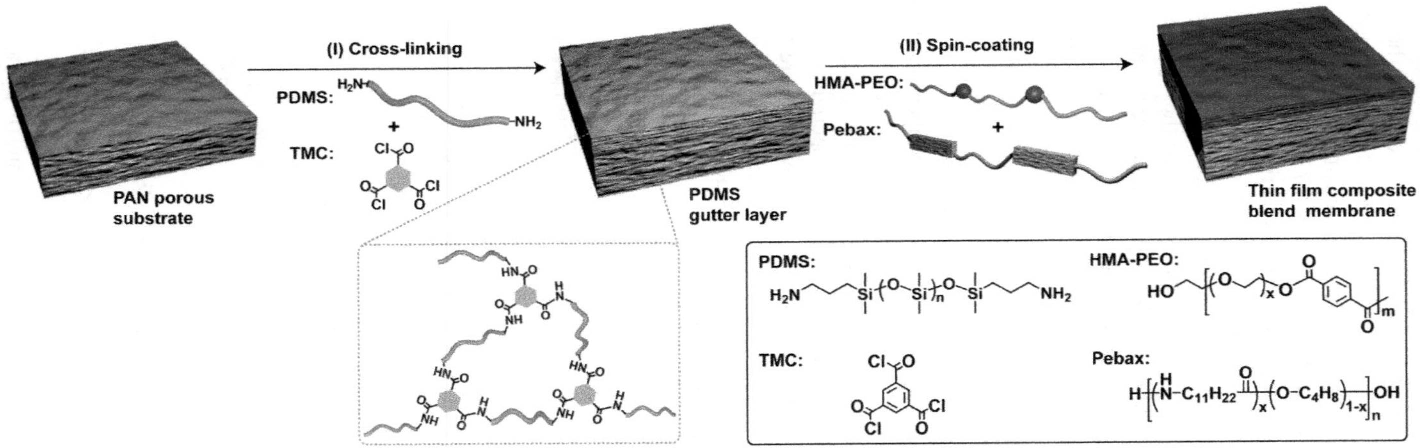

FIGURE 9.6 The fabrication scheme of a thin film composite membrane. (I) The PDMS gutter layer was formed by the cross-linking of amino-terminated PDMS and 1,3,5-benzenetricarbonyl trichloride (TMC). (II) The thin film composite blend membrane was prepared by spin-coating the mixture of Pebax™ 2533 and high molecular weight amorphous PEO onto the PDMS gutter layer. (From Fu, Q. et al., Journal of Materials Chemistry A. 1, 13769–13778, 2013.)

rigid and microporous structure, making them promising candidates for gas separation (Park et al., 2007). These are aromatic polymers with heterocyclic rings, such as polybenzoxazoles (PBO), polybenzimidazole (PBI) or polybenzothiazoles (PBT) structures prepared via thermal rearrangement of polyimide or polyamide precursors with an ortho-functional group (Figure 9.7). The thermal rearrangement occurs in the solid state at a temperature and reaction time that depends on the chemical structure of precursors. The rigid chemical structure shows good chemical and thermal resistance. Meanwhile, the gas permeability is improved at least two orders of magnitude over the precursor polymers due to the significant increase of free volume. The gas permeability of TR polymer membranes are lower than that of the most well-known high free volume polymer, poly(1-trimethylsilyl-1-propyne) (PTMSP), but the CO_2 gas selectivity is much higher. One of the great advantages of TR polymers is the able to control the cavity sizes via tuning polymer structures and thermal reaction mechanisms.

There are many reports regarding the excellent separation performance of thick membranes prepared from TR polymers (Kim and Lee, 2015), however, thin film membranes are rather limited. Recently, TR-PBO hollow fiber membranes were prepared using non-solvent induced phase separation from a hydroxyl poly(amic acid) precursor (Kim et al., 2012). The membranes with a skin layer of 1.5–2 μm showed CO_2 permeance of ~2000 GPU and moderate CO_2/N_2 selectivity of 13. Further optimization is needed to improve the selectivity. Other workers are focusing on lowering the temperature for thermal re-arrangement, which might then allow for the development of thin film composite structures using a temperature-tolerant support material (Guo et al., 2013; Tena et al., 2016).

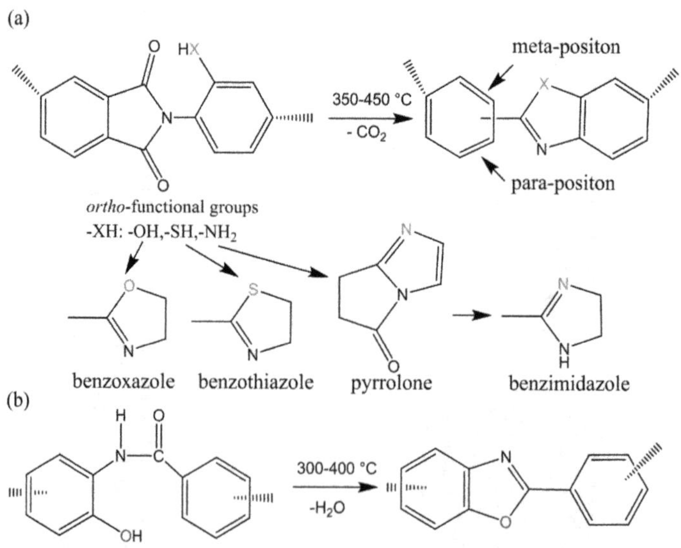

FIGURE 9.7 The thermal rearrangement of (a) polyimides and (b) polyamides. (From Kim, S., Lee, Y.M., Progress in Polymer Science. 43, 1–32, 2015.)

Polymers of intrinsic microporosity (PIMs) are a new class of polymers with interconnected pores, showing very high gas permeability and moderate selectivity. This is attributed to their special ladder-type structures with contorted sites that prevent rotation of polymer chains and effective space packing. The fractional free volume can be more than 20%. The initial PIMs were prepared by a polycondensation reaction of trtrahydroxyl-monomers and tetrafluoro-monomers (or tetrachloro-monomers) with contorted centres, as shown in Figure 9.8.

A few strategies have been developed to further increase the CO_2 permeability and selectivity of these PIM structures: (1) turning the angle of contorted centers, (2) introducing pendant groups and (3) cross-linking polymer chains (Wang et al., 2016). It has been reported that PIM-1 polymer chains tend to undergo self-cross-linking forming triazine rings after the film is treated at 300°C for a period of time, as shown in Figure 9.9 (Li et al., 2012). This cross-linking reaction leads to inefficient chain packing, due to the nature of the contorted structure of the PIM-1 backbone, resulting in an increase in free volume and gas permeability with the increase in thermal treatment time. The best membrane, which was treated at 300°C for 2 days, showed CO_2 permeability of 4000 barrer and an ideal CO_2/CH_4 selectivity of 54.8.

PIMs have good processability as they are readily soluble and thermally stable. They have been formed into thin film composite hollow fibers with an active layer thickness of 1.0–2.9 μm (Gao et al., 2017). However they are glassy polymers and thus physical aging is a challenge due to their high free volume. Gas permeability shows a significant decline with an increase of aging time.

CO_2 facilitated transport membranes have good CO_2 permeability and selectivity as they contain CO_2-reactive carriers, thus enhancing CO_2 transport by both

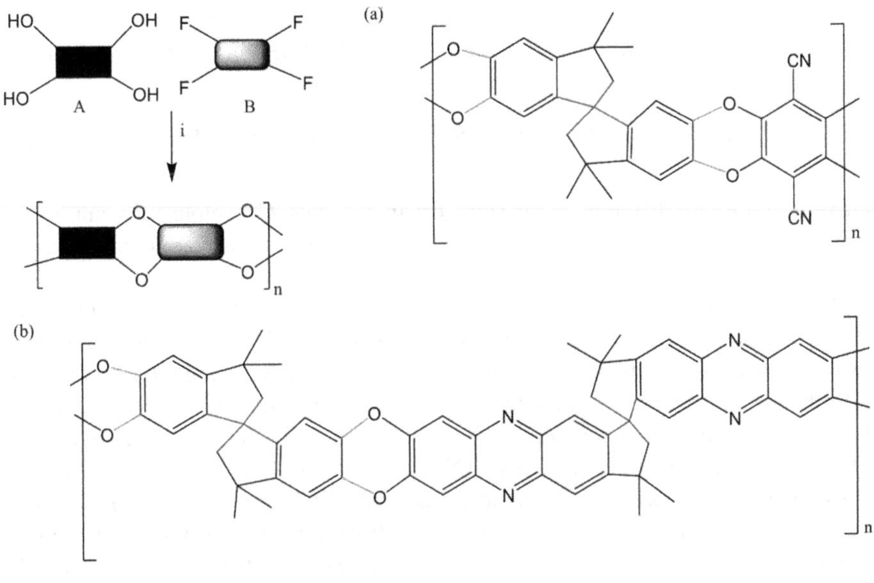

FIGURE 9.8 Synthesis of PIMs (reagents and conditions: i: monomer A, monomer B, K_2CO_3, DMF) and structure of (a) PIM-1 and (b) PIM-7. (From Kim, S., Lee, Y.M., Progress in Polymer Science. 43, 1–32, 2015.)

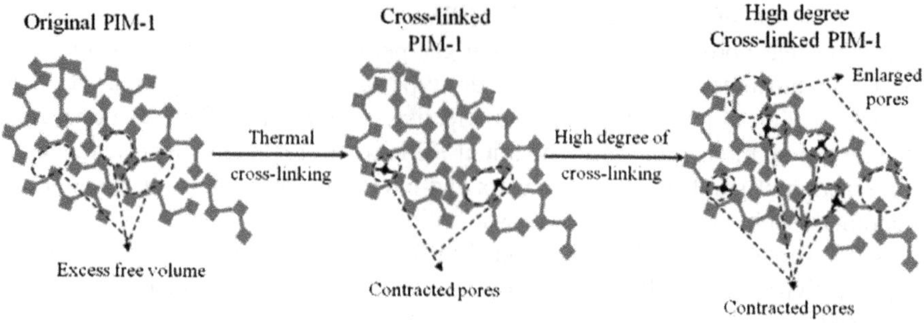

FIGURE 9.9 Two-dimensional representations of the contorted PIM-1 membrane before and after thermal treatment with the formation of triazine rings. (From Li, F.Y. et al., Macromolecules. 45, 1427–1437, 2012.)

reactive species and solution-diffusion transport. Many basic groups such as $-NH_2$, F^-, CO_3^{2-} and $-COO^-$ can be employed as CO_2 carriers. The membranes can also be categorized as containing mobile carriers, where the reactive species are simply dispersed through the matrix or fixed carriers where the reactive species are covalently bonded to the membrane matrix. Although mobile carrier membranes have a high CO_2 transport rate, the loss of the reactive species over time limits their application. Fixed carrier membranes overcome this problem. Polymers with amine groups such as polyvinylamine (PVAm), poly(ethylenimine) (PEI) and chitosan (CS) have been intensively investigated as fixed carrier membranes. Primary and secondary amino groups can react with CO_2 and water as shown in Eqs. 9.7 and 9.8:

$$2CO_2 + 2RNH_2 + H_2O \rightleftharpoons RHNCOOH + RNH_3^+ + HCO_3^- \tag{9.7}$$

$$2CO_2 + 2RR'NH + H_2O \rightleftharpoons RR'NCOOH + RR'NH_2^+ + HCO_3^- \tag{9.8}$$

where R and R′ are different or the same organic groups. CO_2 molecules can transport through the membrane in the form of HCO_3^- ions and the rate is much faster than that of CO_2 molecules.

PVAm is a typical polymer with a high density of primary amine groups. CO_2 readily reacts with the amine groups and water to form HCO_3^-, which dissociates into water and CO_2 at the permeate side. However, PVAm has a high degree of crystallinity due to strong hydrogen bonding between the primary amine groups, which decreases the effective concentration of available carriers. Moreover, it exhibits significant plasticization, thus decreasing CO_2 selectivity. Cross-linking is an effective way to improve membrane performance. Qiao et al. (2012) used a cross-linking agent containing piperazine (PIP) carriers to modify PVAm. A high PIP/PVAm mass ratio of 1.43 was used, leading to a very high CO_2 permeance of 6500 GPU and CO_2/N_2 selectivity of 280 at 110 kPa for a CO_2/N_2 mixture.

A synergistic strategy has been employed to enhance membrane performance via the combination of multiple functional groups. A poly(diallyldimethylammonium carbonate-co-vinylamine) (P(DAD-MACA-co-VAm)) random copolymer containing primary amino groups, carbonate groups and quaternary ammonium groups was synthesized and the membrane was prepared by a casting method (Li et al., 2015b). The membrane exhibited superior CO_2 permeance and selectivity compared to a simple PVAm membrane owing to the synergistic effect of the multiple functional groups, and showed good stability up to 200 hours and resistance to impurities (SO_2, NO_2 and CO) in flue gas. The membrane had low crystallinity, ensuring a high level of utilization of the reactive carriers.

Inspired by carbonic anhydrase (CA), the fastest catalyst for CO_2 hydration, a biomimetic material containing the zinc–poly(N-vinylimidazole) (PVI) complex was synthesized to facilitate the hydration of CO_2, as shown in Figure 9.10. It is worth noting that the crucial component that can efficiently facilitate CO_2 hydration is the complex rather than the Zn(II) ions. By optimizing the synthesis condition, a CO_2 permeance of 1120 GPU and CO_2/N_2 selectivity of 83 was achieved.

Wang and coworkers developed a new membrane by constructing high-speed facilitated transport channels (i.e., hydrotalcite (HT)) in a fixed carrier membrane (e.g., PEI and PVAm based polymers) (Liao et al., 2014, 2015). Hydrotalcite is a type of layered double hydroxide, which consists of positively charged host layers that can readily accommodate hydrated carbonate anions moving through the interlayer channels. These channels thus enhance the CO_2 facilitated transport properties. Figure 9.11 shows a schematic of the formation of the PEIE-HT complex, where PEIE is a polymer synthesized from PEI and epichlorohydrin. The resultant membrane possessed both fixed carrier sites from the polymer matrix and movable carrier sites in the interlayer of the HT. As a result, the CO_2 permeance of the PEIE–HT membrane was 5693 GPU and the CO_2/N_2 selectivity was 268 at 110 kPa, which are 7 and 4 times of those of the PEIE membrane, respectively.

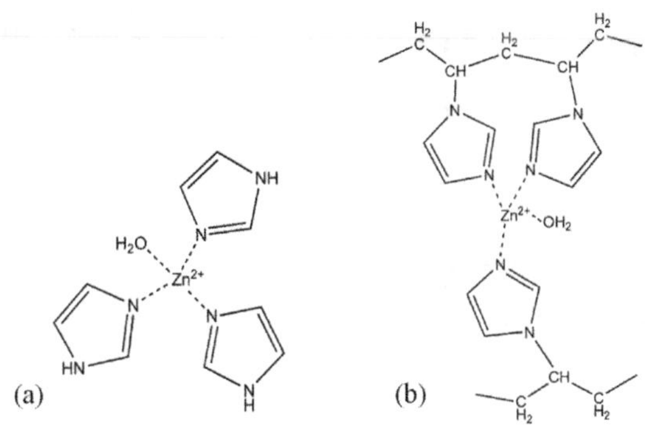

FIGURE 9.10 Model of (a) carbonic anhydrase active site, (b) PVI–Zn (II) complex. (From Yao, K. et al., Chemical Communications. 48, 1766–1768, 2012.)

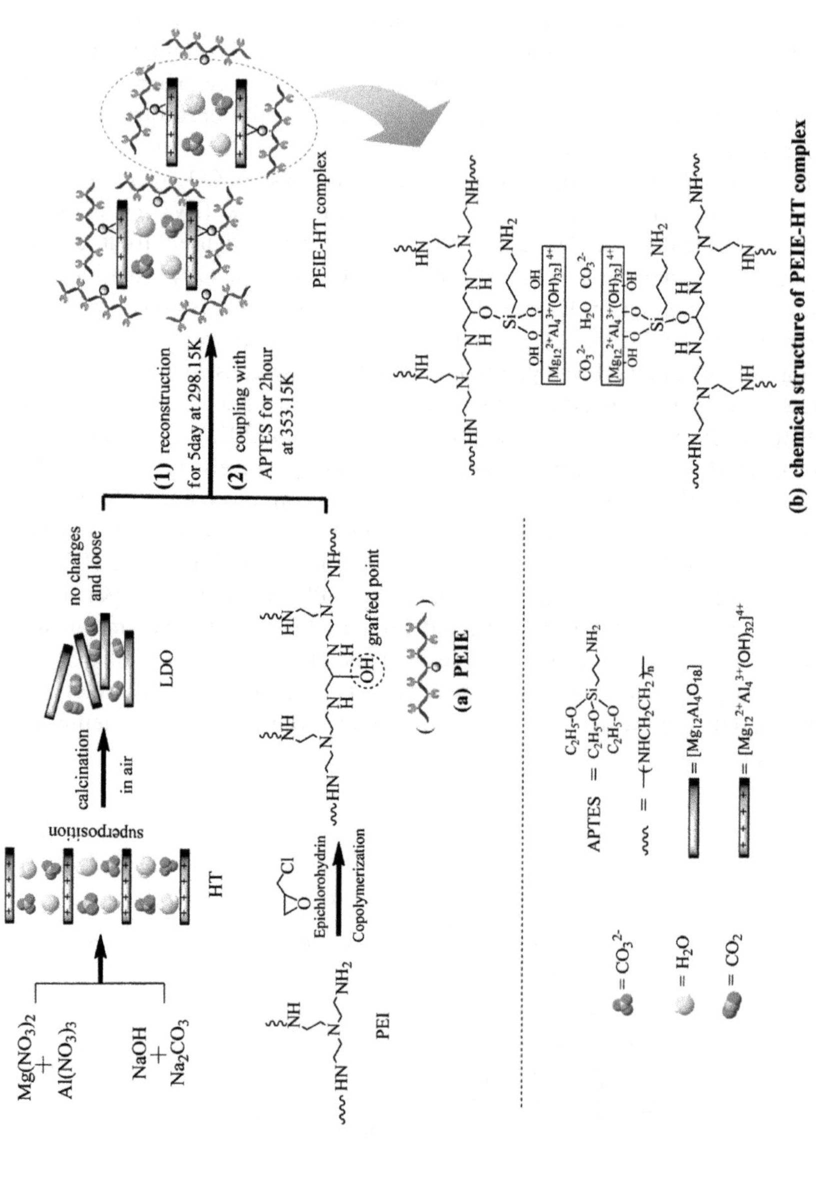

FIGURE 9.11 Schematic of the reaction pathway of a PEIE-HT complex. (From Liao, J. et al., Chemical Science. 5, 2843–2849, 2014.)

9.5 MIXED MATRIX MEMBRANES

Mixed matrix membranes (MMMs), comprising a bulk polymer as the continuous phase and inorganic filler as the dispersed phase, are able to overcome the performance trade-off of polymeric membranes while retaining the advantages of low cost and good processability. Many nanoporous materials have been investigated as fillers such as zeolites, metal organic frameworks (MOFs), carbon nanotubes (CNT) and graphene oxide (GO). They can both tune the free volume elements of polymers to improve diffusional selectivity or provide additional diffusion pathways to increase overall gas diffusivity. Hence, membrane permeability can be enhanced without any reduction of membrane selectivity. The transport properties of MMMs depend to some extent on the interfacial morphology between the polymer and the nanoparticles. Figure 9.12 shows a schematic diagram of various structures at the interface. Among them, Case 1 is an ideal morphology, without any interfacial voids (Case 2), rigidified polymer chains (Case 3) or pore blockage (Case 4), thus enhancing gas permeability and selectivity. While some researchers have clearly shown instances where this morphology is critical to performance (Merkel et al., 2003), there is a tendency from others to overstate the influence of interfacial voids. In particular, such voids can appear in scanning electron microscopy images as an artifact of the sample preparation which have no influence on the membrane performance (Kanehashi et al., 2015). Of greater relevance is the need for the size of the filler to be significantly smaller than the thickness of the ultrathin selective layer, or defects will occur.

Some typical TFC mixed matrix membranes are listed in Table 9.2. The permeance of a PIM-1 membrane was improved by ~60% and the selectivity of CO_2/N_2 slightly increased by incorporation of functionalized multi-walled carbon nanotube (f-MWCNTs). Moreover, the incorporation of these MWCNTS into the PIM-1 reduced the effect of aging. The decline of CO_2 permeance was 17% in 300 days while it was 38% for the original PIM-1 membrane.

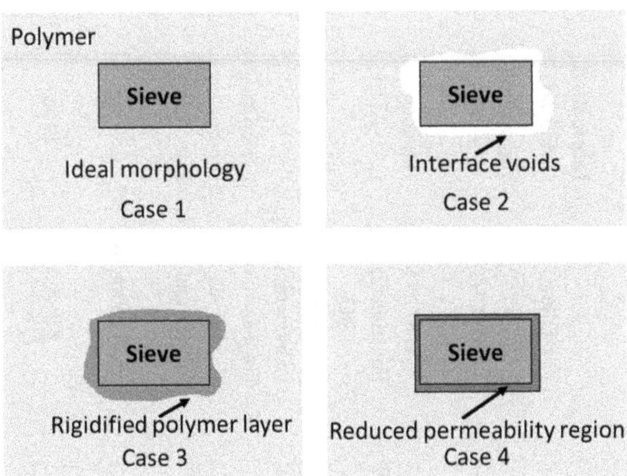

FIGURE 9.12 Schematic of interface morphology in mixed matrix membranes. (From Chung, T.S. et al., Progress in Polymer Science. 32, 483–507, 2007.)

TABLE 9.2
Performance of Mixed Matrix Membranes for CO_2 Separation

Polymer	Filler	Fabrication Method	Operational Conditions	CO_2 Permeance (GPU)	Selectivity		Reference
					CO_2/N_2	CO_2/CH_4	
PDMS	$Cu_3(BTC)_2$	Dip-coating	Pure gas, 25°C, 500 kPa	110		32	(Zulhairun et al., 2015)
Cellulose acetate	MWCNT	Wet phase inversion	Pure gas, room temperature, 300 kPa	740	40		(Ahmad et al., 2014)
PIM-1	MWCNT	Dip-coating		10500	34		(Koschine et al., 2015)
PEGDMA9	Functionalized SiO_2	CAP-ATRP	Pure gas, 35°C, 350 kPa	1290	27		(Kim et al., 2016b)
PEGDMA9	FeDA nanoparticle	CAP-ATRP	Pure gas, 35°C, 350 kPa	1360	30		(Kim et al., 2016a)
Pebax™	Graphene oxide	Dip-coating	Pure gas, 25°C, 200 kPa	410	43		(Zhang et al., 2017)
Pebax™	ZIF-8	Dip-coating	Pure gas, 25°C, 200 kPa	350	32		(Sutrisna et al., 2017)
Pebax™	Ionic liquid	Dip-coating	CO_2/N_2 (50/50 by volume), 25 °C, 200 kPa	350	38		(Chen et al., 2014)
PVAm	PANI nanorods	Casting	CO_2/N_2 (15/85 by volume), 25°C, humidified gas, 110 kPa	3100	250		(Zhao et al., 2013)
PVAm	ZIF-8	Casting	CO_2/N_2 (15/85 by volume), 25°C, humidified gas, 110 kPa	1500	105		(Zhao et al., 2015)
PVAm/PVA	CNT	Casting	CO_2/N_2 (10/90 by volume), 25°C, humidified gas, 200 kPa	81		23	(He et al., 2014)

Kim et al. developed a PEG-based ultra-thin selective layer incorporating PDA-PEI functionalized SiO_2 nanoparticles via the continuous assembly of polymers (CAP) (Kim et al., 2016b). The best results were obtained via codeposition of PEI and PDA onto the silica particles, with a 5 wt% loading. The CO_2 permeance was ~1300 GPU and the CO_2/N_2 selectivity was 27. Moreover, the permeance of the underlying selective layer was over 2000 GPU with a CO_2/N_2 selectivity of 39. They also synthesized PEG-based mixed matrix membrane by incorporating iron dopamine nanoparticles (FeDA NPs). The FeDA NPs were prepared by nano-compexlation between Fe^{3+} and DA, and the particle size varied from 3 to 74 nm by adjusting the molar ratio of DA to Fe^{3+} ions.

Very recently, a high performance graphene oxide/Pebax™ composite hollow fiber membrane with a thin selective layer (~1 μm) was prepared by a facile dip-coating method (Zhang et al., 2017). A 0.1 wt% GO loading significantly improved the CO_2 permeance (~90%) without losing CO_2/N_2 selectivity. The GO laminates provided a highly efficient gas transport pathway for CO_2 through the spaces between the sheets. For this approach to be effective, the GO laminates should be parallel to the support, without any defect formation.

9.6 CHALLENGES AND LIMITATIONS IN INDUSTRIAL APPLICATIONS

Although great efforts are made to test membrane performance, the test time in the lab scale is quite short; varying from a few minutes to days or at a maximum a few months. However, it is expected that gas separation membranes should have a lifetime of three to five years. Many polymers used for gas separation are glassy polymers, which are in a non-equilibrium state and have excess free volume. Over time, this excess free volume will dissipate as the polymer moves towards thermodynamic equilibrium. This physical aging reduces gas permeability and is seen as a reduction of membrane flux. Recently, it has been recognized that physical aging also depends on the thickness of the membrane, with the ultrathin layer within a thin film composite membrane aging more quickly. Figure 9.13 show the impact of film thickness on N_2 permeability and selectivity of Matrimid. Firstly, the bulk values are very different from thin films as the state of the glassy polymer is dependent upon its preparation history. Secondly, it is clear that the aging rate is thickness dependent and the thinner film has a much faster rate. On the other hand, the selectivity increases with aging.

The polymer can also swell when the concentration of gas inside increases, thus increasing its free volume and gas diffusivity. For a rubbery polymer, this swelling occurs rapidly. However, for a glassy polymer, there will be a period of equilibrium swelling, followed by a much slower polymer expansion, termed plasticization, where polymer chains that were not in an equilibrium state re-organize and the polymer moves closer to an equilibrium state. Both swelling and plasticization results in an increase of gas permeability. CO_2 is a well-known plasticizer in gas separation applications. For glassy polymers, the CO_2 permeability generally decreases with increasing feed pressure due to the filling of the larger non-equilibrium free volume voids, until plasticization occurs and gas permeability starts to increase again

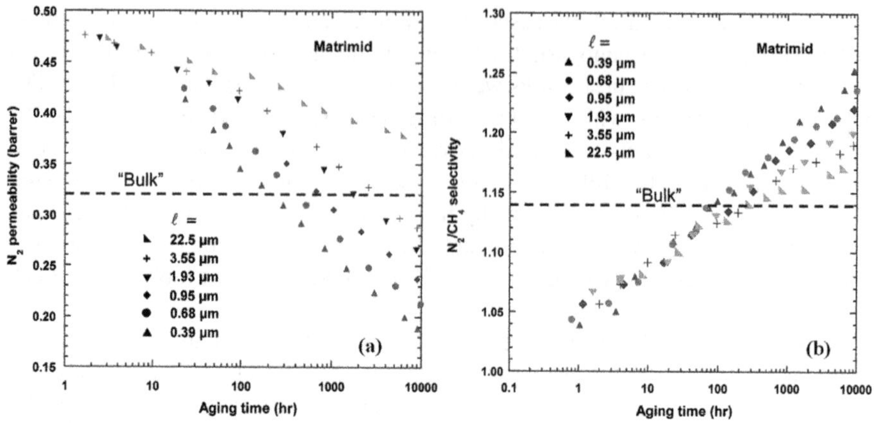

FIGURE 9.13 The influence of film thickness on the aging of Matrimid with: (a) N_2 permeability and (b) N_2/CH_4 selectivity. The membrane was tested at 35°C and 2 bar. (From Huang, Y., Paul, D.R., Industrial and Engineering Chemistry Research. 46, 2342–2347, 2007.)

(Kanehashi et al., 2007). The pressure at which the gas permeability is at a minimum is called the plasticization pressure. The plasticization pressure is also thickness dependent. A Matrimid membrane of 20 μm shows a plasticization pressure of 14 bar while it is ~6 bar for a thin film of 182 nm for CO_2 (Horn and Paul, 2011). The plasticization can also be affected by aging, as shown in Figure 9.14. For the thin films, as aging time increased, the plasticization pressure curve shifts to lower permeability and flattens out due to the loss of excess free volume. The three thick films show little difference in permeability as the pressure changes, which is attributed to the very slow physical aging rate of these bulk glassy polymers. The plasticization effect can also cause selectivity to decrease under mixed gas conditions for CO_2 separation applications. Various strategies have been applied to increase the plasticization resistance of

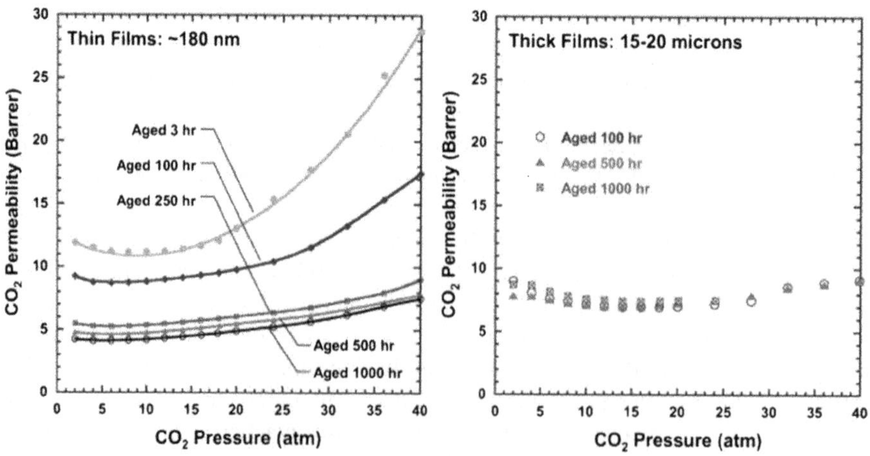

FIGURE 9.14 Effect of aging time on plasticization pressure and CO_2 permeability for thin and thick Matrimid films. (From Horn, N.R., Paul, D.R., Polymer. 52, 1619–1627, 2011.)

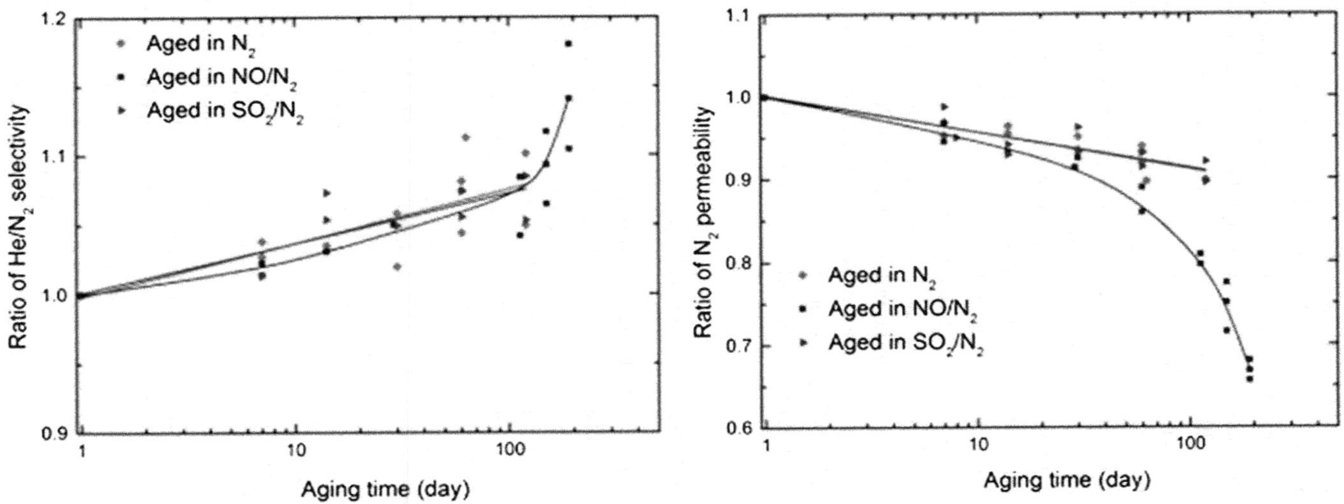

FIGURE 9.15 The influence of N_2, NO and SO_2 on: (left) He/N_2 selectivity and (right) N_2 permeability of a cellulose triacetate membrane. The membranes were tested at 35°C and 7.5 bar after aging in pure N_2, 979 ppm NO in balance N_2 and 1000 ppm SO_2 balance in N_2 at 27.5 bar at 22 ± 2°C. (From Lu, H.T. et al., International Journal of Greenhouse Gas Control. 55, 97–104, 2016.)

glassy membranes, such as cross-linking and thermal-treatment. In principle, both methods reduce the fractional free volume of the polymer, resulting in lower gas permeability, but better plasticization resistance.

Many impurities such as water vapor, SO_x, NO_x, H_2S and hydrocarbons are present as minor components in industrial CO_2-contaning streams. Although the concentration of these components is low, they have a significant influence on membrane performance (Scholes et al., 2009). For example, water vapor can both compete for sorption sites, which reduces the solubility of other gases, and it can cause plasticization, which increases permeability but decreases selectivity. "Anti-plasticization" can also occur, where the water vapour molecules block the pathways of other molecules, reducing diffusion rates (Chen et al., 2011, 2015). The impurities can also chemically react with the polymer materials. For instance, the loss of permeability of cellulose triacetate (CTA) membranes when aged in the presence of 0.74 kPa NO was attributed to reaction of the polymer with trace amounts of NO_2 in the gas (Figure 9.15, (Lu et al., 2016)). NO_2 can oxidize the primary alcohol of cellulosic materials into a carboxyl moiety (Rowen et al., 1947; Yackel and Kenyon, 1942). The permeability was reduced by ~35% of the initial value after 200 days aging in NO_x. Similarly, while metal organic framework materials are often used in mixed matrix membranes, they react with acidic gas impurities, often leading to substantial losses in the permeability of the membranes as structural collapse of the frameworks occur (Kanehashi et al., 2018).

9.7 CONCLUSIONS AND PERSPECTIVES

In this chapter, a review of recent progress of thin film composite membranes for CO_2 separation applications has been provided particularly focusing on polymer-based materials. Two gas transport mechanisms (i.e., solution-diffusion and facilitated transport) and typical membrane preparation methods were briefly explained.

For polymeric membranes, although many novel materials have been investigated for gas separation, most of them are relatively thick films, which are not applicable for industrial deployment. In terms of thin films, PEO-based membranes based on the solution-diffusion mechanism show promising results. On the other hand, TR polymers and PIM polymers exhibit very high gas permeability and moderate gas selectivity due to their high free volume. There are limited reports focusing on fabricating thin film membranes from these materials and this will be need to be a significant focus of future research. Facilitated transport membranes generally show better permeance and selectivity for CO_2/N_2 separation and may be good candidates for CO_2 capture from post-combustion flue gases, considering the low CO_2 partial pressure. However, effort is required to increase their stability over long periods of operation. Mixed matrix membranes provide an alternate option to overcome the performance trade-off of polymeric membranes as they lead to increased permeability with no change in selectivity.

It is well known that the permeance of glassy polymeric membrane will decline with increasing aging time and that these effects are more profound for thin film membranes. The membrane performance can be also affected by plasticization and again these effects are greater in thin films. Impurities in the feed gas stream generally have adverse effects on membrane performance, either through competitive sorption, plasticization, anti-plasticization or by chemical reaction.

Given these issues, the focus of international research needs to shift from generating countless new materials of marginally better performance in idealized conditions, to a more holistic assessment of key target membrane systems. To be deployed in large scale applications and to be cost-effective, the membrane polymers should have good processability and be able to produce a thin film selective layer less than 1 μm. Moreover, the performance of the thin film composite membranes must be evaluated in real industrial conditions or at least simulated conditions (gas mixtures with impurities) and over extended time periods to provide a complete evaluation. Only then will these materials forge the treacherous path from the laboratory into industrial application.

ACKNOWLEDGMENT

The authors would like to acknowledge the funding support from the Australian Research Council through its Discovery program (DP150100977).

REFERENCES

Ahmad, A.L., Jawad, Z.A., Low, S.C., Zein, S.H.S., 2014. A cellulose acetate/multi-walled carbon nanotube mixed matrix membrane for CO_2/N_2 separation. Journal of Membrane Science. 451, 55–66.

Baker, R.W., Low, B.T., 2014. Gas separation membrane materials: A perspective. Macromolecules. 47, 6999–7013.

Baker, R.W., Pinnau, I., He, Z., Amo, K.D., Da Costa A.R., Daniels, R., 2003. Carbon dioxide gas separation using organic-vapor-resistant membranes. U.S. Patent 6,572, 680.

Breck, D.W., 1973. Zeolite Molecular Sieves: Structure, Chemistry and Use. Wiley and Sons, New York.

Brinkmann, T., Naderipour, C., Pohlmann, J., Wind, J., Wolff, T., Esche, E., Müller, D., Wozny, G., Hoting, B., 2015. Pilot scale investigations of the removal of carbon dioxide from hydrocarbon gas streams using poly (ethylene oxide)–poly (butylene terephthalate) PolyActive™) thin film composite membranes. Journal of Membrane Science. 489, 237–247.

Brunetti, A., Scura, F., Barbieri, G., Drioli, E., 2010. Membrane technologies for CO_2 separation. Journal of Membrane Science. 359, 115–125.

Chen, G.Q., Kanehashi, S., Doherty, C.M., Hill, A.J., Kentish, S.E., 2015. Water vapor permeation through cellulose acetate membranes and its impact upon membrane separation performance for natural gas purification. Journal of Membrane Science. 487, 249–255.

Chen, G.Q., Scholes, C.A., Qiao, G.G., Kentish, S.E., 2011. Water vapor permeation in polyimide membranes. Journal of Membrane Science. 379, 479–487.

Chen, H.Z., Thong, Z., Li, P., Chung, T.-S., 2014. High performance composite hollow fiber membranes for CO_2/H_2 and CO_2/N_2 separation. International Journal of Hydrogen Energy. 39, 5043–5053.

Chung, T.S., Jiang, L.Y., Li, Y., Kulprathipanja, S., 2007. 6572, 680Mixed matrix membranes (MMMs) comprising organic polymers with dispersed inorganic fillers for gas separation. Progress in Polymer Science. 32, 483–507.

Deng, L., Kim, T.J., Hägg, M.B., 2009. Facilitated transport of CO_2 in novel PVAm/PVA blend membrane. Journal of Membrane Science. 340, 154–163.

Fu, Q., Halim, A., Kim, J., Scofield, J.M.P., Gurr, P.A., Kentish, S.E., Qiao, G.G., 2013. Highly permeable membrane materials for CO_2 capture. Journal of Materials Chemistry A. 1, 13769–13778.

Fu, Q., Kim, J., Gurr, P.A., Scofield, J.M.P., Kentish, S.E., Qiao, G.G., 2016. A novel cross-linked nano-coating for carbon dioxide capture. Energy & Environmental Science. 9, 434–440.

Fu, Q., Wong, E.H.H., Kim, J., Scofield, J.M.P., Gurr, P.A., Kentish, S.E., Qiao, G.G., 2014. The effect of soft nanoparticles morphologies on thin film composite membrane performance. Journal of Materials Chemistry A. 2, 17751–17756.

Gao, L., Alberto, M., Gorgojo, P., Szekely, G., Budd, P.M., 2017. High-flux PIM-1/PVDF thin film composite membranes for 1-butanol/water pervaporation. Journal of Membrane Science, 529. 207–214.

Gherasim, C.V., Luelf, T., Roth, H., Wessling, M., 2016. Dual-Charged Hollow Fiber Membranes for Low-Pressure Nanofiltration Based on Polyelectrolyte Complexes: One-Step Fabrication with Tailored Functionalities. ACS Applied Materials & Interfaces. 8, 19145–19157.

Guo, R., Sanders, D.F., Smith, Z.P., Freeman, B.D., Paul, D.R., McGrath, J.E., 2013. Synthesis and characterization of Thermally Rearranged (TR) polymers: Influence of ortho-positioned functional groups of polyimide precursors on TR process and gas transport properties. Journal of Materials Chemistry A. 1, 262–272.

Halim, A., Fu, Q., Yong, Q., Gurr, P.A., Kentish, S.E., Qiao, G.G., 2014. Soft polymeric nanoparticle additives for next generation gas separation membranes. Journal of Materials Chemistry A. 2, 4999.

He, X., Kim, T.J., Hägg, M.B., 2014. Hybrid fixed-site-carrier membranes for CO_2 removal from high pressure natural gas: Membrane optimization and process condition investigation. Journal of Membrane Science. 470, 266–274.

Horn, N.R., Paul, D.R., 2011. Carbon dioxide plasticization and conditioning effects in thick vs. thin glassy polymer films. Polymer. 52, 1619–1627.

Huang, Y., Paul, D.R., 2007. Effect of film thickness on the gas-permeation characteristics of glassy polymer membranes. Industrial and Engineering Chemistry Research. 46, 2342–2347.

Jo, E.-S., An, X., Ingole, P.G., Choi, W.-K., Park, Y.-S., Lee, H.-K., 2017. CO_2/CH_4 separation using inside coated thin film composite hollow fiber membranes prepared by interfacial polymerization. Chinese Journal of Chemical Engineering. 25, 278–287.

Kanehashi, S., Aguiar, A., Lu, H.T., Chen, G.Q., Kentish, S.E., 2018. Effects of industrial gas impurities on performance of mixed matrix membranes. Journal of Membrane Science, In Press.

Kanehashi, S., Chen, G.Q., Scholes, C.A., Ozcelik, B., Hua, C., Ciddor, L., Southon, P.D., D'Alessandro, D.M., Kentish, S.E., 2015. Enhancing gas permeability in mixed matrix membranes through tuning the nanoparticle properties. Journal of Membrane Science. 482, 49–55.

Kanehashi, S., Nakagawa, T., Nagai, K., Duthie, X., Kentish, S.E., Stevens, G.W., 2007. Effects of carbon dioxide-induced plasticization on the gas transport properties of glassy polyimide membranes. Journal of Membrane Science. 298, 147–155.

Kim, J., Fu, Q., Scofield, J.M.P., Kentish, S.E., Qiao, G.G., 2016a. Ultra-thin film composite mixed matrix membranes incorporating iron(III)–dopamine nanoparticles for CO_2 separation. Nanoscale. 8, 8312–8323.

Kim, J., Fu, Q., Xie, K., Scofield, J.M.., Kentish, S.E., Qiao, G.G., 2016b. CO_2 separation using surface-functionalized SiO_2 nanoparticles incorporated ultra-thin film composite mixed matrix membranes for post-combustion carbon capture. Journal of Membrane Science. 515, 54–62.

Kim, S., Han, S.H., Lee, Y.M., 2012. Thermally rearranged (TR) polybenzoxazole hollow fiber membranes for CO_2 capture. Journal of Membrane Science. 403–404, 169–178.

Kim, S., Lee, Y.M., 2015. Rigid and microporous polymers for gas separation membranes. Progress in Polymer Science. 43, 1–32.

Kim, T.J., Vrålstad, H., Sandru, M., Hägg, M.B., 2013. Separation performance of PVAm composite membrane for CO_2 capture at various pH levels. Journal of Membrane Science. 428, 218–224.

Koschine, T., Rätzke, K., Faupel, F., Khan, M.M., Emmler, T., Filiz, V., Abetz, V., Ravelli, L., Egger, W., 2015. Correlation of gas permeation and free volume in new and used high free volume thin film composite membranes. Journal of Polymer Science, Part B: Polymer Physics. 53, 213–217.

Li, D., Chung, T.S., Wang, R., 2004. Morphological aspects and structure control of dual-layer asymmetric hollow fiber membranes formed by a simultaneous co-extrusion approach. Journal of Membrane Science. 243, 155–175.

Li, F.Y., Xiao, Y., Chung, T.S., Kawi, S., 2012. High-performance thermally self-cross-linked polymer of intrinsic microporosity (PIM-1) membranes for energy development. Macromolecules. 45, 1427–1437.

Li, P., Chen, H.Z., Chung, T.S., 2013. The effects of substrate characteristics and pre-wetting agents on PAN–PDMS composite hollow fiber membranes for CO_2/N_2 and O_2/N_2 separation. Journal of Membrane Science. 434, 18–25.

Li, P., Wang, Z., Li, W., Liu, Y., Wang, J., Wang, S., 2015a. High-Performance Multilayer Composite Membranes with Mussel-Inspired Polydopamine as a Versatile Molecular Bridge for CO_2 Separation. ACS Applied Materials and Interfaces. 7, 15481–15493.

Li, P., Wang, Z., Liu, Y., Zhao, S., Wang, J., Wang, S., 2015b. A synergistic strategy via the combination of multiple functional groups into membranes towards superior CO_2 separation performances. Journal of Membrane Science. 476, 243–255.

Liao, J., Wang, Z., Gao, C., Li, S., Qiao, Z., Wang, M., Zhao, S., Xie, X., Wang, J., Wang, S., 2014. Fabrication of high-performance facilitated transport membranes for CO_2 separation. Chemical Science. 5, 2843–2849.

Liao, J., Wang, Z., Gao, C., Wang, M., Yan, K., Xie, X., Zhao, S., Wang, J., Wang, S., 2015. A high performance PVAm–HT membrane containing high-speed facilitated transport channels for CO_2 separation. Journal of Materials Chemistry A. 3, 16746–16761.

Lu, H.T., Kanehashi, S., Scholes, C.A., Kentish, S.E., 2016. The potential for use of cellulose triacetate membranes in post combustion capture. International Journal of Greenhouse Gas Control. 55, 97–104.

Merkel, T.C., Freeman, B.D., Spontak, R.J., He, Z., Pinnau, I., Meakin, P., Hill, A.J., 2003. Sorption, transport, and structural evidence for enhanced free volume in poly(4-methyl-2-pentyne)/fumed silica nanocomposite membranes. Chemistry of Materials. 15, 109–123.

Merkel, T.C., Lin, H., Wei, X., Baker, R., 2010. Power plant post-combustion carbon dioxide capture: An opportunity for membranes. Journal of Membrane Science. 359, 126–139.

Ockwig, N.W., Nenoff, T.M., 2007. Membranes for Hydrogen Separation. Chemical Reviews. 107, 4078–4110.

Park, H.B., Jung, C.H., Lee, Y.M., Hill, A.J., Pas, S.J., 2007. Polymers with cavities tuned for fast selective transport of small molecules and ions. Science. 318, 254–259.

Qiao, Z., Wang, Z., Zhang, C., Yuan, S., Zhu, Y., Wang, J., 2012. PVAm–PIP/PS composite membrane with high performance for CO_2/N_2 separation. AIChE Journal. 59, 215–228.

Ren, J., McCutcheon, J.R., 2017. Making thin film composite hollow fiber forward osmosis membranes at the module scale using commercial ultrafiltration membranes. Industrial & Engineering Chemistry Research. 56, 4074–4082.

Robeson, L.M., 2008. The upper bound revisited. Journal of Membrane Science. 320, 390–400.

Robeson, L.M., Freeman, B.D., Paul, D.R., Rowe, B.W., 2009. An empirical correlation of gas permeability and permselectivity in polymers and its theoretical basis. Journal of Membrane Science. 341, 178–185.

Robeson, L.M., Smith, Z.P., Freeman, B.D., Paul, D.R., 2014. Contributions of diffusion and solubility selectivity to the upper bound analysis for glassy gas separation membranes. Journal of Membrane Science. 453, 71–83.

Rowen, J.W., Hunt, C.M., Plyler, E.K., 1947. Absorption spectra in the detection of chemical changes in cellulose and cellulose derivatives. Textile Research Journal. 17, 504–511.

Saeed, M., Deng, L., 2015. CO_2 facilitated transport membrane promoted by mimic enzyme. Journal of Membrane Science. 494, 196–204.

Salih, A.A.M., Yi, C., Peng, H., Yang, B., Yin, L., Wang, W., 2014. Interfacially polymerized polyetheramine thin film composite membranes with PDMS inter-layer for CO_2 separation. Journal of Membrane Science. 472, 110–118.

Schell, W.J., Wensley, C.G., Chen, M.S.K., Venugopal, K.G., Miller, B.D., Stuart, J.A., 1989. Recent advances in cellulosic membranes for gas separation and pervaporation. Gas Separation and Purification. 3, 162–169.

Scholes, C.A., Kentish, S.E., Stevens, G.W., 2009. Effects of minor components in carbon dioxide capture using polymeric gas separation membranes. Separation and Purification Reviews. 38, 1–44.

Scholes, C.A., Smith, K.H., Stevens, G.W., Kentish, S.E., 2010. CO_2 capture from Pre-combustion Processes-Strategies for membrane gas separation. International Journal of Greenhouse Gas Control. 4, 739–755.

Scholes, C.A., Stevens, G.W., Kentish, S.E., 2012. Membrane gas separation applications in natural gas processing. Fuel. 96, 15–28.

Scofield, J.M.P., Gurr, P.A., Kim, J., Fu, Q., Halim, A., Kentish, S.E., Qiao, G.G., 2015. High-performance thin film composite membranes with well-defined poly(dimethylsiloxane)-b-poly(ethylene glycol) copolymer additives for CO_2 separation. Journal of Polymer Science, Part A: Polymer Chemistry. 53, 1500–1511.

Scofield, J.M.P., Gurr, P.A., Kim, J., Fu, Q., Kentish, S.E., Qiao, G.G., 2016a. Development of novel fluorinated additives for high performance CO_2 separation thin-film composite membranes. Journal of Membrane Science. 499, 191–200.

Scofield, J.M.P., Gurr, P.A., Kim, J., Fu, Q., Kentish, S.E., Qiao, G.G., 2016b. Blends of fluorinated additives with highly selective thin-film composite membranes to increase co_2 permeability for co_2/n_2 gas separation applications. Industrial and Engineering Chemistry Research. 55, 8364–8372.

Sutrisna, P.D., Hou, J., Li, H., Zhang, Y., Chen, V., 2017. Improved operational stability of Pebax-based gas separation membranes with ZIF-8: A comparative study of flat sheet and composite hollow fiber membranes. Journal of Membrane Science. 524, 266–279.

Tena, A., Rangou, S., Shishatskiy, S., Filiz, V., Abetz, V., 2016. Claisen thermally rearranged (CTR) polymers. Science Advances. 2, e1501859.

Veríssimo, S., Peinemann, K. V, Bordado, J., 2005. Thin-film composite hollow fiber membranes: An optimized manufacturing method. Journal of Membrane Science. 264, 48–55.

Wang, R., Shi, L., Tang, C., Chou, S., Fane, A.G., 2010. Characterization of novel forward osmosis hollow fiber membranes. Journal of Membrane Science. 355, 158–167.

Wang, S., Li, X., Wu, H., Tian, Z., Xin, Q., He, G., Peng, D., Chen, S., Yin, Y., Jiang, Z., Guiver, M.D., 2016. Advances in high permeability polymer-based membrane materials for CO_2 separations. Energy & Environmental Science. 9, 1863–1890.

Wessling, M., Stamatialis, D., Kopec, K.K., Dutzcak, S.M., 2015. Hollow Fiber Membrane. U.S. Patent 9,067, 180.

Yackel, E.C., Kenyon, W.O., 1942. The oxidation of cellulose by nitrogen dioxide. Journal of the American Chemical Society. 64, 121–127.

Yao, K., Wang, Z., Wang, J., Wang, S., 2012. Biomimetic material—poly(N-vinylimidazole)–zinc complex for CO_2 separation. Chemical Communications. 48, 1766–1768.
Yave, W., Car, A., Funari, S.S., Nunes, S.P., Peinemann, K.V., 2010a. CO_2-Philic polymer membrane with extremely high separation performance. Macromolecules. 43, 326–333.
Yave, W., Car, A., Wind, J., Peinemann, K.V., 2010b. Nanometric thin film membranes manufactured on square meter scale: Ultra-thin films for CO_2 capture. Nanotechnology. 21, 395301.
Yave, W., Huth, H., Car, A., Schick, C., 2011. Peculiarity of a CO_2-philic block copolymer confined in thin films with constrained thickness: 'A super membrane for CO_2-capture'. Energy & Environmental Science. 4, 4656–4661.
Yuan, F., Wang, Z., Li, S., Wang, J., Wang, S., 2012. Formation-structure-performance correlation of thin film composite membranes prepared by interfacial polymerization for gas separation. Journal of Membrane Science. 421–422, 327–341.
Zhang, Y., Shen, Q., Hou, J., Sutrisna, P.D., Chen, V., 2017. Shear-aligned graphene oxide laminate/Pebax ultrathin composite hollow fiber membranes using a facile dip-coating approach. Journal of Materials Chemistry A. 5, 7732–7737.
Zhang, Y., Sunarso, J., Liu, S., Wang, R., 2013. Current status and development of membranes for CO_2/CH_4 separation: A review. International Journal of Greenhouse Gas Control. 12, 84–107.
Zhao, S., Cao, X., Ma, Z., Wang, Z., Qiao, Z., Wang, J., Wang, S., 2015. Mixed-matrix membranes for co_2/n_2 separation comprising a poly(vinylamine) matrix and metal-organic frameworks. Industrial and Engineering Chemistry Research. 54, 5139–5148.
Zhao, S., Wang, Z., Qiao, Z., Wei, X., Zhang, C., Wang, J., Wang, S., 2013. Gas separation membrane with CO_2-facilitated transport highway constructed from amino carrier containing nanorods and macromolecules. Journal of Materials Chemistry A. 1, 246–249.
Zulhairun, A.K., Fachrurrazi, Z.G., Nur Izwanne, M., Ismail, A.F., 2015. Asymmetric hollow fiber membrane coated with polydimethylsiloxane-metal organic framework hybrid layer for gas separation. Separation and Purification Technology. 146, 85–93.

10 Low-Temperature Plasma-Enhanced Chemical Vapor Deposition of Silica-Based Membranes

Synthesis, Characterization, and Gas Permeation Properties

Hiroki Nagasawa and Toshinori Tsuru

CONTENTS

10.1 INTRODUCTION: MICROPOROUS INORGANIC MEMBRANES FOR MOLECULAR SEPARATION

Microporous inorganic membranes have received considerable attention for their potential application in the molecular separation of gaseous or liquid mixtures, because of their excellent stability and high separation performance (Lin et al., 2002; Tsuru, 2008). They are expected to be utilized under harsh operating conditions such as high temperature and high pressure and in the presence of organic vapors and solvents, which conventional polymeric membranes cannot withstand. In microporous inorganic membranes, the pore size plays a significant role in determining separation performance (Lin et al., 2002). Therefore, various microporous inorganic membranes, such as zeolites, carbon, and silica, have been developed to provide attractive permselective characteristics in separation processes (Sommer and Melin, 2005; Ockwig and Nenoff, 2007; Pera-Titsu, 2014).

Amorphous silica is a promising membrane material. It has extremely small pores, approximately 0.3 nm in diameter, formed within siloxane-based networks (Yoshioka et al., 2004; Duke et al., 2008; Hacarlioglu et al., 2008), through which only small molecules such as He and H_2 can permeate; this is shown in Figure 10.1(a). Thus, amorphous silica membranes exhibit excellent permselectivity for H_2, and have been extensively investigated for H_2 separation from gaseous mixtures such as H_2/CO_2, H_2/N_2, and H_2/CH_4, and for applications in membrane reactors for H_2 production (Ockwig and Nenoff, 2007; Dong et al., 2008; Gallucci et al., 2013). Controlling the size of network pores is one of the biggest challenges in the development of silica-based membranes. Various approaches to control the size of network pores have been developed in the past few decades. For example, the size can be controlled by partially replacing siloxane bridges with organic bridges (Kanezashi et al., 2009; Castricum et al., 2011; Kanezashi et al., 2017) as shown in Figure 10.1(b). The incorporation of organic bridges into the siloxane-based networks opens the pore size to be greater than that of ordinary silica membranes. Organosilica membranes with Si-C_2H_4-Si bridges were reported to have a pore size of approximately 0.4–0.7 nm, which is more suitable for the separation of H_2 from

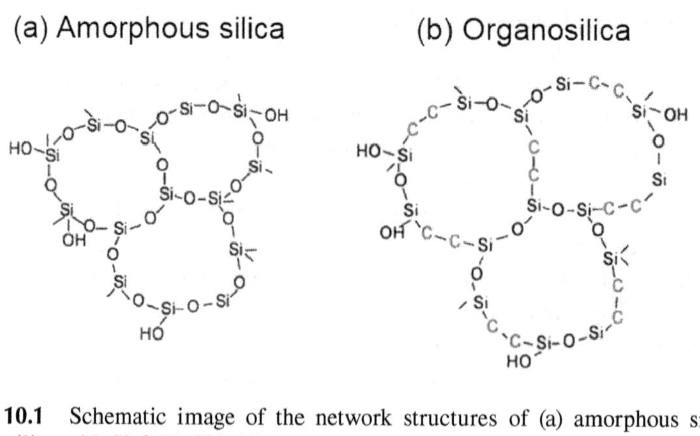

FIGURE 10.1 Schematic image of the network structures of (a) amorphous silica and (b) organosilica with Si-C_2H_4-Si bridges.

organic compounds with large molecular sizes such as methylcyclohexane and toluene (Nagasawa et al., 2014a; Niimi et al., 2014). They have also been demonstrated to exhibit high CO_2 permeance and selectivity for CO_2/N_2 and CO_2/CH_4 separation (Yu et al., 2016).

Typical silica-based membranes consist of a thin separation active layer on a porous support, as shown in Figure 10.2. Such a composite structure is preferable because it provides sufficient mechanical strength and can greatly reduce the thickness of the separation active layer. The methods commonly used for the fabrication of silica-based membranes are thermal chemical vapor deposition (CVD) (Gavalas et al., 1989; Tsapatsis and Gavalas, 1994; Yamaguchi et al., 2000; Nomura et al., 2006; Gu and Oyama, 2007) and sol-gel technique (Kitao and Asaeda, 1990; Brinker et al., 1994; de Vosa et al., 1999; Kanezashi and Asaeda, 2005; Castricum et al., 2008; Kanezashi et al., 2009). The former involves thermal decomposition of volatile precursors to form a thin layer onto a porous substrate (Figure 10.3(a) and (b)), while the latter involves coating of sols onto a porous substrate, followed by calcination, to promote the condensation of silanol into siloxane (Figure 10.3(c)). The thermal decomposition of volatile precursors in CVD and siloxane condensation in sol-gel processing are usually processed at high temperatures above 500°C for pure silica membranes, and at 300–400°C for organosilica membranes. However, the need for high-temperature processing limits the choice of substrates to heat-resistant ceramics such as porous α-alumina. Conventional high-temperature methods are incompatible with the use of polymeric substrates, although their use instead of ceramics may offer great potential for cost saving and large-scale manufacturing (Gong et al., 2014). The incorporation of heat-sensitive functional groups into siloxane-based networks is also difficult to achieve through conventional methods. In limited cases, processing temperatures can be decreased to 100°C or lower by employing acid-vapor-assisted (Wang et al., 2013; Gong et al., 2016) or photoinduced polymerization techniques (Nishibayashi et al., 2015; Nagasawa et al., 2017a); however, the separation performances have not yet reached application level. Therefore, an effective approach to achieve low-temperature fabrication of silica-based membranes is required.

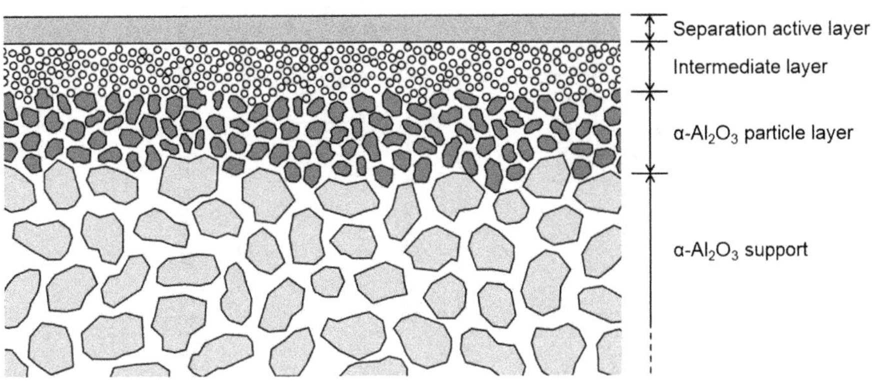

FIGURE 10.2 Schematic image of typical composite structure of silica-based membranes.

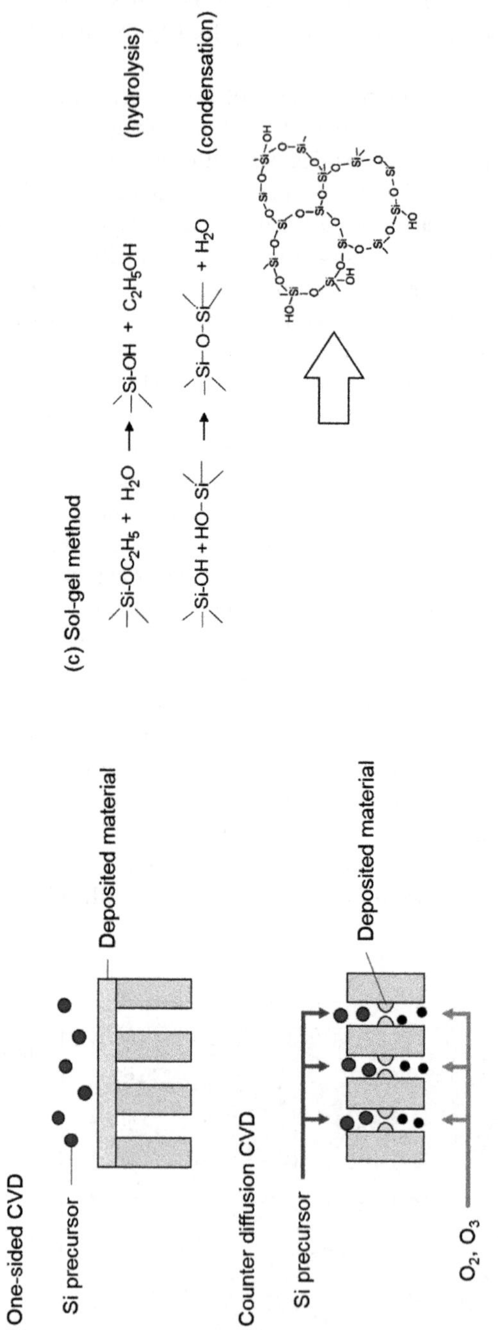

FIGURE 10.3 Methods for the fabrication of amorphous silica and silica-based membranes.

One promising solution for reducing the fabrication temperature is the plasma-enhanced chemical vapor deposition (PECVD) technique. PECVD is one of the commonly used techniques for thin-film deposition, which involves a series of gas phase and surface reactions of volatile precursors induced by plasma discharge. PECVD has been utilized for many different applications. For example, deposition of silicon-based thin films such as silica, silicon nitride, and silicon carbonitride for microelectronics (Maex et al., 2003; Volkesen et al., 2010) and optical devices (Martinu and Poitras, 2000) is a major application of PECVD. An important advantage of PECVD is that the processing temperature can be greatly reduced compared to that in conventional CVD and sol-gel processing (Barranco et al., 2004). Thin films can be deposited even at room temperature, so deposition can be performed onto thermally sensitive substrates. Moreover, because of the high reactivity of chemical species in plasma discharges, the films deposited by PECVD usually have a highly cross-linked structure with strong adhesion on various substrates. These features of PECVD could enable the fabrication of high-performance silica-based membranes on various types of substrates.

In this chapter, fabrication of silica-based membranes using PECVD will be reviewed. First, an overview of the development of PECVD-derived silica-based membranes over the last three decades is presented. Subsequently, the recent efforts to improve separation performance of PECVD-derived silica membranes will be described based on our work. Furthermore, the application of atmospheric-pressure plasma for the PECVD of silica membranes will be presented.

10.2 PECVD AND ITS APPLICATIONS IN MEMBRANE FABRICATION

PECVD is a variant of CVD, which uses a plasma discharge to accelerate the dissociation of volatile precursors into smaller and more reactive species through collisions with high energy electrons. A schematic diagram of PECVD is shown in Figure 10.4. In most PECVD systems, non-equilibrium plasmas such as microwave plasma, radio frequency (RF) plasma, and DC glow plasma, are employed at low-pressure. Non-equilibrium plasma is a plasma whose electron temperature is much

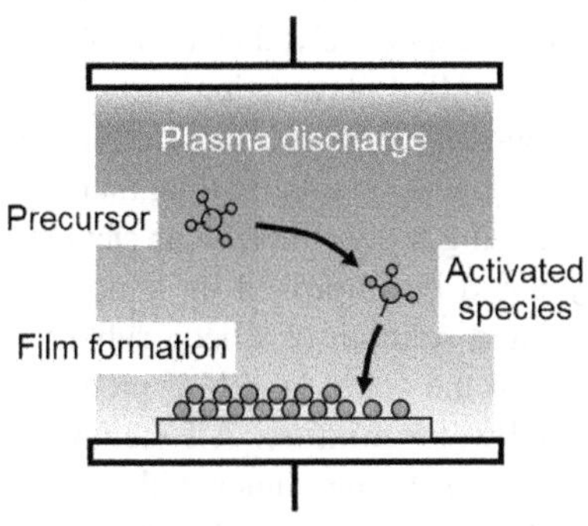

FIGURE 10.4 Schematic diagram of PECVD.

higher than that of heavy particles including ions, atoms, and molecules (Volksen et al., 2010; Bruggeman et al., 2014). In this type of plasma, the temperature of electrons is sufficiently high to make them chemically active to initiate the reaction, while the temperature of background gases including ions, radicals, and excited molecules is close to room temperature. This allows the processing temperature of film deposition to be decreased.

PECVD, sometimes referred to as plasma polymerization, has also been utilized in the fabrication of separation membranes. Membrane fabrication via PECVD was first reported for oxygen/nitrogen separation in the mid-1980s (Yamamoto et al., 1984; Sakata et al., 1987). The application of PECVD-derived membranes was then expanded to gas separation and pervaporation in the 1990s (Weichart and Müller, 1993; Matsuyama et al., 1994). The membranes reported in those early studies were mostly solubility-selective membranes, which are like polymeric membranes, which will be hereafter referred as plasma-polymer membranes. Membranes with a more "inorganic" structure were developed by employing a post-oxidation (Nehlsen et al., 1995) or deposition at high energy input (Roualdes et al., 1999). Membranes that exhibited molecular sieving were then realized by preparing a dense silicon carbonitride as a separation active layer (Kafrouni et al., 2009). In Table 10.1, the gas separation performances of PECVD-derived membranes consisting of silica and related materials reported in the literature are summarized.

In the mid-1980s, Yamamoto and co-workers prepared plasma-polymer membranes from various organic compounds and reported that the membrane prepared from hexamethyldisiloxane (HMDSO) showed permselectivity for O_2 over N_2, owing to the highly cross-linked structure of the plasma-polymer membrane (Yamamoto et al., 1984). They also prepared plasma-polymer membranes on various porous substrates having different pore sizes and pointed out that a plasma-polymer thickness that is approximately five times the pore radius of the porous substrate is needed to plug all pores, so that permselectivity is achieved (Sakata et al., 1987). Weichart and Müller (1993) prepared plasma-polymer membranes by depositing HMDSO and hexamethyldisilazane (HMDSN) onto an anodic porous alumina substrate at 300 K and 5 Pa, using capacitively coupled RF plasma. It was reported that better membranes could be obtained from HMDSO than from HMDSN or HMDSN/O_2 mixtures. The membrane prepared from HMDSO had a higher density compared to silicone rubber membrane; it also exhibited increased values of selectivity for CO_2/N_2 and C_4H_{10}/N_2 of 8.1 and 21, respectively, due to the blocking of non-soluble gases, suggesting that solubility-controlled permeation was dominant for this membrane. The ideal selectivity for non-soluble gases, i.e., He/N_2, which can be regarded as indicative of molecular sieving, was no greater than that expected from the Knudsen diffusion mechanism, suggesting a poor molecular sieving effect. It should also be noted that the selectivity could be improved by simply sealing defects with dip-coating method using a diluted solution of silicon rubber. Matsuyama et al. (1994) investigated the effect of siloxane chain length of precursors on the pervaporation characteristics of an ethanol/water mixture. All membranes obtained were found to be ethanol-permselective, and the monomer with the longer siloxane chain was effective in realizing both high selectivity and high permeation flux. The plasma-polymer membranes developed in early studies were mostly solubility-selective

TABLE 10.1
Gas Separation Performances of PECVD-Derived Membranes

Precursors/ Working Gases	Substrate	Post-Treatment	Thickness [nm]	T [°C]	Permeation Property: Permeance of He [mol m^{-2} s^{-1} Pa^{-1}]	Permeation Property: Permeability of N_2 [mol m^{-1} s^{-1} Pa^{-1}]	Permeance Ratios: He/N_2	Permeance Ratios: He/SF_6	Permeance Ratios: H_2/N_2	Permeance Ratios: CO_2/N_2	Permeance Ratios: Other Gases	Ref.
Low-pressure PECVD												
HMDSO	Anodic alumina	–	500	30		5.7×10^{-14}	2.8			8.1	C_4H_{10}/N_2: 21	Weichar and Müller, 1993
HMDSN	Anodic alumina	–	500	30		1.2×10^{-14}	2.3			3.9	C_4H_{10}/N_2: 9.8	
HMDSO	Anodic alumina	O_2, 350°C	200	20 300	4.9×10^{-7} 5.2×10^{-6}		4.0 15.7			1.8 2.5		Nehlsen et al., 1995
HMDSO	cellulose acetate	–	500	n.a.		4.6×10^{-13}	3.8					Li and Meichsner, 1999
DEDMS	polypropylene	–	–	25		2.5×10^{-15}			8	10.5		Roualdes et al., 1999
HMDSO	cellulose ester	–	–	25		$1.7–5.1 \times 10^{-15}$			4.8–9.9	6.9–11.0		Roualdes et al., 2002
HMDSN + NH_3/Ar	γ-alumina/ α-alumina	–	30	25 150	3.9×10^{-9} 9.1×10^{-8}		2.4 35					Kafrouni et al., 2009
OMCTSO	cellulose ester	–	240	35	1.3×10^{-8} [a]					6		Lo et al., 2010
BDMADMS + NH_3/ Ar	γ-alumina/ α-alumina	–	100	25 150	1.36×10^{-8} 1.64×10^{-7}		~25 20					Kafrouni et al., 2010
2-step PECVD (HMDSO/Ar + HMDSO/O_2)	TiO_2/ α-alumina	–	–	RT	9.7×10^{-9}		7800	28,000				Tsuru et al., 2011

(Continued)

TABLE 10.1 (CONTINUED)
Gas Separation Performances of PECVD-Derived Membranes

				Permeation Property								
							Permeance Ratios					
Precursors/ Working Gases	Substrate	Post-Treatment	Thickness [nm]	T [°C]	Permeance of He [mol m^{-2} s^{-1} Pa^{-1}]	Permeability of N_2 [mol m^{-1} s^{-1} Pa^{-1}]	He/N_2	He/SF_6	H_2/N_2	CO_2/N_2	Other Gases	Ref.
2-step PECVD (HMDSO/Ar+ HMDSO/O_2)	TiO_2/ α-alumina	400°C, vacuum	–	200	1.7×10^{-7}		4900	57,000	1800			Nagasawa et al., 2013
				400	5.2×10^{-7}		4200		1900			
					2.4×10^{-7} [b]							
HMDSO/Ar	SiO_2-ZrO_2/ α-alumina	500°C, vacuum	–	200	1.2×10^{-6}		3.0	220				Nagasawa et al., 2014b
MTMOS/Ar	SiO_2-ZrO_2/ α-alumina	500°C, vacuum	–	200	4.6×10^{-7}		5.0	290				Nagasawa et al., 2015
Propylene	SiO_2-ZrO_2/ α-alumina	–	–	RT	3.7×10^{-9}		23	1750		17		Nagasawa et al., 2016
					Atmospheric-Pressure PECVD							
HMDSO/0.25% N_2 + Ar	SiO_2-ZrO_2/ α-alumina	–	–	50	1.1×10^{-7}		196	820		33		Nagasawa et al., 2017b
		300°C, vacuum	–	50	4.0×10^{-7}		98	770		46	CO_2/CH_4: 166	
HMDSO/1.0% N_2 + Ar	SiO_2-ZrO_2/ α-alumina	–	–	50	4.0×10^{-8}		41	85				Nagasawa et al., accepted
		400°C, vacuum	–	50	1.1×10^{-6}		45	1000		28	CO_2/CH_4: 60	
				200	1.5×10^{-6}		56	1910				
				300	1.7×10^{-6}		58	1890	53			
					1.6×10^{-6} [b]							

[a] Permeance of CO_2.
[b] Permeance of H_2.

membranes, and little research has focused on the molecular sieving characteristics of these membranes.

In the mid-1990s, Nehlsen et al. (1995) investigated the permeation properties of HMDSO-derived plasma-polymer membranes at high temperatures and reported that the ideal selectivity for He/N_2 drastically increased to 15.7 at 300°C after annealing in pure oxygen at 350°C, because of the formation of a molecular sieving layer. Roualdes et al. (1999; 2002) prepared hydrocarbon containing silicon oxide (SiO_xC_y:H) membranes using different organosilicon precursors in RF plasma and investigated the effect of energetic character of the plasma, derived as *V*/*FM*, where *V* is the input voltage, *F* is the monomer flow rate, and *M* is the monomer molecular weight, on the permeation performances of plasma-polymer membranes. It was revealed that gas diffusion through plasma polymers was governed by two different structural factors: chain flexibility and chain cross-linking in polymers. At low *V*/*FM* values, plasma-polymers tended to preserve their monomer structure, and thus membranes prepared from precursors were composed of linear flexible siloxane chains. On the other hand, at high *V*/*FM* values, cross-linking of polymers significantly increased and membranes became denser and more inorganic. Hence, the correlation between gas permeability and *V*/*FM* showed a maximal value. Because the PECVD-derived SiO_xC_y:H membranes were much more cross-linked and had a more inorganic structure than the polydimethylsiloxane (PDMS), the ideal selectivity of 9.9 was achieved at 298 K for H_2/N_2, demonstrating the formation of a molecular sieving layer by using PECVD alone (Roualdes et al., 2002).

Improved permselectivity for small gas molecules such as He and H_2 was achieved by Kafrouni and co-workers by preparing a thin layer of silicon carbonitride (SiC_xN_y:H) as a separation active layer onto an asymmetric α-alumina substrate with a γ-alumina mesoporous layer from HMDSN or bis(dimethylamino)dimethylsilane (BDMADMS), mixed with ammonia in RF plasma power of 100-400 W (Kafrouni et al., 2009; 2010). They reported that the membrane had an inorganic structure, which progressively changed to that of crystalline silicon nitride when both the plasma power and ammonia content were increased. The membranes prepared from HMDSN and BDMADMS with optimized deposition condition showed ideal selectivity for He/N_2 of 35 and 20, respectively, with He permeances of 0.91 and 1.64 × 10^{-7} mol m^{-2} s^{-1} Pa^{-1}, respectively, at 150°C. The activation energies of permeation for HMDSN-derived membrane were 26.2 and 3.6 kJ mol^{-1} for He and N_2, respectively (Kafrouni et al., 2009). Low N_2 activation energy was explained as being due to the local defects of the SiC_xN_y:H layer led by the mesoporous γ-alumina layer defect formation. It should also be noted that these membranes possessed excellent stability at high temperatures in an oxidative atmosphere. The He and N_2 permeances at 25°C showed little change even after 2 hours of annealing at 500°C in air. More recently, Ngamou et al. (2013) prepared hybrid organically-bridged silica membranes on a polyamide-imide substrate using an expanding thermal plasma deposition technique and reported a high water pervaporation flux of ~1.8 kg m^{-2} h^{-1} with a separation factor >1100 in *n*-butanol-water mixture at 95°C, which was comparable with those of conventional sol-gel-derived membranes.

Membrane fabrication via PECVD has been extensively studied in the last three decades, and significant improvements in separation performance have been made.

TABLE 10.2
Characteristics of Low-Pressure and Atmospheric-Pressure PECVD

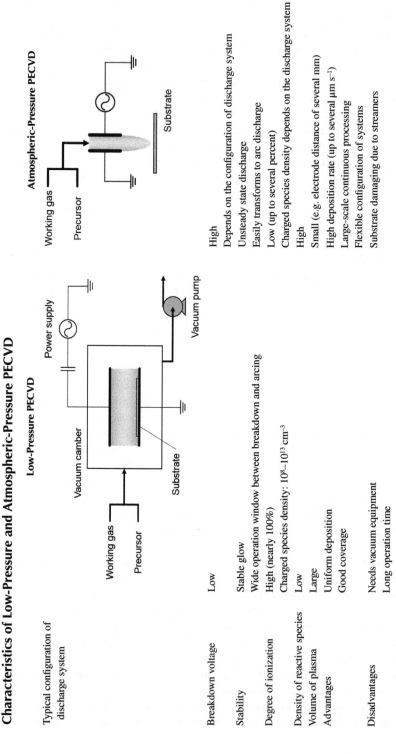

	Low-Pressure PECVD	Atmospheric-Pressure PECVD
Typical configuration of discharge system	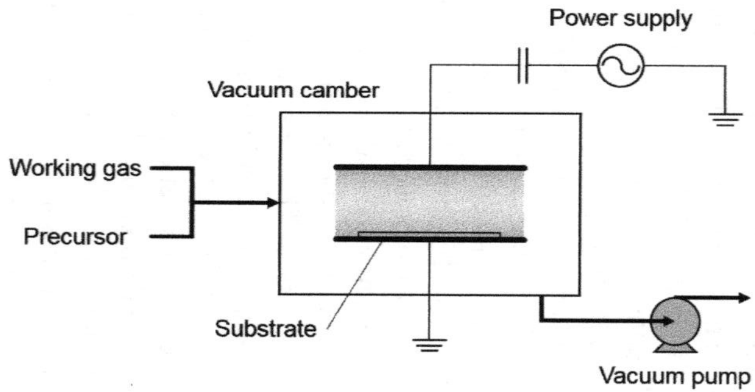 	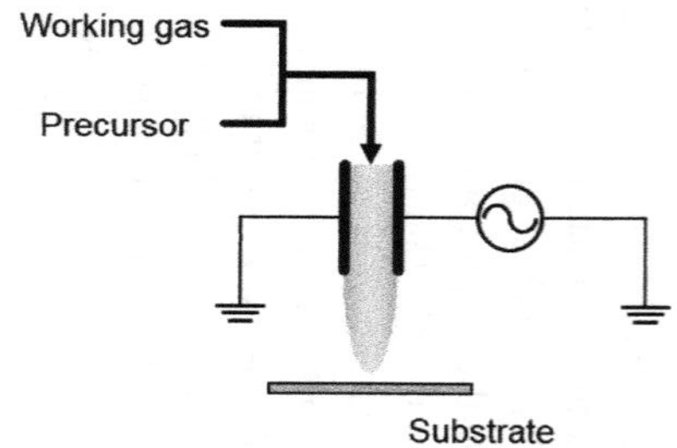
Breakdown voltage	Low	High
Stability	Stable glow Wide operation window between breakdown and arcing	Depends on the configuration of discharge system Unsteady state discharge Easily transforms to arc discharge
Degree of ionization	High (nearly 100%) Charged species density: 10^8–10^{13} cm^{-3}	Low (up to several percent) Charged species density depends on the discharge system
Density of reactive species	Low	High
Volume of plasma	Large	Small (e.g. electrode distance of several mm)
Advantages	Uniform deposition Good coverage	High deposition rate (up to several $\mu m\ s^{-1}$) Large-scale continuous processing Flexible configuration of systems
Disadvantages	Needs vacuum equipment Long operation time	Substrate damaging due to streamers

Nevertheless, improving permselectivity remains a challenge for meeting the requirement for practical applications. In the following sections, recent efforts to improve separation performance of PECVD-derived silica membranes will be described based on our previous work on low-pressure and atmospheric pressure PECVD. The characteristics of low-pressure and atmospheric-pressure PECVD are shown in Table 10.2. First, an overview of the fabrication of silica membranes using low-pressure PECVD is presented (Tsuru et al., 2011; Nagasawa et al., 2013). Subsequently, the effects of the chemical structure of precursors (Nagasawa et al., 2015) as well as the reaction atmosphere on the permeation properties of resultant membranes are discussed (Nagasawa et al., 2013). The thermal stability and permeation characteristics of PECVD-derived silica membranes at elevated temperatures are demonstrated (Nagasawa et al., 2014b). Finally, the use of atmospheric-pressure plasma as a plasma source for PECVD of silica membranes will be presented (Nagasawa et al., 2017b; Nagasawa et al., accepted).

10.3 LOW-PRESSURE PECVD-DERIVED SILICA-BASED MEMBRANES

10.3.1 Membrane Fabrication and Gas Permeation Properties

To obtain a membrane with molecular separation ability, the separation active layer must have a porous structure with an effective pore size in the sub-nanometer scale, because, for example, the size of gaseous molecules is no more than 1 nm. The size of the membrane micropores must be controlled precisely, to within several angstroms, without any defect formation to achieve the desired gas separation. At the same time, the separation active layer should be as thin as possible, for obtaining a high permeance. Our approach for fabricating such a microporous separation active layer via PECVD is to employ a nanoporous intermediate layer with a well-controlled pore size and sharp pore size distribution (Tsuru et al., 2011; Nagasawa et al., 2013). We expected that nanometer-sized pores of the intermediate layer could be plugged rapidly, within a short period of PECVD, leading to a thinner separation active layer. Sharp pore size distribution could prevent accidental defect formation due to the local defect of intermediate layer.

The membrane fabrication procedure is as follows (Nagasawa et al., 2013). Membranes were fabricated using a porous α-alumina capillary tube (average pore size, 150 nm; outer diameter, 3 mm; length, 50 mm) as a substrate, which has a nanoporous intermediate layer consisting of either TiO_2 or SiO_2-ZrO_2 on its outer surface. Intermediate layers were fabricated via sol-gel processing. A detailed fabrication procedure of intermediate layers is reported elsewhere (Tsuru, 2008). Briefly, a TiO_2 sol was coated on the surface of porous substrate, and the substrate was fired at 550°C for 15 min. This procedure was repeated several times to cover large pores. Accordingly, a TiO_2 intermediate layer with a pore size of 5 nm, which was evaluated using the nanopermporometry technique (Tsuru et al., 2001), was obtained. Subsequently, a SiO_2-ZrO_2 (Si/Zr = 1/1) sol was coated onto the TiO_2 intermediate layer and fired at 550°C to form a SiO_2-ZrO_2 intermediate layer with a pore size of 1 nm.

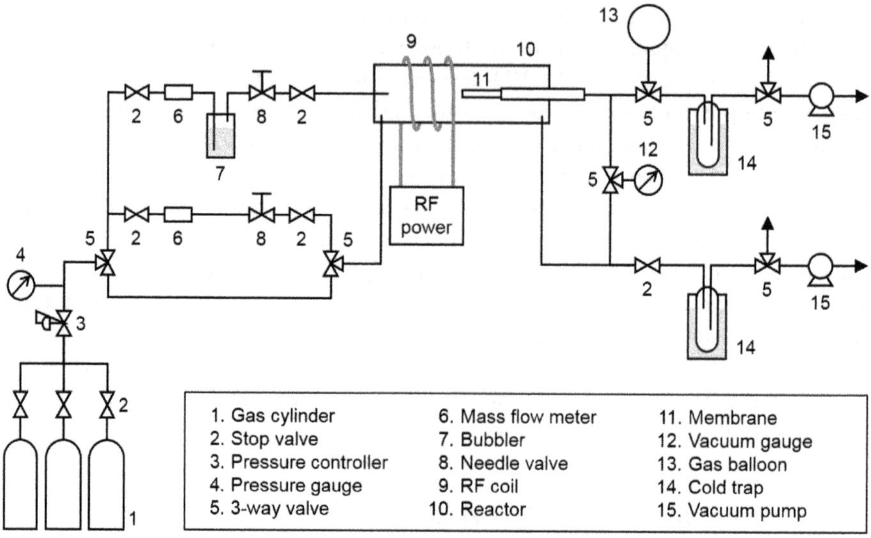

FIGURE 10.5 Schematic diagram of the PECVD experimental apparatus. (From Nagasawa, H. et al., Journal of Membrane Science. 441, 45–53, 2013.)

A silica layer was deposited onto the substrate using an inductively coupled RF plasma reactor (SAMCO Inc., Japan), as shown in Figure 10.5. The reactor consisted of a quartz tube surrounded by an RF coil. The substrate was placed 5 cm downstream from the RF coil, and both the retentate and the permeate of the substrate were connected to each vacuum system. HMDSO was used as the silicon precursor. The silicon precursor was fed to the reactor at a fixed flow rate of 1 sccm, with Ar as the working gas at a flow rate of 10 sccm. Plasma was driven at room temperature using a 13.56 MHz RF generator (AX-300III, AD-TEC, Japan) with RF power of 30 W. During the deposition, the pressure inside the reactor was maintained at 120–150 Pa.

The gas permeation properties of the PECVD-derived membranes were evaluated in situ using the same apparatus that was used for the plasma-deposition. The gas permeances for He (kinetic diameter: 0.26 nm), H_2 (0.289 nm), CO_2 (0.33 nm), Ar (0.346 nm), N_2 (0.364 nm), and SF_6 (0.55 nm) were measured using the constant-volume variable-pressure method. The feed side was maintained at a constant pressure, p_1, of 105 kPa. The permeate side was vacuumed to 10 Pa prior to measurement, and the pressure change in the permeate side, p_2, was measured with a pressure transducer (MKS Baratron type 722A). The single gas permeance, P, was determined by following equation:

$$P = \frac{V_2}{ART\Delta t} \ln \frac{(p_1 - p_{2,1})}{(P_1 - P_{2,2})} \tag{10.1}$$

where A is the surface area of the membrane, R is the universal gas constant, T is the absolute temperature, and $p_{2,1}$ and $p_{2,2}$ are the permeate pressures at $t = 0$ and $t = \Delta t$, respectively.

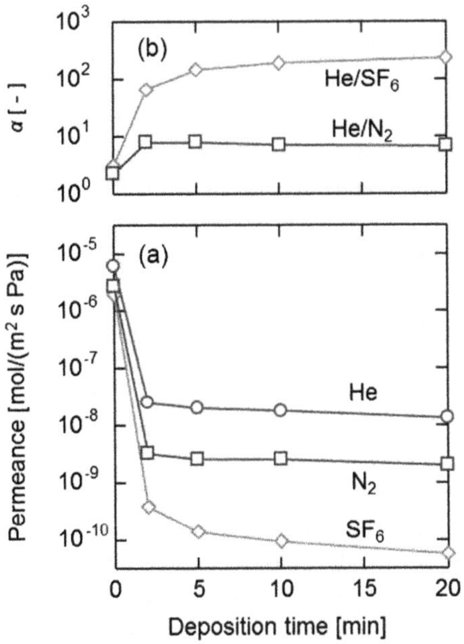

FIGURE 10.6 Deposition time dependence of (a) single gas permeances and (b) permeance ratios at room temperature for an HMDSO/Ar-PECVD-derived membrane. (Modified from Nagasawa, H. et al., Journal of Membrane Science. 489, 11–19, 2015. With permission.)

Figure 10.6(a) shows the time course of single gas permeances of He, N_2, and SF_6 at room temperature for the membrane prepared using HMDSO and Ar, as silicon precursor and working gas, respectively (Nagasawa et al., 2015). The single gas permeances of He and N_2 decreased significantly at the beginning and reached a plateau after 5 min of deposition, while that of larger molecules, namely SF_6, continued to decrease moderately with time. The decrease in the single gas permeance during the first 5 min can be ascribed to the blockage of the pores in the intermediate layer, showing a rapid formation of the plasma-deposited layer on the intermediate layer. The subsequent decrement of the SF_6 permeance can be ascribed to the plugging of a few remaining pinholes. The permeance of He decreased by approximately two orders of magnitude after deposition, while those of N_2 and SF_6 decreased more than that of He. The decreases in single gas permeance during deposition were likely to vary with the molecular size of the permeating species. The permeance ratios for He/N_2 and He/SF_6, as shown in Figure 10.6(b), increased with time and reached steady values of 7.1 and 200, respectively. The observed permeance ratios of He/N_2 and He/SF_6 exceeded those expected from the Knudsen diffusion mechanism (2.64 for He/N_2 and 6.04 for He/SF_6). These results clearly indicated that PECVD led to successful fabrication of a membrane with molecular sieving characteristics.

10.3.2 Effect of Chemical Structure of Precursors

As mentioned in the literature (Weichart and Müller, 1993; Matsuyama et al., 1994; 1999; Roualdes et al., 2002), the permeation properties of PECVD-derived membranes can be controlled by varying the structure of the plasma-deposited layer. One possible approach to change the structure of the plasma-deposited layer is to use a silicon precursor with different chemical structures. We investigated the effect of silicon precursors on the gas permeation properties by using three organosilicons with different O/Si atomic ratios: HMDSO, trimethylmethoxysilane (TMMOS), and methytrimethoxysilane (MTMOS) (Nagasawa et al., 2015). The molecular structures of three organosilicon precursors are shown in Figure 10.7. The chemical structures of the resultant films were determined by FTIR analysis, and the effect of the precursors on the membrane properties was discussed together with the results of gas permeation measurements.

Figure 10.8 shows the FTIR spectra of the plasma-deposited films derived from HMDSO, TMMOS, and MTMOS. The most intense absorption band peaked at approximately 1040 cm^{-1} in the FTIR spectrum of the HMDSO film and was assigned to the asymmetric stretching vibrations related to Si-O-Si bonding. The absorption peaks located at 1410 and 1260 cm^{-1} were due to the asymmetric and symmetric CH_3 bending vibrations in $Si(CH_3)_x$, respectively, and the peak at 1460 cm^{-1} was attributed to the bending vibration of methylene (CH_2) groups (Benitez et al., 2000). The absorption peaks at 870–750 cm^{-1} were assigned to the rocking vibrations of CH_3 (Benitez et al., 2000). More specifically, the peak located at 840 cm^{-1} was from $Si(CH_3)_3$, and that at 800 cm^{-1} was from $Si(CH_3)_2$ (Ngamou et al., 2013). The peaks at 3000–2800 cm^{-1} were due to the stretching vibrations of the saturated C-H bonds (Walkiewicz-Pirtzykowska et al., 2005). The asymmetric and symmetric C-H_3 stretching vibrations at 2960 and 2900 cm^{-1}, respectively, could be identified in the spectrum of the HMDSO film. A broad band located at 2150 cm^{-1} corresponded to the stretching vibrations of Si-H, which is often found in the case of plasma-deposited siloxane-based films (Milella et al., 2007). An absorption peak for Si-OH groups at 900 cm^{-1} was also observed. A substantial level of absorption intensity for the $Si(CH_3)_2$ structure (800 cm^{-1}), which does not exist in the case of the HMDSO monomer structure, indicated that the plasma-deposited HMDSO-film was mainly composed of the dimethyl-siloxane backbone formed in

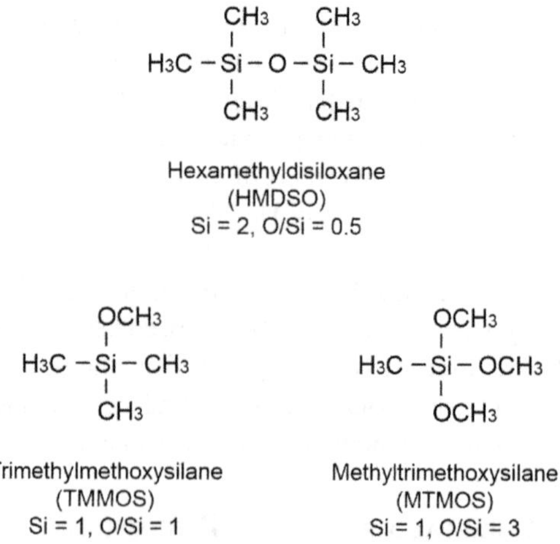

FIGURE 10.7 Molecular structures of hexamethyldisiloxane (HMDSO), trimethylmethoxysilane (TMMOS), and methytrimethoxysilane (MTMOS). (From Nagasawa, H. et al., Journal of Membrane Science. 489, 11–19, 2015.)

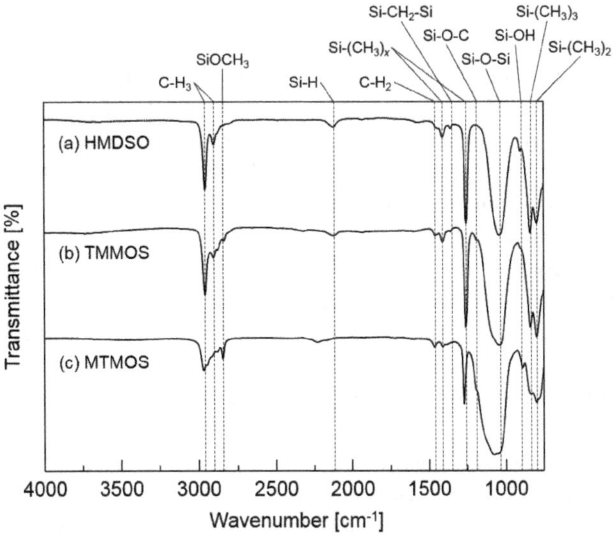

FIGURE 10.8 FTIR spectra of plasma-deposited films prepared using (a) HMDSO, (b) TMMOS, and (c) MTMOS. (From Nagasawa, H. et al., Journal of Membrane Science. 489, 11–19, 2015.)

the plasma-induced cationic polymerization (Steele et al., 2011). The weak peak at 1350 cm^{-1} might be ascribed to the CH_2 scissoring and wagging vibrations in Si-CH_2-Si (Lee et al., 2001). The presence of the Si-CH_2-Si structure indicated that the cross-linking via Si-CH_2-Si contributed as a part of the film deposition in HMDSO/Ar-PECVD.

Compared to the HMDSO-film, the FTIR spectrum of the TMMOS-film showed some new characteristic peaks in addition to all the peaks observed in the HMDSO-film. New peaks at 2840 and 1190 cm^{-1} corresponded to the C-H stretching and the Si-O-C stretching, respectively (Shioya et al., 2008). A high-wavenumber shoulder at around 1100 cm^{-1} on the signal of Si-O-Si asymmetric stretching was also observed. The dominant peak at 1040 cm^{-1} was ascribed to the Si-O-Si chain structure, while the shoulder at 1100 cm^{-1} was often regarded as a contribution of the Si-O-Si cage- and/or ring-type structures (Milella et al., 2007). When compared with the spectra of the HMDSO and TMMOS-films, the FTIR spectrum of MTMOS-film showed a decrease in the intensity of the peaks related to $Si(CH_3)_x$ at 1260 and 800 cm^{-1}, which indicated a lower carbon content. Furthermore, the peak location of the absorption band for Si-O-Si asymmetric stretching was shifted to a higher wavenumber at 1080 cm^{-1}, suggesting that the contribution of the Si-O-Si chain structure decreased relatively, and the Si-O-Si cage- and/or ring-type structure was dominant. It should be also noted that, the peak area ratios of the bands assigned to the symmetric CH_3 bending vibrations in the Si-CH_3 group at 1260 cm^{-1} to the asymmetric stretching vibrations of the Si-O-Si bond at 1000-1150 cm^{-1} (Si-CH_3/Si-O-Si) for plasma-deposited films decreased when the O/Si atomic ratio of the precursor increased. The Si-CH_3/Si-O-Si absorbance ratio for the films prepared using HMDSO, TMMOS and MTMOS were 0.12, 0.09, and 0.05, respectively.

FTIR analysis indicated that the plasma-deposited HMDSO-film was composed of linear dimethylsiloxane chains with a structure that partially cross-linked with the Si-CH_2-Si bridges and was found to be the most organic "silicone-like" structure of the three films studied. Although the FTIR spectrum of the plasma-deposited TMMOS film showed a peak pattern similar to that of the HMDSO-film, the broader Si-O-Si absorption bands of the TMMOS-film suggested that siloxane networks were more branched and denser. The MTMOS-film, on the other hand, had the most inorganic "silica-like" structure, with rigid micropores that resulted from low carbon content, and an abundance of Si-O-Si cage type structures. FTIR analysis revealed that the chemical structure of plasma-deposited films shifted from an organic "silicone-like" structure to an inorganic "silica-like" structure when the O/Si atomic ratio of the precursor was increased. Therefore, the chemical structure of plasma-deposited films could be controlled by changing the chemical structure of the silicone precursor.

Figure 10.9(a) shows the kinetic diameter dependence of the single gas permeances at room temperature for the membranes prepared using HMDSO, TMMOS, and MTMOS with a plasma-deposition time of 20 min. The kinetic diameter dependencies of the relative permeances normalized by He permeance are also represented in Figure 10.9(b). The HMDSO-derived membrane showed the highest He permeance, followed by the TMMOS-derived and MTMOS-derived membranes. The He/N_2 permeance ratio for the TMMOS-derived membrane was 7.7, nearly the same as that of the HMDSO-derived membrane. In contrast, the He/N_2 permeance ratio for MTMOS-derived membranes was 15, which was more than twice that of the HMDSO-derived membrane, suggesting that the MTMOS-derived membrane had smaller pore sizes than the HMDSO-derived membrane. These results suggest that the gas permeation characteristics of plasma-deposited membranes can be controlled by changing the species of the precursors.

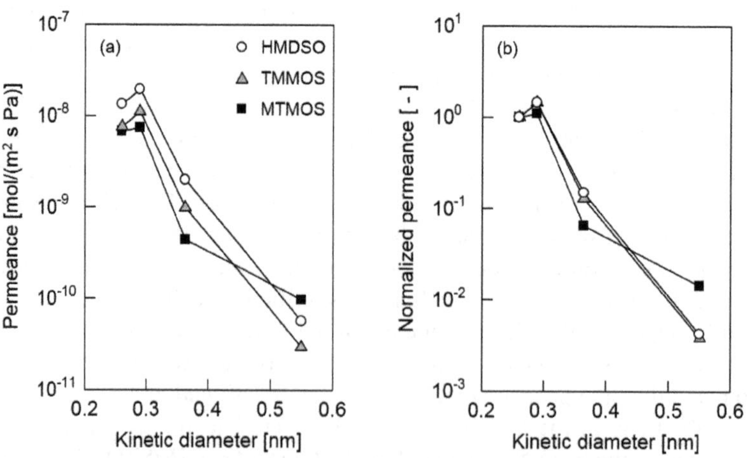

FIGURE 10.9 Kinetic diameter dependence of (a) single gas permeances and (b) normalized permeances for the membranes prepared using HMDSO, TMMOS, and MTMOS with deposition time of 20 min. (From Nagasawa, H. et al., Journal of Membrane Science. 489, 11–19, 2015.)

10.3.3 Effect of Plasma Working Gas and 2-Step PECVD

The reaction atmosphere is another important parameter for controlling the structure of plasma-deposited separation active layer. We investigated the effect of reaction atmosphere on the gas permeation properties of the resultant membranes by using different plasma working gases, namely Ar and O_2, with HMDSO as a silicon precursor (Tsuru et al., 2011; Nagasawa et al., 2013). A 2-step PECVD sequence, which involves plasma-deposition with HMDSO/Ar followed by HMDSO/O_2, is presented because a membrane with an excellent molecular sieving property can be prepared by this method.

Figure 10.10 shows the kinetic diameter dependence of the single gas permeance at 25°C for membranes prepared from HMDSO with Ar (deposition time: 10 min) and O_2 (deposition time: 50 min), respectively, and a 2-step PECVD sequence of plasma-deposition with HMDSO/Ar (deposition time: 10 min) followed by HMDSO/O_2 (deposition time: 5 min). The HMDSO/Ar-derived membrane shows that the permeance of all gases except CO_2 decreased with an increase in kinetic diameters, indicating that the molecular sieving mechanism dominates the permeation mechanism. CO_2 showed a higher permeance than He, the smallest of the molecules, probably because CO_2 had a strong affinity to plasma-polymerized membranes. By contrast, the HMDSO/O_2-derived membrane showed low selectivity for all types of gases, which is approximately equal to the selectivity determined through Knudsen diffusion. For HMDSO/Ar-derived membranes, the PECVD layer formed on the intermediate layer appeared to be homogeneous and crack-free, and the deposited layer seemed to have good adhesion to the porous surface of the intermediate layer (Figure 10.11(a)). In contrast, for HMDSO/O_2-derived membranes, the cross-sectional SEM image (Figure 10.11(b)) shows that many particles with diameters of approximately 100 nm were deposited onto

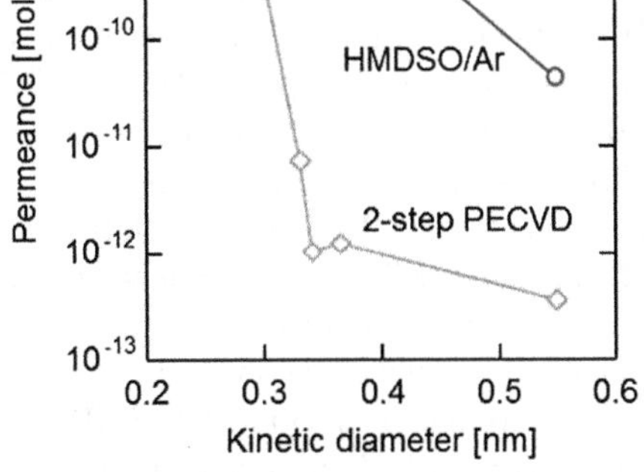

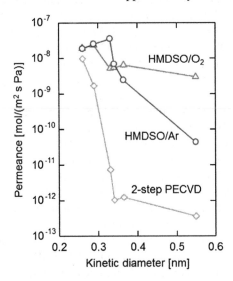

FIGURE 10.10 Kinetic diameter dependency of single gas permeance for the membranes derived from HMDSO/Ar-PECVD (deposition time: 10 min), HMDSO/O_2-PECVD (50 min), and 2-step PECVD (HMDSO/Ar-PECVD 10 min + HMDSO/O_2-PECVD 5 min). (From Tsuru, T. et al., Chemical Communications. 47, 8070–8072, 2011.)

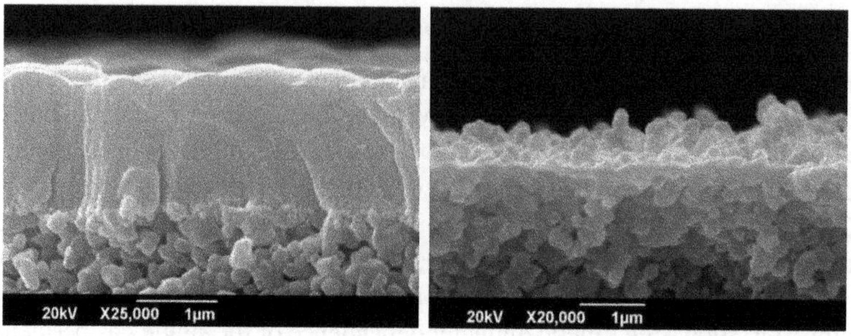

FIGURE 10.11 Cross sectional SEM images of (a) HMDSO/Ar- and (b) HMDSO/O_2-PECVD-derived membranes. (Modified from Nagasawa, H. et al., Journal of Membrane Science. 441, 45–53, 2013. With permission.)

the surface of the intermediate layer and no continuous layer was formed. Accordingly, the lower gas permeation selectivity of the HMDSO/O_2-derived membranes compared with that of the HMDSO/Ar-derived membranes was most likely due to the difficulty in forming a pinhole-free layer via PECVD under this condition.

The 2-step PECVD membrane showed excellent molecular sieving behavior with permeance ratios of 7800 for He/N_2 and 27,000 for He/SF_6. As compared with the HMDSO/Ar-derived membrane, the 2-step PECVD membrane possessed a significantly high molecular sieving properties. Regarding the selectivity for smaller molecules, the permeance ratios for He/H_2 and H_2/CO_2 reached 5 and 200, respectively. The increased selectivity of 2-step PECVD membranes for small molecules suggests that the size of pores formed in the first step HMDSO/Ar-PECVD reduced during the second step. The cross-sectional SEM image of a 2-step PECVD membrane (deposition time: HMDSO/Ar 10 min + HMDSO/O_2 5min) is shown in Figure 10.12(a). After the second step in the 2-step sequence, the membrane had a layer structure, that is, a continuous layer like HMDSO/Ar-PECVD membrane, and the deposition of particles on the top of the membrane similar to HMDSO/O_2-PECVD membrane. The change in the chemical structure of the plasma-deposited layer during the second step HMDSO/O_2-PECVD was elucidated by FTIR measurement, as shown in Figure 10.12(b). The deposition time for the first step HMDSO/Ar-PECVD was 10 min, and the treatment time for the second step was varied between zero and 30 min. In the spectrum of the film without the second-step HMDSO/O_2-PECVD, the characteristic absorption band corresponding to the CH_x group at 2,960–2,900 cm^{-1} was observed. For the films with the second-step PECVD, absorption bands corresponding to the OH group at 3,670 cm^{-1} were observed in addition to a decrease in the absorption at 2,960–2,900 cm^{-1}. As schematically described in Figure 10.13, these results suggest that the organic silicone-like structure formed by the first-step HMDSO/Ar-PECVD gradually converted to a silica-like structure, which became more rigid by oxidation during the second-step HMDSO/O_2-PECVD. A rigid structure allows less vibration of the networks, and results in high selectivity for small molecules such as He and H_2, as shown in Figure 10.10. These results indicate that the 2-step PECVD technique can be used for the low-temperature fabrication of molecular sieving amorphous silica membranes.

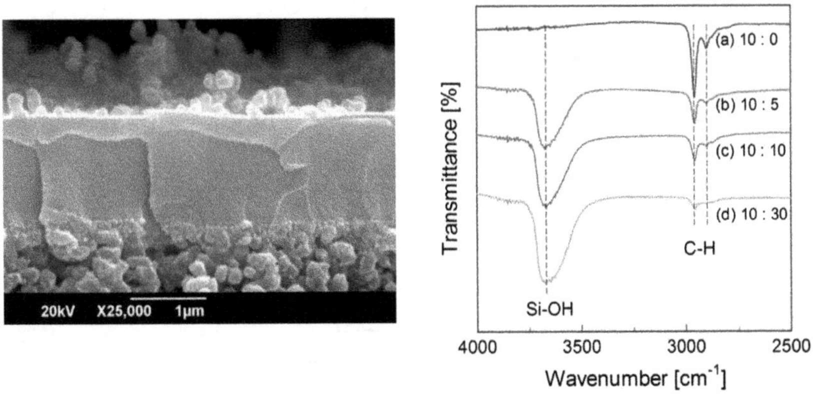

FIGURE 10.12 Cross sectional SEM image of 2-step PECVD membrane (10 min HMDSO/Ar-PECVD + 5 min HMDSO/O_2-PECVD), and FT-IR spectra of 2-step PECVD films with different second step PECVD time: (a) 10 min HMDSO/Ar-PECVD + 0 min HMDSO/O_2-PECVD; (b) 10 min + 5 min; (c) 10 min + 10 min; (d) 10 min + 30 min. (From Nagasawa, H. et al., Journal of Membrane Science. 441, 45–53, 2013.)

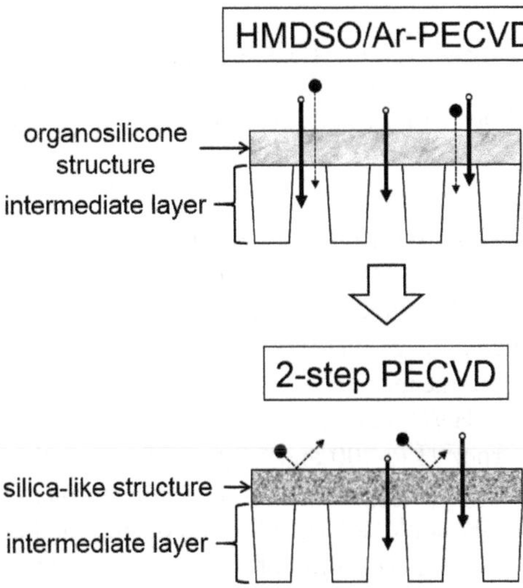

FIGURE 10.13 Schematic of gas transport in PECVD-derived membranes. (From Nagasawa, H. et al., Journal of Membrane Science. 441, 45–53, 2013.)

10.3.4 Thermal Stability and Permeation Properties at Elevated Temperatures

Although the membranes presented in the previous sections were fabricated at room temperature and without additional heating, they exhibited interesting permeation characteristics at elevated temperatures (Nagasawa et al., 2014b). As an example, the permeation behavior of HMDSO/Ar-PECVD-derived membranes at high

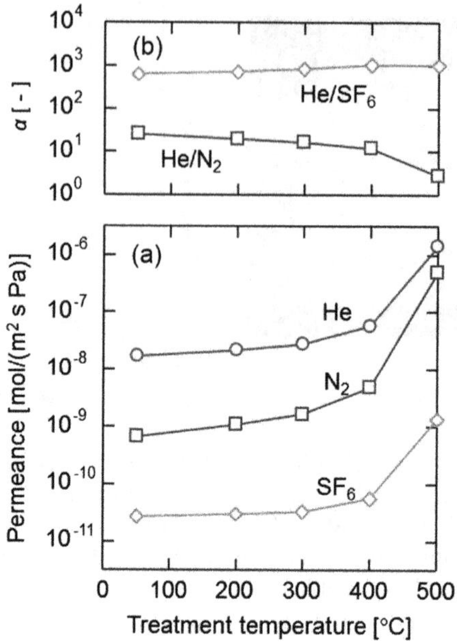

FIGURE 10.14 (a) Single gas permeances and (b) permeance ratios, α, at 50°C for HMDSO/Ar-PECVD-derived organosilica membranes after heat treatments at temperatures from 50 to 500°C. (Modified from Nagasawa, H. et al., Separation and Purification Technology. 13–19, 2014. With permission.)

temperatures is presented in this section. In Figure 10.14, the single gas permeances of He, N_2, and SF_6 at 50°C for HMDSO/Ar-derived membranes before and after heat treatment are plotted as a function of the heat-treatment temperatures. The permeances of He and N_2 increased gradually with heat-treatment temperatures of up to 300°C. In contrast, the permeance of SF_6 was unchanged in this temperature range. The permeance ratio for He/N_2 decreased from 17.4 to 8.2, while those for He/SF_6 and N_2/SF_6 increased from 93 to 200 and 5.3 to 24.5, respectively, after heat treatment at 300°C. These results suggested that micropores through which both He and N_2 could permeate but SF_6 could not were formed during the heat treatment in this temperature range. Since the permeation of SF_6 takes place via large pores such as pinholes, the unchanged SF_6 permeance suggests that no additional pinholes were formed, and that the plasma-deposited layer was unaffected by the heat treatment at this temperature range. The increases in the He and N_2 permeances should, therefore, probably be ascribed to the desorption of small molecules such as unreacted monomers, and/or their oligomers, adsorbed onto the plasma-deposited layer.

Subsequently, when the membranes were heat-treated at 400 and 500°C, the permeances for all types of gases increased more rapidly than during 200 and 300°C-treatments. These increases in the permeance were due to the thermal decomposition of the plasma-deposited layer, which led to a decrease in the thickness of the top layer. However, although the heat-treatment temperature was substantially higher than that of the plasma-deposition, the membrane after heat treatment still exhibited the gas permeation selectivity. While the permeance ratio for He/N_2 decreased to 3.3, the ratio for He/SF_6 remained above 200 even after the treatment at 500°C. More surprisingly, the permeance ratio for N_2/SF_6 increased to 60 after the treatment at 500°C, suggesting that silica networks with larger pore sizes were formed during the heat treatment at 400-500°C. The networks are suitable for the separation of large molecules, such as SF_6, from smaller molecules. The change in the silica network structure in this temperature range was probably due to the thermal degradation of methyl groups in the plasma-deposited layer. These results demonstrate the superior thermal stability of room-temperature-fabricated HMDSO-derived membranes. It is also noteworthy that heat treatment at high temperatures could be a way to modify the porous structure of plasma-deposited organosilica membranes.

The temperature dependence of gas permeances was measured after heat treatments at 200 and 500°C. For the 200°C-treated membrane, as shown in Figure 10.15(a), the permeances of He, H_2 and N_2 increased with increasing temperature, showing activated diffusion behavior. The permeance of CO_2 also slightly increased with increasing temperature, suggesting that the primary mechanism of permeation was activated diffusion. However, the dependence on temperature for the permeance of CO_2 was weaker than it was for the other gases. This might be explained by the additional contribution of the surface or solution diffusion of CO_2 due to its strong affinity for the membrane. The activation energy of gas permeation, E_p, was obtained from the following equation using the experimental single gas permeance data:

$$P = P_0 \exp\left(-\frac{E_p}{RT}\right), \tag{10.2}$$

where P_0 is a pre-exponential factor that reflects the structure of a membrane, R is the gas constant, and T is the temperature. The activation energies of gas permeation for He, H_2, CO_2, and N_2 were calculated to be 13.5, 12.5, 4.8, and 15.4 kJ mol^{-1},

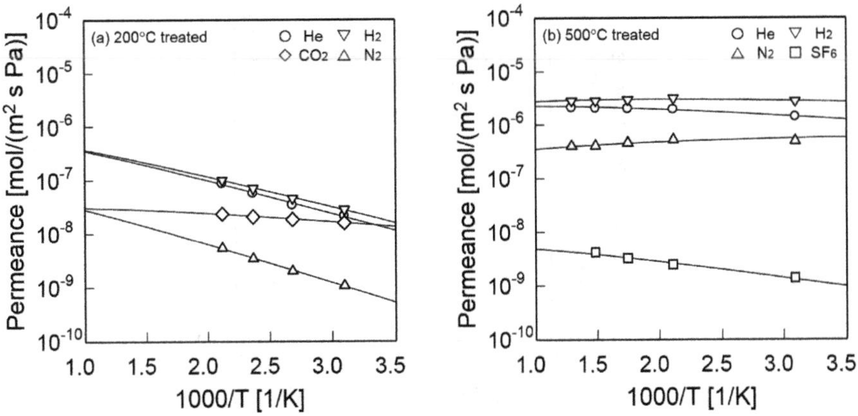

FIGURE 10.15 Temperature dependence of single gas permeances for HMDSO/Ar-PECVD-derived membrane after heat treatment at (a) 200°C and (b) 500°C. (From Nagasawa, H. et al., Separation and Purification Technology. 13–19, 2014.)

respectively. The activation energies of gas permeation for the 200°C-treated membrane were in good agreement with those for the PDMS membranes (Hagg, 2000; Merkel et al., 2001; Clarizia et al., 2004; Sadrzadeh et al., 2009). This indicates that the 200°C-treated membrane retained its organic silicone-like structure. Therefore, the activated diffusion of gas molecules through the 200°C-treated membrane can be attributed to the free volumes between the polymer chains formed by thermal motion.

For the 500°C-treated membrane, as shown in Figure 10.15(b), the permeances of He, H_2, and N_2 were little affected by temperature, showing Knudsen permeation behavior. The activation energies of gas permeation for He, H_2, and N_2 were 3.2, 1.4, and 0.5 kJ mol^{-1}, respectively, all of which were much smaller than those of the 200°C-treated membrane. It was also interesting to note that the permeation of SF_6 showed activated diffusion behavior (E_p = 7.7 kJ mol^{-1}), which suggests the absence of pinholes even after heat treatment. Together with the high gas permeance ratios, it is suggested that the structure of the membrane changed from a PDMS-like structure to a microporous structure with a narrow pore size distribution after heat treatment at 500°C. This structure change might be due to the release of hydrocarbon fragments, which was similar to the results reported in the literature (Lee et al., 2007; Milella at al., 2007).

Figure 10.16 shows the kinetic diameter dependence of dimensionless gas permeances at 50°C, which were normalized by the permeances of He, for the membrane after heat treatments at 200, 400, and 500°C. Although the dimensionless permeances for H_2 and SF_6 did not alter much, a change in the dimensionless permeance for N_2 was clearly observed after heat treatment at 500°C. Therefore, it is strongly suggested that the heat treatment at 500°C formed network pores that allowed the permeation of molecules that were the size of N_2 (0.364 nm) or smaller, but not of larger molecules such as SF_6 (0.55 nm).

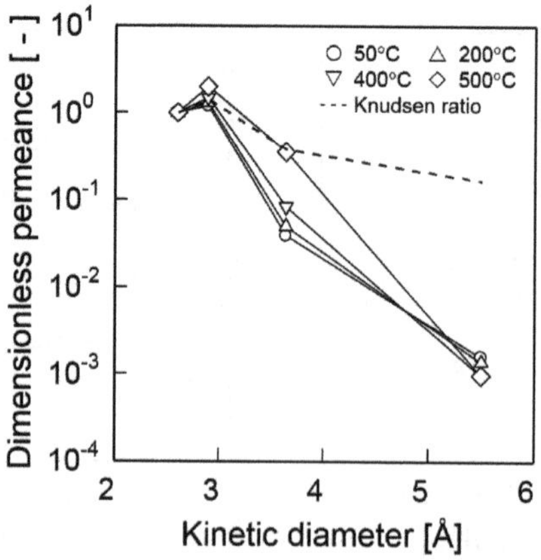

FIGURE 10.16 Kinetic diameter dependence of dimensionless gas permeances at 50°C for an HMDSO/Ar-PECVD-derived membrane before and after heat treatment at different temperatures. (From Nagasawa, H. et al., Separation and Purification Technology. 13–19, 2014.)

The gas separation performances of PECVD-derived membranes before and after heat treatment are also presented in Table 10.1. The membranes after heat treatment displayed an excellent molecular sieving property. It is noteworthy that the membrane derived from the 2-step PECVD sequence, which showed a high H_2/N_2 permeance ratio of 1900 with a H_2 permeance of 2.4×10^{-7} mol m^{-2} s^{-1} Pa^{-1} at 400°C, could be a promising candidate for the H_2 purification application. Furthermore, PECVD has also been applied for the fabrication of amorphous carbon membranes for gas separation (Nagasawa et al., 2016). The PECVD-derived amorphous carbon membranes exhibited molecular sieving characteristics with permeance ratios if 23 and 1750 for He/N_2 and He/SF_6 respectively.

In conclusion, we have successfully prepared silica-based membranes with molecular sieving properties using PECVD and found that it is possible to control the permeation properties of PECVD-derived membranes by using different organosilicon precursors and reaction atmospheres, as well as thermal annealing. The permeation properties of low-pressure PECVD-derived membranes reported in our previous work is summarized in Figure 10.17.

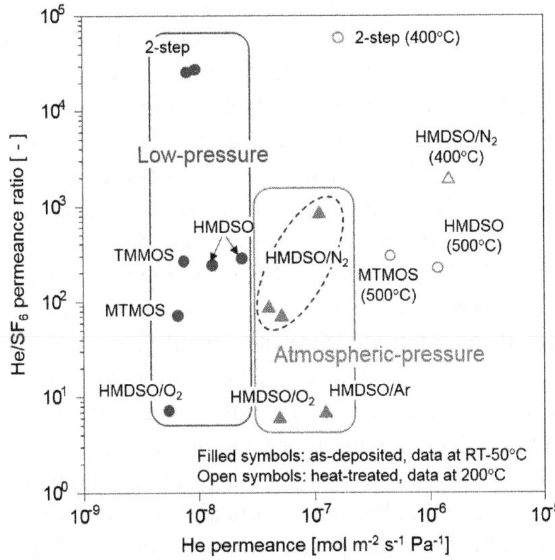

FIGURE 10.17 Trade-off relationship between He permeance and He/SF_6 permeance ratio of PECVD-derived silica membranes prepared on nanoporous intermediate layer. (Data from Tsuru, T. et al., Chemical Communications. 47, 8070–8072, 2011; Nagasawa, H. et al., Journal of Membrane Science. 441, 45–53, 2013; Nagasawa, H. et al., Separation and Purification Technology. 13–19, 2014; Nagasawa, H. et al., Journal of Membrane Science. 489, 11–19, 2015; Nagasawa, H. et al., Journal of Membrane Science. 524. 644–651, 2017; Nagasawa, H. et al., accepted. Atmospheric-pressure plasma-enhanced chemical vapor deposition of hybrid silica membranes. Journal of Chemical Engineering of Japan.)

10.4 ATMOSPHERIC-PRESSURE PECVD-DERIVED SILICA-BASED MEMBRANES

10.4.1 Overview of Atmospheric-Pressure Plasma-Deposition Techniques for Membrane Fabrication

As described in the previous section, PECVD allows low-temperature fabrication of silica-based membranes. It should also be noted that, since an active separation layer can be rapidly formed in a short period of time (~20 min), PECVD affords reduced membrane preparation time (Nagasawa et al., 2013). However, the main challenge associated with this technique is that the vacuum chamber systems used in PECVD are costly and unsuitable for large-scale membrane fabrication.

Recently, atmospheric-pressure plasma has emerged as a new plasma source (Tendero et al., 2006; Massines et al., 2012). The most important advantage of atmospheric-pressure plasma is that a stable discharge can be easily obtained without a vacuum system. This makes atmospheric-pressure processes more versatile than those under a vacuum. It can be operated in open-air as well as in-line processes. Regarding the application in membrane fabrication, atmospheric-pressure plasma processing can contribute to the fabrication of membranes in a continuous process for large-scale manufacturing. One option for preparing inorganic membranes via plasma-based route at atmospheric pressure is the use of atmospheric plasma spraying technique, which uses a high-temperature plasma jet at temperatures of the order of 10,000 K (Tung et al., 2009; Lin et al., 2012). In the atmospheric plasma spraying technique, ceramic and metal particles are fed into and melted in the high-temperature plasma jet, and then precipitated onto a substrate to form a solid alloy or ceramic coating. Tung et al. (2009) prepared ceramic-metallic microfiltration membranes with a pore size of 0.2-0.4 μm using α-Al_2O_3 particles with Ni and Cr as metal binders. They also synthesized photocatalytic TiO_2 membranes with a pore size of 0.35 μm that could be used to treat water through photodegradation of organisms and proteins under UV irradiation (Lin et al., 2012). An attempt to use high-temperature plasma to prepare gas separation membranes was made by Chen and co-workers (2013). They modified poly(dimethylsiloxane) membranes using high-temperature plasma to form a permselective SiO_x layer on the membrane surface. Such high-temperature plasma techniques are suitable for fabricating fully inorganic membranes but cannot be used to produce organic-inorganic hybrid membranes.

Another technique that has received considerable attention in recent years is atmospheric-pressure plasma-enhanced chemical vapor deposition (AP-PECVD), which uses non-equilibrium atmospheric-pressure plasma (Moravej and Hicks, 2005). Reactive species such as ions, radicals, and excited molecules in this type of plasma can induce film-forming reactions both in the gas phase and on the growing film surface. Because of the low temperature of non-equilibrium plasmas, they can greatly lower the processing temperature compared to that using thermal plasma, and thus minimize the thermal damage to the substrate (Starostine et al., 2007; Morent et al., 2009). Therefore, AP-PECVD has emerged as a reliable technique to deposit functional thin films on a wide variety of substrates including thermally sensitive materials. As a result, AP-PECVD can be used to prepare thin films such as dense silica coatings for surface protection layers on plastics (Cui et al., 2012), amorphous silicon nitride thin films for anti-reflective coatings on solar cells (Guruvenket et al., 2012), and titanium nitride films for electronic devices (Dong et al., 2014). The use of AP-PECVD for fabricating membranes for the separation of gas and liquid phases has never been reported. Recently, we reported the synthesis of supported microporous silica membranes for gas separation via AP-PECVD (Nagasawa et al., 2017b; Nagasawa et al., accepted). In the following sections, the fabrication of silica membranes via AP-PECVD and their gas permeation properties are presented.

10.4.2 Experimental Setup

Silica membranes were deposited onto nanoporous substrates in a remote plasma of either pure argon, or a mixture of argon and oxygen or nitrogen, using HMDSO as a silica precursor. A schematic diagram of the AP-PECVD system is shown in Figure 10.18. The system consisted of gas and precursor feed lines, an atmospheric plasma source, and a deposition chamber with a diameter of 100 mm. A damage-free plasma jet (Plasma Concept Tokyo, Inc., Japan), which is a specific type of dielectric barrier discharge plasma jet, was used as the plasma source. The atmospheric plasma was driven by a sinusoidal voltage with a maximum value of 6.0 kV at a frequency of 50 kHz. The plasma working gas was fed through the plasma source at a flow rate of 5.0 L min^{-1} to form reactive species. It was injected from a nozzle with a diameter of 1 mm into the center of the deposition chamber. Pure Ar or a mixture of Ar and O_2 (5.0 vol%) or N_2 (0.25-10.0 vol%) was used as the plasma working gas. The silicon precursor HMDSO was fed into the afterglow region of the plasma from the injection channel located 2 mm downstream of the nozzle exit by bubbling a flow of Ar as a carrier gas through the HMDSO liquid reservoir at 40°C, which corresponds to an HMDSO/carrier gas molar ratio of 1:10. The flow rate of the Ar carrier gas was maintained at 200 mL min^{-1}. The substrate was placed at the center of the deposition chamber. The distance from the nozzle exit to the substrate was 5 mm. The atmosphere in the chamber was prevented from contamination by the surrounding

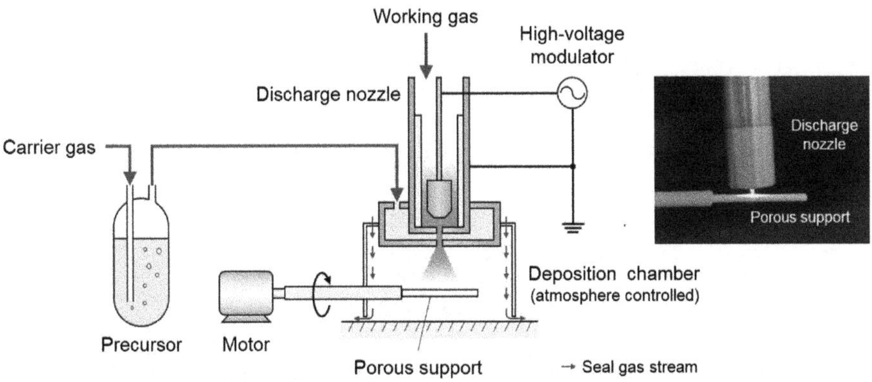

FIGURE 10.18 Schematic illustration of the AP-PECVD membrane fabrication system. (From Nagasawa, H. et al., Journal of Membrane Science. 524, 644–651, 2017.)

atmosphere by using an Ar gas curtain with a flow rate of 5.0 L min^{-1}. All deposition experiments were conducted at room temperature.

A porous α-alumina capillary tube (average pore size, 150 nm; outer diameter, 3 mm; length, 50 mm) was used as a supporting substrate. Prior to the plasma deposition of organosilica layers, SiO_2-ZrO_2 intermediate layers with an average pore size of 1 nm were prepared on the porous α-alumina substrates via a sol-gel method, following the same procedure described in Section 10.3.1. Silica layers were then deposited on the intermediate layers by AP-PECVD. Because the substrates were tubular, they were rotated at 200 rpm and moved toward the axial direction at a rate of 1 mm min^{-1} to ensure that the silica layer could uniformly be deposited on the substrate surface. The gas permeation properties of resultant membranes were evaluated by the constant-volume variable-pressure method, which is also described in Section 10.3.1.

10.4.3 Gas Permeation Properties of AP-PECVD-Derived Silica Membranes

Gas permeances of the membranes were evaluated with He (kinetic diameter, d_i = 0.26 nm), H_2 (0.289 nm), CO_2 (0.33 nm), N_2 (0.364 nm), CH_4 (0.38 nm), and SF_6 (0.55 nm). Figure 10.19 shows gas permeances at 50°C for the silica membranes fabricated by AP-PECVD as a function of the kinetic diameter of the permeating molecules. Considering the difference in the deposition rate among the three types of plasma working gases, the total deposition time was 20 min for the membrane prepared using pure Ar plasma, while that of the membranes deposited using either

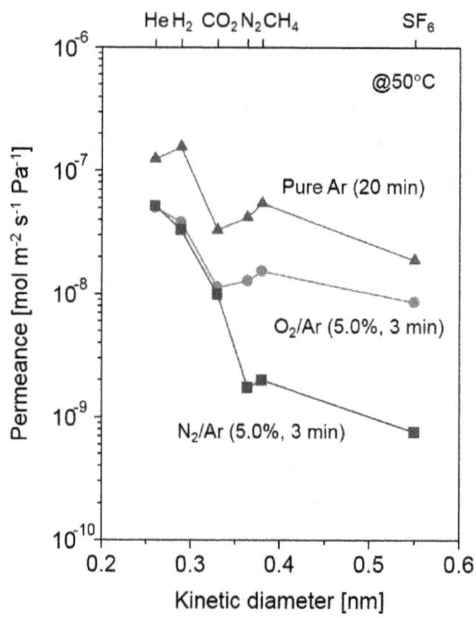

FIGURE 10.19 Single gas permeances at 50°C of membranes prepared by AP-PECVD with different working gases as a function of kinetic diameter of the permeating molecules. (Modified from Nagasawa, H. et al., Journal of Membrane Science. 524. 644–651, 2017. With permission.)

O_2/Ar or N_2/Ar plasma was 3 min, in order to obtain a similar permeance of He. After plasma deposition, the gas permeances decreased by two orders of magnitude from that of a substrate with a nanoporous silica–zirconia intermediate layer (N_2 permeance: ~3×10^{-6} mol m^{-2} s^{-1} Pa^{-1}). Of the three types of membranes, only the membrane prepared using N_2/Ar as the plasma working gas shows selectivity that is much higher than ideal Knudsen selectivity. The membrane fabricated using N_2/Ar plasma exhibits permeance ratios of He/H_2, He/N_2, and He/SF_6 of 1.6, 29.9, and 68.8, respectively, with a He permeance of 0.52×10^{-7} mol m^{-2} s^{-1} Pa^{-1}. Gas permeances decrease with increasing kinetic diameter of the gases, indicating the membrane has a microporous structure with molecular sieving properties. Surprisingly, compared with silica membranes prepared via PECVD under vacuum, which showed He permeances in the range 10^{-9}–10^{-8} mol m^{-2} s^{-1} Pa^{-1}, the membrane prepared using AP-PECVD with N_2/Ar plasma displays permeance that is at least one order of magnitude higher, but similar molecular sieving properties (Nagasawa et al., 2013; 2014b; 2015), as shown in Figure 10.17. These results demonstrate that AP-PECVD can fabricate microporous silica thin layer with molecular sieving properties.

Figure 10.20 shows gas permeances at 50°C for membranes prepared by AP-PECVD using N_2/Ar plasma with different N_2 concentrations as a function of kinetic diameter of the permeating molecules. The total deposition time for the membranes prepared using 0, 0.25, and 5.0 vol% N_2/Ar were 20, 10, and 3 min, respectively, to compare the performance of these membranes with similar He permeances. The gas permselectivity of these membranes is greatly improved by lowering the N_2 concentration in the working gas to 0.25 vol%. The membrane prepared using 0.25 vol% N_2/Ar as the working gas with a deposition time of 10 min shows a

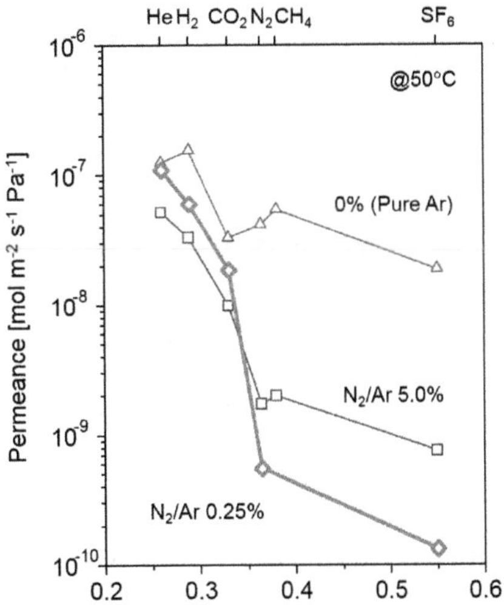

FIGURE 10.20 Single gas permeances at 50°C for membranes prepared by AP-PECVD with different N_2 concentrations as a function of kinetic diameter of the permeating molecules. (Modified from Nagasawa, H. et al., Journal of Membrane Science. 524. 644–651, 2017. With permission.)

high He permeance of 1.1×10^{-7} mol m^{-2} s^{-1} Pa^{-1} with permeance ratios of He/N_2 and He/SF_6 of 196 and 820, respectively. Both the He permeance and permeance ratios are higher than those of the membrane prepared using 5.0 vol% N_2/Ar as a working gas. These results suggest that the deposition of a silica layer using a low N_2 concentration of 0.25 vol% is preferable for membrane fabrication probably because the rate of the gas-phase reaction and the surface reaction are balanced.

10.4.4 Effect of Thermal Annealing and Permeation Property at High Temperature

Finally, we investigated how annealing a membrane at elevated temperatures after plasma deposition affected its properties. A membrane prepared using 0.25 vol% N_2/Ar as the working gas was annealed at 300°C in inert atmosphere for 60 min. This annealing step can help remove physically adsorbed organosilicon monomer and water, which may block the permeation pathway and decrease the membrane performance, from the plasma-deposited layer, and can also rearrange the structure of the plasma-deposited layer by thermally induced reactions. Figures 10.21(a) and (b) display the FTIR spectra of the films fabricated by AP-PECVD before and after annealing at 300°C. For the annealed film, the absorption band attributed to Si-O-Si stretching at 1078 cm^{-1} has the same intensity as that of the as-deposited film. The peak at 1260 cm^{-1} originating from Si-$(CH_3)_x$ bending is also unchanged after annealing. In contrast, the intensities of the absorption bands corresponding to Si-OH and OH stretching at 930 and 3400 cm^{-1}, respectively, decreased after annealing. This decrease in the intensities of Si-OH and OH peaks suggests that thermal

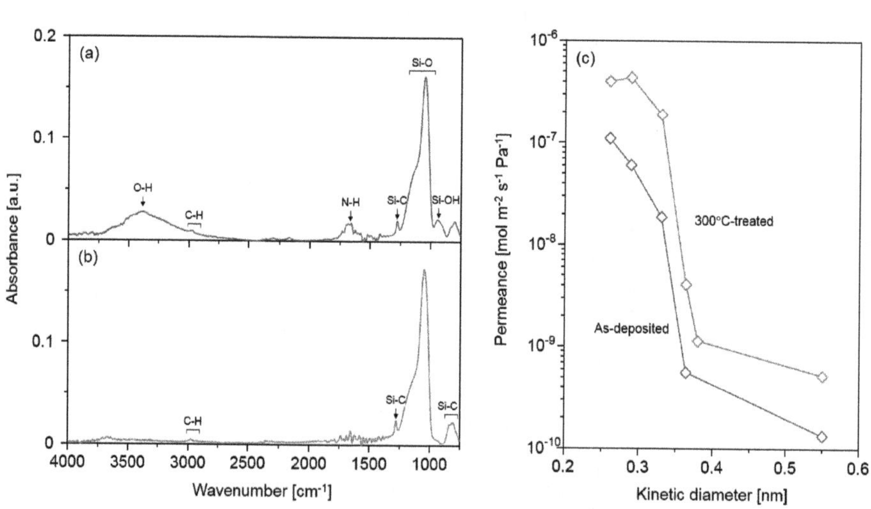

FIGURE 10.21 FTIR spectra of AP-PECVD-derived silica films (a) before and (b) after thermal annealing at 300°C, and (c) single gas permeances at 50°C of membranes before and after thermal annealing at 300°C. Thermal annealing was conducted under vacuum. (Modified from Nagasawa, H. et al., Journal of Membrane Science. 524. 644–651, 2017. With permission.)

treatment causes the condensation of silanol groups, resulting in the formation of the siloxane group. The decrease in the OH peak intensity also suggests the release of adsorbed water from the AP-PECVD-derived layer. Interestingly, as shown in Figure 10.21(c), the thermally annealed membrane shows an improved He permeance of 4.0×10^{-7} mol m^{-2} s^{-1} Pa^{-1} compared with that of the as-deposited membrane, due to the removal of physically adsorbed compounds in the plasma-deposited layer. Even though the He permeance increased 3.6 times, the selectivity of membrane for He/N_2 and He/SF_6 was maintained at 98 and 770, respectively, indicating high thermal stability of the plasma-deposited layer. Furthermore, the membrane also exhibited remarkable permselectivity for CO_2, with permeance ratios for CO_2/N_2 and CO_2/CH_4 of 46 and 166, respectively, and CO_2 permeance of 1.9×10^{-7} mol m^{-2} s^{-1} Pa^{-1} at 50°C. Currently, DDR (Tomita et al., 2004) and SAPO-34 (Carreon et al., 2008) zeolite membranes are reported to be the most effective microporous inorganic membranes developed for CO_2 separation from CO_2/CH_4 mixtures. The membranes fabricated by AP-PECVD have high CO_2 permeance and selectivity comparable with those of zeolite membranes. The high CO_2/N_2 and CO_2/CH_4 permeance ratios of the membranes after thermal annealing can be ascribed to the molecular sieving behavior of the AP-PECVD-derived silica layer with a pore size suitable for CO_2 permeation. CO_2 separation of AP-PECVD-derived silica membranes will be further investigated with a binary mixture in the future work.

To further understand the permeation characteristics of the AP-PECVD-derived membrane, the single gas permeances at high temperatures were investigated. As presented in Table 10.1, the membrane annealed at 400°C shows a remarkable gas permselectivity, with a H_2 permeance of $1.6 \in 10^{-6}$ mol m^{-2} s^{-1} Pa^{-1} and H_2/N_2 and H_2/SF_6 permeance ratios of 53 and 1,800, respectively, at 300°C. The permeation properties of AP-PECVD-derived membranes reported in our previous work are also summarized in Figure 10.17. The AP-PECVD-derived silica membrane exhibited quite high permselectivity for small molecules such as He and H_2, which was comparable with that of silica membranes prepared by conventional methods such as thermal CVD and sol-gel method, suggesting that AP-PECVD has great promise for the fabrication of microporous silica membranes that are highly permselective for gas separation.

10.5 CONCLUDING REMARKS

Room-temperature fabrication of silica-based gas separation membranes using PECVD was reviewed based on our work. The followings are the major conclusions. First, silica-based membranes with high permselectivity were successfully fabricated at room temperature using PECVD techniques. The room-temperature fabrication of silica-based membranes potentially allows the deposition of these membranes onto polymeric supports, which are less expensive but with lower thermal stability than ceramic supports. The membrane prepared via low-pressure 2-step PECVD sequence displayed an excellent molecular sieving property. It is noteworthy that although PECVD was conducted at room temperature, the membrane was stable at elevated temperatures as high as 500°C. The membrane derived from the 2-step PECVD sequence, which showed a high H_2/N_2 permeance ratio of 1900 with a H_2

permeance of 2.4×10^{-7} mol m^{-2} s^{-1} Pa^{-1}, could be a promising candidate for the H_2 purification application.

Second, the use of atmospheric-pressure plasma is the key for developing more versatile deposition systems. We demonstrated for the first time that a silica membrane with a permselective property can be deposited in one step by a remote atmospheric-pressure plasma jet. A membrane deposited using plasma composed of a mixture of argon and nitrogen displayed highly efficient gas separation with ideal selectivity for He/N_2 and He/SF_6 of 196 and 820, respectively, and He permeance of 1.1×10^{-7} mol m^{-2} s^{-1} Pa^{-1} at 50°C. It was also revealed that the membranes after annealing at 300°C exhibited remarkable permselectivity for CO_2, showing ideal selectivity for CO_2/N_2 and CO_2/CH_4 of 46 and 166, respectively, with CO_2 permeance of 1.9×10^{-7} mol m^{-2} s^{-1} Pa^{-1} at 50°C, which was comparable with that of currently available CO_2 permselective zeolite membranes such as DDR and SAPO-34 membranes.

We have also demonstrated that both in low-pressure and atmospheric-pressure PECVD, the structure of the separation active layer, which governs the permeation property of membranes, could be tuned by changing the chemical component of the silicon precursors as well as by changing the plasma reaction atmosphere. The results show great promise for the use of PECVD-derived silica membranes in molecular separation applications.

REFERENCES

Barranco, A., Cotrino, J., Yubero, F., Espinos, J.P., Gonzalez-Elipe, A.R., 2004. Room temperature synthesis of porous SiO_2 thin films by plasma enhanced chemical vapor deposition. Journal of Vacuum Science & Technology A. 22, 1275–1284.

Benitez, F. Martinez, E., Esteve, J., 2000. Improvement of hardness in plasma polymerized hexamethyldisiloxane coatings by silica-like surface modification. Thin Solid Films. 377–378, 109–114.

Brinker, C.J., Sehgal, R., Hietala, S.L., Deshpande, R., Smith, D.M., Loy, D., Ashley, C.S., 1994. Sol-gel strategies for controlled porosity of inorganic materials. Journal of Membrane Science. 94, 85–102.

Bruggeman, P.J., Sadeghi, N., Schram, D.C., Linss, V., 2014. Gas temperature determination from rotational lines in non-equilibrium plasmas: A review. Plasma Sources Science and Technology. 23, 023001.

Carreon, M.A., Li, S., Falconer, J.L., Noble, R.D., 2008. Alumina-supported SAPO-34 membranes for CO_2/CH_4 separation. Journal of American Chemical Society. 130, 5412-5413.

Castricum, H.L., Sah, A., Kreiter, R., Blank, D.H.A, Vente, J.F., ten Elshof, J.E., 2008. Hydrothermally stable molecular separation membranes from organically linked silica. Journal of Materials Chemistry. 18, 2150–2158.

Castricum, H.L., Paradis, G.G., Mittelmeijer-Hazeleger, M.C., Kreiter, R., Vente, J.F., ten Elshof, J.E., 2011. Tailoring the separation behaviour of hybrid organosilica membranes by adjusting the structure of the organic bridging group. Advanced Functional Materials. 21, 2319–2329.

Chen, J.-T., Fu, Y.-J., Tung, K.-L., Huang, S.-H., Hung, W.-S., Lue, S.J., Hu, C.-C., Lee, K.-R., Lai, J.-Y., 2013. Surface modification of poly(dimethylsiloxane) by atmospheric pressure high temperature plasma torch to prepare high-performance gas separation membranes. Journal of Membrane Science. 440, 1–8.

Clarizia, G., Algieri, C., Drioli, E., 2004. Filler–polymer combination: A route to modify gas transport properties of a polymeric membrane. Polymer. 45, 5671–5681.

Cui, L.Y., Ranade, A.N., Matos, M.A., Pingree, L.S., Frot, T.J., Dubois, G., Dauskardt, R.H., 2012. Atmospheric plasma deposited dense silica coatings on plastics. ACS Applied Materials & Interfaces. 4, 6587–6598.

de Vosa, R.M., Maierb, W.F., Verweija, H., 1999. Hydrophobic silica membranes for gas separation. Journal of Membrane Science. 158, 277–288.

Dong, J., Lin, Y.S., Kanezashi. M., Tang, Z., 2008. Microporous inorganic membranes for high temperature hydrogen purification. Journal of Applied Physics. 104, 121301-121317.

Dong, S., Watanabe, M., Dauskardt, R.H., 2014. Conductive transparent TiN_x/TiO_2 hybrid films deposited on plastics in air using atmospheric plasma processing. Advanced Functional Materials. 24, 3075–3081.

Duke, M.C., Pas, S.J., Hill, A.J., Lin, Y.S., da Dosta, J.C.D., 2008. Exposing the molecular sieving architecture of amorphous silica using positron annihilation spectroscopy. Advanced Functional Materials. 18, 3818–3826.

Gallucci, F., Fernandez, E., Corengia, P., van Sint Annaland, M., 2013. Recent advances on membranes and membrane reactors for hydrogen production. Chemical Engineering Science. 92, 40–66.

Gavalas, G.R., Mergiris, C.E., Nam, S.W., 1989. Deposition of H_2-permselective SiO_2 films. Chemical Engineering Science. 44, 1829–1835.

Gong, G., Nagasawa, H., Kanezashi, M., Tsuru, T., 2016. Tailoring the separation behaviour of polymer-supported organosilica layered-hybrid membranes via facile post-treatment using HCl and NH_3 vapors. ACS Applied Materials & Interfaces. 8, 11060–11069.

Gong, G., Wang, J., Nagasawa, H., Kanezashi, M., Yoshioka, T., Tsuru, T., 2014. Synthesis and characterization of a layered-hybrid membrane consisting of an organosilica separation layer on a polymeric nanofiltration membrane. Journal of Membrane Science. 427, 19–28.

Gu, Y., Oyama, S.T., 2007. High molecular permeance in a poreless ceramic membrane. Advanced Materials. 19, 1636–1640.

Guruvenket, S., Andrie, S., Simon, M., Johnson, K.W., Sailer, R.A., 2012. Atmospheric-pressure plasma-enhanced chemical vapor deposition of a-SiCN:H films: Role of precursors on the film growth and properties. ACS Applied Materials & Interfaces. 4, 5293–5299.

Hacarlioglu, P., Lee, D., Gibbs, G.V., Oyama, S.T., 2008. Activation energy for permeation of He and H_2 through silica membranes: An ab initio calculation study. Journal of Membrane Science. 313, 277

Hagg, M., 2000. Membrane purification of Cl_2 gas. I. Permeabilities as a function of temperature for Cl_2, O_2, N_2, H_2 in two types of PDMS membranes. Journal of Membrane Science. 170, 173–190.

Kafrouni, W., Rouessac, V., Julbe, A., Durand, J., 2009. Synthesis of PECVD a-SiC_XN_Y:H membranes as molecular sieves for small gas separation. Journal of Membrane Science. 329, 130–137.

Kafrouni, W., Rouessac, V., Julbe, A., Durand, J., 2010. Synthesis and characterization of silicon carbonitride films by plasma enhanced chemical vapor deposition (PECVD) using bis(dimethylamino)dimethylsilane (BDMADMS), as membrane for a small molecule gas separation. Applied Surface Science. 257, 1196–1203.

Kanezashi, M., Asaeda, M., 2005. Stability of H_2-permeselective Ni-doped silica membranes in steam at high temperature. Journal of Chemical Engineering of Japan. 38, 908–912.

Kanezashi, M., Yada, K., Yoshioka, T., Tsuru, T., 2009. Design of silica networks for development of highly permeable hydrogen separation membranes with hydrothermal stability. Journal of American Chemical Society. 131, 414–415.

Kanezashi, M., Yoneda, Y., Nagasawa, H., Tsuru, T., 2017. Gas permeation properties for organosilica membranes with different Si/C ratios and evaluation of microporous structures. AIChE Journal. 63, 4491–4498.

Kitao, S., Asaeda, M., 1990. Separation of organic acid/water mixtures by thin porous silica membrane. Journal of Chemical Engineering of Japan. 23, 367–370.

Lee, J.H., Jeong, C.H., Lim, J.T., Zavaleyev, V.A., Kyung, S.J., Yeom, G.Y., 2001. SiO_xN_y thin film deposited by plasma enhanced chemical vapor deposition at low temperature using HMDS-O_2-NH_3-Ar gas mixtures. Surface & Coatings Technology. 207, 4957–4960.

Lee, S, Yang, J., Yeo, S., Lee, J., Jung, D., Boo, J., Kim, H., Chae, H., 2007. Effect of annealing temperature on dielectric constant and bonding structure of low-k SiCOH thin films deposited by plasma enhanced chemical vapor deposition. Japanese Journal of Applied Physics. 46, 536–541.

Li, K., Meichsner, J., 1999. Gas-separating properties of membranes coated by HMDSO plamsa polymer. Surface & Coating Technology. 116–119, 841–847.

Lin, Y.-F., Tung, K.-L., Tzeng, Y.-S., Chen, J.-H., Chang, K.-S., 2012. Rapid atmospheric plasma spray coating preparation and photocatalytic activity of microporous titania nanocrystalline membranes. Journal of Membrane Science. 389, 83–90.

Lin, Y.S., Kumakiri, I., Nair, B.N., Alsyouri, H., 2002. Microporous inorganic membranes. Separation and Purification Methods. 31, 229–379.

Lo, C.-H., Lin, M.-H., Liao, K.-S., De Guzman, M., Tsai, H.-A., Rouessac, V., Wei, T.-C., Lee, K.-R., Lai, J.-Y., 2010. Control of pore structure and characterization of plasma-polymerizsed SiOCH films deposited from octamethylsyclotetrasiloxane (OMCTS). Journal of Membrane Science. 365, 418–425.

Maex, K., Baklanov, M. R., Shamiryan, D., Iacopi, F., Brongersma, S.H., Yanovitskaya, Z.S., 2003. Low dielectric constant materials for microelectronics. Journal of Applied Physics. 93, 8793–8841.

Martinu, L., Poitras, D., 2000. Plasma deposition of optical films and coatings: A review. Journal of Vacuum Science & Technology. 18, 2619–2645.

Massines, F., Sarra-Bournet, C., Fanelli, F., Naude, N., Gherardi, N., 2012. Atmospheric pressure low temperature direct plasma technology: Status and challenges for thin film deposition. Plasma Processes and Polymers. 9, 1041–1073.

Matsuyama, H., Kariya, A., Teramoto, M., 1994. Effect of siloxane chain lengths of monomers on characteristics of pervaporation membranes prepared by plasm polymerization. Journal of Applied Polymer Science. 51, 689–693.

Merkel, T., Gupta, R., Turk, B., Freeman, B., 2001. Mixed-gas permeation of syngas components in poly(dimethylsiloxane) and poly(1-trimethylsilyl-1-propyne) at elevated temperatures. Journal of Membrane Science. 191, 85–94.

Milella, A., Palumbo, F., Delattre, J.L., Fracassi, F., d'Agostino, R., 2007. Deposition and characterization of dielectric thin films from allyltrimethylsilane glow discharges. Plasma Processes and Polymers. 4, 425–432.

Moravej, M., Hicks, R.F., 2005. Atmospheric plasma deposition of coatings using a capacitive discharge source. Chemical Vapor Deposition. 11, 469–476.

Morent, R., De Geyter, N., Van Vlierberght, S., Dubruel, P., Leys, C., Gengembre, L., Schacht, E., Payen, E., 2009. Deposition of HMDSO-based coatings on PET substrates using an atmospheric pressure dielectric barrier discharge. Progress in Organic Coatings. 64, 304–310.

Nagasawa, H., Shigemoto, H., Kanezashi, M., Yoshioka, T., Tsuru, T., 2013. Characterization and gas permeation properties of amorphous silica membranes prepared via plasma enhanced chemical vapor deposition. Journal of Membrane Science. 441, 45–53.

Nagasawa, H., Niimi, T., Kanezashi, M., Yoshioka, T., Tsuru, T., 2014a. Modified gas-translation model for prediction of gas permeation through microporous organosilica membranes. AIChE Journal. 60, 4199–4210.

Nagasawa, H., Minamizawa, T., Kanezashi, M., Yoshioka, T., Tsuru, T., 2014b. High-temperature stability of PECVD-derived organosilica membranes deposited on TiO_2 and SiO_2 $-ZrO_2$ intermediate layers using HMDSO/Ar plasma. Separation and Purification Technology. 13–19.

Nagasawa, H., Minamizawa, T., Kanezashi, M., Yoshioka, T., Tsuru, T., 2015. Microporous organosilica membranes for gas separation prepared via PECVD using different O/Si ratio precursors. Journal of Membrane Science. 489, 11–19.

Nagasawa, H., Kanezashi, M., Yoshioka, T., Tsuru, T., 2016. Plasma-enhanced Chemical vapor deposition of amorphous carbon molecular sieve membranes for gas separation. RSC Advances. 6, 59045–59049.

Nagasawa, H., Nishibayashi, M., Kanezashi, M., Yoshioka, T., Tsuru, T., 2017a. Photo-induced sol-gel synthesis of polymer-supported silsesquioxane membranes. RSC Advances. 7, 7150–7157.

Nagasawa, H., Yamamoto, Y., Tsuda, N., Kanezashi, M., Yoshioka, T., Tsuru, T., 2017b. Atmospheric-pressure plasma-enhanced chemical vapor deposition of microporous silica membranes for gas separation. Journal of Membrane Science. 524. 644–651.

Nagasawa, H., Yamamoto, Y., Kanezashi, M., Tsuru, T., accepted. Atmospheric-pressure plasma-enhanced chemical vapor deposition of hybrid silica membranes. Journal of Chemical Engineering of Japan.

Nehlsen, S., Hunte, T., Müller, J., 1995. Gas permeation properties of plasma polymerized thin film siloxane-type membranes for temperatures up to 350°C. Journal of Membrane Science. 106, 1–7.

Ngamou, P., Overbeek, J., Kreiter, R., van Veen, H., Vente, J., Wienk, I., Cuperus, P., Creatore. M., 2013. Plasma deposited hybrid silica membranes with a controlled retention of organic bridges. Journal of Materials Chemistry A. 1, 5567–5576.

Niimi, T., Nagasawa, H., Kanezashi, M., Yoshioka, T., Ito, K., Tsuru, T., 2014. Preparation of BTESE-derived organosilica membranes for catalytic membrane reactors of methylcyclohexane dehydrogenation. Journal of Membrane Science. 455, 375–383.

Nishibayashi, M., Yoshida, H., Uenishi, M., Kanezashi, M., Nagasawa, H., Yoshioka, T., Tsuru, T., 2015. Photo-induced sol-gel processing for low-temperature fabrication of high-performance silsesquioxane membranes for use in molecular separation. Chemical Communications. 51, 9932–935.

Nomura, M., Aida, H., Gopalakrishnan, S., Sugawara, T., Nakao, S., Yamazaki, S., Inada, T., Iwamoto, Y., 2006. Steam stability of a silica membrane prepared by counterdiffusion chemical vapor deposition. Desalination. 193, 1–7.

Ockwig, N.W., Nenoff, T.M., 2007. Membranes for hydrogen separation. Chemical Reviews. 107, 4078–4110.

Pera-Titsu, M., 2014. Porous inorganic membranes for CO_2 capture: Present and prospects. Chemical Reviews. 114, 1413–1492.

Roualdes, S., Sanchez, J., Durand, J., 2002. Gas diffusion and sorption properties of polysiloxane membranes prepared by PECVD. Journal of Membrane Science. 198, 299–310.

Roualdes, S., Van der Lee, A., Berjoan, R., Sanchez, J., Durand, J., 1999. Gas separation properties of organosilicon plasma polymerized membranes. AIChE Journal. 56, 1566–1575.

Sadrzadeh, M., Amirilargani, M., Shahidi, K., Mohammadi, T., 2009. Gas permeation through a synthesized composite PEMS/PES membrane. Journal of Membrane Science. 342, 236–250.

Sakata, J., Hirai, M., Yamamoto, M., 1987. Plasma polymerized membranes and gas permeability III. Journal of Applied Polymer Science. 34, 2701–2711.

Shioya, Y., Ohdaira, T., Suzuki, R., Seino, Y., Omote, K., 2008. Effect of UV anneal on plasma CVD low-k film. Journal of Non-Crystalline Solids. 354, 2973–2982.

Sommer, S., Melin, T., 2005. Performance evaluation of microporous inorganic membranes in the dehydration of industrial solvents. Chemical Engineering and Processing: Process Intensification. 44, 1138–1156.

Starostine, S., Aldea, E., de Vries, H., Creatore, M., van de Sanden, M.C.M., 2007. Atmospheric pressure barrier discharge deposition of silica-like films on polymeric substrates. Plasma Processes and Polymers. 4, S440–S444.

Steele, D.A., Short, R.D., Brown, P., Mayhew, C.A., 2011. On the use of SIFT-MS and PTR-MS experiments to explore reaction mechanisms of volatile organics: Siloxanes. Plasma Processes and Polymers. 8, 287–294.

Tendero, C., Tixier, C., Tristant, P., Desmaison, J., Leprince, P., 2006. Atmospheric pressure plasmas: A review. Spectrochimica Acta Part B: Atomic Spectroscopy. 61, 2–30.

Tomita, T., Nakayama, K., Sakai, H., 2004. Gas separation characteristics of DDR type zeolite membrane. Microporous and Mesoporous Materials. 68, 71–75.

Tsapatsis, M., Gavalas, G., 1994. Structure and aging characteristics of H_2-permselective SiO_2-Vycor membranes. Journal of Membrane Science. 87, 281–296.

Tsuru, T., Hino, T., Yoshioka, T., Asaeda, M., 2001. Permporometry characterization of microporous ceramic membranes. Journal of Membrane Science. 186, 257–265.

Tsuru, T., 2008. Nano/subnano-tuning of porous ceramic membranes for molecular separation. Journal of Sol-Gel Science and Technology. 46, 349–361.

Tsuru, T., Shigemoto, H., Kanezashi, M., Yoshioka, T., 2011. 2-step plasma-enhanced CVD for low-temperature fabrication of silica membranes with high gas-separation performance. Chemical Communications. 47, 8070–8072.

Tung, K.-L., Hsiung, C.-C., Ling, T.-C., Chang, K.-S., Wu, T.-T., Li, Y.-L., Lang, C.-H., Chen, W.-Y., Nanda, D., 2009. Preparation and characterization of aluminium oxide cermet microfiltration membrane using atmospheric plasma spraying. Desalination. 245, 408–421.

Volkesen, W., Miller, R.D., Dubois, G., 2010. Low dielectric constant materials. Chemical Reviews. 110, 56–110.

Wang, J., Kanezashi, M., Yoshioka, T., Ito, K., Tsuru, T., 2013. Pervaporation performance and characterization of organosilica membranes with tuned pore size by solid phase HCl post-treatment. Journal of Membrane Science. 441, 120–128.

Walkiewicz-Pietrzykowska, A., Cotrino, J., Gonzalez-Elipe, A.R., 2005. Deposition of thin films of SiO_xC_yH in a surfatron microwave plasma reactor with hexamethyldisiloxane as precursor. Chemical Vapor Deposition. 11, 317–323.

Weichart, J., Müller, J., 1993. Plasma polymerization of silicon organic membranes for gas separation. Surface and Coating Technology. 59, 342–344.

Yamaguchi, T., Ying, X., Tokimasa, Y., Nair, B.N., Sugawara, T., Nakao, S., 2000. Reaction control of tetraethyl orthosilicate (TEOS)/O_3 and tetramethyl orthosilicate (TMOS)/O_3 counter diffusion chemical vapour deposition for preparation of molecular-sieve membranes. Physical Chemistry Chemical Physics. 2, 4465–4469.

Yamamoto, M., Sakata, J., Hirai, M., 1984. Plasma polymerized membranes and gas permeability. I. Journal of Applied Polymer Science. 29, 2981–2987.

Yoshioka, T., Tsuru, T., Asaeda, M., 2004. Molecular dynamics study of gas permeation through amorphous silica membranes. Molecular Physics. 102, 191–202.

Yu, X., Meng, L., Niimi, T., Nagasawa, H., Kanezashi, M., Yoshioka, T., Tsuru, T., 2016. Network engineering of a BTESE membrane for improved gas performance via a novel pH-swing method. Journal of Membrane Science. 511, 219–227.

11 MOF Membranes for Gas Separation

Hua Jin, Weishen Yang, and Yanshuo Li

CONTENTS

11.1 INTRODUCTION

Metal-organic frameworks (MOFs) are a new class of hybrid crystalline materials built from metal ions and multitopic organic ligands by coordination bonds (Kitagawa et al., 2004; Schoedel et al., 2016). The fine control over the building blocks offers an exceptionally large set of more than 20,000 different MOFs with variable geometry, size or functionality (Furukawa et al., 2013). MOFs typically possess well defined structures, permanent pores, extraordinary surface area and pore volume, which endow them with many potential applications such as gas storage (Suh et al., 2012), chemical catalysis (Lee et al., 2009), adsorption and molecular separation (Li et al., 2009, 2012). In particular, the processing of MOFs into membranes to achieve gas separation is the main focus of this chapter.

As compared to traditional inorganic porous solids with zeolites as the quintessential example, MOFs offer greater structural diversity and chemical varieties owing to the large choice of inorganic and organic components. The features of designability and adjustability are highly significant to tailor MOFs specifically for particular applications. MOFs are usually synthesized at mild conditions, in contrast to the hydrothermal treatment at high temperature and pressure for zeolites. Additionally, activation by calcination that can cause damage to the zeolite structure is not needed for MOFs with no organic structure-directing agents incorporated in the structure. On a fundamental level, MOFs have been an attractive alternative to zeolites as membrane candidates for gas separation. Nevertheless, the poor chemical and hydrothermal stabilities of MOFs due to the lability of coordinate bond have long been criticized, especially when compared to industrially relevant zeolites. The problem has now been resolved to a large extent with the growing number of chemically stable and water-stable MOFs (Howarth et al., 2016; Wang et al., 2016a). Other than the well-known zeolitic imidazolate framework (ZIF) materials, UiO-66, MIL-53, MIL-101, CAU-10 and MOF-74, etc. are typical water-stable MOFs as well. The remarkable development of functional MOFs with excellent hydrothermal stability has been providing essential guarantees for the application of MOF membranes in real industrial conditions.

In recent years, both polycrystalline MOF membranes (Caro, 2011; Qiu et al., 2014; Lin, 2015; Kang et al., 2017a; Liu et al., 2017) and MOF-polymer hybrid membranes (Seoane et al., 2015; Zhang et al., 2016; Dechnik et al., 2017a, 2017b) have been studied intensively, showing tremendous potential for gas separation. In this chapter, we summarize the main developments of these two types of MOF-based membranes, emphasizing the synthetic methods and typical gas separation performance.

11.2 SYNTHESIS OF MOF MEMBRANES

11.2.1 Substrates

Learning from zeolite membranes, the vast majority of MOF membranes have been synthesized on inorganic porous substrates such as a metal net, aluminium oxide and titanium dioxide. Nevertheless, the high cost of inorganic substrates may limit the wide application of MOF membranes, which promotes the search for alternatives. Commercial or customized polymeric substrates, especially polymer hollow fibers associated with a large area per volume, have been identified as a good solution to solve the problem. Since Centrone et al. (2010) reported for the first time the fabrication of MIL-47 directly on polyacrylonitrile using in situ microwave irradiation, extensive studies on the production of MOF membranes on polymer substrates have been reported. Nagaraju et al. (2013) reported the synthesis of HKUST-1 and ZIF-8 on a flat sheet polysulfone (PSF) based porous asymmetric ultrafiltration membrane, by the in situ crystallization followed by layer-by-layer deposition. The resultant continuous and denser MOF@polymer composite membranes showed enhanced hydrogen selectivity compared to the pristine PSF membrane. Brown et al. (2012) prepared ZIF-90 membranes using Torlon hollow fiber as substrate. Cacho-Bailo et al. (2015) grew ZIF-7 and ZIF-8 membranes on the inner-side of a PSF hollow fiber by microfluidic synthesis. Zhang's group has also published fascinating work about MOF membranes on polymer hollow fibers, such as MOF (CuBTC, ZIF-7 and ZIF-8)/PVDF hollow fiber composite membranes, CuBTC/polysulfone (PSF) hollow fiber membranes, CuBTC-MIL-100/ PVDF hollow fiber membranes, etc. (Li et al., 2014a, 2015, 2016c). The obtained membranes demonstrated high permeance and modest selectivity in terms of gas separation. As illustrated by the extensive works published to date, the MOF membranes can be fabricated on polymeric hollow fibers by a technologically scalable process, which will greatly extend the applications of MOF membranes. It is worth noting that

MOFs with harsh synthetic conditions (i.e., high temperatures or strong solvents) rarely formed a continuous membrane on polymeric substrates due to the poor thermal and chemical stability of most polymers. As a result, to attain MOF membranes with excellent gas separation performance, the following two points should be guaranteed: good adhesion between the MOF layer and polymer substrates and the thermal and chemical stability of the polymer substrates.

11.2.2 Polycrystalline MOF Membranes

Polycrystalline MOF membranes associated with high permeability and selectivity have received extensive attention in the last decade. For gas separation application the key is the preparation of continuous and well-intergrown polycrystalline MOF membranes. Drawing on the zeolite membrane synthesis, MOF membranes have been synthesized mainly by in situ growth and seeded-assisted (secondary) growth. Additionally, as a result of the significant difference between coordination chemistry of MOFs and covalent chemistry of zeolites, several innovative techniques such as rapid thermal deposition (RTD) and contra-diffusion have been proposed exclusively for the synthesis of MOF membranes. In the following section, conventional methods accompanied by innovative techniques for the preparation of MOF membranes were discussed at length.

In the in situ growth of MOF membranes, the substrates (normally porous alumina, silica, titania and commercial polymeric porous supports) are immersed in a precursor solution. Subsequently, the nucleation, growth and intergrowth of crystals on the substrate proceeded at a certain temperature over a period of time. Liu et al. (2009) successfully prepared the first continuous and well-intergrown MOF-5 membrane on α-alumina substrate by in situ solvothermal synthesis. Bux et al. (2009) further demonstrated the possibility of preparing highly gas-selective MOF membranes by microwave-assisted in situ solvothermal synthesis. The resultant ZIF-8 membranes exhibited promising H_2 separation abilities. Typically, the preparation of continuous MOF membranes on unmodified substrate is quite challenging due to the poor interaction between MOF membrane and native substrate. To overcome this problem, various techniques for modifying the porous support have been suggested. Huang et al. (2010a) developed a novel covalent functionalization strategy to prepare a hydrogen-selective ZIF-90 membrane by using 3-aminopropyltriethoxysilane (APTES) as covalent linkers between the ZIF-90 layer and the Al_2O_3 support via imines condensation. They also reported the successful formation of ZIF-7, ZIF-8 and ZIF-22 membranes on APTES-modified Al_2O_3 supports (Huang et al., 2010b). In addition to the protocol using organic covalent linkers, bringing a buffering layer between active MOF membranes and supports has proved effective. Zhang et al. (2014) reported that a defect ZIF-8 membrane could be successfully prepared on the modified porous ceramic tube with vertically aligned ZnO nanorods as the buffering layer. The activated nanorods could induce a uniform nucleation of ZIF nuclei on the surface and further initiate and guide the growth of continuous ZIF-8 membranes. Li et al. (2014a) found that the ZIF-7 membranes with excellent H_2 permselectivity could be prepared on PVDF hollow fiber supports using non-activation ZnO array as the buffering layer. Liu et al. (2014b) brought forward a new concept for substrate

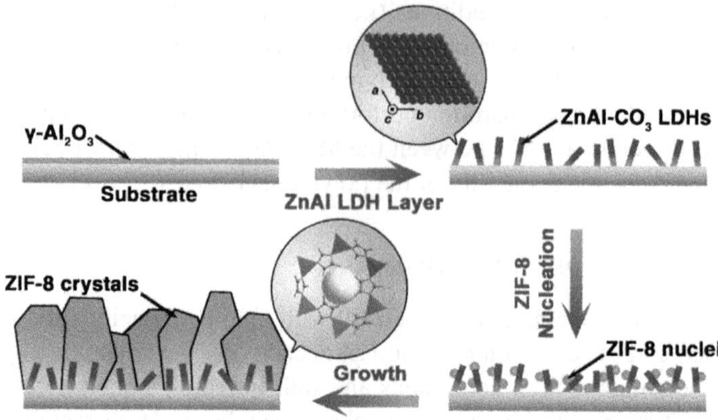

FIGURE 11.1 Schematic illustration of in-situ solvothermal growth of ZIF-8 membrane on a ZnAl-LDH buffer layer-modified γ-Al_2O_3 substrate. (Based on Liu, Y. et al., Journal of the American Chemistry Society. 136, 14353–14356, 2014b.)

modification. As shown in Figure 11.1, ZnAl-CO_3 LDH buffer layers were firstly in situ prepared on porous γ-Al_2O_3 substrates, and then a well-intergrown ZIF-8 membrane was formed in a dilute precursor solution because of the high-affinity interaction between ZnAl-CO_3 LDH grains with the ZIF-8 phase. The developed approach can be applied to prepare other Zn-based MOF membranes such as ZIF-7 and ZIF-90.

On the whole, secondary growth makes continuous MOF membranes without defects more available than in situ growth. Owing to the decoupled nucleation and crystal growth steps, secondary growth is more convenient and effective to control the microstructural of the resultant membranes. For the secondary growth of MOF membranes, the support was first seeded with MOF seed crystals, and then the seeded support went through secondary growth by conventional hydrothermal or solvothermal treatment. Seeding procedure is crucially important due to its great influence on membrane microstructure (membrane thickness, grain boundaries, orientation, etc.). For zeolite membranes, the seed attachment is not an issue because zeolite seeds can be covalently bound into the substrate by calcination treatment. Unfortunately, the approach was not viable for MOF membranes, as MOF cannot withstand high temperatures. So far, several seeding techniques, including dip coating, spin coating, thermal seeding, reactive seeding (RS) and LBL seeding, have demonstrated the effectiveness for achieving uniform and well intergrown MOF membranes. Li et al. (2010a, 2010b, 2010c) have successfully prepared polycrystalline ZIF-7 membranes on α-alumina disks that were first surface-seeded with the ZIF-7 nanoseeds by dip coating. The as-synthesized ZIF-7 membrane exhibited promising H_2/CO_2 separation ability, as well as good thermal and hydrothermal stabilities. Fan et al. (2012) developed a novel approach that introduces an electrospinning technique as a means of seeding. The thickness of seed layer could be precisely tuned, and the achieved continuous and uniform seed layer effectively promoted the synthesis of high-quality MOF membranes. The electrospinning approach described here offers the possibilities of large-area processing. The RS method, in which the a-Al_2O_3 support acts as the inorganic source reacting with 1,4-benzenedicarboxylic acid (H_2BDC) to grow a seed layer, has been developed for the preparation of continuous MIL-53 membranes (Hu et al., 2011). As indicated by the single-gas permeation experiments, the MIL-53 membrane is of very high integrity. The RS method has been shown to be versatile for the fabrication of different types of MOF membranes. Despite the above progress, to develop a versatile seeding method for constructing a seed layer with controllable thickness on various type of substrates is still a challenge and needs more research.

As a result of the fundamental differences between the coordination bond of MOFs and the covalent bond of zeolites, several innovative techniques with low energy requirements (i.e., mild synthetic conditions) or short synthesis time have been developed for the fabrication of MOF membranes. Rapid thermal deposition (RTD), based on the evaporation-induced crystallization, has been used to prepare high-quality membranes of prototypical MOFs, HKUST-1 and ZIF-8, in a relatively short period of time (tens of min) (Shah et al., 2013). The resultant RTD membranes have a notably different microstructure when compared with their conventional counterparts. Results from gas permeation tests indicate that the grain boundary defects were greatly reduced, further confirming the improved microstructure of HKUST-1 and ZIF-8 membranes. RTD with unique properties such as reduced cost of membrane manufacturing, scalability and reproducibility, provides potential for rapid synthesis of MOF membrane for the commercial applications. Aceituno Melgar et al. (2014) reported a simple electrospray deposition technique for synthesizing supported zeolitic imidazolate framework ZIF-7 membranes. Figure 11.2 shows the schematic diagram of electrospray deposition for synthesizing ZIF-7 membranes. The applied voltage to the precursor solution, precursor flow rate and deposition temperature, etc. dominantly affects the crystal integrity and microstructures of the ZIF-7 membrane. In comparison with conventional synthetic routes such as in situ and secondary growth methods, the described electrostatic force-assisted coating approach offers dramatic reduction in synthesis time and precursor consumption,

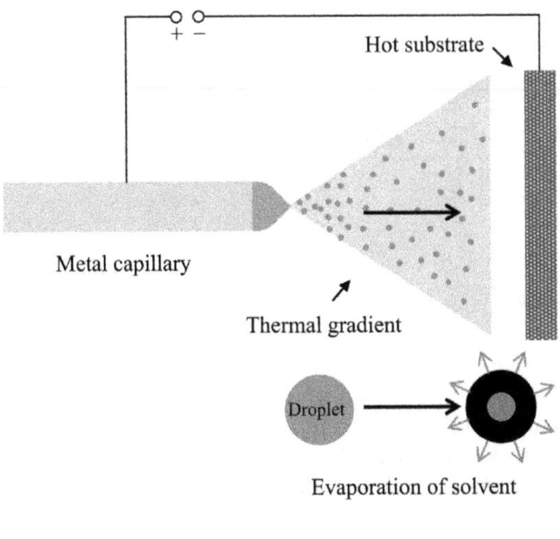

FIGURE 11.2 Schematic diagram of electrospray deposition for synthesizing ZIF-7 membranes. (Based on Aceituno Melgar, V.M. et al., Journal of Membrane Science. 459, 190–196, 2014.)

simplification in the activation process as well as potential scalability. A new method, the contra-diffusion (CD) method in which metal ions and ligand molecules are physically separated and brought into contact by diffusion, has been successfully employed in the facile synthesis of MOF membranes. Yao et al. (2011) successfully prepared ZIF-8 membranes on a flexible nylon substrate and tested them for the separation of H_2 from N_2. As illustrated by Figure 11.3, the zinc nitrate solution and 2-methylimidazole (Hmim) solution were separated by the nylon membrane. After crystallization at room temperature for 72 h, ZIF-8 membranes with a thickness up to 16 μm was formed at the zinc nitrate side and exhibited a H_2/N_2 ideal selectivity of 4.3. Kwon et al. (2013a) further improved the method for preparing ZIF-8 membranes with a significantly enhanced microstructure. As shown in Figure 11.4, the porous α-alumina supports were first soaked in a metal precursor solution, and

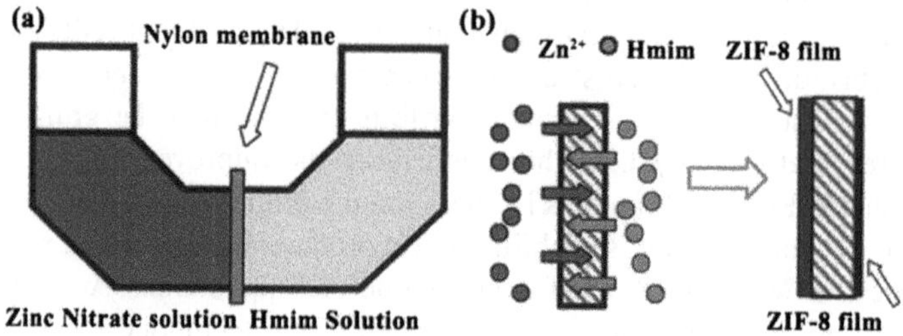

FIGURE 11.3 (a) Diffusion cell for ZIF-8 film preparation and (b) the schematic formation of ZIF-8 films on both sides of the nylon support via contra-diffusion of Zn^{2+} and Hmim through the pores of the nylon support. (From Yao, J. et al., Chemical Communications. 47, 2559–2561, 2011.)

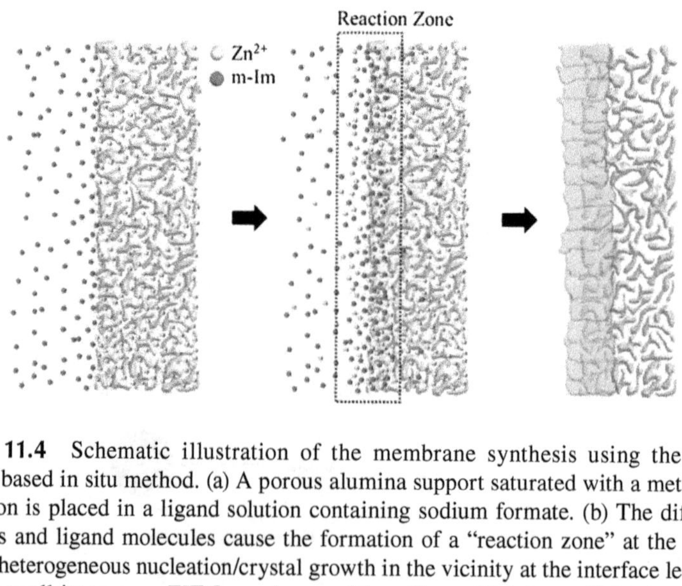

FIGURE 11.4 Schematic illustration of the membrane synthesis using the counter-diffusion-based in situ method. (a) A porous alumina support saturated with a metal precursor solution is placed in a ligand solution containing sodium formate. (b) The diffusion of metal ions and ligand molecules cause the formation of a "reaction zone" at the interface. (c) Rapid heterogeneous nucleation/crystal growth in the vicinity at the interface leads to the continuous well-intergrown ZIF-8 membranes. (From Kwon, H.T., Jeong, H.K., Journal of the American Chemistry Society. 135, 10763–10768, 2013a.)

then followed by a rapid solvothermal reaction in the ligand-containing solution. The obtained ZIF-8 membrane showed remarkable separation performance, suggesting the excellent microstructure of ZIF-8 membrane resulted from the healing nature of contra-diffusion method. The CD-based in situ method has demonstrated general applicability with its successful application for prototypical ZIF-7 and SIM-1 membranes synthesis. However, the approach may prove challenging when the capillary substrates or hollow fibers are used, which could be ascribed to the difficulty of reactant availability and transport in the microscopic confined spaces. Against this backdrop, Brown et al. (2014) developed the interfacial microfluidic membrane processing (IMMP) for in situ fabricating ZIF-8 membranes on Torlon hollow fibers in the module for membrane fabrication as well as permeation. The dilute Zn^{2+}/1-octanol solution flowed through the bore side for the sufficiently replenished supply of Zn^{2+}, while the concentrated 2-mIm aqueous solution was static on the shell side. The ZIF-8 membrane's location was shown to be at the inner surface of the fiber since the transport rate of the limiting Zn^{2+} reactant toward the shell side was hindered by miscibility of water and 1-octanol. The ZIF-8 membrane showed high H_2/C_3H_8 and C_3H_6/C_3H_8 separation factors (~370 at 120°C and ~12 at 25°C, respectively), suggesting the low defect densities. The IMMP approach has successfully applied to prepare three hollow fibers simultaneously, demonstrating the highly promising potential for scalability.

11.2.3 MOF Based Mixed Matrix Membranes

Mixed matrix membranes (MMMs) are generally polymer/ inorganic composites consisting of a primary polymer phase and a secondary phase of dispersed inorganic particles. Hence, MMMs have been expected to overcome the drawbacks of polymeric membranes (i.e., the upper bond limit, plasticization and aging) and inorganic membranes (i.e., lack of reproducibility and high production cost). The field of MMMs is developing with the surprising rapidity, since the rich diversity of polymers and inorganic fillers offer such a wide variety of MMMs.

Figure 11.5 shows two kinds of MMMs in different configurations: symmetric flat dense mixed matrix membrane and asymmetric hollow fiber (Goh et al., 2011). Typically, the manufacture of MMMs is easier and cheaper than that of inorganic membranes. The dense MMMs are normally prepared by casting a solution of polymer and well dispersed inorganic fillers, and then followed by evaporating the solvent in given environment. The researchers in the field of polymeric membranes are inclined to prepare MOF based asymmetric MMMs constructed by a very thin selective layer on the non-selective porous supports. Spiral wound flat sheets and hollow fiber are two frequently used membrane modules. Although the asymmetric structure was first commercialized as flat sheet membranes, Hollow fiber with ten times higher packing density (over 10000 $m^2\,m^{-3}$) and 5 to 20 times lower fabrication costs than spiral wound flat sheets is much more preferred for real industrial application. In this chapter, we mainly focus on asymmetric hollow fiber membranes (Hu et al., 2010; Zhang et al., 2014), but several reports about asymmetric flat membranes (Basu et al., 2010, 2011; Ren et al., 2012) will also be given here for comparison. MOF based asymmetric hollow fiber membranes are usually prepared by the phase

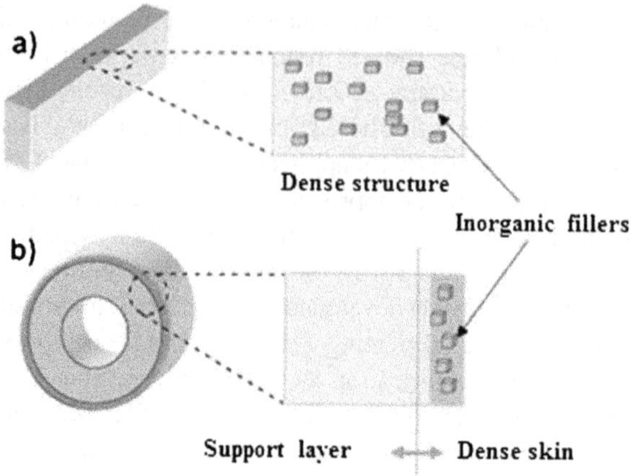

FIGURE 11.5 Mixed matrix membranes in configuration. (a) Symmetric flat dense mixed matrix membrane. (b) Asymmetric hollow fiber. (Adapted from Goh, P.S. et al., Separation and Purification Technology. 81, 243–264, 2011.)

inversion spinning (also called dry jet-wet quench spinning or wet spinning). More specifically, once the dope solution with well dispersed MOF fillers was prepared, the spinning process begins taking place, the spinning solution and a bore fluid are coextruded through a spinneret and precipitated via phase inversion induced by the external coagulation bath to create the asymmetric structure. Zhang et al. (2014) have employed the conventional dry-jet/wet-quench fiber spinning technique for synthesizing dual-layer ZIF-8/6FDA-DAM mixed-matrix hollow fiber membranes that showed significantly enhanced C_3H_6/C_3H_8 selectivity.

Despite their potential for gas separation, MOF based MMMs are still facing many challenges and limitations. MOF based MMMs have reduced permeability compared with their pure MOF counterparts because most of the polymer phase has low gas permeabilities due to its non-porous nature. This limitation can be effectively alleviated by increasing the MOF loadings to form MOF-dominant rather than polymer-dominant composites. Besides, void defects at the particle/polymer interface that are generated from the aggregation of MOF particles and poor dispersion in MMMs will provide no-selective permeation pathways for gas pairs, leading to reduced separation performance. All the two challenges actually involve the same issue, namely two-phase compatibility. Therefore, a good adhesion between polymer phase and inorganic fillers is highly demanded for preparing high performance MMMs. MOFs with partially organic nature offer potential advantages over zeolites as promising additives for MMMs. Furthermore, MOFs display a very high level of tunability, as their size, shape and chemical functionalities can be precisely tuned by the choice of appropriate starting components (metal ions and organic ligands) or post-synthetic modification (Wang and Cohen, 2009), which offers great opportunities to enhance the separation performance of MMMs. Zhang et al. (2012a) prepared a well-performed mixed matrix membrane with 6FDA-DAM and 200 nm BASF ZIF-8 particles without any surface-treating of ZIF-8. The SEM images of as-synthesized MMMs exhibited the absence of "sieve-in-a-cage" morphology, suggesting the good contact of bare ZIF-8 with the 6FDA-DAM matrix. The prepared ZIF-8/6FDA-DAM MMMs with high ZIF-8 loading up to 48.0 wt% exhibited 258% and 150% increases in C_3H_6 permeability and C_3H_6/C_3H_8 ideal selectivity compared to pure 6FDA-DAM membranes, respectively. The results verified that the ZIF-8 particles adhere well with 6FDA-DAM at the polymer-sieve interface. MOFs fillers are expected to provide ideal gas permeation pathways in the MMMs: aspects of particle size and shape control are a key for the filler-polymer integration under certain circumstances. Rodenas et al. (2015) reported three MMMs composed of polyimide (PI)-Matrimid® 5218 and the copper 1,4-benzenedicarboxylate (CuBDC) MOF crystals with different shape. Figure 11.6 shows the CO_2/CH_4 separation performance of the resulting three kinds of CuBDC/PI MMMs. The incorporation of both bulk and nanoparticle CuBDC crystals into the polyimide matrix worsened the separation selectivity as compared with a neat polyimide reference membrane, which might be attributed to the generation of unselective nano- or microvoids at the particle/polymer interface due to the disruption of the polymer chains. Contrarily, the separation selectivity achieved with nanosheet (ns)-CuBDC@PI is 30-80% higher than that with polymeric membrane. What is more interesting is that the CO_2/CH_4 selectivity of ns-CuBDC@PI retained or even increased slightly with the increasing upstream pressure. Obviously, the remarkable performance should be credited with

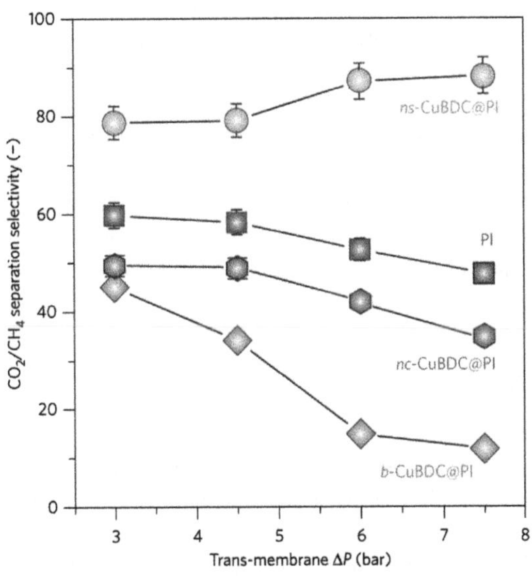

FIGURE 11.6 Separation selectivity, defined as the ratio between the permeability of CO_2 and CH_4, as a function of the pressure difference over the MOF-polymer composite membranes in the separation of CO_2 from an equimolar CO_2/CH_4 mixture at 298K. The MOF-polymer composite membranes are labeled as *x*-CuBDC@PI, where *x* is *ns*, and *b* or *nc* is CuBDC nanosheet, bulk and nanoparticle crystals, respectively. For comparison purposes, results for a neat polyimide membrane (PI) are also presented. (Adapted from Rodenas, T. et al., Nature Materials. 14, 48–55, 2015.)

the beneficial role of the MOF nanosheets as filler material. Additionally, Bachman et al. (2016a) demonstrated that smaller particle sizes are beneficial for minimizing the number of non-selective pathways for gas transport owing to the greater fraction of the polymer at the nanocrystal interface. Postsynthetic polymerization (PSP) of MOF crystals is a promising route to bridge the gap between pure-MOF polymers. The phase interaction is greatly enhanced through the formation of covalent integration between MOF and polymer phases. For example, the PSP method is especially effective in the formation of UiO-66-NH_2/polyurethane PSP membrane by copolymerizing methacrylamide groups functionalized UiO-66-NH_2 microcrystals with butyl methacrylate (Zhang et al., 2015).

11.3 MOF MEMBRANES FOR GAS SEPARATION

Polymer membranes dominated the total gas separation market, partially because of their low production costs and ease of processing and scalability. However, polymers suffer from a fundamental trade-off between permeability and selectivity, which is often referred to as Robeson's "upper bound" (Robeson, 1991, 2008). For decades, substantial numbers of inorganic membranes and mixed matrix membranes have been developed to surpass Robeson's upper bound. MOF membranes with well-defined, regular pore structure have drawn intensive and extensive attention since they may possess high permeability and selectivity simultaneously. In addition, MOF membranes benefit from structural and chemical tailorability for enhancing the separation performance.

During the last 30 years, more than 90% of the gas separation applications has been used for separating mixtures containing noncondensable gases (N_2/O_2, CO_2/CH_4, H_2/N_2, H_2/CH_4, etc.). Baker (2002) discussed the future directions of membrane gas separation technology, and predicted that it will involve the condensable gas separations such as the C_{3+} hydrocarbons from methane or hydrogen, propylene from propane and n-butane from isobutane. In the following section, pure MOF membranes and MOF/polymer composite membranes for relevant gas separation processes will be discussed, aiming at the search for and development of promising candidates with high permeability, selectivity and (hydrothermal) stability.

11.3.1 H_2 Purification

Hydrogen, with the highest energy content per unit weight as compared to other fuels, has the potential to provide a solution to address issues related to energy security and environmental pollution. Generally, many byproducts are associated with the production of hydrogen in industrial process, leading to the high demand of separating hydrogen from other light gases (CO_2, CH_4, C_3H_8, etc.). The commercial separation techniques, mainly pressure swing adsorption (PSA) and cryogenic distillation, are rather energy intensive. As a promising alternative, energy saving membrane technology has currently been studied intensively for the production of high-purity hydrogen. Accordingly, metal-organic frameworks with well-defined pore structures have been employed in the fabrication of hydrogen-selective membranes. Here we mainly focus on the H_2/CO_2 gas pairs. Some of the pioneering research of polycrystalline MOF membranes has been listed in Table 11.1.

TABLE 11.1
Summary of Polycrystalline MOF Membranes for H_2/CO_2 Separation

MOF	Substrate	Temp. (°C)	H_2 Permeance ($mol\ m^{-2}\ s^{-1}\ Pa^{-1}$)	Separation Factor	References
HKUST-1	Copper net	25	1.00×10^{-6}	6.8	Guo et al., 2009
ZIF-90	α-Al_2O_3 disks	200	2.51×10^{-7}	7.3	Huang et al., 2010a
ZIF-22	TiO_2 disks	50	1.60×10^{-7}	7.2	Huang et al., 2010b
ZIF-7	α-Al_2O_3 disks	200	7.60×10^{-8}	6.5	Li et al., 2010a
ZIF-7	α-Al_2O_3 disks	200	9.00×10^{-9}	8.4	Li et al., 2010b
ZIF-7	α-Al_2O_3 disks	220	4.50×10^{-8}	13.6	Li et al., 2010c
HKUST-1	Porous Al_2O_3	25	6.74×10^{-7}	4.6	Nan et al., 2011
MOF-5	α-Al_2O_3 disks	25	4.37×10^{-7}	4.2[a]	Zhao et al., 2011
HKUST-1	Stainless steel net	25	1.13×10^{-6}	9.2	Ben et al., 2012
ZIF-90	Torlon hollow fiber	35	2.00×10^{-7}	1.8[a]	Brown et al., 2012
ZIF-78	Porous ZnO	25	1.00×10^{-7}	9.5	Dong et al., 2012
ZIF-8	Porous SiO_2	25	3.23×10^{-7}	7.3	Fan et al., 2012
ZIF-95	α-Al_2O_3 disks	325	1.95×10^{-6}	25.7	Huang et al., 2012
Ni-MOF-74	α-Al_2O_3 disks	RT	1.27×10^{-5}	9.1[a]	Lee et al., 2012
NH_2-MIL-53(Al)	Porous SiO_2	15	2.00×10^{-6}	30.9	Zhang et al., 2012b
HKUST-1	α-Al_2O_3 hollow fiber	40	4.10×10^{-8}	13.6	Zhou et al., 2012
$Cu(bipy)_2(SiF_6)$	Porous SiO_2	20	2.7×10^{-7}	8.0	Fan et al., 2013
HKUST-1	Free standing	RT	1.50×10^{-6}	6.1	Mao et al., 2013
HKUST-1	PSF	25	7.90×10^{-8}	7.2[a]	Nagaraju et al., 2013
ZIF-9-67	α-Al_2O_3 disks	RT	1.41×10^{-5}	8.89[a]	Zhang et al., 2013
CAU-1	α-Al_2O_3 hollow fiber	25	1.00×10^{-7}	12.3	Zhou et al., 2013
$Zn_2(bim)_4$	α-Al_2O_3 disks	RT	7.74×10^{-7}	230	Peng et al., 2014
ZIF-7	α-Al_2O_3 disks	150	3.05×10^{-7}	18.3	Aceituno Melgar et al., 2014
$Zn(BDC)(TED)_{0.5}$	α-Al_2O_3 disks	180	2.65×10^{-6}	12.1	Huang et al., 2014a
ZIF-8	Stainless steel nets	100	2.10×10^{-5}	8.1	Huang et al., 2014b
ZIF-8@GO	α-Al_2O_3 disks	250	1.30×10^{-7}	14.9	Huang et al., 2014c
JUC-150	Nickel screen	RT	1.83×10^{-7}	38.7	Kang et al., 2014
ZIF-7	PVDF hollow fiber	RT	2.35×10^{-6}	18.43[a]	Li et al., 2014a
ZIF-8	PVDF hollow fiber	RT	2.01×10^{-6}	16.29[a]	Li et al., 2014a
HKUST-1	PAN hollow fiber	20	7.05×10^{-5}	7.14	Li et al., 2014b
HKUST-1	PSF hollow fiber	RT	4.85×10^{-7}	21.03[a]	Li et al., 2014c

(Continued)

TABLE 11.1 (CONTINUED)
Summary of Polycrystalline MOF Membranes for H_2/CO_2 Separation

MOF	Substrate	Temp. (°C)	H_2 Permeance (mol m^{-2} s^{-1} Pa^{-1})	Separation Factor	References
ZIF-8	α-Al_2O_3 tube	100	5.50×10^{-8}	7.8	Drobek et al., 2015
HKUST-1	PVDF hollow fiber	RT	6.01×10^{-6}	7.91	Li et al., 2015
ZIF-8	PVDF hollow fiber	RT	1.90×10^{-6}	12.42	Li et al., 2015
ZIF-7	PVDF hollow fiber	RT	1.02×10^{-6}	15.86	Li et al., 2015
ZIF-8	BPPO	RT	2.05×10^{-6}	12.8[a]	Shamsaei et al., 2015
ZIF-100	α-Al_2O_3 disks	25	5.8×10^{-8}	72	Wang et al., 2015a
MOF-74(Mg)	α-Al_2O_3 disks	25	8.00×10^{-8}	28	Wang et al., 2015b
[COF-300]-[$Zn_2(bdc)_2(dabco)$]	Porous SiO_2	RT	4.48×10^{-7}	12.6	Fu et al., 2016
[COF-300]-[ZIF-8]	Porous SiO_2	RT	3.79×10^{-7}	13.5	Fu et al., 2016
HKUST-1	AAO	RT	1.67×10^{-6}	5.63	Guo et al., 2016
ZIF-8	PVDF hollow fiber	RT	2.01×10^{-5}	7[a]	Hou et al., 2016
ZIF-9	α-Al_2O_3 disks	25	7.43×10^{-6}	14.74[a]	Huang et al., 2016
CAU-10-H	α-Al_2O_3 disks	200	3.80×10^{-9}	10.5	Jin et al., 2016
MIL-96(Al)	α-Al_2O_3 disks	RT	5.30×10^{-7}	9	Knebel et al., 2016
$Ni_2(mal)_2(bpy)$	α-Al_2O_3 disks	25	1.55×10^{-7}	89[a]	Li et al., 2016b
CuBTC/MIL-100	PVDF hollow fiber	85	1.05×10^{-7}	89	Li et al., 2016c
HKUST-1	Ni foam	25	2.72×10^{-6}	6.8	Sun et al., 2016
$Co_3(HCOO)_6$	Ni foam	25	2.09×10^{-6}	6.0	Sun et al., 2016
ZIF-9	Ni hollow fiber	35	2.30×10^{-8}	14.1	Cacho-Bailo et al., 2017
UiO-66	α-Al_2O_3 disks	25	5.30×10^{-7}	5.1	Friebe et al., 2017
HKUST-1	α-Al_2O_3 disks	RT	1.00×10^{-7}	7	Hurrle et al., 2017
$Ni_2(L\text{-}asp)_2(bpe)$	Nickel mesh	25	1.02×10^{-6}	24.3	Kang et al., 2017b
$Zn_2(Bim)_3$	α-Al_2O_3 disks	120	8.00×10^{-7}	166	Peng et al., 2017
MAMS-1	AAO	20	1.88×10^{-7}	235	Wang et al., 2017a

[a] Ideal separation factor.

The membrane material appears to have been favored in classic MOFs such as HKUST-1 and ZIFs. As shown by the early work of Guo et al. (2009), the copper net supported HKUST-1 membrane prepared by means of a "twin copper source" technique exhibited good permeation selectivity for H_2 ($H_2/N_2 = 7$, $H_2/CO_2 = 6.8$ and $H_2/CH_4 = 5.9$). The slightly unsatisfied selectivity might be caused by the larger pore size (9 Å) than common gas molecules. Given that the pore size of ZIF-7 (0.3 nm) is just between the size of H_2 (0.29 nm) and CO_2 (0.33 nm), ZIF-7 is expected as a promising candidate for the development of a H_2-selective membrane. Li et al. (2010a) reported the successful synthesis of ZIF-7 membrane using a microwave-assisted secondary growth technique (Figure 11.7). Before the gas permeation tests, an on-stream activation was carried out to remove the guest molecules within the cavities. The separation factors of H_2/N_2, H_2/CO_2 and H_2/CH_4 binary mixtures were 7.7, 6.5 and 5.9, respectively (at 200°C and 1 bar), i.e., higher than the corresponding Knudsen separation factors (3.7, 4.7 and 2.8, respectively). 2D materials with a layer-like morphology and nanoporous "perforations" spanning the layer thickness may provide unprecedented opportunity for the gas separation process with ultrahigh selectivity and permeability. Peng et al. (2014) constructed 2D layered MOF membranes with $Zn_2(bim)_4$ nanosheets as building blocks, and the ultrathin membrane exhibited H_2/CO_2 selectivity greater than 200, as well as a high permeance up to 2700 gas permeation units (Figure 11.8). $Zn_2(bim)_4$ MOF membranes showed good stability after a 120-hour test under hydrothermal condition (equimolar H_2/CO_2 feed containing ~4 mol% steam, 150°C). Very recently, Peng et al. (2017) prepared a sub-10 nm-thick ultrathin $Zn_2(Bim)_3$ nanosheet based membrane with exceptional H_2/CO_2 separation performance as well.

Limited MOF MMMs showed good H_2/CO_2 selectivity due to the stronger interactions of CO_2 with majorities of polymers (Yang et al. 2011, 2012, 2013a, 2013b; Cao et al., 2013; Kang et al., 2015). Cao et al. (2013) synthesized a thin and compact mixed matrix membrane containing CAU-1-NH_2 and poly(methyl methacrylate) polymer. The as-prepared CAU-1-NH_2/PMMA membrane exhibited H_2/CO_2 selectivity of 13 and high H_2 permeability of 1.1 × 10^4 barrer. The ZIF-8/PBI nanocomposite membrane synthesized by Yang et al. (2013a) exhibited H_2/CO_2 selectivity of 26.3 and high H_2 permeability of 470 barrer. Both the MMMs mentioned here surpassed Robeson's upper bound. In general, the fabrication of MOF MMMs with excellent H_2/CO_2 separation performance is still challenging.

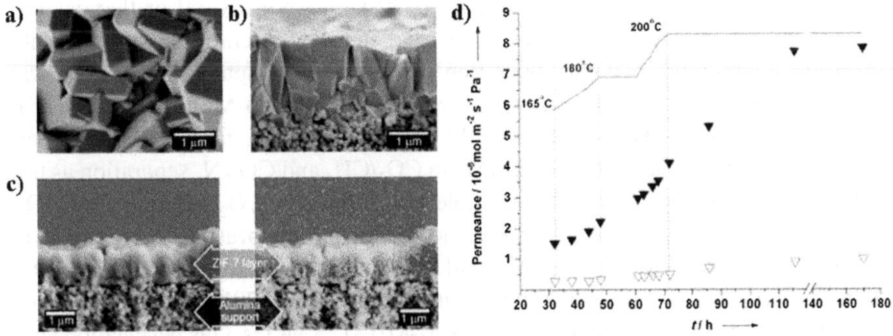

FIGURE 11.7 (a) Top view and (b) cross-section SEM images of the ZIF-7 membrane; (c) EDXS mapping of the ZIF-7 membrane, orange Zn, cyan Al; (d) H_2 (solid triangles) and N_2 (triangles) permeances from the 1:1 mixture through the ZIF-7 membrane during the on-stream activation process with increasing temperature. (From Li, Y. et al., Angewandte Chemie International Edition. 49, 558–561, 2010a.)

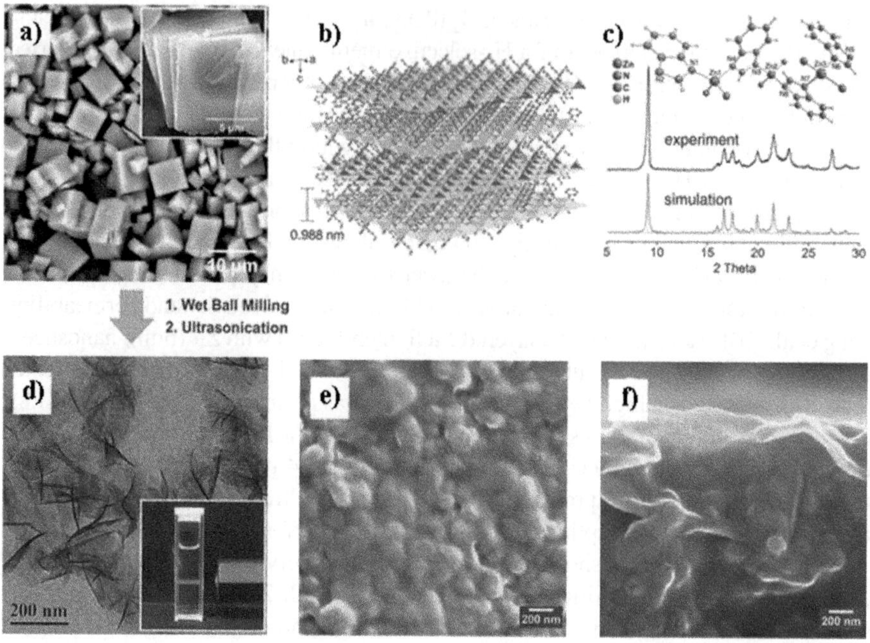

FIGURE 11.8 (a) Scanning electron microscopy (SEM) image of as-synthesized $Zn_2(bim)_4$ crystals. (b) Architecture of the layered MOF precursor. (c) Powder XRD patterns of $Zn_2(bim)_4$. (d) Transmission electron microscopy (TEM) image of $Zn_2(bim)_4$ nanosheets. (e) SEM top view and (f) cross-sectional view of a $Zn_2(bim)_4$ nanosheet layer on α-Al_2O_3 support. (Based on Peng, Y. et al., Science. 346, 1356–1359, 2014.)

11.3.2 CO_2 Separation

In recent years, the purification and recovery of CO_2 from natural or flue gas has attracted worldwide attention in academia and industry. Membrane systems with versatility, adaptability and easy operation are extremely adaptable to CO_2 removal process. MOF based MMMs occupy the vast majority of all MOF membranes for CO_2 separations from natural gas (mainly CH_4) and flue gas (mainly N_2). The performance of different MOF based MMMs in CO_2/CH_4 and CO_2/N_2 separation as has been reported so far is summarized in Table 11.2. Typically, CO_2 separation by MOF materials is adsorption-based rather than molecular sieve separation. Therefore, decoration of MOF structures with functional groups that have strong affinity for CO_2 enables improved CO_2 adsorptive selectivity. Anjum et al. (2016) have synthesized CO_2 selective MMMs with MIL-125 and NH_2-MIL-125 MOFs. Gas permeation tests indicated that NH_2-MIL-125 based MMMs outperformed MIL-125 MMMs both in permeability and CO_2/CH_4 selectivity.

Figures 11.9 and 11.10 can give us a rough idea of the comprehensive separation performance of most reported MMMs with MOF fillers for CO_2/CH_4 and CO_2/N_2 separation, respectively. Obviously, only a few MMMs such as IL@ZIF-8/PSF, ZIF-71/UV-PIM-1, ZIF-71/6FDA-Durene and NH_2-UiO-66/PAO-PIM-1can transcend

TABLE 11.2
Summary of MOF Based MMMs for CO_2/CH_4 and CO_2/N_2 Separation

MOF	Polymer	Temp. (°C)	p (Bar)	P_{CO2} (Barrer)	Separation Factor	References
HKUST-1	PSF			7.5	CO_2/CH_4 (22[a])	Car et al., 2006
				7.5	CO_2/N_2 (25[a])	
HKUST-1	PDMS	–	–	3000	CO_2/N_2 (8.8[a])	
MOF-5	Matrimid®5218	35	2	20.2	CO_2/CH_4 (44.7[a])	Perez et al., 2009
				20.2	CO_2/N_2 (38.8[a])	
ZIF-90	6FDA-DAM	25	2	720	CO_2/CH_4 (37)	Bae et al., 2010
ZIF-8	Matrimid®5218	RT	1	8.0	CO_2/CH_4 (80.8[a])	Ordoñez et al., 2010
CuTPA	PVAc	35	4.5	3.3	CO_2/CH_4(40.4[a])	Adams et al., 2010
				3.3	CO_2/N_2 (35.4[a])	
ZIF-8	PPEES	30	1	50	CO_2/CH_4 (20.8[a])	Díaz et al., 2011
				50	CO_2/N_2 (24.5[a])	
NH_2-MIL-53	PSF	–10	10	2.4	CO_2/CH_4 (117)	Zornoza et al., 2011a
HKUST-1/S1C	PSF	35	2.75	8.9	CO_2/CH_4 (22.4)	Zornoza et al., 2011b
				8.4	CO_2/N_2 (38)	
ZIF-8	PSF	35	2.75	12.1	CO_2/CH_4 (19.8)	
				12.3	CO_2/N_2 (19.5)	
NH_2-MIL-53	6FDA-ODA	35	20	14.5	CO_2/CH_4 (54)	Chen et al., 2012
NH_2-UiO-66	6FDA-ODA	35	10	13.7	CO_2/CH_4 (44.7)	Nik et al., 2012
NH_2-MOF-199				26.6	CO_2/CH_4 (52.4)	
UiO-67				20.8	CO_2/CH_4 (15.0)	
ZIF-8	Matrimid®5218	22	4	16.6	CO_2/CH_4 (35.8[a])	Song et al., 2012
				13.7	CO_2/N_2 (21.6[a])	
ZIF-8	6FDA-Durene/ DABA	35	20	728	CO_2/CH_4 (19.6)	Askari et al., 2013
ZIF-8	PIM-1	RT	1	4270	CO_2/CH_4 (18.6[a])	Bushell et al., 2013
				6300	CO_2/N_2 (18[a])	
$Cu_3(BTC)_2$	PPO	30	–	87	CO_2/CH_4 (28[a])	Ge et al., 2013
				87	CO_2/N_2 (23.5[a])	
ZIF-7	Pebax®1657	25	3.75	111	CO_2/CH_4 (30[a])	Li et al., 2013
				111	CO_2/N_2 (97[a])	
MIL-68(Al)	PSF	35	–	4.7	CO_2/CH_4 (36.5)	Seoane et al., 2013
ZIF-8	6FDA-durene	35	3.5	1153	CO_2/CH_4 (11.0[a])	Wijenayake et al., 2013
				1153	CO_2/N_2 (11.3[a])	
CPO-27(Mg)	XLPEO	25	2	250	CO_2/N_2 (25[a])	Bae and Long, 2013
	PI			850	CO_2/N_2 (23[a])	
NH_2-MIL-53	PMP	30	2	210	CO_2/CH_4 (17)	Abedini et al., 2014
MOF-5	PEI	25	6	5.4	CO_2/CH_4 (23.4[a])	Arjmandi et al., 2014
				5.4	CO_2/N_2 (28.4[a])	
ZIF-8	PBI-BuI	35	20	5.4	CO_2/CH_4 (45[a])	Bhaskar et al., 2014
				5.4	CO_2/N_2 (16[a])	
ZIF-8	DMPBI-BuI	35	20	55	CO_2/CH_4 (15[a])	

(Continued)

TABLE 11.2 (CONTINUED)
Summary of MOF Based MMMs for CO_2/CH_4 and CO_2/N_2 Separation

MOF	Polymer	Temp. (°C)	p (Bar)	P_{CO2} (Barrer)	Separation Factor	References
				55	CO_2/N_2 (12[a])	
ZIF-8	DBzPBI-BuI	35	20	90	CO_2/CH_4 (12[a])	
				90	CO_2/N_2 (14[a])	
HKUST-1	ODPA-TMPDA	35	2	260	CO_2/CH_4 (27.8[a])	Duan et al., 2014
ZIF-71	6FDA-Durene	35	3.5	3435	CO_2/CH_4 (16)	Japip et al., 2014
				4006	CO_2/CH_4 (12.8[a])	
NH_2-MIL-53	Matrimid®5218	0	3	3.6	CO_2/CH_4 (106)	Rodenas et al., 2014
NH_2-MIL-101				3.0	CO_2/CH_4 (99)	
Silica-(ZIF-8) spheres	PSF	35	3.3	26.0	CO_2/CH_4 (31)	Sorribas et al., 2014
MIL-53(Al)	Matrimid®5218	35	2	40.0	CO_2/CH_4 (90.1[a])	Hsieh et al., 2014
				40.0	CO_2/N_2 (95.2[a])	
ZIF-8	Pebax®2533	RT	2	1098	CO_2/N_2 (33)	Nafisi et al., 2014a
ZIF-8	6FDA-durene	25	2	2185	CO_2/CH_4 (17.1[a])	Nafisi et al., 2014b
				2185	CO_2/N_2 (17.0[a])	
IL@ZIF-8	PSF	30	6	291	CO_2/CH_4 (38.3)	Ban et al., 2015
				351	CO_2/N_2 (116)	
$Zn(pyrz)_2(SiF_6)$	XLPEO	25	1	590	CO_2/CH_4 (30)	Gong et al., 2015
				670	CO_2/N_2 (29)	
ZIF-71	UV-PIM-1	35	7	3020	CO_2/CH_4 (28.8)	Hao et al., 2015
				3458	CO_2/N_2 (26.9[a])	
HKUST-1	aBPDA-P1	35	2	4.5	CO_2/CH_4 (43)	Hegde et al., 2015
ZIF-8	PVC-g-POEM	35		623	CO_2/CH_4 (11.2[a])	Hwang et al., 2015
MIL-101	Matrimid®5218	35	10	7.0	CO_2/CH_4 (56)	Naseri et al., 2015
				7.0	CO_2/N_2 (53)	
ns-CuBDC	PI	25	8.5	2.8	CO_2/CH_4 (88)	Rodenas et al., 2015
Ni_2(dobdc)	6FDA-DAM:DAT	35	1	220	CO_2/CH_4 (24)	Bachman and long, 2016b
IL@ZIF-8	Pebax®1657	25	1	105	CO_2/CH_4 (34.8)	Li et al., 2016a
				105	CO_2/N_2 (83.9)	
NH_2-MIL-53	6FDA-DAM	25	3	660	CO_2/CH_4 (28)	Sabetghadam et al., 2016
NH_2-UiO-66	PSF	35	3	45	CO_2/CH_4 (24[a])	Su et al., 2016
				45	CO_2/N_2 (26[a])	
MIL-101/ZIF-8	PSF	35	2	14	CO_2/CH_4 (40)	Tanh Jeazet et al., 2016
ZIF-8@PD	PI	35	1	630	CO_2/CH_4 (27)	Wang et al., 2016b
				695	CO_2/N_2 (21)	
NH_2-MIL-125	Matrimid®9725	35	9	17	CO_2/CH_4 (50)	Anjum et al., 2016
IL@ NH_2-MIL-101(Cr)	PIM-1	25	3	2979	CO_2/N_2 (37[a])	Ma et al., 2016

(*Continued*)

TABLE 11.2 (CONTINUED)
Summary of MOF Based MMMs for CO_2/CH_4 and CO_2/N_2 Separation

MOF	Polymer	Temp. (°C)	p (Bar)	P_{CO2} (Barrer)	Separation Factor	References
NH_2-UiO-66	PEBA	25	3	87	CO_2/N_2 (66)	Shen et al., 2016
MIL-53(Al)	Ultem1000	25	5	24	CO_2/N_2 (41[a])	Zhu et al., 2016
CuBDC-ns	PIM-1	25	1	268	CO_2/CH_4 (15.6)	Cheng et al., 2017
$[Co_4(\mu_4\text{-}O)(Me_2pzba)_3]$	Matrimid®5218	25	3	13	CO_2/CH_4 (60)	Dechnik et al., 2017c
NH_2-UiO-66	PIM-1	RT	1	4000	CO_2/CH_4 (20)	Khdhayyer et al., 2017
PEG-ran-PPG/ CuBTC	Pebax®MH 1657	30	15	1164	CO_2/CH_4 (32.7[a])	Khosravi et al., 2017
Fe-BTC	Pebax1657	RT	12	89	CO_2/CH_4 (26[a])	Mosleh et al., 2017
				89	CO_2/N_2 (48[a])	
ZIF-12	Matrimid®5218	35	4	12.7	CO_2/CH_4 (66.7[a])	Boroglu et al., 2017a
ZIF-11	6FDA-DAM	30	4	258	CO_2/CH_4 (31.0[a])	Boroglu et al., 2017b
ZIF-8	PVC-POEM	RT		103	CO_2/CH_4 (14.4)	Shin et al., 2017
NH_2-UiO-66	PAO-PIM-1	35	1	8500	CO_2/CH_4 (28)	Wang et al., 2017b
				8500	CO_2/N_2 (34)	
NH_2-ZIF-7	XLPEO	35	5	215	CO_2/CH_4 (50)	Xiang et al., 2017
MIL-53(Al)	PDMS	35	5	72	CO_2/CH_4 (30.5[a])	Zhu et al., 2017
				72	CO_2/N_2 (25.8[a])	

[a] Ideal separation factor.

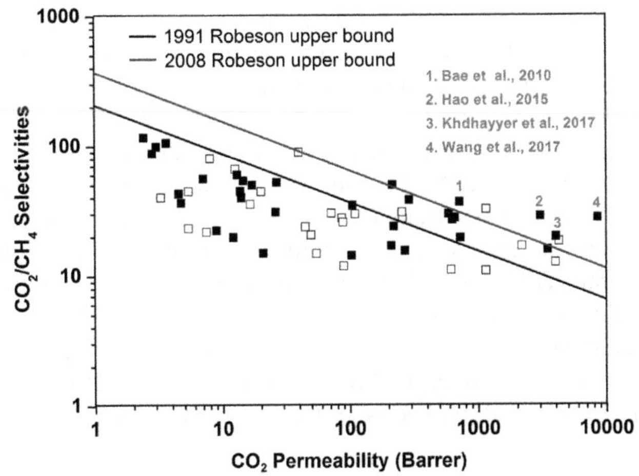

FIGURE 11.9 Summary of comprehensive separation performance of MOF based MMMs for CO_2/CH_4 reported to date. The solid lines are the Robeson upper bound. Open and closed symbols indicate separation data from single and binary gas permeation measurements, respectively.

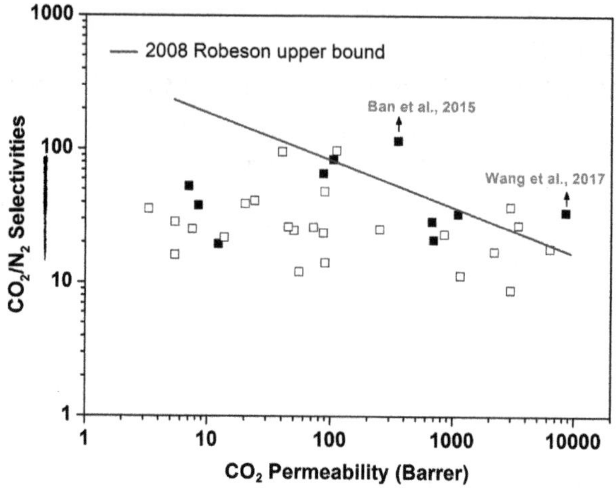

FIGURE 11.10 Summary of comprehensive separation performance of MOF based MMMs for CO_2/N_2 reported to date. The solid line is the Robeson upper bound. Open and closed symbols indicate separation data from single and binary gas permeation measurements, respectively.

the 2008 Robeson upper bond (Japip et al., 2014; Ban et al., 2015; Hao et al., 2015; Khdhayyer et al., 2017; Wang et al., 2017b). The polymer phase is mainly high permeable 6FDA-containg polyimides and microporous polymer PIM-1. From a practical point of view, high permeability is highly crucial to membrane process. In contrast to usually low CO_2 permeability (less than 1000 barrer) of MOF based MMMs, the outstanding PIM-based hybrid membrane with NH_2-UiO-66 as the inorganic filler and amidoxime-functionalized PIM-1 as the polymer matrix fabricated by Wang et al. (2017b) possess very high CO_2 permeability. The well-designed NH_2-UiO-66/PAO-PIM-1 hybrid membranes exhibited excellent gas separation performance with CO_2 permeability of 8500 barrer and mixed selectivities of 28 for CO_2/CH_4 and 34 for CO_2/N_2, which might be attributed to the hydrogen bond network between the two phases.

Polymeric membranes are vulnerable to the plasticization effects that particularly appear with pressurized CO_2 feed steam, which could significantly decrease their performance. Thus, the plasticization of applied MOF MMMs should be mitigated as much as possible for the long-term stability. Fortunately for us, the plasticization effect of membrane at high CO_2 pressure was observed to be partially suppressed upon MOF addition. For example, the NH_2-MIL-53(Al)/PSF MMMs prepared by Zornoza et al. (2011a) showed increased CO_2/CH_4 selectivity with pressure, possibly due to the intrinsic flexibility of the MOF filler. Ban et al. (2015) observed similar phenomena in IL@ZIF-8/PSF MMMs for CO_2/CH_4 separation.

11.3.3 C_3H_6/C_3H_8 Separation

Separation of propylene and propane gases is of fundamental importance to the petrochemical industry. However, the low relative volatility of the two components makes the currently used cryogenic distillation energy intensive, and thus stimulates the energy-efficient and environmentally friendly membrane separation technology. In terms of commercial applications, a minimum propylene permeability of 1 barrer and a propylene selectivity of 35 are required for the propylene-selective separation membrane (Colling et al., 2004). So far, several groups have reported that the ZIF-8 polycrystalline membranes exhibit remarkable propylene/propane separation performance based on the molecular sieving mechanism. Although the size difference between propylene (~4.0 Å) and propane (~4.3 Å) is no more than 0.2 or 0.3 Å, the ratio of their diffusion rate coefficients $D(C_3H_6)/D(C_3H_8)$ in ZIF-8 is 125, indicating that ZIF-8 could effectively separate the two very similar molecules.

Table 11.3 summarizes the polycrystalline ZIF-8 membranes reported to date for propylene/propane separation. The first ZIF-8 membrane that showed excellent propylene/propane separation performance was reported by Pan et al. (2012). The membranes prepared by a hydrothermal seeded growth method showed the propylene permeance of 3.0×10^{-8} mol m^{-2} s^{-1} Pa^{-1} and the propylene/propane separation factor of 35. Later, Jeong's group prepared ZIF-8 membranes with diverse synthesis protocols such as rapid thermal deposition, rapid microwave-assisted seeding and secondary growth, and the resulting continuous well-intergrown membranes display C_3H_6/C_3H_8 selectivities of 30-50 (Shah et al., 2013; Kwon et al., 2013a, 2013b). Very recently, Li et al. (2017) reported the scalable production of ZIF-8 membranes

TABLE 11.3
Summary of Polycrystalline MOF Membranes for C_3H_6/C_3H_8 Separation

MOF	Substrate	Temp. (°C)	C_3H_6 Permeance (mol m^{-2} s^{-1} Pa^{-1})	Separation Factor	References
ZIF-8	α-Al_2O_3 disks	−15	3.00×10^{-8}	35	Pan et al., 2012
ZIF-8	α-Al_2O_3 disks	RT	2.08×10^{-8}	40	Kwon et al., 2013a
ZIF-8	α-Al_2O_3 disks	RT	2.00×10^{-8}	55	Kwon et al., 2013b
ZIF-8	α-Al_2O_3 disks	RT	7.00×10^{-9}	30	Shah et al., 2013
ZIF-8	Torlon hollow fiber	25	1.00×10^{-8}	12	Brown et al., 2014
ZIF-8	α-Al_2O_3 tube	25	2.50×10^{-9}	59[a]	Hara et al., 2014
ZIF-8	α-Al_2O_3 disks	35	1.12×10^{-8}	30	Liu et al., 2014a
ZIF-8	α-Al_2O_3 support	35	6.00×10^{-10}	3.5[a]	Shekhah et al., 2014
ZIF-8	α-Al_2O_3 disks	RT	2.68×10^{-8}	70	Kwon et al., 2015a
ZIF-8/ZIF-67	α-Al_2O_3 support	RT	3.70×10^{-8}	209	Kwon et al., 2015b
ZIF-8	Torlon fiber	25	1.55×10^{-8}	180	Eum et al., 2016a
ZIF-8	Torlon fiber	25	2.28×10^{-8}	65	Eum et al., 2016b
ZIF-8/GO	AAO	25	1.62×10^{-9}	12[a]	Hu et al., 2016
ZIF-8	BPPO	RT	7.50×10^{-9}	28	Shamsaei et al., 2016
ZIF-8	α-Al_2O_3 disks	RT	6.00×10^{-9}	50	Yu et al., 2016
CoZn-ZIF-8	α-Al_2O_3 disks	RT	2.04×10^{-8}	120	Hillman et al., 2017
ZIF-8	PVDF hollow fiber	RT	2.76×10^{-7}	67	Li et al., 2017

[a] Ideal separation factor.

with controllable thickness on PVDF hollow fibers via a gel-vapor deposition (GVD) methodology. The resultant nanometer-thick ZIF-8 membrane showed excellent separation performance for C_3H_6/C_3H_8 mixtures with propylene permeance of 2.76 × 10^{-7} mol m^{-2} s^{-1} Pa^{-1} and selectivity of 73.4. The MOF membrane module with 30 polymeric hollow fibers and effective area of 340 cm^2 can be successfully synthesized with no reduction in selectivities.

Additionally, MMMs comprising ZIF-8 fillers for C_3H_6/C_3H_8 separation have also received researchers' attention with several published studies (Askari et al., 2013; Japip et al., 2014; Zhang et al., 2012a, 2014; Lin et al., 2016). Zhang et al. (2012a) first reported a ZIF-8/6FDA-DAM MMM that lies beyond the reported C_3H_6/C_3H_8 upper bond. The MMM with ZIF-8 loading of 48.0 wt% is superior to the pristine 6FDA-DAM membrane, with C_3H_6 permeability of ~40 and 12 barrer and mixed C_3H_6/C_3H_8 selectivity of 18 and 8, respectively. In the next step, Zhang et al. (2014) prepared highly scalable and high-loading ZIF-8/6FDA-DAM mixed-matrix hollow fiber membranes, bringing the mixed-matrix membrane into an advanced level. Askari et al. (2013) have fabricated a MMM composed of 6FDA-Durene/DABA (9/1) and 40% ZIF-8, which possess a notable ideal C_3H_6/C_3H_8 selectivity of 27.38 and a remarkable C_3H_6 permeability of 47.3 barrer. As can be seen in Figure 11.11, the majority of reported ZIF-8 based MMMs have surpassed the Robeson upper bound, while failed to reach the economically attractive region. Nevertheless, partly polycrystalline ZIF-8 membranes with excellent permeability and selectivity go into the attractive region, carrying a message of hope for the commercialization of polycrystalline ZIF-8 membranes for C_3H_6/C_3H_8 separation.

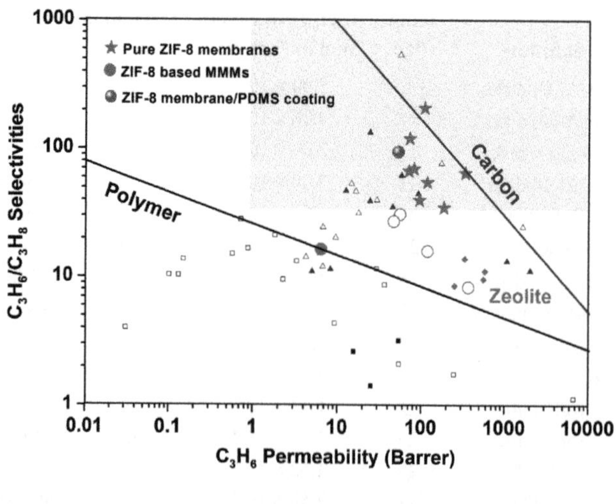

FIGURE 11.11 Comparison of ZIF-8 based membranes with previously reported membranes for the propylene/propane separation performances. The solid lines are the Robeson upper bound. Open and closed symbols indicate separation data from single and binary gas permeation measurements, respectively. Rectangle: polymer membrane; triangle: Carbon membrane; rhombus: zeolite membranes; circle: ZIF-8 based MMMs; star: representative polycrystalline ZIF-8 membranes.

11.4 SUMMARY AND OUTLOOK

The past ten years have witnessed unprecedented prosperity of MOF membranes both in the membrane formation and gas separation processes. Despite the remarkable progress, none of MOF membranes are produced at large scale and used in industrial gas separation applications. Several main challenges that lie ahead in the industrialization of MOF membranes will be described briefly below.

Structural flexibility: MOFs are known for their flexibility as compared to the relatively rigid zeolites. As an example, ZIF-8 can adsorb bulky molecules such as benzene and HMF, despite a pore size estimated from crystallographic analysis to be 3.4 Å (Diestel et al., 2012; Jin et al., 2015). The illuminating work of Ryder et al. (2014) gave an improved understanding of gate-opening and pore-breathing mechanisms of MOF materials by the high-resolution neutron and synchrotron vibrational spectroscopy in combination with ab initio density-functional simulations. However, too much flexibility is undesirable for molecular sieving process. It is well known that most MOF membranes show no clear cut-off in gas separation process owing to their flexible nature. The flexible structure of MOFs is increasingly recognized as a potential drawback for the desired membrane separation process. Researchers have strengthened efforts to control the effective pore size of MOF materials. Thompson et al. (2012) achieved a progressive suppression of the gate-opening phenomena in ZIF-7-8 hybrids by the substitution of ZIF-7 linkers (benzimidazole) in a ZIF-8 framework. Very recently, Sheng et al. (2017) demonstrated an effective manner to hinder the flexibility of ZIF-8 framework by the penetration of polydimethylsiloxane (PDMS) into the underneath ZIF-8 layer. In general, effective control of the gate-opening effects to yield rigid MOF frameworks for desired separation process with high selectivity will be a focus for new technology.

Computational studies: In principle, the number of hypothetical MOF structures is infinitely large due to its tunable property, leading to the difficulty of screening suitable MOF candidates for a specific application. Computational methodologies, as indicated by several excellent reviews, have been a powerful tool for large-scale screening of the best MOFs through identifying structure-property relationships (Düren et al., 2009; Getman et al., 2012; Wilmer et al., 2012). Qiao et al. (2016) reported a computational study to identify the 5 best MOFs for the membrane separation of $CO_2/N_2/CH_4$ mixture by high-throughput screening of 137953 MOFs. To give reliable predictions for various properties of MOFs, the following four issues facing molecular modeling should be addressed in the near future: (1) development of approaches for constructing hypothetical MOFs, (2) method establishment for reliable atomic partial charge estimation, (3) modeling of MOFs with open metal sites or highly flexible frameworks and (4) modeling of membrane-based separation, should be addressed in the near future (Yang et al., 2013c).

Up-scaling of MOF membranes: To date, the scalable fabrication of MOF membranes remains a key issue. In terms of synthesizing large-area membrane with high quality, MOF based MMMs are undoubtedly closer to industrial applications than polycrystalline MOF membranes. MOF-polymer asymmetric hollow fiber membranes with the most scalable membrane geometry, however, have received relatively little attention. In addition, the majority of MOF based MMMs hardly go beyond the upper bound limit, as shown in Figures 11.9 and 11.10, leading to the need for more research. The exploitation of scalable and inexpensive membrane fabrication method is the main barrier facing the commercialization of polycrystalline MOF membranes with promising high permeability and selectivity. Interfacial microfluidic membrane-processing (IMMP) methodology has been identified as a novel strategy for preparing well-defined and continuous ZIF-8 membranes on Torlon hollow fibers. IMMP was successfully applied to the simultaneous processing of three hollow fibers with identical separation behavior to the single-fiber case. This membrane processing approach is a notable step toward realizing scalable molecular sieving MOF membranes. Additionally, several other techniques such as electrospray deposition technique (Aceituno Melgar et al., 2014) and microfluidic synthesis (Cacho-Bailo et al., 2015) facilitate the realization of industrially attractive MOF membranes as well.

In general, molecular design of new MOF materials that combine experimental and computational tools, precise control of membrane microstructure, along with simple and cheap large-scale synthesis methods make the future of MOF membranes in industrial gas separation not so far away.

REFERENCES

Abedini, R., Omidkhah, M., Dorosti, F., 2014. Highly permeable poly(4-methyl-1-pentyne)/NH_2-MIL 53 (Al) mixed matrix membrane for CO_2/CH_4 separation. RSC Advances. 4, 36522–36537.

Aceituno Melgar, V.M., Kwon, H.T., Kim, J., 2014. Direct spraying approach for synthesis of ZIF-7 membranes by electrospray deposition. Journal of Membrane Science. 459, 190–196.

Adams, R., Carson, C., Ward, J., Tannenbaum, R., Koros, W., 2010. Metal organic framework mixed matrix membranes for gas separations. Microporous and Mesoporous Materials. 131, 13–20.

Anjum, M.W., Bueken, B., De Vos, D., Vankelecom, I.F.J., 2016. MIL-125(Ti) based mixed matrix membranes for CO_2 separation from CH_4 and N_2. Journal of Membrane Science. 502, 21–28.

Arjmandi, M., Pakizeh, M., 2014. Mixed matrix membranes incorporated with cubic-MOF-5 for improved polyetherimide gas separation membranes: Theory and experiment. Journal of Industrial and Engineering Chemistry. 20, 3857–3868.

Askari, M., Chung, T.S., 2013. Natural gas purification and olefin/paraffin separation using thermal cross-linkable co-polyimide/ZIF-8 mixed matrix membranes. Journal of Membrane Science. 444, 173–183.

Bachman, J.E., Smith, Z.P., Li, T., Xu, T., Long, J.R., 2016a. Enhanced ethylene separation and plasticization resistance in polymer membranes incorporating metal-organic framework nanocrystals. Nature Materials. 15, 845–849.

Bachman, J.E., Long, J.R., 2016b. Plasticization-resistant Ni_2(dobdc)/polyimide composite membranes for the removal of CO_2 from natural gas. Energy & Environmental Science. 9, 2031–2036.

Bae, T.H., Lee, J.S., Qiu, W., Koros, W.J., Jones, C.W., Nair, S., 2010. A high-performance gas-separation membrane containing submicrometer-sized metal-organic framework crystals. Angewandte Chemie International Edition. 122, 10059–10062.

Bae, T.H., Long, J.R., 2013. CO_2/N_2 separations with mixed-matrix membranes containing Mg_2(dobdc) nanocrystals. Energy & Environmental Science. 6, 3565–3569.

Baker, R.W., 2002. Future directions of membrane gas separation technology. Industrial & Engineering Chemistry Research. 41, 1393–1411.

Ban, Y., Li, Z., Li, Y., Peng, Y., Jin, H., Jiao, W., Guo, A., Wang, P., Yang, Q., Zhong, C., Yang, W., 2015. Confinement of ionic liquids in nanocages: Tailoring the molecular sieving properties of ZIF-8 for membrane-based CO_2 capture. Angewandte Chemie International Edition. 54, 15483–15487.

Basu, S., Cano-Odena, A., Vankelecom, I.F.J., 2010. Asymmetric Matrimid®/[$Cu_3(BTC)_2$] mixed-matrix membranes for gas separations. Journal of Membrane Science. 362, 478–487.

Basu, S., Cano-Odena, A., Vankelecom, I.F.J., 2011. MOF-containing mixed-matrix membranes for CO_2/CH_4 and CO_2/N_2 binary gas mixture separations. Separation and Purification Technology. 81, 31–40.

Ben, T., Lu, C., Pei, C., Xu, S., Qiu, S., 2012. Polymer-supported and free-standing metal-organic framework membrane. Chemistry - A European Journal. 18, 10250–10253.

Bhaskar, A., Banerjee, R., Kharul, U., 2014. ZIF-8@PBI-BuI composite membranes: Elegant effects of PBI structural variations on gas permeation performance. Journal of Materials Chemistry A. 2, 12962–12967.

Boroglu, M.S., Ugur, M., Boz, I., 2017a. Enhanced gas transport properties of mixed matrix membranes consisting of Matrimid and RHO type ZIF-12 particles. Chemical Engineering Research & Design. 123, 201–213.

Boroglu, M.S., Yumru, A.B., 2017b. Gas separation performance of 6FDA-DAM-ZIF-11 mixed-matrix membranes for H_2/CH_4 and CO_2/CH_4 separation. Separation and Purification Technology. 173, 269–279.

Brown, A.J., Brunelli, N.A., Eum, K., Rashidi, F., Johnson, J. R., Koros, W.J., Jones, C.W., Nair, S., 2014. Interfacial microfluidic processing of metal-organic framework hollow fiber membranes. Science. 345, 72–75.

Brown, A.J., Johnson, J.R., Lydon, M.E., Koros, W.J., Jones, C.W., Nair, S., 2012. Continuous polycrystalline zeolitic imidazolate framework-90 membranes on polymeric hollow fibers. Angewandte Chemie International Edition. 51, 1–5.

Bushell, A.F., Attfield, M.P., Mason, C.R., Budd, P.M., Yampolskii, Y., Starannikova, L., Rebrov, A., Bazzarelli, F., Bernardo, P., Carolus Jansen, J., Lanč, M., Friess, K., Shantarovich, V., Gustov, V., Isaeva, V., 2013. Gas permeation parameters of mixed matrix membranes based on the polymer of intrinsic microporosity PIM-1 and the zeolitic imidazolate framework ZIF-8. Journal of Membrane Science. 427, 48–62.

Bux, H., Liang, F., Li, Y., Cravillon, J., Wiebcke, M., Caro, J., 2009. Zeolitic imidazolate framework membrane with molecular sieving properties by microwave-assisted solvothermal synthesis. Journal of the American Chemistry Society. 131, 16000–16001.

Cacho-Bailo, F., Catalán-Aguirre, S., Etxeberría-Benavides, M., Karvan, O., Sebastian, V., Téllez, C., Coronas, J., 2015. Metal-organic framework membranes on the inner-side of a polymeric hollow fiber by microfluidic synthesis. Journal of Membrane Science. 476, 277–285.

Cacho-Bailo, F., Etxeberria-Benavides, M., David, O., Tellez, C., Coronas, J., 2017. Structural contraction of zeolitic imidazolate frameworks: Membrane application on porous metallic hollow fibers for gas separation. ACS Applied Materials & Interfaces. 9, 20787–20796.

Cao, L., Tao, K., Huang, A., Kong, C., Chen, L., 2013. A highly permeable mixed matrix membrane containing CAU-1-NH_2 for H_2 and CO_2 separation. Chemical Communications. 49, 8513–8515.

Car, A., Stropnik, C., Peinemann, K.V., 2006. Hybrid membrane materials with different metal-organic frameworks (MOFs) for gas separation. Desalination. 200, 424–426.

Caro, J., 2011. Are MOF membranes better in gas separation than those made of zeolites. Current Opinion in Chemical Engineering. 1, 77–83.

Centrone, A., Yang, Y., Speakman, S., Bromberg, L., Rutledge, G.C., Hatton, T.A., 2010. Growth of metal-organic frameworks on polymer surfaces. Journal of the American Chemistry Society. 132, 15687–15691.

Chen, X.Y., Vinh-Thang, H., Rodrigue, D., Kaliaguine, S., 2012. Amine-functionalized MIL-53 metal-organic framework in polyimide mixed matrix membranes for CO_2/CH_4 separation. Industrial & Engineering Chemistry Research. 51, 6895–6906.

Cheng, Y., Wang, X., Jia, C., Wang, Y., Zhai, L., Wang, Q., Zhao, D., 2017. Ultrathin mixed matrix membranes containing two-dimensional metal-organic framework nanosheets for efficient CO_2/CH_4 separation. Journal of Membrane Science. 539, 213–223.

Colling, C.W., Huff Jr., G.A., Bartels, J.V., 2004. Processes using solid perm-selective membranes in multiple groups for simultaneous recovery of specified products from a fluid mixture. US Patent, 6,830–691

Dechnik, J., Sumby, C.J., Janiak, C., 2017a. Enhancing mixed-Matrix membrane performance with metal-organic framework additives. Crystal Growth & Design. 17, 4467–4488.

Dechnik, J., Gascon, J., Doonan, C.J., Janiak, C., Sumby, C.J., 2017b. Mixed-matrix membranes. Angewandte Chemie International Edition. 56, 9292–9310.

Dechnik, J., Nuhnen, A., Janiak, C., 2017c. Mixed-matrix membranes of the air-stable MOF-5 analogue $[Co_4(\mu_4\text{-}O)(Me_2pzba)_3]$ with a mixed-functional pyrazolate-carboxylate linker for CO_2/CH_4 separation. Crystal Growth & Design. 17, 4090–4099.

Díaz, K., López-González, M., del Castillo, L.F., Riande, E., 2011. Effect of zeolitic imidazolate frameworks on the gas transport performance of ZIF8-poly(1,4-phenylene ether-ether-sulfone) hybrid membranes. Journal of Membrane Science. 383, 206–213.

Diestel, L., Bux, H., Wachsmuth, D., Caro, J., 2012. Pervaporation studies of n-hexane, benzene, mesitylene and their mixtures on zeolitic imidazolate framework-8 membranes. Microporous and Mesoporous Materials. 164, 288–293.

Dong, X., Huang, K., Liu, S., Ren, R., Jin, W., Lin, Y.S., 2012. Synthesis of zeolitic imidazolate framework-78 molecular-sieve membrane: Defect formation and elimination. Journal of Materials Chemistry. 22, 19222–19227.

Drobek, M., Bechelany, M., Vallicari, C., Abou Chaaya, A., Charmette, C., Salvador-Levehang, C., Miele, P., Julbe, A., 2015. An innovative approach for the preparation of confined ZIF-8 membranes by conversion of ZnO ALD layers. Journal of Membrane Science. 475, 39–46.

Duan, C., Jie, X., Liu, D., Cao, Y., Yuan, Q., 2014. Post-treatment effect on gas separation property of mixed matrix membranes containing metal organic frameworks. Journal of Membrane Science. 466, 92–102.

Düren, T., Bae, Y.S., Snurr, R.Q., 2009. Using molecular simulation to characterise metal-organic frameworks for adsorption applications. Chemical Society Reviews. 38, 1237–1247.

Eum, K., Ma, C., Rownaghi, A., Jones, C.W., Nair, S., 2016a. ZIF-8 membranes via interfacial microfluidic processing in polymeric hollow fibers: Efficient propylene separation at elevated pressures. ACS Applied Materials & Interfaces. 8, 25337–25342.

Eum, K., Rownaghi, A., Choi, D., Bhave, R.R., Jones, C.W., Nair, S., 2016b. Fluidic processing of high-performance ZIF-8 membranes on polymeric hollow fibers: Mechanistic insights and microstructure control. Advanced Functional Materials. 26, 5011–5018.

Fan, L., Xue, M., Kang, Z., Li, H., Qiu, S., 2012. Electrospinning technology applied in zeolitic imidazolate framework membrane synthesis. Journal of Materials Chemistry. 22, 25272–25276.

Fan, S., Sun, F., Xie, J., Guo, J., Zhang, L., Wang, C., Pan, Q., Zhu, G., 2013. Facile synthesis of a continuous thin $Cu(bipy)_2(SiF_6)$ membrane with selectivity towards hydrogen. Journal of Materials Chemistry A. 1, 11438–11442.

Friebe, S., Geppert, B., Steinbach, F., Caro, J., 2017. Metal-organic framework UiO-66 layer: A highly oriented membrane with good selectivity and hydrogen permeance. ACS Applied Materials & Interfaces. 9, 12878–12885.

Fu, J., Das, S., Xing, G., Ben, T., Valtchev, V., Qiu, S., 2016. Fabrication of COF-MOF composite membranes and their highly selective separation of H_2/CO_2. Journal of the American Chemistry Society. 138, 7673–7680.

Furukawa, H., Cordova, K.E., O'Keeffe, M., Yaghi, O.M., 2013. The chemistry and applications of metal-organic frameworks. Science. 341, 974–986.

Ge, L., Zhou, W., Rudolph, V., Zhu, Z., 2013. Mixed matrix membranes incorporated with size-reduced Cu-BTC for improved gas separation. Journal of Materials Chemistry A. 1, 6350–6358.

Getman, R.B., Bae, Y.S., Wilmer, C.E., Snurr, R.Q., 2012. Review and analysis of molecular simulations of methane, hydrogen, and acetylene storage in metal-organic frameworks. Chemical Reviews. 112, 703–723.

Goh, P.S., Ismail, A.F., Sanip, S.M., Ng, B.C., Aziz, M., 2011. Recent advances of inorganic fillers in mixed matrix membrane for gas separation. Separation and Purification Technology. 81, 243–264.

Gong, H., Nguyen, T.H., Wang, R., Bae, T.-H., 2015. Separations of binary mixtures of CO_2/CH_4 and CO_2/N_2 with mixed-matrix membranes containing $Zn(pyrz)_2(SiF_6)$ metal-organic framework. Journal of Membrane Science. 495, 169–175.

Guo. H., Zhu, G., Hewitt, I.J., Qiu, S., 2009. "Twin copper source" growth of metal-organic framework membrane: $Cu_3(BTC)_2$ with high permeability and selectivity for recycling H_2. Journal of the American Chemistry Society. 131, 1646–1647.

Guo, Y., Mao, Y., Hu, P., Ying, Y., Peng, X., 2016. Self-confined synthesis of HKUST-1 membranes from CuO nanosheets at room temperature. ChemistrySelect. 1, 108–113.

Hao, L., Liao, K.S., Chung, T.-S., 2015. Photo-oxidative PIM-1 based mixed matrix membranes with superior gas separation performance. Journal of Materials Chemistry A. 3, 17273–17281.

Hara, N., Yoshimune, M., Negishi, H., Haraya, K., Hara, S., Yamaguchi, T., 2014. Diffusive separation of propylene/propane with ZIF-8 membranes. Journal of Membrane Science. 450, 215–223.

Hegde, M., Shahid, S., Norder, B., Dingemans, T.J., Nijmeijer, K., 2015. Gas transport in metal organic framework-polyetherimide mixed matrix membranes: The role of the polyetherimide backbone structure. Polymer. 81, 87–98.

Hillman, F., Zimmerman, J.M., Paek, S.M., Hamid, M.R.A., Lim, W.T., Jeong, H.K., 2017. Rapid microwave-assisted synthesis of hybrid zeolitic-imidazolate frameworks with mixed metals and mixed linkers. Journal of Materials Chemistry A. 5, 6090–6099.

Hou, J., Sutrisna, P.D., Zhang, Y., Chen, V., 2016. Formation of ultrathin, continuous metal-organic framework membranes on flexible polymer substrates. Angewandte Chemie International Edition. 55, 3947–3951.

Howarth, A.J., Liu, Y., Li, P., Li, Z., Wang, T.C., Hupp, J.T., Farha, O.K., 2016. Chemical, thermal and mechanical stabilities of metal-organic frameworks. Nature Reviews Materials. 1, 1–15.

Hsieh, J.O., Balkus, K.J., Ferraris, J.P., Musselman, I.H., 2014. MIL-53 frameworks in mixed-matrix membranes. Microporous and Mesoporous Materials. 196, 165–174.

Hu, J., Cai, H., Ren, H., Wei, Y., Xu, Z., Liu, H., Hu, Y., 2010. Mixed-matrix membrane hollow fibers of $Cu_3(BTC)_2$ MOF and polyimide for gas separation and adsorption. Industrial & Engineering Chemistry Research. 49, 12605–12612.

Hu, Y., Dong, X., Nan, J., Jin, W., Ren, X., Xu, N., Lee, Y.M., 2011. Metal-organic framework membranes fabricated via reactive seeding. Chemical Communications. 47, 737–739.

Hu, Y., Wei, J., Liang, Y., Zhang, H., Zhang, X., Shen, W., Wang, H., 2016. Zeolitic imidazolate framework/graphene oxide hybrid nanosheets as seeds for the growth of ultrathin molecular sieving membranes. Angewandte Chemie International Edition. 55, 2048–2052.

Huang, A., Dou, W., Caro, J., 2010a. Steam-stable zeolitic imidazolate framework ZIF-90 membrane with hydrogen selectivity through covalent functionalization. Journal of the American Chemistry Society. 132, 15562–15564.

Huang, A., Bux, H., Steinbach, F., Caro, J., 2010b. Molecular-sieve membrane with hydrogen permselectivity: ZIF-22 in LTA topology prepared with 3-aminopropyltriethoxysilane as covalent linker. Angewandte Chemie International Edition. 49, 4958–4961.

Huang, A., Chen, Y., Wang, N., Hu, Z., Jiang, J., Caro, J., 2012. A highly permeable and selective zeolitic imidazolate framework ZIF-95 membrane for H_2/CO_2 separation. Chemical Communications. 48, 10981–10983.

Huang, A., Chen, Y., Liu, Q., Wang, N., Jiang, J., Caro, J., 2014a. Synthesis of highly hydrophobic and permselective metal-organic framework $Zn(BDC)(TED)_{0.5}$ membranes for H_2/CO_2 separation. Journal of Membrane Science. 454, 126–132.

Huang, A., Liu, Q., Wang, N., Caro, J., 2014b. Highly hydrogen permselective ZIF-8 membranes supported on polydopamine functionalized macroporous stainless-steel-nets. Journal of Materials Chemistry A. 2, 8246–8251.

Huang, A., Liu, Q., Wang, N., Zhu, Y., Caro, J., 2014c. Bicontinuous zeolitic imidazolate framework ZIF-8@GO membrane with enhanced hydrogen selectivity. Journal of the American Chemistry Society. 136, 14686–14689.

Huang, Y., Liu, D., Liu, Z., Zhong, C., 2016. Synthesis of zeolitic imidazolate framework membrane using temperature-switching synthesis strategy for gas separation. Industrial & Engineering Chemistry Research. 55, 7164–7170.

Hurrle, S., Friebe, S., Wohlgemuth, J., Woll, C., Caro, J., Heinke, L., 2017. Sprayable, large-area metal-organic framework films and membranes of varying thickness. Chemistry - A European Journal. 23, 2294–2298.

Hwang, S., Chi, W.S., Lee, S.J., Im, S.H., Kim, J.H., Kim, J., 2015. Hollow ZIF-8 nanoparticles improve the permeability of mixed matrix membranes for CO_2/CH_4 gas separation. Journal of Membrane Science. 480, 11–19.

Japip, S., Wang, H., Xiao, Y., Shung Chung, T., 2014. Highly permeable zeolitic imidazolate framework (ZIF)-71 nano-particles enhanced polyimide membranes for gas separation. Journal of Membrane Science. 467, 162–174.

Jin, H., Li, Y., Liu, X., Ban, Y., Peng, Y., Jiao, W., Yang, W., 2015. Recovery of HMF from aqueous solution by zeolitic imidazolate frameworks. Chemical Engineering Science. 124, 170–178.

Jin, H., Wollbrink, A., Yao, R., Li, Y., Caro, J., Yang, W., 2016. A novel CAU-10-H MOF membrane for hydrogen separation under hydrothermal conditions. Journal of Membrane Science. 513, 40–46.

Kang, Z., Xue, M., Fan, L., Huang, L., Guo, L., Wei, G., Chen, B., Qiu, S., 2014. Highly selective sieving of small gas molecules by using an ultra-microporous metal-organic framework membrane. Energy & Environmental Science. 7, 4053–4060.

Kang, Z., Peng, Y., Hu, Z., Qian, Y., Chi, C., Yeo, L.Y., Tee, L., Zhao, D., 2015. Mixed matrix membranes composed of two-dimensional metal-organic framework nanosheets for pre-combustion CO_2 capture: A relationship study of filler morphology versus membrane performance. Journal of Materials Chemistry A. 3, 20801–20810.

Kang, Z., Fan, L., Sun, D., 2017a. Recent advances and challenges of metal-organic framework membranes for gas separation. Journal of Materials Chemistry A. 5, 10073–10091.

Kang, Z., Fan, L., Wang, S., Sun, D., Xue, M., Qiu, S., 2017b. In situ confinement of free linkers within a stable MOF membrane for highly improved gas separation properties. CrystEngComm. 19, 1601–1606.

Khdhayyer, M.R., Esposito, E., Fuoco, A., Monteleone, M., Giorno, L., Jansen, J.C., Attfield, M.P., Budd, P.M., 2017. Mixed matrix membranes based on UiO-66 MOFs in the polymer of intrinsic microporosity PIM-1. Separation and Purification Technology. 173, 304–313.

Khosravi, T., Omidkhah, M., 2017. Preparation of CO_2 selective composite membranes using Pebax/CuBTC/PEG-ran-PPG ternary system. Journal of Energy Chemistry. 26, 530–539.

Kitagawa, S., Kitaura, R., Noro, S., 2004. Functional porous coordination polymers. Angewandte Chemie International Edition. 43, 2334–2375.

Knebel, A., Friebe, S., Bigall, N.C., Benzaqui, M., Serre, C., Caro, J., 2016. Comparative study of MIL-96(Al) as continuous metal-organic frameworks layer and mixed-matrix membrane. ACS Applied Materials & Interfaces. 8, 7536–7544.

Kwon, H.T., Jeong, H.K., 2013a. In situ synthesis of thin zeolitic-imidazolate framework ZIF-8 membranes exhibiting exceptionally high propylene/propane separation. Journal of the American Chemistry Society. 135, 10763–10768.

Kwon, H.T., Jeong, H.K., 2013b. Highly propylene-selective supported zeolite-imidazolate framework (ZIF-8) membranes synthesized by rapid microwave-assisted seeding and secondary growth. Chemical Communications. 49, 3854–3856.

Kwon, H.T., Jeong, H.K., 2015a. Improving propylene/propane separation performance of zeolitic-imidazolate framework ZIF-8 membranes. Chemical Engineering Science. 124, 20–26.

Kwon, H.T., Jeong, H.K., Lee, A.S., An, H.S., Lee, J.S., 2015b. Heteroepitaxially grown zeolitic imidazolate framework membranes with unprecedented propylene/propane separation performances. J Am. Chem. Soc. 137, 12304–12311.

Lee, J., Farha, O.K., Roberts, J., Scheidt, K.A., Nguyen, S.T., Hupp, J.T., 2009. Metal-organic framework materials as catalysts. Chemical Society Reviews. 38, 1450–1459.

Lee, D.J., Li, Q., Kim, H., Lee, K., 2012. Preparation of Ni-MOF-74 membrane for CO_2 separation by layer-by-layer seeding technique. Microporous and Mesoporous Materials. 163, 169–177.

Li, H., Tuo, L., Yang, K., Jeong, H.K., Dai, Y., He, G., Zhao, W., 2016a. Simultaneous enhancement of mechanical properties and CO_2 selectivity of ZIF-8 mixed matrix membranes: Interfacial toughening effect of ionic liquid. Journal of Membrane Science. 511, 130–142.

Li, J.R., Kuppler, R.J., Zhou, H.C., 2009. Selective gas adsorption and separation in metal-organic frameworks. Chemical Society Reviews. 38, 1477–1504.

Li, J.R., Sculley, J., Zhou, H.C., 2012. Metal-organic frameworks for separations. Chemical Reviews. 112, 869–932.

Li, Q., Liu, G., Huang, K., Duan, J., Jin, W., 2016b. Preparation and characterization of $Ni_2(mal)_2(bpy)$ homochiral MOF membrane. Asia-Pacific Journal of Chemical Engineering. 11, 60–69.

Li, T., Pan, Y., Peinemann, K.-V., Lai, Z., 2013. Carbon dioxide selective mixed matrix composite membrane containing ZIF-7 nano-fillers. Journal of Membrane Science. 425-426, 235–242.

Li, W., Meng, Q., Li, X., Zhang, C., Fan, Z., Zhang, G., 2014a. Non-activation ZnO array as a buffering layer to fabricate strongly adhesive metal-organic framework/PVDF hollow fiber membranes. Chemical Communications. 50, 9711–9713.

Li, W., Yang, Z., Zhang, G., Fan, Z., Meng, Q., Shen, C., Gao, C., 2014b. Stiff metal-organic framework-polyacrylonitrile hollow fiber composite membranes with high gas permeability. Journal of Materials Chemistry A. 2, 2110–2118.

Li, W., Zhang, G., Zhang, C., Meng, Q., Fan, Z., Gao, C., 2014c. Synthesis of trinity metal-organic framework membranes for CO_2 capture. Chemical Communications. 50, 3214–3216.

Li, W., Meng, Q., Zhang, C., Zhang, G., 2015. Metal-organic framework/PVDF composite membranes with high H_2 permselectivity synthesized by ammoniation. Chemistry - A European Journal. 21, 7224–7230.

Li, W., Zhang, Y., Zhang, C., Meng, Q., Xu, Z., Su, P., Li, Q.B., Shen, C., Fan, Z., Qin, L., Zhang, G., 2016c. Transformation of metal-organic frameworks for molecular sieving membranes. Nature Communications. 7, 11315.

Li, W., Su, P., Li, Z., Xu, Z., Wang, F., Ou, H., Zhang, J., Zhang, G., Zeng, E., 2017. Ultrathin metal-organic framework membrane production by gel-vapour deposition. Nature Communications. 8, 406.

Li, Y., Liang, F., Bux, H., Feldhoff, A., Yang, W., Caro, J., 2010a. Molecular sieve membrane: Supported metal-organic framework with high hydrogen selectivity. Angewandte Chemie International Edition. 49, 558–561.

Li, Y., Bux, H., Feldhoff, A., Li, G., Yang, W., Caro, J., 2010b. Controllable synthesis of metal-organic frameworks: From MOF nanorods to oriented MOF membranes. Advanced Materials. 22, 3322–3326.

Li, Y., Liang, F., Bux, H., Yang, W., Caro, J., 2010c. Zeolitic imidazolate framework ZIF-7 based molecular sieve membrane for hydrogen separation. Journal of Membrane Science. 354, 48–54.

Lin, R., Ge, L., Diao, H., Rudolph, V., Zhu, Z., 2016. Propylene/propane selective mixed matrix membranes with grape-branched MOF/CNT filler. Journal of Materials Chemistry A. 4, 6084–6090.

Lin, Y., 2015. Metal organic framework membranes for separation applications. Current Opinion in Chemical Engineering. 8, 21–28.

Liu, D., Ma, X., Xi, H., Lin, Y.S., 2014a. Gas transport properties and propylene/propane separation characteristics of ZIF-8 membranes. Journal of Membrane Science. 451, 85–93.

Liu, Y., Ban, Y., Yang, W., 2017. Microstructural engineering and architectural design of metal-organic framework membranes. Advanced Materials. 29, 1606949–1606965.

Liu, Y., Ng, Z., Khan, E.A., Jeong, H.K., Ching, C.B., Lai, Z., 2009. Synthesis of continuous MOF-5 membranes on porous α-alumina substrates. Microporous and Mesoporous Materials. 118, 296–301.

Liu, Y., Wang, N., Pan, J.H., Steinbach, F., Caro, J., 2014b. In situ synthesis of MOF membranes on ZnAl-CO_3 LDH buffer layer-modified substrates. Journal of the American Chemistry Society. 136, 14353–14356.

Ma, J., Ying, Y., Guo, X., Huang, H., Liu, D., Zhong, C., 2016. Fabrication of mixed-matrix membrane containing metal-organic framework composite with task-specific ionic liquid for efficient CO_2 separation. Journal of Materials Chemistry A. 4, 7281–7288.

Mao, Y., Shi, L., Huang, H., Cao, W., Li, J., Sun, L., Jin, X., Peng, X., 2013. Room temperature synthesis of free-standing HKUST-1 membranes from copper hydroxide nanostrands for gas separation. Chemical Communications. 49, 5666–5668.

Mosleh, S., Mozdianfard, M., Hemmati, M., Khanbabaei, G., 2017. Mixed matrix membranes of Pebax1657 loaded with iron benzene-1,3,5-tricarboxylate for gas separation. Polymer Composites. 38, 1363–1370.

Nafisi, V., Hägg, M.B., 2014a. Development of dual layer of ZIF-8/PEBAX-2533 mixed matrix membrane for CO_2 capture. Journal of Membrane Science. 459, 244–255.

Nafisi, V., Hägg, M.B., 2014b. Gas separation properties of ZIF-8/6FDA-durene diamine mixed matrix membrane. Separation and Purification Technology. 128, 31–38.

Nagaraju, D., Bhagat, D.G., Banerjee, R., Kharul, U.K., 2013. In situ growth of metal-organic frameworks on a porous ultrafiltration membrane for gas separation. Journal of Materials Chemistry A. 1, 8828–8835.

Nan, J., Dong, X., Wang, W., Jin, W., Xu, N., 2011. Step-by-step seeding procedure for preparing HKUST-1 membrane on porous alpha-alumina support. Langmuir. 27, 4309–4312.

Naseri, M., Mousavi, S.F., Mohammadi, T., Bakhtiari, O., 2015. Synthesis and gas transport performance of MIL-101/Matrimid mixed matrix membranes. Journal of Industrial and Engineering Chemistry. 29, 249–256.

Nik, O.G., Chen, X.Y., Kaliaguine, S., 2012. Functionalized metal organic framework-polyimide mixed matrix membranes for CO_2/CH_4 separation. Journal of Membrane Science. 413–414, 48–61.

Ordoñez, M.J.C., Balkus, K.J., Ferraris, J.P., Musselman, I.H., 2010. Molecular sieving realized with ZIF-8/Matrimid® mixed-matrix membranes. Journal of Membrane Science. 361, 28–37.

Pan, Y., Li, T., Lestari, G., Lai, Z., 2012. Effective separation of propylene/propane binary mixtures by ZIF-8 membranes. Journal of Membrane Science. 390–391, 93–98.

Peng, Y., Li, Y., Ban, Y., Jin, H., Jiao, W., Liu, X., Yang, W., 2014. Metal-organic framework nanosheets as building blocks for molecular sieving membranes. Science. 346, 1356–1359.

Peng, Y., Li, Y., Ban, Y., Yang, W., 2017. Two-dimensional metal-organic framework nanosheets for membrane-based gas separation. Angewandte Chemie International Edition. 56, 9757–9761.

Perez, E.V., Balkus, K.J., Ferraris, J.P., Musselman, I.H., 2009. Mixed-matrix membranes containing MOF-5 for gas separations. Journal of Membrane Science. 328, 165–173.

Qiao, Z., Peng, C., Zhou, J., Jiang, J., 2016. High-throughput computational screening of 137953 metal-organic frameworks for membrane separation of a $CO_2/N_2/CH_4$ mixture. Journal of Materials Chemistry A. 4, 15904–15912.

Qiu, S., Xue, M., Zhu, G., 2014. Metal-organic framework membranes: From synthesis to separation application. Chemical Society Reviews. 43, 6116–6140.

Ren, H., Jin, J., Hu, J., Liu, H., 2012. Affinity between metal-organic frameworks and polyimides in asymmetric mixed matrix membranes for gas separations. Industrial & Engineering Chemistry Research. 51, 10156–10164.

Robeson, L.M., 1991. Correlation of separation factor versus permeability for polymeric membranes. Journal of Membrane Science. 62, 165–185.

Robeson, L.M., 2008. The upper bound revisited. Journal of Membrane Science. 320, 390–400.

Rodenas, T., van Dalen, M., Serra-Crespo, P., Kapteijn, F., Gascon, J., 2014. Mixed matrix membranes based on NH_2-functionalized MIL-type MOFs: Influence of structural and operational parameters on the CO_2/CH_4 separation performance. Microporous and Mesoporous Materials. 192, 35–42.

Rodenas, T., Luz, I., Prieto, G., Seoane, B., Miro, H., Corma, A., Kapteijn, F., i Xamena, F.X.L., Gascon, J., 2015. Metal-organic framework nanosheets in polymer composite materials for gas separation. Nature Materials. 14, 48–55.

Ryder, M.R., Civalleri, B., Bennett, T.D., Henke, S., Rudic, S., Cinque, G., Fernandez-Alonso, F., Tan, J.C., 2014. Identifying the role of terahertz vibrations in metal-organic frameworks: From gate-opening phenomenon to shear-driven structural destabilization. Physical Review Letters. 113, 215502.

Sabetghadam, A., Seoane, B., Keskin, D., Duim, N., Rodenas, T., Shahid, S., Sorribas, S., Guillouzer, C.L., Clet, G., Tellez, C., Daturi, M., Coronas, J., Kapteijn, F., Gascon, J., 2016. Metal organic framework cystals in mixed-matrix membranes: Impact of the filler morphology on the gas separation performance. Advanced Functional Materials. 26, 3154–3163.

Schoedel, A., Li, M., Li, D., O'Keeffe, M., Yaghi, O.M., 2016. Structures of metal-organic frameworks with rod secondary building units. Chemical Reviews. 116, 12466–12535.

Seoane, B., Coronas, J., Gascon, I., Benavides, M.E., Karvan, O., Caro, J., Kapteijn, F., Gascon, J., 2015. Metal-organic framework based mixed matrix membranes: A solution for highly efficient CO_2 capture. Chemical Society Reviews. 44, 2421–2454.

Seoane, B., Sebastián, V., Téllez, C., Coronas, J., 2013. Crystallization in THF: The possibility of one-pot synthesis of mixed matrix membranes containing MOF MIL-68(Al). CrystEngComm. 15, 9483–9490.
Shah, M.N., Gonzalez, M.A., McCarthy, M.C., Jeong, H.K., 2013. An unconventional rapid synthesis of high performance metal-organic framework membranes. Langmuir. 29, 7896–7902.
Shamsaei, E., Low, Z.X., Lin, X., Mayahi, A., Liu, H., Zhang, X., Zhe Liu, J., Wang, H., 2015. Rapid synthesis of ultrathin, defect-free ZIF-8 membranes via chemical vapour modification of a polymeric support. Chemical Communications. 51, 11474–11477.
Shamsaei, E., Lin, X., Low, Z.X., Abbasi, Z., Hu, Y., Liu, J.Z., Wang, H., 2016. Aqueous phase synthesis of ZIF-8 membrane with controllable location on an asymmetrically porous polymer substrate. ACS Applied Materials & Interfaces. 8, 6236–6244.
Shekhah, O., Swaidan, R., Belmabkhout, Y., du Plessis, M., Jacobs, T., Barbour, L.J., Pinnau, I., Eddaoudi, M., 2014. The liquid phase epitaxy approach for the successful construction of ultra-thin and defect-free ZIF-8 membranes: Pure and mixed gas transport study. Chemical Communications. 50, 2089–2092.
Shen, J., Liu, G., Huang, K., Li, Q., Guan, K., Li, Y., Jin, W., 2016. UiO-66-polyether block amide mixed matrix membranes for CO2 separation. Journal of Membrane Science. 513, 155–165.
Sheng, L., Wang, C., Yang, F., Xiang, L., Huang, X., Yu, J., Zhang, L., Pan, Y., Li, Y., 2017. Enhanced C_3H_6/C_3H_8 separation performance on MOF membranes through blocking defects and hindering framework flexibility by silicone rubber coating. Chemical Communications. 53, 7760–7763.
Shin, H., Chi, W.S., Bae, S., Kim, J.H., Kim, J., 2017. High-performance thin PVC-POEM/ZIF-8 mixed matrix membranes on alumina supports for CO_2/CH_4 separation. Journal of Industrial and Engineering Chemistry. 53, 127–133.
Song, Q., Nataraj, S.K., Roussenova, M.V., Tan, J.C., Hughes, D.J., Li, W., Bourgoin, P., Alam, M.A., Cheetham, A.K., Al-Muhtaseb, S.A., Sivaniah, E., 2012. Zeolitic imidazolate framework (ZIF-8) based polymer nanocomposite membranes for gas separation. Energy & Environmental Science. 5, 8359–8369.
Sorribas, S., Zornoza, B., Téllez, C., Coronas, J., 2014. Mixed matrix membranes comprising silica-(ZIF-8) core-shell spheres with ordered meso-microporosity for natural- and biogas upgrading. Journal of Membrane Science. 452, 184–192.
Su, N.C., Sun, D.T., Beavers, C.M., Britt, D.K., Queen, W.L., Urban, J.J., 2016. Enhanced permeation arising from dual transport pathways in hybrid polymer-MOF membranes. Energy & Environmental Science. 9, 922–931.
Suh, M.P., Park, H.J., Prasad, T.K., Lim, D.W., 2012. Hydrogen storage in metal-organic frameworks. Chemical Reviews. 112, 782–835.
Sun, Y., Yang, F., Wei, Q., Wang, N., Qin, X., Zhang, S., Wang, B., Nie, Z., Ji, S., Yan, H., Li, J.R., 2016. Oriented nano-microstructure-assisted controllable fabrication of metal-organic framework membranes on Nickel foam. Advanced Materials. 28, 2374–2381.
Tanh Jeazet, H.B., Sorribas, S., Román-Marín, J.M., Zornoza, B., Téllez, C., Coronas, J., Janiak, C., 2016. Increased selectivity in CO_2/CH_4 separation with mixed-matrix membranes of polysulfone and mixed-MOFs MIL-101(Cr) and ZIF-8. European Journal of Inorganic Chemistry. 2016, 4363–4367.
Thompson, J.A., Blad, C.R., Brunelli, N.A., Lydon, M.E., Lively, R.P., Jones, C.W., Nair, S., 2012. Hybrid zeolitic imidazolate frameworks: Controlling framework porosity and functionality by mixed-linker synthesis. Chemistry of Materials. 24, 1930–1936.
Wang, C., Liu, X., Demir, N.K., Chen, J.P., Li, K., 2016a. Applications of water stable metal-organic frameworks. Chemical Society Reviews. 45, 5107–5134.

Wang, N., Liu, Y., Qiao, Z., Diestel, L., Zhou, J., Huang, A., Caro, J., 2015a. Polydopamine-based synthesis of a zeolite imidazolate framework ZIF-100 membrane with high H_2/CO_2 selectivity. Journal of Materials Chemistry A. 3, 4722–4728.

Wang, N., Mundstock, A., Liu, Y., Huang, A., Caro, J., 2015b. Amine-modified Mg-MOF-74/CPO-27-Mg membrane with enhanced H_2/CO_2 separation. Chemical Engineering Science. 124, 27–36.

Wang, X., Chi, C., Zhang, K., Qian, Y., Gupta, K.M., Kang, Z., Jiang, J., Zhao, D., 2017a. Reversed thermo-switchable molecular sieving membranes composed of two-dimensional metal-organic nanosheets for gas separation. Nature Communications. 8, 14460.

Wang, Z., Cohen, S.M., 2009. Postsynthetic modification of metal-organic frameworks. Chemical Society Reviews. 38, 1315–1329.

Wang, Z., Wang, D., Zhang, S., Hu, L., Jin, J., 2016b. Interfacial design of mixed matrix membranes for improved gas separation performance. Advanced Materials. 28, 3399–3405.

Wang, Z., Ren, H., Zhang, S., Zhang, F., Jin, J., 2017b. Polymers of intrinsic microporosity/metal-organic framework hybrid membranes with improved interfacial interaction for high-performance CO_2 separation. Journal of Materials Chemistry A. 5, 10968–10977.

Wijenayake, S.N., Panapitiya, N.P., Versteeg, S.H., Nguyen, C.N., Goel, S., Balkus, K.J., Musselman, I.H., Ferraris, J.P., 2013. Surface cross-Linking of ZIF-8/polyimide mixed matrix membranes (MMMs) for gas separation. Industrial & Engineering Chemistry Research. 52, 6991–7001.

Wilmer, C.E., Leaf, M., Lee, C.Y., Farha, O.K., Hauser, B.G., Hupp, J.T., Snurr, R.Q., 2012. Large-scale screening of hypothetical metal-organic frameworks. Nature Chemistry. 4, 83–89.

Xiang, L., Sheng, L., Wang, C., Zhang, L., Pan, Y., Li, Y., 2017. Amino-functionalized ZIF-7 nanocrystals: Improved intrinsic separation ability and interfacial compatibility in mixed-matrix membranes for CO_2/CH_4 separation. Advanced Materials. 29, 1606999.

Yang, T., Xiao, Y., Chung, T.-S., 2011. Poly-/metal-benzimidazole nano-composite membranes for hydrogen purification. Energy & Environmental Science. 4, 4171–4180.

Yang, T., Shi, G.M., Chung, T.-S., 2012. Symmetric and asymmetric zeolitic imidazolate frameworks (ZIFs)/polybenzimidazole (PBI) nanocomposite membranes for hydrogen purification at high temperatures. Advanced Energy Materials. 2, 1358–1367.

Yang, T., Chung, T.-S., 2013a. High performance ZIF-8/PBI nano-composite membranes for high temperature hydrogen separation consisting of carbon monoxide and water vapor. International Journal of Hydrogen Energy. 38, 229–239.

Yang, T., Chung, T.-S., 2013b. Room-temperature synthesis of ZIF-90 nanocrystals and the derived nano-composite membranes for hydrogen separation. Journal of Materials Chemistry A. 1, 6081–6090.

Yang, Q., Liu, D., Zhong, C., Li, J.R., 2013c. Development of computational methodologies for metal-organic frameworks and their application in gas separations. Chemical Reviews. 113, 8261–8323.

Yao, J., Dong, D., Li, D., He, L., Xu, G., Wang, H., 2011. Contra-diffusion synthesis of ZIF-8 films on a polymer substrate. Chemical Communications. 47, 2559–2561.

Yu, J., Pan, Y., Wang, C., Lai, Z., 2016. ZIF-8 membranes with improved reproducibility fabricated from sputter-coated ZnO/alumina supports. Chemical Engineering Science. 141, 119–124.

Zhang, C., Dai, Y., Johnson, J.R., Karvan, O., Koros, W.J., 2012a. High performance ZIF-8/6FDA-DAM mixed matrix membrane for propylene/propane separations. Journal of Membrane Science. 389, 34–42.

Zhang, C., Xiao, Y., Liu, D., Yang, Q., Zhong, C., 2013. A hybrid zeolitic imidazolate framework membrane by mixed-linker synthesis for efficient CO_2 capture. Chemical Communications. 49, 600–602.

Zhang, C., Zhang, K., Xu, L., Labreche, Y., Kraftschik, B., Koros, W.J., 2014. Highly scalable ZIF-based mixed-matrix hollow fiber membranes for advanced hydrocarbon separations. AIChE Journal. 60, 2625–2635.

Zhang, F., Zou, X., Gao, X., Fan, S., Sun, F., Ren, H., Zhu, G., 2012b. Hydrogen selective NH_2-MIL-53(Al) MOF membranes with high permeability. Advanced Functional Materials. 22, 3583–3590.

Zhang, X., Liu, Y., Li, S., Kong, L., Liu, H., Li, Y., Han, W., Yeung, K.L., Zhu, W., Yang, W., Qiu, J., 2014. New membrane architecture with high performance: ZIF-8 membrane supported on vertically aligned ZnO nanorods for gas permeation and separation. Chemistry of Materials. 26, 1975–1981.

Zhang, Y., Feng, X., Li, H., Chen, Y., Zhao, J., Wang, S., Wang, L., Wang. B., 2015. Photoinduced postsynthetic polymerization of a metal-organic framework toward a flexible stand-alone membrane. Angewandte Chemie International Edition. 54, 4259–4263.

Zhang, Y., Feng, X., Yuan, S., Zhou, J., Wang. B., 2016. Challenges and recent advances in MOF-polymer composite membranes for gas separation. Inorganic Chemistry Frontiers. 3, 896–909.

Zhao, Z., Ma, X., Li, Z., Lin, Y.S., 2011. Synthesis, characterization and gas transport properties of MOF-5 membranes. Journal of Membrane Science. 382, 82–90.

Zhou, S., Zou, X., Sun, F., Zhang, F., Fan, S., Zhao, H., Schiestel, T., Zhu, G., 2012. Challenging fabrication of hollow ceramic fiber supported $Cu_3(BTC)_2$ membrane for hydrogen separation. Journal of Materials Chemistry. 22, 10322–10328.

Zhou, S., Zou, X., Sun, F., Ren, H., Liu, J., Zhang, F., Zhao, N., Zhu, G., 2013. Development of hydrogen-selective CAU-1 MOF membranes for hydrogen purification by 'dual-metal-source' approach. International Journal of Hydrogen Energy. 38, 5338–5347.

Zhu, H., Jie, X., Wang, L., Liu, D., Cao, Y., 2017. Polydimethylsiloxane/postmodified MIL-53 composite layer coated on asymmetric hollow fiber membrane for improving gas separation performance. Journal of Applied Polymer Sciences. 134, 44999.

Zhu, H., Wang, L., Jie, X., Liu, D., Cao, Y., 2016. Improved interfacial affinity and CO_2 separation performance of asymmetric mixed matrix membranes by incorporating post-modified MIL-53(Al). ACS Applied Materials & Interfaces. 8, 22696–22704.

Zornoza, B., Martinez-Joaristi, A., Serra-Crespo, P., Tellez, C., Coronas, J., Gascon, J., Kapteijn, F., 2011a. Functionalized flexible MOFs as fillers in mixed matrix membranes for highly selective separation of CO_2 from CH_4 at elevated pressures. Chemical Communications. 47, 9522–9524.

Zornoza, B., Seoane, B., Zamaro, J.M., Tellez, C., Coronas, J., 2011b. Combination of MOFs and zeolites for mixed-matrix membranes. ChemPhyschem. 12, 2781–2785.

Section III

Membranes for Organic Solvent Applications

12 Zeolite Membranes for Pervaporation

Hidetoshi Kita

CONTENTS

12.1 INTRODUCTION

Commercially available zeolite membranes such as LTA, T-type (ERI;OFF), FAU, CHA and DDR zeolite membranes are grown on the surface of porous cylindricalα-alumina tubes by a hydrothermal treatment*†‡§. These membranes are composed of inter-grown zeolite crystal layer, which is ca. 5 to 20 μm in thickness on the surface of the tube. These membranes show excellent water permselective performance in pervaporation (PV) and vapor permeation (VP) for water/organic liquid mixtures.

PV or VP is an attractive separation technique particularly in cases for which ordinary distillation is inapplicable, such as azeotropic mixtures or mixtures of components having close boiling points (Sander and Soukup, 1988; Feng and Huang, 1997; Lipnizki et al., 1999a). Notably, this dehydration of solvents including a small amount of water is energy efficient in comparison with the distillation method (Sander et al., 1988). Zeolite membranes offer significant potential for a PV (VP) agent with high separation factor and permeation flux together with high chemical and thermal stability. In order to purify i-propanol (IPA) from the cleaning process used in industries such as precision machinery and electronics, and also to dehydrate ethanol (EtOH) used in food industry, PV or VP equipment using such zeolite membranes has been put to practical use in Japan and China (Rangnekar et al., 2015) as shown in Figure 12.1. Recently in Europe, as shown in Figure 12.2 from an energy saving perspective, the zeolite membrane has begun to be used industrially to produce bio-ethanol, which is added to gasoline.

* Mitsui E&S, https://www.mes.co.jp/business/environment/environ_plant/detail122.html.
† Mitsubishi Chemical, https://www.m-chemical.co.jp/products/departments/mcc/aquachem/product/1200497_7280.html.
‡ NGK Insulators, Inc., http://www.ngk.co.jp/news/2010/20101216.html.
§ Hitachi Zosen Corporation, https://www.hitachizosen.co.jp/products/products009.html.

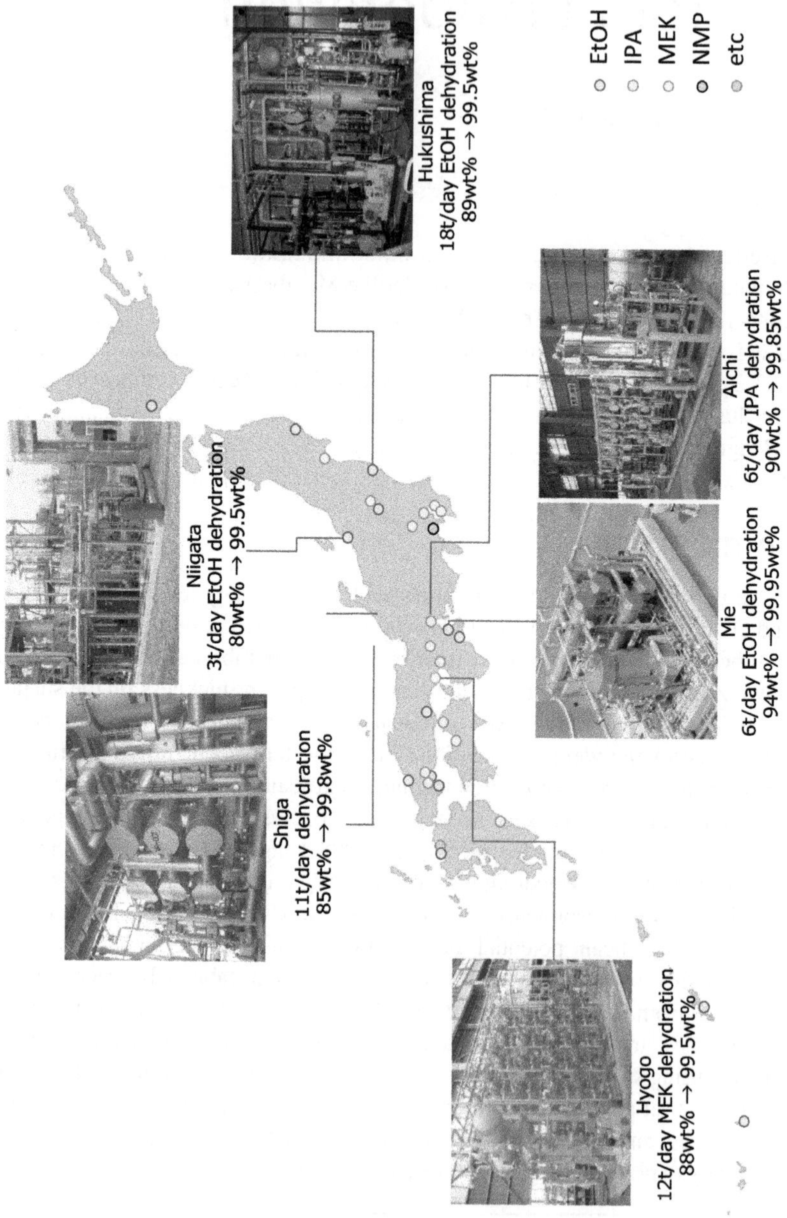

FIGURE 12.1 Examples of commercial PV and VP plants in Japan.

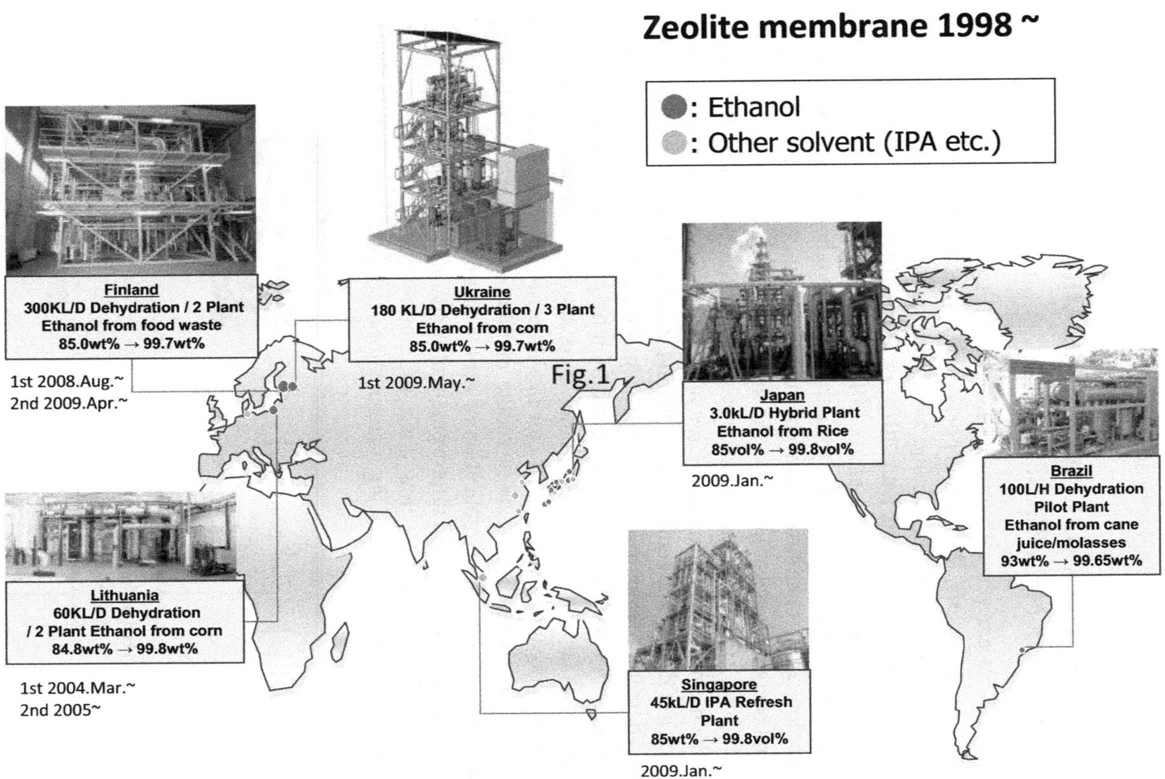

FIGURE 12.2 Examples of large scale commercial PV and VP plants.

In the past decade, the number of publications dealing with zeolite membranes has experienced rapid growth, especially some excellent reviews on zeolite membranes (Yu et al., 2011; Dragomirova and Wohlrab, 2015; Rangnekar et al., 2015; Kosinov et al., 2016). The present chapter is, therefore, focused on important research areas that have potential for industrial use, such as the preparation of long-term acid-stable zeolite membranes for practical dehydration.

12.2 PREPARATION AND CHARACTERIZATION OF ZEOLITE MEMBRANES

Zeolite membranes are generally prepared by in situ growth or secondary growth method with seed crystals. We have reported the synthesis of LTA, FAU, T, MFI, MOR, CHA, AEI, SOD and RHO zeolite membranes using these two methods (Kita, 2006; Li et al., 2009; Kondo and Kita, 2010; Li et al., 2011; Zhou et al., 2012; Zhu et al., 2012; Zhu et al., 2013; Zhu et al., 2014; Qiu et al., 2015; Liu et al., 2016). The formation of zeolite membranes is controlled by synthesis parameters such as gel composition, aging conditions, hydrothermal temperature and time. In addition, supports and seeded crystals may affect the membrane formation (Pina et al., 2004; Noack et al., 2007). Contamination of undesired zeolite phases is one of the causes of defects in the membrane. For example, LTA and FAU type zeolite transforms to other zeolites such as GIS (P-type zeolite) and SOD under certain conditions (Breck, 1974). Achieving a defect-free thin membrane consisting of pure zeolite phase is a challenge. Various methods for zeolite membrane preparation have been implemented in order to eliminate the defects in zeolite membranes. Repeating the hydrothermal synthesis and/or using the organic template that can facilitate zeolite crystallization are two of the ways to eliminate the defects in zeolite membranes (Gu et al., 2005; Zhu et al., 2009). However, industrial application of zeolite membranes requires significant reduction of fabrication costs and improvement of preparation reproducibility.

Figure 12.3 shows a general flow chart of commercialized zeolite membrane preparation (Kita, 2006). The starting materials for preparing the synthesis mixture

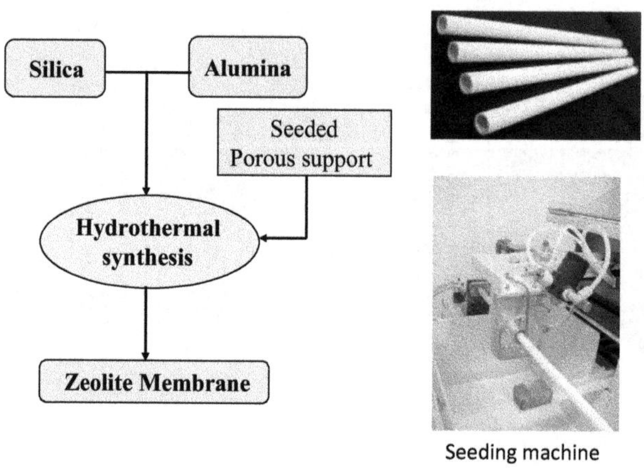

FIGURE 12.3 General flow sheet of commercialized zeolite membrane preparation.

are SiO_2, Al_2O_3 (if necessary), Na_2O (or another alkali metal oxide) and H_2O. Porous single-tubular supports made of α-Al_2O_3 or mullite, or porousα-Al_2O_3 monolith tube containing multiple channels are used. The pore size of the supports is about 0.1–1.5 μm for both supports, and the porosity of the supports is about 30–40%. The surface of the supports is either mechanically rubbed by a seeding machine, as shown in Figure 12.3, with a slurry prepared by mixing zeolite powder with water or dipped into a suspension containing seed crystals and withdrawn.

The formation of the zeolite membranes is confirmed by using X-ray diffraction (XRD). The surface morphology of the membranes is observed by field emission scanning electron microscopy (FESEM). The compositions of the zeolite membrane are analyzed by electron probe microanalysis (EPMA) with wavelength dispersive spectrometer (WDS). An evaluation of non-zeolitic pore size of zeolite membranes is carried out using nanopermporometry, the basic principle of which is based on capillary condensation of vapor and the blocking effect of permeation of a non-condensable gas (Huang et al., 1996; Tsuru et al., 1998; Tsuru et al., 2001). The apparatus and measurement procedure of nanopermporomery is described in detail by Tsuru et al. (2001). Nitrogen is used as non-condensable gas, and the liquid used as a condensable vapor is water. Nanopermporometry is initiated by measuring the permeability of dry nitrogen and checking the steady permeability at the measurement temperature. The vapor pressure is then increased gradually in a stepwise manner to a specific partial pressure, until nitrogen permeation is nearly blocked by capillary condensation. The permeation flux of nitrogen normalized with that of dry nitrogen is shown as a function of Kelvin diameter calculated by the well-known Kelvin equation without correction of the adsorption thickness of water vapor. It should be noted that the Kelvin equation loses physical meaning for pore diameters of less than 2 nm.

Although the present method cannot accurately evaluate the nanopermporometry characterization of zeolite membranes, it is an effective method for the comparison of the non-zeolitic pore distribution of each membrane. For example, average Kelvin diameters, which are defined as the diameters at a dimensionless flux of 50%, of a NaA membrane and a T-type membrane were determined to be 1.4 and 0.8 nm, respectively (Kondo and Kita, 2010). Due to the blocking effect of nitrogen permeation by condensed water, no nitrogen permeation was observed in the conditions that p/p_{sat} is larger than 0.75 for the NaA membrane and 0.61 for the T-type membrane. These results indicate that these zeolite membranes are composed of non-zeolitic and zeolitic pores. The result of the zeolite NaA membrane agrees with a previous paper by Okamoto et al. (2001). Namely, according to the paper, the zeolite NaA membrane consisted of zeolitic and non-zeolitic pores, and the gas molecules permeated through the non-zeolitic pores by Knudsen diffusion mechanism.

A very fine and narrow non-zeolitic pore penetrates from the zeolite crystal layer to the support. The majority of zeolitic pores in the top surface layer are gathered in the non-zeolitic pore. The water molecules in the feed are selectively adsorbed in zeolitic pores in the top surface layer, and then are transported to the non-zeolitic pore through the zeolitic pores. At the narrower space in the non-zeolitic pore, capillary condensation occurs, and then the condensate evaporates and diffuses into the permeation side. Based on theoretical consideration, water flux through both zeolite membranes was correlated in the partial vapor pressure (Wijimans and Baker, 1993; Nomura et al., 1998; Lipnizki et al., 1999b; Sommer and Melin, 2005).

12.3 LONG-TERM ACID-STABLE ZEOLITE MEMBRANES

LTA membranes show very high PV and VP performance for water/organic mixtures even for a water/MeOH mixture. However, a major drawback of LTA membranes is their acid-sensitivity (Hasegawa et al., 2010). Therefore, a chemically inert and acid-stable membrane for dehydration of acidic mixtures is of great interest. One solution to this problem is the use of an acid-resistant zeolite, such as ZSM-5 and MOR zeolites. Due to the medium Si/Al ratio, the high silica CHA (Hasegawa et al., 2012) and T-type zeolite (Tanaka et al., 2002) have been also studied to dehydrate the carboxylic acid mixtures or couple with the esterification reaction.

12.3.1 SDA-Free Preparation of ZSM-5 Membranes

Due to its unique channel structure, thermal stability, acidity and shape-selectivity, MFI zeolite membranes (Silicalite-1 and ZSM-5) have been studied in many fields (Kita, 2006; Kosinov et al., 2016). ZSM-5 membrane, in particular the aluminum rich ZSM-5 membrane, has excellent ion-exchange capability, catalytic activity, hydrophilicity and acid-stability, which have attracted much attention (Hölderich and Bekkum, 1991; Ghobarkar et al., 1999). However, an expensive organic structure-directing agent (SDA), such as tetra-n-propylammonium bromide (TPABr) or tetra-n-propylammonium hydroxide (TPAOH), is necessary to prepare the ZSM-5 membrane. Furthermore, due to the different thermal expansion between the zeolite film and the support and/or by changes in lattice parameters of the zeolite crystals, micro cracks tend to form in the zeolite membranes during the SDA removal process by calcination (Gues and Bekkum, 1995). Small SDA(TPA+) concentrations in the synthesis gel are favorable to prepare aluminum-rich ZSM-5 crystals or membranes (Noack et al., 2005). Arising from the concern about the production cost and post-production waste disposal, it is strongly desirable to prepare ZSM-5 zeolite membranes in the media involving very expensive tetra-n-propylammonium (TPA+) cations' quantity as low as possible or at the absence of TPA+ cations as the SDAs.

Since Mintova et al. (1998) successfully prepared a ZSM-5 film by using an SDA-free synthesis gel, increasing amounts of attention have been paid to synthesize the hydrophilic ZSM-5 membrane without organic SDA (Machado et al., 1999; Lai and Gavalas, 2000; Noack et al., 2000; Lassinantti et al., 2001; Pan and Lin, 2001; Li et al., 2003a; Li et al., 2009). Unfortunately, it was difficult to prepare pure and aluminum-rich ZSM-5 membrane in the absence of organic SDA. Separation performances of the reported ZSM-5 membranes were not high enough for potential industrial interests. As for preparation of the ZSM-5 membranes with low Si/Al in the range of 15–40, the as-synthesized membrane layers were reported to contain a lot of amorphous materials or mordenite phase (Mintova et al., 1998; Machado et al., 1999; Li et al., 2003a). Besides, the synthesis conditions become more rigorous as elevating the aluminum content in the initial synthesis gel (Mintova et al., 1998; Machado et al., 1999; Lai and Gavalas, 2000; Noack et al., 2000; Lassinantti et al., 2001;

Pan et al., 2001; Li et al., 2003a; Li et al., 2009). For example, the crystallization process of the membrane was carried out as long as 3 days under a continuous horizontal rotation at 37.5 rpm, which could limit the high cost to the industrial application of the membrane (Li et al., 2009). Furthermore, there were only a few studies on the PV performance of ZSM-5 membranes for dehydration of organic mixtures (Pan et al., 2001; Li et al., 2009).

In a recent study, static hydrothermal synthesis method was applied to prepare ZSM-5 membranes by using an aluminum rich (Si/Al = 7.5) and template-free precursor gel in a short synthesis time (Zhu et al., 2012). Various synthesis parameters, such as synthesis time, ceramic supports, silica sources, fluoride and alkalinity, were minutely investigated to improve the hydrophilic performance of ZSM-5 membranes and obtain the optimum synthesis condition. The membranes were prepared on the ceramic supports by the static hydrothermal synthesis. Prior to synthesis, the outer surface of the ceramic support was simply rubbed with the water slurry of nanometer silicalite-1 crystals, and then the seeded support was dried at 80°C for 2 h. Figure 12.4 shows the surface and the cross-sectional SEM images of the seeded support. The entirety of the support surface was fully covered with the nanometer silicalite-1 crystals. The molar composition of the synthesis gel for the membrane was SiO_2: 0.04–0.07 Al_2O_3: 0.15–0.25 Na_2O: 0–2 NaF: 30–50 H_2O. The details of preparation have been described in the work of Zhu et al. (2012).

The surface and the cross-sectional SEM images of the membrane are displayed in Figure 12.5. The support surface is fully covered with fine and sand-rose-like zeolite crystals. The thickness of the zeolite layer is about 5 μm. Meanwhile, the molar ratio of Si/Al and Na/Al of the ZSM-5 membrane surface were 13.1 and 1.0, respectively, by the result of EDX characterization. Because the ZSM-5 crystals of the membrane were different from the typical cubic and hexagonal morphology, the ZSM-5 crystals were prepared as that of the membrane. The as-synthesized ZSM-5 crystals were examined by XRD, SEM and TEM. From the high magnification TEM (Figure 12.6), the crystals showed the typically hexagonal morphology. The sand-rose formation of the crystals could be attributed to the very low supersaturation of crystallizing species in the fluoride media; therefore, the nucleation predominates

FIGURE 12.4 Surface and Cross-sectional SEM view of MFI Seeded support.

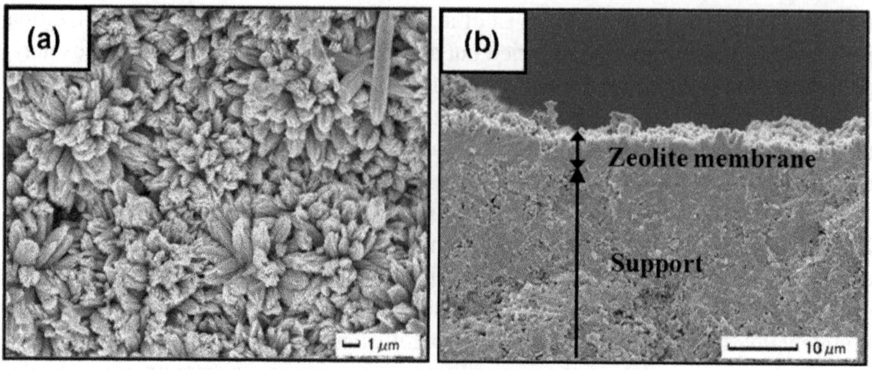

FIGURE 12.5 Surface and Cross-sectional SEM view of Al-rich ZSM-5 membrane.

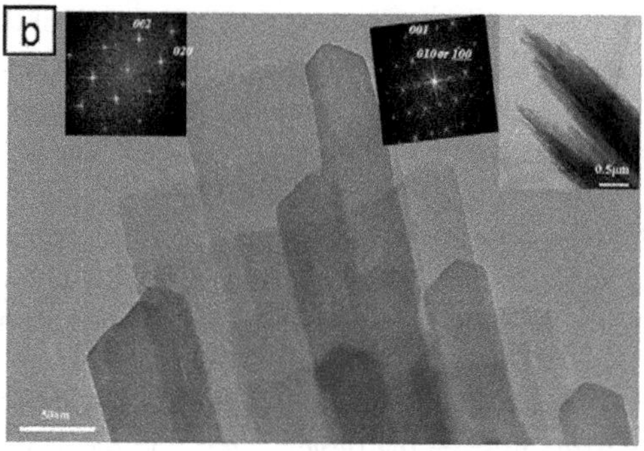

FIGURE 12.6 TEM(b) images of ZSM-5 crystals. (Gel molar composition: SiO_2: $0.067Al_2O_3$: $0.20Na_2O$: NaF:$50H_2O$.)

the crystal growth and induces an intergrowth of the crystals via strong edge-surface platelet interactions (Adachi-Pagano et al., 2003; Arichi et al., 2008). Moreover, the Si/Al ratio of the ZSM-5 crystals was determined to be about 14 by the ICP analysis.

Without organic SDA, appropriate ratios of NaF/SiO_2 and Na_2O/SiO_2 in the precursor gel are crucial for the preparation of a continuous and compact hydrophilic ZSM-5 membrane. Under the optimum synthesis conditions, the ZSM-5 membrane can be used for the dehydration of different H_2O/organic mixtures by PV as shown in Table 12.1.

In addition, the perm-selectivity of the membrane was maintained even after applying pure acetic acid for 77 h and applying esterification product of ethanol and acetic acid for 98 days as shown in Figure 12.7. No change in the membrane structure was observed before and after the dehydration of acidic solutions through XRD, SEM and EDX observations. These results confirm the high acidic stability of the membrane.

TABLE 12.1
Comparison of PV Performance and Synthesis Conditions of ZSM-5 Membranes

Synthesis Conditions			PV Condition		PV Performance	
SiO_2/Al_2O_3	Template	Time	Feed	Temp	Flux	Separation Factor
	–		A/B(A wt%)	°C	Kg/(m²h)	A/B
15	–	16	H_2O/IPA(10)	75	0.85	670
15	–	24	H_2O/IPA(10)	75	3.24	3100
15	–	24	H_2O/IPA(50)	75	6.25	1000
15	–	24	H_2O/Acetone(10)	60	2.40	6400
15	–	24	H_2OEtOH(10)	60	1.30	1700
100	–	19	H_2O/Hac(46)	70	0.63	23
15	–	72	H_2O/EtOH(10)	75	1.48	540
Only SiO_2	TPAOH	8+72	H_2O/EtOH(5)	30	0.17	120
50	TPAOH	24	H_2O/THF(5)	60	0.47	4

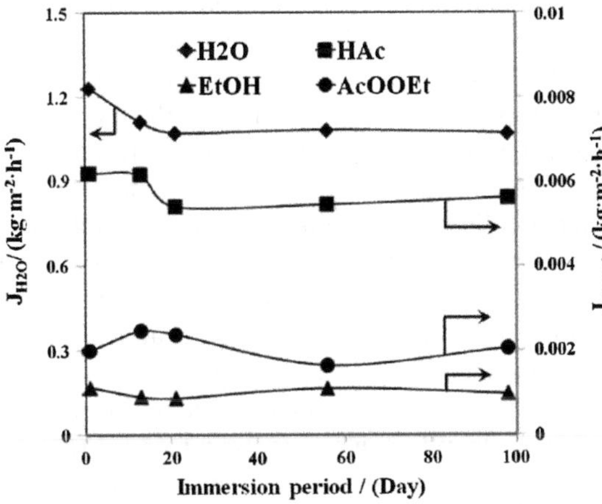

FIGURE 12.7 Long-term stability of ZSM-5 membrane for dehydration of esterification product of ethanol and acetic acid at 75°C.

12.3.2 Synthesis of Fluoride- and Hydroxide-Based MOR Membranes

Mordenite has two types of channels of 0.65 × 0.70 nm parallel to the *c*-axis and 0.26 × 0.57 nm parallel to the *b*-axis, together with a medium Si/Al ratio of 5.0–10.0 in framework. The mordenite membrane is a good candidate for the dehydration of organic solutions containing acids due to its hydrophilicity and acidic resistance.

Preparation, fabrication and application of acid-stable mordenite membranes have drawn many researchers' attention in the last decades. Navajas et al. (2002) reported a mordenite membrane on an alumina tubular support which displayed a flux of 0.20 kg $m^{-2} h^{-1}$ and a water/ethanol selectivity of 150 at a feed temperature of 150°C. In a later work, Navajas et al. (2006) developed a post-synthetic treatment with alkali solutions to increase the flux to 0.91 kg $m^{-2} h^{-1}$and the selectivity to 203. Li et al. (2003b) prepared *c*-oriented mordenite membranes, and the best membrane had a flux of approximately 0.60 kg $m^{-2} h^{-1}$ and a water/isopropanol selectivity of approximately 5000 at 75°C.

Li et al. (2009) fabricated the mordenite membrane from the mixed-mineralizer system of hydroxide and fluoride ions precursor gel, and the membrane had better PV performance for dehydration of HAc/H2O mixture than the nonfluoride-containing mordenite membrane. It is well known that the addition of fluoride into the starting gel favored the crystallization of high-silica zeolites as a mineralizing agent and/or a structure-directing agent (Tavolaro, 2002; Chen et al., 2009; Gualtieri, 2009). Oriented zeolite MFI (Gualtieri, 2009) and beta (Chen et al., 2009) membranes were prepared in fluoride media on seeded alumina supports and stainless steel mesh, respectively. Few studies, however, reported the preparation and especially the role and location of F ion in view of low-silica zeolite using fluoride-mediated gels.

We prepared F-modified mordenite membranes using NH_4F-containing gels by a single hydrothermal synthesis. The role and location of F ion were also clarified. The thin mordenite membrane prepared from the NH_4F-containing gel had a high average flux of 1.50 kg m^{-2} h^{-1} and a high average separation factor of 1380 for a water/ethanol (10/90 wt.%) mixture at 75°C (Zhou et al., 2012). We also reported that the fluorine anion was present in the channels of the mordenite crystal, and the fluoride media could enhance the hydrophilicity of mordenite membranes (Zhou et al., 2012). And the 750 nm thickness mordenite zeolite layer was rapidly prepared by microwave heating and fluoride-containing precursor gel. Either for dehydration of HAc/H_2O mixtures or for dehydration of the esterification mixture online with concentrated H_2SO_4 as catalyst by PV, the membranes showed good acidic durability and dehydration performance (Zhu et al., 2014; Zhu et al., 2016a).

Generally, it is reported that the mordenite membrane prepared from the fluoride-containing precursor gel has a better acid stability and reproducibility than the membrane prepared from non fluoride-containing precursor gel (Li et al., 2009; Zhou et al., 2012). The durable resistance of the mordenite membrane prepared from the optimal synthesis condition (crystallization time and molar composition of the precursor synthesis gel were 5h and SiO_2:0.01 KF:0.08 Al_2O_3:0.2 Na_2O:35 H_2O) for continuously dehydrating a 90 wt% HAc/H_2O mixture at 75°C was studied (Zhu et al., 2016b). The PV performance of the membrane as a function of test time (168 h) is shown in Figure 12.8. Because only the K^+ cations of the membrane surface could be on-exchanged by H^+ cations during the dehydration process, both the fluxes and separation factor are gradually decreased at first with the PV test time and achieved stable values after 72 h. Even if the continuous test time is prolonged to 168 h, the flux and separation factor are 0.53 kg m^{-2} h^{-1} and 620, respectively. Furthermore, Table 12.2 presents the dehydration performance for a 90 wt% HAc/H_2O mixture at 75°C of eight pieces of mordenite membranes prepared independently. All the membranes have good and similar dehydration performances for a 90 wt% HAc/H_2O

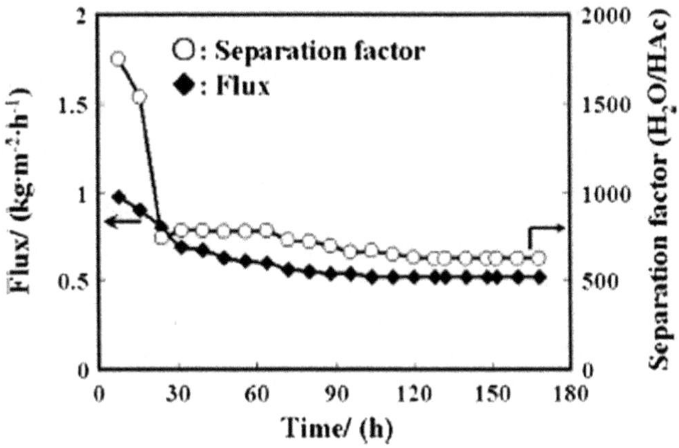

FIGURE 12.8 Long-term PV performance of the mordenite membrane for a 10 wt% H_2O/HAc mixture at 75°C.

TABLE 12.2
Reproducibility of the Mordenite Membranes for a 90 wt% HAc/H_2O Aqueous Mixture at 75°C[a]

No	Flux	Separation Factor
	Kg/(m^2h)	(H_2O/HAc)
1	0.85	760
2	0.76	820
3	0.85	560
4	0.84	580
5	0.77	870
6	0.78	680
7	0.75	810
8	0.84	810

[a] Molar composition of synthesis gel: SiO_2: 0.08, Al_2O_3: 0.2, Na_2O: 0.01, KF: 35 H_2O. Crystallization time: 5 h.

mixture at 75°C, the flux and $\alpha H_2O/HAc$ of the membranes are in the ranges 0.75–0.85 kg m^{-2} h^{-1} and 560–820. Thus, it appears that the mordenite membranes have good reproducibility and long-term acid stability.

Chen et al. (2012) also found that fluoride ions could optimize the distribution of aluminum atoms in mordenite membrane layer and the grain boundaries of mordenite crystals, which could keep the mordenite membrane from long-term acidic corrosion. Li et al. (2016) prepared the mordenite membranes from fluoride-containing dilute solution (H_2O/SiO_2 = 120) and microwave-assisted heating in a short synthesis time.

12.4 CONCLUSION

Separation processes play critical roles in manufacturing and their proper application can significantly reduce costs and increase profits. Alternative energy-saving and high efficiency separation process are strongly expected to be applied to many industries. Membrane separation appears to be a promising candidate because of low energy consumption, compact unit, simple operation and low environmental impact. Therefore, strong interest exists in the synthesis of membranes that exhibit both higher permeabilities and higher selectivities than presently available polymers. Membranes made from inorganic materials are generally superior to organo-polymeric materials in thermal and mechanical stability, and chemical resistance.

The introduction of microporous properties in inorganic membranes such as zeolite membranes appears to have strong potential application with respect to high temperature gas separation, pervaporation and vapor permeation for liquid mixtures and catalysis. Although disadvantages of zeolite membranes in comparison with polymer membranes are the presently higher production costs of the membranes and modules, significant progress has been made in synthesis and permeation performance of zeolite membranes in the past decade. Long-term acid-stable zeolite membranes for the dehydration process will certainly advance the acceptance of zeolite membranes in industrial processes.

ACKNOWLEDGMENTS

This work was supported by JSPS KAKENHI Grant Number 15H04174.

REFERENCES

Adachi-Pagano, M., Forano, C., Besse, J.P., 2003. Synthesis of Al-rich hydrotalcite-like compounds by using the urea hydrolysis reaction-control of size and morphology. Journal of Materials Chemistry. 13, 1988–1993.

Arichi, J., Louis, B., 2008. Toward microscopic design of zeolite crystals: Advantages of the fluoride-mediated synthesis. Crystal Growth & Design. 8, 3999–4005.

Breck, D.W., 1974. *Zeolite Molecular Sieves.* New York: Wiley.

Chen, Y.L., Zhu, G.S. Peng, Y., Yao, X.D., Qiu, S.L., 2009. Synthesis and characterization of (h0l) oriented high-silica zeolite beta membrane. Microporous and Mesoporous Materials. 124, 8–14.

Chen, Z., Li, Y.H., Yin, D.H., Song, Y.M., Ren, X.X., Lu, J.M., Yang, J.H., Wang, J.Q., 2012. Microstructural optimization of mordenite membrane for pervaporation dehydration of acetic acid. Journal of Membrane Science. 411–412, 182–192.

Dragomirova, R., Wohlrab, S., 2015. Zeolite membranes in catalysis – From separate units to particle coatings. Catalysts, 5, 2161–2222.

Feng, X., Huang, R.Y.M., 1997. Liquid separation by membrane pervaporation: A review. Industrial & Engineering Chemistry Research. 36, 1048–1066.

Ghobarkar, H., Schaf, O., Guth, U., 1999. Zeolites-from kitchen to space. Progress in solid state. Chemistry. 27, 29–73.

Gu, X., Dong, J., Nenoff, T.M., 2005. Synthesis of defect-free fau-type zeolite membranes and separation for dry and moist co2/n2 mixtures. Industrial & Engineering Chemistry Research. 44, 937–944.

Gualtieri, M.L., 2009. Synthesis of MFI films on α-alumina at neutral pH. Microporous and Mesoporous Materials. 117, 508–510.
Gues, E.R., Bekkum, H.V., 1995. Calcination of large MFI-type single crystals, part 2: Crack formation and thermomechanical properties in view of the preparation of zeolite membranes. Zeolites. 15, 333–341.
Hasegawa, Y., Nagase, T., Kiyozumi, Y., Hanaoka, T., Mizukami, F., 2010. Influence of acid on the permeation properties of NaA-type zeolite membranes. Journal of Membrane Science. 349, 189–194.
Hasegawa, Y., Abe, C., Mizukami, F., Kowata, Y., Hanaoka, T., 2012. Application of a CHA-type zeolite membrane to the esterification of adipic acid with isopropyl alcohol using sulfuric acid catalyst. Journal of Membrane Science. 415–416, 368–374.
Hölderich, W.F., Bekkum, H.V., 1991. Zeolites in organic syntheses. Studies in Surface Science and Catalysis. 58, 631–726.
Huang, P., Xu, N., Shi, J., Lin, Y.S., 1996. Characterization of asymmetric ceramic membranes by modified permporometry. Journal of Membrane Science. 116, 301–305.
Kita, H., 2006. Zeolite membranes for pervaporation and vapor permeation. In *Materials Science of Membranes for Gas and Vapor Separation,* Yampolskii, Yu., Pinnau, I., Freeman, B.D., (Ed.), p. 373. New York: Wiley.
Kondo, M., Kita, H., 2010. Permeation mechanism through zeolite NaA and T-type membranes for practical dehydration of organic solvents. Journal of Membrane Science. 361, 223–231.
Kosinov, N., Gascon, J., Kapteijn, F., Hensen, E.J.M., 2016. Recent developments in zeolite membranes for gas separation. Journal of Membrane Science. 499, 65–79.
Lai, R., Gavalas, G.R., 2000. ZSM-5 membrane synthesis with organic-free mixtures. Microporous and Mesoporous Materials. 38, 239–245.
Lassinantti, M., Jareman, F., Hedlund, J., Creaser, D., Sterte, J., 2001. Preparation and evaluation of thin ZSM-5 membranes synthesized in the absence of organic template molecules. Catalysis Today. 67, 109–119.
Li, G., Kikuchi, E., Matsukata, M., 2003a. A study on the pervaporation of water–acetic acid mixtures through ZSM-5 zeolite membranes. Journal of Membrane Science. 218, 185–194.
Li, G., Kikuchi, E., Matsukata, M., 2003b. Separation of water–acetic acid mixtures by pervaporation using a thin mordenite membrane. Separation and Purification Technology. 32, 199–206.
Li, L.Q., Yang, J.H., Li, J.J., Han, P., Wang, J.X., Zhao, Ye., Wang, J.Q., Lu, J.M., Yin, D.H., Zhang, Y., 2016. Synthesis of high performance mordenite membranes from fluoride-containing dilute solution under microwave-assisted heating. Journal of Membrane Science. 512, 83–92.
Li, X., Kita, H., Zhu, H., Zhang, Z., Tanaka, K., 2009. Synthesis of long-term acid-stable zeolite membranes and their potential application to esterification reactions. Journal of Membrane Science. 339, 224–232.
Li, X., Kita, H., Zhu, H., Zhang, Z., Tanaka, K., Okamoto, K., 2011. Influence of the hydrothermal synthetic parameters on the pervaporative separation performances of CHA-type zeolite membranes. Microporous and Mesoporous Materials. 143, 270–276.
Lipnizki, F., Field, R.W., Ten, P.-K., 1999a. Pervaporation-based hybrid process: A review of process design: Applications and economics. Journal of Membrane Science. 153, 183–210.
Lipnizki, F., Hausmanns, S., Ten, P.-K., Field, R.W., Laufenberg, G., 1999b. Organophilic pervaporation prospects and performance. Chemical Engineering Journal. 73, 113–129.
Liu, B., Kumakiri, I., Tanaka, K., Chen, X., Kita, H., 2016. Preparation of Rho zeolite membranes on tubular supports. Membrane. 41, 81–86.

Machado, F.J., Lópeza, C.M., Centeno, M.A., Urbina, C., 1999. Template-free synthesis and catalytic behavior of aluminum-rich MFI-type zeolites. Applied Catalysis A: General. 181, 29–38.

Mintova, S., Hedlund, J., Valtchev, V., Schoeman, B.J., Sterte, J., 1998. ZSM-5films prepared from template free precursors. Journal of Materials Chemistry. 8, 2217–2221.

Navajas, A., Mallada, R., Tellez, C., Coronas, J., Menendez, M., Santamaria, J., 2002. Preparation of mordenite membranes for pervaporation of water-ethanol mixtures. Desalination. 148 (2002), 25–29.

Navajas, A., Mallada, R., Tellez, C., Coronas, J., Menendez, M., Santamaria, J., 2006. The use of post-synthetic treatments to improve the pervaporation performance of mordenite membranes. Journal of Membrane Science. 270, 32–41.

Noack, M., Kölsch, P., Caro, J., Schneider, M., Toussaint, P., Sieber, I., 2000. MFI membranes of different Si/Al ratios for pervaporation and steam permeation. Microporous and Mesoporous Materials. 35–36, 253–265.

Noack, M., Kölsch, P., Dittmar, A., Stohr, M., Georgi, G., Schneider, M., Dingerdissen, U., Feldhoff, A., Caro, J., 2007. Proof of the ISS-concept for LTA and FAU membranes and their characterization by extended gas permeation studies. Microporous and Mesoporous Materials. 102, 1–20.

Noack, M., Kölsch, P., Seefeld, V., Toussaint, P., Georgi, G., Caro, J., 2005. Influence of the Si/Al-ratio on the permeation properties of MFI-membranes. Microporous and Mesoporous Materials. 79, 329–337.

Nomura, M., Yamaguchi, T., Nakao, S., 1998. Ethanol/water transport through silicalite membranes. Journal of Membrane Science. 144, 161–171.

Okamoto, K., Kita, H., Horii, K., Tanaka, K., Kondo, M., 2001. Zeolite NaA membrane: Preparation, single-gas permeation, and pervaporation and vapor permeation of water/organic liquid mixtures. Industrial & Engineering Chemistry Research. 40, 163–175.

Pan, M., Lin, Y.S., 2001. Template-free secondary growth synthesis of MFI type zeolite membranes. Microporous and Mesoporous Materials. 43, 319–327.

Pina, M.P., Arruebo, M., Felipe, M., Fleta, F., Bernal, M.P., Coronas, J., Menendez, M., Santamaria, J., 2004. A semi-continuous method for the synthesis of NaA zeolite membranes on tubular supports. Journal of Membrane Science. 244, 141–150.

Qiu, L., Kumakiri, I., Tanaka, K., Kita, H., 2015. Dehydration performance of sodalite membranes prepared by secondary growth method. Membrane. 40, 349–354.

Rangnekar, N., Mittal, N., Elyassi, B., Caro, J., Tsapatsis, M., 2015. Zeolite membranes – A review and comparison with MOFs. Chemical Society Reviews. 44, 7128–7154.

Sander, U., Soukup, P., 1988. Design and operation of a pervaporation plant for ethanol dehydration. Journal of Membrane Science. 36, 463–475.

Sommer, S., Melin, T., 2005. Influence of operation parameters on the separation of mixtures by pervaporation and vapor permeation with inorganic membranes. Part 1: Dehydration of solvents. Chemical Engineering Science. 60, 4500–4523.

Tanaka, K., Yoshikawa, R., Cui, Y., Kita, H., Okamoto, K., 2002. Application of Zeolite T Membrane to Vapor-permeation aided Esterification of Lactic Acid with Ethanol. Chemical Engingeering Science. 57, 1577–1584.

Tavolaro, A., 2002. VS-1 composite membrane: Preparation and characterization. Desalination. 147, 333–338.

Tsuru, T., Wada, S., Izumi, S., Asaeda, M., 1998. Silica–zirconia membranes for nanofiltration. Journal of Membrane Science. 149, 127–135.

Tsuru, T., Hino, T., Yoshioka, T., Asaeda, M., 2001. Permporometry characterization of microporous ceramic membranes. Journal of Membrane Science. 186, 257–265.

Wijimans, J.G., Baker, R.W., 1993. A simple predictive treatment of the permeation process in pervaporation. Journal of Membrane Science. 79, 101–113.

Yu, M., Noble, R.D., Falconer, J.L., 2011. Zeolite membranes: Microstructure characterization and permeation mechanism. Accounts of Chemical Research. 44, 1196–1206.

Zhou, R., Hu, Z., Hu, N., Duan, L., Chen, X., Kita, H., 2012. Preparation and microstructural analysis of high-performance mordenite membranes in fluoride media. Microporous and Mesoporous Materials. 156, 166–170.

Zhu, G., Li, Y., Zhou, H., Liu, J., Yang, W., 2009. Microwave synthesis of high performance FAU-type zeolite membranes: Optimization, characterization and pervaporation dehydration of alcohols. Journal of Membrane Science. 337, 47–54.

Zhu, M., Lu, Z., Kumakiri, I., Tanaka, K., Chen, X., Kita, H., 2012. Preparation and characterization of high water perm-selectivity ZSM-5 membrane without organic template. Journal of Membrane Science. 415, 57–65.

Zhu, M., Kumakiri, I., Tanaka, K. Kita, H., 2013. Dehydration of acetic acid and esterification products by acid-stable ZSM-5 membranes. Microporous and Mesoporous Materials. 181, 47–53.

Zhu, M., Xia, S.L., Hua, X.M., Feng, Z.J., Hu, N., Zhang, F., Kumakiri, I., Lu, Z.H., Chen, X., Kita, H., 2014. Rapid preparation of acid-stable and high dehydration performance mordenite membranes. Industrial & Engineering Chemistry Research. 53, 19168–19174.

Zhu, M., Feng, Z.J., Hua, X.M., Hu, H.L., Xia, S.L., Hu, N., Zhen, Y., Kumakiri, I., Chen, X., Kita, H., 2016a. Application of a mordenite membrane to the esterification of acetic acid and alcohol using sulfuric acid catalyst. Microporous and Mesoporous Materials. 233, 171–176.

Zhu, M., Hua, X., Liu, Y., Hu, H., Li, Y., Hu, N., Kumakiri, I., Chen, X., Kita, H., 2016b. Influences of synthesis parameters on preparation of acid-stable and reproducible mordenite membrane. Industrial & Engineering Chemical Research. 55, 12268–12275.

13 Solvent Resistant Nanofiltration Membranes Prepared via Phase Inversion

Maarten Bastin and Ivo F.J. Vankelecom

CONTENTS

13.1 INTRODUCTION

Solvent resistant nanofiltration (SRNF) is the branch of membrane technology that focuses on the separation of products smaller than 1000 Da from organic solvents. SRNF is a fairly new field, with the earliest publications around the 1990s. Taking into account the diversity of applications in (petro-)chemical and pharmaceutical industry and the impact of separation on the overall costs and energy requirements of a finished product, interest in SRNF has grown substantially ever since (Marchetti et al., 2014).

Membranes can either be of organic (polymeric) or inorganic (ceramic) nature or a mix of both (mixed matrix membrane). When developing solvent resistant membranes, inorganic or ceramic membranes have the advantage that they have a high chemical and thermal stability, often a longer lifetime and are easier to clean (Mulder, 1996). However, they also have high capital costs making polymeric membranes often more cost-effective to implement.

When it comes to preparation methods for polymeric membranes for SRNF, phase inversion is one of the most commonly used techniques. Sometimes, only a support layer is prepared via phase inversion, and subsequently, an interfacial polymerization (IFP) step is introduced to create the selective layer. In this case, the support layer is impregnated with a monomer dissolved in a solvent (usually water). On this support, a second solution is poured, containing a second monomer that is dissolved in another solvent, immiscible and of lower density than the first solvent. In this way, the monomers can only react on the interface, forming, often through a polycondensation reaction, an ultrathin polymeric layer, on top of the support (Raaijmakers and Benes, 2016). This chapter will only deal with SRNF membranes prepared via phase inversion.

Phase inversion, introduced as early as 1918 by Zsigmondy (Zsigmondy and Bachmann, 1918), is based on the principle of transforming a polymer from a liquid to a solid state in a controlled manner. This solidification process can be approached via different methods of which immersion precipitation is the most commonly used in the production of commercial membranes. This technique was introduced in the 1960s by Loeb and Sourirajan (Loeb and Sourirajan, 1963), and can be considered the most important breakthrough in membrane development. This method can result in asymmetric membranes, with a thin and dense top-layer, and an underlaying, more open, support layer, as shown in Figure 13.1. All layers originate from the same material in such membranes, hence the term Integrally Skinned Asymmetric (ISA) membranes (Marchetti et al., 2014).

When the process of phase inversion is described in this chapter, it will mostly refer to immersion precipitation. There are other phase inversion techniques, like thermally induced or vapour induced phase separation, but they are outside the scope of this chapter, as they have not yet been applied for SRNF.

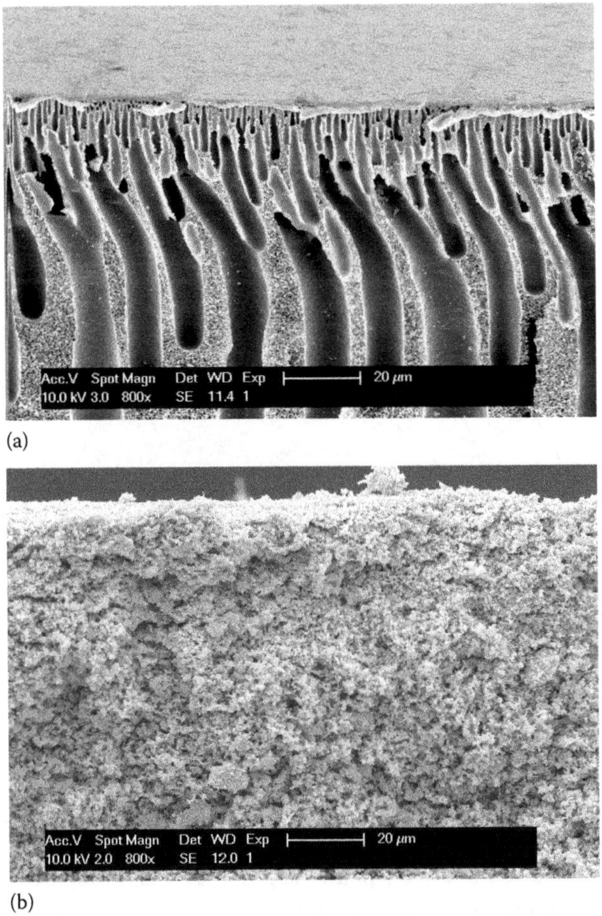

FIGURE 13.1 SEM pictures of the cross section of two typical phase inversion membranes. (a) Membrane with presence of macrovoids. (b) Membrane with a sponge-like structure.

13.2 BASIC PRINCIPLES

Phase inversion is a process in which at least three, but sometimes up to nine or ten, components interact with each other. Two key components are a polymer dissolved in a solvent to form a single phase. Such homogeneous solution is cast as a thin film to produce a flat-sheet membrane, or as a hollow fiber, capillary or tubular membrane when cast from a spinneret. After casting and possible exposure to a short evaporation stage, the actual process of phase inversion will take place. This happens when

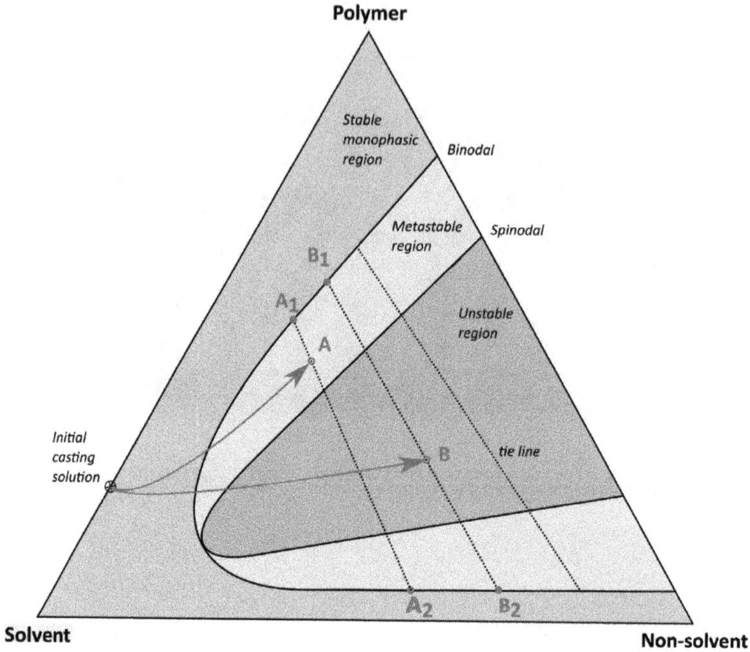

FIGURE 13.2 Ternary phase diagram and the pathways that can be followed during phase-inversion. A: binodal demixing. B: spinodal decomposition.

a third component is introduced to this system, which is (a) miscible with the solvent and (b) not capable of dissolving the polymer. The thermodynamic equilibrium then gets disturbed, demixing takes place, and the polymer solidifies. This process can best be described by a ternary phase diagram, as seen in Figure 13.2. This figure is often a simplification of a real phase inversion process, where it is possible that several other components, like co-solvents or additives, are present in the cast membrane solution or in the non-solvent bath (Baker, 2004; Hołda and Vankelecom, 2015; Mulder, 1996).

13.2.1 Thermodynamic Considerations

Initially, a homogeneous polymer solution is prepared. This solution is thermodynamically stable and mostly located on the polymer/solvent axis of the diagram. However, it is possible to add a small portion of non-solvent directly to the initial polymer solution, while still keeping it in the stable monophasic region. This solution is then located in the region between the polymer/solvent axis, the solvent/non-solvent axis and the binodal. The binodal is defined as the curve where two distinct phases can co-exist (Jansen et al., 2005). It can be determined by cloudpoint measurements. These involve addition of non-solvent to a polymer/solvent solution untill turbidity in this solution remains permanently. Between the binodal and spinodal curve, a metastable region is present. In this region, the polymer solution is unstable but will not precipitate, unless well nucleated. When a polymer film is cast and immersed into the coagulation bath, the composition inside the film will change due to the increase in non-solvent concentration. As the cast film loses more and more solvent during the solvent/non-solvent exchange, a region is entered where the cast solution is thermodynamically unstable and it will spontaneously separate in two phases (Baker, 2004; Vandezande et al., 2008). The spinodal curve is the border between this unstable and metastable region, and can be determined by Pressure Pulsed Induced Critical Scattering (PPICS) (Wells et al., 1993) or by small angle neutron scattering (SANS) (Lefebvre et al., 2002).

During coagulation, two situations can occur, as described in Figure 13.2 by end-points A and B, representing the so-called binodal demixing and spinodal decomposition, respectively. When point A is the endpoint, the composition of the cast polymer film will end up in the metastable region between the binodal and the spinodal. As more non-solvent enters the composition, liquid-liquid demixing occurs, and phase separation will cause the solution to form a polymer-rich phase and a polymer-lean phase. The polymer-lean phase will start to form nuclei and will withdraw more and more liquid from the polymer-rich phase. This process will continue until thermodynamic stability is re-established. The polymer-rich phase will solidify and form the polymer matrix of the resulting membrane. If during this process the polymer-lean nuclei grow to such an extent that they can merge together, a contineous porous structure is formed. The composition of these two phases is indicated on the phase diagram by the ends of the so-called tie-lines, connecting point A_1 (composition of the polymer-rich phase) and A_2 (composition of the polymer-lean phase), i.e., lines that connect two phases in equilibrium with each other (Hołda and Vankelecom, 2015).

13.2.2 Kinetic Considerations and Pore Formation

The phase diagram can give information on the composition of the membrane and how the membrane is formed. It will not always give full information about the final morphology of the formed phases, since the kinetic aspects of the phase inversion process also need to be taken into account. The structure of the overall membrane is mostly dependent on the exchange rate of the solvent out and the non-solvent into the casting solution. The mass transfer of both components (i.e., solvent and non-solvent) over the two phases is driven by the chemical potential gradient. Young and Chen (1995) clarified this process with a diffusion-controlled model. In this model, the cast solution can be divided in N hypothetical inner layers, parallel to the surface of the cast solution. If the thickness of each layer is made small enough, the diffusion ratio of solvent to non-solvent can be considered as a constant value for each layer. This ratio can be written as:

$$k = \frac{\bar{n}_2}{\bar{n}_1} \tag{13.1}$$

where $\bar{n}_1$ and $\bar{n}_2$ are the diffusive flux of non-solvent and solvent, respectively.

The non-solvent penetrating the cast film will interact with the polymer, while the diffusion of solvent into the coagulation bath is related to its heat of mixing with the non-solvent (Young and Chen, 1995). As it is not easy to find these diffusion parameters for each layer, the model was simplified towards a two-step mechanism model where formation of a membrane is a process of the formation of a top-layer and a sublayer, both with different k-values. This makes it possible to make a distinction between an asymmetric and a homogeneous membrane.

With a large k-value of the top-layer, the solvent diffuses faster out of the cast solution than that the non-solvent diffuses in. This will result in a dense top-layer because no solvent is left to supply the pores to grow. The dense top-layer will control the diffusion in and out of the sublayer, hindering the diffusion of solvent out and non-solvent into the sublayer. When the sublayer has a small k-value, the top-layer mainly limits solvent out-diffusion. As a result, the solvent is still present when non-solvent diffuses into the sublayer, and the pores are allowed to grow. After a longer time, the composition path will still cross the binodal, resulting in binodal demixing to form a porous sublayer underneath the dense top-layer (Figure 13.2, endpoint A). However, when the dense top-layer mainly limits non-solvent in-diffusion, i.e., with a large k-value of the sublayer, the solvent will leave the sublayer, not giving the pores the chance to grow. At the same time, the non-solvent will diffuse slowly into the sublayer, giving the polymer-rich phase more time to solidify with a dense, homogenous membrane as endresult. Solvent and non-solvent fluxes are large when driving forces are large (i.e., chemical potential gradient) and/or when molecular size is small.

With a small k-value of the top-layer, the solvent diffuses more slowly out of the cast film than that the non-solvent diffuses in. The composition of the top-layer will cross the binodal very fast to cause liquid-liquid binodal demixing. The diffusion in and out of the sublayer is only slightly affected by the top-layer and the same morphology is obtained throughout the membrane: a homogeneous porous structure is then formed. Figure 13.3 gives a schematic overview of the effect of the k-values on the overall membrane morphology.

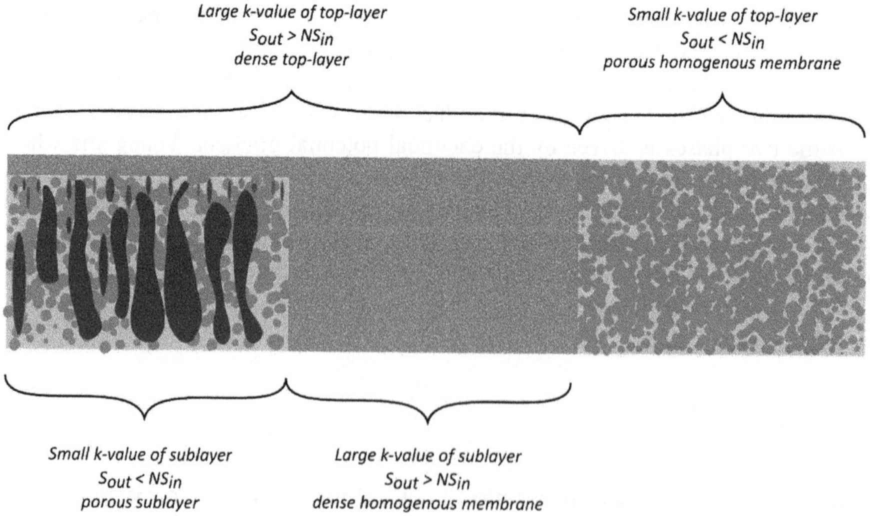

FIGURE 13.3 Kinetics of Phase Inversion Membrane Formation.

Not only can the difference between dense and porous or homogenous and asymmetric membrane formation be partly derived from a ternary phase diagram, so too can the morphology of the pores. The pores are formed by the polymer-lean phase. This phase contains a mixture of solvent and non-solvent, the composition of which can be seen when following the tie-lines. From a thermodynamic point of view, the main driving force for pore growth is a decrease in the free energy on mixing solvent and non-solvent in the pores. The pore will only stop growing when solvent diffusion into the polymer-lean phase has stopped. As long as a sufficient supply of solvent, and a large enough potential gradient remains available, pores can continue to grow. When the solution reaches the solidification region, the pore wall is formed. When the mutual affinity between solvent and non-solvent is large, this pore wall formation will happen late and the pore is able to grow into finger-like macrovoid structures. Another condition for getting finger-like pores, is that the top-layer limits the in-diffusion of large amounts of non-solvent. The non-solvent is responsible for nuclei formation. This way, the formation of many small nuclei is inhibited.

When, on the other hand, the affinity between solvent and non-solvent is smaller, the mixing tendency and the liquid-liquid phase separation of the system are both slower. As a consequence, many nuclei are initiated later. The growth of the nuclei is limited by the solvent consumption of every other nuclei, and growth of macrovoids is impossible: only smaller pores are formed and a sponge-like structure is obtained (Young and Chen, 1995).

13.3 SYNTHESIS OF AN INTEGRALLY SKINNED ASYMMETRIC PHASE INVERSION MEMBRANE VIA IMMERSION PRECIPITATION

It has become clear in the previous sections that membrane formation is greatly dependent on getting control over the amount of non-solvent that diffuses into the cast film versus the amount of solvent that diffuses out. In this section, different parameters that have an influence on these diffusion rates (as shown in Table 13.1) will be discussed. Most of these will influence solvent and non-solvent diffusion in the same direction but to a different extent. Sometimes very specific interactions with either solvent or non-solvent will only impact the diffusion of one of them.

13.3.1 Composition of the Casting Solution

Making a polymeric membrane always starts from dissolving a polymer in a solvent. It is obvious that its physical and chemical properties will be a dominant factor in the final membrane morphology. However, besides polymer and solvent, other components like co-solvents, pore formers or non-solvents can be added to the casting solution and can change the phase diagram completely.

13.3.1.1 Polymer

The selection of the polymer is one of the key factors in preparing a good membrane for a certain application. The thermal, chemical and mechanical properties of the polymer will strongly affect the membrane performance. These properties

TABLE 13.1
Overview of Parameters Influencing Phase Inversion Membrane Formation

Composition of the Casting Solution	
	Polymer
	Solvent
	Additives
Casting conditions	
	Evaporation time and temperature
	Relative humidity (RH)
	Membrane thickness
	Casting speed
Coagulation conditions	
	Composition and temperature of the coagulation bath
Post treatment	
	Annealing
	Drying via solvent exchange and use of conditioning agents
	Solvent treatment
	Crosslinking

are directly derived from the chemical structure, the polymer molecular weight, glass transition temperature and polydispersity index. Also the purity of the polymer sample is a parameter that may not be overlooked. Often, commercially available polymers are used because they are easily accessible. However, companies usually produce these polymers for other applications on a larger scale than for merely membrane preparation. Additives, like flame retardants, are then sometimes present in the polymer sample, which will obviously influence the membrane preparation process. A pretreatment, like dissolving the polymer in different solvents and precipitating it, can be necessary in these cases (Hołda and Vankelecom, 2015).

A variety of polymers has already been used for the preparation of membranes via phase inversion for use in liquid filtrations. The most important ones are shown in Table 13.2. With these polymers, membranes have been made both for the NF and UF range. UF membranes can later also be used as support for TFC NF membranes (Marchetti et al., 2014).

Besides polymer type, the concentration of polymer in the casting solution is also important. The higher the concentration, the more viscous the solution becomes. This will have an effect on the non-solvent in-diffusion and the solvent out-diffusion, with a more viscous solution producing denser and thicker skinlayers (Hołda and Vankelecom, 2015).

13.3.1.2 Solvent

The solvent has to be chosen as a function of the polymer, since it may not undergo any reactions with it while being able to dissolve it. The final membrane morphology depends on the affinity of the solvent for the non-solvent for a given polymer.

TABLE 13.2
Polymers Used for ISA Membranes

Polymer	Structure
Polyamide-imide (Torlon®)	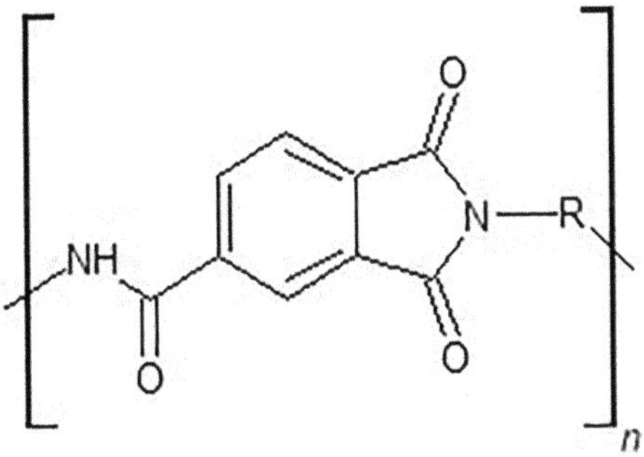
Polyaniline (PANI)	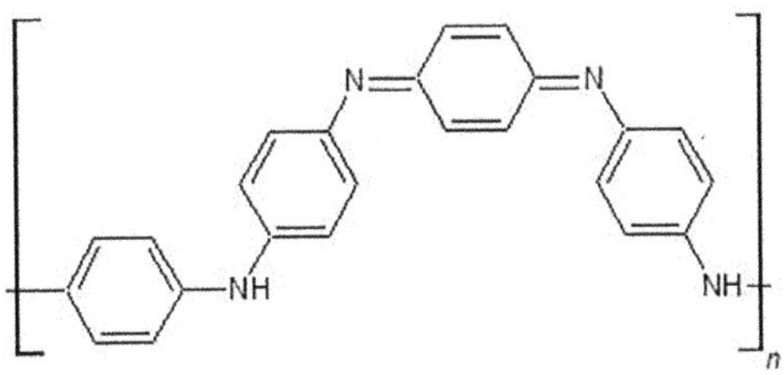
Polybenzimidazole (PBI)	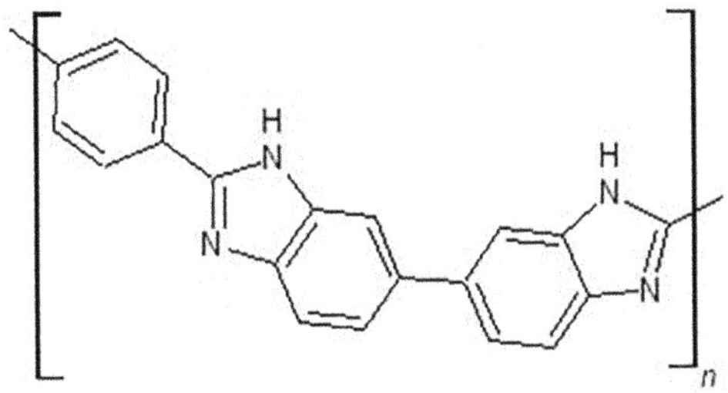
Poly (ether-ether-ketone) (PEEK)	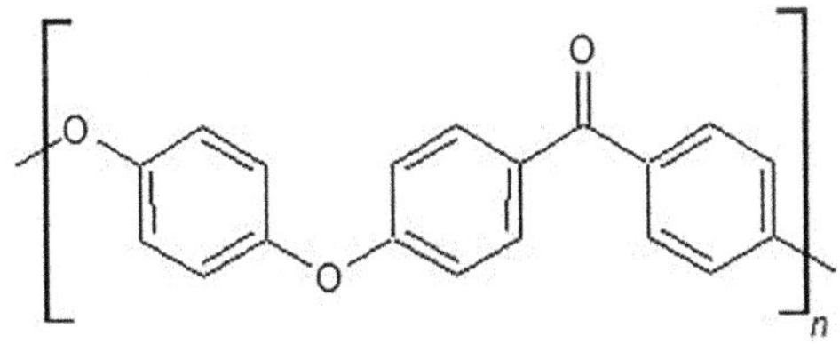
BPAPEEK	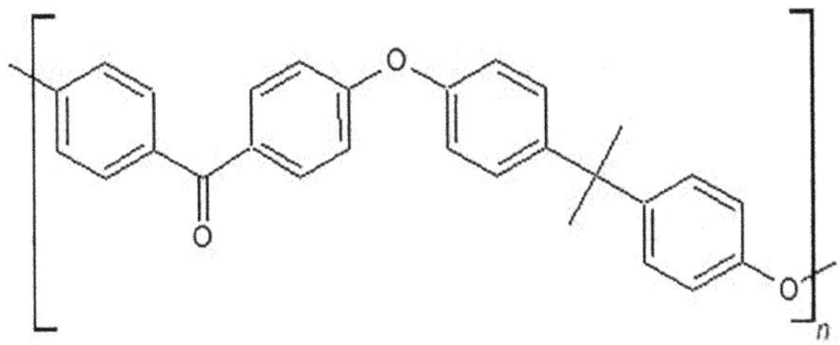

(Continued)

TABLE 13.2 (CONTINUED)
Polymers Used for ISA Membranes

Polymer	Structure
TBPEEK	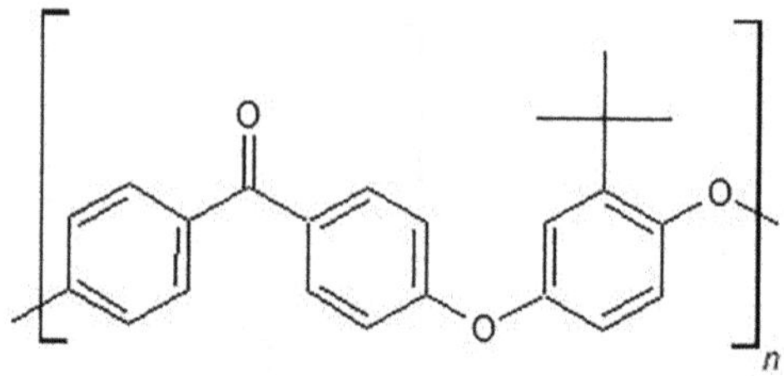
VAPEEK	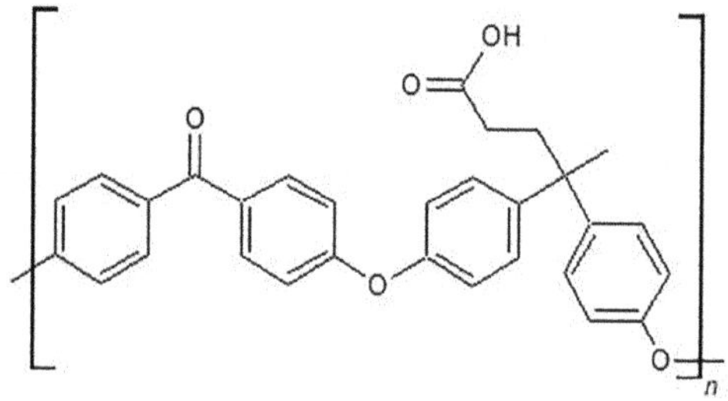
Poly (ether imide) (PEI)	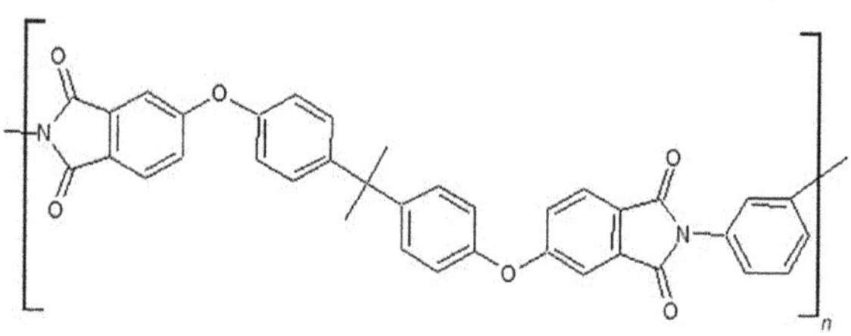
Polyimide (PI) Matrimid®	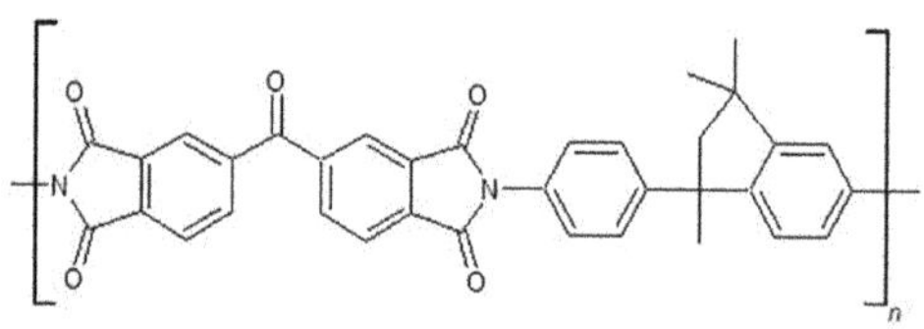
Polyimide (PI) P84®	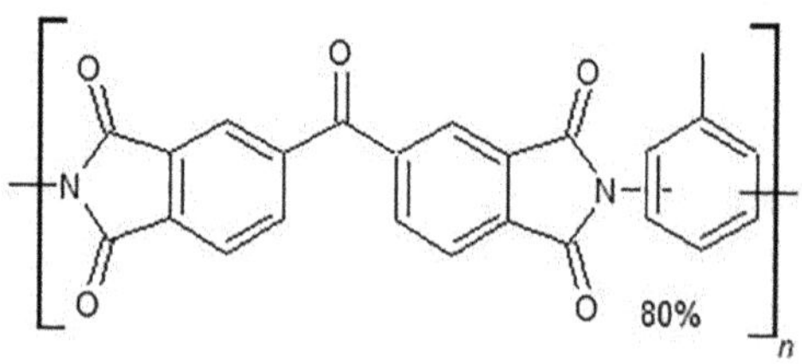

(*Continued*)

TABLE 13.2 (CONTINUED)
Polymers Used for ISA Membranes

Polymer	Structure
	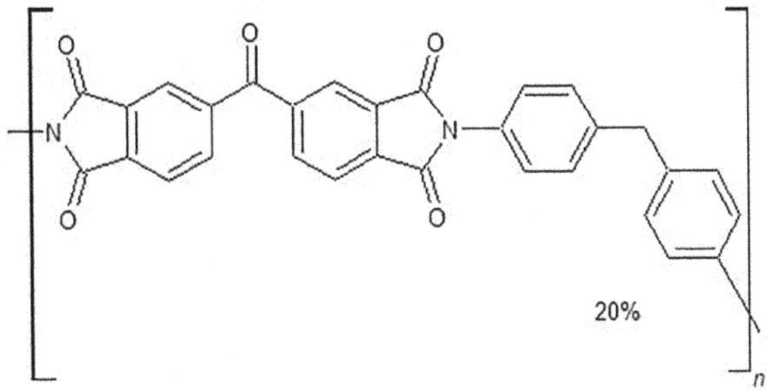
Polyphenylsulfone (PPSf)	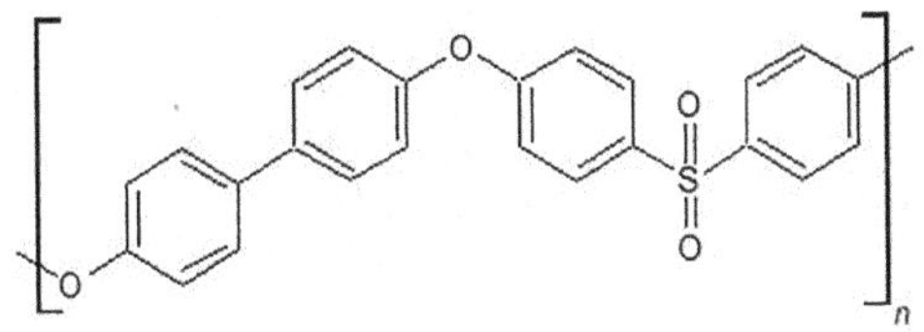
Polysulfone (PSf)	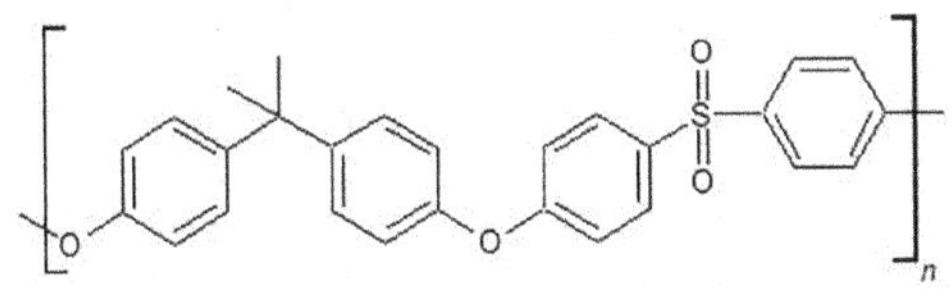

As described in Figure 13.4, a different solvent can completely change the ternary phase diagram. The binodal curve will be at a different place, and the tie-lines will be different. Depending on the miscibility with the non-solvent, delayed demixing or instantaneous demixing can occur and more porous or more dense membranes will be obtained. Typical solvents used for SRNF polymers are NMP, DMF, THF, DMSO and DCM. Many of these solvents are highly toxic, carcinogenic or hazardous (Szekely et al., 2014). In the last decade, a lot of effort has going into replacing them with greener alternatives, such as ionic liquids (Mariën et al., 2016). A useful tool to help select the correct solvent for a given polymer is by calculating the Hansen solubility parameters, which should be as similar as possible for polymer and solvent (Guillen et al., 2011).

13.3.1.3 Additives

To optimize membrane performance, different components can be added to the casting solution, like co-solvents, non-solvents or inorganic and organic additives. They can be present as a dispersion or fully dissolved and are added to influence the phase diagram and membrane morphology, to have a sharper or lower MWCO-curve, to change hydrophilicity or to increase permeance and/or retention. In general, the additive properties that cause these changes are their concentration, viscosity,

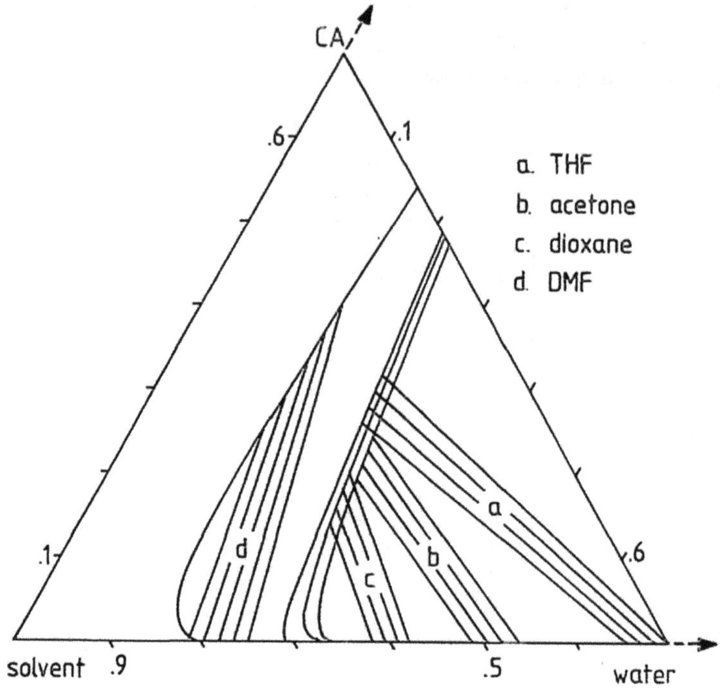

FIGURE 13.4 Binodals and tie-lines for ternary CA/solvent/water systems. (From Reuvers, A.J., 1987. Membrane formation: Diffusion induced demixing processes in ternary polymeric systems (Doctoral dissertation). Retrieved from https://research.utwente.nl/en/publications/membrane-formation-diffusion-induced-demixing-processes-in-ternar.)

volatility, non-solvent power and their specific interactions with polymer, solvent and non-solvent (Hołda and Vankelecom, 2015).

Volatile co-solvents can be added to create a denser top-layer by introducing an evaporation step just before the film is moved into the coagulation bath. After casting, they will selectively evaporate from the cast film, creating locally a higher polymer concentration. The dense top-layer will function as a barrier between the non-solvent and the solvent, slowing down the demixing process (Vandezande et al., 2009). In this way, volatile co-solvents promote delayed demixing and can avoid macrovoid formation.

When adding a non-solvent to the casting solution, the opposite will be accomplished. In the phase diagram, the composition of the polymer film will be closer to the binodal curve and after immersing the film in the coagulation bath, more instantaneous demixing will occur and a more porous structure will form. The maximal concentration of non-solvent in the casting solution is determined by the position of the binodal. The composition may not yet loose its thermodynamic stability before casting the membrane, or demixing would already happen.

Surfactants affect the interfacial properties between the polymer solution and the coagulant and influence the phase inversion. The addition of surfactants can enhance the affinity between solvent and non-solvent, resulting in a shift from delayed demixing to instantaneous demixing (Guillen et al., 2011).

Organic or inorganic components are added to the casting solution to enhance pore formation, to improve pore interconnectivity, to increase hydrophilicity or to prevent macrovoid formation (Guillen et al., 2011). For example, LiCl is used to reduce macrovoid formation and increase the mechanical stability of the membranes (Fontananova et al., 2006). When using polyvinylpyrrolidone (PVP) as an additive, macrovoid formation can increase or decrease, depending on the molecular weight of the PVP (Yoo et al., 2004). Another example of a polymer that has been extensively studied in its function as additive is polyethylene glycol (PEG) (Kim and Lee, 1998; Ma et al., 2011; Zheng et al., 2006). With an increase in molecular weight of PEG, not only the pore size increased but also the number of pores (Chakrabarty et al., 2008).

Metal oxides can be introduced to the polymer solution as another macrovoid suppressor, as they increase the viscosity of the polymer solution, slow down the solvent/non-solvent exchange rate and impose delayed demixing (Li et al., 2007; Soroko and Livingston, 2009). Noble metal nanoparticles have been added to the polymer solution to increase the permeance without compromising retention as their photothermal heating properties can convert light into heat. Gold, silver and copper especially exhibit unique and tunable optical properties due to their surface plasmon resonance (Li et al., 2013).

13.3.2 Casting Conditions

Besides the chemical composition, there are a wide range of parameters that influence the structure and performance of the finished membrane. These parameters are more related to the process and the environment in which the membrane is prepared, like the casting thickness and speed, or the temperature and relative humidity in the casting room. In this section, the effects of these parameters will be discussed.

13.3.2.1 Evaporation Time and Temperature

When using a volatile (co-)solvent, the time between casting and coagulation will influence the membrane structure. As long as (co-)solvent evaporates out of the film, the concentration of polymer at the top will increase and a kind of "skin-layer" is formed which increases in thickness with longer evaporation times (See-Toh et al., 2007b). This layer will act as a barrier and will prevent to a certain extent the diffusion of non-solvent into the bulk of the membrane and conversely also slows down solvent out-diffusion. The rate of evaporation is a function of the temperature of the casting environment.

13.3.2.2 Relative Humidity (RH)

In most cases, water is a non-solvent for the casting solution. Water from the air can be sorbed in the cast polymer film, and a partial demixing can already take place before the aimed coagulation has started. This problem is of a bigger concern when hygroscopic solvents are used, like NMP or DMF, or when longer evaporation times are used. Usually, a certain critical RH exists below which RH will be of no influence on the membrane formation (Boussu et al., 2006).

13.3.2.3 Membrane Thickness

Membrane casting thickness is a parameter that is not often taken into account because ideally only the resistance of the selective top-layer should determine the permeance. However, when the top-layer gets thin enough while remaining defect-free, the transport through the support layer can become rate limiting. There is a critical structure-transition thickness, indicating a transition of the membrane morphology from a sponge-like to a finger-like structure with an increase in membrane thickness. The thickness is different for each cast solution and depends on its viscosity, surface energy, phase diagram and many other characteristics (Li et al., 2004). Thinner membranes can also lower the production costs, because less polymer and solvents are used and less solvents will contaminate the coagulation bath, which thus needs less frequent replacement.

13.3.2.4 Casting Speed

The casting speed is another factor that is often overlooked. It had a direct effect on the membrane morphology in polyethersulfone (PES) membranes where membranes cast at higher shear rates formed thinner selective skins. The casting speed also affected pore radius and macrovoid formation. Hence, an optimum casting speed can be found for a certain viscosity and casting solution (Ali et al., 2010).

13.3.3 Coagulation Conditions

Composition and temperature of the coagulation bath will influence the membrane morphology, as they influence the exchange rate of solvent and non-solvent during phase inversion. When upscaling is considered, heating up the coagulation bath, or using other non-solvents than water, can seriously raise the production costs.

13.3.3.1 Composition and Temperature of the Coagulation Bath

Similar to adding non-solvent to the casting solution, solvent can be added to the coagulation bath. When solvent is already present in the coagulation bath, instantaneous demixing can become delayed demixing, and a less porous membrane will be formed (Chun et al., 2000). The maximum amount of solvent that can be added depends on the binodal of the phase diagram and on the temperature. For systems with an upper critical solution temperature, increasing temperatures will decrease the demixing area, as described in Figure 13.5. When the temperature is high enough, the components can become miscible in all proportions (Mulder, 1996).

13.3.4 Post Treatment

Post treatment can be performed to increase the performance of the membranes, but also to make them more stable in different environments, like harsh solvents, extreme pH and high temperatures. Often as a last step, a conditioning treatment is executed to handle the membrane more easily.

13.3.4.1 Annealing

By annealing, the membrane structure will densify and the number of defects will be reduced. The retention of the membrane will increase while the permeance goes

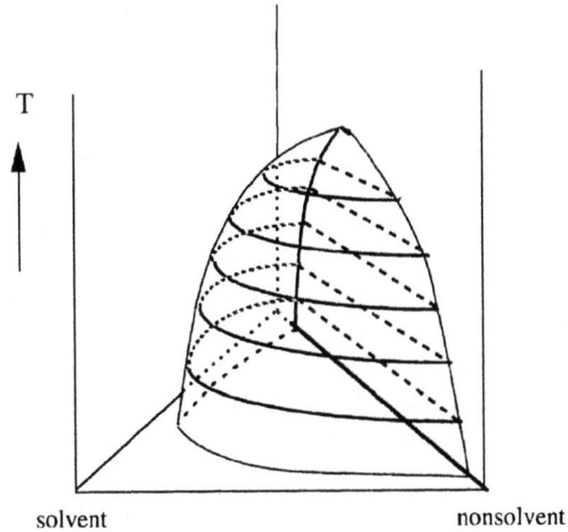

FIGURE 13.5 The shape of binodals at various temperatures for a ternary system consisting of polymer, solvent and a non-solvent. (From Mulder, M., 1996. *Basic Principles of Membrane Technology* (Second ed.). Dordrecht, The Netherlands: Kluwer Academic Publishers.)

down. To avoid excessive pore collapse, wet thermal annealing is done by immersing the membrane in a hot bath, typically filled with water, around 70–90°C, depending on the polymer (Hołda and Vankelecom, 2015; Su et al., 2010).

13.3.4.2 Drying via Solvent Exchange and Use of Conditioning Agents

Dry membrane storage can be a solution for practical issues, like transport or ease of handling. Unfortunately, capillary pressure can cause pore collapse when water evaporates from the pores (Siddique et al., 2014), and a conditioning step needs to be taken before storage is possible without damaging the membrane structure. This procedure often involves multiple liquid exchanges. First, the residual non-solvent needs to be removed from the membrane with a liquid that is miscible with the non-solvent and not able to dissolve the polymer. Second, this liquid needs to be replaced with a (series of) more volatile solvent(s), that eventually can be removed via evaporation without damaging the membrane structure (Macdonald et al., 1974).

13.3.4.3 Solvent Treatment

When applying dry membranes, it is often recommended to presoak them before starting the filtration in the feed solution. This treatment will swell the polymer, resulting in increased permeation (Hołda and Vankelecom, 2015). Presoaking the membrane in solvents like DMF or DMSO, can increase the permeance of certain membranes even further. These so called "activating solvents" are believed to cause temporary swelling of the top-layer, so that unreacted oligomers will dissolve in the activating solvent. This process is not fully understood yet, but as a consequence creates additional free volume in the top-layer, and thus increasing the permeance, without compromising retention (Hermans et al., 2015; Jimenez Solomon et al., 2012).

13.3.4.4 Crosslinking

Polymeric membranes prepared via phase inversion are usually not stable enough to withstand the harsh conditions of many feed solutions and need an additional crosslinking step before they can be used as SRNF membrane. Possible methods are thermal, UV, electron beam (EB) or chemical crosslinking. Heat treatment can turn ionically crosslinked membranes into a covalent crosslinked network structure (Joseph et al., 2017) or create radicals that form interchain bonds (Sairam et al., 2010). With UV-crosslinking, it is necessary to add a photo-initiator and crosslinker to the casting solution. After radiation, a radical polymerization reaction is initiated and a crosslinked network is formed (Strużyńska-Piron et al., 2014). When the membrane is cured with an electron beam, the polymerization reaction can be initiated without the need for a photo-initiator, but it is accompanied with a drastic decrease of permeance (Altun et al., 2016). Chemical crosslinking is extensively studied on PI-membranes, where the imide-group is transformed in an amide, with addition of amines (Hendrix et al., 2012; See-Toh et al., 2007a). Some polymers are stable enough in most solvents, so they do not require crosslinking, however they often are very hard to dissolve for membrane preparation (e.g., Torlon® and PEEK) (Szekely et al., 2014).

13.4 APPLICATIONS AND OVERVIEW OF RECENTLY DEVELOPED SOLVENT RESISTANT ISA MEMBRANES

Although the principle of making membranes through phase inversion has already been known for almost a century, the development of solvent resistant polymeric membranes is relatively new. It took some time to go from the pioneering work of Sourirajan (Sourirajan, 1964) to the first commercial SRNF membrane producers in the 1990s, like Grace Davison and Koch (Marchetti et al., 2014). SRNF is a versatile unit operation and can be applied in almost any separation process: solute enrichment (Priske et al., 2010), solvent recovery (Bastin et al., 2017), solvent exchange (Sheth et al., 2003), purification of a product (Székely et al., 2011) and in chemical synthesis (Xiong et al., 2009).

Probably one of the most famous SRNF applications is the MAX-DEWAX® process, for the recovery of solvents used in the refinery of lubricants (Gould et al., 2001). Other petrochemical applications for which membranes have been developed are the enrichment of aromatics in refening streams (White, 2002) and the desulfurization of gasoline (White, 2006).

SRNF can also be of use in the food industry, for example in edible oil processes, where the crude vegetable oils are extracted from the seeds often with hexane or acetone as solvent. In order to recover these solvents, SRNF membrane modules can replace the more energy consuming evaporators and destillation columns (Hendrix et al., 2014b). Also in further processing of edible oil, solvent resistant membranes can be of use: ultrafiltration membranes have been developed for the phospholipid removal with phosphoric acid (degumming), and SRNF membranes have been developed for the deacidification step (Firman et al., 2013). Another potential SRNF application in the food industry is found in the recovery of unreacted amino acids in the synthesis of aspartame (Reddy et al., 1996).

For the pharmaceutical sector, the purification of compounds like antibiotics, enzymes and peptides is another field in which SRNF can be the solution. Not only is purification through membrane separation often more environmently friendly, it can also be performed at ambient temperatures, which is necessary for bio-active compounds. With regards to catalytic applications, SRNF can be a solution for the separation of reaction products from their catalysts (Priske et al., 2010; Vankelecom, 2002).

The review by Marchetti et al. (2014) gives an overview of polymeric ISA membranes. Table 13.3 is largely adopted from this, supplemented with more recent polymeric ISA membranes and with a brief summary in the column "highlights." A list of the abbreviations used under the column "membrane material" is listed below Table 13.3 in Table 13.4.

13.5 FUTURE PERSPECTIVES AND CONCLUSIONS

SRNF is slowly finding its way into large industrial applications. In some places not to completely substitute conventional separation methods, but more as a technique to complement them and enhance their efficiency. For instance, membrane pretreatment can be an ideal way to concentrate a product so that more energy demanding techniques only need to perform the final (i.e., reduced) part of a process. Over the last decade, a lot of progress has been made when it comes to material development. This will often result into well retaining and highly permeable membranes on lab-scale. Still, long-term and pilot-scale testing of these membranes are required for most, but those studies are so far absent in open literature.

When it comes to membrane development for SRNF, a lot of attention nowadays goes towards stability in harsh environments, like extreme pH or high temperatures or require regular cleaning under such conditions. When it comes to membrane performance, optimizing permeance together with the rejection within a given environment will always be the main challenge, but focusing on a steep MWCO-curve can open pathways to new applications. When a clear distinction can be made in terms of rejection between two molecules that only slightly differ in size, a big step is made towards a versatile technique in separation and purification technology.

TABLE 13.3
Summary of Recently Developed ISA Membranes and Their Performance

Mem. Material	Highlights	P ($L\ m^{-2}\ h^{-1}\ bar^{-1}$)	Solvent	Marker (Da)	R (%)	Ref.
Cellulose	Cellulose membranes made from ionic liquid + volatile co-solvent	0.3	EtOH	Bromothymol Blue (BB) (624)	94	(Sukma and Çulfaz-Emecen, 2018)
CA	CA-membranes with gold nanoparticles incorporated, to heat membranes during filtration through light irradiation	0.4	EtOH	BB (624)	82	(Vanherck et al., 2011)
PAI Torlon®	PAI-membranes crosslinked with di-isocyanates for applications at elevated temperatures	1.2	acetone	Styrene oligomers (236–1200)	MWCO ~ 260 Da	(Dutczak et al., 2013)
PANI	PANI-membranes crosslinked, either with α,α'-dichloro-p-xylene or with glutaraldehyde, membranes stable up to 70°C	0.6	DMF	Styrene oligomers (236–1200)	95 (300 Da)	(Loh et al., 2009)
		0.8	DCM		100 (236 Da)	
		0.8	MeOH		97 (236 Da)	
		1.0	EtOAc		95 (236 Da)	
		1.4	acetone		99 (236 Da)	
PBI	Crosslinked PBI (patented)	0.4	DMF	Styrene oligomers (236–1200)	99 (236 Da)	(Livingston and Bhole, 2013)
		3.5	THF		87 (236 Da)	
		9.6	DCM		87 (236 Da)	
		15.6	acetone		80 (236 Da)	
	PBI crosslinked with gluteraldehyde or 1,2,7,8-diepoxyoctane, starting from a phase inversion membrane where PBI was dissolved in an IL	3.7	EtOH	Remazol Briljant Blue R (627)	99	(Xing et al., 2014)
		5.2	EtOAc	Eosin Y (648)	85	

(Continued)

TABLE 13.3 (CONTINUED)
Summary of Recently Developed ISA Membranes and Their Performance

Mem. Material	Highlights	P ($L\ m^{-2}\ h^{-1}\ bar^{-1}$)	Solvent	Marker (Da)	R (%)	Ref.
	PBI crosslinked with either an aliphatic (1,4-dibromobutane) or an aromatic (α,α'-dibromo-p-xylene) bifunctional crosslinker, good resistance towards extreme pH	6–11 1–7	ACN DMF	PEG (400–8000)	90 (2000 Da) 90 (2000 Da)	(Valtcheva et al., 2014)
	PBI membranes acting as shape-specific adsorbent or size exclusion membrane	4	ACN	Roxythromycin (837)	99	(Székely et al., 2015)
PEEK	PEEK phase inversion SRNF membranes, suitable for high temperatures and extreme pH, for NF performance, a drying step is necessary	0.22 0.07	THF DMF	Styrene oligomers (236–1200)	90 90	(da Silva Burgal et al., 2015)
PEEK-WC	Phenolphtalein based poly (ether-ether-ketone) used to prepare NF membranes	0.9 1.7	IPA MeOH	Rose Bengal (RB) (1017)	99.8 90	(Buonomenna et al., 2011)
BPA-PEEK	Phase inversion parameter study with BPAPEEK	0.1	IPA	RB (1017)	87	(Hendrix et al., 2014a)
TB-PEEK	Phase inversion parameter study with TBPEEK	0.4–1.0	IPA	RB (1017)	90	(Hendrix et al., 2013a; Hendrix et al., 2014a)
VA-PEEK	Crosslinked PEEK by introducing carboxylic acid groups which can crosslink diamines	0.1	IPA	RB (1017)	90	(Hendrix et al., 2013b)
PEI	PEI membranes with polyvinyl pyrrolidone (PVP-K30) as additive to increase rejection	4–8	EtOH	Methylene blue (MB) (373)	95	(Zhang et al., 2017)
PI	Crosslinked P84 PI with jeffamine and diamines. Study on pore preservation by post-treatment with PEG 400 and PEG 400/IPA	0.2 0.2 0.3	toluene DMF acetone	Styrene oligomers (236–1200)	90 (236 Da) 90 (236 Da) 90 (236 Da)	(Siddique et al., 2014)

(Continued)

TABLE 13.3 (CONTINUED)
Summary of Recently Developed ISA Membranes and Their Performance

Mem. Material	Highlights	P ($L\ m^{-2}\ h^{-1}\ bar^{-1}$)	Solvent	Marker (Da)	R (%)	Ref.
	Starmem 240 commercial membrane (now Duramem and Puramem) from Evonik	0.1 0.7	n-heptane toluene	Styrene oligomers (236–1200)	90 (400 Da) 90 (380 Da)	(Fritsch et al., 2012)
	Effect of polymer/solvent/non-solvent system on permeance and rejection, polymer dissolved in DMF and 1,4 dioxane	0.5 1.0 1.6	DMF DMF DMF	Styrene oligomers (236–1200)	90 (236 Da) 90 (500 Da) 92 (236 Da)	(Soroko et al., 2011a)
	SRNF membrane by applying solvent treatment to crosslinked PI UF membranes (DMF)	1.1	EtOH	Sudan Black B (327)	95	(Mariën and Vankelecom, 2017)
	UV-cured polyimide NF membrane	1.4	IPA	RB (1017)	96	(Strużyńska-Piron et al., 2013)
	Control over MWCO by varying DMF/1,4-dioxane ratio in casting solution	1.6 3.6	DMF toluene	Styrene oligomers (236–1200)	95 (236 Da) 95 (236 Da)	(See-Toh et al., 2008)
	Study on the effect of polymer chain length and use of random vs. block co-polymers in casting solution, block co-polymers to give higher rejections and lower flux	2.8	DMF	Styrene oligomers (236–1200)	92 (236 Da)	(Soroko et al., 2011b)
MMM with CNTs	Carboxyl-functionalized multi-walled carbon nanotubes incorporated in a P84 XL-PI membrane increases porosity, after thermal annealing, this membrane has improved rejections, but also a lower permeance	9.6	EtOH	RB (1017)	85	(Farahani et al., 2017)

(Continued)

TABLE 13.3 (CONTINUED)
Summary of Recently Developed ISA Membranes and Their Performance

Mem. Material	Highlights	P ($L\ m^{-2}\ h^{-1}\ bar^{-1}$)	Solvent	Marker (Da)	R (%)	Ref.
PPSf	Hollow fiber membranes from PPSf	0.02	IPA	RB (1017)	98.6	(Darvishmanesh et al., 2011b)
	Study on the impact of solvent exposure of PPSf membranes	0.4	MeOH	RB (1017)	88	(Darvishmanesh et al., 2011a)
sPPSf	Crosslinking reaction between sulfonated polyphenylsulfone (sPPSU) and hyperbranched polyethyleneimine (HPEI)	1.5	EtOH	RB (1017)	99.9	(Asadi Tashvigh et al., 2018)
blend PPSf/PI	Membranes prepared from a blend of PPSf with PI in different ratios and the effect of immersion in different solvents	2	MeOH	Sudan II (276)	95	(Jansen et al., 2013)
PSf	Effect of polymer concentration and evaporation time	0.07	IPA	RB (1017)	92	(Hołda et al., 2013)
	Influence of PEG as additive: high concentration gives few macrovoids, low MW PEG increases flux, high MW PEG makes denser	<0.1	IPA	RB (1017)	95	(Hołda and Vankelecom, 2014a)
	High concentrations of low MW additives are necessary to create SRNF membranes from PSf	4.5	IPA	RB (1017)	>90	(Hołda and Vankelecom, 2014b)
	UV-crosslinking of PSf via acrylate crosslinkers	0.2 8.1	IPA EtOAc	RB (1017)	94 91	(Strużyńska-Piron et al., 2013)
	UV-crosslinking of PSf via acrylate crosslinkers	1.2	IPA	RB (1017)	94	(Strużyńska-Piron et al., 2014)

TABLE 13.4
Abbreviations of Polymers from Table 13.3

Abbreviation	Polymer/Membrane Material
CA	Cellulose acetate
PAI	Polyamide-imide
PANI	Polyaniline
PBI	Polybenzi-midazole
PEEK	Poly (ether ether ketone)
PEEKWC	Phenolphthalein based poly(ether ether ketone)
BPAPEEK	Bisphenol A Poly (ether ether ketone)
TBPEEK	*tert*-butyl Poly (ether ether ketone)
VAPEEK	Valeric acid Poly (ether ether ketone)
PEI	Poly (ether imide)
PI	Polyimide
MMM with CNTs	Mixed matrix membranes with Carbon Nanotubes
PPSf	Polyphenylsulfone
sPPSf	sulfonated PPSf
PSf	Polysulfone

REFERENCES

Ali, N., Halim, N.S.A., Jusoh, A., and Endut, A., 2010. The formation and characterisation of an asymmetric nanofiltration membrane for ammonia–nitrogen removal: Effect of shear rate. *Bioresource Technology*. 101(5), 1459–1465.

Altun, V., Bielmann, M., and Vankelecom, I.F.J., 2016. EB depth-curing as a facile method to prepare highly stable membranes. *RSC Advances*. 6(60), 55526–55533.

Asadi Tashvigh, A., Luo, L., Chung, T.-S., Weber, M., and Maletzko, C., 2018. A novel ionically cross-linked sulfonated polyphenylsulfone (sPPSU) membrane for organic solvent nanofiltration (OSN). *Journal of Membrane Science*. 545, 221–228.

Baker, R.W., 2004. *Membrane Technology and Applications* (Second ed.). West Sussex, England: John Wiley & Sons, Ltd.

Bastin, M., Hendrix, K., and Vankelecom, I., 2017. Solvent resistant nanofiltration for acetonitrile based feeds: A membrane screening. *Journal of Membrane Science*. 536(Supplement C), 176–185.

Boussu, K., Vandecasteele, C., and Van der Bruggen, B., 2006. Study of the characteristics and the performance of self-made nanoporous polyethersulfone membranes. *Polymer*. 47(10), 3464–3476.

Buonomenna, M.G., Golemme, G., Jansen, J.C., and Choi, S.-H., 2011. Asymmetric PEEKWC membranes for treatment of organic solvent solutions. *Journal of Membrane Science*. 368(1), 144–149.

Chakrabarty, B., Ghoshal, A.K., and Purkait, M.K., 2008. Effect of molecular weight of PEG on membrane morphology and transport properties. *Journal of Membrane Science*. 309(1), 209–221.

Chun, K.-Y., Jang, S.-H., Kim, H.-S., Kim, Y.-W., Han, H.-S., and Joe, Y., 2000. Effects of solvent on the pore formation in asymmetric 6FDA–4,4′ODA polyimide membrane: Terms of thermodynamics, precipitation kinetics, and physical factors. *Journal of Membrane Science*. 169(2), 197–214.

Darvishmanesh, S., Jansen, J.C., Tasselli, F., Tocci, E., Luis, P., Degrève, J., Drioli, E. et al., 2011a. Novel polyphenylsulfone membrane for potential use in solvent nanofiltration. *Journal of Membrane Science*. 379(1), 60–68.

Darvishmanesh, S., Tasselli, F., Jansen, J.C., Tocci, E., Bazzarelli, F., Bernardo, P., Luis, P. et al., 2011b. Preparation of solvent stable polyphenylsulfone hollow fiber nanofiltration membranes. *Journal of Membrane Science*. 384(1), 89–96.

Dutczak, S.M., Cuperus, F.P., Wessling, M., and Stamatialis, D.F., 2013. New crosslinking method of polyamide–imide membranes for potential application in harsh polar aprotic solvents. *Separation and Purification Technology*. 102(Supplement C), 142–146.

Farahani, M.H.D.A., Hua, D., and Chung, T.-S., 2017. Cross-linked mixed matrix membranes consisting of carboxyl-functionalized multi-walled carbon nanotubes and P84 polyimide for organic solvent nanofiltration (OSN). *Separation and Purification Technology*. 186(Supplement C), 243–254.

Firman, L.R., Ochoa, N.A., Marchese, J., and Pagliero, C.L., 2013. Deacidification and solvent recovery of soybean oil by nanofiltration membranes. *Journal of Membrane Science*. 431(Supplement C), 187–196.

Fontananova, E., Jansen, J.C., Cristiano, A., Curcio, E., and Drioli, E., 2006. Effect of additives in the casting solution on the formation of PVDF membranes. *Desalination*. 192(1), 190–197.

Fritsch, D., Merten, P., Heinrich, K., Lazar, M., and Priske, M., 2012. High performance organic solvent nanofiltration membranes: Development and thorough testing of thin film composite membranes made of polymers of intrinsic microporosity (PIMs). *Journal of Membrane Science*. 401–402(Supplement C), 222–231.

Gould, R.M., White, L. S., and Wildemuth, C.R., 2001. Membrane separation in solvent lube dewaxing. *Environmental Progress*. 20(1), 12–16.

Guillen, G.R., Pan, Y., Li, M., and Hoek, E.M.V., 2011. Preparation and characterization of membranes formed by nonsolvent induced phase separation: A review. *Industrial & Engineering Chemistry Research*. 50(7), 3798–3817.

Hendrix, K., Vanherck, K., and Vankelecom, I.F.J., 2012. Optimization of solvent resistant nanofiltration membranes prepared by the in-situ diamine crosslinking method. *Journal of Membrane Science*. 421–422(Supplement C), 15–24.

Hendrix, K., Van Eynde, M., Koeckelberghs, G., and Vankelecom, I.F.J., 2013a. Synthesis of modified poly(ether ether ketone) polymer for the preparation of ultrafiltration and nanofiltration membranes via phase inversion. *Journal of Membrane Science*. 447, 96–106.

Hendrix, K., Van Eynde, M., Koeckelberghs, G., and Vankelecom, I.F.J., 2013b. Crosslinking of modified poly(ether ether ketone) membranes for use in solvent resistant nanofiltration. *Journal of Membrane Science*. 447(Supplement C), 212–221.

Hendrix, K., Koeckelberghs, G., and Vankelecom, I.F.J., 2014a. Study of phase inversion parameters for PEEK-based nanofiltration membranes. *Journal of Membrane Science*. 452, 241–252.

Hendrix, K., Vandoorne, S., Koeckelberghs, G., and Vankelecom, I.F.J., 2014b. SRNF membranes for edible oil purification: Introducing free amines in crosslinked PEEK to increase membrane hydrophilicity. *Polymer*. 55(6), 1307–1316.

Hermans, S., Dom, E., Mariën, H., Koeckelberghs, G., and Vankelecom, I.F.J., 2015. Efficient synthesis of interfacially polymerized membranes for solvent resistant nanofiltration. *Journal of Membrane Science*. 476(Supplement C), 356–363.

Hołda, A.K., Aernouts, B., Saeys, W., and Vankelecom, I.F.J., 2013. Study of polymer concentration and evaporation time as phase inversion parameters for polysulfone-based SRNF membranes. *Journal of Membrane Science*. 442, 196–205.

Hołda, A.K., and Vankelecom, I.F.J., 2014a. Integrally skinned PSf-based SRNF-membranes prepared via phase inversion—Part A: Influence of high molecular weight additives. *Journal of Membrane Science*. 450(Supplement C), 512–521.

Hołda, A.K., and Vankelecom, I.F.J., 2014b. Integrally skinned PSf-based SRNF-membranes prepared via phase inversion—Part B: Influence of low molecular weight additives. *Journal of Membrane Science*. 450, 499–511.

Hołda, A.K., and Vankelecom, I.F.J., 2015. Understanding and guiding the phase inversion process for synthesis of solvent resistant nanofiltration membranes. *Journal of Applied Polymer Science*. 132(27), n/a–n/a.

Jansen, J.C., Darvishmanesh, S., Tasselli, F., Bazzarelli, F., Bernardo, P., Tocci, E., Friess, K. et al., 2013. Influence of the blend composition on the properties and separation performance of novel solvent resistant polyphenylsulfone/polyimide nanofiltration membranes. *Journal of Membrane Science*. 447(Supplement C), 107–118.

Jansen, J.C., Macchione, M., and Drioli, E., 2005. High flux asymmetric gas separation membranes of modified poly(ether ether ketone) prepared by the dry phase inversion technique. *Journal of Membrane Science*. 255(1), 167–180.

Jimenez Solomon, M.F., Bhole, Y., and Livingston, A.G., 2012. High flux membranes for organic solvent nanofiltration (OSN)—Interfacial polymerization with solvent activation. *Journal of Membrane Science*. 423–424(Supplement C), 371–382.

Joseph, D., Krishnan, N.N., Henkensmeier, D., Jang, J.H., Choi, S.H., Kim, H.-J., Han, J. et al., 2017. Thermal crosslinking of PBI/sulfonated polysulfone based blend membranes. *Journal of Materials Chemistry A*. 5(1), 409–417.

Kim, J.-H., and Lee, K.-H., 1998. Effect of PEG additive on membrane formation by phase inversion. *Journal of Membrane Science*. 138(2), 153–163.

Lefebvre, A.A., Lee, J.H., Balsara, N.P., and Vaidyanathan, C., 2002. Determination of critical length scales and the limit of metastability in phase separating polymer blends. *The Journal of Chemical Physics*. 117(19), 9063–9073.

Li, D., Chung, T.-S., Ren, J., and Wang, R., 2004. Thickness Dependence of macrovoid evolution in wet phase-inversion asymmetric membranes. *Industrial & Engineering Chemistry Research*. 43(6), 1553–1556.

Li, J.-B., Zhu, J.-W., and Zheng, M.-S., 2007. Morphologies and properties of poly(phthalazinone ether sulfone ketone) matrix ultrafiltration membranes with entrapped TiO2 nanoparticles. *Journal of Applied Polymer Science*. 103(6), 3623–3629.

Li, Y., Verbiest, T., and Vankelecom, I., 2013. Improving the flux of PDMS membranes via localized heating through incorporation of gold nanoparticles. *Journal of Membrane Science*. 428(Supplement C), 63–69.

Livingston, A.G., and Bhole, Y.S., 2013. Asymmetric membranes for use in nanofiltration. Retrieved from http://www.google.ch/patents/US20130118983

Loeb, S., and Sourirajan, S., 1963. Sea water demineralization by means of an osmotic membrane. *Saline Water Conversion—II*, Advances in Chemistry (Vols. 1-0, Vol. 38, 117–132).

Loh, X.X., Sairam, M., Bismarck, A., Steinke, J.H.G., Livingston, A.G., and Li, K., 2009. Crosslinked integrally skinned asymmetric polyaniline membranes for use in organic solvents. *Journal of Membrane Science*. 326(2), 635–642.

Ma, Y., Shi, F., Ma, J., Wu, M., Zhang, J., and Gao, C., 2011. Effect of PEG additive on the morphology and performance of polysulfone ultrafiltration membranes. *Desalination*. 272(1), 51–58.

MacDonald, W., and Pan, C., 1974. Method for drying water-wet membranes. Retrieved from http://www.google.com/patents/US3842515

Marchetti, P., Jimenez Solomon, M.F., Szekely, G., and Livingston, A.G., 2014. Molecular separation with organic solvent nanofiltration: A critical review. *Chemical Reviews*. 114(21), 10735–10806.

Mariën, H., Bellings, L., Hermans, S., and Vankelecom, I.F.J., 2016. Sustainable process for the preparation of high-performance thin-film composite membranes using ionic liquids as the reaction medium. *ChemSusChem*. 9(10), 1101–1111.

Mariën, H., and Vankelecom, I.F.J., 2017. Transformation of cross-linked polyimide UF membranes into highly permeable SRNF membranes via solvent annealing. *Journal of Membrane Science*. 541(Supplement C), 205–213.

Mulder, M., 1996. *Basic Principles of Membrane Technology* (Second ed.). Dordrecht, The Netherlands: Kluwer Academic Publishers.

Priske, M., Wiese, K.-D., Drews, A., Kraume, M., and Baumgarten, G., 2010. Reaction integrated separation of homogenous catalysts in the hydroformylation of higher olefins by means of organophilic nanofiltration. *Journal of Membrane Science*. 360(1), 77–83.

Raaijmakers, M.J.T., and Benes, N.E., 2016. Current trends in interfacial polymerization chemistry. *Progress in Polymer Science*. 63, 86–142.

Reddy, K.K., Kawakatsu, T., Snape, J.B., and Nakajima, M., 1996. Membrane concentration and separation of L-Aspartic acid and L-Phenylalanine derivatives in organic solvents. *Separation Science and Technology*. 31(8), 1161–1178.

Reuvers, A.J., 1987. Membrane formation: Diffusion induced demixing processes in ternary polymeric systems (Doctoral dissertation). Retrieved from https://research.utwente.nl/en /publications/membrane-formation-diffusion-induced-demixing-processes-in-ternar

Sairam, M., Loh, X.X., Bhole, Y., Sereewatthanawut, I., Li, K., Bismarck, A., Steinke, J.H.G. et al., 2010. Spiral-wound polyaniline membrane modules for organic solvent nanofiltration (OSN). *Journal of Membrane Science*. 349(1), 123–129.

See-Toh, Y.H., Lim, F.W., and Livingston, A.G., 2007a. Polymeric membranes for nanofiltration in polar aprotic solvents. *Journal of Membrane Science*. 301(1), 3–10.

See-Toh, Y.H., Ferreira, F.C., and Livingston, A.G., 2007b. The influence of membrane formation parameters on the functional performance of organic solvent nanofiltration membranes. *Journal of Membrane Science*. 299(1), 236–250.

See-Toh, Y.H., Silva, M., and Livingston, A., 2008. Controlling molecular weight cut-off curves for highly solvent stable organic solvent nanofiltration (OSN) membranes. *Journal of Membrane Science*. 324(1), 220–232.

Sheth, J.P., Qin, Y., Sirkar, K.K., and Baltzis, B.C., 2003. Nanofiltration-based diafiltration process for solvent exchange in pharmaceutical manufacturing. *Journal of Membrane Science*. 211(2), 251–261.

Siddique, H., Bhole, Y., Peeva, L.G., and Livingston, A.G., 2014. Pore preserving crosslinkers for polyimide OSN membranes. *Journal of Membrane Science*. 465(Supplement C), 138–150.

da Silva Burgal, J., Peeva, L.G., Kumbharkar, S., and Livingston, A., 2015. Organic solvent resistant poly(ether-ether-ketone) nanofiltration membranes. *Journal of Membrane Science*. 479(Supplement C), 105–116.

Soroko, I., and Livingston, A., 2009. Impact of TiO2 nanoparticles on morphology and performance of crosslinked polyimide organic solvent nanofiltration (OSN) membranes. *Journal of Membrane Science*. 343(1), 189–198.

Soroko, I., Lopes, M.P., and Livingston, A., 2011a. The effect of membrane formation parameters on performance of polyimide membranes for organic solvent nanofiltration (OSN): Part A. Effect of polymer/solvent/non-solvent system choice. *Journal of Membrane Science*. 381(1), 152–162.

Soroko, I., Sairam, M., and Livingston, A.G., 2011b. The effect of membrane formation parameters on performance of polyimide membranes for organic solvent nanofiltration (OSN). Part C. Effect of polyimide characteristics. *Journal of Membrane Science*. 381(1), 172–182.

Sourirajan, S., 1964. Separation of hydrocarbon liquids by flow under pressure through porous membranes. *Nature*. 203(4952), 1348.

Strużyńska-Piron, I., Bilad, M.R., Loccufier, J., Vanmaele, L., and Vankelecom, I.F.J., 2014. Influence of UV curing on morphology and performance of polysulfone membranes containing acrylates. *Journal of Membrane Science*. 462(Supplement C), 17–27.

Strużyńska-Piron, I., Loccufier, J., Vanmaele, L., and Vankelecom, I.F.J., 2013. Synthesis of solvent stable polymeric membranes via UV depth-curing. *Chemical Communications.* 49(98), 11494.

Su, J., Zhang, S., Chen, Hangzheng, Chen, Hongmin, Jean, Y.C., and Chung, T.-S., 2010. Effects of annealing on the microstructure and performance of cellulose acetate membranes for pressure-retarded osmosis processes. *Journal of Membrane Science.* 364(1), 344–353.

Sukma, F.M., and Çulfaz-Emecen, P.Z., 2018. Cellulose membranes for organic solvent nanofiltration. *Journal of Membrane Science.* 545(Supplement C), 329–336.

Székely, G., Bandarra, J., Heggie, W., Sellergren, B., and Ferreira, F.C., 2011. Organic solvent nanofiltration: A platform for removal of genotoxins from active pharmaceutical ingredients. *Journal of Membrane Science.* 381(1), 21–33.

Szekely, G., Jimenez-Solomon, M.F., Marchetti, P., Kim, J.F., and Livingston, A.G., 2014. Sustainability assessment of organic solvent nanofiltration: From fabrication to application. *Green Chemistry.* 16(10), 4440–4473.

Székely, G., Valtcheva, I.B., Kim, J.F., and Livingston, A.G., 2015. Molecularly imprinted organic solvent nanofiltration membranes – Revealing molecular recognition and solute rejection behaviour. *Reactive and Functional Polymers.* 86(Supplement C), 215–224.

Valtcheva, I.B., Kumbharkar, S.C., Kim, J.F., Bhole, Y., and Livingston, A.G., 2014. Beyond polyimide: Crosslinked polybenzimidazole membranes for organic solvent nanofiltration (OSN) in harsh environments. *Journal of Membrane Science.* 457(Supplement C), 62–72.

Vandezande, P., Gevers, L.E.M., and Vankelecom, I.F.J., 2008. Solvent resistant nanofiltration: separating on a molecular level. *Chemical Society Reviews.* 37(2), 365–405.

Vandezande, P., Li, X., Gevers, L.E.M., and Vankelecom, I.F.J., 2009. High throughput study of phase inversion parameters for polyimide-based SRNF membranes. *Journal of Membrane Science.* 330(1), 307–318.

Vanherck, K., Hermans, S., Verbiest, T., and Vankelecom, I., 2011. Using the photothermal effect to improve membrane separations via localized heating. *Journal of Materials Chemistry.* 21(16), 6079.

Vankelecom, I.F.J., 2002. Polymeric membranes in catalytic reactors. *Chemical Reviews.* 102(10), 3779–3810.

Wells, P.A., de Loos, T.W., and Kleintjens, L.A., 1993. Pressure pulsed induced critical scattering: spinodal and binodal curves for the system polystyrene + methylcylcohexane. *Fluid Phase Equilibria.* 83(Supplement C), 383–390.

White, L.S., 2002. Transport properties of a polyimide solvent resistant nanofiltration membrane. *Journal of Membrane Science.* 205(1), 191–202.

White, L.S., 2006. Development of large-scale applications in organic solvent nanofiltration and pervaporation for chemical and refining processes. *Journal of Membrane Science.* 286(1), 26–35.

Xing, D.Y., Chan, S.Y., and Chung, T.-S., 2014. The ionic liquid [EMIM]OAc as a solvent to fabricate stable polybenzimidazole membranes for organic solvent nanofiltration. *Green Chemistry.* 16(3), 1383–1392.

Xiong, W.-W., Wang, W.-F., Zhao, L., Song, Q., and Yuan, L.-M., 2009. Chiral separation of (R,S)-2-phenyl-1-propanol through glutaraldehyde-crosslinked chitosan membranes. *Journal of Membrane Science.* 328(1), 268–272.

Yoo, S.H., Kim, J.H., Jho, J.Y., Won, J., and Kang, Y.S., 2004. Influence of the addition of PVP on the morphology of asymmetric polyimide phase inversion membranes: Effect of PVP molecular weight. *Journal of Membrane Science.* 236(1), 203–207.

Young, T.-H., and Chen, L.-W., 1995. Pore formation mechanism of membranes from phase inversion process. *Desalination.* 103(3), 233–247.

Zhang, Y., Zhong, M., Luo, B., Li, J., Yuan, Q., and Yang, X.J., 2017. The performance of integrally skinned polyetherimide asymmetric nanofiltration membranes with organic solvents. *Journal of Membrane Science*. 544(Supplement C), 119–125.

Zheng, Q.-Z., Wang, P., and Yang, Y.-N., 2006. Rheological and thermodynamic variation in polysulfone solution by PEG introduction and its effect on kinetics of membrane formation via phase-inversion process. *Journal of Membrane Science*. 279(1), 230–237.

Zsigmondy, R., and Bachmann, W., 1918. Über neue Filter. *Zeitschrift für anorganische und allgemeine Chemie*. 103(1), 119–128.

14 Ceramic-Supported Composite Membranes for Pervaporation

Gongping Liu and Wanqin Jin

CONTENTS

14.1 INTRODUCTION

Pervaporation is a membrane process that could realize molecular separation for liquid mixtures in which a feed solution is passed over a membrane surface and some of the components are able to preferentially pass through the membrane and be concentrated as vapors in the permeate (Shao and Huang, 2007). The vapor pressure difference between the feed solution and permeate vapor provides the driving force of the pervaporation process, which is usually maintained by applying a vacuum on the downstream side. The key component of the pervaporation process is the membrane. An ideal pervaporation membrane allows for a high separation factor (the ratio of two components on the permeate side divided by the ratio of two components on the feed side of the membrane) and a high permeate flux (the mass flow rate per unit membrane area). Generally, the separation factor is primarily determined by the intrinsic transport property of the membrane material, while the permeate flux depends on the membrane structure. For practical application, a composite membrane comprising a thin separation layer and a porous support layer is often required to offer sufficient flux and mechanical strength (Koros and Zhang, 2017).

Compared with polymeric supports, inorganic (e.g., ceramic) supports exhibit higher thermal and mechanical stability, showing great potential in developing composite membranes for pervaporation application (Liu et al., 2012). Various membrane materials have been used as the separation layer of ceramic-supported composite pervaporation membranes, including polymers, metal-organic frameworks (MOFs), mixed-matrix, and graphene oxide (GO). This chapter discusses recent progresses in ceramic-supported composite membranes for pervaporation, focusing on nanostructures and separation performance of the membranes derived from the above-three kinds of materials. Zeolite membranes also often use ceramic supports, and they are discussed in depth in Chapter 12.

14.2 MEMBRANE STRUCTURE AND FABRICATION

Ceramic-supported composite membranes consist of a porous ceramic support layer and a thin separation layer made of polymer, MOF, mixed-matrix, or GO. Depending on the material of separation layer, the fabrication method for the membranes is varied from dip-coating, crystal growth to filtration. Meanwhile, some methods such as the layer-by-layer approach can be generally applied for all kinds of ceramic-supported composite membranes. The membrane configuration is another important feature for designing the composite membrane for practical application, which depends on the geometry of ceramic support. Tubular and hollow fiber ceramic-supported composite membranes are industrially preferred owing to the advantages in scalable fabrication and module packing (Koros and Zhang, 2017), while the ceramic disk is mainly used for fundamental study and generally limited to lab-scale fabrication.

14.2.1 Polymer/Ceramic Composite Membranes

Like most membrane processes, polymers dominate the membrane materials for pervaporation, owing to their wide variety and ease of processing (Shao and Huang, 2007). Polymeric membrane for pervaporation is considered a dense membrane in which the separation mechanism follows the sorption-diffusion model (Lee et al., 1989). Namely, the transport of components through a polymeric pervaporation membrane is governed by both the sorption in the polymers and the diffusion through the polymeric layer.

The dip-coating method is often used for depositing a thin polymer separation layer on top of a porous ceramic substrate to fabricate polymer/ceramic composite membranes (Liu et al., 2012). The polymer precursor is dissolved in solvent to form a polymer solution in which the polymer concentration can be tuned and/or a crosslinker is added to produce a desirable viscosity. Meanwhile, the ceramic support with suitable pore size is selected and pre-treated with polishing and/or solvent immersion to obtain favorable surface properties for polymer coating. After dipping the ceramic support into the polymer solution and subsequent solvent evaporation and thermal treatment, a polymer layer with typical thickness of approximately 5–10 μm is formed on the surface of ceramic support. The structure, photo, and SEM images of the polymer/ceramic composite membrane are shown in Figure 14.1.

Polymers for the separation layer of ceramic-supported composite membranes can be divided into two kinds: hydrophobic polymers and hydrophilic polymers. The former includes polydimethylsiloxane (PDMS) and polyether block amide (PEBA), and the latter includes polyvinyl alcohol (PVA), chitosan (CS), and polyelectrolytes. Jin and co-workers (Xiangli et al., 2007) developed an array of ceramic-supported composite membranes based on the above-mentioned polymers via applying the facile dip-coating method on a macroporous ZrO_2/Al_2O_3 tube (average pore size ~200 nm). To optimize the preparation parameters for PDMS/ceramic composite membranes, response surface methodology (RSM) was applied to study the effect of three dominant factors on the separation performance (flux and separation factor): polymer concentration, cross-linking concentration, and dip-coating time (Xiangli et al., 2008). The regression equations between the preparation variables and the separation performance were established by the RSM. It was found that PDMS concentration was the most significant variable affecting the membrane performance.

Recent works (Shu et al., 2012; Wang et al., 2009) showed that the hollow fiber supported zeolite membranes showed much higher flux than the conventional tubular zeolite membranes, highlighting the important role that support configuration plays in the separation performance of ceramic-supported composite membrane.

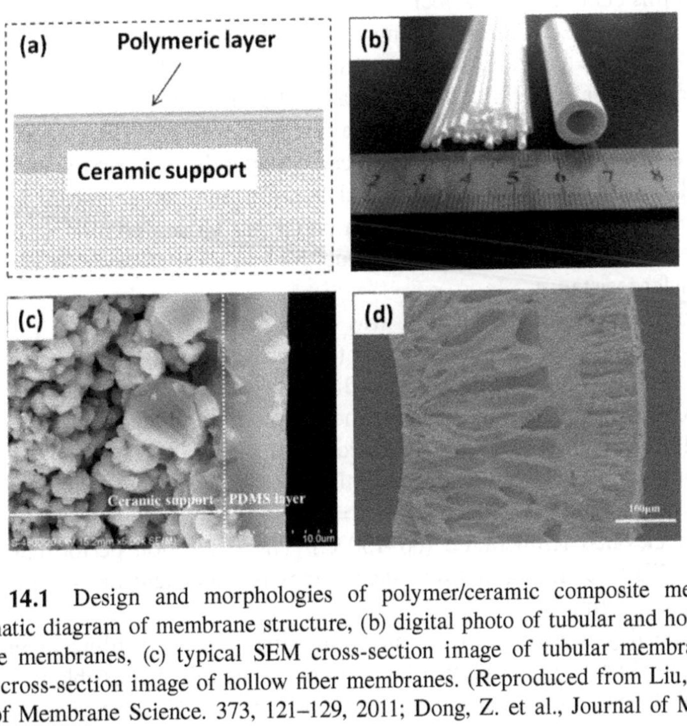

FIGURE 14.1 Design and morphologies of polymer/ceramic composite membranes: (a) schematic diagram of membrane structure, (b) digital photo of tubular and hollow fiber composite membranes, (c) typical SEM cross-section image of tubular membranes, and (d) SEM cross-section image of hollow fiber membranes. (Reproduced from Liu, G. et al., Journal of Membrane Science. 373, 121–129, 2011; Dong, Z. et al., Journal of Membrane Science. 450, 38–47, 2014.)

Besides the low transport resistance, the ceramic hollow fiber could achieve 1-2 orders of magnitude higher packing density than the tubular configuration (Shu et al., 2012). Despite these advantages, the fabrication of a polymer layer on the hollow fiber ceramic support was rarely reported, probably due to the challenges in controlling the delicate micro-structures of the ceramic hollow fiber.

Jin and co-workers (Dong et al., 2014) realized high-performance ceramic hollow fiber supported PDMS composite membranes via the facile dip-coating method. In addition to optimizing the polymer layer thickness, the pore structure of the ceramic hollow fibers was tuned for the deposition of thin and defect-free polymer layer. Generally, there are two kinds of structure (finger-like and sponge-like) in the ceramic hollow fiber. The sponge-like layer contributes to mechanical strength enhancement and provides a relatively smooth surface that is often controlled as the outer layer for depositing additional thin membrane layers, while the finger-like layer reduces resistance for molecule permeation. Moreover, three kinds of ceramic hollow fibers were spun, with average pore size of 300 nm (S1), 1100 nm (S2) and 1500 nm (S3). It was found that although the supports with large pores have low transport resistance, the large pores would lead to more PDMS solution penetrated into the ceramic support to form a thick transition layer, resulting in higher transport resistance in the PDMS composite membrane. By selecting the ceramic hollow fiber support S2 with 1100 nm pore size and the thinnest sponge-like structure, the resulting PDMS/ceramic composite membrane showed the highest flux and good separation factor. Likewise, the PEBA pervaporation membrane was also dip-coated on the surface of this ceramic hollow fiber.

For fabricating PEBA/ceramic composite membranes, the temperature-dependent viscosity of PEBA solution was quite important since PEBA is a thermoplastic elastomer polymer (Li et al., 2016). Because the viscosity of PEBA solution was inversely proportional to the temperature, the thickness of PEBA layer increased with the viscosity of coating solution. Recent study indicated that the appropriately enhanced viscosity was favor of forming defect-free and thin composite membrane. Excessively low viscosity PEBA solution could not form a continuous separation layer, but too high viscosity would form too thick of a PEBA layer, leading to high mass transfer resistance.

Sometimes, the ceramic support was functionalized to build or enhance the chemical bonding with the polymer layer to improve the interfacial adhesion of the polymer/ceramic composite membranes. Ji and co-workers (Wu et al., 2015) modified the Al_2O_3 tube by 3-aminopropyl-trimethoxysilane and then dip-coated the substrate in PEBA solution to form a separation layer. A thermal treatment was applied to crosslink the PEBA chains on the amino-modified ceramic surface. Hydrogen bonding was formed between the H–N–C=O from PEBA and –N–H– from ceramic surface at elevated temperature (60–150°C). The separation performance of the resulting PEBA/ceramic composite membranes is dependent on the crosslinking temperature and time.

Permeate flux is inverse to the membrane thickness, according to Fick's law. Thus, the thickness of the polymer layer should be as thin as possible to maximize the flux of the ceramic-supported composite membranes. However, the reduction in thickness would be accompanied by sacrifice of selectivity due to the formation of non-selective defects in thinner membrane. To fabricate a thin and dense polymeric layer, Peters et al. (2006) prepared 4 intermediate γ-Al_2O_3 layers 3–4 μm thick on top of the α-Al_2O_3 hollow fiber to provide a sufficiently smooth surface for coating a thin PVA layer with 300–800 nm thickness. Moreover, the pore size of the modified ceramic hollow fiber is only 4 nm, which is much smaller than the Flory radius of the PVA polymer (56 nm, calculated from its monomer size of 0.37 nm by $R_F = N^{0.6}a$, where N is the number of monomer units per chain and a is the monomer size).

Polymer blending is another way to improve the pervaporation performance of a separation layer supported by the ceramic substrate. PVA is the most commonly used hydrophilic polymers for pervaporation, although it has to be crosslinked to inhibit the excessive swelling when applied in water solution. The crosslinked PVA membrane showed lower permeate flux due to the reduced free volumes. Another hydrophilic polymer, CS was blended with PVA to increase the amorphous region of the pristine PVA membrane, thereby advancing the permeate flux of PVA/ceramic composite membrane (Zhu et al., 2010). It was proposed that intramolecular hydrogen bonding within PVA is substituted by intermolecular hydrogen bonding between CS and PVA, so the membrane structure becomes less compact and the free volume of the membrane increases, thus the permeation flux increases with CS content in the PVA-CS/ceramic composite membrane.

Importantly, there is a transition layer between the separation layer and support layer, resulting from the penetration of polymer coating solution into the ceramic pores. On the one hand, the transition layer greatly enhances the interfacial adhesion of the polymer layer onto the ceramic support; on the other hand, it would increase the transport resistance and reduce the permeate flux (Wei et al., 2011). Particular attention should be paid to optimize the polymer solution and coating process as well as the pore size and surface roughness of ceramic support in order to form a transition layer with appropriate thickness that minimizes the transport resistance meanwhile maintains sufficient interfacial adhesion. Moreover, with this transition layer, the rigid ceramic support could inhibit the excessive swelling of the polymeric layer caused by elevated operating temperature or feed concentration. As shown in Figure 14.2, the proposed "confinement effect" offered higher structural stability for the ceramic-supported composite membranes than that of the composite membranes using polymeric support (Wei et al., 2011).

Layer-by-layer (LBL) assembly technique is used for the fabrication of polyelectrolyte separation layer on the ceramic support (Chen et al., 2007), in which the ceramic support is coated in cationic and anionic polyelectrolyte solutions alternatively for a certain time and then the membrane is rinsed to get cycles of the above processes. Initially, the driving force for the LBL process is electrostatic force and now has been extended to hydrogen bonding, halogen atom, coordinate bonding and even chemical bonding by rationally designing the molecular structure of the polyelectrolyte pair. Several modified LBL methods have been developed to increase the fabrication efficiency of traditional static LBL method, such as dynamic LBL, pressure-driven LBL, electric field-enhanced LBL, and spray coating LBL. Also, crosslinking was proposed to improve the structural stability of the assembled polyelectrolyte layer. As shown in Figure 14.3, Zhang and co-workers (Wang et al., 2012a) applied thermal treatment to crosslink the 3-aminopropyl-trimethoxysilane

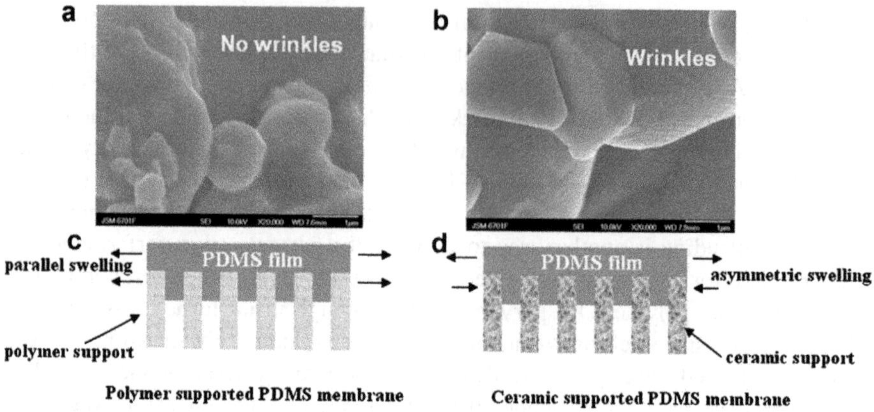

FIGURE 14.2 FESEM images of the penetrated PDMS in the pore of two supports: (a) BCA (blend cellulose acetate) support and (b) ceramic support. Schematic drawings of the parallel swelling and asymmetric swelling: (c) polymer-supported PDMS membrane and (d) ceramic-supported PDMS membrane. (Reproduced from Wei, W. et al., Journal of Membrane Science. 375, 334–344, 2011. With permission.)

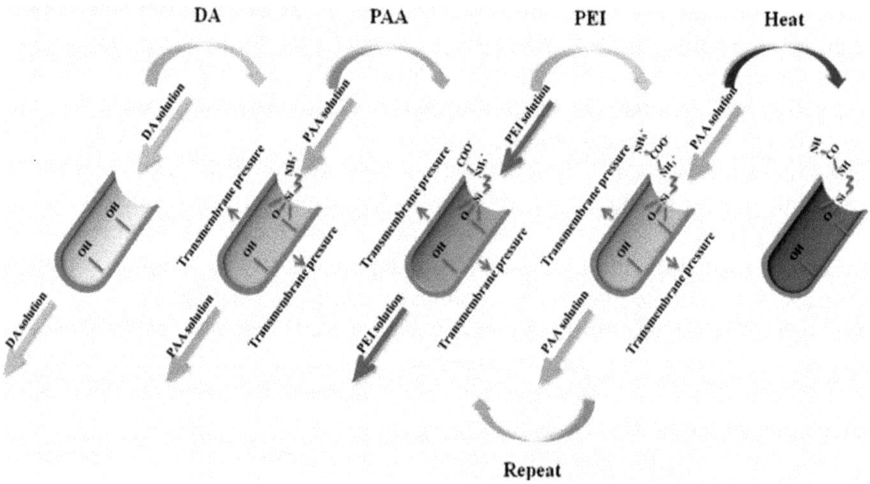

FIGURE 14.3 Schematic illustration of the preparation of a multilayer on a ceramic hollow fiber substrates membrane by dynamic pressure-driven LbL assembly. (Reproduced from Wang, N. et al., AIChE Journal. 58, 3176–3182, 2012. With permission.)

modified ceramic-supported poly(acrylic acid)/poly(ethyleneimine) (PAA/PEI) multilayer via heat-induced amide formation from carboxylate-ammonium complexes.

In addition, surface grafting techniques can be used to form a polymeric layer on the porous ceramic support. Cohen and co-workers (Jou et al., 1999) deposited polyvinyl acetate (PVAc) separation layer on vinyltrimethoxysilane modified silica support with pore size of 50 nm via graft polymerization. They also used asymmetric γ-alumina/α-alumina tubes with pore size of ~5 nm as the support for grafting poly(vinyl pyrrolidone) (PVP) or PVAc layer (Yoshida and Cohen, 2003), in which the polymer chains had an estimated radius of gyration a factor ~4.5–6.8 larger than the membrane pore radius. Wang et al. (2015) introduced hyperbranched polymer (Boltorn W3000) monomers into the Al_2O_3 pores by a simple immersion process, and applied thermal treatment to crosslink the monomers via the reaction between –OH and –COOH in hyperbranched polymer (HBP) to form the HBP/ceramic composite membrane.

14.2.2 MOF/Ceramic Composite Membranes

Metal-organic frameworks (MOFs) or porous coordination polymers have emerged as a new class of microporous crystalline materials composed of a metal-containing cluster bridged by organic ligands. The wide topological varieties and customizable chemistry render MOFs as superior materials for membrane separation (Qiu et al., 2014). Until now, the majority of MOF membranes have been focused on gas separation while only a few attempts were reported on the pervaporation application, partially due to the challenge of water stability for MOF materials (Wang et al., 2016). Recently, the development of water stable MOFs such as zeolitic imidazolate frameworks (ZIFs) (Park et al., 2006) and zirconium(IV) MOFs (Zr-MOFs) (Cavka et al., 2008) opens the door to using MOF membranes for pervaporation separation.

Although there are more diverse pore structures and functionalities for MOFs than zeolite, the fabrication of high-quality MOF membranes is challenging compared with zeolite membranes (Rangnekar et al., 2015). This is because MOF membranes often show (Venna and Carreon, 2015): (i) low heterogeneous nucleation density; (ii) relative weakness of the coordination bonds and intergrowth of MOF crystals; and (iii) differences in thermal expansion coefficients between the MOF layer and the support causing cracks. Thus, some of the fabrication methods (e.g., in-situ growth, secondary growth) were borrowed from zeolite membranes while some new methods (e.g., reactive seeding, contra diffusion) were developed for MOF membranes.

The in-situ growth method forms the MOF membrane directly on a support that is not modified with any seeding crystals, in which the crystal nucleation and growth occurs at the same time. Li and co-workers (Liu et al., 2017) fabricated UiO-66 (a prototypical Zr-MOF, UiO stands for University of Oslo) on the Yttria-Stabilized Zirconia (YSZ) hollow fiber support by an in-situ solvothermal growth approach via careful control of the heating duration, composition, and temperature of the synthetic mother solutions. A very thin amorphous gel layer was formed on the surface of the substrate during the initial heating process (2 h) where the gel particles were transferred from the mother solution onto the substrate driven by Brownian motion and chemical interaction between the ligands and substrate. Heterogeneous nucleation occurred probably at the interface of the gel and the solution in the consequent synthesis. After this nucleation period, crystals propagated through the gel network and then sank to the substrate by consuming the gel around them. With prolonged heating, crystal growth took place by acquisition of nutrients from bulk solution, nearby unreacted amorphous gel and small UiO-66 crystals. Finally, a well-intergrown membrane layer was formed after 48 h continuous heating after narrowing inter-crystalline gaps. It is worth noting that the selected YSZ support provided Zr to chemically bond with BDC (1,4-benzene-dicarboxylate) ligands thus promoting the heterogeneous nucleation of UiO-66 on the ceramic support.

In contrast to in-situ growth method, the secondary growth method separates the nucleation and growth steps, thereby easier controlling the membrane microstructure. In the in-situ method, MOF crystals are more easily crystallized and grown in the bulk solution rather than on the support surface, thus resulting in a defective MOF layer. It is much more popular to use the secondary growth method to fabricate MOF membranes. Lin and co-workers (Zhao et al., 2011) prepared continuous and crack-free MOF-5 membranes of ~14 μm thickness on alumina support by seeding size-reduced (via physical grinding) MOF-5 crystals followed by the secondary growth synthesis. Caro and co-workers (Bux et al., 2011) fabricated a highly oriented ZIF-8 membrane on the ceramic disk by seeding and secondary growth. ZIF-8 nanocrystals were attached to the ceramic surface using polyethyleneimine as the coupling agent by dip-coating. The solvothermal synthesis produced a continuous and well-intergrown ZIF-8 layer with ~12 μm thickness. Time-dependent investigations by scanning and transmission electron microscopy as well as X-ray diffraction indicated that the preferred orientation of the "100" plane parallel to the support develops during an evolutionary growth process.

MOFs are synthesized by reacting a hydrated metal salt with an organic linker using hydrothermal or solvothermal methods. It is also possible to link the seed layer and the support by chemical interaction, that is, let the support participate in a reaction to generate a seed layer. To achieve this end, Jin and co-workers (Hu et al., 2011) designed a novel reactive seeding (RS) method, in which the inorganic support acts as the inorganic source reacting with the organic precursor to grow a seed layer for secondary growth of the MOF membrane. The feasibility of the proposed RS method was demonstrated by the preparation of MIL-53 ($Al(OH)[O_2C–C_6H_4–CO_2]\cdot[O_2C–C_6H_4–CO_2]_{0.7}$) membrane on a porous alumina support. A schematic of

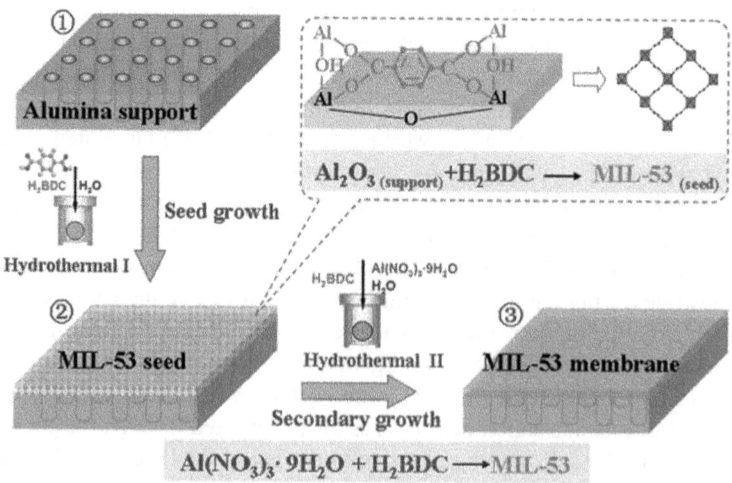

FIGURE 14.4 Schematic diagram of preparation of the MIL-53 membrane on alumina support via the reactive seeding method. (Reproduced from Hu, Y. et al., Chemical Communications. 47, 737–739, 2011. With permission.)

the RS method is shown in Figure 14.4. Al_2O_3 support rather than $Al(NO_3)_3{\cdot}9H_2O$ acted as the Al precursor, which reacted with 1,4-benzenedicarboxylic acid (H_2BDC) under mild hydrothermal conditions to produce a seed layer. This was followed by a secondary growth process in which $Al(NO_3)_3{\cdot}9H_2O$ and H_2BDC formed the MIL-53 membrane under hydrothermal conditions. By this method, MIL-53 (Hu et al., 2011), MIL-96 (Hu et al., 2011), Zn-BLD (Wang et al., 2012b), ZIF-78 (Dong et al., 2012), and ZIF-71 (Dong and Lin, 2013) membranes have been successfully prepared on porous Al_2O_3 or ZnO support.

For some kinds of MOFs, like ZIF-71, it was found to be difficult to prepare high-quality MOF membranes by the secondary growth method because the MOF growth is too fast to be controlled (Huang et al., 2015a). In this context, the contra-diffusion method develop by Yao et al. (2011) could be useful to fabricate this kind of MOF membranes. In this method, two synthesis solutions, metal salt solution and organic ligand solution, are separated by the porous support, and the crystallization takes place on the membrane surface through solution contra-diffusion. Jin and co-workers (Huang et al., 2015a) used a modified contra-diffusion method to prepare ZIF-71 membrane on ceramic hollow fiber. The end-sealed ceramic hollow fiber was immersed in a $Zn(Ac)_2{\cdot}MeOH$ solution while a 4,5-dichloroimidazole MeOH solution was filled in the bore size of the fiber. Thus, the metals Zn^{2+} and the organic links imidazole would meet and react on the outer surface of ceramic hollow fiber through diffusion to form an intergrowth ZIF-71 membrane with thickness of ~2.5 μm well bonded to the ceramic hollow fiber support.

14.2.3 Mixed-Matrix/Ceramic Composite Membranes

Generally, polymeric membranes suffer the performance trade-off between permeate flux and selectivity, and it remains challenging for scalable fabrication of inorganic membranes despite their high separation performance. Alternatively, a mixed-matrix membrane (MMM) is prepared by incorporating inorganic fillers into polymeric matrix, which combines the respective advantages of good processability from polymeric membranes and high performance from inorganic membranes well (Koros and Zhang, 2017). There are some reports fabricating mixed-matrix separation layer on the ceramic support to form mixed-matrix/ceramic composite membranes for pervaporation (Liu et al., 2011c; Liu et al., 2011d; Liu et al., 2015a). Homogenous dispersion of the filler and well-adhered filler-polymer interface are two important points for the development of high-performance MMMs.

To fabricate ceramic-supported ZSM-5 (a type of MFI zeolite) filled PDMS MMMs, Jin and co-workers (Liu et al., 2011c) proposed a surface graft/coating approach to modify the ZSM-5 fillers to achieve homogenous dispersions of ZSM-5 particles in PDMS matrix with filler loading up to 40 wt%. Specifically, octyl group was attached onto the zeolite surface via the silylation using n-octyltriethoxysilane, thereby providing interactions with PDMS chain in the following coating process by adding 10% of the total amount of PDMS. Molecular dynamic simulation study indicates that the interaction energy between zeolite and PDMS was greatly improved (from −9.4 to −921.8 kcal/mol), owing to the entangling of grafted long n-octyl chains on zeolite surface with PDMS chains.

Compared with the purely inorganic filler such as zeolite, MOFs with organic linkers exhibit much higher compatibility and interfacial adhesion with polymers and thus have been widely used as high-performing fillers for MMMs in recent years. Yang and co-workers (Liu et al., 2011d) introduced ZIF-8 nanoparticles (particle size ~40 nm) into the polymethylphenylsiloxane (PMPS) and dip-coated the resulting ZIF-8/PMPS solution on the inside surface of alumina capillary support. As shown in Figure 14.5, the thickness of the separation layer on the ceramic support is ~2.5 μm with 9 wt% ZIF-8 uniformly dispersed in the PMPS matrix.

Besides the well-defined transport channels, the incorporated inorganic fillers can affect the packing and mobility of the polymer chains, thereby tuning the membrane free volumes for molecular separation. Polyhedral oligomeric silsesquioxane (POSS) filled PDMS mixed-matrix membranes were fabricated on a ceramic tube (Liu et al., 2015a). The molecular interactions between POSS and PDMS not only were beneficial for the dispersion of POSS in PDMS and the POSS-PDMS interfacial morphology, but also finely regulated the free volumes of the MMMs. By increasing the POSS loading, the small free volume elements declined, whereas the large volume elements gradually increased, which was profitable for selective permeating larger-sized molecules, for instance: butanol recovery from water solution.

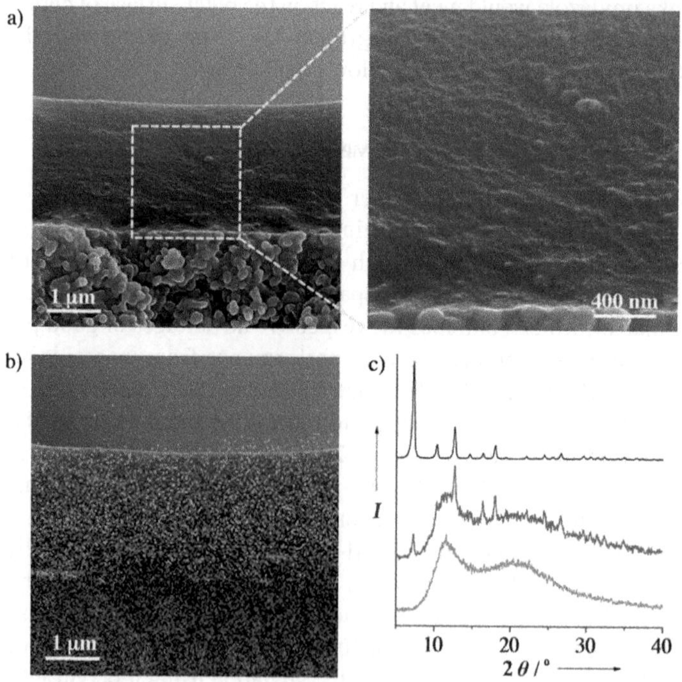

FIGURE 14.5 (a) Cross-sectional SEM images and (b) EDXS mapping of the ZIF-8-PMPS membrane (ZIF-8 loading: 9wt%; Zn signal: white; Al signal: dark gray; Si signal: medium gray); (c) XRD patterns of ZIF-8 nanoparticles (black line), pure PMPS membrane (dark gray line) and ZIF-8-PMPS membrane (medium gray line). (Reproduced from Liu, X.-L. et al., Angewandte Chemie International Edition. 50, 10636–10639, 2011. With permission.)

Instead of dispersing the filler in the polymer, Yang and co-workers (Liu et al., 2011e) proposed another way to combine the inorganic filler and polymer as the separation layer on top of a ceramic tube. First, silicalite-1 (a type of MFI zeolite) nanocrystals were deposited onto a porous alumina capillary support via dip-coating (packing). Second, the interspaces within the calcined silicalite-1 nano-crystal layer were filled with PDMS phase by using the capillary condensation effect (filling). The so-called "packing-filling" method produced a thin (300 nm) and uniform silicalite-PDMS MMM that was fabricated on the ceramic support.

14.2.4 GO/Ceramic Composite Membranes

The discovery of graphene triggered the research on two-dimensional (2D) materials in related fields of physics, materials science, and chemistry. The unique feature of 2D materials, atomic thickness, enabled them as emerging building blocks for developing high-performance separation membranes by minimizing the membrane thickness to maximize the permeate flux (Liu et al., 2016). As an important derivative of graphene, GO is regarded as the most promising candidate for 2D-material membranes owing to its easy fabrication and processing, as well as abundant functional groups. Generally, the GO nanosheets are assembled into laminar membranes consisting of interlayer galleries for molecular separation. These GO membranes have shown excellent separation performance for water purification, solvent dehydration, ions separation, and gas separation (Liu et al., 2015b).

Several methods have been developed to fabricate GO membranes, including filtration, coating (drop-casting, dip-coating, spin-coating, or spray-coating), and LBL assembly. Among them, filtration is the most common and straightforward route. Ruoff and co-workers (Dikin et al., 2007) first fabricated free-standing GO papers by filtration GO nanosheets dispersed in water. Tightly packed interlocking nanosheets subtly undulate along the paper surface. The laminar structure is mainly controlled by the slowly flowing water in the confined galleries and the electrostatic and van der Waals attractive forces between the GO nanosheets. Typically, the thickness of the GO membranes can be tuned by the filtrated amount of GO nanosheets that is often varied by the concentration and volume of GO suspension.

The drop-casting or dip-coating process forms relatively heterogeneous GO laminates caused by the electrostatic repulsion between the GO edges (Shen et al., 2016). In contrast, spin-coating often prepares a highly interlocked laminate because the capillary interactions between the faces of GO sheets could overcome the electrostatic forces between the GO edges (Kim et al., 2013). The application of spray-coating approach is not limited by the configuration of substrates, not only flat sheet, but also tubular and hollow fiber forms. In the LBL approach, the membrane thickness can be precisely controlled by varying the number of LBL deposition cycles of GO nanosheets (Hu and Mi, 2013). Moreover, this approach is ideal for uniformly introducing molecules or nanomaterials during each layer deposition to functionalize or intercalate GO membranes (Mi, 2014). When integrated with spin-coating, the LBL method shows higher efficiency and structural homogeneity for preparing GO membrane (Shen et al., 2016).

Likewise, a porous support is often used to fabricate a thin-film composite GO membrane to afford high permeate flux and good mechanical strength. The pore size, porosity, and surface roughness of the support would affect the formation of uniform GO membrane. Surface modification of the substrate by introducing some functional groups can improve the interfacial adhesion between GO membrane and substrate (Lou et al., 2014). Most GO membranes were prepared on the flat-sheet polymeric supports or aluminum oxide substrate (Zhang et al., 2017). Compared with these supports, ceramic hollow fibers exhibit advantages in high packing density and chemical and structural stabilities (Wang et al., 2009). However, conventional coating approaches are difficult to deposit GO membranes on the ceramic hollow fiber support due to the high curvature and elongated geometry.

In view of above-mentioned challenges, Jin and co-workers (Huang et al., 2014) proposed a vacuum suction (modified filtration) method to construct the GO/ceramic hollow fiber composite membranes. As shown in Figure 14.6, one side of the ceramic hollow fiber is sealed, and the other side is connected to a vacuum pump. Thus, GO nanosheets can be easily stacked on the curved surface of the ceramic hollow fiber with pressure as the driving force. SEM images of the resulting membrane indicate a smooth surface without visible defects and a laminar structure with ~1.5 μm thickness of the GO separation layer firmly deposited on the porous ceramic hollow fiber support. The thickness of the GO layer can be easily tuned by controlling the concentration of the GO aqueous suspension or by altering the operation time of the vacuum suction operation.

Furthermore, they designed a bio-inspired membrane that couples an ultrathin surface water-capturing polymeric layer (<10 nm) and GO laminates consisting of water channels, which was constructed by simply by vacuum filtrating the polymer onto the above-discussed GO/ceramic hollow fiber composite membrane (Huang et al., 2015b). In the prepared CS@GO membrane, the hydrophilic CS coating acts as the surface layer to preferentially capture a large number of water molecules from feed mixtures (e.g., alcohol/water), resulting in an increase of the driving force across the GO laminates. As a result, the 2D water channels within the interlayers of GO nanosheets would be fully used, realizing fast and selective water permeation through the integrated membrane.

14.2.5 Interfacial Adhesion

Interfacial adhesion between the separation layer and support layer plays an important role in determining the structural stability of the composite membranes. Thus, it is necessary to directly measure the interfacial adhesion force of the active layer on the support. In contrast to the conventional tensile and peeling test, nano-indentation/scratch technique provides a simple, accurate, versatile, and rapid means to assess the mechanical strength and interfacial adhesion of small-volume samples on nano-scale (Fischer-Cripps, 2006). In particular, nano-scratch test moves sample perpendicular to the scratch probe to determine interfacial adhesion strength through film delamination caused by the accumulation of interfacial shear stress. Based on this advanced technique, Jin and co-workers developed an efficient characterization method to

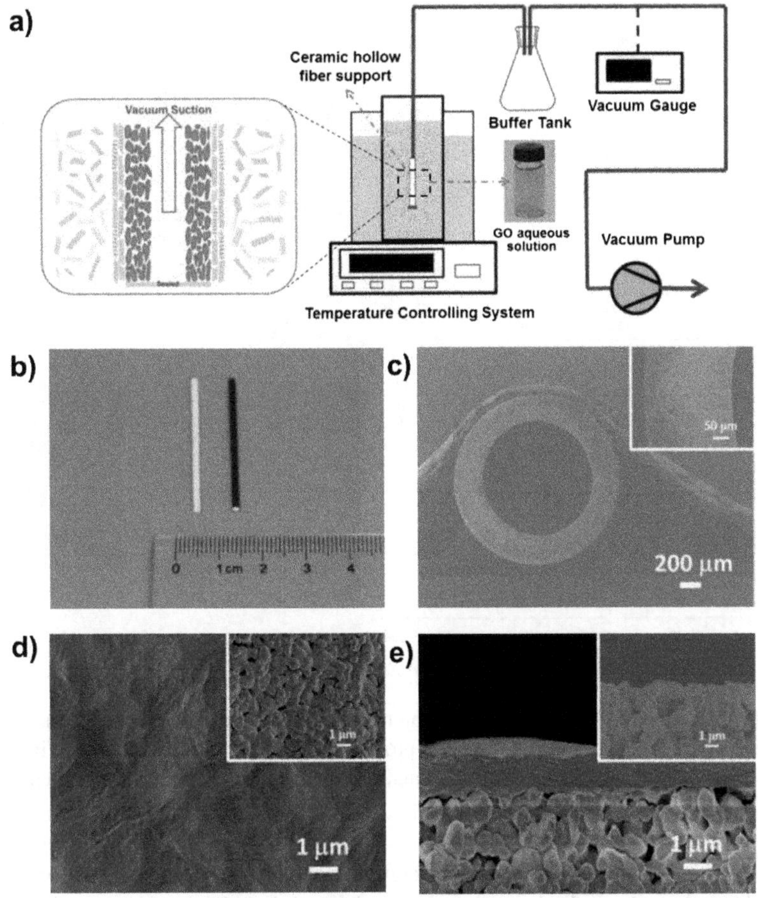

FIGURE 14.6 GO/ceramic hollow fiber composite membrane: (a) Schematic diagram of fabrication setup and forming process of the GO membrane; (b) Photographs of the blank hollow fiber (white) and the GO membrane (black); SEM images of (c) a ceramic hollow fiber (insert: an enlarged cross-section of the ceramic hollow fiber), (d) surface, and (e) cross-section of the blank hollow fiber (inset) and the GO membrane. (Reproduced from Huang, K. et al., Angewandte Chemie International Edition. 53, 6929–6932, 2014. With permission.)

measure interfacial adhesion of various ceramic-supported composite membranes (Hang et al., 2015; Wei et al., 2010).

The interfacial adhesive properties of PDMS/ceramic composite membranes were studied by using the nano-indentation/scratch technique (Hang et al., 2015). Figure 14.7 gives the typical scratch profiles. Two clear transition points can be observed in scratch profile (Figure 14.7(a)) and friction profile (Figure 14.7(b)): (i) At 121 μm scan displacement for the onset of edge cracking (Lc1 = 15.9 mN). (ii) At 337 μm scan displacement for the total film failure (Lc2 = 50.5 mN). These transitions can be confirmed by SEM characterization of the scratch morphology. In the range before 121 μm, the PDMS layer was fully recovered under elastic contact, and no obvious

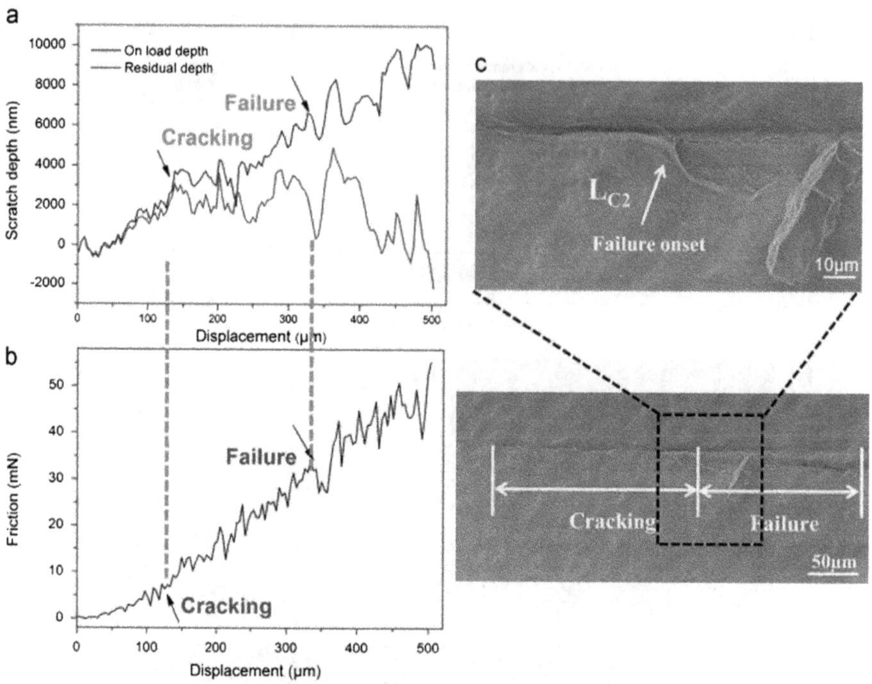

FIGURE 14.7 Typical results of nano-scratch test of PDMS/ceramic composite membranes (PDMS layer thickness: 12 μm): (a) scratch profile, (b) friction profile and (c) SEM images of scratch morphology. (Reproduced from Hang, Y. et al., Journal of Membrane Science. 494, 205–215, 2015. With permission.)

scratch can be found. When the displacement was over 121 μm, fluctuations in the on-load depth and residual depth are found, implying the beginning of cracking formation. The PDMS layer continued to crack with the increase of applied load until it peeled from the ceramic support, indicated by the onset of larger fluctuations in the scratch depth at displacement of 337 μm. In the SEM images (Figure 14.7(c)), porous ceramic support was exposed within the scratch track after Lc2 on the PDMS/ceramic composite membrane. Thus, such critical load (Lc2) at the failure onset of PDMS layer is considered the adhesive strength of the PDMS layer onto the ceramic support layer, namely, the interfacial adhesion force of the PDMS/ceramic composite membrane. The total film removal after Lc2 was accompanied with broad scratch trace and the scratch damage was remarkable at the end of trace.

It was found that the interfacial adhesion of the composite membranes can be tuned by varying the roughness of ceramic support and the thickness of the PDMS layer. As the surface roughness of ceramic support increased from 0.440 to 1.510 μm, the interfacial adhesion force (critical load) was improved from 27.6 to 50.7 mN. Moreover, there is a linear advance in the critical load from 10 mN to >50 mN, as increasing the PDMS layer thickness from 3 μm to 14 μm. The interfacial adhesion of PDMS separation layer onto the porous ceramic support was attributed into three parts: adsorption, chemical bonding, and mechanical interlocking (Wei et al., 2010).

First, the ceramic support has higher surface energy than the PDMS solution, thus the coated PDMS layer could adhere to the surface of ceramic support. Second, the hydrogen bonding occurs between –OH group in the ceramic support and oxygen atom of PDMS. Third, the ceramic pores form the enchased structure between the ceramic support and PDMS layer. The nano-indentation/scratch characterization suggests that the mechanical interlocking dominates the interfacial adhesion of PDMS/ceramic composite membrane.

The nano-indentation/scratch platform is also useful to probe the interfacial adhesion of GO/ceramic composite membranes (Huang et al., 2015b). In the above-discussed CS@GO membrane, the ultrathin polymeric layer was demonstrated to be helpful for increasing the mechanical stability of the GO/ceramic hollow fiber composite membrane. The CS@GO membrane began to fail with a drastic depth change at 233.7 μm, which was larger than that of the pristine GO laminates (141.6 μm). The load-displacement curve revealed that the critical load for CS@GO membranes (27.5 mN) is higher than that of pristine GO membrane as well. The enhanced structural stability is beneficial for the application in cross-flow operation of practical pervaporation process.

14.3 MEMBRANE PERVAPORATION APPLICATIONS

14.3.1 Solvent Recovery

Hydrophobic or organophilic pervaporation membranes, preferentially permeating organic compounds over water, are often used for solvent recovery from aqueous solution. In this area, one of the intensive studies is of the removal of biofuel from a dilute water system (Peng et al., 2010). Bio-alcohols produced by biomass fermentation, such as ethanol or butanol, are critical for the development of renewable energy, due to the emerging scarcity of oil resources and the demand for environmental protection (Vane, 2005). The bio-alcohols produced by traditional fermentation process have low concentrations, typically ~10 wt% for ethanol and ~1 wt% for butanol, thereby requiring a high energy cost to concentrate the biofuels by using distillation. Alternatively, pervaporation is an attractive separation technology to reduce the energy consumption. Moreover, if integrating pervaporation with fermentation process, the bio-alcohol can be in-situ removed from the fermentation broth as it is produced. Thus, the inhibitory effect of the solvent on the microbial growth would be relieved, thereby realizing continuous fermentation to improve the efficiency of bio-alcohol production (Liu et al., 2014). In the past decade, different kinds of ceramic-supported composite membranes have been investigated for bio-alcohol recovery.

PDMS, the most representative hydrophobic membrane material for pervaporation, has been widely used for separating ethanol or butanol from aqueous solution. Jin and co-workers (Liu et al., 2011a; Xiangli et al., 2007; Xiangli et al., 2008) studied the pervaporation performance of tubular PDMS/ceramic composite membranes in ethanol/water mixtures and n-butanol/water mixtures. It was found that the ceramic-supported PDMS composite membrane exhibited a higher flux than the other PDMS composite membranes using polymeric supports, while kept the separation factor at the same level. This high performance might be owing to the dip-coated thin and defect-free PDMS layer, as well as the macroporous ceramic support with low transport resistance. Furthermore, the separation factor can be well maintained at increased operating temperature or feed concentration, which is usually unable to be achieved in the composite membranes using polymeric supports. This favorable result can be attributed to the "confinement effect" of the rigid ceramic support restricting the excessive swelling of PDMS layer (Wei et al., 2011).

The performance of the tubular PDMS/ceramic composite membrane was further enhanced when replacing the ceramic tube with ceramic hollow fiber as the support (Dong et al., 2014). By rationally designing the nanostructures, the selected ceramic hollow fiber exhibited larger average pore size resulting from the finger-like pores; meanwhile, sponge-like pores induced smoother surface for polymer layer deposition, achieving lower transport resistance and thinner defect-free separation layer at the same time. Thus, for 1 wt% n-butanol/water mixtures at 40°C, 181% higher total flux (1282 g/m^{2}h) and 65% higher separation factor (42.9) were obtained in the PDMS composite membranes supported by the ceramic hollow fiber than in those supported by ceramic tube. In view of this, another typical organophilic material, PEBA was also fabricated on the ceramic hollow fiber support via the facile dip-coating method (Li et al., 2016). By optimizing the PEBA concentration, coating viscosity and membrane thickness, total flux of ~2000 g/m^{2}h and separation factor of 21 was obtained in the resulting PEBA/ceramic hollow fiber composite membrane (feed: 1 wt% n-butanol/water mixtures at 40°C). Moreover, during long-term (100–200 h) continuous operation in model or real butanol fermentation broth, the ceramic hollow fiber supported PDMS or PEBA composite membranes exhibited high and stable butanol recovery performance that is fairly beyond the performance of state-of-the-art membranes for butanol/water separation, as shown in Figure 14.8.

Hydrophobic fillers have been introduced into PDMS/ceramic composite membranes to further improve the ethanol or butanol removal performance. Two types of MFI zeolites were used: ZSM-5 and silicalite-1. In the first case (Liu et al., 2011c), the incorporation of ZSM-5 particles modified by the surface grafting/coating approach significantly enhanced the separation factor of the pristine PDMS membrane for ethanol/water separation owing to the increased hydrophobicity, while the resulting

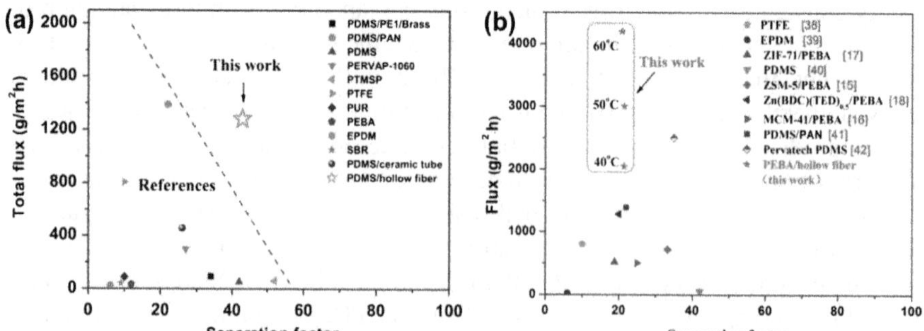

FIGURE 14.8 Performance comparison of hollow fiber ceramic-supported polymer composite membranes with reported membranes for butanol recovery: (a) PDMS membrane; (b) PEBA membrane. (Reproduced from Dong, Z. et al., Journal of Membrane Science. 450, 38–47, 2014; Li, Y. et al., Journal of Membrane Science. 510, 338–347, 2016. With permission.)

thicker ZSM-5/PDMS mixed-matrix layer reduced the total flux. The second case (Liu et al., 2011e) demonstrated ultra-thin PDMS phase filled silicalite-1/ceramic composite membranes exhibiting high flux (5.0–11.2 kg/m^{2}h) and good separation factor (25.0–41.6) for the recovery of i-butanol from aqueous solution (0.2–3 wt%) at 80°C.

In contrast to the hydrothermal synthesis of zeolite crystals at high temperature and autogenous pressure, the preparation of ZIF-8 nanoparticles (using Zinc nitrate and 2-methylimidazole) can be conducted at room temperature in solution, making it lower cost and time saving. Meanwhile, the flexible window aperture (minimum size: ~0.34 nm) and superhydrophobicity of ZIF-8 lead high sorption capacity and excellent selectivity towards biofuels over water. These features enabled ZIF-8 as high-performing filler to improve the alcohol/water separation performance of polymeric membranes. By incorporating 9 wt% ZIF-8 with particle size of ~40 nm into the PMPS matrix, the resulting ceramic-supported ZIF-8/PMPS mixed-matrix composite membrane showed a total flux of 6.4 kg/m^{2}h and separation factor of 40.1 for 1 wt% i-butanol/water at 80°C (Liu et al., 2011d). This separation factor and the high permeance clearly transcend the upper limit of state-of-the-art membranes (Figure 14.8), and reach the economically attractive region. Furthermore, both the butanol permeability and separation factor increased simultaneously as the ZIF-8 loading (up to 9 wt%), suggesting that the ZIF-8 nanoparticles can create preferential pathways for butanol molecules based on its ultrahigh butanol/water adsorption selectivity.

A few pure MOF membranes showed potential in pervaporation recovery of solvent from water solution. Although the hydrophobic ZIF-8 highly improved the butanol/water performance of the polymeric membranes, the pure ZIF-8 membranes have not been reported for use in this application yet. Alternatively, ZIF-71 (Morris et al., 2010), another ZIF family member, having a RHO topology with small windows (0.48 nm) and big cages (1.68 nm), and intrinsic hydrophobicity, has been prepared as pure MOF membranes for alcohol/water separation. The ZIF-71 membrane fabricated on ZnO disk via reactive seeding method showed low flux (332 g/m^{2}h for 5 wt% ethanol/water at 25°C) due to the high transport resistance of the support (Dong and Lin, 2013). Predictably, the flux was highly improved (to 2601 g/m^{2}h, with the same separation factor of ~6–7) by using the low-resistance ceramic hollow fiber as the support (Huang et al., 2015a).

14.3.2 Solvent Dehydration

The widest application of pervaporation is the dehydration of organic solvents, especially for azeotropes or the close-boiling mixtures for which distillation processing is energy intensive (Chapman et al., 2008). In the literature, dehydration of alcohol/water mixtures is still the most widely studied system because of the great potential of bio-alcohols and the separation of alcohol/water azeotropes consumes high energy.

Benes and co-workers (Peters et al., 2008) investigated the long-term performance and thermal stability of the PVA/ceramic hollow fiber composite membrane over a period of 6 months and operating temperature up to 100°C. The separation performance of the membranes was tested for pervaporation dehydration of n-propanol, 2-propanol, and n-butanol. After a period of 3 days in which the water flux decreased, both water flux and selectivity were stable at a value of 1000 g/m^{2}h and 450, respectively, during a period of 8 days at 80°C for dehydration of n-butanol. After 6 months, the PVA/ceramic composite membrane showed a higher water flux and kept a high selectivity despite lower than the fresh membrane. It is attributed to the looseness of the crosslinked PVA networks due to the removal of maleic acid in water. It is worth noting that by increasing temperature or feed water concentration, both the water flux and selectivity improved simultaneously. This remarkable behavior is in contrast to the trade-off generally observed for polymer membranes, i.e., an increase in flux is typically combined with a decrease in separation factor (Shao and Huang, 2007). Jin and co-workers (Xia et al., 2011) observed similar results when testing the tubular PVA/ceramic composite membranes at higher feed temperature and alcohol concentration. A possible explanation is the so-called "confinement effect" discussed above: a reduced three-dimensional swelling of the PVA membrane in the region of the PVA-ceramic interface due to the rigid structure of ceramic support.

Pure UiO-66 membranes supported by ceramic hollow fiber were applied for dehydration of i-butanol/water mixtures, exhibiting a very high flux of up to ~6.0 kg/m^{2}h and excellent separation factor (>45,000) for 5 wt% water in the feed at 70°C (Liu et al., 2017). This separation factor is 1–2 orders of magnitude higher than that of commercially available polymeric and silica membranes with equivalent flux. This performance is also comparable to that of commercial NaA zeolite membranes. Moreover, the UiO-66 membrane remained robust during a long-term stability test for ~300 h, including exposure to aggressive conditions such as boiling benzene, boiling water, and sulfuric acid, which some commercial membranes like NaA zeolite membranes cannot survive.

The potential of ceramic-supported graphene-based membranes for dehydration of bio-alcohols was demonstrated, as well. Li et al. (2014) used positively charged ethanediamine to modify GO nanosheets to reverse the surface charge of GO (negative) in order to enhance the interaction between GO and the negatively charged surface of tubular Al_2O_3 support. As a result, good separation performance (flux, selectivity, and stability) was achieved in the pervaporation dehydration of ethanol, n-propanol, iso-propanol, ethyl acetate or butanol isomers aqueous solution. Jin and co-workers (Huang et al., 2015b) applied the ceramic hollow fiber supported CS@GO composite membrane for removing water from 90 wt% n-butanol/water mixtures at 70°C via pervaporation, showing highly improved water flux (>10 kg/m^{2}h) with separation factor of over 1000. This excellent performance, resulting from a synergistic effect of the hydrophilic polymer surface coating and the 2D interlayer channels of graphene oxide laminates, exceeds the upper bound of state-of-the-art membranes for butanol dehydration.

Besides the bio-alcohols, ceramic-supported composite membranes were also used for dehydration of other organic solvents, such as ethyl acetate (EtAc). EtAc is widely used in manufacture of varnishes, thinners, nitrocellulose lacquers, and various drugs owing to its low toxicity, good volatility, and solubility. Current azeotropic distillation or extractive distillation for producing EtAc is highly energy intensive, and thus can be replaced by pervaporation-based process.

Jin and co-workers (Xia et al., 2011; Xia et al., 2012) studied the pervaporation dehydration performance of the tubular PVA/ceramic membrane for both EtAc/water binary mixtures (Xia et al., 2011) and EtAc/ethanol (EtOH)/water ternary mixtures (Xia et al., 2012). In sorption measurement of binary mixtures, the increase of water concentration in the mixtures enhanced the swelling degree of the PVA layer, owing to the favorable interaction between water molecules and hydrophilic PVA chains. Meanwhile, the swollen PVA layer with enlarged free volumes also adsorbed more EtAc molecules, thus resulting in a declined water/EtAc sorption selectivity. As a result, in the pervaporation test, the PVA/ceramic composite membrane showed an improvement in flux while a decrease in selectivity when increasing the water concentration in the feed. Moreover, the Flory-Huggins theory was applied to analyze the interactions of the membrane and penetrants (water, EtAc, and EtOH). It was found that the interaction of the membrane and the penetrant was in the order of water>EtOH>EtAc, suggesting that the interaction between water and membrane dominates the sorption process. This result agreed well with the sorption measurement, which is also caused by the hydrophilic characters of the PVA. In the 82.6 wt% EtAc/8.4 wt% EtOH/9 wt% water mixtures at 60°C, the PVA/ceramic composite membrane exhibited a total flux of 2.1 kg/m^{2}h and permeate water concentration of 94.9 wt%, and this performance changed slightly during 110 h continuous operation.

In addition, the ceramic hollow fiber supported UiO-66 membrane was used for dehydration of different solvents, including ethanol, n-propanol, i-propanol, furfural, tetrahydrofuran (THF), and acetone (Liu et al., 2017). The pervaporation performance for the furfural and THF were as good as the membrane used for dehydration of i-butanol/water mixtures discussed above: water flux of ~6.0 kg/m^{2}h and separation factor >45,000 for 5 wt% water in the feed at 70°C. Moreover, it was found that the selectivity of water over solvent increased with the increase of kinetic diameter of these solvents, along with a lower flux, suggesting a size-selective diffusion in the UiO-66 membrane. Meanwhile, the selectivity and water permeance varied with the feed concentration and temperature, indicating that water adsorption also affects the membrane performance (Figure 14.9). Thus, the UiO-66 membrane-based pervaporation separation follows the classic sorption-diffusion model.

GO membrane prepared on the ceramic hollow fiber showed an excellent dehydration performance for dimethyl carbonate (DMC)/water mixtures (Huang et al., 2014). The reactor for producing DMC typically contains 1–3 wt% water that requires further purification. Thus, water concentration ranging from 1–2.6 wt% and temperature of 25–40°C was used to study the pervaporation performance. The results indicated that the total flux was enhanced by increasing the feed concentration or temperature caused by the improved driving force and the swelling of the GO laminate, while the separation factor was lowered at higher temperatures. At 25°C and 2.6 wt% water in the feed, the water concentration in the permeate was more than 95 wt% (separation factor: 740) and the total flux was 1.7 kg/m^{2}h, which are better than the reported values. The sorption measurement using a quartz crystal microbalance technique (QCM) revealed that sorption capacity of the GO membrane followed the order of water>methanol>DMC. It is believed that both the sorption and diffusion contributed to this excellent dehydration performance achieved in GO/ceramic hollow fiber composite membrane.

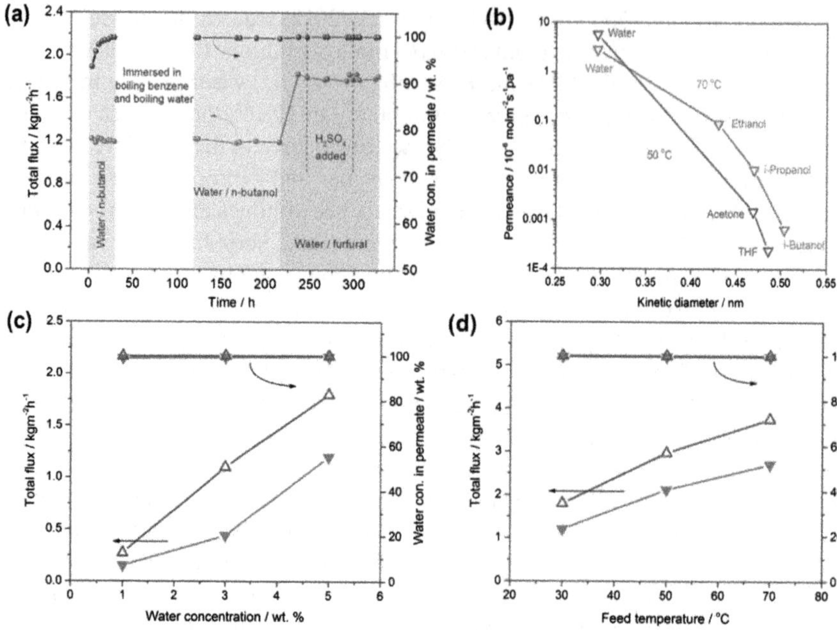

FIGURE 14.9 Pervaporation performance of the ceramic hollow fiber supported UiO-66 membrane: (a) Separating water from n-butanol and furfural during on-stream activation and stability test processes at 30°C. The concentration of water in each feed was kept at 5 wt%. (b) Water and organic permeance versus their kinetic diameters. The concentration of water in each feed was set at 5 wt%. (c) Effect of water concentration in the feed on the pervaporation performance at 30°C. (d) Effect of feed temperature on the pervaporation performance using 5 wt% aqueous solutions as feed. For (c) and (d), filled and open triangles represent membrane performance for separating water from n-butanol and furfural, respectively. (Reproduced from Liu, X. et al., Advanced Functional Materials. 27, 1604311, 2017. With permission.)

14.3.3 Separation of Organic Mixtures

Separation of close-boiling organic/organic mixtures such as aromatic/aliphatic hydrocarbon is one of the most significant yet challenging processes in chemical industry. Currently, distillations (e.g., azeotropic, extractive) dominate this kind of separation, but are very energy intensive. Several advanced membrane materials have been developed for pervaporation separation of organic/organic mixtures, which are mostly organophilic membranes owing to the favorable interactions with organics. But, these favorable interactions also cause large swelling of the polymeric membrane during the pervaporation process, and thus might decrease the membrane selectivity and structural stability. Crosslinking (Wu et al., 2015) or improving the crosslinking degree (Xu et al., 2010) is often used to enhance the performance and stability of the polymeric membrane in the organic/organic mixtures. On the other hand, MOF crystalline membrane, possessing a more rigid structure than polymeric membrane, is another promising candidate for separating organic mixtures.

Ji and co-workers (Wu et al., 2015) used the crosslinked PEBA/ceramic composite membrane for separation of 50/50 (wt%) aromatic/aliphatic mixed hydrocarbons. In toluene/n-heptane solution, the feed temperature was gradually increased from 40 to 80°C, and each temperature was operated for 6 h. It was found that the permeate flux is gradually improved with the feed temperature when the toluene content in the permeate is kept at ~80 wt%, suggesting a stable membrane selectivity during the total 30 h continuous pervaporation process. The pervaporation performance in other hydrocarbons, including toluene/cyclohexane and toluene/iso-octane, were also investigated. The results indicated that the physical and chemical properties of the organics affected the membrane performance.

Sulfur removal from gasoline has received special attention due to environmental concerns. In 2002, W.R. Grace & Co's Davison Company proposed a pervaporation membrane-based S-Brane process for the gasoline desulfurization. Jin and co-workers (Xu et al., 2010) demonstrated the potential of PDMS/ceramic composite membranes for removal of sulfur impurities out of model gasoline (n-octane/thiophene mixtures). They found that increasing the concentration of crosslinker from 10 wt% to 20 wt% significantly improved the membrane selectivity (the enrichment factor was increased from 1.4 to 4.3), although the flux was decreased to some extent. Compared with the reported membranes that use polymeric supports, the ceramic-supported PDMS membrane exhibited a higher permeation flux of 5.37 kg/m^{2}h and an acceptable sulfur enrichment factor of 4.22 for 400 μg/g sulfur in feed at 30°C.

Only a few studies report pervaporation performance of MOF membranes for separating organic mixtures. Lin and co-worker (Ibrahim and Lin, 2016) tested the pervaporation of MOF-5 membranes for pure toluene, o-xylene, and 1,3,5-triisopropylbenzene (TIPB), and their binary mixtures. They found that MOF-5 membrane was fouled by these organics, indicated by a continuous decrease of flux during 10 h of pervaporation. Activating the fouled membrane at 100°C in vacuum could not restore the flux, suggesting that the fouling was not fully reversible in the MOF-5 membrane. A higher activation temperature was not tried due to the limited thermal stability of the MOF membrane layer. Caro and co-workers (Diestel et al., 2012) measured the permeation of pure n-hexane (critical diameter σ = 0.43 nm), benzene (σ = 0.58 nm), and TIPB (σ = 0.84 nm), as well as the pervaporation separation of n-hexane/benzene and n-hexane/TIPB liquid mixtures. The high quality of the ZIF-8 membrane was ensured by the measured H_2/CH_4 mixed-gas selectivity of 15. Although the window aperture is estimated to be 0.34 nm from crystallographic data, ZIF-8 has a flexible framework structure, leading the observed passage of methane (σ = 0.38 nm), n-hexane, or even benzene through the integrated ZIF-8 membrane. Benzene showed a lower flux than *n*-hexane, whereas TIPB showed only a very small leakage rate through the O-ring gasket. Based on the difference in molecular size, medium separation factors were found for the pervaporation separation of n-hexane/benzene mixtures.

14.3.4 Pervaporation-Based Hybrid Process

Pervaporation process is often integrated with another process such as distillation or reaction to further improve the energy efficiency. The reversible chemical equilibrium limitation can be overcome by in-situ removing the product as soon as it is produced via pervaporation (or vapor permeation), leading to a higher yield and conversion rate. In this regard, Jin and co-workers applied their ceramic-supported polymer composite membranes for the hybrid processes built on integrating pervaporation with fermentation (Liu et al., 2011b), reactive distillation (Lv et al., 2012), or hydrolysis reaction (Li et al., 2010).

The hydrophobic pervaporation membranes were also integrated with fermentation to in-situ recover bio-alcohol as it is produced in the broth. A tubular PDMS/ceramic composite membrane was coupled with acetone-butanol-ethanol (ABE) fermentation process (Liu et al., 2011b). It was found that the inorganic salts in the ABE fermentation broth are in favor of increasing the membrane selectivity, owing to the increased ABE activity in the feed, while the microbial cells in the broth reduced the pervaporation performance because the attached bio-components reduced the membrane hydrophobicity and hindered the molecular transport. During the fermentation-PV coupled process under 37°C, although a fluctuation of membrane performance was observed due to the occurrence of membrane fouling, the PDMS/ceramic composite membrane exhibited a high flux of 0.670 kg/m^2h and good ABE/water separation factor of 16.7. The SEM, AFM, and IR characterizations confirmed that the membrane fouling was caused by the adsorption of active cells on the PDMS membrane surface. Nevertheless, the fouled membrane can be almost fully recovered after a simple water rinse, indicating the membrane fouling from the coupled process was reversible. The solvent productivity of the ABE fermentation was 0.303 g/L h, which was ~33% higher than that of fermentation without in-situ recovery by pervaporation.

The PDMS/ceramic composite membrane was also integrated with hydrolysis conversion of ethyl lactate, which is a typical reversible reaction controlled by thermodynamic equilibrium. Ethanol in the products was in-situ removed via vapor permeation (Li et al., 2010). Reaction temperature and ethanol concentration were varied to test the membrane performance. It was shown that the membrane flux increased with temperature, and the separation factor varied between 7 and 9. The total flux was 2.54 kg/m^2h with a separation factor of 7.5 for the vapor permeation of 5 wt% ethanol/water mixtures at 85°C. As the ethanol concentration increased to 10 wt%, the flux was more than 4.0 kg/m^2h and the separation factor was ~11. By coupling with vapor permeation, the conversion of the ethyl lactate was increased from 77.1% to 98.2% with the initial water/ethyl lactate of 10:1 at 85°C.

Another example for the reversible reactions is esterification (e.g., ethanol + acetic acid → ethyl acetate), in which the equilibrium limitation is a key issue and now is often performed by reactive distillation (RD). Compared with the distillate, much more water is created with the unreacted acetic acid (HAc) in the reboiler, significantly increasing the energy consumption of the reboiler. Thus, a new RD-pervaporation hybrid process (Lv et al., 2012) was proposed for the EtAc production, where PVA/ceramic composite membrane was applied to in-situ remove the water from the reboiler and recycle the HAc into the feed (Figure 14.10). In this new process, not only was the water removed from the products to promote the esterification, but also the unreacted HAc was reused that provided feedback to the

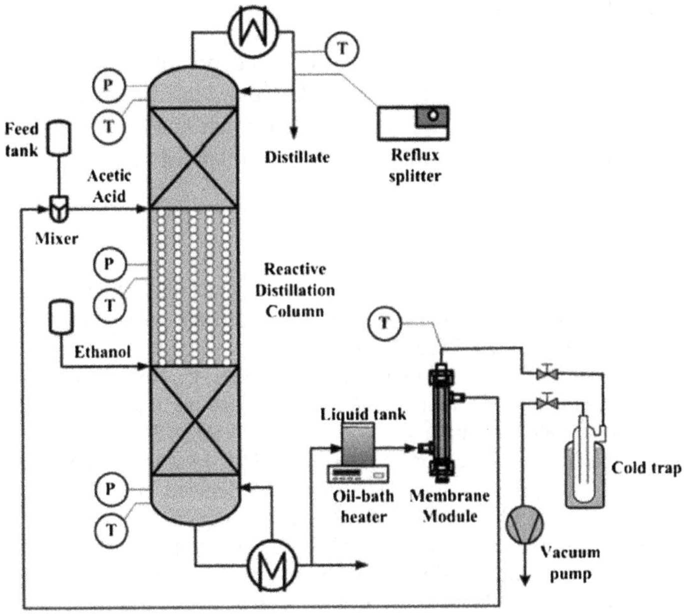

FIGURE 14.10 Experiment setup for reactive distillation-pervaporation hybrid process. (Reproduced from Lv, B. et al., Industrial & Engineering Chemistry Research. 51, 8079–8086, 2012. With permission.)

RD process. The PVA/ceramic composite membrane for dehydration of HAc/water was first studied. It showed a good stability with total flux of 600 g/m^2h and separation factor of 14 in 90 wt% HAc/water at 70°C. In the hybrid process, the removal of water and recycling of bottom withdrawal, realized by integrating a hydrophilic pervaporation membrane, significantly enhanced both the product purity and the reaction conversion of the RD process. The EtAc purity and EtOH conversion simultaneously increased from 82.4 to 85.6 wt% and from 81.3 to 84.8%, respectively, at the low HAc/EtOH molar ratio of 1.1.

14.4 CONCLUSION AND PERSPECTIVES

In conclusion, this chapter summarized the fabrication and application of ceramic-supported (polymer, MOF, mixed-matrix, GO) composite membranes for pervaporation. These membranes, especially polymer- and GO-based membranes, have shown great potential in solvent recovery, solvent dehydration, separation of organic mixtures, and hybrid process, demonstrating that the ceramic support is a promising candidate for developing high-performance composite membranes. Moreover, the interfacial adhesion is a key component for this kind of composite membrane. Future work in this area may focus on fundamental understanding of the membrane formation and transport mechanism, extending the application spectra, and scalable fabrication.

ACKNOWLEDGMENTS

We acknowledge the National Natural Science Foundation of China (Nos. 21490585, 21476107, 21776125, 51861135203), Innovative Research Team Program by the Ministry of Education of China (No. IRT17R54), and the Topnotch Academic Programs Project of Jiangsu Higher Education Institutions (TAPP) for financial support.

REFERENCES

Bux, H., Feldhoff, A., Cravillon, J., Wiebcke, M., Li, Y.-S., Caro, J., 2011. Oriented zeolitic imidazolate framework-8 membrane with sharp h2/c3h8 molecular sieve separation. Chemistry of Materials. 23, 2262–2269.

Cavka, J.H., Jakobsen, S., Olsbye, U., Guillou, N., Lamberti, C., Bordiga, S., Lillerud, K.P., 2008. A new Zirconium inorganic building brick forming metal organic frameworks with exceptional stability. Journal of the American Chemical Society. 130, 13850–13851.

Chapman, P.D., Oliveira, T., Livingston, A.G., Li, K., 2008. Membranes for the dehydration of solvents by pervaporation. Journal of Membrane Science. 318, 5–37.

Chen, Y., Xiangli, F., Jin, W., Xu, N., 2007. Organic–inorganic composite pervaporation membranes prepared by self-assembly of polyelectrolyte multilayers on macroporous ceramic supports. Journal of Membrane Science. 302, 78–86.

Diestel, L., Bux, H., Wachsmuth, D., Caro, J., 2012. Pervaporation studies of n-hexane, benzene, mesitylene and their mixtures on zeolitic imidazolate framework-8 membranes. Microporous and Mesoporous Materials. 164, 288–293.

Dikin, D.A., Stankovich, S., Zimney, E.J., Piner, R.D., Dommett, G.H.B., Evmenenko, G., Nguyen, S.T., Ruoff, R.S., 2007. Preparation and characterization of graphene oxide paper. Nature. 448, 457–460.

Dong, X., Huang, K., Liu, S., Ren, R., Jin, W., Lin, Y.S., 2012. Synthesis of zeolitic imidazolate framework-78 molecular-sieve membrane: Defect formation and elimination. Journal of Materials Chemistry. 22, 19222–19227.

Dong, X., Lin, Y.S., 2013. Synthesis of an organophilic ZIF-71 membrane for pervaporation solvent separation. Chemical Communications. 49, 1196–1198.

Dong, Z., Liu, G., Liu, S., Liu, Z., Jin, W., 2014. High performance ceramic hollow fiber supported PDMS composite pervaporation membrane for bio-butanol recovery. Journal of Membrane Science. 450, 38–47.

Fischer-Cripps, A.C., 2006. Critical review of analysis and interpretation of nanoindentation test data. Surface and Coatings Technology. 200, 4153–4165.

Hang, Y., Liu, G., Huang, K., Jin, W., 2015. Mechanical properties and interfacial adhesion of composite membranes probed by in-situ nano-indentation/scratch technique. Journal of Membrane Science. 494, 205–215.

Hu, M., Mi, B., 2013. Enabling graphene oxide nanosheets as water separation membranes. Environmental Science & Technology. 47, 3715–3723.

Hu, Y., Dong, X., Nan, J., Jin, W., Ren, X., Xu, N., Lee, Y.M., 2011. Metal-organic framework membranes fabricated via reactive seeding. Chemical Communications. 47, 737–739.

Huang, K., Li, Q., Liu, G., Shen, J., Guan, K., Jin, W., 2015a. A ZIF-71 hollow fiber membrane fabricated by contra-diffusion. ACS Applied Materials & Interfaces. 7, 16157–16160.

Huang, K., Liu, G., Lou, Y., Dong, Z., Shen, J., Jin, W., 2014. A graphene oxide membrane with highly selective molecular separation of aqueous organic solution. Angewandte Chemie International Edition. 53, 6929–6932.

Huang, K., Liu, G., Shen, J., Chu, Z., Zhou, H., Gu, X., Jin, W., Xu, N., 2015b. High-efficiency water-transport channels using the synergistic effect of a hydrophilic polymer and graphene oxide laminates. Advanced Functional Materials. 25, 5809–5815.

Ibrahim, A., Lin, Y.S., 2016. Pervaporation separation of organic mixtures by MOF-5 membranes. Industrial & Engineering Chemistry Research. 55, 8652–8658.

Jou, J.-D., Yoshida, W., Cohen, Y., 1999. A novel ceramic-supported polymer membrane for pervaporation of dilute volatile organic compounds. Journal of Membrane Science. 162, 269–284.

Kim, H.W., Yoon, H.W., Yoon, S.-M., Yoo, B.M., Ahn, B.K., Cho, Y.H., Shin, H.J., Yang, H., Paik, U., Kwon, S., Choi, J.-Y., Park, H.B., 2013. Selective gas transport through few-layered graphene and graphene oxide membranes. Science. 342, 91–95.

Koros, W.J., Zhang, C., 2017. Materials for next-generation molecularly selective synthetic membranes. Nature Materials. 16, 289–297.

Lee, Y.M., Bourgeois, D., Belfort, G., 1989. Sorption, diffusion, and pervaporation of organics in polymer membranes. Journal of Membrane Science. 44, 161–181.

Li, G., Shi, L., Zeng, G., Zhang, Y., Sun, Y., 2014. Efficient dehydration of the organic solvents through graphene oxide (GO)/ceramic composite membranes. RSC Advances. 4, 52012–52015.

Li, W., Zhang, X., Xing, W., Jin, W., Xu, N., 2010. Hydrolysis of ethyl lactate coupled by vapor permeation using polydimethylsiloxane/ceramic composite membrane. Industrial & Engineering Chemistry Research. 49, 11244–11249.

Li, Y., Shen, J., Guan, K., Liu, G., Zhou, H., Jin, W., 2016. PEBA/ceramic hollow fiber composite membrane for high-efficiency recovery of bio-butanol via pervaporation. Journal of Membrane Science. 510, 338–347.

Liu, G., Hou, D., Wei, W., Xiangli, F., Jin, W., 2011a. Pervaporation separation of butanol-water mixtures using polydimethylsiloxane/ceramic composite membrane. Chinese Journal of Chemical Engineering. 19, 40–44.

Liu, G., Hung, W.-S., Shen, J., Li, Q., Huang, Y.-H., Jin, W., Lee, K.-R., Lai, J.-Y., 2015a. Mixed matrix membranes with molecular-interaction-driven tunable free volumes for efficient bio-fuel recovery. Journal of Materials Chemistry A. 3, 4510–4521.

Liu, G., Jin, W., Xu, N., 2015b. Graphene-based membranes. Chemical Society Reviews. 44, 5016–5030.

Liu, G., Jin, W., Xu, N., 2016. Two-dimensional-material membranes: A new family of high-performance separation membranes. Angewandte Chemie International Edition. 55, 13384–13397.

Liu, G., Wei, W., Jin, W., 2014. Pervaporation membranes for biobutanol production. ACS Sustainable Chemistry & Engineering. 2, 546–560.

Liu, G., Wei, W., Jin, W., Xu, N., 2012. Polymer/ceramic composite membranes and their application in pervaporation process. Chinese Journal of Chemical Engineering. 20, 62–70.

Liu, G., Wei, W., Wu, H., Dong, X., Jiang, M., Jin, W., 2011b. Pervaporation performance of PDMS/ceramic composite membrane in acetone butanol ethanol (ABE) fermentation–PV coupled process. Journal of Membrane Science. 373, 121–129.

Liu, G., Xiangli, F., Wei, W., Liu, S., Jin, W., 2011c. Improved performance of PDMS/ceramic composite pervaporation membranes by ZSM-5 homogeneously dispersed in PDMS via a surface graft/coating approach. Chemical Engineering Journal. 174, 495–503.

Liu, X.-L., Li, Y.-S., Zhu, G.-Q., Ban, Y.-J., Xu, L.-Y., Yang, W.-S., 2011d. An organophilic pervaporation membrane derived from metal–organic framework nanoparticles for efficient recovery of bio-alcohols. Angewandte Chemie International Edition. 50, 10636–10639.

Liu, X., Li, Y., Liu, Y., Zhu, G., Liu, J., Yang, W., 2011e. Capillary supported ultrathin homogeneous silicalite-poly(dimethylsiloxane) nanocomposite membrane for bio-butanol recovery. Journal of Membrane Science. 369, 228–232.

Liu, X., Wang, C., Wang, B., Li, K., 2017. Novel organic-dehydration membranes prepared from zirconium metal-organic frameworks. Advanced Functional Materials. 27, 1604311-n/a.

Lou, Y., Liu, G., Liu, S., Shen, J., Jin, W., 2014. A facile way to prepare ceramic-supported graphene oxide composite membrane via silane-graft modification. Applied Surface Science. 307, 631–637.

Lv, B., Liu, G., Dong, X., Wei, W., Jin, W., 2012. Novel reactive distillation–pervaporation coupled process for ethyl acetate production with water removal from reboiler and acetic acid recycle. Industrial & Engineering Chemistry Research. 51, 8079-8086.

Mi, B., 2014. Graphene oxide membranes for ionic and molecular sieving. Science. 343, 740–42.

Morris, W., Leung, B., Furukawa, H., Yaghi, O.K., He, N., Hayashi, H., Houndonougbo, Y., Asta, M., Laird, B.B., Yaghi, O.M., 2010. A combined experimental–computational investigation of carbon dioxide capture in a series of isoreticular zeolitic imidazolate frameworks. Journal of the American Chemical Society. 132, 11006–11008.

Park, K.S., Ni, Z., Côté, A.P., Choi, J.Y., Huang, R., Uribe-Romo, F.J., Chae, H.K., O'Keeffe, M., Yaghi, O.M., 2006. Exceptional chemical and thermal stability of zeolitic imidazolate frameworks. Proceedings of the National Academy of Sciences. 103, 10186-10191.

Peng, P., Shi, B., Lan, Y., 2010. A review of membrane materials for ethanol recovery by pervaporation. Separation Science and Technology. 46, 234–246.

Peters, T.A., Benes, N.E., Keurentjes, J.T.F., 2008. Hybrid ceramic-supported thin PVA pervaporation membranes: Long-term performance and thermal stability in the dehydration of alcohols. Journal of Membrane Science. 311, 7–11.

Peters, T.A., Poeth, C.H.S., Benes, N.E., Buijs, H.C.W.M., Vercauteren, F.F., Keurentjes, J.T.F., 2006. Ceramic-supported thin PVA pervaporation membranes combining high flux and high selectivity; contradicting the flux-selectivity paradigm. Journal of Membrane Science. 276, 42–50.

Qiu, S., Xue, M., Zhu, G., 2014. Metal-organic framework membranes: From synthesis to separation application. Chemical Society Reviews 43, 6116–6140.

Rangnekar, N., Mittal, N., Elyassi, B., Caro, J., Tsapatsis, M., 2015. Zeolite membranes - a review and comparison with MOFs. Chemical Society Reviews. 44, 7128–7154.

Shao, P., Huang, R.Y.M., 2007. Polymeric membrane pervaporation. Journal of Membrane Science 287, 162–179.

Shen, J., Liu, G., Huang, K., Chu, Z., Jin, W., Xu, N., 2016. Subnanometer two-dimensional graphene oxide channels for ultrafast gas sieving. ACS Nano. 10, 3398–3409.

Shu, X., Wang, X., Kong, Q., Gu, X., Xu, N., 2012. High-flux MFI zeolite membrane supported on YSZ hollow fiber for separation of ethanol/water. Industrial & Engineering Chemistry Research. 51, 12073–12080.

Vane, L.M., 2005. A review of pervaporation for product recovery from biomass fermentation processes. Journal of Chemical Technology & Biotechnology. 80, 603–629.

Venna, S.R., Carreon, M.A., 2015. Metal organic framework membranes for carbon dioxide separation. Chemical Engineering Science. 124, 3–19.

Wang, C., Liu, X., Keser Demir, N., Chen, J.P., Li, K., 2016. Applications of water stable metal-organic frameworks. Chemical Society Reviews. 45, 5107–5134.

Wang, N., Wang, L., Zhang, R., Li, J., Zhao, C., Wu, T., Ji, S., 2015. Highly stable "pore-filling" tubular composite membrane by self-crosslinkable hyperbranched polymers for toluene/n-heptane separation. Journal of Membrane Science. 474, 263–272.

Wang, N., Zhang, G., Ji, S., Fan, Y., 2012a. Dynamic layer-by-layer self-assembly of organic–inorganic composite hollow fiber membranes. AIChE Journal. 58, 3176–3182.

Wang, W., Dong, X., Nan, J., Jin, W., Hu, Z., Chen, Y., Jiang, J., 2012b. A homochiral metal-organic framework membrane for enantioselective separation. Chemical Communications. 48, 7022–7024.

Wang, Z., Ge, Q., Shao, J., Yan, Y., 2009. High performance zeolite LTA pervaporation membranes on ceramic hollow fibers by dipcoating–wiping seed deposition. Journal of the American Chemical Society. 131, 6910–6911.

Wei, W., Xia, S., Liu, G., Dong, X., Jin, W., Xu, N., 2011. Effects of polydimethylsiloxane (PDMS) molecular weight on performance of PDMS/ceramic composite membranes. Journal of Membrane Science. 375, 334–344.

Wei, W., Xia, S., Liu, G., Gu, X., Jin, W., Xu, N., 2010. Interfacial adhesion between polymer separation layer and ceramic support for composite membrane. AIChE Journal. 56, 1584–1592.

Wu, T., Wang, N., Li, J., Wang, L., Zhang, W., Zhang, G., Ji, S., 2015. Tubular thermal crosslinked-PEBA/ceramic membrane for aromatic/aliphatic pervaporation. Journal of Membrane Science. 486, 1–9.

Xia, S., Dong, X., Zhu, Y., Wei, W., Xiangli, F., Jin, W., 2011. Dehydration of ethyl acetate–water mixtures using PVA/ceramic composite pervaporation membrane. Separation and Purification Technology. 77, 53–59.

Xia, S., Wei, W., Liu, G., Dong, X., Jin, W., 2012. Pervaporation properties of polyvinyl alcohol/ceramic composite membrane for separation of ethyl acetate/ethanol/water ternary mixtures. Korean Journal of Chemical Engineering. 29, 228–234.

Xiangli, F., Chen, Y., Jin, W., Xu, N., 2007. Polydimethylsiloxane (PDMS)/Ceramic Composite Membrane with High Flux for Pervaporation of Ethanol–Water Mixtures. Industrial & Engineering Chemistry Research. 46, 2224–2230.

Xiangli, F., Wei, W., Chen, Y., Jin, W., Xu, N., 2008. Optimization of preparation conditions for polydimethylsiloxane (PDMS)/ceramic composite pervaporation membranes using response surface methodology. Journal of Membrane Science. 311, 23–33.

Xu, R., Liu, G., Dong, X., Wanqin, J., 2010. Pervaporation separation of n-octane/thiophene mixtures using polydimethylsiloxane/ceramic composite membranes. Desalination. 258, 106–111.

Yao, J., Dong, D., Li, D., He, L., Xu, G., Wang, H., 2011. Contra-diffusion synthesis of ZIF-8 films on a polymer substrate. Chemical Communications. 47, 2559–2561.

Yoshida, W., Cohen, Y., 2003. Ceramic-supported polymer membranes for pervaporation of binary organic/organic mixtures. Journal of Membrane Science. 213, 145–157.

Zhang, M., Guan, K., Shen, J., Liu, G., Fan, Y., Jin, W., 2017. Nanoparticles@rGO membrane enabling highly enhanced water permeability and structural stability with preserved selectivity. AIChE Journal. 63, 5054–5063.

Zhao, Z., Ma, X., Li, Z., Lin, Y.S., 2011. Synthesis, characterization and gas transport properties of MOF-5 membranes. Journal of Membrane Science. 382, 82–90.

Zhu, Y., Xia, S., Liu, G., Jin, W., 2010. Preparation of ceramic-supported poly(vinyl alcohol)–chitosan composite membranes and their applications in pervaporation dehydration of organic/water mixtures. Journal of Membrane Science. 349, 341–348.

[illegible]

15 Ceramic Nanofiltration for Organic Solvents

Anita Buekenhoudt

CONTENTS

15.1 INTRODUCTION

In aqueous streams, molecular separation with membranes has been possible since the second half of the 20th century thanks to the emergence of reverse osmosis (RO) and nanofiltration (NF) membranes. NF and RO have since long proven their efficient applicability in a broad range of water treatment processes. In contrast, membrane separation at the molecular level in organic solvents emerged as a new area in membrane technology at the end of the 20th century, triggered by the trend towards a more sustainable chemistry. The technology was first developed in the area of refining, and a large scale plant has been constructed in the USA (Exxon Mobil) for the energy efficient recovery of solvents in lube oil refining (White, 2006). The progress in this area of so-called solvent resistant NF (SRNF) or organic solvent nanofiltration (OSN) has accelerated since 2000, mainly driven by European research, due to its enormous potential in a wide range of industrial applications. Possible applications vary from bulk production processes such as (bio)refineries where high-volume streams need to be separated with low energy use, to low-volume pharmaceutical and fine-chemical processes where better and greener separations of high-value products are required. OSN gives interesting alternative options not only for downstream purification, but also for process intensification by coupling membranes to reactions. Main drivers are the important possible savings in energy (compared to distillation), drastic reduction of solvent waste (e.g., compared to extractions, crystallization or preparative chromatography), enhanced product quality (e.g., higher purity, functionality or "naturalness"), in addition to optimization of raw material usage (e.g., loop closure by recovery of spent solvent or homogeneous catalysts; valorization of underutilized side streams). In a recent review, Marchetti et al. (2014) gave a good overview of the field. After the installation in 1999 and several years of trouble-free operation of the Wax-dewax process in Exxon-Mobil, it took many years before new OSN-plants appeared at industrial scale (Vandezande et al., 2008). This has changed drastically over the past 2-3 years with some large-scale (up to 25 ton/h) installations by the most important polymeric membrane suppliers Evonik, GMT-Borsig and Solsep (Franke et al., 2010; Priske et al., 2010; Buekenhoudt et al., 2014; Cuperus, 2015; Haverkamp, 2017; Shnitzer, 2017).

While the market of aqueous filtrations is dominated by polymeric membranes (mainly due to price), the superior chemical and structural (no swelling) stability of ceramics opens a specific potential for ceramic membranes (typically multilayer, tubular membranes as shown in Figure 15.1) in solvent-based streams. Therefore, alongside the extensive research on polymeric membrane development and applications for OSN (see Chapter 13), increasing efforts are also dedicated to modification of ceramic membranes or ceramic membrane synthesis for improved performance in solvents. Numerous successful OSN results for ceramic membranes have been communicated ranging from lab scale up to pilot and implementation scale. This chapter intends to give an overview of past and new achievements in this exciting R&D field. It starts with an overview of the Strengths, Weaknesses, Opportunities and Threats (SWOT analysis) for ceramic membranes in OSN.

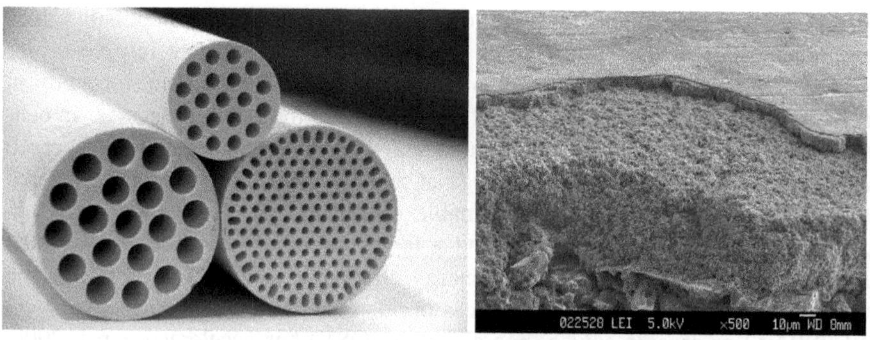

FIGURE 15.1 Ceramic membranes: samples of different commercial-scale multichannel tubes (left), SEM picture of the typical multilayer structure for a NF membrane (right).

15.2 SWOT ANALYSIS OF CERAMIC MEMBRANES FOR OSN

The trump card of ceramic membranes is their exceptionally wide chemical, structural and temperature stability. Polymeric OSN membranes, on the contrary, are often only stable in a range of solvents with similar characteristics, and aprotic solvents remain challenging. Swelling issues hamper performance comprehension and prediction. Moreover, stability issues do not only relate to the membrane itself, but also to the potting materials used to mount polymeric membranes in modules. Users have found different situations where the membrane material is sufficiently stable, but the potting material is not. Leaking and leaching are not uncommon. Here, ceramic membranes have a clear advantage as they are easily and universally mounted in modules using solvent-resistant glass or Teflon coatings on the tubular membrane ends in combination with chemically stable O-rings. Another important asset of ceramic membranes is the fact that they do not contain any pore fillers, and thus can be stored dry. Consequently, their implementation is easier, as no preconditioning to remove the pore fillers and/or to compact the membrane is needed, and no wet (solvent) conditions need to be foreseen for their storage after use.

The price is currently less of a discriminating factor, since the procedure to prepare good solvent-stable polymeric membranes is generally more complex than for a standard one (e.g., cross-linking), and therefore prices are very similar.

The most important drawback of the commercialized ceramic membranes is their limited pore size. Currently, the tightest ceramic membrane commercially available (from the German company Inopor) is a TiO_2 membrane with a pore size of 0.9 nm, leading to a molecular weight cut-off (MWCO) of ~450 Da in water (Puhlfürss et al., 2000). However, recent research is tackling this issue, and some interesting new developments are described further in this chapter. The limited suppliers of sufficiently tight membranes might become a threat.

Of concern is also the inherent hydrophilic nature of the native ceramic membranes, currently mainly consisting of metal oxides. This is not a limitation for applications in water-solvent mixtures or in solvents with relative high polarity. However, considering implementation in rather apolar solvents, this is an issue. This aspect has received already a lot of attention, and good hybrid membranes have been developed by grafting organic groups on the surface of existing ceramic membranes, or by incorporating organic groups in the ceramic matrix using modified synthesis procedures (see further sections). Grafted membranes also open the path to affinity-based separations.

As in water, operation of ceramic membranes is somewhat different than for polymeric membranes. Trans membrane pressures (TMPs) for ceramic membranes are typically somewhat lower (up to 20–25 bar) without jeopardizing the flux, while cross-flows are higher due to the tubular nature of most ceramic membranes. For the treatment of high volume streams, the spiral wound character of most polymeric membranes might be advantageous over the multichannel (typically 19 channels)

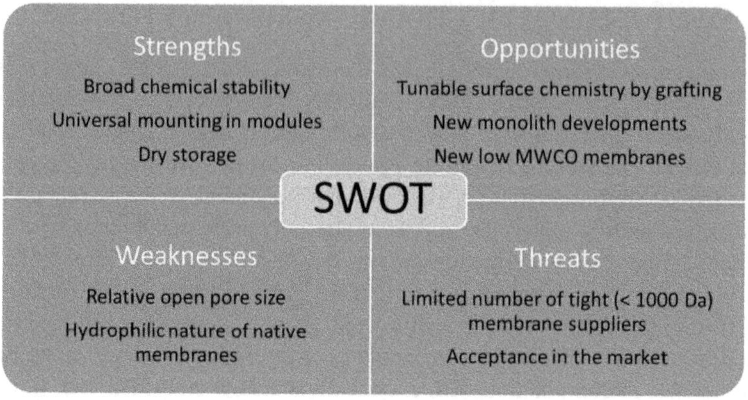

FIGURE 15.2 SWOT analysis of ceramic membranes for OSN.

tubular design of the ceramic membranes with its somewhat lower surface to volume ratio. However, Inopor is currently developing NF monoliths (163 to 500 channel elements, and beyond; a picture of a 163 channel elements is included in Figure 15.1, left) with a surface to volume ratio similar to the one of spiral wound elements (Puhlfürss et al., 2016). This new development will also strongly reduce the price of the ceramic membranes (goal is 200 Euro/m^2).

Despite the clearly growing interest of the market for the extensive potential of membranes in solvent related streams and despite the increasing number of pilot-scale testing, the acceptance of OSN in industry is still low. More success stories on pilot and demonstration scale will be needed to remediate this, especially for (hybrid) ceramic membranes.

Figure 15.2 gives an overview of this SWOT analysis.

15.3 NATIVE CERAMIC MEMBRANES FOR OSN

In the 1990s, several groups worldwide succeeded in a proper control of the sol-gel chemistry and synthesized good microporous ceramic NF membranes of different metal oxide materials, first from the rather unstable γ-alumina, and later from robust titania, zirconia, hafnia and silica-zirconia (Peterson et al., 1990; Larbot et al., 1994; Sarrade et al., 1994; Sarrade et al., 1996; Palmeri et al., 1998; Richter et al., 1998; Tsuru et al., 1998; Voigt et al., 1998). At the start of the 21st century, two high-quality titania membranes with pore sizes in the order of 1 nm, developed by the group of Voigt (Puhlfürss et al., 2000), have been commercialized via the German company Inopor. The tightest membrane has a MWCO in water of ~450 Da and a pure water permeability of ~20 L/m^2 h bar.

More recent membrane developments have led to tighter NF membranes also interesting for OSN. By properly choosing the additive reducing the reactivity of the sol precursor prior to hydrolysis, Van Gestel et al. (2006; 2008) developed defect-free, lab-scale ZrO_2 and TiO_2 NF membranes with a dense pore structure, a MWCO in water of about 200 to 300 Da, but relatively low water permeabilities of ~2.5 L/m^2 h bar. These membranes have merely been used for pervaporation. Voigt et al. (2012) used a mixture of titania and zirconia precursors in a ratio of 12 to 1, to prevent crystallization of the amorphous metal oxide during sintering, leading again to extra dense membranes with a MWCO in water of ~200 Da and very decent water permeabilities of ~10 L/m^2 h bar. These membranes are on the verge of being commercialized (code name LC1), but currently, no results in solvent streams are available in the literature. Since 2011 in the USA, the company Cerahelix has commercialized titania NF membranes with a unique pore structure created by a DNA templating process (Bishop et al., 2011). After membrane deposition, removal of the DNA leaves behind continuous, linear pores with a very narrow size distribution and high porosity. Membranes with different pore sizes (created by different DNA templates) are available as multichannel tubes. The tightest membrane shows a MWCO of 400 Da and a permeability of ~7 L/m^2 h bar in water. As far as we know these membranes have only been used in aqueous NF.

Very recently, Chen et al. (2017) created ultra-thin (~10 nm) titania NF membranes, using atomic layer deposition of titanicone layers on tubular ceramic UF membranes with a pore size of ~5 nm, and subsequent calcination for micropore formation. Tests of the resulting membranes in aqueous NF showed a quality very similar to the existing commercially available sol-gel titania membranes: permeabilities of 30 L/m^2 h bar, MWCO in water of 680 Da. However, the flexibility of this synthesis route and the possible avoidance of some costly high temperature calcination steps may make this synthesis route advantageous over the traditional procedures.

Table 15.1 gives an overview of different types of ceramic NF membranes developed over the last few decades. The ones commercially available are highlighted in

TABLE 15.1
Overview of Ceramic NF Membranes Developed over the Last Decades. The Membranes Mentioned in Bold, Have Been Commercialized or Will Be Commercialized Soon

Toplayer Material	MWCO[a] (Da)	Water Permeability (L/m2 h bar)	Applied for OSN	Reference[b]
γ-Al_2O_3	2000	2.5	no	Peterson et al. (1990)
γ-Al_2O_3	400	1 to 2	no	Larbot et al. (1994)
γ-Al_2O_3	600	2.5	no	Sarrade et al. (1994)
TiO_2	900	75	no	Voigt et al. (1998)
TiO_2 and ZrO_2	900	80	no	Richter et al. (1998)
SiO_2/ZrO_2 mix	200 to 1000	0.5 to 6	yes	Tsuru et al. (1998)
HfO_2	500	–	no	Palmeri et al. (1998)
TiO_2	**450**	**20**	**yes**	**Puhlfürss et al. (2000)**
TiO_2 and ZrO_2	200 to 300	205	no	Van Gestel et al. (2006)
TiO_2	500 to 1000	7 to 15	yes	Tsuru et al. (2008)
TiO_2	**400**	**7**	**no**	**Bishop et al. (2011)**
TiO_2/ZrO_2 mix	**200**	**10**	**yes**	**Voigt et al. (2012)**
TiO_2	680	30	no	Chen et al. (2017)

[a] MWCO values measured in water. Most values are approximate values.
[b] First publication for this type of membrane.

bold. Note the systematically low water permeability for the γ-Al_2O_3 membranes (thick membrane layer) and the correlation between permeability and MWCO.

15.3.1 Flux Performance

The group of Tsuru et al. (2000, 2003) was the first to test their NF membranes (silica-zirconia) in non-aqueous solvents in a very systematic way. Their experiments with different alcohols revealed that the performance of fine-porous ceramic membranes deviates from the viscous flow behavior observed for open-porous polymeric and ceramic membranes. This is concluded from permeability × viscosity values dependent on solvent and temperature (i.e., deviation from the Hagen-Poiseuille equation: permeability ~1/viscosity). Particularly, permeability × viscosity decreases for larger and less polar alcohols (methanol > ethanol > isopropanol), and increases with temperature with an activation energy larger than the one of the solvent viscosity. This points to a more hindered solvent transport in the case of larger molecules, lower solvent-membrane affinity and lower temperatures. The extra high activation energy signifies that higher temperatures can alleviate the hindering. Consistently, solvent and temperature effects are more substantial for smaller pore sizes and more hydrophilic membranes. The deviations from viscous flow are also obvious for solvent mixtures of water, methanol and ethanol (Van der Bruggen et al., 2007). Similar effects were observed for polymeric NF membranes as well (Machado et al., 1999; Yang et al., 2001; Geens, 2006); however, due to their non-rigid structure (swelling) and a variable pre-conditioning step, the interpretation was not so straightforward.

If no specific precautions are taken, fluxes of apolar solvents as hexane and toluene through different metal oxide NF membranes are much lower than for more polar solvents and show a stronger activated temperature dependence (Guizard et al., 2002; Chowdhury et al., 2003; Van Gestel et al., 2003; Tsuru et al., 2004; Dobrak et al., 2010). However, in explicit dry conditions (for membranes and solvents), fluxes of apolar solvents can be high, and no deviating temperature dependence is found (viscous flow). Tsuru et al. clarified this difference by proving a systematic decrease of the hexane flux as the water content of the solvent increases (Tsuru et al. 2004, 2006, 2008). This effect was again more substantial for smaller pores, lower temperatures and more hydrophilic membranes, and could be assigned to the adsorption of water molecules at the pore walls of the hydrophilic membranes, and thus to a decrease of the available pore size and the hexane-membrane affinity. Other "history" effects (effects of the sequence of experimentation) as a threshold pressure for apolar solvents have also been observed (Chowdhury et al., 2003).

From all these flux results, the significant influence of solvent-membrane affinity for the native hydrophilic ceramic NF membranes is undeniable: the lower the affinity, the more hindered solvent transport, and thus the more deviation from viscous flow.

Subsequently, numerous researchers measured fluxes for the commercially available titania NF membranes using a wide variety of solvents, and proceeded to model their results quantifying the solvent-membrane affinity in different ways. Darvishmanesh et al. (2009) developed a model based on parallel convection-diffusion transport that also included a resistance to the surface, defined by the ratio of the surface tension of solvent and membrane. Marchetti et al. (2012) used a complex correction factor in

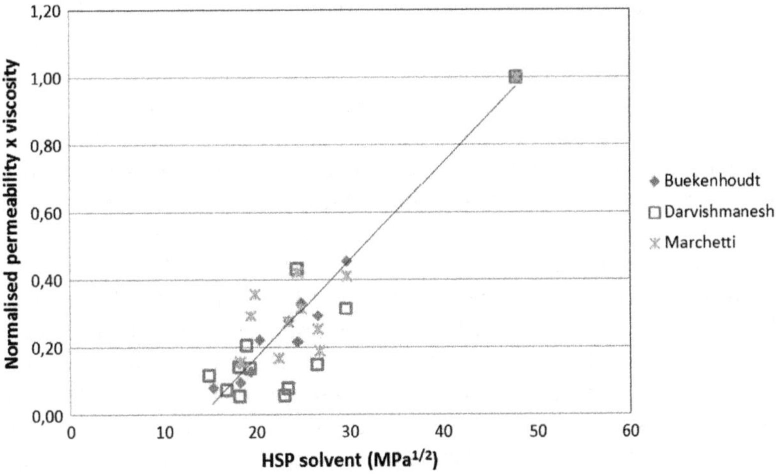

FIGURE 15.3 Normalized permeability x viscosity values as function of solvent polarity (quantified by solvent HSP) for tight titania NF membranes, as measured by three different studies: Buekenhoudt et al. (From Buekenhoudt, A. et al., Journal of Membrane Science. 439, 36–47, 2013.) (0.9 nm membrane, batch 1092); Darvishmanesh et al. (From Darvishmanesh, S. et al., Separation and Purification Technology. 70, 46–52, 2009.) (HITK275) and Marchetti et al. (From Marchetti, P. et al., Journal of Membrane Science. 415/416, 444–458, 2012.) (Inopor Nano 450 Da). The line is a linear fit through the results of Buekenhoudt et al. (From Buekenhoudt, A. et al., Journal of Membrane Science. 439, 36–47, 2013.)

the Hagen-Poiseuille equation including different solvent and membrane parameters, together accounting for capillary effects, dipole interactions and steric hindrance. Both researchers used one single membrane to determine all fluxes, soaking the membrane in the next solvent between measurements. To avoid any history effects this procedure might create, Buekenhoudt et al. (2013) measured each flux on a new membrane (used as received, without any prior drying) as one would do starting the investigation of a specific application. Measured as such, the results of permeability x viscosity decrease almost linearly with the decreasing solvent polarity when characterized by its total Hansen solubility parameter (HSP) (note that the HSP is strongly correlated to the dielectric constant). The slope of the straight line was dependent on the pore size and the membrane hydrophilicity. Figure 15.3 compiles the results of the three groups for tight titania NF membranes, and shows a very similar general trend.

15.3.2 Solute Retentions

Tsuru et al. (2001) were also the first to investigate the retention behavior of native ceramic NF membranes. They observed for their silica-zirconia membranes high retentions for sufficiently large ethylene glycols (molecular size close to pore size) in relatively polar solvents such as alcohols. But the retentions were strongly dependent on the type of alcohol and on TMP. Around the same time, similar solvent dependencies were also shown for tight polymeric membranes (Koops et al., 2001; Yang et al., 2001).

The pressure dependence observed is well known for aqueous NF: the retention increases with applied TMP and then reaches a constant value. This behavior can be well described by the Spiegler-Kedem theory (Spiegler et al., 1966) defining the solute transport through the membrane, J_s, as partly due to diffusion (governed by the solute diffusion permeability P_{diff}) and partly by convection (governed by the volume flux J and the reflection coefficient σ, function of the ratio of solute size to pore size). In this equation (15.1), Δx is the membrane thickness, and C the solute concentration. From J_s, the solute retention R and its observed evolution as function of J (and thus TMP) can be derived, Equation (15.2). At sufficiently high TMP or high volume flux J, when convection overrules diffusion (i.e., $J/P_{diff} >> 1$), R equals the reflection coefficient σ and is solely determined by size exclusion (no TMP and no temperature dependence). When diffusion cannot be discarded, R is always smaller than σ and dependent on J/P_{diff} and thus on TMP (via J) and solvent (via J and P_{diff}). The limiting case of size exclusion is expected when solvent-membrane affinity is high (J large), and at the same time solute-membrane affinity is low (P_{diff} small). The limiting case of dominant diffusion is expected when solvent-membrane affinity is low (J small) and at the same time solute-membrane affinity is high (P_{diff} high).

$$J_s = P_{diff}\,\Delta x\, dC/dx + C(1-\sigma)J \tag{15.1}$$

$$R = (1-F)\sigma/(1-\sigma F) \text{ with } F = \exp(-(1-\sigma)J/P_{diff}) \tag{15.2}$$

$$\text{If } J/P_{diff} >> 1 \rightarrow F \approx 0 \rightarrow R \approx (1-F)\sigma \rightarrow R \approx \sigma \tag{15.3}$$

$$\text{If } J/P_{diff} << 1 \rightarrow F \approx 1 \rightarrow R \approx \sigma J/P_{diff} << \sigma \rightarrow R \approx f(\text{TMP}, \text{solvent})$$

The effect of temperature on the performance was investigated by Tsuru et al. (2006a). In this study, Tsuru expanded his previous work including other solutes as alkanes (hexane, decane and tetradecane) and alcohols (hexanol, octanol, decanol and hexanediol) in the solvent ethanol. The fluxes of all mixtures showed activated temperature behavior as for pure ethanol. For all retentions, the typical Spiegler-Kedem pressure dependence was found. However, no temperature dependence was observed for the alkanes, while the alcohol retentions decreased slightly with temperature. The results were analyzed with the Spiegler-Kedem theory, although the effect of the MW of the different solutes complicates the interpretation and should be taken into account. Regardless, the results point to P_{diff} values for both solutes increasing with temperature stronger than the solvent viscosity (activated diffusive transport as for the solvent), and to J/P_{diff} values independent of temperature for alkanes, while slightly decreasing for alcohols. The effect of the different solute-membrane affinities is bound to play a role.

From the results mentioned above, it is already clear that retentions for native hydrophilic membranes, as for polymeric membranes, are very solvent, solute and temperature dependent, and governed by the complex competition of interactions between solvent-membrane and solute-membrane, further influenced by the solvent-solute affinity. The two extreme situations where solvent-membrane affinity or solute-membrane affinity dominates are visually represented in a simplified way in Figure 15.4. Both situations correspond well to the two limits of the Spiegler-Kedem theory, Equation (15.3).

This method of visualizing the phenomena governing OSN has proved very valuable in interpreting other results of ceramic NF membranes, including those in solvents more apolar than alcohols. For instance, Voigt et al. (2003) pointed to the low retentions of polystyrene (PS) molecules in apolar toluene, consistent with the low toluene-membrane affinity and corresponding to Figure 15.4, right. Dobrak et al. (2010) compared retentions of the polar solute brilliant blue (826 Da) and the apolar bromothymol blue (624 Da) in the polar solvent ethanol and the apolar toluene for titania NF membranes. High, temperature independent retentions, consistent with size exclusion, were found for the mixtures ethanol-brilliant blue (826 Da) and toluene-bromothymol blue (624 Da), corresponding to Figure 15.4, left. This work shows that even for an apolar solvent as toluene, an hydrophilic ceramic NF membrane can work well, as long as the solute is sufficiently apolar, i.e., at sufficiently low solute-membrane affinity (making $P_{diff} \ll J$). Adequate fluxes are then guaranteed when working at higher temperatures. On the contrary, the retentions of the bromothymol blue in ethanol were very low, most likely due to the low solute-solvent affinity (low solubility) increasing the importance of solute-membrane interactions. The study of Rezaei Hosseinabadi et al. (2015) included retention measurements for native titania NF membranes for polyethylene glycol (PEG, polar) and polystyrene (PS, apolar) molecules of about the same size in toluene and acetone. All retentions are smaller than size exclusion, but retentions in acetone are systematically larger than in toluene, consistent with the higher solvent-membrane affinity for acetone. Moreover, retentions for PEG are systematically lower than for PS, caused by the higher solute-membrane affinity for PEG. In toluene the PEG retentions are even negative (~–20%).

It was actually Tsuru's group that observed the first negative retentions for titania NF membranes when filtrating linoleic acid in hexane (Tsuru et al., 2006b; 2008). The retentions were more negative for denser membranes. The phenomenon of negative

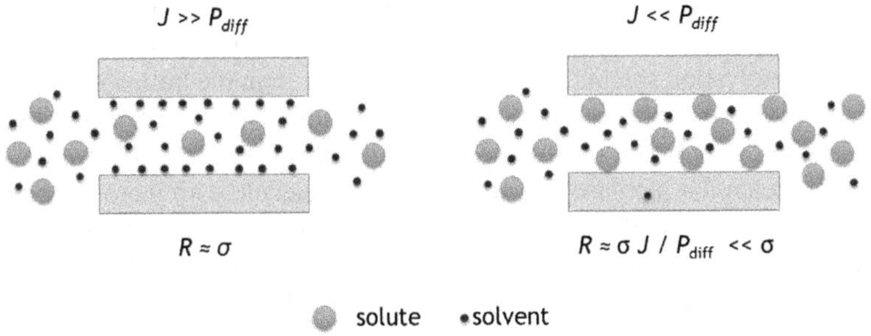

FIGURE 15.4 Simplified visual representation of the two extreme situations of the Spiegler-Kedem theory: dominating solvent-membrane affinity (left) or dominating solute-membrane affinity (right). (Adapted from Marchetti, P. et al., Journal of Membrane Science. 444, 101–115, 2013.)

retentions was already encountered for polymeric (cellulose acetate) NF membranes, in case of a mixture of carboxylic acids in hexane (Koops et al., 2001). In both situations, the negative retentions are explained by a strong solute-membrane affinity, overruling the solvent-membrane affinity, and further supported by a limited solubility of the solute in the solvent. This interpretation is supported by adsorption measurements. Negative retentions appear regularly in similar conditions (for ceramic membranes: Vandezande et al., 2009; Marchetti et al., 2013; for polymeric membranes: Postel et al., 2013).

Due to the broad chemical resistance of ceramic NF membranes, these membranes work as well in aprotic solvents or in other solvents difficult for polymeric membranes. For example, Buekenhoudt et al. (2004) showed high retentions governed by size exclusion in dicholoromethane (DCM) and n-methyl-2-pyrrolidone (NMP), and Marchetti et al. (2013) proved good stable performance in solvents as acetonitrile (ACN) and dimetyl sulfoxide (DMSO). Although DCM is a rather apolar solvent, the high solution power of this solvent for a broad range of solutes could be the cause of the overall good results with the native ceramic NF membranes.

Different researchers have tried to describe the ruling solvent-solute-membrane affinities in different ways and to model their experimental results gathered for a clever choice of solvents (or solvent mixtures) and solutes. Most modeling work has been performed for polymeric membranes (e.g., the recent studies of Darvishmanesh et al., 2010; 2011; Zeidler et al., 2013; Blumenschein et al., 2017), showing also a clear influence of solvent dependent swelling on polymeric OSN performance. The Hildebrand or HSP proved to be very helpful in describing the mutual affinities of solvent, solutes and membranes. Similar modeling has shown to be valid for grafted ceramic membranes (see Section 15.4.2). Less modeling work has, however, been dedicated to retentions of native ceramic NF membranes. Geens et al. (2006) investigated modeling of their retention results in 4 solvents and 6 solutes for both ceramic and polymeric membranes, taking into account solvation of solutes and of rigid pores (no swelling). Retention prediction was reasonably successful, although solute-membrane affinity was not incorporated in the model. Solvation was also seen as a powerful tool to clarify the retention of salts and small peptide molecules in water/solvent mixtures for titania NF membranes (Marchetti et al., 2013a). For water/ACN mixtures the salt retentions strongly decrease as the amount of ACN increases. Marchetti et al. (2013a) was able to explain this using the preferential solvation theory determining the composition of the solvation shell around the ions. For ACN this solvation shell is always richer in water than the bulk solvent, leading to increased solute-membrane affinity, and thus to lower retentions. Similar effects, though smaller, were noted in acetone/water mixtures. Preferential solvation could also explain: retention increase of a small peptide molecule with increasing ACN in the mixture (solvation shell contained more ACN, thus lower solute-membrane affinity), further retention enhancement when adding HCl or NaCl to the mixture, and retention decrease when adding trifluoroacetic acid (TFA).

15.3.3 Applications

As can be concluded from the results described in the previous sections, native ceramic NF membranes are especially applicable in sufficiently polar solvents or water/solvent mixtures, when fluxes and retentions are high due to high solvent-membrane affinity and for a wide range of solutes (see Figures 15.3 and 15.4).

One of the first OSN applications of the newly commercialized ceramic NF membranes, commissioned by a Belgian company, Beneo-Orafti, involved the purification of chemically modified sugars as Inutec® (Buekenhoudt et al., 2004, Levecke, 2006). Inutec® (MW ~1000 Da) was synthesized in pure NMP, and this NMP needed to be removed down to 0.1 wt%. While traditional methods such as evaporation and extraction (even supercritical extraction) led to rest NMP contents of a few wt% due to strong solute-NMP interactions, diafiltration with water, using ceramic NF membranes, proved to be an efficient purification method (at the start of the project in 2002, polymeric OSN membranes were insufficiently stable). Screenings tests were followed by a study determining the optimal pre-dilution (before diafiltration), minimizing water use and process time. Subsequently, pilot trials were performed during ~6 months, using 5.25 m^2 of titania NF membranes (3 modules, each with 7 × 19 channel tubes with a length of 1,20 m). The piloting confirmed the success of the method at process fluxes of ~200 L/h, and re-use of part of the permeate as diafiltration fluid was demonstrated, diminishing the NMP-containing wastewater to ~10%. Later, this application was successfully implemented at industrial scale, first using 36 m^2 of membranes, then expanded with another 36 m^2 (see Figure 15.5).

Another successful application described by Buekenhoudt et al. (2004) is the concentration of an active pharmaceutical ingredient (API) with a MW of 530 Da in

FIGURE 15.5 Implementation of ceramic NF membranes at the Belgian company Beneo-Orafti (membrane surface 72 m^2).

DCM. High retentions of ~98 to 92% combined with high permeabilities of ~2.5 L/m^2 h bar were found for titania NF membranes using mixtures with concentrations of 0.2 to 10 g/L, allowing efficient concentration with at least a factor of 20. Further, efficient catalyst recovery was demonstrated for the transition metal Co-Jacobsen catalyst (~600 Da) in the hydrolytic kinetic resolution of an epoxide (Aerts et al., 2006). When using diethyl ether as the solvent, the titania membrane flux was too low. However, in isopropanol (IPA) the membrane showed fluxes and retentions far superior than most of the tested polymeric membranes. Membrane reactor tests, with subsequent reaction and filtration steps, proved the success of the homogeneous catalyst recovery and its re-use in further reaction steps, and thus the increase of the turnover number (TON) of the catalyst. Similar work was performed by Chowdhury et al. (2006) showing the size exclusion of the more bulky hydrophobic Polyoxometallate ($Q_{12}POM$) catalyst over 2 to 5 nm γ-alumina membranes in toluene. The catalyst activity in the epoxidation reaction of cyclooctene proved to be maintained for up to 6 recycles.

At VITO in collaboration with different pharmaceutical companies, the potential of the commercial ceramic NF membranes was also shown in a variety of other applications in different lab-scale projects from 2005 to 2010 (Buekenhoudt 2010). Removal of small genotoxins (~150 Da) from an API (~700 Da) in DCM proved possible, as far as no genotoxin-API interaction prevailed. An API recovery >99% was reachable in a 2-stage diafiltration process. Further, diafiltration with a 3 nm zirconia membrane allowed successful removal of unwanted polymers and oligomers from an API (~675 Da) again in DCM, de-bottlenecking the purification with chromatography. Moreover, a second denser titania NF membrane allowed

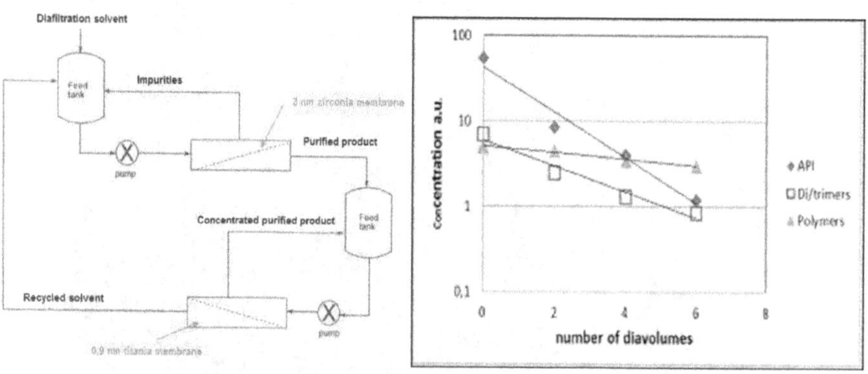

FIGURE 15.6 Two-step membrane process to remove oligomers and polymers from a monomer API (left). Evolution of the impurity removal during the diafiltration in the first step (right). The final purified product quality (permeate first step) was 91% API with 8.5% di/trimers and 0.6% polymers. (Data from Buekenhoudt, A., The potential of ceramic membranes in different industrial separation problems. Presentation at the 3th International Conference on Solvent Filtration, London, UK, 2010.)

for in situ recycling of the solvent needed for the diafiltration. The process scheme and the results during the first membrane step are shown in Figure 15.6. Several other opportunities were also demonstrated in peptide purification, where residual organic solvents (as acetonitrile, DMSO or others) need to be removed before the chromatography or the lyophilisation step. For example, diafiltration with titania NF membranes proved successful to remove the acetonitrile from an acid water/acetonitrile peptide mixture (containing TFA) from 50% down to 0.1 wt%. The process ran for different peptide molecules with molecular weight varying from ~800 to 2000 Da. The loss of peptide was minimal, even for the smallest peptides, and high purity was maintained during the process (no impurities were introduced). Here OSN is a relative easy and low-temperature alternative for the standard tedious evaporation of the thermolabile peptide molecules, requiring high manpower input for quality assurance. Hence, pay back times of the NF process were calculated to be below 3 years. A similar peptide concentration and salt/solvent exchange with ceramic NF membranes was successfully performed by Marchetti et al. (2013b), again with minimal losses of peptide. They also showed that the design of experiments can be a very valuable tool to choose for the optimal process. In their case, pre-concentration followed by constant volume diafiltration was identified as the best one.

More recent studies of Ormerod et al. (2013; 2015; 2016) showed the potential to use native ceramic NF membranes for different ways of process intensification. In a continuous reaction-filtration set-up, different metathesis catalysts were efficiently retained by lab-scale titania NF membranes, while the model ring-closing product was well transported through the membrane (Ormerod et al., 2013). A clever choice of solvent added to the process success (acetone better than DCM). In contrast, an alternative polymeric membrane showed either non-stable retentions (in DCM), or too high retentions (in acetone) for both catalyst and product. A later similar study (Ormerod et al., 2016) also showed good results for the recovery of Pd cross-coupling catalysts in a model Suzuki reaction performed in ethanol; however, it also proved that high catalyst stability is a prerequisite to obtain high yields in a continuous membrane-filtration coupling. Very high rejection of Pd species and consequently very low Pd contamination of reaction products (<10 ppm) has been demonstrated with a single OSN step with a lab-scale titania NF membrane, avoiding losses of product encountered when using more conventional adsorbents. In a proof of concept study in collaboration with the company Polypeptide, Ormerod et al. (2015) showed the successful use of a 50 cm long 19 channel titania membrane for the in-line solvent recycling during the formation of the cyclic peptide Desmopressin. This membrane-assisted process results in a significant reduction in the solvent load (>80%) of this high-dilution reaction, while allowing a similar reaction yield and even somewhat higher product quality as in the standard high-dilution batch process. Direct coupling to existing reactors is feasible but has not yet been implemented.

Table 15.2 gives an overview of the OSN applications with native ceramic NF (or tight UF) membranes described above.

TABLE 15.2
Overview of OSN Applications Tested with Native Ceramic NF (or Tight UF) Membranes. The Performance Lists the Retention of the Solute and the Permeability of the Tested Mixture. The Applications Mentioned in Bold Have Been Tested Up to Pilot Scale or Have Been Implemented

Membrane Producer[a]	Solvent	Solute MW[b]	Process Objective	Performance R and J	Reference
TiO_2 Inopor	**NMP/water mixtures**	**Inutec® 1000 Da**	**Diafiltration to remove NMP**	**>99% ~0.4 L/m²hbar**	**Buekenhoudt et al. (2004); Levecke (2006)**
TiO_2 VITO	DCM	API 530 Da	Concentration	92 to 98% ~2.5 L/m²hbar	Buekenhoudt et al. (2004)
TiO_2 VITO	IPA	Co-catalyst 600 Da	Catalyst recovery in reaction	83 to 89% ~0.15 L/m²hbar	Aerts et al. (2006)
γ-Al_2O_3 UTwente	Toluene	POM-catalyst 9300 Da	Catalyst recovery in reaction	93 to 100% 0.2 to 0.6 L/m²hbar	Chowdhury et al. (2006)
TiO_2 Inopor	DCM	API 700 Da	Diafiltration to remove genotoxin	98% 4 L/m²hbar	Buekenhoudt (2010)
ZrO_2 Inopor	DCM	API 675 Da	Diafiltration to remove polymers	50% 25 L/m²hbar	Buekenhoudt (2010)
TiO_2 Inopor	DCM	API 675 Da	Solvent recycling	95% 0.5 L/m²hbar	Buekenhoudt (2010)
TiO_2 Inopor	Acetonitrile/ water/TFA	Peptides 800 to 2000 Da	Diafiltration to remove acetonitrile	>99% 2 to 5 L/m²hbar	Buekenhoudt (2010)
TiO_2 Inopor	Acetonitrile/ water/TFA	Peptide 3000 Da	Concentration[b] diafiltration to remove acetonitrile	95 to 99 % 1 to 4 L/m²hbar	Marchetti et al. (2013b)
TiO_2 Inopor	DCM and Acetone	Ru-catalysts 500 Da	Catalyst recovery in reaction	90 to 100% 0.3 to 2 L/m²hbar	Ormerod et al. (2013)
TiO_2 Inopor	Ethanol	Pd catalysts 630 Da	Catalyst recovery in reaction	97 to 98% 0.7 to 16 L/m²hbar	Ormerod et al. (2016)
TiO_2 Inopor	Ethanol	Pd catalyst Nanoparticles	Metal removal after reaction	>99.99 % 0.7 to 16 L/m²hbar	Ormerod et al. (2016)
TiO_2 Inopor	Water/ ethanol mixtures	Cyclic peptide 1070 Da	In-line solvent recycling during cyclization	96 to 99% 0.05 to 0.2 L/m²hbar	Ormerod et al. (2015)

[a] Producer means commercial supplier, or research group making the membrane used.
[b] Most values are approximate values.

15.4 HYBRID ORGANIC-INORGANIC MEMBRANES FOR OSN

In order to widen the application potential of tight ceramic membranes towards more apolar solvents, hybrid organic-inorganic membranes have been developed to decrease the membrane surface polarity. The organic groups can be incorporated during the sol-gel synthesis using appropriate hybrid precursors (the in situ method is mainly limited to silica), or the groups can be grafted onto existing microporous or tight mesoporous ceramic membranes. Different techniques have been used, and the procedures as well as the advantages and disadvantages of each technique can be found in the recent review by Meynen et al. (2014). For the grafting methods, it is important to modify the complete pore surface and not only the outer membrane surface using grafting moieties that fit into the micropores (Van Gestel et al., 2003) and/or to avoid pore blockage by possible polymerization reactions (Bothun et al., 2007). Table 15.3 gives an overview of the hybrid organic-inorganic membranes developed and used for OSN. Their performance is described in this and the following sections.

Voigt et al. (2003) developed a good hydrophobic NF membrane by silanating their commercially available 3 nm zirconia membrane with long hydrophobic chains. This silanated membrane is semi-commercially available on the market (product code HOC), and has been used in different research projects and in some applications. Buekenhoudt et al. (2010) developed an innovative method for membrane grafting based on Grignard chemistry, leading to a unique direct metal-C bond (other than the metal-O-C bond of the other grafting methods) and a partial surface coverage. Samples of amphiphilic, alkyl/phenyl-grafted membranes are available even at commercial membrane scale (multichannel tubes up to 120 cm length). Moreover,

TABLE 15.3
Overview of Hybrid Organic-Inorganic NF Membranes Developed and Used for OSN Over the Last 15 Years. The Membranes Mentioned in Bold, Have Been Commercialized or Will Be Commercialized Soon

Membrane Description	Polarity	Reference[a]
Silanated mix SiO_2/ZrO_2 membranes	Hydrophobic	Tsuru et al. (2003)
Silanated TiO_2 membranes	Hydrophobic	Van Gestel et al. (2003)
Silanated 3 nm ZrO_2 membrane of Inopor (HOC)	**Hydrophobic**	**Voigt et al. (2003)**
Silanated TiO_2 and γ-Al_2O_3 membranes	Hydrophobic	Bothun et al. (2007)
Grignard grafted TIO_2 and ZrO_2 membranes	**Amphiphilic**	**Buekenhoudt et al. (2010)**
Hybrid SiO_2 made from methylated precursors	Hydrophobic	Tsuru et al. (2011)
Mix TiO_2/ZrO_2 membranes with extra complexing agent and sintered under $N_2(LC_2)$	**Hydrophobic**	**Zeidler et al. (2014)**
γ-Al_2O_3 membranes grafted with PDMS brushes	Hydrophobic	Pinheiro (2013)
γ-Al_2O_3 membranes grafted with PI brushes	Amphiphilic	Pinheiro (2013)
γ-Al_2O_3 membranes grafted with longer PEGs	Hydrophilic	Tanardi et al. (2016b)

[a] First publication for this type of membrane.

the Grignard method is quite flexible, opening the possibility for a wide variety of different membranes with different functional moieties (under development).

Zeidler et al. (2014) also tested other innovative routes to incorporate carbon in the NF membrane layer. In particular, they investigated sintering under nitrogen of 3 different sol-gel layers: (1) the titania/zirconia layer used to prepare the hydrophilic 200 Da membranes described in Section 15.3, (2) adding a complexation agent as diethanol amine to this sol, (3) adding extra carbon to sol (1) in the form of a phenolic resin. The sintering under nitrogen leads to pyrolysis of remaining alkoxide groups or the extra carbon in the gel. Membranes formed by route (1) were tight but not hydrophobic enough, while membranes formed by route (3) were too open. In contrast, the membranes formed by route (2) showed a narrow pore distribution, and proved interesting for OSN due to their sufficient hydrophobic character. By combining 2 layers (route (1) + route (2)), an excellent, quite well reproducible membrane with a low cut-off of 350 to 450 Da as measured with PS oligomers in THF could be formed. Hence, the commercialization of these membranes has started (product code LC2).

15.4.1 Flux and Retention Performance

Tsuru et al. (2003) was the first to show that the hydrophobic nature of a silica-zirconia NF membrane grafted with trimethylchlorosilane leads to solvent transport according to viscous flow. In contrast to the strong deviations encountered for the native membranes (see Section 15.3.1), permeability x viscosity values were almost solvent and temperature independent, attributed to the increased wide solvent-membrane affinity for the grafted membranes. Moreover, the influence of water concentration on the hexane flux was minimal (Tsuru et al., 2004). All this was also observed for a hybrid silica membrane synthesized using methylated silica precursors (Tsuru et al., 2011). Dobrak et al. (2010) and Buekenhoudt et al. (2013) measured similar viscous flow behavior for the silanated 3 nm ZrO_2 membrane of Voigt et al. (2003). They mentioned also a zero water flux for this strongly hydrophobic membrane. Rezaei Hosseinabadi et al. (2014) showed the evolution towards viscous flow for 1 nm TiO_2 membranes grafted with Grignard reagents with increasing alkyl chains (C1 to C12). The water flux for these amphiphilic membranes with water contact angles of 70 to 80° is still high and in line with the fluxes of other solvents. Bothun et al. (2007) also tested liquid CO_2 and observed a particularly high permeability of this fluorophilic solvent in a fluoroalkyl grafted titania membrane, consistent with the high solvent-membrane affinity.

Pinheiro et al. (2013; 2014) and Tanardi et al. (2015; 2016a; 2016b) grafted longer polymer brushes of polydimethylsiloxane (PDMS)-, polyimide (PI)-based polymers or longer polyethylene glycol (PEG) molecules into the pores of 5 nm γ-alumina membranes. They used a 2-step process, with and without a coupling agent allowing additional growth of the organic chain. Logically, the membranes with the thickest organic layer in the pores show the lowest fluxes (Tanardi et al., 2016a). However, Tanardi et al. (2015; 2016a) also observed that the hydrophobic PDMS polymer brushes show swelling effects governed by solvent-membrane affinity, similar to non-crosslinked PDMS material. This swelling into the rigid ceramic pores, especially by the more apolar solvents, leads to a reduction of the permeable volume in each pore. When taking these solvent dependent pore changes into account, fluxes do meet viscous flow behavior. For the apolar solvents, the swelling is temperature independent, but for a more polar solvent as IPA, the swelling increases with temperature, leading to a decrease of the IPA flux at higher temperatures. For the hydrophilic PEG-grafted membrane, similar swelling effects were also observed for the more polar solvents (Tanardi et al., 2016b). For the intermediate polarity PI-grafted membrane, the permeability of hexane showed the typical decrease characteristic for the water adsorption on the native membranes (Pinheiro, 2013). For all these membranes with an extensive organic layer, flux evolution as function of TMP is linear, indicating the absence of compaction and of shear-flow influence of the grafted layer.

Voigt et al. (2003) investigated also the retention of their silanated 3 nm zirconia membranes using PS oligomers dissolved in the apolar solvent toluene. The most hydrophobic membrane showed a cut-off value of 600 Da, compared to the ~1200 Da of the unmodified membrane. Also the hydrophobic methylated silica membranes of Tsuru et al. (2011) showed high retentions for polyolefin oligomers in apolar hexane. MWCO determined from these measurements was 1000 to 2000 Da, corresponding well to the measured pore sizes i.e., the size exclusion behavior of Figure 15.4 (left). Permeabilities in the mixtures were similarly high as for the pure solvent, even at high oligomer concentrations. This points to low fouling most likely due to the hydrophobic surface chemistry and its low affinity towards most solutes.

Similar retention increase compared to native membranes, has been observed for Grignard grafted titania membranes using PS in acetone, and assigned to the increased solvent-membrane affinity, and thus the evolution from Figure 15.4, right, to Figure 15.4, left (Rezaei Hosseinabadi et al., 2014). Consistently, the PS retentions in acetone are close to what is expected for size exclusion of the non-grafted membrane ($R = \sigma$). Rezaei Hosseinabadi et al. (2015) also compared the retentions of PS and PEG molecules of similar size, in acetone and in toluene. As for native NF membranes (see Section 15.3.2), PS retentions were systematically higher than PEG retentions in the same solvents, and retentions in toluene were always lower than retentions in acetone, consistent with the changing solute-solvent-membrane affinities. The influence of solute-membrane interactions was also underlined by the filtration of a mixture of a phosphine cyclophane (690 Da) in IPA. When in contact with air, this mixture contains also the mono (~15 wt%) and dioxide (~1.3 wt%) of this phosphine. All three molecules have about the same MW, but different polarity. Retention results over a C8 Grignard grafted membrane show high retentions for the phosphine and its mono-oxide (99 and 90%, respectively), while negative retention for the di-oxide (–160%). This negative retention is attributed to the high solute-membrane affinity of this molecule (referring to Figure 15.4, right).

The PDMS- and PI-grafted membranes of Pinheiro et al. (2013; 2014) showed a MWCO of 500 and 830 Da, respectively, using a mixture of polyisobuthylene in toluene. For their different PDMS grafted membranes, Tanardi et al. (2016a) measured Sudan Black B (456 Da) retentions in toluene varying from 45% to 95%, correlated to the thickness of the organic layer. The retentions of the same dye for the more polar IPA were systematically lower, ranging from 20 to 80%, due to the lower solvent-membrane affinity combined with the larger effective pore size (lower swelling of the PDMS brushes inside the rigid ceramic pores). This behavior is different than for unconfined, free swelling PDMS polymeric membranes, as they show lower rejection in apolar solvents such as toluene, and higher rejections in the more polar IPA (higher free swelling, is lower rejections), see e.g., Vankelecom et al. (2004). Sudan Black retentions were also measured for the hydrophilic PEG-grafted membranes (Tanardi et al., 2016b) and showed again the effect of solvent-membrane affinity and changing effective pore size: retentions decreased from 89% in polar ethanol, to 54% in apolar hexane.

Zeidler et al. (2014) thoroughly investigated the OSN performance of their new LC2-type OSN membranes. As mentioned before, the membranes show a steep cut-off curve for PS in THF with MWCO 350 to 450 Da, retentions >98.5% for large PS > 1000 Da, and an economically valuable permeability of ~3.5 L/hm^{2}bar. However, the authors also proved the importance of avoiding defects in the membrane sub-structure: e.g., in the case of remaining defects of ~5% as measured with permporometry, the retention of big PS oligomers (~6000 Da) stagnated at a value of ~95%. In contrast, in aqueous filtrations the influence of defects was observed to be negligible: in water, a dense hydrophilic NF membrane with about the same amount of defects, does show retentions up to 100%, even for (PEG) molecules of only 600 Da. In OSN, transport through defects is most likely more important due to the lower flux in the nanopores (lower solvent-membrane affinity). The LC2-type of membranes were also tested with PS in other solvents, showing some lower performance in ethanol but especially low retentions in n-heptane. However, these results have to be treated with care, as the influence of the cyclohexane treatment used in the permporometry pre-characterisation of the membranes was clearly demonstrated, an effect that is most probably strengthened by the small size of the pores of these membranes.

15.4.2 Modeling of Retentions

From the above sections it is clear that the Spiegler-Kedem theory, and the interpretation of Figure 15.4, can give a good qualitative explanation of most retention results of native and hybrid ceramic NF membranes. Some groups have also tried to use this theory for more quantitative modeling.

To this end, Rezaei Hosseinabadi et al. (2016) have performed a more elaborated performance study of Grignard grafted membranes, including a wide range of solvents, four solutes with different polarity of about the same size (~600 Da) and 3 different TMPs. The solutes studied were 3 PEGs (normal PEG, partially methylated PEG and fully methylated PEG) and PS. The polarity of these molecules was quantified by their total Hansen solubility parameter (HSP in MPa$^{1/2}$): 25.2 for normal PEG, 22.8 for partially methylated PEG, 18.6 for fully methylated PEG and 17.0 for PS. Solvent polarities ranged from methylcyclohexane (HSP = 16.0) to water (HSP = 47.8). TMPs varied from 5 to 11 bar. This work confirmed again the existence of 2 performance regions, correlated to Figure 15.4 left and right. For the amphiphilic Grignard grafted membranes, in the region at high solvent polarity, retentions are high, solvent and pressure independent, and consistent with size exclusion of the non-grafted membranes ($R = \sigma$, Figure 15.4, left). In the other region at low solvent polarity, retentions are lower than size exclusion, and solvent and pressure dependent (R decreases with decreasing solvent HSP and decreasing TMP). Similar retention evolutions are also observed during diafiltrations: in this study shown for diafiltration over 1 diavolume from ethylacetate (HSP = 18.1) to ethanol (HSP = 26.5). Figure 15.7 combines the results of PS and PEG for one type of Grignard grafted membrane. From this figure

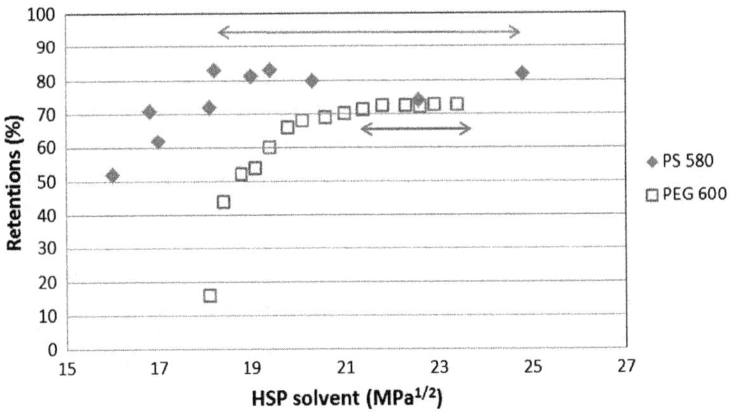

FIGURE 15.7 Retentions of 2 solutes PEG 600 Da (HSP 22.8 $MPa^{1/2}$), and PS 580 Da (HSP 17.0 $MPa^{1/2}$) as function of the polarity of the solvents (quantified by the solvent HSP) for 1 nm titania membranes with MWCO of 750 to 800 Da, Grignard grafted with a C8 group. The arrows indicate the extension of the size exclusion region ($R = \sigma$). (Re-interpretation of data from Rezaei Hosseinabadi, S. et al., Journal of Membrane Science. 513, 177–185, 2016.)

it is also clear that the switch from one region to the other depends on the polarity of the solute: the more apolar the solute (lower HSP), the lower the inflection point or, in other words, the wider the size exclusion region. Moreover, for another grafted membrane all the PS retentions (different solvents, different TMPs) were successfully fitted using the Spiegler-Kedem theory with two fixed parameters σ, and P_{diff}. Note that the Spiegler-Kedem theory accurately describes the general trend of the retentions, but that small systematic deviations from this general trend occur for specific solvents. Maybe these deviations can be clarified when working with the three partial Hansen solubility parameters (hydrogen HSP, polar HSP and dispersion HSP). Andrecochea Saiz et al. (2017) are investigating this route, looking at the distances of solute, solvent and membrane points when visualized in a ternary diagram. Hereto, they developed also a method to calculate the 3 HSP parameters of Grignard grafted membranes.

The work of Rezaei Hosseinabadi et al. (2016) also showed some indication of the existence of another performance region with retentions $< \sigma$, when the solute-solvent affinity becomes low. This led, e.g., to retentions $< \sigma$ for the apolar fully methylated PEG (HSP = 18.6) for solvents with relative high polarity (HSP > 22.5). More substantial evidence for the existence of this 3rd region was provided in Ormerod et al. (2017). This study investigated the solvent exchange for a mixture of an API intermediate from the reaction solvent (THF) to the solvent (ethanol) that will be used in the subsequent crystallization step. The obvious low solubility of the API in ethanol, leads to low retentions for all polymeric and ceramic membranes tested e.g., for a C1 Grignard grafted 0.9 nm titania membrane the retentions changes from 97% in THF to 50% in ethanol. Interpretation with the Spiegler-Kedem theory is possible with a varying P_{diff}, i.e., larger P_{diff} in case of ethanol than for THF. Further investigation of the TMP dependence of the retentions in this region is planned (there are some primary indications of reverse TMP dependence: retentions decreasing with increasing pressure).

Merlot et al. (2017) utilized the Spiegler-Kedem theory to interpret the retention performance of the PDMS grafted membranes of Tanardi et al. (2015; 2016a). In that article, they extended their previous investigation, using again Sudan Black B in a range of solvents from cyclooctane to IPA, but also including other neutral hydrophobic dyes with different sizes and varying TMP. As observed previously (Tanardi et al., 2016a), the Sudan black retentions increased with decreasing solvent polarity due to swelling of the PDMS brushes into the rigid ceramic pores (see Section 15.4.1). As a function of TMP, they observed the typical Spiegler-Kedem behavior, and the parameters σ and P_{diff} were derived by fitting Equation (15.2) for each solvent. Subsequently, from the σ-values for each solvent, an effective pore size applicable for convection was calculated using the Ferry model, the Steric hindrance pore model or the Verniory model (all providing different equations for σ as function of the ratio solute size to pore size). To predict the retentions of the other dyes (different solute sizes d_s), P_{diff} values were calculated using the Stokes-Einstein relation (free diffusion), the Renkin equation (ratio of hindered diffusion in a pore to free diffusion), the solvent viscosity μ and an estimated pore size applicable for diffusion d_{diff} (Equation 15.4).

$$P_{diff} = D_{s,hindered}/\Delta x \tag{15.4}$$

with

$$D_{s,hindered} = D_{s,free}\,(1\text{-}d')^2\,(1\text{-}2.104\,d' + 2.09\,d'^3 - 0.95\,d'^5)$$

$$d' = d_s/d_{diff}$$

and

$$D_{s,free} = kT/(3\pi\mu d_s)$$

The experimental retentions fit well when the pore size for diffusion (~3 nm) was bigger than the one for convection (~0.8 nm). This is physically possible as solutes can diffuse through the grafted PDMS brushes. It is not clear that this way of working can also predict retentions for solutes with other polarity, as solute-membrane affinity is not accounted for. Note also that, in contrast to the Grignard grafted membranes, all retention data (different solvents and different TMPs) for one specific PDMS grafted membrane cannot be fitted with the Spiegler-Kedem theory using fixed σ and P_{diff} parameters. This underlines that the same kind of swelling effects, most likely, do not dominate the transport through the Grignard grafted membranes.

15.4.3 Applications

As clarified in the previous sections, hydrophobic hybrid organic-inorganic membranes work very well in apolar solvents where native membranes are unfit. In addition, amphiphilic grafted membranes are still useful in polar solvents and expand their applicability to a wider range of medium to low polarity solvents. As most of the hybrid membranes are less mature, they have, to-date, only been tested in a more limited number of applications.

Voigt et al. (2003) tested their hydrophobic silanated 3 nm membrane (HOC) for homogeneous catalyst recovery using the Pd-BINAP catalyst in toluene. The retention of the full catalyst (~1000 Da) was observed to be >99%, while the retention of the BINAP ligand (623 Da) was about 65%. Economically high permeabilities from ~6 L/m^2hbar (pure toluene) to ~1 L/m^2h bar (with catalyst) were obtained. The retention of the BINAP ligand in toluene through Grignard grafted membranes was only <40%, most likely due to the lower toluene-membrane affinity for these amphiphilic membranes. However, Ormerod et al. (2013; 2016) showed that these grafted membranes work as well as the native membranes for the recovery and re-use of different metathesis and cross-coupling catalysts in continuous reaction-filtration conditions in a variety of relevant solvents. An advantage is their higher and/or more stable flux in some conditions. As for the native membranes, very low Pd contamination of reactions products (<10 ppm) has been reached in a one-step process.

So et al. (2010) used the silanated membrane of Voigt et al. (2003) for membrane enhanced peptide synthesis (MEPS). The concept of coupling membranes to solution phase peptide synthesis, offers major advantages over the traditional solid state synthesis, by combining the high mass transfer and smaller excess of reagents for synthesis in solution, and the ease of non-thermal purification with membranes. In the MEPS concept, amide coupling and deprotection steps are alternated by diafiltrations removing by-products and excess reagents, and for each amino acid added to the peptide (chain propagation), the cycle is repeated. The choice of membrane is not evident, as long-term robustness is required in the coupling solvent DMF and during deprotection when the organic base piperidine (20%) is added to DMF. Moreover, the membrane requires high selectivity between the growing peptide and the by-products and excess reagents. To guarantee this selectivity, in this work, the peptide was grown on a soluble PEG support (~5000 Da). In 2010, the silanated 3 nm membrane (HOC) turned out to be the most optimal membrane. Different peptides were produced in this way, and the yield and purity proved to be very good (~92 and ~94%, respectively) and better than the corresponding solid phase product (purity ~77%). During the whole MEPS process, membrane performance was very stable, and fluxes remained high, pointing to low fouling.

For several years at VITO, Grignard grafted membranes have been successfully scaled up to commercial scale and are extensively used in different bilateral projects with a variety of industrial companies (Buekenhoudt, 2017). Applications involve, e.g., diafiltrations for purification or for solvent operations, concentrations, fractionations and catalyst recovery. Solutes used were mainly different APIs, peptides, fragrance molecules and other fine chemicals in a variety of solvents as methanol, butanol, IPA, acetone, ethylacetate, DMSO, THF, DMF, DCM or mixtures thereof. In most cases the performance of the Grignard grafted membranes beats the one of the native membranes and corresponds well to size exclusion. Good results with these membranes have also been obtained in the purification and fractionation of hydrolysed lignine streams (based on water, ethanol, ethylacetate, DMSO or mixtures thereof). In a recent Dutch ISPT project "Energy efficient membrane-assisted recovery of acetone", the long-term

FIGURE 15.8 Pilot tests in a food-oil company using Grignard grafted ceramic membranes for energy-efficient recovery of acetone (From Buekenhoudt, A. Presentation at the 6th International Conference on Solvent Filtration, St Petersburg, Russia, June 5–6, 2017.)

performance of Grignard grafted membranes is demonstrated at pilot scale (0.75 m^2) in a food-oil company (see Figure 15.8). The membranes show good retentions (~90%) for different types of edible oils and economic fluxes of 20 to 40 L/hm^2 dependent on the oil. Comparison with alternative polymeric OSN membranes is in progress.

Table 15.4 gives an overview of the current OSN applications tested with hybrid organic-inorganic membranes.

15.5 RECENT CERAMIC MEMBRANE DEVELOPMENT FOR OSN

Carbon-based membranes such as carbon nanotubes, graphene and graphene oxide show promise as next-generation separation membranes. Among those membranes, graphene oxide (GO) membranes have recently been tested for OSN. A GO membrane consists of a stacking of impermeable graphene sheets (~1 μm in size) with hydroxyl, carboxyl and other oxidized groups at their outer edges. The oxidized regions act as spacers to keep adjacent graphene sheets apart and help water or other polar molecules to intercalate between the sheets. The pristine graphene sheets provide a network of capillaries that allow nearly frictionless flow of water, similarly to flow through carbon nanotubes. Huang et al. (2015) showed that the graphene sheets not only accommodate water but also other polar solvents, such as ethanol or acetone, creating nanochannels of about 1 nm, in the right range for molecular sieving. The GO nanochannels can further be tuned by thermal annealing, creating narrowed nanochannels due to reduction of the oxygen-containing groups. However,

TABLE 15.4
Overview of OSN Applications Tested with Hybrid Organic-Inorganic NF (or Tight UF) Membranes. ThePerformance Lists the Retention of the Solute and the Permeability of the Tested Mixture. The Applications Mentioned in Bold Are Being Tested at Pilot Scale

Membrane Producer[a]	Solvent	Solute MW[b]	Process Objective	Performance R and J	Reference
ZrO_2 HOC Fraunhofer	**Toluene**	**Pd catalyst 1000 Da**	**Catalyst recovery in reaction**	**>99% ~1 L/m²hbar**	**Voigt et al. (2003)**
ZrO_2 HOC Fraunhofer	DMF DMF+ piperidine	Growing peptide 5000 Da	Membrane enhanced peptide synthesis	99% ~4 L/m²hbar	So et al. (2010)
Grafted TiO_2 VITO	DCM and Acetone	Ru-catalysts 500 Da	Catalyst recovery in reaction	83 to 99% 0.6 to 5 L/m²hbar	Ormerod et al. (2013)
Grafted TiO_2 VITO	Ethanol	Pd catalysts 630 Da	Catalyst recovery in reaction	97 to 99% 1 to 6.5 L/m²hbar	Ormerod et al. (2016)
Grafted TiO_2 VITO	Ethanol mixtures	Pd catalyst nanoparticles	Metal removal after reaction	>99.99% 1 to 3 L/m²hbar	Ormerod et al. (2016)
Grafted TiO_2 VITO	**Acetone**	**Edible oil 600 Da**	**Concentration for acetone recovery**	**~90% 1 to 2 L/m²hbar**	**Buekenhoudt (2017)**
Grafted TiO_2 VITO	Various	Various	Purification, concentration, diafiltration, solvent switch	Better than native membranes	Buekenhoudt (2017)

[a] Producer means commercial supplier, or research group making the membrane used
[b] Most values are approximate values

the first OSN results in GO membranes show complex behavior, questioning a transport mechanism solely governed by the nanochannels. Indeed, Aher et al. (2017) and Akbari et al. (2018) showed a clear difference in performance for dry or wet GO membranes. Dry membranes show steady state organic solvent fluxes higher than for water, while retentions are systematically lower in organic solvents than in water. On the other hand, organic solvent fluxes are much lower in wet membranes, while the retentions increase. The authors suggest that microstructural defects created during membrane formation and/or water immobilized around the oxidized functional groups could play a role in the OSN transport. A comparison of the results of both authors also points to very big differences in fluxes and selectivities, most likely related to the different way of membrane synthesis. More R&D will be required to clarify if GO membranes could be good OSN membranes.

Recently other novel carbon-based membranes with interesting properties for OSN have been developed by Koh et al. (2016). These hollow fiber carbon molecular sieve membranes have been prepared by dual-layer, dry-wet spinning of poly(vinylidene fluoride) (PVDF). In order to avoid the pore collapse during the subsequent pyrolysis, the porous structure is fixed by cross-linking before the high temperature treatment. The thin toplayer (<100 nm) and the slit-like nanopores (bimodal distribution with peaks at 0.6 and 0.8 nm) guarantee a high mass transport and high selectivities, fit for organic solvent reverse osmosis (OSRO). The thin hollow fiber membranes were subjected to first OSRO tests with a para-xylene/ortho-xylene mixture at room temperature and at TMPs of 50 to 120 bar. The results proved the excellent molecular sieving character of these membranes: a 50/50 (90/10) mixture at the feed side (shell side) resulted in a 95/5 (55/45) mixture at the permeate side. The retention further increased sigmoidally with increasing pressure. Consequently, these membranes open the door for the performance of important bulk separations in petrochemistry as the separation of xylene isomers at low operating temperatures, removing the necessity for energy-intensive phase changes.

15.6 CONCLUSION

The wide chemical robustness, universal mounting in modules and possibility of dry storage of ceramic NF membranes gives them a great potential for applications in organic solvents. Although less dense than most of their polymeric counterparts, the pore size of ~1 nm of the current commercially available membranes has proved to be sufficient for a broad range of OSN separations. Native membranes are particularly fit for use in more polar solvents, while hybrid organic-inorganic membranes widen (for amphiphilic membranes) or shift (for hydrophobic membranes) the application potential to more apolar solvents.

Solvent permeabilities are mainly governed by solvent-membrane affinity, and lead to strong deviations from viscous flow, especially for the hydrophilic native membrane. Retentions are very solvent, solute and temperature dependent, and are governed by the complex competition of solvent-membrane and solute-membrane affinity, further influenced by solvent-solute affinity. The Spiegler-Kedem theory has proven to be a great help in explaining the retention results of both native and hybrid membranes in a qualitative way.

A vast variety of OSN applications have been tested with ceramic NF membranes and have confirmed their potential. More and more feasibility testing evolves towards long-term pilot trials, and some rare cases have already been implemented.

Recent new developments of denser ceramic membranes (possibly with increased intrinsic hydrophobicity) and monolith design with higher surface to volume ratio will further enlarge the opportunities of ceramics for OSN. In short, ceramic NF membranes are becoming a mature alternative for the existing polymeric OSN membranes.

REFERENCES

Aerts, S., Buekenhoudt, A., Weyten, H., Gevers, L.E.M., Vankelecom, I.F.J., Jacobs, P.A., 2006. The use of solvent resistant nanofiltration in the recylcing of the Co-Jacobsen catalyst in the hydrolytic kinetic resolution (HKR) of epoxides, Journal of Membrane Science. 280, 245–252.

Aher, A., Cai, Y., Majumder, M., Bhattacharyya, D., 2017. Synthesis of graphene oxide emmbranes and their behavior in water and isopropanol. Carbon. 116, 145–153.

Akbari, A., El Meragawi, S., Martin, S.T., Corry, B., Shamsaei, E. Bhattacharrya, D., Majumder, M., 2018. Solvent transport behavior of shear aligned graphene oxide membranes and implications in organic solvent nanofiltration. Submitted to ACS Applied Materials & Interfaces, August 2018, 10, 2067–2074.

Andrecochea Saiz, C., Darvishmanesh, S., Buekenhoudt, A., Van der Bruggen, B., 2017. Shortcut applications of the Hansen solubility parameter for organic solvent nanofiltration. Journal of Membrane Science, 546, 120–127.

Baumgarten, G., 2010. Presentation at the BMG evening at Janssen, Beerse, Belgium, October 14, 2010.

Bishop, K.D., Kirkmann, T.J., 2011. Structure for molecular separation, US 8,426,333.

Blumenschein, S., Kätzel, U., 2017. An heuristic-based selection process for organic solvent nanofiltration membranes. Separation and Purification Technology. Accepted Manuscript. 183, 83–95.

Bothun, G., Peay, K., Ilias, S., 2007. Role of tail chemistry on liquid and gas transport through organosilane-modified mesoporous ceramic membranes. Journal of Membrane Science. 301, 162–170.

Buekenhoudt, A., Dotremont, C., Aerts, S., Vankelecom, I., Jacobs, P.A., 2004. Successful applications of ceramic nanofiltration in non-aqueous solvents. Proceedings of the 8th International Conference on Inorganic Membranes, Cincinnati, OH.

Buekenhoudt, A., Wyns, K., Meynen, V., Maes, B., Cool, P., 2010. Surface modified inorganic matrix and method for preparation thereof, WO2010/106167.

Buekenhoudt, A., 2010. The potential of ceramic membranes in different industrial separation problems. Presentation at the 3th International Conference on Solvent Filtration, London, UK.

Buekenhoudt, A., Bisignano, F., De Luca, G., Vandezande, P., Wouters, M., Verhulst, K., 2013. Unravelling the solvent flux behaviour of ceramic nanofiltration and ultrafiltration membranes. Journal of Membrane Science. 439, 36–47.

Buekenhoudt, A., Bulut, M., Beckers, H., Vleeschouwers, R., van Zanten, D., 2014. OSN: Successful API recovery from a distillation residue at Sitech-DSM. Proceedings of the Aachener Membrane Colloquium, AMK2014, Aachen, Germany.

Buekenhoudt, A., 2017. Presentation at the 6th International Conference on Solvent Filtration, St Petersburg, Russia, June 5–6, 2017.

Chen, H., Jia, X., Wei, M., Wang, Y., 2017. Ceramic tubular nanofiltration membranes with tunable performances by atomic layer deposition and calcination. Journal of Membrane Science. 528, 95–102.

Chowdhury, S., Schmuhl, R., Keizer, K., ten Elshof, J., Blank, D., 2003. Pore size and surface chemistry effects on the transport of hydrophobic and hydrophilic solvents through mesoporous γ-alumina and silica MCM-48. Journal of Membrane Science. 225, 177–186.

Chouwdhury, S.R., Witte, P.T., Blank, D.H.A., Alsters, P.L., ten Elshof, J.E., 2006. Recovery of homogeneous polyoxometallate catalysts from aqueous and organic media by a mesoporous ceramic membrane without loss of catalytic activity. Chemistry A European Journal. 12, 3061–3066.

Cuperus, P., 2015. Presentation at the 5th International Conference on Solvent Filtration, Antwerp, Belgium, November 18–19, 2015.

Darvishmanesh, S., Buekenhoudt, A., Degrève, J., Van der Bruggen, B., 2009. Coupled series-parallel resistance model for transport of solvent through inorganic nanofiltration membranes. Separation and Purification Technology. 70, 46–52.

Darvishmanesh, S., Degreve, J., Van Der Bruggen, B., 2010. Mechanisms of solute rejection in solvent resistant nanofiltration: The effect of solvent on solute rejection. Physical Chemistry and Chemical Physics. 12, 13333–13342.

Darvishmanesh, S., Vanneste, J., Tocci, E., Jansen, J., Tasseli, F., Degreve, J., Drioli, E., Van Der Bruggen, B., 2011. Physicochemical characterization of solute retention in solvent resistant nanofiltration: The effect of solute size, polarity, dipole moment, and solubility parameter. Journal of Physical Chemistry B. 115, 14507–14517.

Dobrak, A., Verrecht, B., Van den Dungen, H., Buekenhoudt, A., Vankelecom, I.F.J., Van der Bruggen B., 2010. Solvent flux behavior and rejection characteristics of hydrophilic and hydrophobic mesoporous and microporous TiO_2 and ZrO_2 membranes. Journal of Membrane Science. 346, 344–352.

Franke, R., Rudek, M., Baumgarten, G., 2010. Nanofiltration development to application readiness. Evonik Science Newsletter. 30, 6–11.

Geens, J., 2006. Nanofiltration in organic solvents using polymeric and ceramic membranes, PhD thesis, KULeuven, Belgium.

Geens, J., Boussu, K.,Vandecasteele, C., Van der Bruggen, B., 2006. Modelling of solute transport in non-aqueous nanofiltration. Journal of Membrane Science. 281, 139–148.

Guizard, C., Ayral, A., Julbe, A., 2002. Potentiality of organic solvents filtration with ceramic membranes, A comparison with polymer membranes. Desalination. 147, 275–280.

Haverkamp M.-S. (Borsig), 2017. Presentation at the 6th International Conference of Solvent Filtration, St Petersburg, Russia, June 5–6, 2017.

Huang, L., Li, Y., Zhou, Q., Yuan, W., Shi, G., 2015. Graphene oxide membranes with tunable semipermeability in organic solvents. Advanced Materials. 27, 3797–3802.

Koh, D.-Y., McCool, B.A., Deckman, H.W., Lively, R.P., 2016. Reverse osmosis molecular differentiation of organic liquids using carbon molecular sieve membranes. Science. 353, 804–807.

Koops, G.H., Yamada, S., Nakao, S., 2001. Separation of linear hydrocarbons and carboxylic acids from ethanol and hexane solutions by reverse osmosis. Journal of Membrane Science. 189, 241–254.

Larbot, A., Alami-Younssi, A., Persin, M., Sarrazin, J., Cot, L., 1994. Preparation of γ-alumina nanofiltration membrane. Journal of Membrane Science. 97, 167–173.

Levecke, B., 2006. Successful industrial application of diafiltration with ceramic NF. Presentation at the International Workshop on Membranes in Solvent Filtration, Leuven, Belgium, March 23–24, 2006.

Machado, D.R., Hasson, D., Semiat, R., 1999. Effect of solvent properties on permeate flow through nanofiltration membranes. Part I. Investigation of parameters affecting solvent flux. Journal of Membrane Science. 163, 93–102.

Marchetti, P., Butté, A., Livingston, A.G., 2012. An improved phenomenological model for solvent permeation through NF and UF membranes. Journal of Membrane Science. 415/416 444–458.

Marchetti, P., Butté, A., Livingston, A.G., 2013a. NF in organic solvent/water mixtures: Role of preferential solvation. Journal of Membrane Science. 444, 101–115.

Marchetti, P., Butté, A., Livingston, A.G., 2013b. Quality by design for peptide nanofiltration: Fundamental understanding and process selection. Chemical Engineering Science. 101, 200–212.

Marchetti, P., Jimenez Solomon, M., Szekely, G., Livingston, A., 2014. Molecular separation with Organic Solvent Nanofiltration: A critical review. Chemical Reviews. 114, 10735–10806.

Merlot, R.B., Tanardi, C.R., Vankelecom, I.F.J., Nijmeijer, A., Winnubst, L., 2017. Interpreting rejection in SRNF across grafted ceramic membranes through the Spiegler-Kedem model. Journal of Membrane Science. 525, 359–367.

Meynen, V., Castricum, H.L., Buekenhoudt, A., 2014. Class II hybrid organic-inorganic membranes creating new versatility in separations. Current Organic Chemistry. 18, 2334–2350.

Ormerod, D., Bongers, B., Porto-Carrero, W., Siegas, S., Vijt, G., Lefevre, N., Lauwers, D., Brusten, W., Buekenhoudt, A., 2013. Separation of metathesis catalysts and reaction product in flow reactors using organic solvent nanofiltration. RSC Advances. 3, 21501–21510.

Ormerod, D., Buekenhoudt, A., Bongers, B., Baramov, T., Hassfeld, J., 2017. From reaction solvent to crystallization solvent, membrane assisted reaction work-up and interpretation of membrane performance results by application of Spiegler – Kedem theory. Organic Process Research & Development. Accepted for publication November 2017, 21, 2060–2067.

Ormerod, D., Lefevre, N., Dorbec, M., Eyskens, I., Vloemans, P., Duyssens, K., Diez de la Torre, V., Kaval, N., Merkul, E., Sergeyev, S., Maes, B.U.W., 2016. Potential of homogeneous Pd catalyst separation by ceramic membranes. Application to downstream and continuous flow processes. Organic Process Research & Development. 20, 911–920.

Ormerod, D., Noten, B., Dorbec, M., Andersson, L., Buekenhoudt, A., Goetelen, L., 2015. Cyclic peptide formation in reduced solvent volumes via in-line solvent recycling by organic solvent nanofiltration. Organic Process Research & Development. 19, 841–848.

Palmeri, J., Blanc, P., Larbot, A., David, P., 1998. Hafnia ceramic nanofiltration membranes: Modeling of pressure-driven transport of netural solutes and ions. Proceedings of the Fifth International Conference on Inorganic Membranes, Nagoya, Japan.

Peterson, R.A., Anderson, M.A., Hill, C.G., 1990. Permselectivity characteristics of supported ceramic alumina membranes. Separation Science and Technology. 25, 1281–1285.

Pinheiro, A.F.M., 2013. Development and characterization of polymer-grafted ceramic membranes for solvent nanofiltration, PhD thesis, University of Twente.

Pinheiro, A.F.M., Hoogendoorn, D., Nijmeijer, A., Winnubst, L., 2014. Development of a PDMS-grafted alumina membrane and its evaluation as solvent resistant nanofiltration membrane. Journal of Membrane Science. 463, 24–32.

Postel, S., Spalding, G., Chirnside, M., Wessling, M., 2013. On negative retentions in organic solvent nanofiltration. Journal of Membrane Science. 447, 57–65.

Priske, M., Wiese, K.-D., Drews, A., Kraume, M., Baumgarten, G., 2010. Reaction integrated separation of a homogenous catalysts in hydroformylation of higher olefins by means of organophilic nanofiltration. Journal of Membrane Science. 360, 77–83.

Puhlfürss, P., Voigt, A., Weber, R., Morbé, M., 2000. Microporous TiO_2 membranes with a cut off < 500 Da. Journal of Membrane Science. 174, 123–133.

Puhlfürss, P., Richter, H., Weyd, M., Voigt, I., 2016. NF membranes for the cleaning of "recycle water" in oil sand extractions. Annual Report 2015/2016 of Fraunhofer IKTS, 50–51.

Rezaei Hosseinabadi, S., Wyns, K., Buekenhoudt, A., Van der Bruggen, B., Ormerod, D., 2015. Performance of Grignard functionalized ceramic nanofiltration membranes. Separation and Purification Technology. 147, 320–328.

Rezaei Hosseinabadi, S., Wyns, K., Meynen, V., Buekenhoudt, A., Van der Bruggen, B., 2016. Solvent-membrane-solute interactions in organic solvent nanofiltration (OSN) for Grignard functionalised membranes: Explanation via Spiegler-Kedem theory. Journal of Membrane Science. 513, 177–185.

Rezaei Hosseinabadi, S., Wyns, K., Meynen, V., Carleer, R., Adriaensens, P., Buekenhoudt, A., Van der Bruggen, B., 2014. Organic solvent nanofiltration with Grignard functionalised ceramic nanofiltration membranes. Journal of Membrane Science. 454, 496–504.

Richter, H., Tomandl, G., Siewert, S., Piorra, A., 1998. Ceramic nanofiltration membranes made of ZrO_2 and TiO_2. Proceedings of the Fifth International Conference on Inorganic Membranes, Nagoya, Japan.

Sarrade, S., Rios, G.M., Carlès, M., 1994. Dynamic characterisation and transport mechanism of two inorganic membranes for nanofiltration. Journal of Membrane Science. 97, 155–166.

Sarrade, S., Rios, G.M., Carlès, M., 1996. Nanofiltration membrane behavior in a supercritical medium. Journal of Membrane Science. 114, 81–91.

Shnitzer, C. (Evonik), 2017. Presentation at the 6th International Conference of Solvent Filtration, St Petersburg, Russia, June 5, 2017.

So, S., Peeva, L.G., Tate, E.W., Leatherbarrow, R.J., Livingston, A., 2010. Organic solvent nanofiltration: A new paradigm in peptide synthesis. Organic Process Research & Development. 14, 1313–1325.

Spiegler, K.S., Kedem, O., 1966. Thermodynamics of hyperfiltration (reverse osmosis): Criteria for efficient membranes. Desalination. 1, 311–326.

Tanardi, C.R., Vankelcom, I.F.J., Pinheiro, A.F.M., Tetala, K.K.R., Nijmeijer, A., Winnubst, L., 2015. Solvent permeation behavior of PDMS grafted γ-alumina membranes. Journal of Membrane Science. 495, 216–225.

Tanardi, C.R., Nijmeijer, A., Winnubst, L., 2016a. Coupled-PDMS grafted mesoporous γ-alumina membranes for solvent nanofiltration. Separation and Purification Technology. 169, 223–229.

Tanardi, C.R., Catana, R., Barboiu, M., Ayral, A., Vankelecom, I.F.J., Nijmeijer, A., Winnubst, L., 2016b. Polyethyleneglycol grafting of γ-alumina membranes for solvent resistant nanofiltration. Micorporous and Mesoporous Materials. 229, 106–116.

Tsuru, T., Wada, S., Izumi, S., Asaeda, M., 1998. Preparation of microporous silica-zirconia membranes for nanofiltration. Journal of Membrane Science. 149, 127–135.

Tsuru, T., Sudoh, T., Kawahara, S., Yoshioka, T., Asaeda, M., 2000. Permeation of liquids through inorganic nanofiltration membranes. Journal of Colloid and Interface Science. 228, 292–296.

Tsuru, T., Sudoh, T., Yoshioka, T., Asaeda, M., 2001. Nanofiltration in non-aqueous solutions by porous silica-zirconia membranes. Journal of Membrane Science. 185, 253–261.

Tsuru, T., Miyawaki, M., Kondo, H., Yoshioka, T., Asaeda, M., 2003. Inorganic porous membranes for nanofiltration of nonaqueous solutions. Separation and Purification Technology. 32, 105–109.

Tsuru, T., Kondo, H., Yoshioka, T., Asaeda, M., 2004. Permeation of nonaqueous solution through organic/inorganic hybrid nanoporous membranes. AIChE Journal. 50, 1080–1087.

Tsuru, T., Miyawaki, M., Yshioka, T., Asaeda, M., 2006a. Reverse Osmosis of nonaqueous solutions through porous silica-zirconia membranes. AIChE Journal. 52, 522–531.

Tsuru, T., Narita, M., Shinagawa, R., Yoshioka, T., Asaeda, M., 2006b. Nanoporous ceramic membranes for permeation and filtration of organic solutions. Proceedings of the 9th International Conference on Inorganic Membranes, Lillehammer, Norway.

Tsuru, T., Narita, M., Shinagawa, R., Yoshioka, T., 2008. Nanoporous titania membrane for permeation and filtration of organic solutions. Desalination. 233, 1–9.

Tsuru, T., Nakasuji, T., Oka, M., Kanezashi, M., Yoshioka, T., 2011. Preparation of hydrophobic nanoporous methylated SiO_2 membranes and application to nanofiltration of hexane solutions. Journal of Membrane Science. 384, 149–156.

Van der Bruggen, B., Buekenhoudt, A., Verrecht, B., Van den Dungen, H., Geens, J., Leysen, R., 2007. Preparation and performance of nanoporous hydrophilic and hydrophobic ceramic membrane for use in non-aqueous nanofiltration. in Ceramic Materials Research Trends, Nova Publishers, Hauppauge NY, Editor P.B. Lin, 209–225.

Vandezande, P., Gevers, L.E.M., Vankelecom, I.F.J., 2008. Solvent Resistant Nanofiltration: Separating on a molecular level. Chemical Society Reviews. 37, 365–405.

Vandezande, P., Wyns, K., Meynen, V., Buekenhoudt, A., 2009. Solvent filtration "matrix": Assessing the potential of ceramic nanofiltration membranes in solvent environments, Proceedings of Euromembrane 2009, Montpellier, France.

Van Gestel, T., Van der Bruggen, B., Buekenhoudt, A., Dotremont, C., Luyten, J., Vandecasteele, C., Maes, G., 2003. Surface modification of γ-Al_2O_3/TiO_2 multilayer membranes for applications in non-polar organic solvent streams. Journal of Membrane Science. 224, 3–10.

Van Gestel, T., Kruidhof, H., Blank, D.H.A., Bouwmeester, H.J.M. 2006. ZrO_2 and TiO_2 membranes for nanofiltration and pervaporation Part 1. Preparation and characterization of a corrosion-resistant ZrO_2 nanofiltration membrane with a MWCO < 300. Journal of Membrane Science. 284, 128–136.

Van Gestel, T., Sebold, D., Kruidhof, H., Bouwmeester, H.J.M., 2008. ZrO_2 and TiO_2 membranes for nanofiltration and pervaporation Part 2. Development of ZrO_2 and TiO_2 toplayers for pervaporation. Journal of Membrane Science. 318, 413–421.

Vankelecom, I.F.J., De Smet, K., Gevers, L.E.M., Livingston, A., Nair, D., Aerts, S., Kuypers, S., Jacobs, P.A., 2004. Physico-chemical interpretation of the SRNF transport mechanism for solvents through dense silicone membranes. Journal of Membrane Science. 231, 99–108.

Voigt, I., Fisher, G., Puhlfürss, P., Seifert, D., 1998. New filtration ceramics from support to NF membrane completely of TiO_2. Proceedings of the Fifth International Conference on Inorganic Membranes, Nagoya, Japan.

Voigt, I., Puhlfürss, P., Holborn, T., Dudziak, D., Mutter, M., Nickel, A., 2003. Ceramic nanofiltration membranes for applications in organic solvents. Proceedings of the Aachener Membrane Colloquium, Aachen, Germany.

Voigt, I., Puhlfürss, P., Richter, H., Endter, A., Duscher, S., Herrmann, K., 2012. Ceramic NF-membranes with a cut-off of 200 Da. Presentation at the International Conference on Inorganic Membranes, Eschede, The Netherlands.

White, L.S., 2006. Development of large-scale applications in organic solvent nanofiltration and pervaporation for chemical and refining processes. Journal of Membrane Science. 286, 26–35

Yang, X.J., Livingston, A.J., Freitas dos Santos, L., 2001. Experimental observations of nanofiltration with organic solvents. Journal of Membrane Science. 190, 45–55.

Zeidler, S., Kätzel, U., Kreis, P., 2013. Systematic investigation on the influence of solutes on the separation behavior of a PDMS membrane in organic solvent nanofiltration. Journal of Membrane Science. 429, 295–303.

Zeidler, S., Puhlfürss, P., Kätzel, U., Voigt, I., 2014. Preparation and characterization of new low MWCO ceramic nanofiltration membranes for organic solvents. Journal of Membrane Science. 470, 421–430.

Section IV

Membranes for Energy Applications

16 Membrane Materials for Ion Exchange Membrane Fuel Cell Applications

Yubin He, Jianqiu Hou, and Tongwen Xu

CONTENTS

16.1 INTRODUCTION

Ion exchange membrane fuel cells (IEMFC), which can directly convert chemical energy to electric power in a highly efficient and environmental-friendly manner, are recognized as a promising technology to solve the challenges of energy shortage and environmental pollution (Zhang and Shen, 2012a). Ion exchange membranes (IEMs) generally work as the ion conductor and fuel separator in a fuel cell (Ran et al., 2017).

The performance of a fuel cell, especially its power density and service time, are directly affected by the properties of IEMs. For example, Nafion® as the benchmark cation exchange (CEM) membrane shows high ion conductivity (around 70 mS/cm at room temperature), good mechanical strength and suitable water swelling. Therefore, it was most frequently adopted in the fuel cell applications despite its high-cost and high methanol permeability.

IEMs are typically composed of hydrophobic substrates, immobilized ion conducting groups and movable counter ions. Depending on the type of ionic groups, IEMs are broadly classified into CEMs and anion exchange membranes (AEMs). Naturally, the ion pair between counter ions and ion conducting groups will dissociate after the penetration of sufficient water molecules, releasing cations or anions for the transfer of corresponding ions. The common functional moiety in CEMs is a sulfonic acid group, while that in AEMs is quaternary ammonium cation.

Good ion conductivity, low swelling ratio, high mechanical strength and sufficient chemical stability are the essential properties of IEM. However, the trade-off effect among these properties makes it difficult to prepare an ideal IEM. For example, high ion conductivity requires high ion exchange group density, which will reduce the dimensional and chemical stability at the same time. This dilemma can be disentangled by developing advanced materials for IEM.

To date, numerous efforts have been devoted to modifying the pristine membrane structures or exploiting novel structures of IEMs. A wide variety of IEMs were prepared by either post-functionalization of pre-existing polymers or direct polymerization of functionalized monomers. In this chapter, novel IEM structures as well as the corresponding synthetic procedures will be classified and reviewed.

16.2 PROTON EXCHANGE MEMBRANE FOR FUEL CELLS (PEMFCs)

Fuel cell requires that proton exchange membranes (PEMs) possess high proton conductivity as well as superior dimensional stability. Thus, the Nafion® membrane is always chosen as a benchmark membrane for PEMFCs. However, the high cost of Nafion® membranes is the main problem blocking their practical application, especially in PEMFC, which employs expensive Pt catalysts. Besides, the capacity to retain water retention at high temperature and low humidity, which is not possessed by the Nafion® membranes, is critical to fuel cell applications due to the high operating temperature (>80°C). And for direct methanol fuel cells (DMFCs), the Nafion® membrane cannot exhibit good resistance for methanol leakage, leading to decreased cell performance in long-term operation (Smitha et al., 2005). Therefore, the exploitation of cheap materials, such as polymers with aliphatic and aromatic backbone, is necessary for PEMFC applications. Besides, it is necessary for the membrane materials to have good water retention capacity and low fuel permeability.

The PEMs are generally composed of polymer backbones and functional groups. The common polymer materials which have the potential of replacing the Nafion® membrane for fuel cells, are poly(ether sulfone) (PES), poly(ether ketone) (PEK), polybenzimidazole (PBI), polyimide (PI), polystyrene (PS) and polyvinylidene fluoride (PVDF). However, directly attaching the functional groups (sulfonic acid group or carboxylic group) on these polymer backbones can hardly create PEM with comparable performance to Nafion®. In order to improve the properties of low-cost PEMs, it is essential to figure out the structure-morphology-property relationship of Nafion® membrane. In the past decades, the Nafion® membrane, which consists of a polytetrafluoroethylene (PTFE) backbone and pendant sulfonic acid groups, was investigated by small angle X-ray scattering (SAXS) (Gierke et al., 1981), transmission electron spectroscopy (TEM) (Fujimura et al., 1982) and atomic force microscopy (AFM) (Lehmani et al., 1998) to demonstrate its nanoscale-ionic-cluster morphology. When the membrane is hydrated, the ionic clusters become well connected, resulting in proton conductive channels (PCCs) and fast proton conduction. Therefore, the generation and alignment of PCCs is the main reason for the high proton conductivity of Nafion® membranes.

To improve the proton conductivity of PEMs made from low-cost materials, it is crucial to construct well-connected PCCs within the membrane matrix. In this section, the corresponding polymer structures that can facilitate the formation of ion channels were summarized and discussed (Debe, 2012; Elabd and Hickner, 2011; Hou et al., 2012; Ran et al., 2017; Wu et al., 2013; Zhang and Shen, 2012b).

16.2.1 The Distribution of Functional Groups

It is difficult to improve the proton conductivity only by increasing the content of functional groups due to the concomitantly decreased dimensional stability. Tuning the distribution of ion-exchange groups to facilitate the aggregation of ionic cluster is thus critical to improve the performance of PEMs. Inspired by the benchmarked Nafion® membranes, the side-chain type and the densely sulfonated PEMs were designed. Their obviously enhanced properties were due to the aggregation of hydrophilic moiety, which is in favor of constructing well-connected PCCs.

16.2.1.1 Side-Chain Type Functional Groups

As inspired by Nafion® membrane, the side-chain type functional groups can improve the mobility of the sulfonic acid group in a hydrated membrane matrix, which is critical to access high proton conductivity. The most frequently adopted method to prepare side-chain type PEM is by reacting the phenol groups on the polymer main chain with 1,3-propanesultone or 1,4-butanesultone (Na et al., 2012). Additionally, the azide-alkyne click reaction (Qi et al., 2015) and potassium carbonate mediated nucleophilic substitution reaction (Lafitte et al., 2005) have been developed for preparing side-chain type PEMs. To precisely control the content of sulfonated groups, the side-chain type PEMs can be synthesized by the polymerization of pre-sulfonated monomers (Ding et al., 2002). Nakabayashi et al. (2011) synthesized poly(phenylene ether) containing pendant perfluoroalkyl sulfonic acids for PEMFCs (Figure 16.1). The proton conductivity is comparable to that of the Nafion® 117 membrane, while the water uptake is significantly lower (19.3–28.0%) due to the high fluorine content. Jutemar and Jannasch (2010) studied the effects of rigid side chains, including poly(arylene ether sulfone)s (PAES) bearing sulfobenzoyl, sulfonapthoxybenzoyl, disulfonathoxybenzoyl or trisulfopyrenoxyben-zoyl moieties. SAXS analysis reveals that the longer side chains are favorable for ionic clustering, whereas the ionic clustering in the shorter side chain SPAEs is suppressed

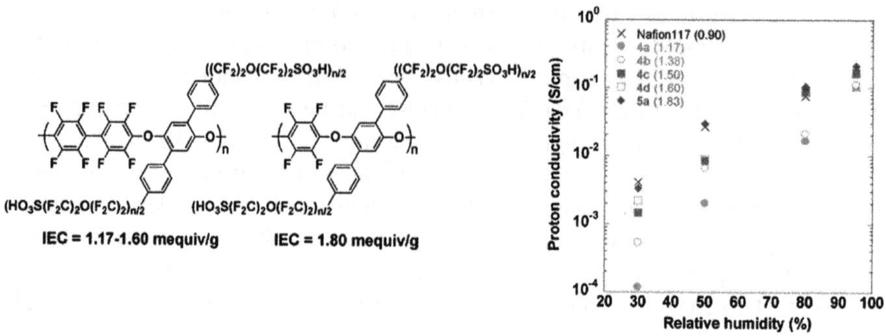

FIGURE 16.1 Polymer structures (left) and proton conductivities (right) of the PEMs with pendant perfluoroalkyl side chains. (Reprinted from Nakabayashi, K. et al., Macromolecules. 44, 1603–1609, 2011.)

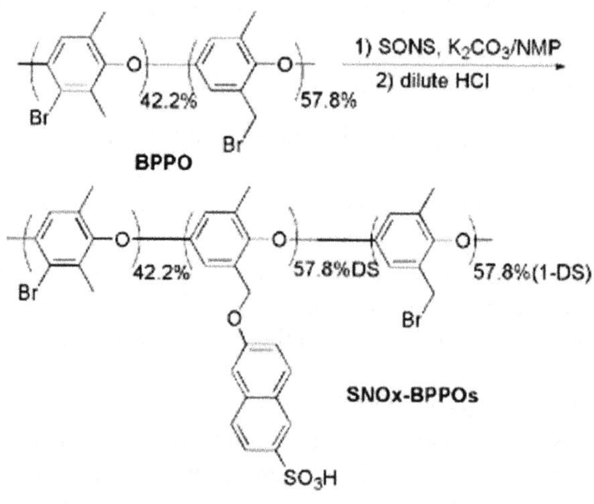

FIGURE 16.2 Schematic synthesis of side-chain type PEM via etherification reaction of BPPO. (Adapted from Zhang, Z. et al., Journal of Membrane Science. 373, 160–166, 2011.)

(Jutemar and Jannasch, 2010). The aggregation of pendant side chains is beneficial for enhancing the phase separated micro-morphology due to the polarity difference between the sulfonated hydrophilic moiety and the hydrophobic polymer backbone. For example, our group synthesized side-chain type PEM based on the synthetic route described in Figure 16.2 (Zhang et al., 2011). UV spectrometry suggests that p-p stacking between pendant naphthalene rings can assist in the formation of ionic channels. High proton conductivities ranging from 38 to 71 mS/cm (25°C, 100% RH) could be achieved at relatively low ion exchange capacity (IEC) (0.93–1.62 mmol/g).

16.2.1.2 Densely Sulfonated PEMs

Attaching densely sulfonated moieties onto the polymer backbones is anther strategy to improve the proton conductivity of PEMs. The aggregation of sulfonic acid

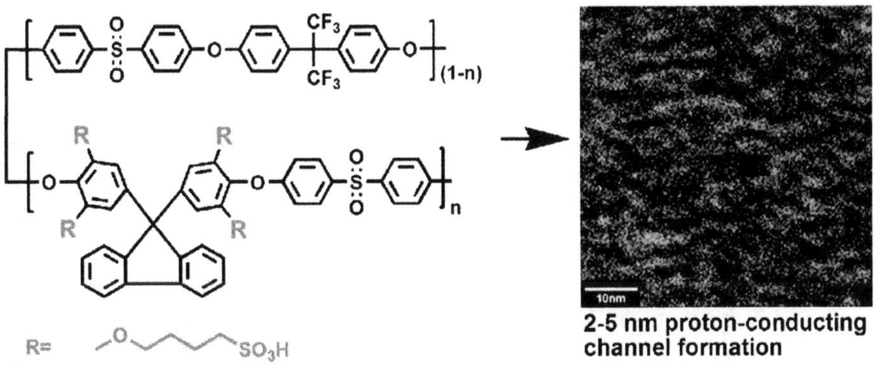

FIGURE 16.3 Molecular structure and TEM image of the densely sulfonated PAES. (Reprinted from Wang, C. et al., Macromolecules. 44, 7296–7306, 2011.)

groups can lead to good micro-phase separation ability as well as high-efficiency PCCs. Wang et al. (2011) designed the densely sulfonated PAES with one single moiety attached by four sulfonated side chains. TEM images revealed well-connected PCCs with the diameter of 2–5 nm (Figure 16.3). The resultant membrane exhibits excellent proton conductivities in the range of 61–209 mS/cm at 30°C. In addition, a branched poly(ether ketone) bearing higher density of sulfonic acid groups was synthesized by postsulfonation. SAXS analysis reveals a distinct phase separation and resulting high proton conductivity (>80 mS/cm at room temperature) at low IEC of 1.12 mmol/g (Chen et al., 2011).

To further improve the proton conductivity of PEMs, SPAES with two or four sulfonic acid groups in one side chain were designed and synthesized (Figure 16.4). The resulting PEMs exhibit higher proton conductivities of 63 mS/cm at room temperature (Kim et al., 2008). To attach more sulfonic acid groups on one side chain, the anionic polymerization and atom transfer radical polymerization is adopted. The polymers bearing reactive C-Cl or C-Br bonds were employed as the macro-initiator and a special comb-shaped PEM was constructed. Comparable proton conductivity to that of the Nafion® membrane can be achieved due to the distinct phase separation as observed by transmission electron microcopy (TEM) (Figure 16.5) (Tsang et al., 2009).

In summary, the PEMs with dense functional groups in one side chain are good candidates to disentangle the dilemma between conductivity and other membrane properties like swelling and fuel permeability due to efficiently using sulfonic acid groups. This can increase mobility of ion-exchange groups, enhance micro-phase separation and construct well-connected PCCs. The further works will focus on building aligned and ordered ion channels in PEM matrix.

16.2.2 PEMs Based on Block Copolymers

Besides the distribution of functional groups, the structure of polymer backbones also determines the properties of PEMs. The conventional PEMs are usually based on a random copolymer. It is thus difficult to form extensive phase separation morphology even with a long side chain. Conversely, in a blocked PEM, the hydrophilic

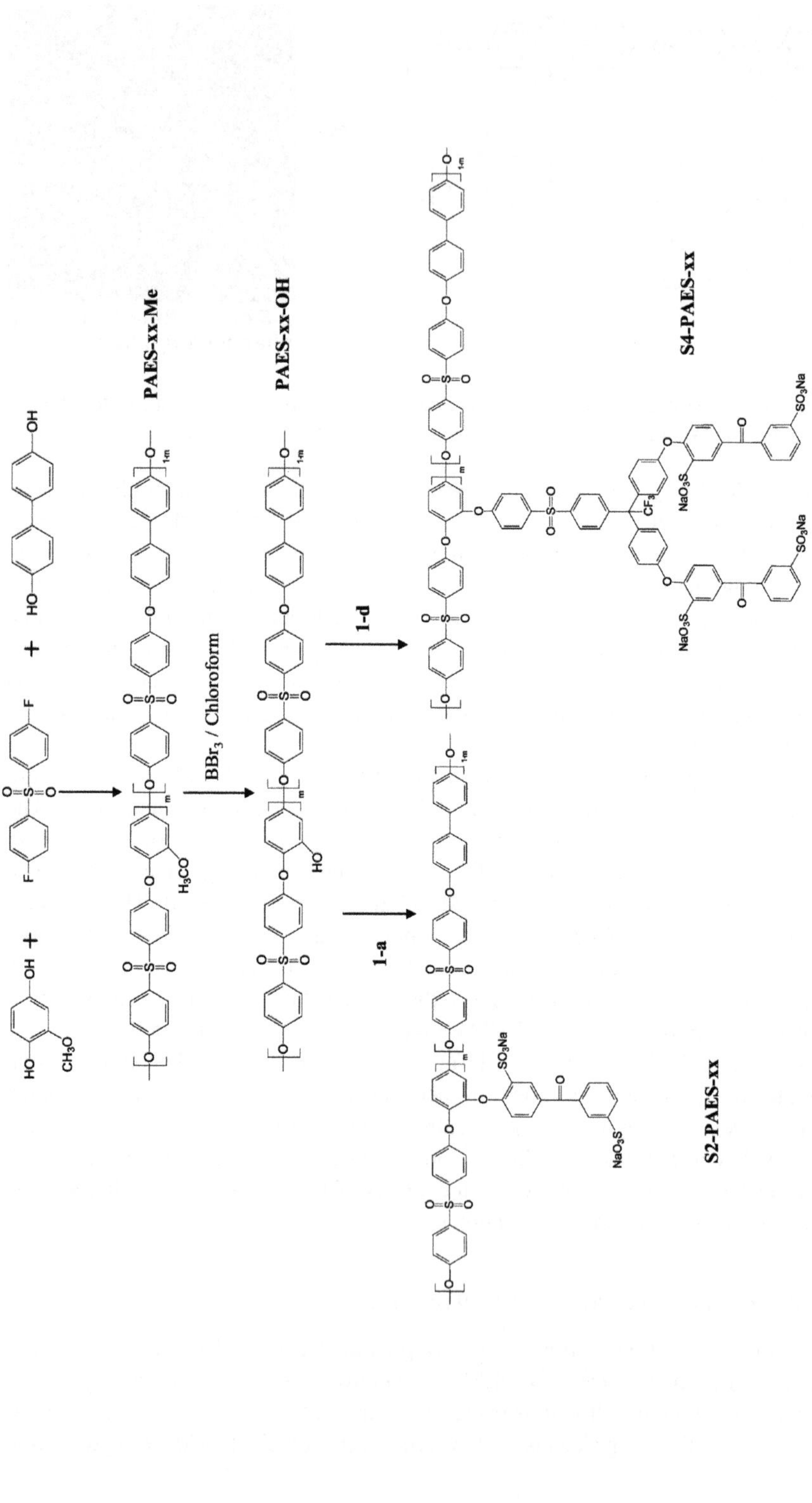

FIGURE 16.4 Synthesis of SPAES with two or four sulfonic acid groups in one side chain. (Reprinted from Kim, D.S. et al., Macromolecules. 41, 2126, 2008.)

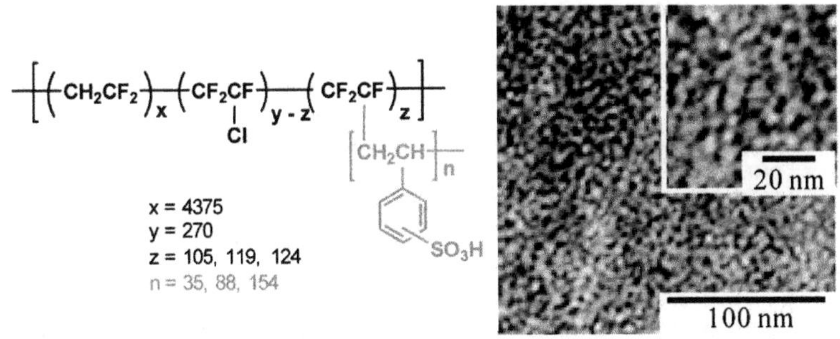

FIGURE 16.5 Molecular structure and TEM image of the comb-shaped PEM. (Reprinted from Tsang, E.M.W. et al., Macromolecules. 42, 9467–9480, 2009.)

segments and the hydrophobic segments are completely separated due to their chemical incompatibility; this can lead to higher phase-separation degree through self-assembly.

The synthesis of block copolymers for PEM originates from polystyrene and its derivatives due to the well-established synthesis procedure of controlled radical polymerization. The resulting copolymers with low poly-dispersity indices (PDI) are in favor of self-assembly forming various morphologies, such as disordered, lamellar, hexagonal packed cylinders, hexagonally perforated lammellar and gyroid phases (Park and Balsara, 2008). However, the rigid polystyrene backbone leads to suppressed mechanical strength, and the IEC is difficult to reach more than 1 meq/g in order to maintain highly ordered morphology (Storey et al., 1997). To solve this problem, the aromatic block copolymers have been synthesized by step-growth polymerization. Despite a higher PDI that is above 2.0, phase separated morphologies were observed by controlling the chemical composition and the reaction conditions (Figures 16.6 and 16.7) (Bae et al., 2010). These membranes generally show better chemical stability than aliphatic blocked copolymers when the formation of ordered morphology is hampered. For example, Miyatake et al. (2009) investigated the effects of random and block copolymers on the membrane morphology and properties. The scanning force microscopy (SFM) images indicate that block copolymers exhibit more obvious phase separated morphology with better long-range connected PCCs, whereas random copolymers have smaller and quite isolated phase domains.

The annealing process is necessary for forming highly ordered morphology of conventional block copolymers. However, annealing sulfonated block copolymers is difficult due to their high glass transition temperature (T_g), which is close to the degradation temperature of the sulfonic acid group (Hegedus,1990; Markó, 1991). When annealing in solution and vapor, the aggregation of ion-exchange groups makes it difficult to process due to the limited solubility of sulfonated block copolymers (Gromadzki et al., 2006). Therefore, the thermal or vapor annealing strategies have not been widely employed for PEMs.

By introducing block copolymers into PEMs, the proton conductivity increases obviously due to the better-developed phase separated morphology. Nakabayashi et al. (2008) synthesized SPAEs block copolymer and investigated the effect of the

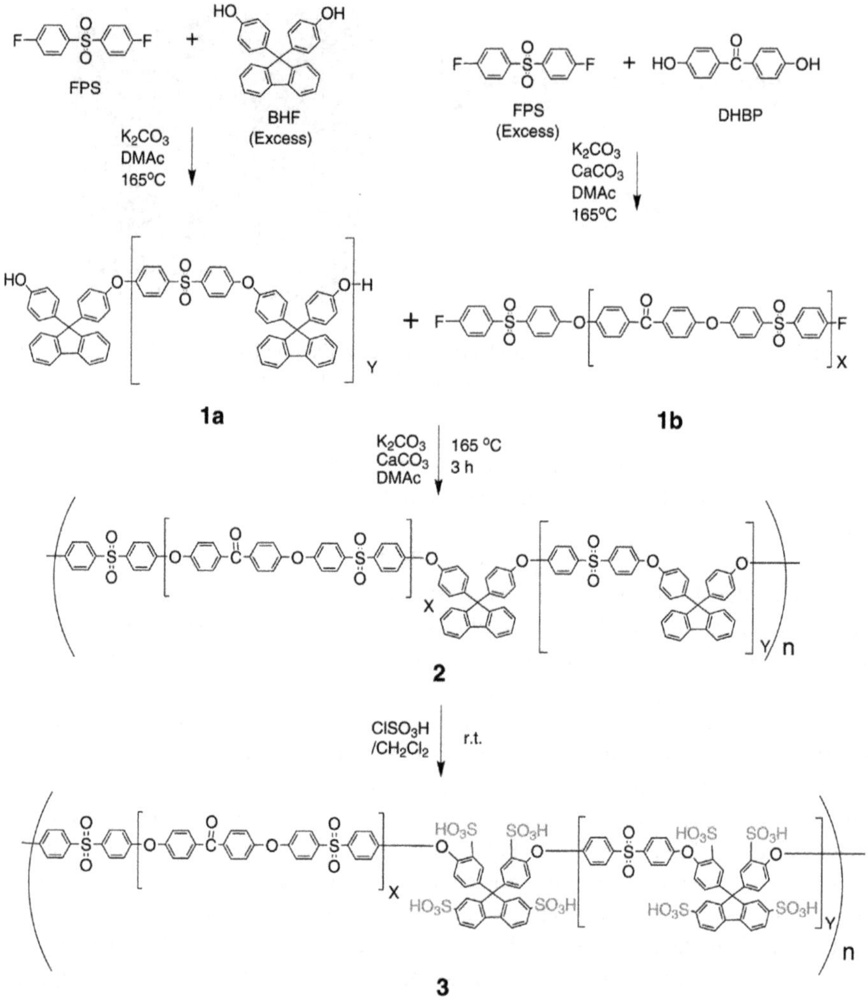

FIGURE 16.6 Synthesis of sulfonated block copolymers with various chemical compositions. (Reprinted from Bae, B. et al., Macromolecules. 43, 2684–2691, 2010.)

block length and the chemical composition. They found that proton conductivity increased as a function of block lengths due to increased phase separation degree. By optimizing the length of hydrophilic and hydrophobic blocks, the proton conductivity was higher than that of Nafion® 117 (Nakabayashi et al., 2008). The more complex and ordered morphology, such as hexagonally arranged cylinders, hexagonally perforated lamellae and lamellae, can be achieved by altering the length of hydrophobic blocks in a rod-coil triblock copolymer. A fully aromatic triblock copolymer has been designed by Qu et al. (2012). This sulfonated triblock copolymer exhibit scomparable conductivity to Nafion® 112 even at a low IEC of 0.97 mmol/g due to the unique phase separated morphology, i.e., PCCs.

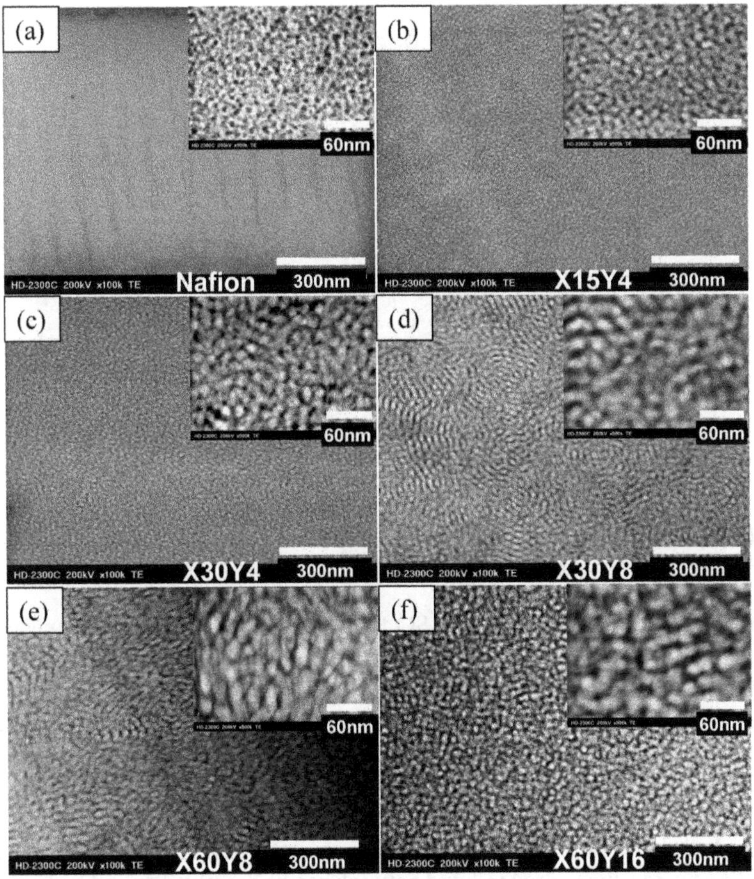

FIGURE 16.7 Scanning transmission electron microscope (STEM) images of lead ion-exchanged (a) Nafion® NRE 212, (b) X15Y4; IEC = 2.04 mequiv/g, (c) X30Y4; IEC = 1.34 mequiv/g, (d) X30Y8; IEC = 1.86 mequiv/g, (e) X60Y8; IEC = 1.07 mequiv/g, and (f) X60Y16; IEC = 1.74 mequiv/g membranes, where X and Y are referred to the degree of polymerization of hydrophilic and hydrophobic blocks, respectively. (Reprinted from Bae, B. et al., Macromolecules. 43, 2684–2691, 2010.)

16.2.3 Cross-Linked PEMs

Cross-linking is an effective method to decrease water swelling and improve mechanical strength of IEMs. It can be readily processed by thermal treatment and/or adding cross-linking agents. For PEMs, cross-linking is beneficial for decreasing methanol permeability, i.e., increasing the selectivity, and improving water-retaining capacity. Typically, three types of interactions, hydrogen bonding, ionic bonding and covalent bonding, are involved in PEM cross-linking.

16.2.3.1 Hydrogen Bond Cross-Linking

Hydrogen bonds can be formed between a hydrogen atom and a highly electronegative atom such as nitrogen (N), oxygen (O), or fluorine (F). To construct a hydrogen-bond

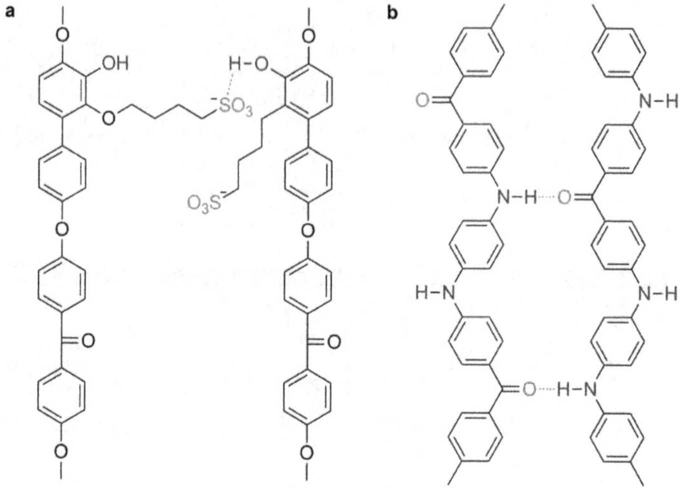

FIGURE 16.8 Hydrogen-bonded interactions between –OH, $-SO_3H$ (a) and –NH–, –CO– (b) in PEMs.

network in PEMs, –OH, –COOH, $-SO_3H$, –CO– or $-NH_2$ groups are necessary. Bonding these groups with polymer material is the main challenge when preparing cross-linked PEMs. Among these functional groups, the –COOH and $-SO_3H$ groups can be distributed on the side chain to form a side-chain-like network or at the end of the polymer chain to form an end-capped network (Zhu et al., 2010). The –OH group can be obtained by the demethylation of the methoxy group (Figure 16.8a) (Li et al., 2015). And the –NH- and –CO– groups could be introduced by using various functional monomers during the polymerizing reaction (Figure 16.8b) (Chang et al., 2015). Afterwards, the hydrogen-bond network can be readily obtained by heat treatment during the membrane preparation process.

Although crosslinking with hydrogen bonds can construct a flexible polymer network to enhance the dimensional stability as well as the selectivity of PEMs, the weak hydrogen bonds can be easily broken above 80°C during long-term operation. Therefore, more stable cross-linking strategies are necessary.

16.2.3.2 Ionic Bond Cross-Linking

Ionic bond cross-linking, i.e., acid-base cross-linking, is usually processed by forming acid-base pairs between polymer chains. The acid part is segment with –COOH or $-SO_3H$ groups, and the base part is segment with imine, imidazole, pyridine or another nitrogen-containing heterocycle group. For example, 5,5′-bis[2,4-(hydroxyphenyl) benzimidazole] (BHPB) as an intermolecular cross-linking moiety was polymerized with other monomers to synthesized a poly(arylene ether sulfone) copolymer (Ko et al., 2012). The ionic bonds between sulfonic acid and amine groups can improve the mechanical and thermal stability, reduce water swelling and decrease methanol permeability. Additionally, the ionic cross-linking network can be constructed by blending two polymers. Compared to direct polymerization with acid and base moieties, blending two types of polymers is easily executed and

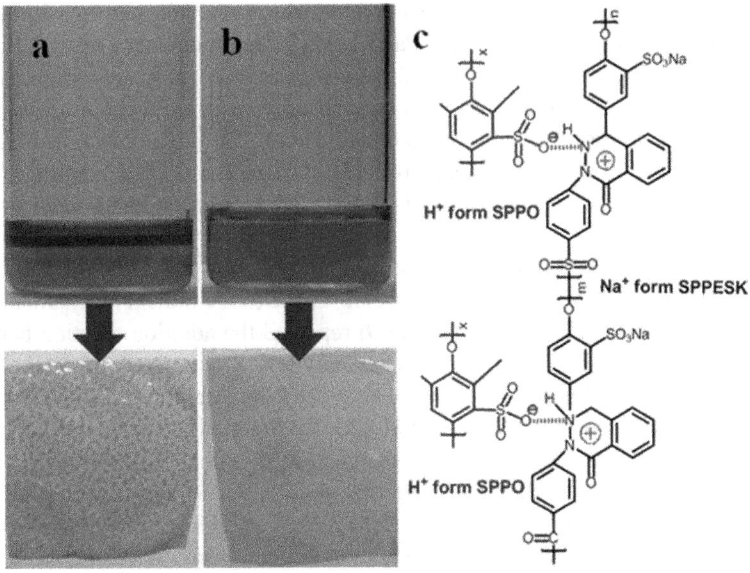

FIGURE 16.9 Physical observations of the casting solutions and their corresponding membranes, which are prepared from (a) Na^+-form SPPO and the Na^+-form SPPESK and (b) 100% protonated SPPO and the Na^+-form SPPESK, respectively; (c) Ionically acid-base interactions between the H^+-form SPPO and the Na^+-form SPPESK. (Adapted from Wu, L. et al., Fuel Cells. 15, 189–195, 2015.)

much more cost effective. However, the poor compatibility between blended polymers is observed. To solve this problem, Wu et al. (2015) fabricated cross-linked proton exchange membranes by blending sulfonated poly(phthalazinone ether sulfone ketone) (SPPESK) with sulfonated poly(2,6-dimethyl-1,4-phenylene oxide) (Figure 16.9). They found that increasing the protonation degree of SPPO can make the membrane translucent and flexible due to the hydrogen bond formation between hydrogen ion in SPPO and N atom in SPPESK (Wu et al., 2015).

In comparison with hydrogen bond cross-linking, an acid-base ionic bond is much more stable. It was reported that PEM cross-linked via ionic bonds can exhibit hydrolytic stability in boiling water for 1000 h (Zhang et al., 2008). However, the decrease of proton exchange capacity was generally observed due to the consumption of functional groups after ionic bond cross-linking. In 2010, Holdcroft et al. found that when blending sulfonated poly-benzimidazole (SuPBI) with high IEC SPEEK, the PEM exhibited significantly enhanced dimensional stability. However, the IEC of the blend membrane was close to zero due to the consumption of sulfonic acid groups by intermolecular ionic bond cross-linking, leading to decreased proton conductivity in comparison with the pure SPEEK membrane (Thomas et al., 2010).

16.2.3.3 Covalent Bond Cross-Linking

Covalent bond cross-linking is the most widely used cross-linking method due to its excellent stability. The general process for covalent cross-linking is reserving potential cross-linking sites on polymer chains, which can then self-crosslinking or react with other cross-linking agents. Particularly, using monomers with multifunctional groups will construct cross-linked networks via direct polymerization, such as the divinylbenzene (DVB) for radical polymerization and the 3,3-diaminobenzidine (DAB) for condensation polymerization (Bhadra et al., 2010). The common reactions for PEM covalent cross-linking can be roughly classified as six types: addition reaction, Friedel-Crafts reaction, amidation, esterification reaction, ring-opening reaction and click reaction.

The addition reaction usually occurs between two C=C bonds on the side chain of polymers. It requires temperature above 80°C due to the radical mechanism (Yao et al., 2015). For example, Lee et al. (2009) reported the addition reaction between ethyl groups at 250°C to form an end-capped network. After cross-linking, the water swelling and methanol permeability were both decreased obviously while the oxidative resistance was enhanced. Zhang et al. (2010b) reported the Friedel-Crafts reaction between carboxylic acid group and the nucleophilic phenyl rings, which can be processed at 160°C. The sulfonic acid groups in the PEM act as a benign solid catalyst, i.e., acid sources for the Friedel-Craft reaction. Copolymerizing monomers with –COOH, $–NH_2$ and $–OCH_3$ groups can lead to amide or ester bonds between polymer chains. In 2015, Han et al. (2015) reported cross-linked sulfonated poly(arylene ether ketone sulfone) polymers with the amide bonds. After cross-linking, the water retention capacity increased without apparent membrane swelling. TEM shows the formation of a continuous proton transport channel that is attributable to the high cross-linking degree. The ring-opening reaction of epoxy was usually used to cross-link PBI membrane for high temperature PEMFCs (Wang et al., 2012). In particular, the click reaction has attracted more and more attention due to its high reactivity and nearly 100% reaction yield. Some attempts to introduce click reaction into the preparation PEMs are as follows. In 2013, Na et al. synthesized a water-soluble sulfonated poly (ether ether ketone) containing dipropenyl groups. Cross-linked membranes could be obtained employing 1,2-ethanedithiol and 1,6-hexanedithiol as the cross-linker (Jiang et al., 2013). In 2015, Ko et al. synthesized cross-linked PEM by simultaneously casting the polymer solutions of azido functionalized sulfonated poly(arylene ether sulfone), diethynylbenzene crosslinker and the CuBr catalyst (Ko et al., 2015). Because the cross-linker was incorporated by two sulfonic acid groups, the IEC and proton conductivity increased with increasing cross-linking degree, solving the trade-off effect between the mechanical and the electrochemical properties of PEMs.

The choice and design of cross-linking agents are critical for the properties of PEMs in many cases (see Figure 16.10). Besides the common cross-linking agents like divinylbenzene (DVB) and 3,3-diaminobenzidine (DAB), there are two special types of cross-linker, namely the silane coupling agent and the macromolecular cross-linking agent. By silane-crosslinking, a hybrid PEM can be prepared. It has been proved that the Si–O–Si cross-linking bonds can efficiently enhance the stability of the PEM. Meanwhile, the methanol permeability is nearly three times lower than that of Nafion® (Lin et al., 2011b). On the other hand, polymer blending is an effective approach to improve the properties of PEMs because it can combine advantages of different polymer materials. However, the compatibility of two

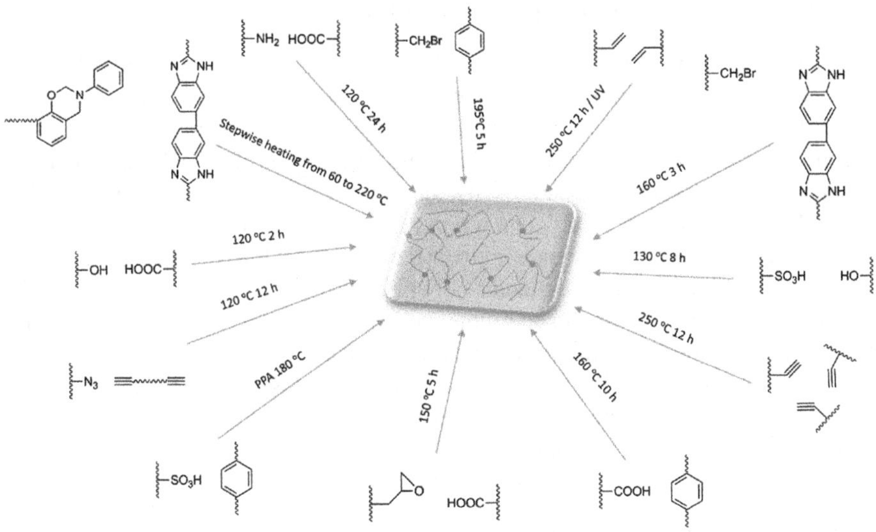

FIGURE 16.10 Reactions for covalently cross-linked PEMs.

components is always a problem for practical application. Thus, the macromolecular cross-linking strategy was developed. For example, Zhang et al. (2014) reported the crosslinking of brominated poly(ether ether ketone) with PBI as a macromolecular cross-linker. The highest proton conductivity was 0.081 S cm^{-1} at 200°C when the content of PBI was 20%. In addition, the stability and tensile strength were improved when the content of PBI increased.

16.3 ANION EXCHANGE MEMBRANE FOR FUEL CELLS (AEMFCs)

Traditional ion exchange membrane fuel cell (IEMFC) adopted a PEM, typically Nafion® (Zakil et al., 2016) as the gas separator and a proton conductor, namely proton exchange membrane fuel cell (PEMFC). However, the dependence on Pt catalyst has severely hindered its further development and applications. On the other hand, the emerging anion exchange membrane fuel cells (AEMFCs), which adopt AEM instead, enable the use of a non-precious metal catalyst (Ni, Ag, Fe, etc.) due to the faster oxygen reduction kinetics under alkaline conditions (Lu et al., 2008). In addition, AEMFCs bear other distinct advantages like more facile water management (Zhang and Shen, 2012b), wide choice of fuels (Zakaria et al., 2016), lowered chance of catalyst poisoning (Varcoe and Slade, 2005), etc. But there are still remaining problems to be solved in the field of AEMFCs. The power density of AEMFCs is generally lower compared to PEMFCs due to the lack of a high performance AEM. In addition, the life time of AEMs in fuel cell is still unsatisfying, especially when compared with the state of art Nafion® membrane for PEMFCs. Thus, the development of AEM with high conductivity as well as good stability has received increasing attention, and strategies to enhance the conductivity as well as the stability will be introduced in this section.

16.3.1 Strategies to Improve the Conductivity of AEM

Conventional AEMs are prepared by the quaternization of randomly bromo-methylated (or chloro-methylated) polymer precursors. However, the resulting "main-chain" type AEMs generally exhibit low hydroxide ion conductivity and poor chemical stability, and thus can hardly fulfil the requirement of AEMFCs. In order to enhance their hydroxide ion conductivity, various strategies have been developed by researchers. One of the most efficient ways is to construct high-efficient ion conducting channels within the membrane matrix (Pan et al., 2014). By tuning the location of ion conducting groups, a particular polymer structure composed of hydrophilic segments as well as a hydrophobic segment could be constructed. During the membrane formation process, the phase separation arising from the polar discrimination between the hydrophilic segments and the hydrophobic segment results in an inter-connected hydrophilic phase. Due to the containment of high density of ion conducting groups, this hydrophilic phase is considered to be the main pathway for hydroxide ion conduction, namely the ion conducting channel.

16.3.1.1 Side-Chain Type AEMs

Aiming to maximum the nano-phase separation ability of anion exchange membranes, researchers firstly designed a polymer architecture that was composed of a hydrophobic main chain and hydrophilic side chains, as shown in Figure 16.11 (Yang et al., 2015). The ion conducting groups in the side chain bear high mobility because

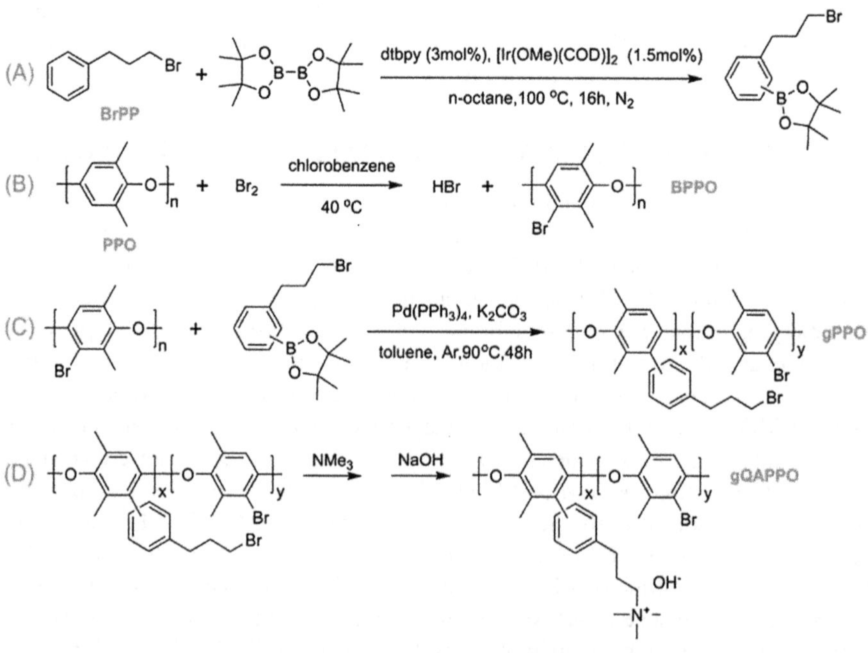

FIGURE 16.11 The synthetic procedure of a typical side-chain type AEM. (Reproduced from Yang, Z. et al., Journal of Materials Chemistry A. 3, 15015–15019, 2015.)

of the flexibility of the side chain and thus can aggregate to form the interconnected ion conducting channels. The side-chain type AEMs generally exhibit hydroxide ion conductivity of 30–45 mS/cm and their nano-phase separated morphology could be clearly observed by atomic force microscopy (AFM), TEM or small angle X-ray scattering (SAXS) (Gao et al., 2016; Guo et al., 2016; Jannasch and Weiber, 2016; Lin et al., 2016; Liu et al., 2016; Qi et al., 2015; Yan et al., 2016; Yang et al., 2015; Zhang et al., 2016; Zheng et al., 2017; Zuo et al., 2016). Although enhanced hydroxide ion conductivity was observed compared to the traditional main-chain type AEM, the performance of the side-chain type AEMS can hardly match that of Nafion® (a commercial CEM with proton conductivity ~80 mS/cm). This could be mainly ascribed to the following reasons: (1) The inferior migration rate of OH^- compared with H^+ (Varcoe et al., 2014), (2) the lower disassociation ability of OH^- from quaternary ammonium groups and (3) the need to further enhance the nano-phase separation ability of AEM.

16.3.1.2 Densely Functionalized AEMs

Afterwards, researchers developed another strategy to obtain AEMs with improved nano-phase separation ability. As generally known, the nature of ion conducting channels is the aggregation of ion clusters. The side-chain type AEMs exhibit good nano-phase separation ability because of the increased mobility of their ion conducting groups. On the other hand, decreasing the distance between the ion conducting groups can facilitate their aggregation and thus improve the nano-phase separation ability. To verify this hypothesis, ion conducting groups were selectively allocated on particular segments of the polymer main chain of AEMs, namely the "main chain densely functionalized" AEMs (Takaba et al., 2017). Tanaka et al. (2011) successfully synthesized AEMs with two ion conducting groups on a same fluorene unit. But the resulting AEM did not exhibit the expected high conductivity (20 mS/cm at 25°C) and the morphology of the AEM was not investigated. Later, Patric Jannasch synthesized densely functionalized AEM by bromination and quaternization of the duroquinol units in the polymer backbone (Weiber and Jannasch, 2014). Four closely located ion conducting groups were proved by nuclear magnetic resonance (NMR) spectrum. Although distinct micro-phase separation was observed by small angle X-ray scattering, the researchers did not report the hydroxide ion conductivity. In 2016, Lin et al. (2016) reported another AEM with four ion conducting groups on a same structural unit. The resulting AEMs exhibited clear phase separated morphology as well as high hydroxide conductivity of 36 mS/cm (20°C).

Based on those prior works, one can conclude that the "main chain densely functionalization" strategy can effectively improve the nano-phase separation ability and hydroxide ion conductivity of anion exchange membrane. But due to the restricted mobility of ion conducting groups by the rigid polymer main chain, the performance of densely functionalized AEMs can hardly exceed the side-chain type AEMs. Additionally, because of the electro-withdrawing effect of positively charged ion conducting groups, the alkaline stability of polymer main chains is still worth studying (Arges and Ramani, 2013).

By analyzing the structural features of the side-chain type AEMs and the densely functionalized AEMs, it can be inferred that the mobility and the density of ion

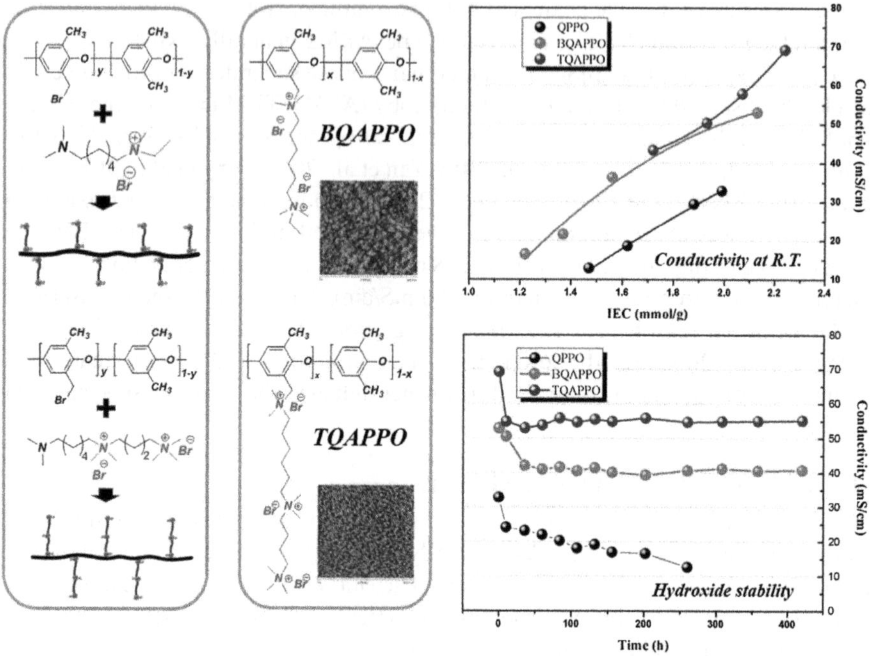

FIGURE 16.12 Chemical structure, hydroxide conductivity and alkaline stability of the "side chain densely functionalized" AEM. (Reproduced from He, Y. et al., Scientific Reports. 5, 13417, 2015.)

conducting groups in the ion conducting channels are two key factors that decide the nano-phase separation ability and the hydroxide ion conductivity. In a side-chain type AEM, the flexible spacer between the ion conducting group and the polymer main chain have rendered high mobility to the ion conducting group. On the other hand, the decreased distance between ion conducting groups is the main reason for the enhanced performance of densely functionalized AEMs. To combine the advantages of these two strategies, we have designed "side chain densely functionalized" AEMs with two or more ion conducting groups in one same side chain (He et al., 2015). As depicted in Figure 16.12, the flexibility of the side chains ensured the high mobility of ion conducting groups. The increased cation number in the side chains can enhance the hydrophilicity of side chains and facilitate the aggregation of ionic segments. As a result, this type of AEM with densely functionalized side chains shows improved nano-phase separation ability and high hydroxide ion conductivity of 69 mS/cm at room temperature.

16.3.1.3 Functional Side Chain AEMs

Different from the side-chain type AEMs and the densely functionalized AEMs, another strategy to enhance the nano-phase separation ability is to introduce a functional side chain onto the backbones of AEMs. The functional side chain can either facilitate the nano-phase separation or assist the conduction of hydroxide ions. In 2013, Li et al. (2013) reported AEMs with a comb-shaped structure. By directly reacting brominated poly phenylene oxide with hexadecyldimethylamine, highly conductive AEM (30 mS/cm, 25°C) with obvious nano-phase separated morphology could be easily prepared. Its high performance depends on the construction of unique polymer architecture with a hydrophilic main chain and the hydrophobic side chains. The main drawback of this design is the hindered interaction between the hydroxide ions and the ion conducting groups because of the long aliphatic tails. Therefore, Pan et al. (2014) reported another comb-shaped AEM with separated ion conducting groups and aliphatic side chains (Figure 16.13). After optimizing the length of aliphatic side chains, a maximum hydroxide conductivity of 40 mS/cm was achieved. More recently, with the goal of further enhancing the hydrophobicity of side chains, researchers have developed AEMs with ionic main chain and fluorinated side chains (Zeng and Zhao, 2016). But unfortunately, despite observed excellent nano-phase separation, the hydroxide conductivity was not further improved.

Although this "functional side chain" strategy did not lead to AEM with obviously enhanced performance compared with the previous two strategies, the synthetic procedure of this type of AEM is quite simple and convenient. By optimizing the structure and length of functional side chains, the performance of resulting AEMs may be further enhanced.

16.3.1.4 Hyper-Branched AEMs

In addition to the numerous works based on the modification of main chain and side chain structure, we have developed AEMs based on polymers with novel topology (Figure 16.14). It is widely accepted that the enhancement of the AEM performance depends on the construction of high-efficient ion conducting channels. In addition to the insufficient mobility and density of ion conducting groups, the entanglement of polymer chain will hinder the aggregation of ionic segments, and thus lower the nano-phase separation ability of AEMs. Therefore, AEMs with new polymer topology, namely the hyper-branched anion exchange membrane, were designed. The structural feature of this type of AEM is that the ion clusters mainly allocate on the edge of branched macromolecule, and the weakened chain entanglement effect can promote the aggregation of the ion conducting groups. Experimentally, a hyper-branched oligomer was synthesized by the ATRP polymerization of 1-(chloromethyl)-4-vinylbenzene. Afterwards, the hyper-branched oligomer will be crosslinked and quaternized by a multi-tertiary amine containing molecule. The resulting membrane exhibits high conductivity of 40 mS/cm at room temperature due to its unique hyper-branched polymer architecture (Ge et al., 2016).

16.3.2 Strategies to Enhance the Chemical Stability of AEMs

16.3.2.1 Polymer Architecture Optimization

Insufficient chemical stability is another obstacle blocking the development and applications of AEMs. In the field of PEMs, perfluorosulfonic acid (PFSA) membranes represented by Nafion® demonstrates excellent long-term stability against radical attack and outstanding acid tolerance. On the other hand, compared to PEMs, chemical stability of AEMs is much less reliable when exposed to alkaline media at

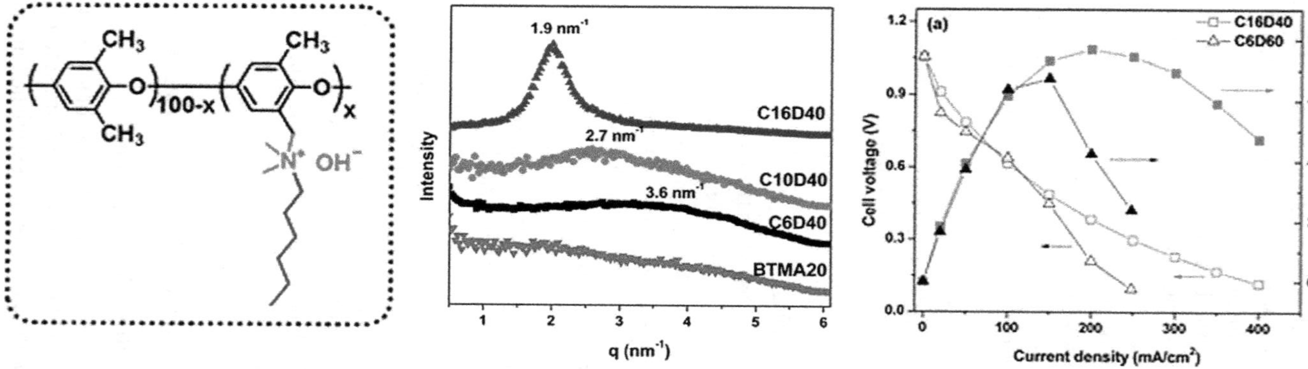

FIGURE 16.13 Chemical structure, small angle X-ray scattering and fuel cell performance of a typical comb-shaped AEM. (Reproduced from Li, N. et al., Journal of the American Chemical Society. 135, 10124–10133, 2013.)

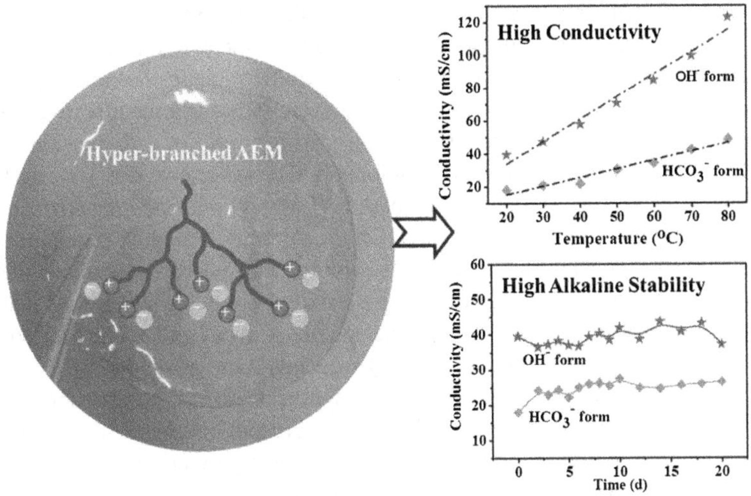

FIGURE 16.14 Hydroxide conductivity and alkaline stability of the hyper-branched AEM. (Reproduced from Ge, Q. et al., Chemical Communications. 52, 10141–10143, 2016.)

elevated temperatures, which should be one of the first considerations when developing new high performance AEMs. The well-known degeneration pathway of quaternary ammonium in basic environments includes β-hydrogen Hofmann elimination, direct nucleophilic substitution at α-carbon or nitrogen ylide formation. Edson et al. (2012) investigated the thermal decomposition pathway of various alkyltrimethyl ammonium cations in hydroxide form. Increased importance of nucleophilic attack on N-connected methyl groups was observed when β hydrogen was replaced by methyl groups. The stability of benzyltrimethyl ammonium and phenyltrimethyl ammonium was examined as they are representative cations for AEMs materials. Phenyltrimethyl ammonium displays much inferior alkaline stability because of the strong electro-withdrawing effect of phenyl rings, which accelerates the nucleophilic attack of hydroxide on α-carbon. Likewise, inadequate base resistance was observed for phenyltrimethyl ammonium based AEMs (Einsla et al., 2007).

On the other hand, degradation of the polymer main chain is another factor that restricts the chemical stability of AEMs. Conventional AEMs are prepared either by chloromethylation or bromomethylation followed by quaternization, which leads to anion exchange membranes with quaternary ammonium groups closely attached to the polymer backbones. Positively charged quaternary ammonium groups can inevitably damage the chemical inertness of polymer backbones resulting in accelerated degeneration. For example, pristine poly (phenyl oxide) (PPO) exhibits excellent alkaline stability in hash test conditions, however, degeneration of polymer backbones was observed after quaternization as characterized by two-dimensional NMR spectra (Arges et al., 2013).

In order to reduce the damage of anion exchanging moieties to polymer backbones, a lower grafting ratio of quaternary ammonium groups is desired while maintaining enough IEC values. AEMs containing two cations in one side chain, as synthesized by Pan et al. (2013), demonstrated enhanced alkaline stability over mono-QA functionalized anion exchange membranes because of their reduced grafting ratio. Similarly, PPO functionalized by tri-cation containing moieties reported by Li et al. (2014) exhibits superior chemically stability as well, indicating the effectively alleviated damage to polymer backbones by lower grafting ratio of cationic side groups. The influence of various backbones on cationic functional groups was investigated by Nuñez and Hickner (2013). Polystyrene based AEMs exhibit best long term stability among the selected polymeric materials. On the contrary, anion exchange polyelectrolytes based on polysulfone degenerates rapidly, induced by the reduced electron density of benzyl carbon because of strong electro-withdrawing sulfone moieties.

Long side-chain type AEMs were designed to avoid the reciprocal induced degeneration of polymer backbones and QA groups. Our group reported aromatic polyelectrolytes with long pendent spacer synthesized via polycondensation of functionalized monomer (Zhang et al., 2013); outstanding chemical stability was observed because of the weakened interaction between polymer main chain and QA groups. This is attributable to the superior chemically stability of long side-chain type imidazolium functionalized AEMs synthesized by Lin et al. (2011a). Despite excellent alkaline stability, the long side-chain type still suffers from degeneration through Hoffman elimination of β hydrogen. The alkaline stability of various cations was investigated by Price et al. (2014). Alkyltrimethylammonium cations with two methyl groups replacing the β hydrogen exhibit significantly improved chemically stability than conventional benzyl connected quaternary ammonium cations, make them a promising candidate for the AEM design community.

Introducing steric crowding groups around the cationic quaternary ammonium is considered effective to reduce the chance of radical attack by hydroxide. Comb-shaped AEMs with long aliphatic chains attached to a nitrogen atom were synthesized by Li et al. (2013); although β-hydrogen Hoffman elimination may be a possible degeneration pathway, the steric hindrance effect of long aliphatic chain is recognized to be essential for the dramatically improved chemical stability of the comb shaped AEMs.

Enhanced chemical stability through steric hindrance protection was verified by Thomas et al. (2012) (Figure 16.15). Chemically robust poly-benzimidazolium hydroxide was synthesized by installation of adjacent bulky groups to C2 positions. Computer simulation was employed to further demonstrate the effective shielding of C2 position from nucleophilic attack from OH^-.

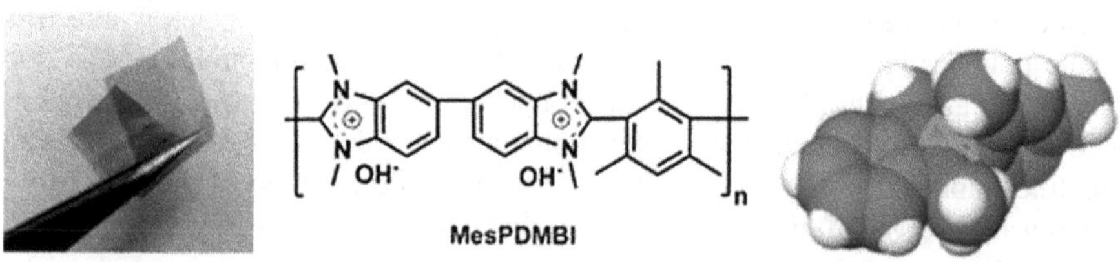

FIGURE 16.15 Chemical stable poly-benzimidazolium via steric hindrance protection on the C2 position. (Reproduced from Thomas, O.D. et al., Journal of the American Chemical Society. 134, 10753–10756, 2012.)

16.3.2.2 Designing Ion Conducting Groups

Conventional quaternary ammonium group has long been criticized for its insufficient chemically stability in basic media. To date, various hydroxide conductive groups have been designed and applied in the preparation of AEM including ammonium, guanidinium (Lin et al., 2012; Liu et al., 2014; Qu et al., 2012; Wang et al., 2010; Zhang et al., 2010a), imidazolium (Deavin et al., 2012; Gu et al., 2014; Lin et al., 2013a), sulfonium (Wang et al., 2012), and phosphonium (Gu et al., 2009; Gu et al., 2011).

Guanidinium based AEMs (Figure 16.16), first synthesized by Wang et al. (2010), exhibits significantly higher hydroxide conductivity than quaternary ammonium based ones (67 mS/cm at 20°C). However, the stability of guanidinium in alkaline media is quite a problem for applications in AEMFCs. Despite continuous modification by numerous researchers (Jiang et al., 2012), guanidinium based AEMs can hardly survive more than 48 h in 1 M KOH at 60°C.

Other cationic heterocyclic compounds such as pyridinium (Kim et al., 2017) and benzimidazolium (Lin et al., 2013b) were investigated as functional groups of AEMs, but poor alkaline stability was observed. C2 unsubstituted imidazolium was reported to be unstable under basic conditions (Deavin et al., 2012), and the prospective degradation pathway was revealed by Holloczki et al. (2011). However, significantly enhanced stability was achieved by simply introducing a methyl group to the C2 position. The relationship between C2 substitute and alkaline stability was investigated by Price et al. (2014); they observed that imidazolium with more bulky C2 substitute like tert-butyl group is quite unstable in tested conditions, which is obviously different from the stereotypical interpretation. It was proposed that the steric hindrance effect is a less important stabilizing factor than the ability to provide alternative deprotonation reactions with hydroxide (Table 16.1). The effect of N-substitute on the chemical stability of imidazolium and prospective degeneration mechanism was revealed by ^{1}H-NMR spectra and density functional theory (GGA-BLYP) calculations (Gu et al., 2014). N3-isopropyl substitute imidazolium exhibits the highest alkaline stability, which is in line with the results of theoretical calculations. AEMs based on N3-propyl substitute were fabricated and exhibited superior alkaline stability than conversional QA type anion exchange membranes (Si et al., 2014).

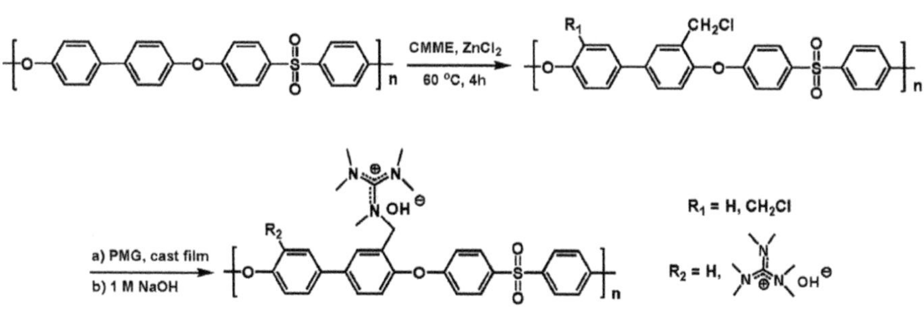

FIGURE 16.16 Synthetic procedure for the guanidinium based anion exchange membrane. (Reproduced from Wang, J. et al., Macromolecules. 43, 3890–3896, 2010.)

TABLE 16.1
Chemical Structure and Alkaline Stability of Selected Imidazolium Cations

Cation	Test Condition	Stability Test Method	Stability
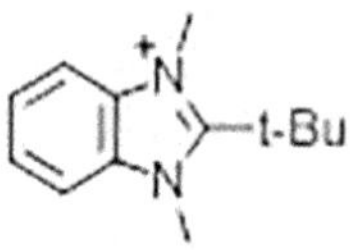	0.5 M NaOH, RT	NMR	$t_{1/2}$ = too fast to measure
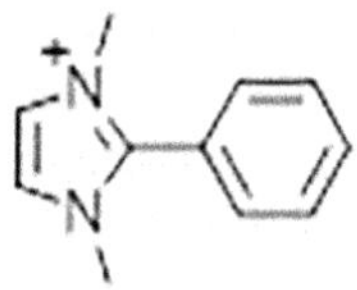	1 M NaOH, 80°C	NMR	$t_{1/2}$ = 6400 min
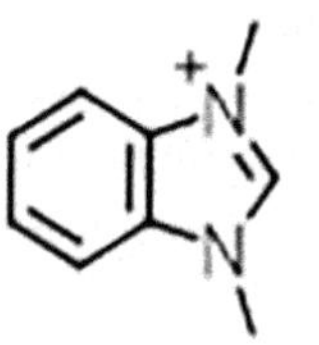	1 M NaOH, 80°C	NMR	$t_{1/2}$ = 46 min
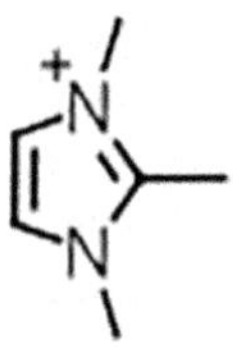	1 M NaOH, 88°C	NMR	$t_{1/2} \geq$ 1000 h
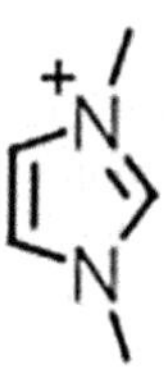	1 M NaOH, 80°C	NMR	$t_{1/2}$ = 800 min
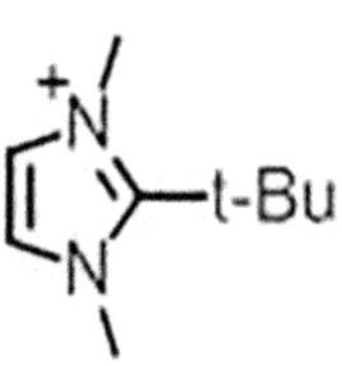	1 M NaOH, 80°C	NMR	$t_{1/2}$ = 170 min

Source: Reproduced from Price, S.C. et al., ACS Macro Letters. 3, 160–165, 2014.

Phosphonium is another promising candidate to be employed as anion exchanging groups for AEMs. However, AEMs functionalized by trialkyl benzyl phosphonium suffer from severe nucleophilic substitution when exposed to an alkaline environment (Ye et al., 2013). The first base-stable phosphonium based AEMs was synthesized by quaternary phosphorization of chloromethylated polysulfone with tris(2,4,6-trimethoxyphenyl)phosphine (Gu et al., 2009). Strong electro-donating methoxy groups and conjugation of phenyl rings is recognized to be critical for the stabilization of positively charged phosphonium groups: hydroxide attack was minimized because of the steric hindering trimethoxyphenyl groups. Such bulky trimethoxyphenyl groups were utilized as C2 substitute of imidazolium, and improved chemical stability was observed compared to C2-unsubstituted imidazolium (Wang et al., 2013). Another category of tetrakis(dialkylamino) phosphonium based AEMs were synthesized by Noonan et al. (2012) (Figure 16.17). It is believed that highly resonance system and delocalization of charge density is responsible for the extraordinary base resistance of the tetrakis(dialkylamino) phosphonium. It should be noted that the phosphonium cations prepared from readily available inexpensive materials exhibit adequate hydroxide conductivity at relatively low IEC values, which makes them promising candidates as functional moieties for AEMs.

Sulfonium is also evaluated as a prospective hydroxide conductive functional moiety for AEMs. However, tirphenyl and trialkyl-sulfonium was proved to hold poor base resistance which undergoes nucleophilic degeneration pathway. The first base-stable sulfonium functionalized AEMs was synthesized by Wang et al. (2012)

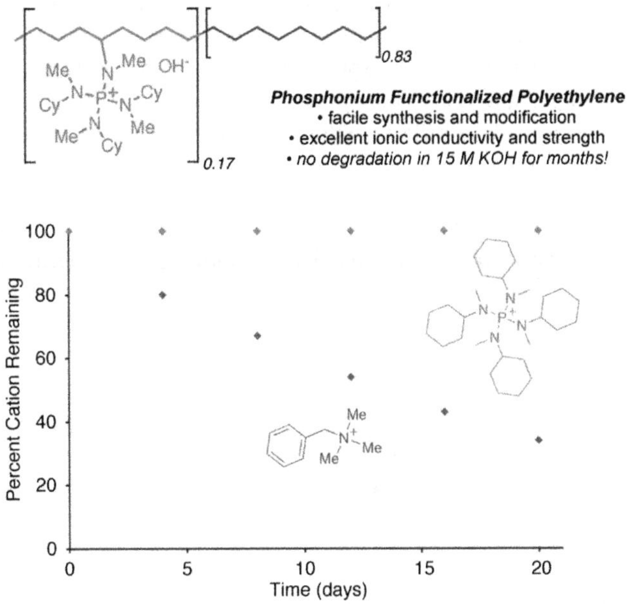

FIGURE 16.17 Membrane material based on tetrakis(dialkylamino) phosphonium cations and its alkaline stability. (Reproduced from Noonan, K.J.T. et al., Journal of the American Chemical Society. 134, 18161–18164, 2012.)

by attaching methoxyl group onto the phenyl ring of tirphenyl-sulfonium, the electro-donating effect is recognized to be crucial to stabilize the positively charged sulfur atom. Beside the above-mentioned organic cations, metal cations were explored as the ion conducting groups for AEMs. Zha et al. (2012) reported the first metal cation based AEM employing bis(terpyridine)ruthenium(II) complex as the ion conducting groups. The resulting membrane was alkaline stable when soaking in 1 M KOH at 80°C for around 48 hours. Afterwards, Gu et al. (2015) also synthesized Co^+ based AEMs with excellent alkaline stability. The hydroxide conductivity maintained constant after soaking in 1M KOH at 80°C solution for 2000 h. This inspired the exploration of AEMs based on alkaline-stable metal complex.

16.4 CONCLUSIVE SUMMARIES AND PERSPECTIVES

In summary, exploiting inexpensive materials to prepare high performance PEMs is the main challenge for PEMFC. Inspired by the structure of Nafion® membranes, constructing the PCCs in PEMs will efficiently improve the proton conductivity without decreasing the stability. The strategies for constructing PCCs can be divided into two types: densely functionalized side chains and block copolymers. Although both can significantly increase the properties of hydrocarbon-backbone PEMs, they still have a long way to go for practical application due to their complex synthesis. The novel microporous materials are gradually attracting more and more attention for application in many fields, such as catalyst, separation and energy storage. For PEMs, the potential to directly construct well-connected PCCs is an interesting topic that deserves further investigation.

On the other hand, the major problems of AEMs are their insufficient hydroxide conductivity and alkaline stability. In the past decades, some promising ion conducting groups like imidazolium and phosphonium show superior stability or conductivity than the traditional quaternary ammonium groups. And AEMs with new polymer architecture have been developed, including the side chain AEM, densely functionalized AEM and the hyper-branched AEM. However, the performance of AEMs is still below the standard of the Nafion® membrane. In the future, the conduction mechanism of OH^- and the degeneration pathway of AEMs need to be further studied to help design new AEM materials. The combination of novel polymer architecture and stable ion conducting groups also requires an extensive research effort.

ACKNOWLEDGMENTS

This project has been supported by the National Natural Science Foundation of China (Nos. 21720102003, 91534203), the K.C. Wong Education Foundation (2016-2011), and International Partnership Program of Chinese Academy of Sciences (No. 21134ky5b20170010).

REFERENCES

Arges, C.G., and Ramani, V., 2013. Two-dimensional NMR spectroscopy reveals cation-triggered backbone degradation in polysulfone-based anion exchange membranes. Proceedings of the National Academy of Sciences. 110, 2490–2495.

Arges, C.G., Wang, L., Parrondo, J., and Ramani, V., 2013. Best practices for investigating anion exchange membrane suitability for alkaline electrochemical devices: Case study using quaternary ammonium poly(2,6-dimethyl 1,4-phenylene)oxide anion exchange membranes. Journal of the Electrochemical Society. 160, F1258–F1274.

Bae, B., Miyatake, K., and Watanabe, M., 2010. Sulfonated poly(arylene ether sulfone ketone) multiblock copolymers with highly sulfonated block. Synthesis and Properties. Macromolecules. 43, 2684–2691.

Bhadra, S., Kim, N.H., and Lee, J.H., 2010. A new self-cross-linked, net-structured, proton conducting polymer membrane for high temperature proton exchange membrane fuel cells. Journal of Membrane Science. 349, 304–311.

Chang, G., Shang, Z., and Yang, L., 2015. Hydrogen bond cross-linked sulfonated poly(imino ether ether ketone) (PIEEK) for fuel cell membranes. Journal of Power Sources. 282, 401–408.

Chen, D., Wang, S., Xiao, M., Meng, Y., and Hay, A.S., 2011. Novel polyaromatic ionomers with large hydrophilic domain and long hydrophobic chain targeting at highly proton conductive and stable membranes. Journal of Materials Chemistry. 21, 12068–12077.

Deavin, O.I., Murphy, S., Ong, A.L., Poynton, S.D., Zeng, R., Herman, H., and Varcoe, J.R., 2012. Anion-exchange membranes for alkaline polymer electrolyte fuel cells: Comparison of pendent benzyltrimethylammonium- and benzylmethylimidazolium-head-groups. Energy & Environmental Science. 5, 8584–8597.

Debe, M.K., 2012. Electrocatalyst approaches and challenges for automotive fuel cells. Nature. 486, 43–51.

Ding, J., Chuy, C., and Holdcroft, S., 2002. Solid polymer electrolytes based on ionic graft polymers: Effect of graft chain length on nano-structured, ionic networks. Advanced Functional Materials. 12, 389–394.

Edson, J.B., Macomber, C.S., Pivovar, B.S., and Boncella, J.M., 2012. Hydroxide based decomposition pathways of alkyltrimethylammonium cations. Journal of Membrane Science. 399–400, 49–59.

Einsla, B.R., Chempath, S., Pratt, L., Boncella, J., Rau, J., Macomber, C., and Pivovar, B., 2007. Stability of cations for anion exchange membrane fuel cells. ECS Transactions. 11, 1173–1180.

Elabd, Y.A., and Hickner, M.A. 2011. Block copolymers for fuel cells. Macromolecules. 44, 1–11.

Fujimura, M., Hashimoto, T., and Kawai, H., 1982. Small-angle X-ray scattering study of perfluorinated ionomer membranes. 2. Models for ionic scattering maximum. Macromolecules. 15, 136–144.

Gao, L., He, G., Pan, Y., Zhao, B., Xu, X., Liu, Y., Deng, R., and Yan, X., 2016. Poly(2,6-dimethyl-1,4-phenylene oxide) containing imidazolium-terminated long side chains as hydroxide exchange membranes with improved conductivity. Journal of Membrane Science. 518, 159–167.

Ge, Q., Liu, Y., Yang, Z., Wu, B., Hu, M., Liu, X., Hou, J., and Xu, T., 2016. Hyper-branched anion exchange membranes with high conductivity and chemical stability. Chemical Communications. 52, 10141–10143.

Gierke, T.D., Munn, G.E., and Wilson, F.C., 1981. The morphology in Nafion* perfluorinated membrane products, as determined by wide- and small- angle x-ray studies. Journal of Polymer Science Part B: Polymer Physics. 19, 1687–1704.

Gromadzki, D., Černoch, P., Janata, M., Kůdela, V., Nallet, F., Diat, O., and Štěpánek, P., 2006. Morphological studies and ionic transport properties of partially sulfonated diblock copolymers. European Polymer Journal. 42, 2486–2496.

Gu, F., Dong, H., Li, Y., Si, Z., and Yan, F., 2014. Highly stable N3-substituted imidazolium-based alkaline anion exchange membranes: Experimental studies and theoretical calculations. Macromolecules. 47, 208–216.

Gu, S., Cai, R., Luo, T., Chen, Z., Sun, M., Liu, Y., He, G., and Yan, Y., 2009. A soluble and highly conductive ionomer for high-performance hydroxide exchange membrane fuel cells. Angewandte Chemie International Edition. 48, 6499–6502.

Gu, S., Cai, R., and Yan, Y., 2011. Self-crosslinking for dimensionally stable and solvent-resistant quaternary phosphonium based hydroxide exchange membranes. Chemical Communications. 47, 2856–2858.

Gu, S., Wang, J., Kaspar, R.B., Fang, Q., Zhang, B., Bryan Coughlin, E., and Yan, Y., 2015. Permethyl cobaltocenium (Cp*2Co+) as an ultra-stable cation for polymer hydroxide-exchange membranes. Scientific Reports. 5, 11668.

Guo, D., Lai, A.N., Lin, C.X., Zhang, Q.G., Zhu, A.M., and Liu, Q.L., 2016. Imidazolium-functionalized poly(arylene ether sulfone) anion-exchange membranes densely grafted with flexible side chains for fuel cells. ACS Applied Material Interfaces. 8, 25279–25288.

Han, H., Liu, M., Xu, L., Xu, J., Wang, S., Ni, H., and Wang, Z., 2015. Construction of proton transport channels on the same polymer chains by covalent crosslinking. Journal of Membrane Science. 496, 84–94.

He, Y., Pan, J., Wu, L., Zhu, Y., Ge, X., Ran, J., Yang, Z., and Xu, T., 2015. A novel methodology to synthesize highly conductive anion exchange membranes. Scientific Reports. 5, 13417.

Hegedus, L.S., 1990. Transition metals in organic synthesis annual survey covering the year 1989. Journal of Organometallic Chemistry. 392, 285–607.

Holloczki, O., Terleczky, P., Szieberth, D., Mourgas, G., Gudat, D., and Nyulaszi, L., 2011. Hydrolysis of imidazole-2-ylidenes. Journal of the American Chemical Society. 133, 780–789.

Hou, H., Di Vona, M.L., and Knauth, P., 2012. Building bridges: Crosslinking of sulfonated aromatic polymers—A review. Journal of Membrane Science. 423–424, 113–127.

Jannasch, P., and Weiber, E.A., 2016. Configuring anion-exchange membranes for high conductivity and alkaline stability by using cationic polymers with tailored side chains. Macromolecular Chemistry and Physics. 217, 1108–1118.

Jiang, H., Guo, X., Zhang, G., Ni, J., Zhao, C., Liu, Z., Zhang, L., Li, M., Xu, S., and Na, H., 2013. Cross-linked high conductive membranes based on water soluble ionomer for high performance proton exchange membrane fuel cells. Journal of Power Sources. 241, 529–535.

Jiang, L., Lin, X., Ran, J., Li, C., Wu, L., and Xu, T., 2012. Synthesis and Properties of Quaternary Phosphonium-based Anion Exchange Membrane for Fuel Cells. Chinese Journal of Chemistry. 30, 2241–2246.

Jutemar, E.P., and Jannasch, P., 2010. Locating sulfonic acid groups on various side chains to poly(arylene ether sulfone)s: Effects on the ionic clustering and properties of proton-exchange membranes. Journal of Membrane Science. 351, 87–95.

Kim, D.S., Robertson, G.P., and Guiver, M.D., 2008. Comb-shaped poly(arylene ether sulfone)s as proton exchange membranes. Macromolecules. 41, 2126.

Kim, S., Yang, S., and Kim, D., 2017. Poly(arylene ether ketone) with pendant pyridinium groups for alkaline fuel cell membranes. International Journal of Hydrogen Energy. 42, 12496–12506.

Ko, H.-N., Yu, D.M., Choi, J.-H., Kim, H.-J., and Hong, Y.T., 2012. Synthesis and characterization of intermolecular ionic cross-linked sulfonated poly(arylene ether sulfone)s for direct methanol fuel cells. Journal of Membrane Science. 390–391, 226–234.

Ko, T., Kim, K., Jung, B.-K., Cha, S.-H., Kim, S.-K., and Lee, J.-C., 2015. Cross-linked sulfonated poly(arylene ether sulfone) membranes formed by in situ casting and click reaction for applications in fuel cells. Macromolecules. 48, 1104–1114.

Lafitte, B., Puchner, M., and Jannasch, P., 2005. Proton conducting polysulfone ionomers carrying sulfoaryloxybenzoyl side chains. Macromolecular Rapid Communications. 26, 1464–1468.

Lee, K.-S., Jeong, M.-H., Lee, J.-P., and Lee, J.-S., 2009. End-group cross-linked poly(arylene ether) for proton exchange membranes. Macromolecules. 42, 584–590.

Lehmani, A., Durand-Vidal, S., and Turq, P., 1998. Surface morphology of Nafion 117 membrane by tapping mode atomic force microscope. Journal of Applied Polymer Science. 68, 503–508.

Li, G., Zhao, C., Li, X., Qi, D., Liu, C., Bu, F., and Na, H., 2015. Novel side-chain-type sulfonated diphenyl-based poly(arylene ether sulfone)s with a hydrogen-bonded network as proton exchange membranes. Polymer Chemistry. 6, 5911–5920.

Li, N., Leng, Y., Hickner, M.A., and Wang, C.Y., 2013. Highly stable, anion conductive, comb-shaped copolymers for alkaline fuel cells. Journal of the American Chemical Society. 135, 10124–10133.

Li, Q., Liu, L., Miao, Q., Jin, B., and Bai, R., 2014. A novel poly(2,6-dimethyl-1,4-phenylene oxide) with trifunctional ammonium moieties for alkaline anion exchange membranes. Chemical Communications. 50, 2791–2793.

Lin, B., Dong, H., Li, Y., Si, Z., Gu, F., and Yan, F., 2013a. Alkaline stable C2-substituted imidazolium-based anion-exchange membranes. Chemistry of Materials. 25, 1858–1867.

Lin, B., Qiu, L., Qiu, B., Peng, Y., and Yan, F., 2011a. A soluble and conductive polyfluorene ionomer with pendant imidazolium groups for alkaline fuel cell applications. Macromolecules. 44, 9642–9649.

Lin, C.X., Huang, X.L., Guo, D., Zhang, Q.G., Zhu, A.M., Ye, M.L., and Liu, Q.L., 2016. Side-chain-type anion exchange membranes bearing pendant quaternary ammonium groups via flexible spacers for fuel cells. Journal of Materials Chemistry A. 4, 13938–13948.

Lin, H., Zhao, C., Jiang, Y., Ma, W., and Na, H., 2011b. Novel hybrid polymer electrolyte membranes with high proton conductivity prepared by a silane-crosslinking technique for direct methanol fuel cells. Journal of Power Sources. 196, 1744–1749.

Lin, X., Liang, X., Poynton, S.D., Varcoe, J.R., Ong, A.L., Ran, J., Li, Y., Li, Q., and Xu, T., 2013b. Novel alkaline anion exchange membranes containing pendant benzimidazolium groups for alkaline fuel cells. Journal of Membrane Science. 443, 193–200.

Lin, X., Wu, L., Liu, Y., Ong, A.L., Poynton, S.D., Varcoe, J.R., and Xu, T., 2012. Alkali resistant and conductive guanidinium-based anion-exchange membranes for alkaline polymer electrolyte fuel cells. Journal of Power Sources. 217, 373–380.

Liu, L., Ahlfield, J., Tricker, A., Chu, D., and Kohl, P.A., 2016. Anion conducting multiblock copolymer membranes with partial fluorination and long head-group tethers. Journal of Materials Chemistry A. 4, 16233–16244.

Liu, L., Li, Q., Dai, J., Wang, H., Jin, B., and Bai, R., 2014. A facile strategy for the synthesis of guanidinium-functionalized polymer as alkaline anion exchange membrane with improved alkaline stability. Journal of Membrane Science. 453, 52–60.

Lu, S., Pan, J., Huang, A., Zhuang, L., and Lu, J., 2008. Alkaline polymer electrolyte fuel cells completely free from noble metal catalysts. Proceedings of the National Academy of Sciences. 105, 20611–20614.

Markó, L., 1991. Transition metals in organic synthesis: Hydroformylation, reduction, and oxidation: Annual survey covering the year 1989. Journal of Organometallic Chemistry. 404, 325–504.

Miyatake, K., Shimura, T., Mikami, T., and Watanabe, M., 2009. Aromatic ionomers with superacid groups. Chemical Communications. 42, 6403–6405.

Na, T., Shao, K., Zhu, J., Sun, H., Liu, Z., Zhao, C., Zhang, Z., Lew, C.M., and Zhang, G., 2012. Block sulfonated poly(arylene ether ketone) containing flexible side-chain groups for direct methanol fuel cells usage. Journal of Membrane Science. 417–418, 61–68.

Nakabayashi, K., Higashihara, T., and Ueda, M., 2011. Polymer electrolyte membranes based on poly(phenylene ether)s with pendant perfluoroalkyl sulfonic acids. Macromolecules. 44, 1603–1609.

Nakabayashi, K., Matsumoto, K., Higashihara, T., and Ueda, M., 2008. Influence of adjusted hydrophilic-hydrophobic lengths in sulfonated multiblock copoly(ether sulfone) membranes for fuel cell application. Journal of Polymer Science Part A: Polymer Chemistry. 46, 7332–7341.

Noonan, K.J.T., Hugar, K.M., Kostalik IV, H.A., Lobkovsky, E.B., Abruña, H.D., and Coates, G.W., 2012. Phosphonium-functionalized polyethylene: A new class of base-stable alkaline anion exchange membranes. Journal of the American Chemical Society. 134, 18161–18164.

Nuñez, S.A., and Hickner, M.A., 2013. Quantitative 1H NMR analysis of chemical stabilities in anion-exchange membranes. ACS Macro Letters. 2, 49–52.

Pan, J., Chen, C., Li, Y., Wang, L., Tan, L., Li, G., Tang, X., Xiao, L., Lu, J., and Zhuang, L., 2014. Constructing ionic highway in alkaline polymer electrolytes. Energy & Environmental Science. 7, 354–360.

Pan, J., Li, Y., Han, J., Li, G., Tan, L., Chen, C., Lu, J., and Zhuang, L., 2013. A strategy for disentangling the conductivity–stability dilemma in alkaline polymer electrolytes. Energy & Environmental Science. 6, 2912–2915.

Park, M.J., and Balsara, N.P., 2008. Phase behavior of symmetric sulfonated block copolymers. Macromolecules. 41, 3678–3687.

Price, S.C., Williams, K.S., and Beyer, F.L., 2014. Relationships between structure and alkaline stability of imidazolium cations for fuel cell membrane applications. ACS Macro Letters. 3, 160–165.

Qi, Z., Gong, C., Liang, Y., Li, H., Wu, Z., Feng, W., Wang, Y., Zhang, S., and Li, Y., 2015. Side-chain-type clustered sulfonated poly(arylene ether ketone)s prepared by click chemistry. International Journal of Hydrogen Energy. 40, 9267–9277.

Qu, C., Zhang, H., Zhang, F., and Liu, B., 2012. A high-performance anion exchange membrane based on bi-guanidinium bridged polysilsesquioxane for alkaline fuel cell application. Journal of Materials Chemistry. 22, 8203–8207.

Ran, J., Wu, L., He, Y., Yang, Z., Wang, Y., Jiang, C., Ge, L., Bakangura, E., and Xu, T., 2017. Ion exchange membranes: New developments and applications. Journal of Membrane Science. 522, 267–291.

Si, Z., Sun, Z., Gu, F., Qiu, L., and Yan, F., 2014. Alkaline stable imidazolium-based ionomers containing poly(arylene ether sulfone) side chains for alkaline anion exchange membranes. Journal of Materials Chemistry A. 2, 4413–4421s.

Smitha, B., Sridhar, S., and Khan, A.A., 2005. Solid polymer electrolyte membranes for fuel cell applications—A review. Journal of Membrane Science. 259, 10–26.

Storey, R.F., Chisholm, B. J., and Lee, Y., 1997. Synthesis and mechanical properties of poly(styrene-lsobutylene-&styrene) block copolymer lonomers. Polymer Engineering and Science. 37, 73–80.

Takaba, H., Hisabe, T., Shimizu, T., and Alam, M.K., 2017. Molecular modeling of OH– transport in poly(arylene ether sulfone ketone)s containing quaternized ammonio-substituted fluorenyl groups as anion exchange membranes. Journal of Membrane Science. 522, 237–244.

Tanaka, M., Koike, M., Miyatake, K., and Watanabe, M., 2011. Synthesis and properties of anion conductive ionomers containing fluorenyl groups for alkaline fuel cell applications. Polymer Chemistry. 2, 99–106.

Thomas, O.D., Peckham, T.J., Thanganathan, U., Yang, Y., and Holdcroft, S., 2010. Sulfonated polybenzimidazoles: Proton conduction and acid-base crosslinking. Journal of Polymer Science Part A: Polymer Chemistry. 48, 3640–3650.

Thomas, O.D., Soo, K.J., Peckham, T.J., Kulkarni, M.P., and Holdcroft, S., 2012. A stable hydroxide-conducting polymer. Journal of the American Chemical Society. 134, 10753–10756.

Tsang, E.M.W., Zhang, Z., Yang, A.C.C., Shi, Z., Peckham, T.J., Narimani, R., Frisken, B.J., and Holdcroft, S., 2009. Nanostructure, morphology, and properties of fluorous copolymers bearing ionic grafts. Macromolecules. 42, 9467–9480.

Varcoe, J.R., Atanassov, P., Dekel, D.R., Herring, A.M., Hickner, M.A., Kohl, P.A., Kucernak, A.R., Mustain, W.E., Nijmeijer, K., and Scott, K., 2014. Anion-exchange membranes in electrochemical energy systems. Energy & Environmental Science. 7, 3135–3191.

Varcoe, J.R., and Slade, R.C., 2005. Prospects for alkaline anion-exchange membranes in low temperature fuel cells. Fuel Cells. 5, 187–200.

Wang, C., Li, N., Shin, D.W., Lee, S.Y., Kang, N.R., Lee, Y.M., and Guiver, M.D., 2011. Fluorene-based poly(arylene ether sulfone)s containing clustered flexible pendant sulfonic acids as proton exchange membranes. Macromolecules. 44, 7296–7306.

Wang, J., Gu, S., Kaspar, R.B., Zhang, B., and Yan, Y., 2013. Stabilizing the imidazolium cation in hydroxide-exchange membranes for fuel cells. ChemSusChem. 6, 2079–2082.

Wang, J., Li, S., and Zhang, S., 2010. Novel hydroxide-conducting polyelectrolyte composed of an poly(arylene ether sulfone) containing pendant quaternary guanidinium groups for alkaline fuel cell applications. Macromolecules. 43, 3890–3896.

Wang, S., Zhao, C., Ma, W., Zhang, G., Liu, Z., Ni, J., Li, M., Zhang, N., and Na, H., 2012. Preparation and properties of epoxy-cross-linked porous polybenzimidazole for high temperature proton exchange membrane fuel cells. Journal of Membrane Science. 411–412, 54–63.

Weiber, E.A., and Jannasch, P., 2014. Ion distribution in quaternary-ammonium-functionalized aromatic polymers: Effects on the ionic clustering and conductivity of anion-exchange membranes. ChemSusChem. 7, 2621–2630.

Wu, L., Zhang, Z., Ran, J., Zhou, D., Li, C., and Xu, T., 2013. Advances in proton-exchange membranes for fuel cells: An overview on proton conductive channels (PCCs). Physical Chemistry Chemical Physics. 15, 4870–4887.

Wu, L., Zhou, D., Wang, H., Pan, Q., Ran, J., and Xu, T., 2015. Ionically cross-linked proton conducting membranes for fuel cells. Fuel Cells. 15, 189–195.

Yan, X., Gao, L., Zheng, W., Ruan, X., Zhang, C., Wu, X., and He, G., 2016. Long-spacer-chain imidazolium functionalized poly(ether ether ketone) as hydroxide exchange membrane for fuel cell. International Journal of Hydrogen Energy. 41, 14982–14990.

Yang, Z., Zhou, J., Wang, S., Hou, J., Wu, L., and Xu, T., 2015. A strategy to construct alkali-stable anion exchange membranes bearing ammonium groups via flexible spacers. Journal of Materials Chemistry A. 3, 15015–15019.

Yao, H., Zhang, Y., Liu, Y., You, K., Song, N., Liu, B., and Guan, S., 2015. Pendant-group cross-linked highly sulfonated co-polyimides for proton exchange membranes. Journal of Membrane Science. 480, 83–92.

Ye, Y., Stokes, K.K., Beyer, F.L., and Elabd, Y.A., 2013. Development of phosphonium-based bicarbonate anion exchange polymer membranes. Journal of Membrane Science. 443, 93–99.

Zakaria, Z., Kamarudin, S.K., and Timmiati, S., 2016. Membranes for direct ethanol fuel cells: an overview. Applied Energy. 163, 334–342.

Zakil, F.A., Kamarudin, S.K., and Basri, S., 2016. Modified Nafion membranes for direct alcohol fuel cells: An overview. Renewable and Sustainable Energy Reviews. 65, 841–852.

Zeng, L., and Zhao, T.S., 2016. An effective strategy to increase hydroxide-ion conductivity through microphase separation induced by hydrophobic-side chains. Journal of Power Sources. 303, 354–362.

Zha, Y., Disabb-Miller, M.L., Johnson, Z.D., Hickner, M.A., and Tew, G.N., 2012. Metal-cation-based anion exchange membranes. Journal of the American Chemical Society. 134, 4493–4496.

Zhang, F., Li, N., Cui, Z., Zhang, S., and Li, S., 2008. Novel acid–base polyimides synthesized from binaphthalene dianhydrie and triphenylamine-containing diamine as proton exchange membranes. Journal of Membrane Science. 314, 24–32.

Zhang, H., and Shen, P.K., 2012a. Advances in the high performance polymer electrolyte membranes for fuel cells. Chemical Society Reviews. 41, 2382–2394.

Zhang, H., and Shen, P.K., 2012b. Recent development of polymer electrolyte membranes for fuel cells. Chemical Reviews. 112, 2780–2832.

Zhang, M., Shan, C., Liu, L., Liao, J., Chen, Q., Zhu, M., Wang, Y., An, L., and Li, N., 2016. Facilitating anion transport in polyolefin-based anion exchange membranes via bulky side chains. ACS Applied Materials & Interfaces. 8, 23321–23330.

Zhang, N., Zhao, C., Ma, W., Wang, S., Wang, B., Zhang, G., Li, X., and Na, H., 2014. Macromolecular covalently cross-linked quaternary ammonium poly(ether ether ketone) with polybenzimidazole for anhydrous high temperature proton exchange membranes. Polymer Chemistry. 5, 4939.

Zhang, Q., Li, S., and Zhang, S., 2010a. A novel guanidinium grafted poly(aryl ether sulfone) for high-performance hydroxide exchange membranes. Chemical Communications. 46, 7495–7497.

Zhang, Y., Wan, Y., Zhang, G., Shao, K., Zhao, C., Li, H., and Na, H., 2010b. Preparation and properties of novel cross-linked sulfonated poly(arylene ether ketone) for direct methanol fuel cell application. Journal of Membrane Science. 348, 353–359.

Zhang, Z., Wu, L., Varcoe, J., Li, C., Ong, A.L., Poynton, S., and Xu, T., 2013. Aromatic polyelectrolytes via polyacylation of pre-quaternized monomers for alkaline fuel cells. Journal of Materials Chemistry A. 1, 2595.

Zhang, Z., Wu, L., and Xu, T., 2011. Synthesis and properties of side-chain-type sulfonated poly(phenylene oxide) for proton exchange membranes. Journal of Membrane Science. 373, 160–166.

Zheng, J., Zhang, Q., Qian, H., Xue, B., Li, S., and Zhang, S., 2017. Self-assembly prepared anion exchange membranes with high alkaline stability and organic solvent resistance. Journal of Membrane Science. 522, 159–167.

Zhu, J., Shao, K., Zhang, G., Zhao, C., Zhang, Y., Li, H., Han, M., Lin, H., Xu, D., Yu, H., and Na, H., 2010. Novel side-chain-type sulfonated hydroxynaphthalene-based Poly(aryl ether ketone) with H-bonded for proton exchange membranes. Polymer. 51, 3047–3053.

Zuo, P., Su, Y., and Li, W., 2016. Comb-like poly(ether-sulfone) membranes derived from planar 6,12-diaryl-5,11-dihydroindolo[3,2-b]carbazole monomer for alkaline fuel cells. Macromolecular Rapid Communications. 37, 1748–1753.

17 Anion Exchange Membranes for Electrodialysis through Layer-by-Layer Deposition

Ryosuke Takagi, Sri Mulyati, and Hideto Matsuyama

CONTENTS

17.1 INTRODUCTION

The shortage of freshwater has been a serious problem in the world due to the increase in population and the contamination of fresh water resources. There is an increasing demand for producing potable water by treating ground, surface and seawater. Electrodialysis (ED) is one of the useful methods to make drinking water from surface water or groundwater. ED is an electrochemical desalination process that utilizes the ion permselectivity of cation and anion exchange membranes (Strathmann, 2010).

Figure 17.1 shows the fundamental system of ED. In ED process, a feed is separated by a paired cation exchange membrane (CEM) and anion exchange membrane (AEM). In a practical ED system, more than several tens of pairs are stacked between electrodes. Then, the external voltage is applied to the system. Anions migrate to the anode through the AEM and cations migrate to the cathode through the CEM under the driving force mainly due to the external potential gradient. However, anions cannot permeate through the CEM and cations cannot permeate through the AEM because of the electric repulsion between ionic charge and membrane charge. The CEM has a negative charge and the AEM a positive charge. Thus, ions are selectively transported from one compartment (dilute compartment) to another compartment (concentrate compartment) through ion exchange membranes. So, the ion concentration in a dilute compartment is decreased. This is a principle of ED. One advantage of ED compared to other processes, such as reverse osmosis (RO), is that a higher brine concentration can be achieved because there are no osmotic pressure limitations (Strathmann, 2010). ED is especially useful for removing harmful ions like fluoride, F^- and nitrate, NO_3^- (Kesore et al., 1997; Kabay et al., 2008). It can also be used to soften water (Kabay et al., 2002) and to remove heavy metals (Smara et al., 2007).

Unfortunately, fouling, especially fouling of AEM, is a serious problem in ED as well as RO (Lindstrand et al., 2000; Wang et al., 2017). Fouling is generally divided into three categories depending on types of foulants, such as inorganic, organic and biofouling. Most foulants included in the feed, like proteins, surfactants, humic acid, bacteria and so on, have a negative charge (Noble et al., 1995; Strathmann, 2004; Persico et al., 2016), while AEM has a positive charge. The fouling of AEM takes place easily and decreases the system performance. Thus, a high antifouling property is demanded for AEMs and researchers have made a continuing effort to

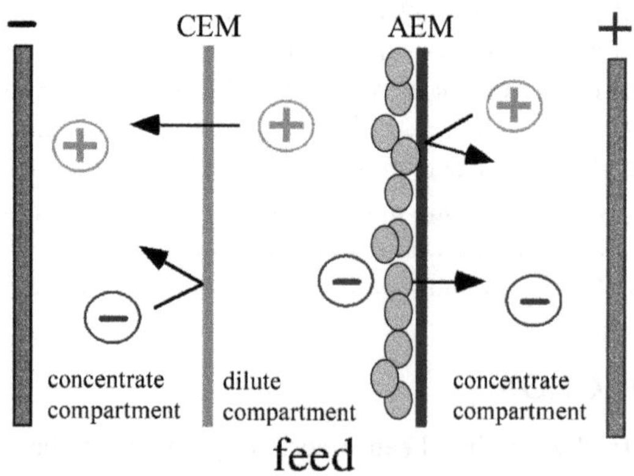

FIGURE 17.1 Fundamental setup of ED and fouling of AEM. CEM: cation exchange membrane, AEM: anion exchange membrane, –: negatively charged foulants such as protein, humic acid, surfactant, etc.

improve their antifouling property (Mulyati et al., 2012; Vaselbehagh et al., 2014; Zhao et al., 2016b; Fernandez-Gonzalez et al., 2017b).

Another important property of ion exchange membranes is the permselectivity towards specific ions. In particular, the monovalent anion selectivity of AEM is very important because most of the harmful ions have to be removed from feed water are monovalent anions, such as F^- and NO_3^-. Infants younger than 6 months old are susceptible to nitrate poisoning, which interferes with the blood's ability to carry oxygen (Mahler et al., 2007). A high concentration of fluoride in drinking water results in fluorosis (dental/skeletal abnormalities) and several types of neurological damage (Kabay et al., 2008). In addition, AEMs must be impermeable to sulfate ions to prevent the precipitation of calcium sulfate during the ED process (Sata, 2000). Calcium sulfate precipitates as a scale in the concentrate compartment, causing an inorganic fouling of CEM. Thus, high monovalent anion selectivity is also demanded for AEMs and the research has been continued (Mulyati et al., 2013a; Takagi et al., 2014; Liu et al., 2017; Zhao et al., 2017).

Figure 17.2 shows one concept for simultaneously improving the antifouling potential and a monovalent anion selectivity of AEMs. It has been widely recognized that increasing the hydrophilicity of the membrane surface and imparting a negative surface charge are very effective methods to prevent fouling (Ba et al., 2010; Zou et al., 2011). A high hydrophilicity prevents the adsorption of foulants by reducing hydrophobic interactions. A negative surface charge prevents the adsorption of negatively charged foulants through electrostatic repulsion. In addition, the partition of multivalent anions into AEM will also be decreased by Donnan exclusion due to the negative surface charge (Takagi et al., 2014), and the monovalent selectivity will be increased simultaneously.

The common way to give a negative charge to AEM is a surface modification. One of the easy and simple ways for surface modification is layer-by-layer (LbL) deposition (Krasemann et al., 1999; Hoffmann et al., 2009). LbL deposition is a thin

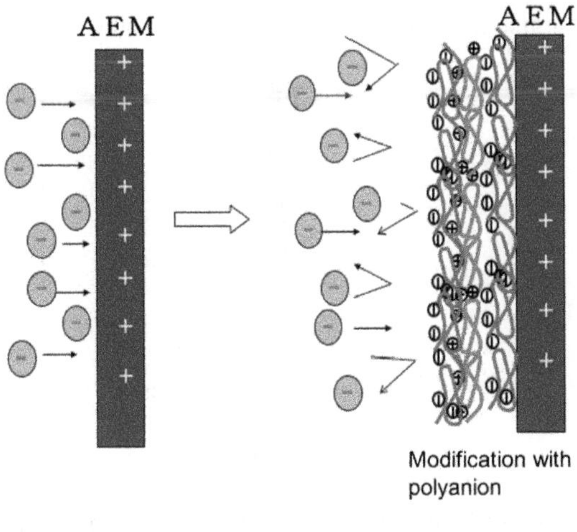

FIGURE 17.2 Concept of improvement of antifouling property and monovalent selectivity of AEM.

film multilayer fabrication technique and is also used in the surface modification of reverse osmosis membranes (Xu et al., 2015). The aim of this chapter is to review the recent improvement of antifouling property and monovalent anion selectivity of AEMs by surface modification, mainly by LbL deposition. At first, we will review the mechanism of LbL multilayer formation and the effect of polyelectrolyte layer on surface properties. Then, we review the effect of polyelectrolyte layer on the antifouling property and the monovalent anion selectivity including the mechanism.

17.2 LAYER-BY-LAYER DEPOSITION

17.2.1 LbL Multilayer Formation

Table 17.1 shows polyelectrolytes commonly used in LbL deposition. The multilayer is formed by alternately dipping a membrane into a polyelectrolyte solution with an opposite charge. Generally, the polyelectrolyte solution contains an additional electrolyte as a supporting electrolyte. Figure 17.3 shows the poly(sodium 4-styrene

TABLE 17.1
Polyelectrolytes Commonly Used in LbL Deposition

Polyelectrolytes	Examples
Polyanion	poly(sodium 4-styrene sulfonate) (PSS)
	poly(acrylic acid) (PAA)
	poly(vinylsulfonic acid) (PVS)
Polycation	poly(allylamine hydrochloride) (PAH)
	poly(diallydimethyl ammonium chloride) (PDADMAC)
	2-hydroxypropyltrimethyl ammonium chloride chitosan (HACC)

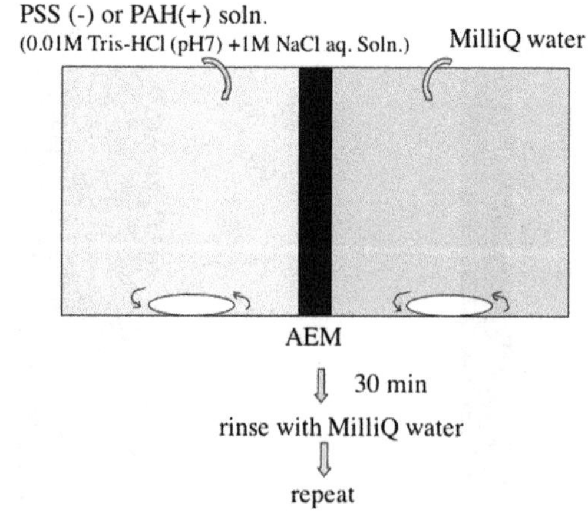

FIGURE 17.3 Formation of polyelectrolyte multilayer (LbL deposition). PSS and PAH are an example of polyanion and polycation, respectively.

sulfonate) (PSS)/poly(allylamine hydrochloride) (PAH) multilayer formation on one surface of AEM reported by Mulyati et al. (2013a) as an example. AEM was clamped between two cells tightly, and first the PSS solution was put on one side of the membrane and the Milli-Q water was put on the other side. After 30 min, the membrane was removed from the cells and rinsed with Milli-Q water. After resetting the AEM, the PAH solution was put on the same side of the membrane as before and the Milli-Q water was put on the other side. The PSS/PAH multilayer was formed by repeating this procedure.

17.2.2 Factors That Affect a Polyelectrolyte Deposition

The supporting electrolyte affects the morphology and the amount of adsorbed polyelectrolyte (Steeg et al., 1992; Shubint et al., 1995; Stanton et al., 2003), since the adsorption of polyelectrolyte depends on a balance between electrostatic and non-electrostatic (hydrophobic interactions) attractions. In the case of pure electrostatic attraction, adsorption always decreases with increasing supporting electrolyte concentration. In the case where both the electrostatic and non-electrostatic attractions affect adsorption, the adsorbed amount shows a maximum with the increase of supporting electrolyte. Mulyati et al. (2012) investigated the factors that affect a polyelectrolyte deposition on a charged surface. They measured the adsorbed amount of PSS on an Au-coated quartz crystal surface modified with a cationic poly(diallyldimethylammonium chloride) using a quartz crystal microbalance (QCM). Figure 17.4 shows the amount of PSS adsorbed to the positively charged surface as a function of NaCl concentration as a supporting electrolyte. The amount of adsorbed PSS measured by QCM showed a maximum. Thus, it was found that the adsorption of polyelectrolyte on a charged surface was affected by both electrostatic and non-electrostatic attraction.

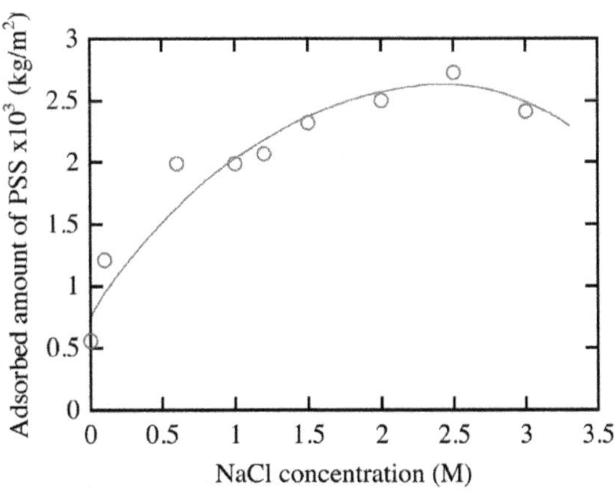

FIGURE 17.4 Amount of PSS adsorbed as a function of supporting electrolyte NaCl concentration. (From Mulyati, S. et al., Journal of Membrane Science. 417–418, 137–143, 2012.)

17.2.3 Effect of Supporting Electrolyte and Polymer Concentration

Mulyati et al. (2012) investigated the effect of supporting electrolyte and polymer concentration on surface properties of a modified AEM by measuring the water contact angle and the ζ-potential of the modified AEM surface with polyelectrolyte. Figure 17.5 shows the water contact angle and the ζ-potential of the surface of a commercial AEM, AMX (Astom Corp., Tokyo, Japan), modified with single PSS layer as a function of supporting NaCl for PSS concentration of 0.3 kg/m^3. As seen in Figure 17.5(a), the modified membrane surface was more hydrophilic than the unmodified membrane in the region of 0.8–1.0 M NaCl, since the contact angle for the unmodified membrane surface was 69°. The membrane surface was more hydrophobic at other NaCl concentrations. On the other hand, as seen in Figure 17.5(b), the ζ-potential initially became negative and decreased with NaCl concentration and then increased after showing a minimum (maximum absolute value) at 1.0 M NaCl. The ζ-potential of the unmodified membrane surface was 0.6 mV.

The phenomenon shown in Figure 17.5 is discussed as follows. The PSS molecule is extended in a solution at low supporting electrolyte concentration, and the charge density of PSS is low because there is insufficient electrolyte to shield intramolecular electrostatic repulsion. PSS will be adsorbed onto the AMX membrane surface in an extended string-like form (mostly train form) from the PSS solution at low supporting NaCl concentrations, and the aromatic rings of PSS will have a strong hydrophobic interaction with the membrane surface (Bukhovets et al., 2010). Thus, at very low supporting NaCl concentrations, the membrane surface was mainly covered with the hydrophobic segments of PSS, and the modified AMX membrane surface became more hydrophobic than the unmodified membrane surface, as shown in Figure 17.5(a). At low supporting NaCl concentrations, the ζ-potential did not become as highly negative relative to the unmodified membrane, as shown in Figure 17.5(b), because the charge density of adsorbed PSS was low. The surface area covered with the hydrophobic segments of PSS would not be expected to increase

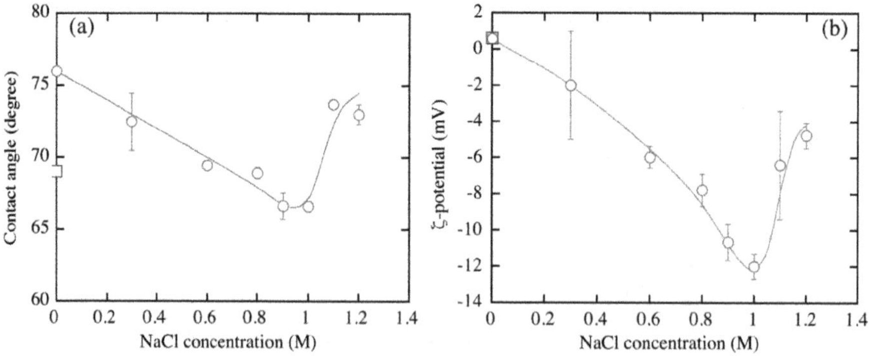

FIGURE 17.5 The effect of supporting electrolyte (NaCl) on (a) the water contact angle and (b) the ζ-potential of a modified membrane surface. Circles represent the modified membrane and squares represent the unmodified membrane. In these experiments, the PSS concentration of the PSS solution (modification solution) was 0.3 kg/m^3. (From Mulyati, S. et al., Journal of Membrane Science. 417–418, 137–143, 2012.)

significantly with increasing supporting NaCl concentration because the concentration of PSS was constant. However, the portion of tails or loops of adsorbed PSS would increase at the higher NaCl concentrations. Thus, the amount of adsorbed PSS increases with increasing supporting NaCl concentration (Shubint et al., 1995; Stanton et al., 2003). The negative charge density on the membrane surface also increases, and the absolute value of the ζ-potential increases. In addition, the membrane surface becomes more hydrophilic, and the contact angle decreases. In the region where the supporting NaCl concentration was higher than 1.0 M, the amount of PSS adsorbed decreased through shielding of the electrostatic attraction between PSS and the membrane surface. Thus, the contact angle increased and the absolute value of ζ-potential decreased with increasing supporting NaCl concentration at concentrations higher than 1.0 M.

Figure 17.6 shows the water contact angle and the ζ-potential of the modified AMX surface as a function of PSS concentration for an NaCl concentration of 1.0 M. It can be seen from Figure 17.6(a) that the contact angle of the modified membrane was lower than that of the unmodified membrane at PSS concentrations lower than 0.6 kg/m^3. A further increase in PSS concentration caused an increase in the contact angle. On the other hand, the absolute value of the ζ-potential increased with PSS concentration, as shown in Figure 17.6(b).

As described above, when the supporting NaCl concentration is 1.0 M, the morphology of the PSS molecule mainly exists in loop or tail conformations (Steeg et al., 1992; Stanton et al., 2003). PSS is adsorbed on the AMX surface by the electrostatic interaction between the negative charge of PSS and the positive charge of AMX at PSS concentration less than 0.1 kg/m^3. Thus, the hydrophobic segments of the PSS molecule are not extensively exposed on the membrane surface. In this situation, the membrane surface becomes more hydrophilic. The contact angle decreases and the absolute value of ζ-potential increases.

On the other hand, at PSS concentration ≥ 0.1 kg/m^3, some PSS chains are adsorbed on the membrane surface by hydrophobic interaction (non-electrostatic

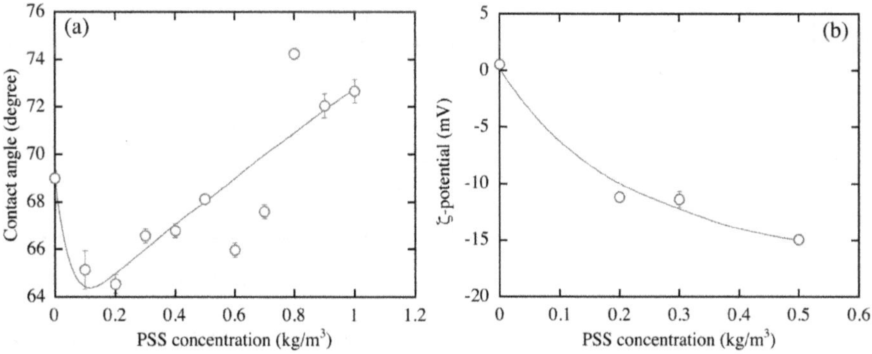

FIGURE 17.6 (a) Water contact angle and (b) ζ-potential of the modified membrane surface as a function of PSS concentration. The NaCl concentration of the PSS solution was 1.0 M. Data at PSS = 0 correspond to those of the unmodified membrane. (From Mulyati, S. et al., Journal of Membrane Science. 417–418, 137–143, 2012.)

attraction) because the positive surface charge of the membrane is neutralized by adsorbed PSS. Since the amount of PSS adsorbed increases, the negative surface charge density increases, which leads to a further increase in the absolute value of the ζ-potential. The hydrophilicity of the membrane surface is determined by the balance between the number of charged sites (hydrophilic part) and the surface area covered by hydrophobic segments of PSS. When the PSS concentration exceeds 0.1 kg/m^3, the membrane surface area covered with hydrophobic segments increases and thus, the membrane surface becomes more hydrophobic, even though the ζ-potential is more negative when the PSS concentration is higher than 0.1 kg/m^3.

17.2.4 LbL Multilayer Formation

Mulyati et al. (2013a) discussed the mechanism of PSS/PAH multilayer formation considering the effect of each factor described above. They modified AMX based on the procedure outlined in Section 17.2.1 using polyelectrolyte solutions prepared by dissolving 0.3 kg/m^3 of PSS or PAH with 1.0 M supporting NaCl in 10 mM Tris-HCl buffer (pH 7). Figure 17.7 shows the water contact angle and the ζ-potential of the modified AMX membrane surface as a function of the number of layers. As shown in Figure 17.7(a), the contact angles on the membrane surface initially decreased with the number of layers, then increased after showing a minimum at seven or eight layers, and became constant for both surface with PSS or PAH top layer. The contact angles for the membrane surface with PAH top layer were larger than those for the membrane surface with PSS top layer. This is because the hydrohobicity of PAH layer is higher than PSS layer (Kolasińska et al., 2005).

On the other hand, as seen in Figure 17.7(b), the ζ-potential showed a weak dependence on the number of layers and became almost constant above 15 layers. It is known that the inversion of surface charge produced by sequential adsorption steps is

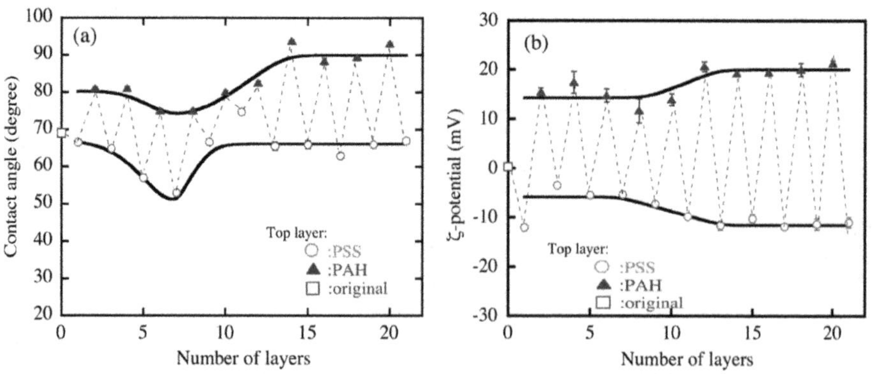

FIGURE 17.7 (a) Water contact angle and (b) ζ-potential of the AMX membrane surface modified using LbL deposition. Squares represent the original membrane, circles the membrane with PSS top layer, and triangles the membrane with PAH top layer. The modification solutions contained 0.3 kg/m^3 of PSS or PAH, and 1.0 M NaCl. All error bars in Figure 17.7(a) are within symbols. (From Mulyati, S. et al., Journal of Membrane Science. 431, 113–120, 2013a.)

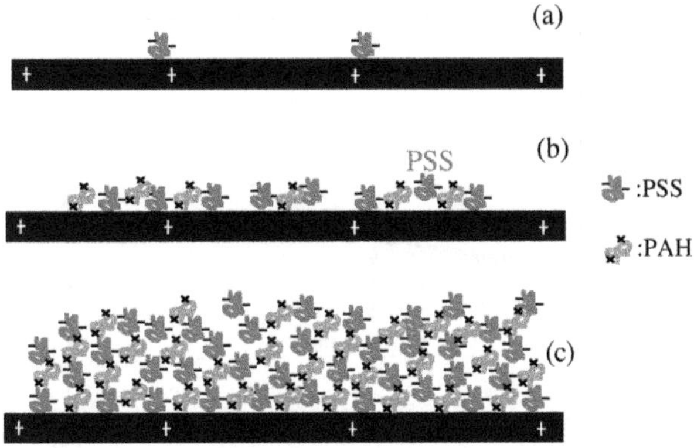

FIGURE 17.8 Schematic representation of multilayer fabrication. (a) Representation of the first layer, (b) representation of approximately seven layers, and (c) representation of more than 15 layers. (From Mulyati, S. et al., Journal of Membrane Science. 431, 113–120, 2013a.)

due to the charge overcompensation at each step (Schlenoff et al., 1998; 2001). Thus, it is considered that the absolute value of the change in the ζ-potential after each step (negative-to-positive or positive-to-negative) reflected the amount of polymer adsorbed at each step. From Figure 17.7(b), the absolute value of the change in the ζ-potential for the first seven or eight layers was approximately 20 mV. The change in the ζ-potential then increased to approximately 32 mV and became constant above 15 layers. This indicated that the amount of PAH and PSS adsorbed in each of the first eight layers was smaller than that absorbed in each of the layers above 15 layers.

Mulyati et al. (2013a) suggested a schematic speculation of the multilayer fabrication as shown in Figure 17.8. The total charge in Figure 17.8(b) is the same as in Figure 17.8(a), but the number of ionic groups in Figure 17.8(b) is larger than in Figure 17.8(a). Thus, in the first seven to eight layers deposition, the ζ-potential was the constant, while the water contact angle decreased, as shown in Figure 17.7. After eight layers, the amount of polyelectrolyte adsorbed per layer increased, and the surface properties shifted to those determined by PSS or PAH layers. The membrane surface was fully covered with LbL layers above 15 layers, and then the contact angle and the ζ-potential became constant. This is consistent with previous reports which show that several layer pairs are required to fully cover the substrate using LbL deposition (Stanton et al., 2003; Ouyang et al., 2008).

17.3 IMPROVEMENT OF ANTIFOULING POTENTIAL

17.3.1 Single Layer Deposition

Mulyati et al. (2012) quantitatively evaluated the antifouling property of AEM using the transition time. The transition time is defined as the time elapsed until the occurrence of fouling. They measured the time course of voltage across the membrane, ΔE, under the constant electric current density ED operation and obtained the transition time from the point at which ΔE sharply increased. Figure 17.9 shows an example of measuring the transition time. The transition time is large when little fouling takes

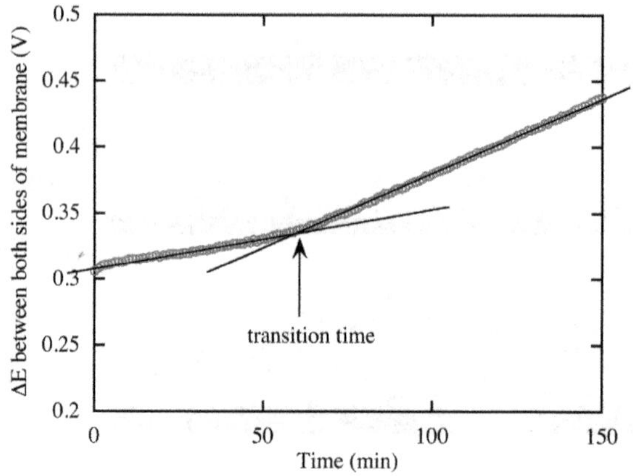

FIGURE 17.9 Determination of transition time. The feed solution contained 0.05 M NaCl as an electrolyte and 0.015 kg/m^3 sodium dodecylbenzene sulfonate (SDBS) as a foulant. The membrane was unmodified AMX. (From Mulyati, S. et al., Journal of Membrane Science. 417–418, 137–143, 2012.)

place, while the transition time is short when fouling occurs easily. Thus, the transition time is used as a quantitative parameter that reflects the antifouling potential of AEM.

Figure 17.10 shows the transition time for AMX modified with PSS single layer under the constant electric current density of 20 A/m^2 (Mulyati et al., 2012). Figure 17.10(a) shows the transition time as a function of the supporting NaCl concentration for a PSS concentration of 0.3 kg/m^3. Figure 17.10(b) shows the transition time as a function of the PSS concentration for a supporting NaCl concentration of 1.0 M. The feed solution contained 0.05 M NaCl and 0.052 kg/m^3 sodium dodecylbenzene sulfonate (SDBS) as a model foulant. This condition is very prone to fouling because the

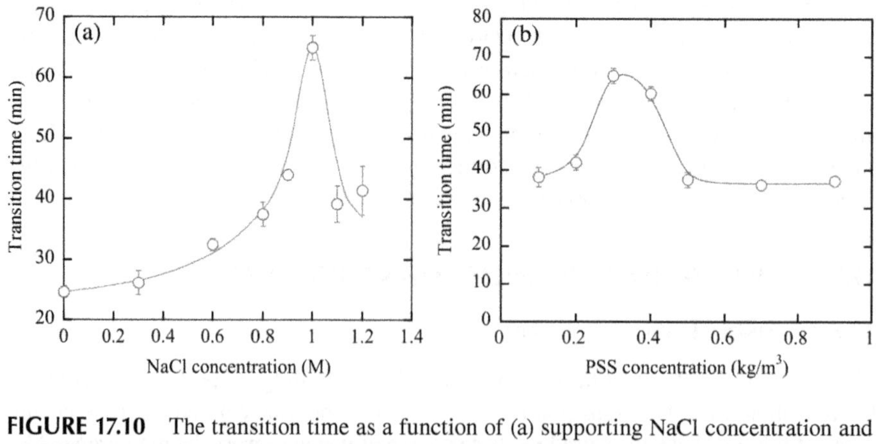

FIGURE 17.10 The transition time as a function of (a) supporting NaCl concentration and (b) PSS concentration. The current density was 20 A/m^2. The feed solution contained 0.05 M NaCl and 0.052 kg/m^3 SDBS. In Figure 17.10(a), the PSS concentration was 0.3 kg/m^3 and the supporting NaCl concentration was1.0 M in Figure 17.10(b). (From Mulyati, S. et al., Journal of Membrane Science. 417–418, 137–143, 2012.)

SDBS concentration is about 3 times higher than the critical micelle concentration (CMC). It has been reported that the formation of SDBS micelles tends to promote fouling (Lee et al., 2008). It can be seen from Figure 17.10 that the transition time (i.e., the antifouling property) strongly depends on the supporting electrolyte and polyelectrolyte concentration. It is clear that there is an optimum condition for surface modification with polyelectrolyte. In the case of surface modification with PSS, the optimum supporting NaCl concentration is 1.0 M and the optimum PSS concentration is 0.3 kg/m^3 because the transition time shows a maximum value. Comparing Figure 17.10 with Figures 17.5 and 17.6 reveals that the optimum condition for antifouling property is determined by the combination of hydrophilicity (contact angle) and negative charge density (ζ-potential) of membrane surface.

Figure 17.11 shows the transition time of modified and unmodified AMX membranes as a function of SDBS concentration in the feed solution. In Figure 17.11, an AMX membrane was modified under the optimum condition that is PSS solution containing 0.3 kg/m^3 PSS and 1.0 M supporting NaCl. The transition time of the modified membrane was longer than that of the unmodified membrane in whole SDBS concentration. Thus, it is clear that surface modification with polyanion is quite useful for enhancing the antifouling potential of AEMs.

Other works reported the improvement of antifouling potential of AEM using single layer surface modification other than LbL. Vaselbehagh et al. (2014) modified AMX with polydopamine and measured the transition time under the constant electric current density of 20 A/m^2 using the feed solution containing 0.05 M NaCl and 0.052 kg/m^3 SDBS. This condition was the same as that in Figure 17.10. They reported the maximum transition time was about 290 min, while it was about 68 min for AMX modified with single PSS layer as shown in Figure 17.10. The reason why AMX modified with polydopamine could show longer transition time than AMX modified

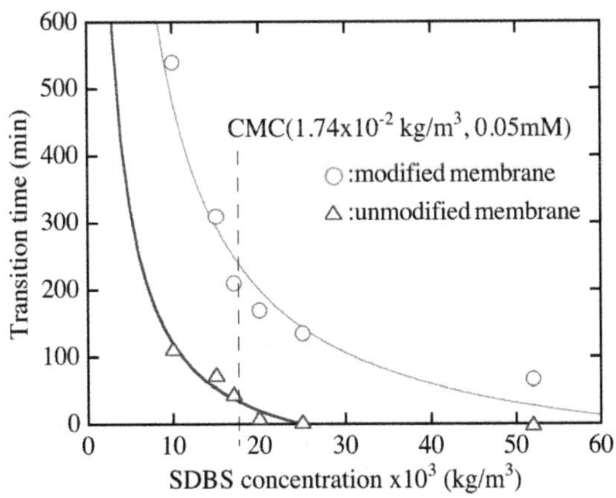

FIGURE 17.11 Transition time of modified and unmodified AMX membranes. Circles and triangles show the modified and unmodified AMX membranes, respectively. The feed solution contained 0.05 M NaCl and SDBS. The PSS and supporting NaCl concentrations in the PSS solution were 0.3 kg/m^3 and 1.0 M, respectively. (From Mulyati, S. et al., Journal of Membrane Science. 417–418, 137–143, 2012.)

with PSS is likely due to the increase in surface hydrophilicity. AMX modified with polydopamine still showed a long transition time, about 130 min, after keeping it in Milli-Q water for 3 months. Zhao et al. (2016b) investigated the effect of polyanion with different functional groups using SDBS as a model foulant under the constant electric current density. They modified the commercial AEM, JAM-II-07 (Yanrun, China), with PSS, PVS and poly(sodiumacrylate) by electrodeposition. They evaluated the antifouling potential from the increase rate of the voltage across AEM and showed that PVS was most effective to improve the antifouling potential. Fernandez-Gonzalez et al. (2017b) modified commercial polyethylene AEMs with sulfonated poly(2,6-dimethyl-1,4-phenylene oxide) (sPPO) and oxidized multi-walled carbon nanotubes (CNTs-COO$^-$) or sPPO and sulfonated iron oxide nanoparticles (Fe_2O_3-SO_4^{2-}) by physical coating. They evaluated the antifouling potential from the increase rate of the voltage across AEM using sodium dodecyl sulfate as a model foulant and showed that the optimized loading of Fe_2O_3-SO_4^{2-} and CNTs-COO$^-$ at 0.4 wt% and 0.6 wt% improved membrane fouling resistance by 45% and 53%, respectively.

17.3.2 Multilayer Layer Deposition

Mulyati et al. (2013a) evaluated the antifouling potential of AMX modified with PSS/PAH multilayer based on the method described in Section 17.2.4. Figure 17.12 shows the transition time as a function of number of multilayers. It can be seen in Figure 17.12 that the transition times for an even number of layers (PAH top layer) were almost zero. When PAH formed the top layer of the membrane, the membrane surface charge was positive. Foulants would thus easily deposit and adhere to the membrane surface due to the electrostatic attraction between the negatively charged foulants and the positively charged membrane surface.

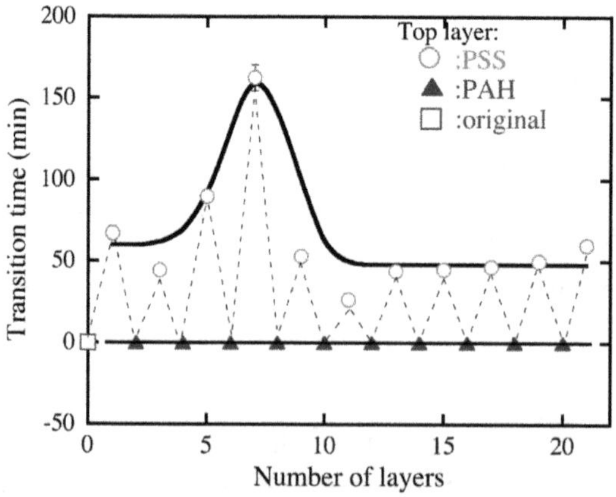

FIGURE 17.12 Effect of the number of layers on the transition time. The feed solution contained 0.05 M NaCl and 0.052 kg/m^3 SDBS. The square represents the original membrane, circles represent the membrane with PSS top layer, and triangles represent the membrane with PAH top layer. The modification solution contained 0.3 kg/m^3 of PSS or PAH and 1.0 M NaCl. (From Mulyati, S. et al., Journal of Membrane Science. 431, 113–120, 2013a.)

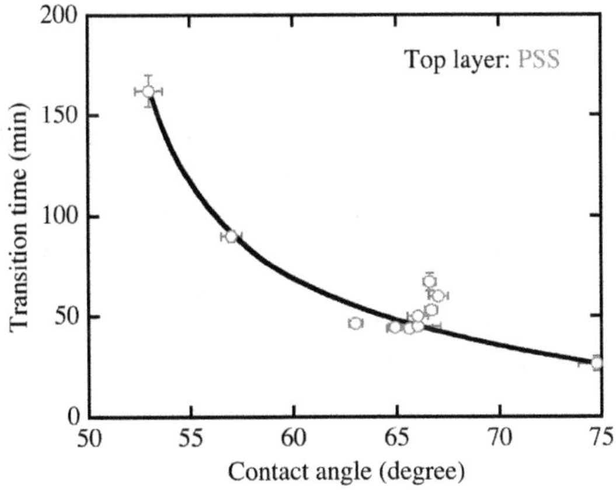

FIGURE 17.13 Transition time as a function of the contact angle for the membrane with PSS top layer. (From Mulyati, S. et al., Journal of Membrane Science. 431, 113–120, 2013a.)

In contrast, for odd numbers of layers (PSS top layer), the transition time increased with the number of layers, and showed a maximum at seven layers. Figure 17.7 shows that the contact angle is the lowest at seven layers, but the ζ-potential is not at its highest negative value. Figure 17.13 shows the transition time as a function of contact angle. It is clear that the antifouling potential increased with the increase of membrane surface hydrophilicity (reduced contact angle). As seen in Figure 17.7(b), the ζ-potential was negative for all odd numbers of layers and there were no large differences in the absolute values. Thus, it was concluded that increased hydrophilicity was the dominant factor in improving the antifouling potential of the AMX membranes, if AMX had a sufficient negative surface charge. In addition, as clearly shown in Figure 17.12, the surface modification via multilayer deposition improved the antifouling potential more than the single-layer deposition shown in Figure 17.10. The maximum transition time was about 68 min in Figure 17.10, while that in Figure 17.12 was about 160 min.

17.4 IMPROVEMENT OF MONOVALENT ANION SELECTIVITY

17.4.1 Evaluation of Monovalent Anion Selectivity of AEM

The permselectivity between anions passing through AEMs during ED is governed by the affinity of the anions with the membrane (ion exchange equilibrium constant) and differences in the migration speed of the respective anions (mobility ratios for the anions) (Sata, 2000). The monovalent anion selectivity of the modified AEM is generally evaluated by the transport number ratio between sulfate and chloride ions. The transport number of ion k, t_k is defined by Eq. 17.1(a) and rewritten by Eq. 17.1(b) using Faraday's law:

$$t_k = I_k / \sum I_s \qquad (17.1(a))$$

$$t_k = [J_k] / \sum [J_s] \quad (17.1(b))$$

where I_k denotes the electric current (A) carried by ion k, ΣI_s the total electric current (A) carried by ions, $[J_k]$ the ion flux (eq/m²s) of ion k and $\Sigma[J_s]$ the total ion flux (eq/m²s). The transport number depends on the mole fraction since the ion flux depends on the mole fraction. Thus, to evaluate the permselectivity of ion exchange membrane, the transport number divided by its concentration is generally treated to avoid the mole fraction effect. The transport number ratio between SO_4^{2-} and Cl^- ions (P_{Cl}^{SO4}) is defined by Eq. 17.2 (Sata et al., 1998):

$$P_{Cl}^{SO4} = \left(t_{SO4} / [SO_4]_B\right) / \left(t_{Cl} / [Cl]_B\right) = \left([J_{SO4}] / [J_{Cl}]\right) / \left([Cl]_B / [SO_4]_B\right) \quad (17.2)$$

where t_{SO4} and t_{Cl} denote the transport numbers of SO_4^{2-} and Cl^- ions, respectively; $[SO_4]_B$ and $[Cl]_B$ are the concentrations (eq/m³) of SO_4^{2-} and Cl^- in the dilute compartment, respectively; and $[J_{SO4}]$ and $[J_{Cl}]$ denote the fluxes (eq/m²s) of SO_4^{2-} and Cl^-, respectively. P_{Cl}^{SO4} larger than unity means the high multivalent anion selectivity of AEM, while P_{Cl}^{SO4} less than unity means the high monovalent anion selectivity. The lower P_{Cl}^{SO4} is, the higher monovalent anion selectivity is.

17.4.2 Effect of LbL Multilayer

Mulyati et al. (2013a) evaluated the monovalent anion selectivity of AMX modified with PSS/PAH multilayer formed by LbL deposition using the feed solution containing 0.01 M NaCl and 0.01 M Na_2SO_4. They obtained the ion flux from the time course of anion concentrations in the dilute compartment during ED under the constant electric current density, 20 A/m². The transport number ratio depends on the time elapsed during the ED process (Takagi et al., 2011), since the ion flux depends on the time. Thus, the authors used the ion flux at time zero, immediately after the beginning of ED operation, and obtained the transport number ratio P_{Cl}^{SO4} by substituting the ion flux and ion concentration into Eq. 17.2 for unmodified and modified AMXs with a PSS top layer.

Figure 17.14 shows P_{Cl}^{SO4} as a function of number of layers. As can be seen, the P_{Cl}^{SO4} decreased with increasing numbers of layers, until 15 layers, after which it became almost constant. This indicated that the monovalent anion selectivity of the AMX membrane increased with increasing numbers of LbL layers. Above 15 layers, P_{Cl}^{SO4} remained constant at approximately 0.55. Although this value was slightly higher than the value of 0.4 quoted for a commercially available monovalent anion-selective membrane (Neosepta ACS, Astom Corp.), it is good enough for practical use.

Mulyati et al. (2013a) consider that the dependency of P_{Cl}^{SO4} on number of layers is due to the excess negative charge within the multilayer. The LbL layers are quite disordered and interpenetrating (Decher, 1997). The polymer in a particular layer is dispersed up to three or four layers away from its normal location within the thin film. The interpenetration between layers with same charge causes the excess charge

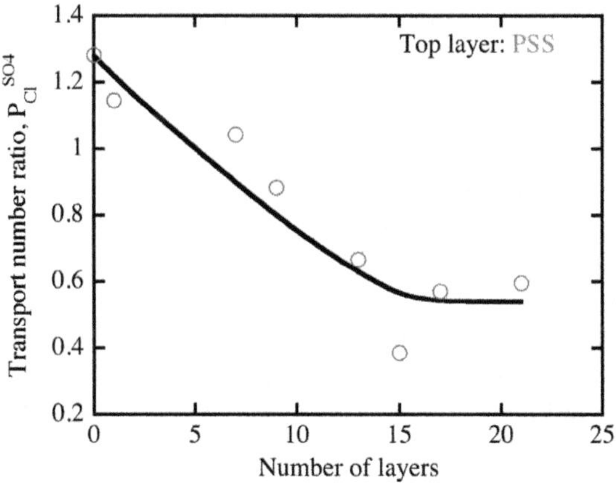

FIGURE 17.14 Transport number ratio for the modified AMX with PSS top layer as a function of the number of layers. (From Mulyati, S. et al., Journal of Membrane Science. 431, 113–120, 2013a.)

within the LbL layer. The excess charge is balanced by small counterions derived from the supporting electrolytes in the polymer solution used to construct the LbL layer (Schlenoff et al., 1998; 2001; Dubas et al., 1999). The excess negative charge within the multilayer is given by Eq. 17.3 (Klitzing et al., 1995; Schlenoff et al., 2001):

$$c^{-}_{ex.l} = c^{-}_{ex.0}\exp(-l/lcp) \tag{17.3}$$

where $c^{-}_{ex,l}$ denotes the excess negative charge (kg m^{-3}) at position l, l is the distance (m) between the surface and the multilayer position, $c^{-}_{ex,O}$ denotes the excess negative charge (kg m^{-3}) at the multilayer surface and l_{cp} is the decay length (m) where the excess charge becomes $1/e$ times $c^{-}_{ex,O}$.

It is expected that a high negative charge will produce a larger repulsive electrostatic force against divalent anions than against monovalent anions (Krasemann et al., 1999; Takagi et al., 2014). The total excess areal negative charge (kg m^{-2}), $c^{-}_{ex,total}$, is obtained by integrating Eq. 17.3 over the multilayer thickness, and is given by Eq. 17.4:

$$c^{-}_{ex.total} = c^{-}_{ex.0} l_{cp}(1 - \exp(-l/l_{cp})) \tag{17.4}$$

where $c^{-}_{ex,total}$ indicates the amount of excess charge included in an LbL layer of volume 1 m^2 × l m. It is clear from Eq. 17.4 that the total excess areal negative charge increases with an increase in l and becomes constant when $l >> l_{cp}$. As can be expected from Eq. 17.4, the monovalent anion selectivity increases with increasing multilayer thickness, up to l_{cp}, and becomes constant above l_{cp}. Mulyati et al. (2013a) considered that l satisfied the condition $l >> l_{cp}$ above 15 layers and then the monovalent anion selectivity increased with increasing the number multilayers, up to 15 layers.

17.4.3 Effect of Electrodeposition Multilayer

Recently, the electrodeposition of polyelectrolytes has been used as an option to form a polyelectrolyte multilayer on an AEM surface (Zhao et al., 2016a; 2016b). Liu et al. (2017) combined the cross-linking with the electrodeposition and formed PSS/2-hydroxypropyltrimethyl ammonium chloride chitosan (HACC) multilayer on commercial AEM using 4,4-diazostilbene-2,2-disulfonic acid disodium salt (DAS) as a cross-linker. Figure 17.15 shows the diagram of cross-linked electrodeposition multilayer formation.

They applied the constant current density of 50 A/m^2 for the feed solution containing 0.05 M NaCl and 0.05 M Na_2SO_4 and found that P_{SO4}^{Cl} (= $1/P_{Cl}^{SO4}$) was increased by cross-linking between layers. The highest value of P_{SO4}^{Cl} (= $1/P_{Cl}^{SO4}$) was 4.36 (0.23 of P_{Cl}^{SO4}) for the membrane modified with cross-linked electro-deposition $(PSS/HACC)_5PSS$ layer.

Table 17.2 shows some examples of monovalent anion selectivity reported in the recently published works. The effects of polyethyleneimine (PEI) are considered as a sieving effect of dense PEI layer for SO_4^{2-} ion and the hydrophobic effect of PEI layer (Wang et al. 2015; Pan et al., 2017). The monovalent selectivity P_{Cl}^{SO4} depends on the experimental condition, such as feed composition, the way to determine the ion flux and so on. P_{Cl}^{SO4} is also affected by the electric resistances of ion exchange membranes and dilute compartment (Takagi et al., 2014). Thus, it is difficult to compare the data reported in literature. Nevertheless, it can be said that the monovalent anion selectivity of AEMs is improved upon surface modification.

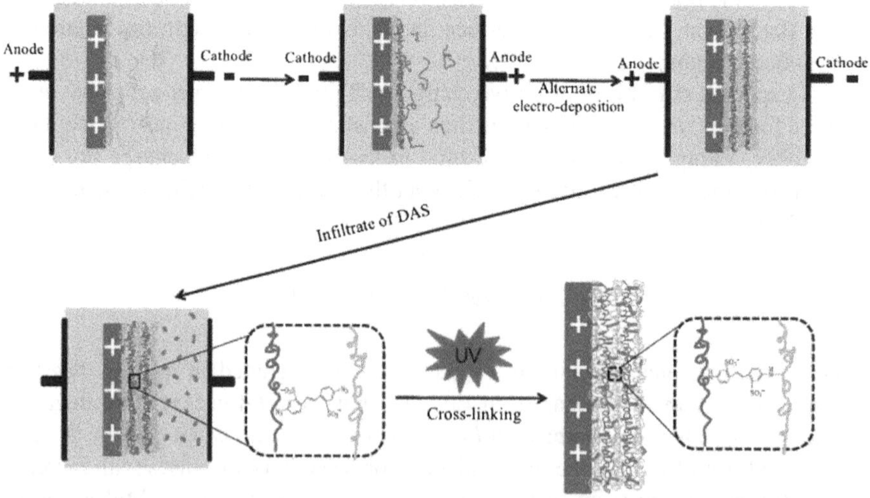

FIGURE 17.15 Schematic diagram of cross-linked electrodeposition multilayer membranes modification. (From Liu, H. et al., Journal of Membrane Science. 543, 310–318, 2017.)

TABLE 17.2
Example of Monovalent Anion Selectivity Reported in Recent Years

Membrane	Layer	P_{Cl}^{SO4}	Note	References
homogenous AMX (Astom, Japan)	PSS/PAH	0.55	LbL	Mulyati et al., 2013a
homogenous JAM-II-07 (Yanrun, China)	sulfonated reduced graphene oxide	0.43[a]	adsorption	Zhao et al., 2017
homogenous (Fujifilm, Japan)	PSS/HACC	0.34[a]	electrodeposition	Zhao et al., 2016b
homogenous (Fujifilm, Japan)	PSS/HACC	0.23[a]	cross-linked electrodeposition	Liu et al., 2017
homogenous (Fujifilm, Japan)	polyethyleneimine	0.23[a]	electrodeposition	Pan et al., 2017
homogenous AM-PP (Mega, Czech Republic)	carbon nanotube	0.806	coating	Fernandez-Gonzalez et al., 2017a
homogenous AM-PP (Mega, Czech Republic)	iron oxide	0.81	coating	Fernandez-Gonzalez et al., 2017a
homogenous AMX (Astom, Japan)	polydopamine	0.22	adsorption	Vaselbehagh et al., 2015
heterogeneous[b]	polyethyleneimine	0.35	grafting	Wang et al. 2015

[a] $1/P_{SO4}^{Cl}$

[b] Zhe-jiang Qian-qiu Environmental Protection & Water Treatment, China.

17.5 SIMULTANEOUS IMPROVEMENT OF ANTIFOULING PROPERTY AND MONOVALENT ANION SELECTIVITY

Most researchers have focused on the improvement of either antifouling property or monovalent anion selectivity. Few research works have been conducted to simultaneously improve both the antifouling property and the monovalent anion selectivity. Mulyati et al. (2013a) showed that the antifouling property and the monovalent anion selectivity of AEM could be simultaneously improved by the surface modification with $(PSS/PAH)_nPSS$, owing to the improvement in the negative membrane surface charge density and hydrophilicity. The dominant factor determining the antifouling property is the hydrophilicity of the membrane surface with sufficient negative charge (refer to Section 17.3.2), while the dominant factor determining the monovalent anion selectivity is the total excess negative charge within the LbL layer (refer to Section 17.4.2). Thus, the optimum condition for antifouling potential is different from that for monovalent anion selectivity. But, Mulyati et al. (2013a) reported that it was possible to improve both properties simultaneously by optimizing the surface modification conditions. For example, in the case of AMX modified with PSS/PAH multilayer, AMX modifed with more than 15 layers (PSS top layer) had a high monovalent anion selectivity (P_{Cl}^{SO4} = 0.55, Figure 17.14). This membrane also had an adequate antifouling property (transition time ≈ 50 min., Figure 17.12) even though it was not a mixmum.

Vaselbehagh et al. (2014; 2015) reported that the antifouling potential and the monovalent anion selectivity were improved by the surface modification with polydopamine. Fernandez-Gonzalez et al. (2017a; 2017b) also reported that these properties were improved by the surface modification with carbon nanotubes or sulfonated iron oxide nanoparticles. Thus, it is clear that the antifouling property and the monovalent anion selectivity are improved simultaneously by the surface modification of AEM.

17.6 EFFECT OF MEMBRANE SURFACE ON THE OPTIMUM CONDITION OF LbL DEPOSITION

As discussed above, the multilayer surface modification is a very useful method to improve the antifouling potential and the monovalent anion selectivity of AEM. However, Mulyati et al. (2013b) reported that the optimum condition of surface modification depended on the membrane surface. They modified the surface of commercial monovalent anion selective AEM, ACS (Astom Corp., Tokyo, Japan), with PSS/PAH multilayer and compared with AMX modified with PSS/PAH (see Section 17.3.2).

Figure 17.16 shows the transition time as a function of the number of layers for the original and modified ACS membranes. The transition times for an even number of layers (PAH top layer) were almost zero. This is the same for AMX with PAH top

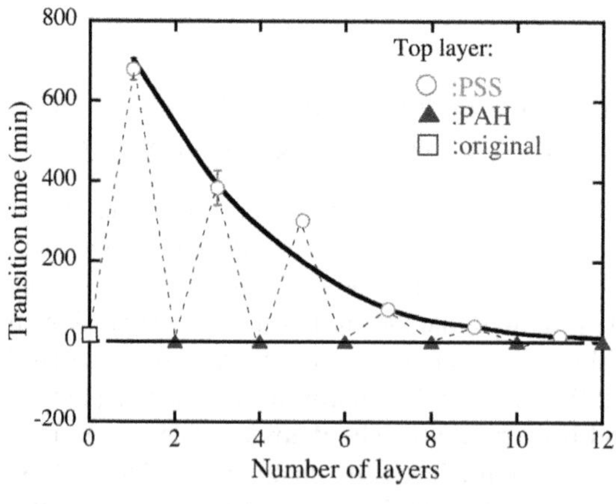

FIGURE 17.16 The transition time of modified ACS. The feed solution contained 0.05 M NaCl and 0.052 kg/m³ SDBS. The square represents the original membrane, circles the membrane with PSS top layer, and triangles the membrane with PAH top layer. The modification solution contained 0.3 kg/m³ of PSS or PAH, and 1.0 M NaCl. (From Mulyati, S. et al., Membrane. 38, 137–144, 2013b.)

layer shown in Figure 17.12. In contrast, for odd numbers of layers (PSS top layer), the transition time showed a maximum with the first layer and decreased with the number of layers and approached zero, while the transition time showed maximum at seven layers and then decreased with the number of layers for AMX. Mulyati et al. (2013b) explained this phenomenon as follows: the ACS surface had a negative charge and the ACS surface was fully covered by negative charge layer with first PSS layer deposition. Thus, the first layer on ACS corresponded to the seven layers on AMX. The water contact angle and the ζ-potential of modified ACS surface also showed the variation with the number of layers that were well explained by considering the first layer corresponded to seven layers of AMX. This phenomenon clearly shows that the optimum condition of surface modification depends on the membrane surface.

17.7 CONCLUSION

The recent research on the improvement of antifouling property and monovalent anion selectivity of AEMs by surface modification, mainly polyelectrolyte LbL deposition, was reviewed in this chapter. It is clear that the antifouling property and the monovalent anion selectivity are improved upon surface modification as a result of improved surface negative charge and hydrophilicity.

The dominant factor determining the antifouling property is the hydrophilicity of the membrane surface with negative charge, while the factor determining the monovalent anion selectivity is the total excess negative charge within the LbL layer. Thus, the optimum condition for antifouling property is different than that for monovalent anion selectivity. However, it is possible to improve both properties simultaneously by optimizing the surface modification condition.

In conclusion, the surface modification with polyelectrolyte deposition by LbL is simple and easy in process. Meanwhile, it is a very effective method to simultaneously improve the antifouling property and the monovalent anion selectivity of AEM. However, the important point to notice is that the optimum condition depends on an original membrane surface.

REFERENCES

Ba, C., Ladnerb, D.A., Economy, J., 2010. Using polyelectrolyte coatings to improve fouling resistance of a positively charged nanofiltration membrane. Journal of Membrane Science. 347, 250–259.

Bukhovets, A., Eliseeva, T., Oren, Y., 2010. Fouling of anion-exchange membranes in electrodialysis of aromatic amino acid solution. Journal of Membrane Science. 364, 339–343.

Decher, G., 1997. Fuzzy nanoassemblies: Toward layered polymeric multicomposites. Science. 277, 1232–1237.

Dubas, S.T., Schlenoff, J.B., 1999. Factors controlling the growth of polyelectrolyte multilayers. Macromolecules. 32, 8153–8160.

Fernandez-Gonzalez, C., Kavanagh, J., Dominguez-Ramos, A., Ibañez, R., Irabien, A., Chen, Y., Coster. H., 2017a. Electrochemical impedance spectroscopy of enhanced layered nanocomposite ion exchange membranes. Journal of Membrane Science. 541, 611–620.

Fernandez-Gonzalez, C., Zhang, B., Dominguez-Ramos, A., Ibañez, R., Irabien, A., Chen, Y., 2017b. Enhancing fouling resistance of polyethylene anion exchange membranes using carbon nanotubes and iron oxide nanoparticles. Desalination. 411, 19–27.

Hoffmann, K., Tieke, B., 2009. Layer-by-layer assembled membranes containing hexacyclen-hexaacetic acid and polyethyleneimine N-acetic acid and their ion selective permeation behaviour. Journal of Membrane Science. 341, 261–267.

Kabay, N., Arra, O., Samatya, S., Yuksel, U., Yuksel, M., 2008. Separation of fluoride aqueous solution by electrodialysis: Effect of process parameters and other ionic species. Journal of Hazardous Materials. 153, 107–113.

Kabay, N., Demircioglu, M., Ersiiz, E., Kurucaovali, I., 2002. Removal of calcium and magnesium hardness by electrodialysis. Desalination. 149, 342–349.

Kesore, K., Janowski, F., Shaposhnik, V.A., 1997. Highly effective electrodyalisis for selective elimination of nitrate from drinking water. Journal of Membrane Science. 127, 17–24.

Klitzing, R., Möhwald, H., 1995. Proton concentration profile in ultrathin polyelectrolyte films. Langmuir. 11, 3554–3559.

Kolasińska, M., Warszyński, P., 2005. The effect of support material and conditioning on wettability of PAH/PSS multilayer films. Bioelectrochemistry. 66, 65–70.

Krasemann, L., Tieke, B., 1999. Selective ion transport across self-assembled alternating multilayers of cationic and anionic polyelectrolytes. Langmuir. 16, 287–290.

Lee, H.-J., Hong, M.-K., Han, S.-D., Shim, J., Moon, S.-H., 2008. Analysis of fouling potential in the electrodialysis process in the presence of an anionic surfactant foulant. Journal of Membrane Science. 325, 719–726.

Lindstrand, V., Soundstrom, G., Jonsson, A., 2000. Fouling of electrodialysis by organic substances. Desalination. 128, 91–102.

Liu, H., Ruan, H., Zhao, Y., Pan, J., Sotto, A., Gao, C., Bruggen, B., Shen, J., 2017. A facile avenue to modify polyelectrolyte multilayers on anion exchange membranes to enhance monovalent selectivity and durability simultaneously. Journal of Membrane Science. 543, 310–318.

Mahler, R.L., Colter A, Hirnyck R., 2007. Quality Water for Idaho; Nitrate and Groundwater. University of Idaho Extention. (http://www.cals.uidaho.edu/edcomm/pdf/CIS/CIS0872.pdf.)

Mulyati, S., Takagi, R., Fujii, A., Ohmukai, Y., Maruyama, T., Matsuyama, H., 2012. Improvement of the antifouling potential of an anion exchange membrane by surface modification with a polyelectrolyte for an electrodialysis process. Journal of Membrane Science. 417–418, 137–143.

Mulyati, S., Takagi, R., Fujii, A., Ohmukai, Y., Maruyama, T., Matsuyama, H., 2013a. Simultaneous improvement of the monovalent anion selectivity and antifouling properties of an anion exchange membrane in an electrodialysis process, using polyelectrolyte multilayer deposition. Journal of Membrane Science. 431, 113–120.

Mulyati, S., Takagi, R., Ohmukai, Y., Maruyama, T., Matsuyama, H., 2013b. Effect of the membrane surface on performance improvements to anion exchange membranes for electrodialysis through Layer-by-Layer deposition. Membrane. 38, 137-144.

Noble, R.D., Stern, S.A., 1995. *Membrane Separations Technology: Principles and Applications* (1st Ed.). Elsevier Science.

Ouyang, L., Malaisamy, R., Bruening, M.L., 2008. Multilayer polyelectrolyte films as nanofiltration membranes for separating monovalent and divalent cations. Journal of Membrane Science. 310, 76–84.

Pan, J., Ding, J., Tan, R., Chen, G., Zhao, Y., Gao, C., Van der Bruggen, B., Shen, J., 2017. Preparation of a monovalent selective anion exchange membrane through constructing a covalently crosslinked interface by electro-deposition of polyethyleneimine. Journal of Membrane Science. 539, 263–272.

Persico, M., Mikhaylin, S., Doyen, A., Firdaous, L., Bazinet, L., 2016. How peptide physico-chemical and structural characteristics affect anion-exchange membranes fouling by a tryptic whey protein hydrolysate. Journal of Membrane Science. 520, 914–923.

Sata, T., 2000. Studies on anion exchange membranes having permselectivity for specific anions in electrodialysis—Effect of hydrophilicity of anion exchange membranes on permselectivity of anions. Journal of Membrane Science. 167, 1–31.

Sata, T., Mine, K., Tagami, Y., Higa, M., Matsusaki, K., 1998. Changing permselectivity between halogen ions through anion exchange membranes in electrodialysis by controlling hydrophilicity of the membranes. Journal of the Chemical Society, Faraday Transactions. 94, 147–153.

Schlenoff, J.B., Dubas, S.T., 2001. Mechanism of polyelectrolyte multilayer growth: charge overcompensation and distribution. Macromolecules. 34, 592–598.

Schlenoff, J.B., Ly, H., Li, M., 1998. Charge and mass balance in polyelectrolyte multilayers. Journal of the American Chemical Society. 120, 7626–7634.

Shubint, V., Linse, P., 1995. Effect of electrolytes on adsorption of cationic polyacrylamide on silica: Ellipsometric study and theoretical modeling. Journal of Physical Chemistry. 99, 1285–1291.

Smara, A., Delimi, R., Chainet, E., Sandeaux, J., 2007. Removal of heavy metals from diluted mixtures by a hybrid ion-exchange/electrodialysis process. Separation and Purification Technology. 57, 103–110.

Stanton, B.W., Harris, J. J., Miller, M.D., Bruening, M.L., 2003. Ultrathin, multilayered polyelectrolyte films as nanofiltration membranes. Langmuir. 19, 7038–7042.

Steeg, H.G.M., Stuart, M.A.C., Keizer, A., Bijsterbosch, B.H., 1992. Polyelectrolyte adsorption: A subtle balance of forces. Langmuir. 8, 2538–2546.

Strathmann, H., 2004. *Ion-exchange Membrane Separation Processes* (1st Ed.). Elsevier Ltd.

Strathmann, H., 2010. Electrodialysis, a mature technology with a multitude of new applications. Desalination. 264, 268–288.

Takagi, R., Mulyati, S., Arahman, N., Ohmukai, Y., Maruyama, T., Matsuyama, H., 2011. Time dependence of transport ratio during electrodialysis process. Desalination and Water Treatment. 34, 25–31.

Takagi, R., Vaselbehagh, M., Matsuyama. H., 2014. Theoretical study of the permselectivity of an anion exchange membrane in electrodialysis. Journal of Membrane Science. 470, 486–493.

Vaselbehagh, M., Karkhanechi, H., Mulyati, S., Takagi, R., Matsuyama, H., 2014. Improved antifouling of anion-exchange membrane by polydopamine coating in electrodialysis process. Desalination. 332, 126–133.

Vaselbehagh, M., Karkhanechi, H., Takagi, R., Matsuyama, H., 2015. Surface modification of an anion exchange membrane to improve the selectivity for monovalent anions in electrodialysis – Experimental verification of theoretical predictions. Journal of Membrane Science. 490, 301–310.

Wang, W., Fu, R., Liu, Z., Wang, H., 2017. Low-resistance anti-fouling ion exchange membranes fouled by organic foulants in electrodialysis. Desalination. 417, 1–8.

Wang, X., Wang, M., Jia, Y., Wang, B., 2015. Surface modification of anion exchange membrane by covalent grafting for imparting permselectivity between specific anions. Electrochimica Acta. 174, 1113–1121.

Xu, R., Wang, S., Zhao, H., Wu, S., Xu, J., Li, L., Liu, X., 2015. Layer-by-layer (LBL) assembly technology as promising strategy for tailoring pressure-driven desalination membranes. Journal of Membrane Science. 493, 428–443.

Zhao, Y., Tang, K., Liu, H., Bruggen, B.V., Díaz, A.S., Shen, J., Gao, C., 2016a. An anion exchange membrane modified by alternate electro- deposition layers with enhanced monovalent selectivity. Journal of Membrane Science. 520, 262–271.

Zhao, Y., Tang, K., Ruan, H., Xue, L., Bruggen, B.V., Gao, C., Shen, J., 2017. Sulfonated reduced graphene oxide modification layers to improve monovalent anions selectivity and controllable resistance of anion exchange membrane. Journal of Membrane Science. 536, 167–175.

Zhao, Z., Cao, H., Shi, S., Li, Y., Yao, L., 2016b. Characterization of anion exchange membrane modified by electrodeposition of polyelectrolyte containing different functional groups. Desalination. 386, 58–66.

Zou, L., Vidalis, I., Steele, D., Michelmore, A., Low, S.P., Verberk, J.Q.J.C., 2011. Surface hydrophilic modification of RO membranes by plasma polymerization for low organic fouling. Journal of Membrane Science. 369, 420–428.

18 Membranes for Bioenergy Production and Wastewater Treatment in Microbial Fuel Cells

María Jose Salar-García, Víctor Manuel Ortiz-Martínez, Antonia Pérez de los Ríos, and Francisco José Hernández-Fernández

CONTENTS

18.1 INTRODUCTION

Sustainable energy sources are needed to face current pressing environmental problems, e.g., global warming and climate change, as well as future challenges such as the depletion of fossil fuels. In this context, microbial fuel cell (MFC) technology has drawn increasing attention due to the possibility of converting the chemical energy present in a substrate directly into electrical energy by action of microorganisms (Santoro et al., 2017). The types of substrates that can be used in these devices range from simple organic substrates to biomass and wastes. The use of wastewater as fuel is of special interest since energy recovery and water treatment take place simultaneously (Gude, 2016).

Ideally, in a MFC device, oxidation is performed at the anode by bacteria producing protons and electrons. While electrons are transferred to the cathode through an external circuit, protons typically pass through a semipermeable separator to the cathode to balance the cell. The cathode reaction involves the reduction of an electron acceptor such as oxygen, in which electrons and protons from the anode are consumed. However, in practice, when complex substrates are used, not only proton exchange is involved, but also the exchange of other ion species (Hernández-Fernández et al., 2015b). The ions transferred through the membrane can take part in secondary reactions at the cathode depending on the exchange mechanism of the separator employed. Figure 18.1 shows a schematic representation of two typical arrangements in MFCs. While double-chamber MFCs comprise both anodic and cathodic chambers with electrodes submerged in respective solutions, single-chamber MFCs only have an anodic chamber, and the cathode electrode is generally assembled with the membrane and exposed to the air. In recent years, many works have focused on developing new materials for the main components of MFC devices as well as on designing new configurations for the optimization of this technology. The components that have received more attentions are the anode and cathode electrodes and the separator (Rinaldi et al., 2008; Santoro et al., 2017).

The separator is a key element in an MFC system that keeps the anode and the cathode physically separated. It has important implications on the overall performance since crucial operational factors directly depend on the nature and properties of this component. A good separator must display selective properties in order to allow proton movement from the anode to the cathode (high proton conductivity) and, at the same time, must limit oxygen diffusion to the anode. If oxygen penetrates into the anodic chamber, it acts as electron acceptor in this compartment, preventing electrons from reaching the cathode and thus reducing the current output in the

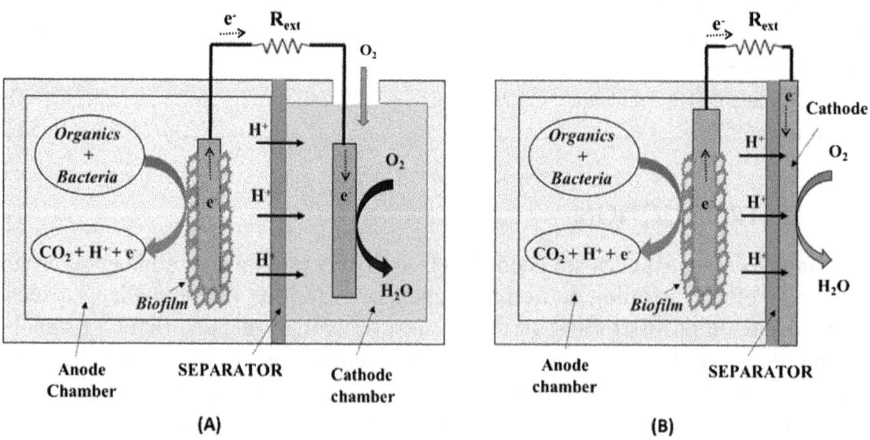

FIGURE 18.1 Schematic representation of double-chamber (A) and single chamber (B) MFCs.

system. The separator in a MFC also has an important influence on other electrical performance parameters such as coulombic efficiency and internal resistance. This component is one of the main contributors to the total internal resistance of the device. High values of internal resistance lead to the decrease of power output (Logan, 2007). On the other hand, the separator should prevent the occurrence of biofouling phenomenon, which implies the accumulation of biomaterial over the separator surface with a consequent blocking effect on proton/ions exchange mechanisms (Zhang et al., 2009).

In addition, the cost of the separator is a critical aspect when it comes to up-scaling this technology. Nafion is a reference material as a proton exchange membrane (PEM) in MFCs due to its outstanding properties such as high proton conductivity and mechanical stability but, by contrast, it is very expensive. Therefore, efficient and low costs separators are needed for the practical implementation of MFC technology (Leong et al., 2013). Another design approach consists of the construction of membraneless MFCs. However, this type of configuration brings on operational problems such as substrate and oxygen crossover through the cathode and the anode, respectively, which greatly reduces coulombic efficiency (Hernández-Fernández et al., 2015b).

This chapter focuses on the developments of new materials for the construction of MFC separators. It will cover the different types of separators that have been employed in MFC systems and discuss in detail the advantages and drawbacks of each separator design. Among the separators included in this chapter are cation and anion exchange membranes, polymer/composite membranes and porous material based separators. Table 18.1 summarizes the types of separators that will be analyzed in this chapter.

TABLE 18.1
Types of MFC Separators

Separator	Advantages	Drawbacks
Ion exchange membranes (CEM, AEM, BEM)	• High ion selectivity including proton conductivity • Long term chemical and thermal stability	• High cost • Substrate and oxygen crossover • Biofouling • pH splitting
Salt bridge	• Simple set-up • Low cost • Low oxygen diffusion	• High internal resistance • Low performance
Porous size-selective (MF, UF, J-cloth, glass fibre, ceramic materials, natural rubber, etc)	• High proton selectivity • Low pH spliting • Low cost • Low biofouling • Long term stability	• High internal resistance • J-cloth exhibits high oxygen permeation • Ceramic materials are brittle

18.2 ION EXCHANGE MEMBRANES (IEMs)

The most widespread separators employed in MFCs are ion exchange membranes (IEMs). They can be grouped in cation exchange membranes (CEMs), anion exchange membranes (AEMs) and bipolar exchange membranes (BEMs). This classification is made according the type of ions that can be transported through them. Figure 18.2 represents ion transfer across IEMs (Li et al., 2011). As seen, cations are selectively transported from the anode to the cathode when CEMs are used as separator, while in the case of AEMs, anions are transferred from the cathode to the anode. BEMs incorporate the assembly of the two mentioned types of monopolar membranes, a CEM and an AEM, allowing the simultaneous movement of cations and anions to the two MFC chambers. When a CEM and an AEM are assembled with a separation between them forming a middle chamber, they can be used for desalination purposes (see Figure 18.2(C)), as will be discussed later.

18.2.1 Cation Exchange Membranes

Among the different types of ion exchange membranes, cation exchanges membranes are commonly used as separators in MFCs. Some of them are specifically designed to transfer protons from the anode to the cathode. The perfluorosulfonic acid membrane Nafion(DuPont) is one of the most widespread commercial separators. It contains hydrophilic sulfonate groups attached to a hydrophobic fluorocarbon backbone. The negatively charged sulfonate groups enable the transfer of several cations (Oh and Logan, 2006). Nafion membranes come in different thickness and types. For instance, Nafion-112 is thinner than Nafion-117, and its performance as MFC separator is higher in terms of power production despite displaying higher oxygen permeability (Rahimnejad et al., 2010). As mentioned above, the main drawback of this type of membranes is its high cost, which has promoted the development of other membranes with lower price (Ghasemi et al., 2013).

Ultrex CMI 7000 is another type of commercial CEM commonly used in MFC technology. It is a strong acid polymer membrane consisting of polystyrene cross-linked with divinylbenzene and has sulphonic groups responsible for cation exchange.

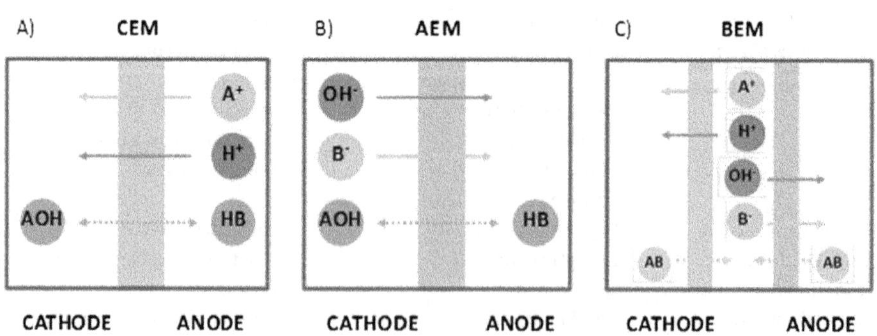

FIGURE 18.2 Transference across: (A) cation; (B) anion and (C) bipolar (for desalination purposes) exchange membranes.

Despite its relatively high-ohmic resistance, this type of membrane is as widely used as Nafion since both exhibit similar cation conductivity and mechanical stability (Harnisch et al., 2008).

There are other materials suitable as cation exchangers such as Hyflon, which consists of short-side-chain perfluorosulfonic acid membranes. This material, as well as Ultrex, exhibits higher ohmic resistance than Nafion but shows better chemical stability and ionic conductivity. By contrast, Zirfon, a membrane made of zirconium dioxide (ZrO_2) and polysulfone, offers lower resistance than Nafion but its permeability to the oxygen is higher (Arico' et al., 2006; Ieropoulos et al., 2010; Pant et al., 2010).

All these CEMs show several limitations. First, the above-discussed membranes can offer higher selectivity values towards Na^+, K^+ or Ca^+ versus protons under neutral conditions. This fact promotes pH splitting between the anodic and cathodic chambers. The accumulation of protons in the anode can negatively affect the bioelectrocatalytic activity of bacteria. By contrast, the performance of air-cathodes may be affected by the development of a carbonate salt layer over their surface. The high internal resistance of these CEMs also represents a constraint. Finally, the growth of biofilm around CEMs during long operation times tends to reduce MFC performance.

The modification of conventional perfluorosulfonic acid polymer membranes, the use of sulfonated and non-sulfonated aromatic hydrocarbon and polymer composites are among the usual approaches to obtain CEMs with enhanced properties to be employed in fuel cell devices (Vaghari et al., 2013). For example, several polymer composites including Nafion have been developed as MFC separators in order to reduce the above-mentioned drawbacks of this material. This is the case of Nafion-112 membranes impregnated with polyaniline. The interaction of polyaniline with sulfonic groups from the Nafion-112 matrix significantly affects membrane performance by increasing proton conductivity while decreasing biofouling phenomenon. The enhanced properties make it possible to increase the level of power density by nine times in the presence of polyaniline (124 $mW.m^{-2}$ in comparison with 14 $mW.m^{-2}$ when using neat Nafion-112) and to achieve almost the same level of power output achieved with Nafion-117 (Mokhtarian et al., 2013).

Nanomaterial-based composites are a new pathway for the development of alternative MFC membranes. The inclusion of nanoparticles into a polymer matrix changes its chemical and physical properties. For instance, they can enhance the selectivity and permeation of the polymer membrane towards certain chemical species boosting power performance (Rahimnejad et al., 2012). Activate carbon nanofibers and Nafion composites have shown to generate higher levels of power production than neat Nafion-112 and Nafion-117, with a higher associated coulombic efficiency (18%). The presence of carbon nanofibers changes the physical properties of neat Nafion such as roughness and porosity so that oxygen crossover to the anode decreases while power performance improves (Ghasemi et al., 2012a). Rahimnejad et al. (2012) developed polyethersulfone composite membranes with magnetite (Fe_3O_4) nanoparticles at different concentrations (from 5 to 20 wt%) by casting technique and compared their behaviour in double chamber MFC devices with Nafion-117 as a conventional PEM. The nanocomposite with an intermediate concentration of 15 wt% of

Fe_3O_4 outperformed the Nafion-117, offering maximum values of power densities of 20 $mW.m^{-2}$ and 148 $mA.m^{-2}$, respectively, which represent almost 30% more energy in comparison with the Nafion membrane. It was also found that more than 15% of Fe_3O_4 is an observable aggregation state over the membrane surface that diminishes power performance. As seen above, polyaniline has been employed with Nafion to synthesize MFC polymer composites. In line with this approach, polysulfone membranes doped with polyaniline have been proposed as new class of PEM. The interest of using this low cost polymer as composite material lies on its conductive and electroactive properties. However, the nature of the matrix polymer greatly influences on the suitability of the doping with nanoparticles. Ghasemi et al. (2012b) found that the doping of polysulfone with polyaniline nanoparticles disfavors power performance versus the undoped membrane since polyaniline nanoparticles can block the proton exchange pathways. In the case of the undoped polysulfone membrane, a significant percentage out of the power density generated with a Nafion-117 membrane (more than 80%) can be achieved.

Low cost polymers such as sulfonated polyether ether ketones (SPEEK) are also an alternative to commercial CEMs. In SPEEKs, the degree of sulfonation (DS) can be modified by varying the conditions of the sulfonation reaction (Yee et al., 2013). SPEEK shows a similar structure to Nafion, but the aromatic groups contained in SPEEKs make them more rigid, which increases their internal resistance (Rinaldi et al., 2008). On the other hand, their low conductivity and high activation losses negatively affect electrical MFC performance (up to 20% less compared with Nafion). However, the wastewater treatment capacity of MFCs using these membranes can be comparable to that displayed by the Nafion-based devices. The most important benefit of SPEEK versus Nafion is its lower price, which can fall by as much as two-thirds (Ghasemi et al., 2013).

Advanced membrane materials based on SPEEK include composites prepared with nanoparticles of metal oxides such as titanium dioxide both in the forms of pure oxide (TiO_2) or as acidic additive with a sulfonic acid group (TiO_2-SO_3H) (Ayyaru and Dharmalingam, 2013). The introduction of metal oxides into a SPEEK matrix has clearly shown to improve the membrane performance in comparison with the neat SPEEK membrane by increasing maximum power performance from 11% to over 75% when tested in single-chamber MFCs. However, the use of SPEEK/TiO_2-SO_3H is preferable versus SPEEK/TiO_2. Ayyaru and Dharmalingam (2013) achieved a significant power density level of ~1200 $mW.m^{-2}$ by using a SPEEK/TiO_2-SO_3H (at a percentage of 7.5%w/w of TiO_2-SO_3H), while in the same system the membrane SPEEK/TiO_2 only achieved around 770 $mW.m^{-2}$. The neat SPEEK offered a power density of 676 $mW.m^{-2}$. The presence of the acidic active leads to a higher water uptake capacity, which in turn improves the proton conductivity of the membrane up to $1.382 \cdot 10^{-2}$ Scm^{-1}, while this value remains at $0.5391 \cdot 10^{-2}$ Scm^{-1} for the SPEEK/TiO_2 and at $0.78 \cdot 10^{-2}$ Scm^{-1} for the neat SPEEK membrane. Moreover, comparing the values with other reported results, SPEEK exhibited a lower oxygen permeability of up to one order of magnitude when compared with Nafion-117 (Ayyaru and Dharmalingam, 2013; Rhee et al., 2006).

Another low cost polymer used as cation exchange membrane in MFCs is sulfonated polystyrene-ethylene-butylene-polystyrene (PSEBS). The structure of this thermoplastic elastomer consists of styrene blocks dispersed in an ethylene-butylene matrix. It is a long-term stable material because of the absence of fluorinated groups in its aromatic backbone combined with a hydrocarbon network. SPSEBS membranes allow MFCs to reach higher power densities than Nafion. SPSEBS-based separators exhibit higher proton conductivity, which enhances overall MFC performance and prevents proton accumulation in the anode, reducing pH gradient between anode and cathode (Ayyaru et al., 2012; Mishra et al., 2012).

Polyethersulfone has been researched in combination with SPEEK as MFC membranes. Specifically, polyethersulfone membrane incorporated with 5% of SPEEK was capable of producing 140 $mW.m^{-2}$, 50% more than the power output achieved by the Nafion membranes, when used in a membrane-electrode assembly in which the membrane was hot-pressed onto the cathode. This composite membranes offered higher conductivity in comparison with Nafion (and versus neat polyethersulfone membranes) (Daud et al., 2011).

18.2.2 Anion Exchange Membranes

The second most widely used separator type in MFCs is anion exchange membranes. This type of material overcomes the inefficient proton transfer of CEMs. In this case, AEMs facilitate the proton transport by using functionalized groups such as carbonate or phosphate that act as anion carriers. In addition, these anions also favour the balance of pH between anodic and cathodic chambers by avoiding pH splitting issues (Kim et al., 2007). This fact improves MFC performance in comparison with CEMs. Kim et al. (2007) compared the performance of AEMs and CEMs, demonstrating that the former are capable of offering higher power density, reaching a maximum power output of 0.48 $W.m^{-2}$. Similar results have been obtained in other studies that employed membrane cathode assemblies (MCAs) based on CEMs and AEMs. In such cases, MFCs set up with CEMs achieved almost half of the power output reached by MFCs using AEMs (Kim et al., 2007; Zuo et al., 2008). In addition to higher power performance, AEMs can display higher coulombic efficiency and better long-term stability versus CEMs. This can be explained by the enhanced properties of AEMs, which present: (i) low oxygen permeability, ensuring anaerobic condition in the anodic chamber, (ii) low pH gradient between anodic and cathodic chambers, (iii) less cathode resistance, (iv) less amount of precipitates on the cathode-catalyst and (v) higher ionic conductivity due to their low internal ohmic resistances (Varcoe et al., 2014). On the contrary, the two main drawbacks of AEMs are: (i) their high permeability to the substrate, which promotes biofouling phenomenon and (ii) they can be easily deformed in single-chamber MFCs. Thus, further efforts are still needed to improve the practical applications of AEMs (Zhang et al., 2010).

To address the biofouling issue, Elangovan and Dharmalingam (2017) have recently proposed the modification of the surface of an AEM with ethanolamine (AEOH). This compound creates hydrophilic functional groups such as amide (CONH) and hydroxyl groups (OH) over the surface of the membrane that does not affect its morphology. These authors synthesized a hydrophilic quaternized poly(ether imide) supported membrane and evaluated the effect of the modification time on the bacteria adhesion among other properties. They reported that the optimum time is 30 min, which allows MFCs to reach high performance. In their previous work, Elangovan and Dharmalingam (2016) also developed an easy method to prevent the biofouling on AEMs. The technique consists of modifying the surface of functionalized graphene oxide (FGO) sheets with poly(-diallyldimethylammonium)chloride (PDDA) employing a physical absorption followed by a dispersion step into a quaternized polysulphone (QPSU) anion exchange membrane. This process increases the density of the anion exchange zones of the separator. Among the ratio investigated, they observed that the modified membranes QPSU with a proportion of 1.0 wt% of FGO outperform commercial AMI-7001 membranes in terms of power output and long term stability, doubling the power density obtained by the commercial membrane. While the membrane AMI-7001 was able to achieve up to 576 $mW.m^{-2}$ (and 1800 $mA.m^{-2}$), the maximum power output of the QPSU/FGO membrane was 1036 $mW.m^{-2}$ (and 2880 $mW.m^{-2}$). The enhanced performance of QPSU/FGO in comparison with the commercial separator is mainly due to due its tailored anti-biofouling properties.

Poly(vinyl alcohol) (PVA) is often used as a polymer matrix for the fabrication of membranes due its advantageous properties such as anti-fouling properties, physicochemical properties and low cost. This polymer has been functionalized to form AEMs for their application in MFC with other compounds including metal oxides. Tao et al. (2015) improved the performance of MFCs using novel organic-inorganic hybrid anion exchange membranes. They compared the performance of MFCs working with Nafion-117, a hybrid membrane based on titanium dioxide (TiO_2)-quaternized poly(vinyl alcohol) (QAPVA/TiO_2) and LeHoAM-III, a commercial AEM. The second material type showed better oxygen resistance. Furthermore, the hybrid membrane QAPVA/TiO_2 allowed MFCs to reach a maximum power output of approximately 125 $W.m^{-2}$ versus 65 $W.m^{-2}$ in the case of QAPVA/TiO_2, which represents over 92% increase in power output in comparison to the commercial AEM.

18.2.3 Bipolar Exchange Membranes

Bipolar exchange membranes (BPMs) consist of two monopolar membranes assembled together. Unlike CEMs and AEMs, BPMs allow protons and hydroxide ions generated from water molecule breakdown to be simultaneously transported. In this case, there is no competition between the transport of cation and anions, mitigating the acidification of the anodic chamber. In this type of membranes, the ionic conductivity is caused by water splitting instead of ion flux across the membrane. Very recently, several authors have proven that BEMs can offer higher power performance in MFCs when compared with AEMs and CEMs since this configuration would provide an increase in the ion conductivity of the assembled membrane (Khera and Chandra, 2017). However, BPMs have also shown to increase both the polarization potential of the membrane and the internal resistance of the system (Harnisch et al., 2008; Hurwitz and Dibiani, 2001). On the contrary, this type of membrane offers many advantages for water treatment, especially for desalination processes in a modified configuration of a MFC known as desalination microbial cell (DFC), which includes a middle chamber (See Figure 18.2(C)). When the CEM and the AEM that form a BEM are assembled separated, the ions present in the waster trapped between these monopolar membranes move to the electrode chambers while the desalination of the water takes place (see Figure 18.2(C)) (Cao et al., 2009; Li et al., 2011).

Bipolar membranes can be also useful to maintain the optimal conditions for cathodic reactions that involve the reduction of metallic species assembled with graphite electrodes. BEMs in combination with graphite electrodes assembled in the form of flat plates have been investigated to operate MFCs that include the reduction of ferric iron (redox couple Fe^{3+}/Fe^{2+}) as an electron mediator for oxygen reduction. In such case, a low pH in the cathode chamber is required to ensure the solubility of ferric iron, which can be achieved by the use of a bipolar membrane. Using this configuration as an efficient cathode system, Heijne et al. (2006) have reported a maximum power density of 0.86 W/m^2 with a high associated coulombic efficiency of up to 95%.

18.2.4 Salt Bridges

Salt bridges are an inexpensive option for proton exchange in MFCs. Salt bridges are used in double-chamber configurations in which the anode and the cathode compartments are connected through an electrolyte solution. The electrolyte solution is often placed in a glass tube and thus a membrane is not required (Sevda and Sreekrishnan, 2012). Typical electrolyte solutions include potassium chloride and phosphate buffer solutions in the presence of agar to separate these electrolytes and the anodic and cathodic bulk solutions (Daud et al., 2015). Several studies have shown that salt bridges can offer high coulombic efficiency values due to low oxygen diffusion rates to the anode chamber. In contrast, high values of internal resistance have been found for MFCs that employ salt bridges, which hinder the spread of their use as separator for this technology (Liu and Li, 2007; Min et al., 2005).

Min et al. (2005) compared the performances of a salt bridge and a conventional proton exchange membrane Nafion-117, respectively, in a double chamber system. The salt bridge consisted of a U-tube connecting the two chambers and filled with phosphate buffered solution. In the first case, the level of power density achieved was 2.2 $mW.m^{-2}$, which is clearly lower compared to that obtained when using the proton exchange membrane, 38 $mW.m^{-2}$. Oxygen diffusion to the anode chamber was no detected in the first case, while oxygen diffused to the anode at a rate of 0.014 $mg.h^{-1}$ in the case of the membrane based system. However, the key factor for the differences observed in terms of power density achieved was the internal resistance of the system, which was clearly higher in the salt bride based configuration.

Other works have studied the use of agar salt bridges as media for proton conduction in MFCs (Sevda and Sreekrishnan, 2012). This option consisted of a water solution of 10 wt% of agar and varying concentrations of sodium chloride (NaCl). The production of power densities ranges from 0.32 to 16.02 $mW.m^{-2}$, depending on the NaCl concentration employed (1 to 10 wt%). The results showed the maximum value of power density was obtained at 5% NaCl concentration with 10% agar. Under the optimal conditions, the system offered a removal percentage of chemical oxygen demand (COD) over 88% using synthetic wastewater. Figure 18.3 shows a schematic representation of a double chamber MFC system containing an agar salt bridge.

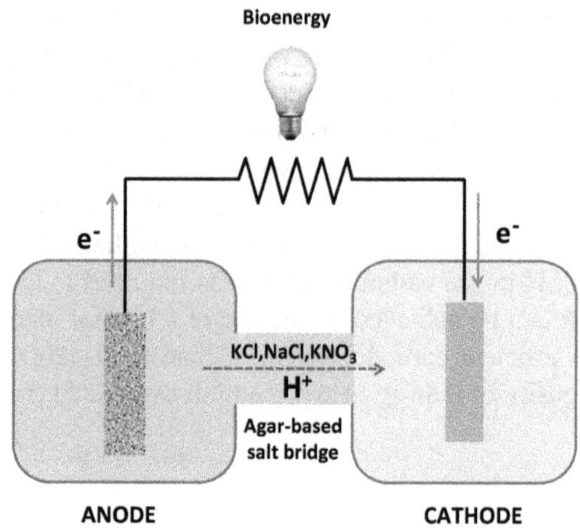

FIGURE 18.3 Double chamber MFC system employing an agar salt bridge.

18.2.5 Other Types of Ionic Exchange Membranes

Recently, novel membranes based on ionic liquids (ILs) have been employed as a separator in microbial fuel cells to replace conventional membranes. Ionic liquids are organic molten salts whose melting point is below 100°C. In recent years, these materials have attracted much attention in different research fields. Their unique properties such as long term chemical and thermal stability, high conductivity, near-zero vapor pressure and the possibility of modifying their properties by varying the anion and the cation of its structure have earned them the consideration of "green solvent". ILs have been employed in synthesis, catalysis, polymer science and as extraction media (Hallett and Welton, 2011; Han and Armstrong, 2007; Lu et al., 2009). Ionic liquids can be supported in a polymer matrix by ultrafiltration or immobilized by casting method. They can also be polymerized or gelled to form a membrane (Armand et al., 2009; Hernández-Fernández et al., 2015a; 2016).

Salar-García et al. (2015) employed a spectroscopy impedance technique to characterize the behavior of different IL-based polymer inclusion membranes as separators in MFCs. Among the ILs tested, triisobutyl-(methyl)-phosphonium tosylate showed the best performance in batch mode, reaching a maximum power output of 794.7 $mW.m^{-3}$. This value outperformed the same type of MFC system working with Nafion membranes (see Figure 18.4).

Later, a scaled-up MFC assembly was designed with a novel IL embedded type membrane. In this case, phosphonium and ammonium-based ionic liquids were used to prepared novel polymer inclusion membranes embedded into the cathode. This set up is more compact compared with a traditional assembly in which the membrane is placed between the electrodes after being manufactured separately. Its internal resistance is also lower, which favors the energy performance of MFCs. The up-flow MFC design working with the embedded phosphonium-based membrane reached a maximum power output of 12.3 $W.m^{-3}$ with a simultaneous wastewater treatment

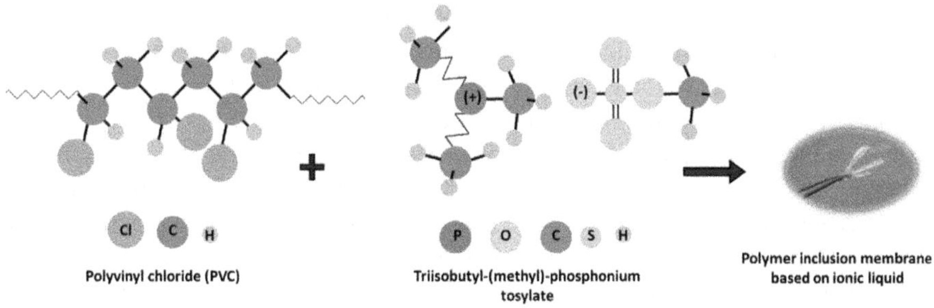

FIGURE 18.4 Synthesis of polymer inclusion membranes based on ionic liquids.

capacity of 60% in terms of COD removal at a feed flow rate of 0.25 $mL.min^{-1}$ (Ortiz-Martínez et al., 2015; Salar-García et al., 2016b).

Gohil and Karamanev (2016) also immobilized imidazolium-based IL into a modified poly(vinyl alcohol) matrix for its use as separator in a stacked MFC. They optimized the properties of the membranes synthesized by varying the cross-link time and achieved to increase the proton conductivity of the membranes up to 3.6 $mS.cm^{-1}$, with an associated maximum power output of 245 $mW.cm^{-2}$.

More recently, Koók et al. (2017) employed non-water soluble ILs based on imidazolium cations to prepare supported membranes. 1-butyl-3-methylimidazolium bis{(trifluoromethyl)sulfonyl}imide and 1-hexyl-3-methylimidazolium hexafluorophosphate were immobilized into an hydrophobic support based on polyvinylidene difluoride and were used as separators in double-chambered MFCs. The results in terms of power output were similar to those reached by commercial Nafion when MFCs were fed with glucose. However, when the devices were fuelled with low acetate loading, MFCs with the membrane containing the cation bis{(trifluoromethyl) sulfonyl}imide showed better performance than that of commercial membrane (77.2 $mW.m^{-3}$ *vs* 60.2 $mW.m^{-3}$). Although further work is still needed for a better understanding of the role of ILs as electrolytes in separators for MFCs, the recent research work has already showed the potential of ILs to replace commercial membranes.

18.3 POROUS SIZE-SELECTIVE SEPARATORS

This section covers porous material used as separators in MFC systems. This type of separators is not ion-selective but size-selective. This implies that the exchange mechanism is based on the presence of porous that enables the movement of electrolytes from the anode to the cathode. Porous separators are generally categorized into two groups, microporous and pore filter materials (Gugliuzza et al., 2013; Li et al., 2011).

Among microporous separators are microfiltration (MF) membranes and ultrafiltration (UF) membranes, which are suitable for proton exchange. Some works studying MF membranes have proven that this type of separators can reduce pH splitting across the membrane in comparison to the Nafion while offering similar values of internal resistance, similar power performance and high COD removal rates in double chamber systems (Tang et al., 2010). Moreover, they can attenuate membrane biofouling phenomena due to enhanced hydrophilicity and the presence of electrostatic repulsive forces (Huang et al., 2017). Considerable power density levels in the order of 18 $W.m^{-3}$ and coulombic efficiency over 70% have been reported for up-scaled MFCs equipped with a tubular UF membrane made of polysulfone on a composite polyester carrier (Zuo et al., 2007), showing the efficiency of microporous materials. UF membranes have also been employed as bifunctional components serving as both membrane and cathode for wastewater filtration in MFCs. In this case, power output reached over 14.5 $W.m^{-3}$ (Malaeb et al., 2013). Nevertheless, when microporous membranes are used, oxygen and substrate crossover to the anode and the cathode, respectively, needs to be addressed in order to avoid the decrease in power performance (Li et al., 2011).

Among the pore filter materials used as MFC separators are fabrics (J-cloth, glass fiber, nylon mesh, etc.), ceramic materials, natural rubber and cellulose (Daud et al., 2015). Both J-cloth and glass fiber have been investigated by Zhang et al. (2009) as effective separators, allowing similar values of power density to be obtained (46 $W.m^{-3}$). Both classes of materials outperformed CEMs in single-chamber MFCs. However, these two materials behaved differently in terms of coulombic efficiency. This parameter clearly improves with the use of glass fiber up to 80% versus 40% for J-Cloth since glass fiber is less permeable to oxygen. Moreover, the lower biodegradability of glass fiber enables a higher stability of the separator in the long term and largely prevents the accumulation of biomass over it surface.

Non-woven fabrics are another type of low cost separators that can be used both as raw material or treated with proton conductive compounds to form composites (Daud et al., 2015). These materials have shown comparable performance when compared with Nafion membranes due to their high transfer ratio of chemical species including protons. Choi et al. (2013) compared the performance of a separator based on a non-woven fabric filter to that provided by Nafion-117 and found that non-woven separator exhibited higher power efficiency and higher stability over a period of 300 days. The performance of the non-woven separator was also higher versus Nafion when assembled with a membrane of 2,5-benzimidazole as ion-specific conductor. This hydrogel can decrease impedance while improving the contact between the membrane and the cathode.

Ceramic materials such as earthen, terracotta, earthware and clayware have recently gained increasing attention as low cost separator materials in MFCs (Winfield et al., 2016). Ceramics are effective at conducting proton (Park and Zeikus, 2003) while offering important advantages such as chemical and thermal stability, low maintenance requirements and mechanical strength (Yousefi et al., 2017). Several studies have shown the good performance of ceramics in terms of power generation and wastewater treatment in MFCs. Behera et al. (2010) proved that earthen pot could offer higher levels of power generation versus Nafion with lower associated values of internal resistance. Furthermore, ceramic-based MFCs have proven to be capable of offering higher COD removal rates versus Nafion membranes when synthetic wastewater was used as fuel (Jana et al., 2010).

Due to the varying degrees of permeability of these materials, air-cathode configurations can present problems related to anolyte leakage and losses due to evaporation

(Winfield et al., 2016). In this regard, the cathode electrode can be constructed with hydrophobic components (e.g., by applying polytetrafluoroethylene on the gas diffusion layer) to alleviate this problem (Santoro et al., 2011; Yousefi et al., 2017).

Various types of MFC configuration based on ceramics materials have been developed in both single and double chamber arrangements. For example, several interesting cylindrical MFC configurations have been recently reported in the literature. Jana et al. (2010) designed an up-flow MFC containing an earthen cylinder that serves as both anodic chamber and separator. In such configuration, the overall COD removal efficiency was over 90% for organic loading rates from 0.6 to 2.0 kg COD $m^{-3}d^{-1}$, thus showing its high efficiency in terms of wastewater treatment.

Gajda et al. (2015) developed another cylindrical configuration consisting of a terracotta cave assembled with carbon electrodes. The cave serves as separator and as cathode chamber simultaneously. This simple and practical design can be directly immersed in wastewater, offering operation ease (Figure 18.5). During MFC operation, this configuration generates a caustic catholyte, which is housed inside the caves, allowing cations to be extracted from the anolyte as a result an electro osmosis process (Winfield et al., 2016). Among other applications, this by-product can be reused in the same device for the pre-treatment of complex substrates (Salar-García et al., 2016a).

Several properties of the ceramic materials, from the nature of the material itself to the physical properties of the separators employed, can greatly influence their efficiency in MFC devices. Among other parameters, the thickness of this type of separators can play a key role in MFC performance. Some authors have shown that power generation in MFCs using earthen pot separators decreases as the thickness of the membrane increases (Behera and Ghangrekar, 2011). Winfield et al. (2013a) also compared the performance of earthenware and terracotta clays by analyzing properties such as porosity. They found that earthenware, which presented higher

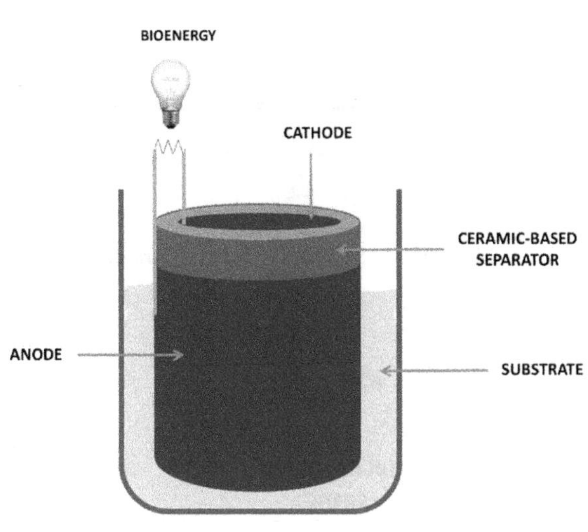

FIGURE 18.5 Schematic representation of ceramic microbial fuel cells.

porosity versus terracotta, was also more efficient in terms of power production. In contrast, earthenware separators can present some drawbacks such as oxygen crossover to the anode, which in turns reduces the coulombic efficiency of the system (Daud et al., 2015).

Although there are some issues that need to be addressed, ceramic separators are a promising option comparable to conventional ion exchange membranes and offer advantageous properties for the construction of inexpensive MFC devices (Winfield et al., 2016).

Other alternative options to standard IEMs are separators made out of natural rubber, which is a biodegradable material. An interesting feature of this material is that it can improve its performance over time. The enhancement in power performance with time has been explained by the formation of microspores due to its biodegradable nature. In fact, natural rubber can offer higher levels of power output in comparison with conventional AEMs and CEMs for the long term (over 11 months). These microspores have been reported to be the pathways for proton exchange (Winfield et al., 2013b). Furthermore, the use of biodegradable materials points to the possibility of creating disposable MFC devices (Winfield et al., 2014). Figure 18.6 shows a stack MFC system employing a separator made out of rubber.

Other materials such as nylon and polycarbonate have been explored as separator materials in MFCs as potential alternatives to the Nafion membranes. The performance of such options is comparable to Nafion-117 membranes in terms of power production and durability according to previous studies (Biffinger et al., 2007). More recently, it has been proven that mixed cellulose ester filters can also compete with

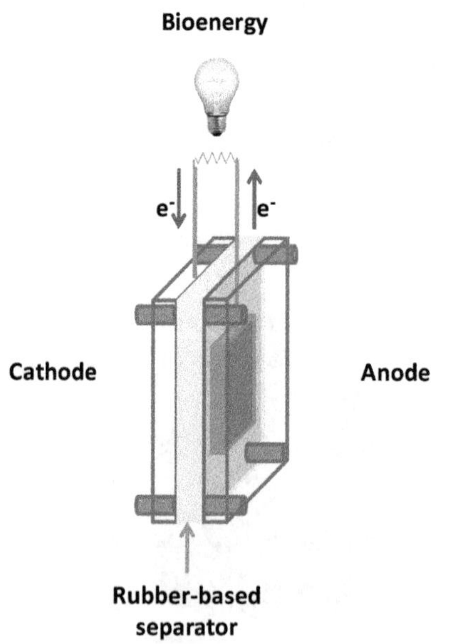

FIGURE 18.6 Stack MFC based on a rubber type separator.

Nafion membranes, showing higher stability over time. Cellulose ester filters are also more efficient at preventing biofouling when compared with the conventional proton exchange membranes (Wang and Lim, 2017).

Similarly, Zhang et al. (2015) compared the performance of several types of separators, including a syntheic fabric membrane made out of cellulose acetate and cellulose nitrate, a polycarbonate filter (Nuclepore) and a convencional Nafion-117 employed as standard in a double chamber MFC system. Among these materials, the first type of membrane showed the highest power performance, with a total power output over 24% of that obtained with the Nafion membrane. The polycarbonate filter membrane also exhibited a good performance, with a level of power density comparable to that of Nafion (675 $mW.m^{-3}$ versus 657 $mW.m^{-3}$, respectively). In contrast, the celullose based membrane also offered the highest oxygen difussion rates from the cathodic chamber to the anodic chamber, since this type of material displays a larger pore aperture, promoting oxygen transport. Thus, the coulombic efficiency was affected by this fact, with an associated value of 23%, which was lower compared to Nafion, with a coulombic efficiency of over 27%.

As seen above, due to their low cost and functionality, there is special interest in size-selectives materials as a new class of separator in MFC devices in order to spread this technology.

18.4 CONCLUSIONS

Microbial fuel cell technology is a very promising option for wastewater treatment with simultaneous bioenergy production. However, the practical implementation of MFCs requires the use of low cost materials for the construction of their main components. The separator is a key factor in ensuring a high performance in these devices. Conventional materials such as Nafion and Ultrex are effective as proton exchange membranes but also have a high cost. Thus, over the recent years, much effort has been devoted to develop alternative membranes that exhibit high performance while offering lower costs. Important parameters such as mechanical and chemical stability, biofuling, oxygen permeability and internal resistance need to be taken into consideration to select a suitable material for the fabrication of MFC separators. Several types of membranes have been developed as effective separator options for these devices. They include alternative cation and anion conductive materials, polymer membranes, composites, nanocomposites, ionic liquid-based membranes and porous size-selective separators. Among these options, inexpensive ceramic membranes have become a reality for the practical implementation of microbial fuel cell technology. Among the separator materials reviewed in this chapter, many types can indeed offer similar levels of power density when compared with Nafion membranes and also represent cost effective options. These novel materials can be tailored to prevent or reduce biofouling phenomena, minimize oxygen transport from the cathode to the anode and increase coulombic efficiency. Moreover, the new class of advanced separators developed tends to display more biodegradable and environmentally friendly features. The development of low cost and effective ion conductive separators will contribute considerably to the practical implementation of MFC technology.

REFERENCES

Arico', A.S., Baglio, V., Di Blasi, A., Antonucci, V., Cirillo, L., Ghielmi, A., Arcella, V., 2006. Proton exchange membranes based on the short-side-chain perfluorinated ionomer for high temperature direct methanol fuel cells. Desalination. 199, 271–273.

Armand, M., Endres, F., MacFarlane, D.R., Ohno, H., Scrosati, B., 2009. Ionic-liquid materials for the electrochemical challenges of the future. Nature Materials. 8, 621–629.

Ayyaru, S., Dharmalingam, S., 2013. Improved performance of microbial fuel cells using sulfonated polyether ether ketone (SPEEK) TiO_2–SO_3H nanocomposite membrane. RSC Advances. 3, 25243–25251.

Ayyaru, S., Letchoumanane, P., Dharmalingam, S., Stanislaus, A.R., 2012. Performance of sulfonated polystyrene–ethylene–butylene–polystyrene membrane in microbial fuel cell for bioelectricity production. Journal of Power Sources. 217, 204–208.

Behera, M., Ghangrekar, M.M., 2011. Electricity generation in low cost microbial fuel cell made up of earthenware of different thickness. Water Science & Technology. 64, 2468–2473.

Behera, M., Jana, P.S., Ghangrekar, M.M., 2010. Performance evaluation of low cost microbial fuel cell fabricated using earthen pot with biotic and abiotic cathode. Bioresource Technology. 101, 1183–1189.

Biffinger J.C., Ray R., Little B., Ringeisen B.R., 2007. Diversifying biological fuel cell designs by use of nanoporous filters. Environmental Science & Technology. 41, 1444–1449.

Cao, X., Huang, X., Liang, P., Xiao, K., Zhou, Y., Zhang, X., Logan, B.E., 2009. A new method for water desalination using microbial desalination cells. Environmental Science Technology. 43, 7148–7152.

Choi, S., Kim, J.R., Cha, J., Kim, Y., Premier, G.C., Kim, C., 2013. Enhanced power production of a membrane electrode assembly microbial fuel cell (MFC) using a cost effective poly [2,5-benzimidazole] (ABPBI) impregnated non-woven fabric filter. Bioresource Technology. 128, 14–21.

Daud, S.M., Kim, B.H., Ghasemi, M., Daud, W.R.W., 2015. Separators used in microbial electrochemical technologies: Current status and future prospects. Bioresource Technology. 195, 170–179.

Daud, W.R.W., Ghasemi, M., Chong, P.S., Jahim J.M., Lim, S.S., Ismail, M., 2011. SPEEK/PES composite membranes as an alternative for proton exchange membrane in microbial fuel cell (MFC). IEEE Conference on Clean Energy and Technology (CET). 400–403.

Elangovan, M., Dharmalingam, S., 2017. Anti-biofouling anion exchange membrane using surface modified quaternized poly(ether imide) for microbial fuel cells. Journal of Applied Polymer Science. 134, 1–9.

Elangovan, M., Dharmalingam, S., 2016. A facile modification of a polysulphone based anti biofouling anion exchange membrane for microbial fuel cell application. RSC Advances. 6, 20571–20581.

Gajda, I., Stinchcombe, A., Greenman, J., Melhuish, C., Ieropoulos, I., 2015. Ceramic MFCs with internal cathode producing sufficient power for practical applications. International Journal of Hydrogen Energy. 40, 14627–14631.

Ghasemi, M., Daud, W.R.W., Ismail, A.F., Jafari, Y., Ismail, M., Mayahi, A., Othman, J., 2013. Simultaneous wastewater treatment and electricity generation by microbial fuel cell: Performance comparison and cost investigation of using Nafion 117 and SPEEK as separators. Desalination. 325, 1–6.

Ghasemi, M., Shahgaldi, S., Ismail, M., Yaakob, Z, Daud, W., 2012a. New generation of carbon nanocomposite proton exchange membranes in microbial fuel cell systems. Chemical Engineering Journal. 184, 82–89.

Ghasemi, M., Rahimnejad, M., Esmaeili, C., Daud, W.R.W., Masdar, M.S., Majlan, E.H., Hassan S.H.A., Alam, J., Ismail M., Alhoshan, M.S., 2012b. Polysulfone composed of polyaniline nanoparticles as nanocomposite proton exchange membrane in microbial fuel cell. American Journal of Biochemistry and Biotechnology. 8, 311–319.

Gohil, J.M. and Karamanev, D.G., 2016. Novel approach for the preparation of ionic liquid/ imidazoledicarboxylic acid modified poly(vinyl alcohol) polyelectrolyte membranes. Journal of Membrane Science. 513, 33–39.

Gude, V.G., 2016. Wastewater treatment in microbial fuel cells: an overview. Journal of Cleaner Production. 122, 287–307.

Gugliuzza, A., Basile, A. (Eds), 2013. *Membranes for Clean and Renewable Power Applications*. Philadelphia, USA: Woodhead Publishing Limited.

Hallett, J.P., Welton, T., 2011. Room-temperature ionic liquids. Solvents for synthesis and catalysis, 2. Chemical Reviews. 111, 3508–3576.

Han, X., Armstrong, D.W., 2007. Ionic liquids in separations. Accounts of Chemical Research. 40, 1079–1086.

Harnisch, F., Schröder, U., Scholz, F., 2008. The suitability of monopolar and bipolar ion exchange membranes as separators for biological fuel cells. Environmental Science Technology. 42, 1740–1746.

Heijne, A.T., Hamelers, H.V.M., Wilde, V.D., Rozendal, R.A., Buisman, C.J.N., 2006. A bipolar membrane combined with ferric iron reduction as an efficient cathode system in microbial fuel cells. Environmental & Science Technology. 40, 5200–5205.

Hernández-Fernández, F.J., Pérez de los Ríos, A., Mateo-Ramírez, F., Juarez, M.D., Lozano-Blanco, L.J., Godínez, C., 2016. New application of polymer inclusion membrane based on ionic liquids as proton exchange membrane in microbial fuel cell. Separation and Purification Technology. 160, 51–58.

Hernández-Fernández, F.J., Pérez de los Ríos, A., Mateo-Ramírez, F., Godínez, C., Lozano-Blanco, L.J., Moreno, J.I., Tomás-Alonso, F., 2015a. New application of supported ionic liquids membranes as proton exchange membranes in microbial fuel cell for waste water treatment. Chemical Engineering Journal. 279, 115–119.

Hernández-Fernández, F.J., Pérez de los Ríos, A., Salar-García, M.J., Ortiz-Martínez, V.M., Lozano-Blanco, L.J., Godínez, C., Tomás-Alonso, F., Quesada-Medina, J., 2015b. Recent progress and perspectives in microbial fuel cells for bioenergy generation and wastewater treatment. Fuel Processing Technology. 138, 284–297.

Huang, L., Li, X., Ren, Y., Wang, X., 2017. Preparation of conductive microfiltration membrane and its performance in a coupled configuration of membrane bioreactor with microbial fuel cell. RCS Advances. 7, 20824–20832.

Hurwitz, H., Dibiani, R., 2001. Investigation of electrical properties of bipolar membranes at steady state and with transient methods. Electrochimica Acta. 47, 759–773.

Ieropoulos, I., Greenman, J., Melhuish, C., 2010. Improved energy output levels from small-scale microbial fuel cells. Bioelectrochemistry. 78, 44–50.

Jana, P.S., Behera, M., Ghangrekar, M.M., 2010. Performance comparison of up-flow microbial fuel cells fabricated using proton exchange membrane and earthen cylinder. International Journal of Hydrogen Energy. 35, 5681–5686.

Khera, J., Chandra, A., 2017. Use of an alternated cation–anion exchange membrane assembly for improved microbial fuel cell performance. Proceedings of the National Academy of Sciences, India Section A: Physical Sciences. 87, 297–301.

Kim, J.R., Cheng Ch., Oh, S.-E., Logan B.E., 2007. Power generation using different cation, anion, and ultrafiltration membranes in microbial fuel cells. Environmental Science & Technology. 41, 1004–1009.

Koók, L., Nemestóthy, N., Bakonyi, P., Zhen, G., Kumar, G., Lu, X., Su., L., Saratale, G.D., Kim, S.-H., Gubicza, L., 2017. Performance evaluation of microbial electrochemical systems operated with Nafion and supported ionic liquid membranes. Chemosphere. 175, 350–355.

Leong, J.X., Daud, W.R.W., Ghasemi, M., Liew, K. Ben, Ismail, M., 2013. Ion exchange membranes as separators in microbial fuel cells for bioenergy conversion: A comprehensive review. Renewable & Sustainable Energy Reviews. 28, 575–587.

Li, W.-W., Sheng, G.-P., Liu, X.-W., Yu, H.-Q., 2011. Recent advances in the separators for microbial fuel cells. Bioresource Technology. 102, 244–252.

Liu, Z.-D. and Li, H.-R., 2007. Effects of bio- and abio-factors on electricity production in a mediatorless microbial fuel cell. Biochemical Engineering Journal. 36, 209–214.

Logan, B.E., 2007. *Microbial Fuel Cells.* Hoboken, NJ, USA: John Wiley Sons, Inc.

Lu, J., Yan, F., Texter, J., 2009. Advanced applications of ionic liquids in polymer science. Progress in Polymer Science. 34, 431–448.

Malaeb, L., Katuri, K.P., Logan, B.E., Maab, H., Nunes, S.P., Saikaly, P.E., 2013. A hybrid microbial fuel cell membrane bioreactor with a conductive ultrafiltration membrane biocathode for wastewater treatment. Environmental Science & Technology. 47, 11821–11828.

Min, B., Cheng, S., Logan, B.E., 2005. Electricity generation using membrane and salt bridge microbial fuel cells. Water Research. 39, 1675–1686.

Mishra, A.K., Bose, S., Kuila, T., Kim, N.H., Lee, J.H., 2012. Silicate-based polymer-nanocomposite membranes for polymer electrolyte membrane fuel cells. Progress in Polymer Science. 37, 842–869.

Mokhtarian, N., Ghasemi, M., Daud, W.R.W., Ismail, M., Najafpour, G., Alam, J., 2013. Improvement of microbial fuel cell performance by using nafion polyaniline composite membranes as a separator. Journal of Fuel Cell Science and Technology. 10, 1–5.

Oh, S.-E., Logan, B.E., 2006. Proton exchange membrane and electrode surface areas as factors that affect power generation in microbial fuel cells. Applied Microbiology and Biotechnology. 70, 162–169.

Ortiz-Martínez, V.M., Salar-García, M.J., Hernández-Fernández, F.J., de los Ríos, A.P., 2015. Development and characterization of a new embedded ionic liquid based membrane-cathode assembly for its application in single chamber microbial fuel cells. Energy. 93, 1748–1757.

Pant, D., Van Bogaert, G., De Smet, M., Diels, L., Vanbroekhoven, K., 2010. Use of novel permeable membrane and air cathodes in acetate microbial fuel cells. Electrochimica Acta. 55, 7710–7716.

Park, D.H., Zeikus, J.G., 2003. Improved fuel cell and electrode designs for producing electricity from microbial degradation. Biotechnology and Bioengineering, 81, 348–355.

Rahimnejad M., Jafary T., Haghparast F., Najafpour G., Ghoreyshi A.A., 2010. Nafion as a nanoproton conductor in microbial fuel cells. Turkish Journal of Engineering and Environmental Sciences. 34, 289–292.

Rahimnejad, M., Ghasemi, M., Najafpour, G., Ismail M., Mohammad, A., Ghoreyshi, A., Hassan, S.H.A., 2012. Synthesis, characterization and application studies of self-made Fe3O4/PES nanocomposite membranes in microbial fuel cell. Electrochimica Acta. 85, 700–706.

Rhee, C.H., Kim, Y., Lee, J.S., Kim, H.K., Chang, H., 2006. Nanocomposite membranes of surface-sulfonated titanate and Nafion® for direct methanol fuel cells. Journal of Power Sources. 159, 1015–1024.

Rinaldi, A., Mecheri, B., Garavaglia, V., Licoccia, S., Di Nardo, P., Traversa, E., 2008. Engineering materials and biology to boost performance of microbial fuel cells: A critical review. Energy & Environmental Science. 1, 417–429.

Salar-García, M.J., Gajda, I., Ortiz-Martínez, V.M., Greenman, J., Hanczyc, M.M., de los Ríos, A.P., Ieropoulos, I.A., 2016a. Microalgae as substrate in low cost terracotta-based microbial fuel cells: Novel application of the catholyte produced. Bioresource Technology. 209, 380–385.

Salar-García, M.J., Ortiz-Martínez, V.M., Baicha, Z., de los Ríos, A.P., Hernández-Fernández, F.J., 2016b. Scaled-up continuous up-flow microbial fuel cell based on novel embedded ionic liquid-type membrane-cathode assembly. Energy. 101, 113–120.

Salar-García, M.J., Ortiz-Martínez, V.M., de los Ríos, A.P., Hernández-Fernández, F.J., 2015. A method based on impedance spectroscopy for predicting the behavior of novel ionic liquid-polymer inclusion membranes in microbial fuel cells. Energy. 89, 648–654.

Santoro, C., Agrios, A., Pasaogullari, U., Li, B., 2011. Effects of gas diffusion layer (GDL) and micro porous layer (MPL) on cathode performance in microbial fuel cells (MFCs). International Journal of Hydrogen Energy. 36, 13096–13104.

Santoro, C., Arbizzani, C., Erable, B., Ieropoulos, I., 2017. Microbial fuel cells: From fundamentals to applications. A review. Journal of Power Sources. 356, 225–244.

Sevda, S. and Sreekrishnan, T.R., 2012. Effect of salt concentration and mediators in salt bridge microbial fuel cell for electricity generation from synthetic wastewater. Journal of Environmental Science and Health, Part A. 47, 878–886.

Tang, X., Guo, K., Li, H., Du, Z., Tian, J., 2010. Microfiltration membrane performance in two-chamber microbial fuel cells. Biochemical Engineering Journal. 52, 194–198.

Tao, H.-C., Sun, X.-N., Xiong, Y., 2015. A novel hybrid anion exchange membrane for high performance microbial fuel cells. RSC Advances. 5, 4659–4663.

Vaghari, H., Jafarizadeh-Malmiri H., Berenjian A., Anarjan N., 2013. Recent advances in application of chitosan in fuel cells. Sustainable Chemical Processes. 1, 1–12.

Varcoe, J.R., Atanassov, P., Dekel, D.R., Herring, A.M., Hickner, M.A., Kohl, P.A., Kucernak, A.R., Mustain, W.E., Nijmeijer, K., Scott, K., Xu, T., Iwashita, N., Zhuang L., 2014. Anion-exchange membranes in electrochemical energy systems. Energy and Environmental Science. 7, 3135–3191.

Wang, Z., Lim, B., 2017. Mixed cellulose ester filter as a separator for air-diffusion cathode microbial fuel cells. Environmental Technology. 38, 979–984.

Winfield, J., Gajda, I., Greenman, J., Ieropoulos, I., 2016. A review into the use of ceramics in microbial fuel cells. Bioresource Technolog. 215, 296–303.

Winfield, J., Chambers, L.D., Rossiter, J., Greenman, J., Ieropoulos, I., 2014. Towards disposable microbial fuel cells: Natural rubber glove membranes. International Journal of Hydrogen Energy. 39, 21803–21810.

Winfield, J., Greenman, J., Huson, D., Ieropoulos, I., 2013a. Comparing terracotta and earthenware for multiple functionalities in microbial fuel cells. Bioprocess and Biosystems Engineering. 36, 1913–1921.

Winfield, J., Ieropoulos, I., Rossiter, J., Greenman, J., Patton, D., 2013b. Biodegradation and proton exchange using natural rubber in microbial fuel cells. Biodegradation. 24, 733–739.

Yee, R., Zhang, K., Ladewig, B., 2013. The effects of sulfonated poly(ether ether ketone) ion exchange preparation conditions on membrane properties. Membranes. 3, 182–195.

Yousefi, V., Mohebbi-Kalhori, D., Samimi, A., 2017. Ceramic-based microbial fuel cells (MFCs): A review. International Journal of Hydrogen Energy. 42, 1672–1690.

Zhang, S., Hui, Y., Han, B., 2015. Effects of three types of separators membranes on the microbial fuel cells performance. International conference on Mechatronics, Electronic, Industrial and Control Engineering. 2015, 1592–1595.

Zhang, X., Cheng, S., Huang, X., Logan, B.E., 2010. Improved performance of single-chamber microbial fuel cells through control of membrane deformation. Biosensors and Bioelectronics. 25, 1825–1828.

Zhang, X., Cheng, S., Wang, X., Huang, X., Logan, B.E., 2009. Separator characteristics for increasing performance of microbial fuel cells. Environmental Science Technology. 43, 8456–8461.

Zuo, Y., Cheng, S., Call D., Logan, B.E., 2007. Tubular membrane cathodes for scalable power generation in microbial fuel cells. Environmental Science & Technology. 41, 3347–3353.

Zuo, Y., Cheng, S., Logan, B.E., 2008. Ion exchange membrane cathodes for scalable microbial fuel cells. Environmental Science & Technology. 42, 6967–6972.

Section V

Membranes for Biomedical Applications

19 Recent Development of Hemodialysis Membrane Materials

Muhammad Irfan, Masooma Irfan,
Ani Idris, and Ghani ur Rehman

CONTENTS

19.1 INTRODUCTION

The healthy human kidney controls the emission of body waste products including hormones, uremic solutes, unchanged drugs, and hydrophilic metabolites; regulates the water level; and exhibits a glomerulus filtration rate (GFR) greater than 90 mL/min (1.73 m^2) with no signs of proteins in urine. When the filtration efficiency of kidney lowers to 15 mL/min, and its 90% working capability is lost, a state of renal failure called chronic kidney disease (CKD) is reached. Vascular disease (e.g., diabetes) or chronic glomerulonephritis and hypertension are the major causes of CKD. At this level, the waste products of the blood begin to accumulate and cause many health issues including gastrointestinal and cardiovascular issues and bone diseases. The progressive increment of CKD can lead to kidney failure, called end-stage renal disease (ESRD), because CKD is generally considered an irreversible disease if it is not cured properly (Daugirdas et al., 2012; Daugirdas et al., 2013).

Hemodialysis (HD) is a commonly adopted clinical therapy for ESRD patients that consist of a proportioning system for blood and electrolyte solutions, delivery and safety monitoring treatment, and, most importantly, a hemodialyzer. The dialyzer works on diffusive and convective processes where semi permeable membranes (dialysis membranes) are housed in the form of a compact bundle and provide a counter current flow path for blood and dialysate solutions across the membrane surface. The central part of dialyzer is the membrane, and the dialyzer itself is the core of the extracorporeal treatment. This concept clearly emphasizes the importance of the dialyzer's membrane, whose performance is determined by the biocompatibility and the efficiencies of solute clearances. Under normal conditions, the compounds that are removed by the healthy kidney into the urine are named uremic toxins or uremic solutes. Vanholder et al. (2003) and Duranton et al. (2012) reported more than 95% of uremic solutes have the potential to develop toxicity when their amount increases beyond a certain limit. The passage of the uremic solutes via the membrane surface is strongly affected by differences in the inter-membrane diffusion rate, solubility of solutes, electrical charge density, and polarity on the surface of the intact membrane

Technological advances in membrane chemical compositions; dialyzer design and sterilization techniques have led to improved performance of dialyzer and prolong survival time of ESRD patients. Currently, different types of dialyzers are commercially available in the market, and they can be classified in terms of membrane chemical composition, membrane surface area and membrane permeability to water or different blood proteins such as middle molecules (ß2-microglobulin) (Chowdhury et al., 2012; Daugirdas et al., 2012; Nissenson and Fine, 2016).

The total global expenditures for ESRD patients have reached more than $435.6 billion, and it is also estimated that the number of kidney patients will increase in the future due to different diseases, poor quality lifestyles, and improper medical treatments. Thus, the application of HD is also expected to increase in the future, and it is considered a multi-million dollar industry where formulations of commercial grade dialyzers are also considered trade secrets (Saran et al., 2017; Young-Hyman, 2013). Keeping these points in consideration, this chapter critically reviews the advanced materials used for the fabrication and modification of dialysis membranes in research and commercial scale, including their basic.

19.2 FABRICATION OF MEMBRANES FOR BLOOD-FILTERING UNIT

Generally, HD membranes can be fabricated in two different forms: flat sheet and cylindrical form. Flat sheet is normally used in the production of spirally wound, disc shaped, plate, and frame modules. The cylindrical membranes are employed in capillary and tubular, or hollow fiber (HF) modules. These membranes can be formed by one of the following techniques: (i) thermally-induced phase separation, (ii) vapor-induced phase separation, (iii) immersion precipitation (wet phase inversion), or (iv) dry phase inversion.

The dry-wet spinning method is generally used for the preparation of HF membranes, which is based on liquid-liquid phase separation. The dope composition, spinneret dimension, bore and dope flow rates, coagulation bath composition, and temperature are the important parameters in fabricating HF membranes (Roy and De, 2017). The medical use of coil and parallel-plate dialyzers have recently declined in commercial use, but the use of HF dialyzers is increasing as they are effective, small, and reusable (Yamashita and Tomisawa, 2009).

In a typical HF dialyzer, a fiber bundle is fixed in plastic casing where both ends are potted in polyurethane. Usually, four end caps are present on each HF dialyzer; the vertical end caps are used for the blood inlet and outlet ports, whereas the parallel end caps are used for the dialysate inlet and outlet ports. The HF membranes utilized in clinical dialyzers consist of ~1,000 to 15,000 fibers that have an internal diameter of ~175–250 μm with a surface area of 1 to 2.5 m^2. The molecular weights of proteins that should be removed from dialysis patients are in the range of 10–55 kDa, which includes removal of middle (ß2-microglobulin) and low molecular weight uremic toxins (urea, uric acid, creatinine, etc.). However, proteins such as albumin (66 kDa) should be retained in the blood (Daugirdas et al., 2012; Nissenson and Fine, 2016).

19.3 HISTORY, BACKGROUND AND CURRENT COMMERCIAL DIALYSIS MEMBRANES

The first purification of blood in terms of HD was performed for canines using collodion based dialyzer material as reported by Abel et al. in 1914. Collodion was very fragile for dialysis therapy, and so other materials were also examined. Finally, cellophane was used to replace the collodion. Kolff et al. (1943) performed the first clinical trial using a rotating drum dialyzer that consisted of cellophane. Stuart and Lipps (1967) developed the first cellulosic HF membrane based dialyzer in 1967 and Cordis-Dow Co. (Miami, FL) offered them as the first commercial product in 1972. The cellulose dialyzer was compact, had a large surface area, and looked similar to a multi-tube heat exchanger.

West Germany registered the trade name of a dialysis membrane (Cuprophan®) synthesized by cuprammonium rayon, also known as regenerated cellulose (RC), that consisted of cellulose liquefied in cuprammonium solution. Later, Asahi-Kasei Co. (Tokyo, Japan) registered another RC membrane by the trade name Bemberg®, followed by Terumo (Tokyo, Japan). The RC membranes, however, exhibited very poor biocompatibility because of their hydroxyl groups. Cellulose acetate, di-acetate, and triacetate brands were then formed.

Currently, different brands of cellulose acetate-base products are available for commercial use, but the commercial availability of RC dialyzers are discontinued. Baxter offers a wide range of dialyzers composed of cellulose diacetate (DICEA®), cellulose triacetate (Exeltra®), polyarylethersulfone/polyvinylpyrrolidone/polyimide (PAES/PVP/PA) (Polyflux®), PAES/PVP (Ravaclear®) and polyethersulfone (PES) (Xenium XPH®). Its two brands, Evodial® and Nephral®, contain the modified AN69® membrane that has acrylonitrile and sodium methallyl sulfonate copolymer-polyethyleneimine-Heparin grafted based formulation in its composition. These products have higher biocompatibility, prevent blood contact activation for longer time, and reduce cardiovascular issues.

A broad spectrum of hemodialyzers is also offered by Fresenius Medical Care for general (Optiflux and Hemoflow) and cardio-protective (FX-class) dialysis. Those hemodialyzers use Fresenius polysulfone® and Helixone® dialysis membranes. The Helixone® is also a polysulfone (PSf) based membrane but is fabricated by nano-controlled spinning technology that is able to produce a distinct pore structure and pore pattern of the inner membrane layer (Fresenius, 2017). The core membrane materials for the Asahi-Kasei Co. company is PSf, but they also sell ethylene vinyl alcohol copolymer (KF-201C series) and vitamin E-coated PSf (ViE series) based hemodialyzers (Kasei, 2017). The chemical formulation of Nikkiso Co. Ltd. is different from the other membrane producers as it sells polyester-polymer alloy/PVP based dialyzers that demonstrate good antithrombogenicity properties (GmbH, 2017).

Commercially, it appears that synthetic polymer based dialyzers have a strong edge on cellulose based membranes due to their excellent biocompatibility performances, higher sieving coefficients, better uremic solute dialysis, and higher protein retention. Table 19.1 shows the general specifications from in vitro experiments of a number of commercially available dialyzers; Table 19.2 compares the main membrane materials of only commercially available dialyzers. The commercial dialyzer companies generally followed the ISO EN 1283:1996 test methods, which were then replaced with BS EN ISO 8637:2014 and BS EN ISO 8638:2014 to register the evaluation reports of their dialysis performances.

19.3.1 Permeation and Clearance Rate of Commercial Dialysis Membranes

In Table 19.1, K_{uf} represents the ultrafiltration (UF) coefficient that is the volume of plasma filter water in mm/h at mmHg transmembrane pressure (TMP). The synthetic membranes showed higher K_{uf} (>10 to 100 mL/hr/mmHg) values, whereas the cellulose based membranes showed lower K_{uf} results; thus, cellulose materials represent the lower permeability of plasma water. The dialyzer company reported the vitro results of K_{uf} values, and the values are normally 5–30% lower in the vivo dialysis therapy. The solute clearance data of commercial dialyzers are generally measured at the blood flow (Qb) rates of 200, 300, and 400 mL/min with dialysate flow rate (Qd) fixed at 500 mL/min.

The creatinine (MW: 113.12 g/mol) and urea (MW: 60.06 g/mol) levels are the conventional identifiers used to evaluate the severity of the uremic syndrome (Mavroidis, 2006). The amount of creatinine in normal serum is 0.8–1.4 mg/dl and 0.56–1 mg/dl in men and women, respectively. When the creatinine level reaches 5.0 mg/dl in blood serum, it indicates that the kidney functions have been severely damaged and the patient needs kidney therapy (Ejaz et al., 2016).

Renal insufficiency is also represented by the blood urea nitrogen (BUN) concentration. Healthy adult individuals show BUN levels up to 7–20 mg/dl, which will be greatly increased to 40–60 mg/dl in case of kidney failure (Daugirdas et al., 2007; Mavroidis, 2006). The phosphate (MW: 1355 g/mol), insulin (MW: 5808 g/mol), and β2-microglobin (β2-m) (MW: 11800 g/mol) clearance results are utilized to estimate the dialyzer performance for the passage of larger molecular weight solutes. In vitro experiments of the dialysis of β2-m are difficult and rarely reported by commercial

TABLE 19.1
The General Specification of Main Hollow Fiber Brands of Commercially Available Dialyzers

						Min-Max Clearance %Age at Qb = 300 mL/min, Qd = 500 mL/min				HF Membrane		
Company	Brand Name	Membrane Main Material	Type	Surface Area (m^2)	K_{Uf}[a] (mL/h/mmHg)	Urea	Cr.	Ph.	Vit- B12	Inner dia. (um)	Thickness (um)	Ref.
Fresenius medical care	Optiflux	PSf	High flux series	1.5–2.0	45–107	265–286	238–271	230–271	152–199	–	–	Fresenius, 2017
			LF series	1.5–2.0	10–14	261–271	229–243	175–199	95–121	–	–	
	Hemoflow	PSf	High flux series	0.7–1.8	20–55	200–246	165–220	158–216	86–150	–	–	
			LF series	0.4–1.8	1.7–11	183–241	145–215	88–165	34–76	–	–	
	FX-class	PSf (Helixone® plus)	High flux series	0.6–2.2	20–73	250–278	210–261	201–248	130–192	185	35	
			LF series	1.0–1.8	8–14	228–261	200–231	164–210	94–138	185	35	
Baxter Inc. & Baxter (Gambro)	DICEA	Cellulose Diacetate		0.9–2.1	6.8–15.5	214–268	177–240	129–202	61–121	200	15	Baxter, 2017; Gambro, 2017
	Exeltra	CTA	(150–210)	1.5–2.1	3.1–4.7	262–287	242–277	227–252	152–202	200	15	
	Revaclear	PAES/PVP	All series	1.4–1.8	50–54	196–376	250–348	239–326	170–228	190	35	
	Polyflux R	PAES/PVP/ PA	(17R-24R)	1.7–2.4	71–83	254–274	229–255	223–249	159–192	–	–	
	Polyflux H	PAES/PVP/ PA	6H	0.6	33	(QB = 200) 167	146	136	90	215	50	
	Xenium	PES	XPH	1.1–2.1	59–89	257–291	233–275	213–265	148–206	200	40	
	Evodial	AN 69.[b]	(1.0–2.2)	1.05–2.15	33–65	216–265	187–237	156–207	92–143	210	42	
Asahi Kasei	APS series	PSf	APS, APS H, APS U	1.1–2.1	69–100	226–261	210–245	180–218	145–192	200	45	Asahi Kasei, 2017
	Rexeed	PSf	A & L series	1.5–2.5	66–90	264–284	247–272	227–257	155–194	185	45	
	LEOceed	PSf	16H-21H	1.6–2.1	68–88	271–278	254–311	241–296	173–221	185	35	

(Continued)

TABLE 19.1 (CONTINUED)
The General Specification of Main Hollow Fiber Brands of Commercially Available Dialyzers

Company	Brand Name	Membrane Main Material	Type	Surface Area (m²)	K_{Uf}[a] (mL/h/mmHg)	Min-Max Clearance %Age at Qb = 300 mL/min, Qd = 500 mL/min: Urea	Cr.	Ph.	Vit- B12	HF Membrane: Inner dia. (um)	Thickness (um)	Ref.
			16N-21N	1.6–2.1	14–17	256–263	238–253	196–209	120–133	185	35	
	ViE series	Vitamin E-coated Psf	ViE, ViE A, ViE L series	1.3–2.1	51–81	235–257	213–246	200–235	136–175	200	45	
	KF-201- C Series	EVAL	1.0C–1.8C	1–1.8		218–250	183–220	149–188	102–120	175	25	
Nikkiso	FDX & FDY series	Polyester-Polymer Alloy/ PVP[c]	–	1.2–2.1	47–64	242–260	221–244	206–231	142–170	210	30	GmbH, 2017
Nipro medical	ELISIO®	PES (Polynephron™)	(9H–25H)	0.9–2.5	53–93	246–294	218–285	200–276	134–224	200	40	Nipro, 2017
	Solacea series	Asymmetric CTA (ATA™)	15H–25H	1.5–2.5	61–87	QB 200 ml/min 196–352	191–331	185–318	150–246	200	25	
Allmed	Polypure	PSf	High flux (10H–20H)	1–2	33–68	238–279	215–266	193–251	125–186	200	40	Allmed, 2017
			LF (P 10–P 20)	1–2	8.1–16.2	234–275	199–249	162–212	92–137	200	40	
	Biorema	PES	Low and high flux series	1–2.2	8.4–86	217–286	192–272	140–253	64–199	200	35	

[a] K_{UF}; bovine blood, TP = 60 ± 5 g/L, Hct = 32 ± 2%, Qb = 300 mL/min.

[b] Acrylonitrile and sodium methallyl sulfonate copolymer–polyethyleneimine–heparin grafted.

[c] Polyarrylate/PES/PVP.

TABLE 19.2
The Brand Name of Core Membrane Materials of Some Leading Vendors

Company	Brand Name	Main Material	Company	Brand Name	Main Material
B. Braun	Dicap®	PSf	Toray Medical	Filtryzer BG, NF, BK-F, BK-U, B1, B3	PMMA
Bain Medical Equipment	DORA®	PES		Toraylight NS	PSf
Membrana GmbH, Germany	PUREMA®	PES/PVP		Toraysulfone TS	PSf
	DIAPES®	PES/PVP		ETRF TE-12R	PSf
	SYNPHAN®	PES/PVP	Silvermed	FSF low and high flux series	PSf
Farmasol	PUREMA®	PES/PVP		FES low and high flux series	PES
Hemoclean	Hemopex	PES		FPM low and high flux Purema series	PES/PVP
Medica	Smatflux	PES/PVP			

dialyzer companies. Higher β2-m removal rate corresponds to poor protein retention. Since the commercial dialyzing companies are producing different types of dialyzers on the basis of surface area of hollow fibers under the same main brand name, thus for ease of comparison, the solute dialysis efficiencies organized from minimum to maximum clearance percentage are shown in Table 19.1.

The commercial dialyzers are also divided into high and low flux membranes. The high flux membranes demonstrate a K_{uf} greater than 20 mL/h/mm, high efficiency (mass transfer coefficient/area > 600 mL/min), and good permeability of β2-m clearance > 20 mL/min (Ambalavanan et al., 1999; Finelli et al., 2005; Tokars et al., 1994). The larger surface area coupled with fast blood (Qb ≥ 350 mL/min) and dialysate (Qd ≥ 500 mL/min) flow could lead to higher efficiency dialyzers. The CKD patients treated with high flux dialyzers have a lesser margin of safety and can suffer from dialysis disequilibrium syndrome, potential vascular access damage, and hemodynamic instability. Although low treatment time is used with a high flux dialyzer, but it is more expensive than low flux dialyzer as the patients require additional dosage (Slinin et al., 2015).

19.4 HIGH PERFORMANCE HEMODIALYSIS MATERIALS

Since the HD is a general therapeutic technique for ESRD patients, the quality and composition of dialysis membranes are very important. The criteria for a high-performance membrane (HPM) include biocompatibility, good balance in water and solutes permeability, sufficient mechanical strength at different trans-membrane pressures, ability to be sterilized, and low cost (Daugirdas et al., 2012). Other attributes, such as larger pore size and higher molecular weight cut-off (MWCO) than the conventional HD membranes, are also important, since bigger pore size promotes elimination of protein-bound uremic solutes with β2-m type. The HPM of superior flux normally exhibits MWCO closer to 65 kDa, whereas the other class of dialyzers has MWCO in the range 3 kDa to >15 kDa (Roy and De, 2017; Saito, 2011).

On the basis of transport properties, composition, and biocompatibility, HD hollow fiber membranes can be synthesized from various types of materials: regenerated cellulose, modified cellulose, and synthetic polymers. Figure 19.1 shows the list of commonly used polymeric materials with their chemical structures for HD membranes fabrication. The regenerated cellulose based HD membranes are extremely hydrophilic and are not expensive. They exhibit excellent clearances of uremic solutes and can prevent platelet adhesion but possess poor mechanical strength. They have limited capacity to eliminate the middle molecules (β2-m) and their nucleophilic groups, which subsequently promotes the activation of the complement system and result in bio-incompatibility.

Besides cellulose, the engineered thermoplastics based polymeric membranes, which are highly hydrophobic and contain asymmetric structures, can also be used. Polyacrylonitrile (PAN), ethylene vinyl alcohol copolymers (EVAL), polyester-polymer alloy (PEPA), polymethylmethacrylate (PMMA), PES, and PSf are some of the commonly used thermoplastics. Currently, most of the HD membranes are composed of synthetic polymers with variable compositions, thus, exhibiting their own advantages and disadvantages in terms of biocompatibility, uremic solute filtration, and adsorption characteristics (Sakai and Matsuda, 2011; Sunohara and Masuda, 2017; Yamashita and Sakurai, 2015).

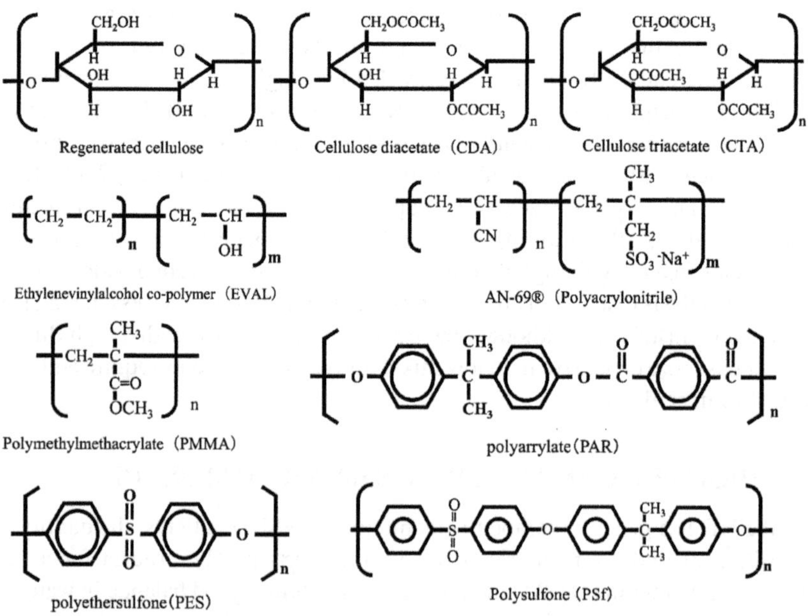

FIGURE 19.1 Common polymeric materials with their chemical formulations used for fabrication of HD membranes.

When HD membranes comes into contact with blood, the dialyzer materials are considered foreign surfaces. If the biocompatibility of the polymeric membrane is not satisfactory, proteins will be adsorbed and accumulated onto the membrane surface when they encounter blood. A number of factors are responsible for adsorption of proteins onto the polymer surfaces. These include surface charge, surface chemistry, and hydrophilicity of polymers, instrument operating conditions, and the size and shape of adsorbed protein. Among these factors, the size and shape of the adsorbed protein are the most important. The globulins, albumin, and fibrinogen are the plasma proteins in blood whose accumulation on the membrane surface causes not only membrane fouling but also invokes the body's immune system against the membrane and activates the inflammatory response, coagulation cascade activation, fibrinolysis and complement reactions, adhesion of platelets and red blood cells. These phenomena alter the polymer surface and gradually result in the decline of membrane flux, selectivity, and permeability of uremic solutes (Irfan and Idris, 2015; Kokubo et al., 2015).

To improve biocompatibility, the polymeric membranes need to be modified using various techniques such as blending or surface modification. Blending is normally referred to as bulk modification and can be performed during dope solution mixing. Sulfonation and carboxylation are also considered bulk modification (Irfan and Idris, 2015). In addition, the choice of membrane material with specific formulation is also equally important (Bowry et al., 2011). In order to understand why several compositions of HD membranes are commonly chosen, some of the important synthetic polymers and their characteristics are briefly discussed.

19.4.1 Polyarylsulfone Family

At present, 93% of dialyzer materials are derived from the parent polyarylsulfone family, of which PSf and PES are 71% and 22% of their share, respectively. PSf and PES are also the primary membrane materials in the new super high-flux dialyzers (Bowry et al., 2011). The PSf polymers contain alkyl- or aryl- and sulfone chemical groups in their composition. Additionally, the Asahi-PSf, α-PSf, Helixone, and Fresenius-PSf contain isopropylidene groups (Table 19.1). The PSf dialyzers are preferred because they exhibit low cytotoxicity and less anticoagulation requirements, offer superior intrinsic biocompatibility, retain endotoxins, and also eliminate a broad range of uremic solutes compared to cellulose based membranes. Increased hydraulic permeability and sieving capability of PSf stimulates efficient transport through the convection process (Bowry et al., 2011). Most of the current commercial dialyzers brands are made of PSf.

19.4.1.1 Enhancement of PSf

Mahmoudi et al. (2017) immobilized a peptoid with 2-methoxyethyl chain on a PSf based membrane and reported that the modified membranes demonstrated significant resistant against lysozyme and BSA with less fibrinogen adsorption in comparison to other published low-fouling surfaces. Zhu et al. (2017) fabricated the hemocompatible membrane by cross-linked polymerization of vinyltriethoxysilane (VTEOS) and vinyl pyrrolidone (VP) with PSf. They reported that the modified membrane showed 60–80% less protein adhesion (BSA, Fbg) and 14–20% high blood clotting factors (APTT, TT, Fibronogen) than the pristine PSf membrane. Tian and Qiu (2017) grafted the sulfonated hydroxypropyl chitosan on the PSf and evaluated the biocompatibility of formulating membranes. After modification, the BSA adsorption of membrane was reduced (87%) and whole blood clotting time and plasma recalcification time were enhanced up to 67% and 24%, respectively, whereas the hemolysis ratio significantly decreased (92%) (Liu et al., 2017).

Kaleekkal et al. (2015) sulfonated the PES and blended it with PSf and evaluated the hemodialysis performance of fabricated membranes. After alteration, platelet adhesion reduced from 14.2 to 8.75 (10^3 cells/mm), and clotting times were prolonged. Furthermore, lower complement activation and less protein adsorption were also observed than in the pristine PSf membrane. The dialysis experiments revealed that diffusive permeability of urea, creatinine, and cytochrome were also improved, and these details are charted in Table 19.3.

PES is considered a new generation material, widely utilized in a number of blood purification devices such as plasma collectors, hemofiltration, plasmapheresis, hemodiafiltration, and hemodialyzers (Kokubo et al., 2015; Nissenson and Fine, 2016). The PES based membranes in dialyzers hold consistent, larger, densely and distributed pores, which improves the selectivity of low molecular weight proteins and promotes the removal of B2-m with nominal albumin loss. Moreover, PES dialysis membranes showed the highest standard of biocompatibility and endotoxin retaining features upon modifications (Krieter and Lemke, 2011). Although the commercial share of PSf based dialyzers is currently higher than PES dialyzers, this trend is predicted to discontinue because nephrologists recently detected the presence of BPA in the blood from the PSf based dialyzer. Unlike PSf, PES is a BPA free material (Huang et al., 2012; Vandentorren et al., 2011). However, PES membranes must also be modified with mono-, di-, and triblock of additives to improve the targeted characters as shown in Table 19.3. Figures 19.2 and 19.3 represent the chemical formulation of some modifiers and their schematic attachment to different polymers.

19.4.1.2 Optimization of PES

Polyethylene glycol (PEG) and polyethylene oxide (PEO) based blocks (co-polymer) of chemicals when used with hydrophobic base polymers in the dialysis membranes could exhibit steric repulsion and work like molecular cilia. Such modified structures improved the membrane capability to repel cell adhesion, reduce protein adsorption and platelet adhesion, and enhance the blood clotting times in body environment. The protein resistance ability of PEG and PEO based blocks depend on the surface density and chain length. PEG and PEO are highly hydrophilic, and their leaching from the base polymer is unavoidable when mixed alone. However, if they are used in the form of amphiphilic, zwitterionic, or neutral co-polymers as a membrane modifier, the leaching problem is highly minimized and biocompatibility improved (Harris, 2013).

He et al., (2013) prepared an amphiphilic block of PEG (terpolymer) consisting of poly (ethylene glycol)-poly(sodium styrene sulfonate-co-methyl methacrylate) by atom transfer radical polymerization. Blending of terpolymer with PES reduced BSA adsorption by 40% and platelet adhesion by 97% and enhanced the APTT time up to 28% more than pristine PES membrane (He et al., 2013). Xie et al. (2015)

TABLE 19.3
Modification of PES and PSf Membranes by Different Additives and Their Outcomes

Core Composition	Min. Contact Angle (θ)	Max. UF Rate (mL/m²mmHg)	Protein Adsorption Reduced (ug/cm²)	Platelet Adhesion Reduced	Clotting Time Enhanced (TT, PT, APTT, WBCT, PRT)	Other Tests (Com. A, Z.P)	Ref.
				Polymer-PES			
MPEG-P(SSNa-co-MMA	67	618	12.2 to 7.3 (BSA)	85 to 2 ($\times 10^5$ cells/cm^2)	APTT 43 to 60s		(He et al., 2013)
Nanozeolite +TPGS	–	265	40% (BSA)	Highly reduced	TT- 27 to 30s PT- 18 to 19s APTT- 60 to 65s	Hemolysis < 0.25%	(Verma et al., 2017)
PES-PGMA	43	60	5.3 to 0.5 (BSA) 5.7 to 0.4 (Fbg)	12 to 0.1 (× 105 cells/cm^2)	APTT 40–56s TT 17–19s		(Xie et al., 2015)
2-hydroxyethl methacrylate and AA	44	32	14 to 3 (BSA) 13.5 to 2.8 (Fbg)	13.5 to 0.5 ($\times 10^5$ cells/cm^2)	WBCT 90 to 370s APTT 38 to 68s		(Qin et al., 2014)
P(AN-AA-VP)	53		12 to 3 (BSA)				(Li et al., 2010)
CPES/SPES/PES (phase inversion)	45		11.8 to 5 (Fbg)	128 to 5 ($\times 10^3$ cells/cm^2)	APTT 38 to 85s PT 14 to 17.8s	(Com. A): C3a 42.5–34 C5a 1.8–1.15 ng/mL TAT III 11–13.8%	(Nie et al., 2014)
CPES/SPES/PES (evaporation)	60		5 to 2.5 (Fbg)	88 to 19 ($\times 10^3$ cells/cm^2)	APTT 42 to 80s PT 14.5 to 16s	(Com. A): C3a 42.5–34 C5a 1.8–1.15 ng/mL TAT III 12–14%	

(Continued)

TABLE 19.3 (CONTINUED)
Modification of PES and PSf Membranes by Different Additives and Their Outcomes

Core Composition	Min. Contact Angle (θ)	Max. UF Rate (mL/m²mmHg)	Protein Adsorption Reduced (ug/cm²)	Platelet Adhesion Reduced	Clotting Time Enhanced (TT, PT, APTT, WBCT, PRT)	Other Tests (Com. A, Z.P)	Ref.
P(AN-co-AA-co-VP)	60		24 to 13.5 (BSA) 19 to 11 (Fbg)		APTT 32 to 72s PT 19% higher TAT III generation 11.76%.	(Com. A) C3a and C5a decreased 4% and 37.5%	(Tang et al., 2012)
P(AN-co-AA-co-VP) + S-PES	38		24 to 15 (BSA) 11 to 7 (Fbg)		APTT 32 to 90s PT 18 to 24s, TAT III 9% high.	(Com. A): C3a and C5a decreased 19.23% and 40.62%	
P(St-co-AA)-b-PVP-b-P(St-co-AA))	60.7	86.2	26 to 5 (BSA) 17 to 7 (Fbg)	7% than control	APTT 35 to 95s TAT III generation 9%.		(Nie et al., 2012)
P(PVP-b-PMMA-b-PVP)	59	95	48% (BSA)		APTT 55 to 95s		(Ran et al., 2011)
P(VP-co-AN-co-VP)	53.7	42	58% (BSA) 60% (Fbg)	Reduced 100% to 0.5%	APTT 42 to 70s PT showed 12s.		(Yin et al., 2012)
PVP (12.1 × 10,000 to 0.8 × 10,000 mol. wt.)	54	–	88% (BSA)	17.5 to 1 (× 10^7 cells/cm²)	APTT 38 to 90s		(Qin, Nie, et al., 2014)
PVP (12.1 × 10,000 mol. wt.)	62	–	90 % (BSA)	17.5 to 0.2 (× 10^7 cells/cm²)	APTT 35 to 78s		
PVP-PES	–	–	–	96% reduction	APTT 58 to 93s		(Ran et al., 2014)

(Continued)

TABLE 19.3 (CONTINUED)
Modification of PES and PSf Membranes by Different Additives and Their Outcomes

Core Composition	Min. Contact Angle (θ)	Max. UF Rate (mL/m²mmHg)	Protein Adsorption Reduced (ug/cm²)	Platelet Adhesion Reduced	Clotting Time Enhanced (TT, PT, APTT, WBCT, PRT)	Other Tests (Com. A, Z.P)	Ref.
Dopamine–g-(P(sodium 4-vinylbenzenesulfonate) -*co*-P (sodium methacrylate))			26.3 to 3 (BSA) 17.5 to 3.25 (Fbg)	14.8 to 0.8 (× 105 cells/cm²)	APTT 58 to 104s TT 18 to 27s		(Ma et al., 2014)
Heparinized-PES	66		179 to 36 (BSA)	Highly reduced (SEM images)	PRT 7 to 14 min	(Z.P at pH 3) Changed 25 to –16 mv.	(Wang et al., 2014)
Chitosan derivatives	25		23 to 11 (BSA) 17.5 to 8.3 (Fbg)	67 to 21 (× 105 cells/cm²)	APTT 50 to 84s TT 19–22s	(Com. A): C3a changed 17.2 to 18.9 and C5a 5.5 to 5.3 ng/ml	(Xue et al., 2013)
PCA-g-MWCNT	56	95 (L/h)	11 to 4.5 (BSA) 40 to 16 (Fbg)		APTT 34–35s PT 11 to 13s		(Abidin et al., 2017)
CNT–P(Na SS-co-EGMA)	56	136	15.5 to 7.5 (BSA) 14 to 7 (Fbg)	57.3 to 3.07 (× 10^4 cells/cm²)	APTT 40 to75s TT 19 to 21s	TAT III decreased 10.7 to 10.2 ng/ml creatinine absorb. increased 180 to 1200 ug/g.	(Nie et al., 2015)
PVPK90/f-MWCNT composites	51	72.20 (L/m².h)	–	–	–	Dialysis clearance Urea 10.3 to 49% Creatinine 8 to 55%, Lysozyme 0.93 to 28.04%	(Irfan et al., 2014)

(Continued)

TABLE 19.3 (CONTINUED)
Modification of PES and PSf Membranes by Different Additives and Their Outcomes

Core Composition	Min. Contact Angle (θ)	Max. UF Rate (mL/m²mmHg)	Protein Adsorption Reduced (ug/cm²)	Platelet Adhesion Reduced	Clotting Time Enhanced (TT, PT, APTT, WBCT, PRT)	Other Tests (Com. A, Z.P)	Ref.
S-PES+ PVPK90/f-MWCNT	50.1	93.4 (L/m².h)	17.7 to 7.7 (BSA) 12.7 to 4.5 (lysozyme)	–	APTT 36 to 67 TT 13 to 19 PT 10 to 16.5	Dialysis clearance Urea 10.3 to 57% Creatinine 8 to 59%, Lysozyme 0.93 to 32%	(Irfan et al., 2016)
Polymer- PSf and PAN							
PSF-g-PMEA	59.5		31 to 3.5 (BSA)	Reduced (SEM images)	APTT 40 to 42s PT 14 to 15s TT 20.5 to 20.6s	–	(Tian and Qiu, 2017)
PSf-PDA-NMEG5	75		(Normalized) 0.9 to 0.65 (lysozyme) 0.95–0.45 (Fbg)			20%, 43%, and 64% less fouled with BSA, lysozyme, and Fbg proteins	(Mahmoudi et al., 2017)
PSf/P(VP-VTEOS)	44	186 (L/m².h)	60 to 12 (BSA) 20 to 8 (Fbg)	Reduced (SEM images)	APTT 44 to 55.6s PT 15.2 to 18.5s	Fibrinogen reduced 100 to 86 mg/dl.	(Zhu et al., 2017)
Psf/Sulfonated hydroxypropyl chitosan	20		400 to 50	Reduced	APTT 31 to 36s PT 11 to 12s TT 19 to 20s WBCT 66 to 200s PRT 152 to 201s.	Hemolysis reduced 8.5 to 0.8%	(Liu et al., 2017)

(*Continued*)

TABLE 19.3 (CONTINUED)
Modification of PES and PSf Membranes by Different Additives and Their Outcomes

Core Composition	Min. Contact Angle (θ)	Max. UF Rate (mL/m^2mmHg)	Protein Adsorption Reduced (ug/cm^2)	Platelet Adhesion Reduced	Clotting Time Enhanced (TT, PT, APTT, WBCT, PRT)	Other Tests (Com. A, Z.P)	Ref.
PSf/S-PES		1.04 (mL/ m^2·h·Pa)	3 to 1 (BSA)	14.2 to 8.753 (10*3 cells/ mm)	APTT 35 54.4s, PTT 13.5 to 22.7s	C3a increased 98.5 to 122 and C5a 14.3 to 22.4 mg/dl. Diffusive permeability increased (× 104 cm/min): Urea 32 to 96 Creatinine 25 to 52 Cytochrome C 8 to 18	(Kaleekkal et al., 2015)
PAN/ Carboxylated-polyetherimide	50.4	285	46 to 27 (BSA) 29.6 to 16.3 (Fbg)	4.87 to 1.96 (10*3 cells/ mm)	APTT 32.3 to 41.2 PT 14.7 to 17.1	Solute permeability improved (× 104/cm min); Urea 74 to 77 Creatinine 44 to 51 Cytochrome C 11 to 14	(Senthilkumar et al., 2013)

Abbreviations: Ultrafiltration (UF), whole blood clotting time (WBCT), activated partial thrombin time (APTT), thrombin time (TT), prothrombin time (PT), plasma recalcification time (PRT), complement activation (Com. A), poly(ethylene glycol)-poly(sodium styrene sulfonate-co-methyl methacrylate) (MPEG-P(SSNa-co-MMA)), zeta potential (Z.P), vitamin E D-a-Tocopherol polyethylene glycol succinate (TPGS), acrylonitrile (AN), acrylic acid (AA), vinyl pyrrolidone (VP), bovine serum albumin (BSA), fibrinogen (Fbg), carboxylic polyethersulfone (CPES), sulfonated polyethersulfone (SPES), polystyrene-co-acrylic acid (St-co-AA), functionalized multiwall carbon nanotube (f-MWCNT), polyurethane (PU), citric acid (CA), sodium styrene sulfonate(Na-SS) and methyl ether methacrylate (EGMA) units grafted CNT (CNT–P(Na SS-co-EGMA)), polyglycidyl methacrylate (PGMA), poly (citric acid)-grafted-MWCNT (PCA-g-MWCNT), poly(2-methoxyethylacrylate) (PMEA), 5-merpeptoid with 2-methoxyethyl (NMEG5), polydopamine (PDA), vinyl pyrrolidone (VP), vinyltriethoxysilane. (VTEOS).

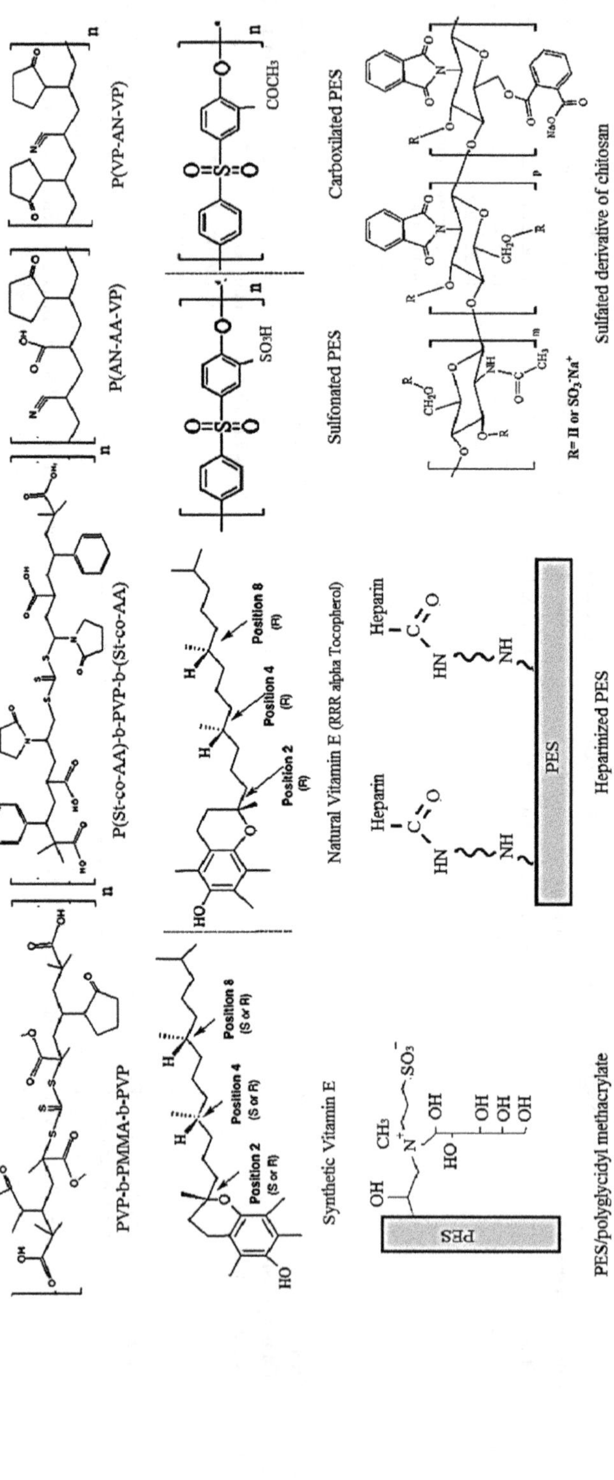

FIGURE 19.2 Chemical structures and schematic representation of some additives used with PES.

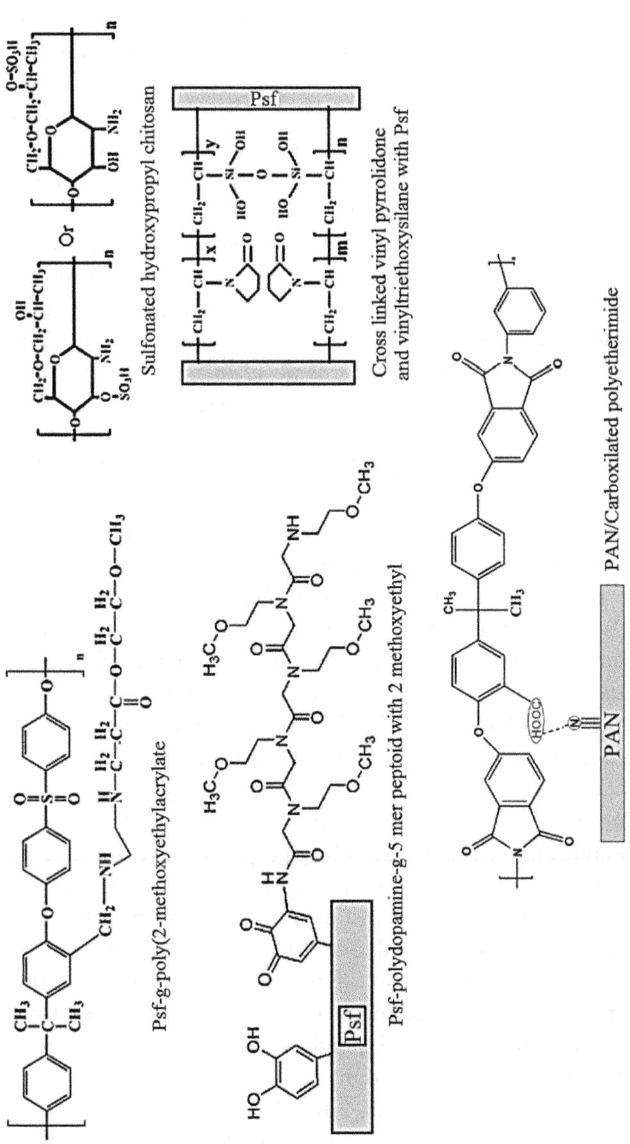

FIGURE 19.3 Chemical structures of the modifiers used with PSf and PAN membranes.

developed a zwitterionic polyglycidyl methacrylate (PGMA) block via cross-linking polymerization and modified the PES membrane and concluded that the modified membranes exhibited 99% lesser platelet adhesion and 28% higher clotting time (APTT) (Xie et al., 2015).

Polyvinylpyrrolidone (PVP) is an amphiphilic, hydrophilic, and inert water-soluble polymer, widely used in dialysis membrane composition. It promotes the macrovoid-free structure in the HFs and enhances the membrane biocompatibility. It is also believed that PVP could form poor miscible blends with hydrophobic polymers due to lack of attractive forces between them. However, as PVP leaching is likely to occur during dialysis, the issue needs to be solved. Researchers prepared binary, ternary, and tetra-blocks or copolymer by composite and covalent methods and blended them with a polymer such as PES (Nissenson and Fine, 2016).

Ran et al. (2011) and Nie et al. (2012) used reversible addition fragmentation chain transfer (RAFT) polymerization process and prepared amphiphilic triblocks of "poly(vinyl pyrrolidone)–b-poly(methyl methacrylate)–b-poly(vinyl pyrrolidone)" and "poly(styrene–co-acrylic acid)–b-poly(vinyl pyrrolidone)–b-poly(styrene–co-acrylic acid)", respectively. Similarly, Li et al. (2010) and Yin et al. (2012) utilized solution polymerization method to develop terpolymers blocks consist of poly(acrylonitrile–acrylic acid–vinyl pyrrolidone) and poly(vinylpyrrolidone–acrylonitrile–vinylpyrrolidone), respectively. The structures of PVP blocks are illustrated in Figure 19.3.

In these blocks, methyl methacrylate, styrene, and acrylonitrile are hydrophobic in nature and tend to form hydrophobic-hydrophobic interactions with hydrophobic PES. They also acted as anchoring molecules. The other parts of the block (PVP and acrylic acid) stretched out on the polymer surface and acted as cilia (polymer brush). After blending these blocks with PES, the leaching problem of PVP could be minimized and further improved membrane biocompatibility (Li et al., 2010; Nie et al., 2012; Ran et al., 2011; Yin et al., 2012). Qin et al. (2014) synthesized different molecular weights of PVP by RAFT process and fabricated PES/PVP composite HD membranes by blending process. The results suggested that low molecular weight PVPs showed additional leaching compared to high molecular weight PVPs. Moreover, the antithrombotic and antifouling properties of composite membranes (PES/PVP) were highly improved as compared to pristine PES (Qin et al., 2014).

Protein and lipid oxidation by reactive oxygen species of blood through neutrophils and monocyte produced the oxidative stress, which is responsible for chronic inflammatory diseases, atherosclerosis, hypertension, chronic inflammatory diseases, and nephritis during dialysis treatment. In the HD treatment, the vitamin E composite or coated membrane proved to reduce the oxidative stress with very little change in hydrophobic nature. The resulting membrane showed good biocompatibility, but poor UF coefficient (Dahe et al., 2011). Verma at al. (2017) used vitamin E (D-α-Tocopherol polyethylene glycol succinate) (TPGS) and nanozeolite (NZ) to improve the biocompatibility of PES HF membranes. They used NZ as a filler and TPGS as an additive. The fabricated membranes exhibited lower (0.25%) hemolysis, higher coagulation time, and 40% lower protein adsorption than commercial (F60S) HF dialysis membrane. Moreover, the uremic solute clearance of goat blood was five times higher than F60S dialyzers.

The blood protein cofactors (fibrin and thrombin) cause blood to clot and vitamin K controls the formation of these cofactors. Injection of an anticoagulant (warfarin or heparin) to increase the blood clotting time is commonly used on kidney patient. Heparin averts the working of cofactors, and warfarin prevents the precise working of vitamin K (Irfan and Idris, 2015). Want et al. (2014) introduced amino groups onto the PES surface followed by covalent immobilization of heparin via amide bond (Figure 19.3, heparinized PES). These modifications of PES caused a slight influence on the membrane morphology but enhanced the hydrophilicity and hemocompatibility behavior.

The heparin molecules contain a number of carboxylate (COO-) and sulfonate (SO_3H-) groups which create steric repulsion on most of the blood proteins, thus reduces protein adsorption onto the membrane surface. The blending of carboxylated or sulfonated (anionic) or both functional groups with hydrophobic polymer developed heparin like interface on the membrane surface. Nie at al. (2014) fabricated heparin-like structured on membrane surface by two different methods (phase inversion and evaporation process) using carboxylic polyethersulfone (CPES), sulfonated polyethersulfone (SPES), and PES with variable ratios. The results showed that both types of membranes exhibited almost the same biocompatible behavior, but the membranes made by phase inversion process were more hydrophilic (refer to Table 19.3) (Nie et al., 2014).

Tang at al. (2012) developed a heparin-like surface by blending SPES and triblock of acrylonitrile, acrylic acid and vinylpyrrolidone (P(AN--AA-VP)), in which sulfonate and carboxylate groups were provided by the SPES and triblock, respectively. The results showed that SPES/triblock/PES based membranes exhibited higher clotting time and lesser complement activation than triblock/PES and pristine PES membranes. The sulfonated membrane (heparin-like surface) showed 64% and 9% higher APTT and TAT III results and 40.6% lesser complement activation of C5a than the pristine PES membrane (Tang et al., 2012).

Functionalized and modified nanoparticles (NPs) and nanotubes have a broad spectrum of usage in the dialysis field. The major problems with nanomaterial are poor dispersion and agglomeration on the polymeric matrix. Different factors such as nanomaterial morphology, surface charge, pH of solution, surface chemistry, and presence of functional groups might affect the dispersion and stability of NPs. Generally, the use of surfactants, longer blending time with high speed mixer, proper chemical functionalization, and change in pH could improve the dispersion and stability of NPs to a certain degree.

Irfan et al., (2014) prepared composite additive consisting of functionalized multiwall carbon nanotubes (f-MWCNT) and PVP via non-covalent bonding and blended it with PES to fabricate new type of HD membranes. The dialysis results of model solution of uremic solutes showed that modified membrane enhanced the clearance ratio of creatinine, urea and lysozyme up to 55%, 56% and 28.4%, respectively. In other research work, Irfan et al. (2016) fabricated the composite membranes (f-MWCNT/PVP/PES) with variable amounts of SPES polymer and determined the biocompatibility and dialysis properties. The results demonstrated that upon addition of SPES to the composite membranes, the biocompatibility of PES membrane was greatly enhanced, reducing protein adsorption by 56% (BSA) and 64% (lysozyme). Moreover, the SPES based membrane enhanced the APTT by 46%, TT by 31%, and PT by 40% with higher dialysis clearance of uremic solutes.

Meanwhile, the PAES is a 3rd member of the polyarylsulfone family and has been used in the manufacturing of commercial dialyzers for more than a decade. The brand products of dialyzers named Baxter, Revaclear®, Polyflux® R, and Polyflux® H are made of PAES (Baxter, 2017).

19.4.2 Polyacrylonitrile (PAN) Improvements

The copolymer of acrylonitrile (AN) and PAN have been effectively used as the main membrane material for HD application because they are moderately hydrophilic in nature, form hydrogel structures with water that favors hydraulic permeability, and have high diffusive process and outstanding membrane forming properties. The pristine surface of PAN exhibits relatively poor biocompatibility, brittleness, and adsorbs middle molecules sized proteins during dialysis. Thus, PAN is modified prior to its use to improve its biocompatibility and productivity of solute removal.

The commercial brand Evodial® (Baxter) contains AN69®acrylonitrile based membrane composition that has acrylonitrile and sodium methallyl sulfonate copolymer-polyethyleneimine-heparin grafted composition. These dialyzers provide heparin-free dialysis and are suitable for patients that have high risk of bleeding (Baxter, 2017; Laville et al., 2014). The AN based membrane can also be coated with PEG or vitamin E materials so as to increase its biocompatibility and reduce the leukocyte activation (Frascà et al., 2015).

Yu et al. (2017) fabricated a new type of two-tier composite thin-film nanofibrous membrane that consisted of cross-linked PVA and a supporting layer of an electrospun PAN nanofiber. The resulting membrane exhibited good mechanical strength and biocompatibility and demonstrated 45.8% and 82.6% for lysozyme and urea clearance, respectively, and 98.8% BSA retention (Yu et al., 2017). Wei et al. (2016) grafted maleic anhydride onto a PAN membrane surface by UV irradiation followed by chemically attaching hyper-branched polyester with anhydride groups of maleic anhydride. The modified membrane showed better antifouling, biocompatibility, and hydrophilic properties than the non-modified membranes (Wei et al., 2016). Roy et al. (2015) fabricated three different types of membranes with dialysis grade MWCO (6–16 kDa) and examined the hemocompatibility, cytocompatibility, and permeation performance of urea and creatinine solutes. These membranes consisted of PSf blended with PVP and PEG, chemically modified PSf with PVP, and dialysis grade of PAN. The surface charge study showed that the PAN membrane was slightly negative charged (–0.03 mV), the most hydrophobic (80° contact angle), and demonstrated higher levels of protein adsorption (32.91 ug/cm^2), platelet adhesion, thrombus formation and dialysis permeation as compared to the other two sets of membranes (Roy et al., 2015). Senthilkumar et al. (2013) modified the PAN membrane with carboxylated polyetherimide (CPEI) and examined its biocompatibility and solutes permeability. They concluded that the modified membrane not only reduced the platelet adhesion, plasma protein adsorption, and thrombus formation but also demonstrated high solute permeability (see Table 19.3) (Senthilkumar et al., 2013).

19.4.3 Polymethylmethacrylate (PMMA) Amendments

Toray Industries, Inc. developed the PMMA synthetic membrane that is widely used in their commercial grade dialyzers (Filtryzer BG, Filtryzer NF, Filtryzer BK-F, Filtryzer BK-U, FiltryzerB1, and Filtryzer B3) (Toray, 2017). The PMMA membrane contains symmetric microporous uniform structures all over its surface, which contribute to its higher adsorptive properties of endotoxins. PMMA membrane is superior at removing the middle (vitamin B12, β-m) to high (albumin) molecular weight chemicals compared to cellulose based membranes. Moreover, improvement in neutropenia, suppression of complement activation, high blood-coagulant time, and low cell-adhesive during dialysis make them highly biocompatible membrane materials (Kokubo et al., 2015; Nubé et al., 2015).

Masakane et al. (2017) studied the protein adsorption and platelet adhesion characters of commercial PMMA based membrane dialyzers (Filtryzer® BG and Filtryzer® NF) through in vitro experiments. They found the same pattern of protein adsorption on both BG and NF dialyzers, but the activation and adhesion of platelets were much less on NF. The accumulation of erythrocytes, leukocytes, and platelets were started by activation of platelets during the HD therapy and caused side effects such as reduction in blood circulation. Thus, lesser activation favors a good dialysis therapy and NF dialyzer proved better than BG dialyzer (Masakane et al., 2017).

19.4.4 Cellulose Triacetate (CTA)

The brand names Solacea® from Nipro medicals and Exeltra® from Baxter contain CTA as a base membrane material and are currently available in the market (Baxter, 2017; Nipro, 2017). The CTA membranes usually exhibit homogenous surface structures with thick separation layer, whereas the asymmetrical membrane contained a thin separation layer (Yamazaki et al., 2011). CTA fibers are thin with moire structures, exhibit high solute permeability and good diffusive efficiency that can eliminate β2-m by diffusion process. The biocompatibility of the CTA membrane is slightly better than diacetate or acetate or other cellulosic dialysis membranes. Enhancement in lipid metabolism, high antithrombogenicity, lower activation of the coagulation cascade, and lessening of end products of biomarkers (glycation and homocysteine) are the clinical benefits of CTA based dialyzers. But CTA membranes suffer from high adsorption of albumin that may affect the membrane chemistry and flow conditions of blood (Sunohara and Masuda, 2011; Urbani et al., 2012). Takouli et al. (2010) used vitamin E-coated CTA membranes in HD patients and measured some of the inflammation biomarkers, oxidative stress biomarkers, total antioxidant capacity, and reactive oxygen metabolites. They concluded that a coating of vitamin E on CTA membranes improved the biocompatibility and suppressed inflammation and oxidative stress (Takouli et al., 2010).

19.4.5 Ethylene Vinyl Alcohol Copolymer (EVAL)

Ethylene vinyl alcohol copolymer (EVOH) is recognized by the trade name "EVAL." The commercial dialyzer of Ashi-Kasei (KF-201 Series) is EVAL based. It is a low flux dialyzer that exhibits enhanced biocompatibility and good removal capability and is ideal for patients who are allergic to PSf dialyzers. The EVAL membranes are smooth, neutral and hydrophilic in nature, and absorb more water and fewer blood proteins. They exhibit minimal activation of platelets and are weakly interlined with blood components. Pro-inflammatory cytokines such as monocyte chemo-attractant protein and interleukin-6 are produced in negligible amounts in blood with EVAL based dialyzers, thus a minimal level of allergic response of patients may lead to better peripheral circulation. It is also reported that the oxidative stress and inflammatory response may be reduced with the use of EVAL based dialyzers (Kasei, 2017; Nakano, 2011).

19.5 CONCLUSION

In the fabrication of HD membranes, biocompatibility is very important because the cell system of the dialysis patient treats the membrane material as an external invader that may result in a number of interactions between the membrane and blood element. Poor biocompatible membrane material may adsorb plasma proteins, activate the complement system, and promote adherence of platelets upon exposure to blood. These then invoke thrombosis reactions against membrane materials. Poor biocompatibility can cause higher rates of morbidly and mortality of HD patients. Synthetic polymers are broadly utilized in biomedical application, especially in the HD field, but they are mostly hydrophobic and deficient in biocompatible factors and cannot be used alone in membrane fabrications. In order to optimize the membrane performance, different anticoagulant molecules, non-ionic, zwitterionic, biomimetic, amphoteric, and hydrophilic brushes were blended or immobilized with base polymers to achieve the target properties in the resulting membranes. These additives not only enhanced the hemocompatibility, but also improved the dialysis results of uremic toxins. The clotting inhibition rates and complex formation of blood slowed down when anticoagulant (heparin) based additives were used. For instance, the combination of SPES and CPES or SPES with carboxylate based chemicals help to develop the heparin-like membrane surface. These modified surfaces caused steric repulsion to the blood proteins and platelet, thus improved membranes' biocompatibility. The incorporation of PVP and PEG blocks is also commonly practiced to improve the hydrophilicity of membranes. These block additives are effective to reduce the elution problems and their hydrophobic part could act as an anchoring material with the main polymer, improving membrane stability. Generally, the addition of additives in small quantities was reported to be ideal to remarkably reduce the platelet adhesion and protein adsorption. With respect to the use of NPs for HD membrane fabrication, there are currently no standard and safety guidelines to evaluate their impacts on human health. Nevertheless, it is necessary to entirely assess the biocompatibility of NPs-based membranes though in vivo and in vitro tests prior to clinical trial.

REFERENCES

Abidin, M.N.Z., Goh, P.S., Ismail, A.F., Othman, M.H.D., Hasbullah, H., Said, N., Kadir, S.H.S.A., Kamal, F., Abdullah, M.S., Ng, B.C., 2017. Development of biocompatible and safe polyethersulfone hemodialysis membrane incorporated with functionalized multi-walled carbon nanotubes. Materials Science and Engineering: C. 77, 572–582.

Allmed, 2017. Products - ADVENTA Health. http://www.adventa-health.com/products-single.php?id=150

Asahi Kasei, 2017. Product Line | Asahi Kasei Medical Co., Ltd. http://www.asahi-kasei.co.jp/medical/en/dialysis/product/

Baxter, 2017. Dialyzers for Hemodialysis | Baxter. https://www.baxter.com/products-expertise/renal-failure-treatments/hemodialysis-products/dialyzers.page

Bowry, S.K., Gatti, E., Vienken, J., 2011. Contribution of polysulfone membranes to the success of convective dialysis therapies. High-Performance Membrane Dialyzers. 173, 110–118.

Chowdhury, N., Islam, F., Zafreen, F., Begum, B., Sultana, N., Perveen, S., Mahal, M., 2012. Effect Of Surface Area Of Dialyzer Membrane On The Adequacy Of Haemodialysis. Journal of Armed Forces Medical College, Bangladesh. 7(2), 9–11.

Dahe, G.J., Teotia, R.S., Kadam, S.S., Bellare, J.R., 2011. The biocompatibility and separation performance of antioxidative polysulfone/vitamin E TPGS composite hollow fiber membranes. Biomaterials. 32(2), 352–365.

Daugirdas, J.T., Blake, P.G., Ing, T.S., 2012. *Handbook of Dialysis*. Lippincott Williams & Wilkins.

Daugirdas, J.T., Greene, T., Rocco, M.V., Kaysen, G.A., Depner, T.A., Levin, N.W., Chertow, G.M., Ornt, D.B., Raimann, J.G.,Larive, B, Kliger, A.S., 2013. Effect of frequent hemodialysis on residual kidney function. Kidney International. 83(5), 949–958.

Frascà, G. M., Sagripanti, S., D'arezzo, M., Oliva, S., Francioso, A., Mosconi, G., Zambianchi, L., Sopranzi, F., Boggi, R., Fattori, L., Rigotti, A., Maldini, L., Gattiani, A., Del Rosso, G., Federico, A., Da Lio, L., Ferrante, L., 2015. Post-dilution hemodiafiltration with a heparin-grafted polyacrylonitrile membrane. Therapeutic Apheresis and Dialysis. 19(2), 154–161.

Fresenius, 2017. Fresenius Polysulfone Dialyzers. http://fmcna-dialyzers.com/dialyzers-site/products.html

Gambro, 2017. Dialyzers - Gambro. http://www.gambro.at/en/global/Products/Hemodialysis/-Dialyzers/index.html

GmbH, N.-M., 2017. NIKKISO: Dialyzers. https://www.nikkiso-europe.eu/en/products/dialyzers/

Harris, J.M., 2013. *Poly (ethylene glycol) Chemistry: Biotechnical and Biomedical Applications*. Springer Science & Business Media.

He, C., Nie, C.X., Zhao, W.F., Ma, L., Xiang, T., Cheng, C.S., Sun, S.-D., Zhao, C.S., 2013. Modification of polyethersulfone membranes using terpolymers engineered and integrated antifouling and anticoagulant properties. Polymers for Advanced Technologie. 24(12), 1040–1050.

Huang, Y., Wong, C., Zheng, J., Bouwman, H., Barra, R., Wahlström, B., Neretin, L., Wong, M.H., 2012. Bisphenol A (BPA) in China: A review of sources, environmental levels, and potential human health impacts. Environment International. 42, 91–99.

Irfan, M., Idris, A., 2015. Overview of PES biocompatible/hemodialysis membranes: PES–blood interactions and modification techniques. Materials Science and Engineering: C. 56, 574–592.

Irfan, M., Idris, A., Nasiri, R., Almaki, J.H., 2016. Fabrication and evaluation of polymeric membranes for blood dialysis treatments using functionalized MWCNT based nanocomposite and sulphonated-PES. RSC Advances. 6(103), 101513–101525.

Irfan, M., Idris, A., Yusof, N.M., Khairuddin, N.F.M., Akhmal, H., 2014. Surface modification and performance enhancement of nano-hybrid f-MWCNT/PVP90/PES hemodialysis membranes. Journal of Membrane Science. 467, 73–84.

Kaleekkal, N.J., Thanigaivelan, A., Tarun, M., Mohan, D., 2015. A functional PES membrane for hemodialysis—Preparation, characterization and biocompatibility. Chinese Journal of Chemical Engineering. 23(7), 1236–1244.

Kokubo, K., Kurihara, Y., Kobayashi, K., Tsukao, H., Kobayashi, H., 2015. Evaluation of the biocompatibility of dialysis membranes. Blood Purification. 40(4), 293–297.

Krieter, D.H., Lemke, H.-D. (2011). Polyethersulfone as a high-performance membrane. In A. Saito, H. Kawanishi, A.C. Yamashita, M. Mineshima (Eds.), *High-Performance Membrane Dialyzers* (Vol. 173, pp. 130–136). Karger Publishers.

Laville, M., Dorval, M., Ros, J.F., Fay, R., Cridlig, J., Nortier, J.L., Juillard, L., Dębska-Ślizień, A., Fernández Lorente, L., Thibaudin, D., Franssen, C., Schulz, M., Moureau, F., Loughraieb, N., Rossignol, P., 2014. Results of the HepZero study comparing heparin-grafted membrane and standard care show that heparin-grafted dialyzer is safe and easy to use for heparin-free dialysis. Kidney International. 86(6), 1260–1267.

Li, L., Yin, Z., Li, F., Xiang, T., Chen, Y., Zhao, C., 2010. Preparation and characterization of poly (acrylonitrile-acrylic acid-N-vinyl pyrrolidinone) terpolymer blended polyethersulfone membranes. Journal of Membrane Science. 349(1), 56–64.

Lipps, B.J., Stewart, R.D., Perkins, H.A., Holmes, G.W., McLain, E.A., Rolfs, M.R., Oja, P.D., 1967. The Hollow Fibr Artificial Kidney. ASAIO Journal. 13(1), 200–207.

Liu, T.-M., Xu, J.-J., Qiu, Y.-R., 2017. A novel kind of polysulfone material with excellent biocompatibility modified by the sulfonated hydroxypropyl chitosan. Materials Science and Engineering: C. 79, 570–580.

Ma, L., Qin, H., Cheng, C., Xia, Y., He, C., Nie, C., Wang, L., Zhao, C., 2014. Mussel-inspired self-coating at macro-interface with improved biocompatibility and bioactivity via dopamine grafted heparin-like polymers and heparin. Journal of Materials Chemistry B. 2(4), 363–375.

Mahmoudi, N., Reed, L., Moix, A., Alshammari, N., Hestekin, J., Servoss, S.L., 2017. PEG-mimetic peptoid reduces protein fouling of polysulfone hollow fibers. Colloids and Surfaces B: Biointerfaces. 149, 23–29.

Masakane, I., Esashi, S., Yoshida, A., Chida, T., Fujieda, H., Ueno, Y., Sugaya, H., 2017. A new polymethylmetacrylate membrane improves the membrane adhesion of blood components and clinical efficacy. Renal Replacement Therapy. 3(1), 32.

Nakano, A., 2011. Ethylene vinyl alcohol co-polymer as a high-performance membrane: An EVOH membrane with excellent biocompatibility. In A. Saito, H. Kawanishi, A.C. Yamashita, M. Mineshima (Eds.), *High-Performance Membrane Dialyzers* (Vol. 173, pp. 164–171). Karger Publishers.

Nie, C., Ma, L., Xia, Y., He, C., Deng, J., Wang, L., Cheng, C., Sun, S., Zhao, C., 2015. Novel heparin-mimicking polymer brush grafted carbon nanotube/PES composite membranes for safe and efficient blood purification. Journal of Membrane Science. 475, 455–468.

Nie, S., Tang, M., Yin, Z., Wang, L., Sun, S., Zhao, C., 2014. Biologically inspired membrane design with a heparin-like interface: Prolonged blood coagulation, inhibited complement activation, and bio-artificial liver related cell proliferation. Biomaterials Science. 2(1), 98–109.

Nie, S., Xue, J., Lu, Y., Liu, Y., Wang, D., Sun, S., Ran, F., Zhao, C., 2012. Improved blood compatibility of polyethersulfone membrane with a hydrophilic and anionic surface. Colloids and Surfaces B: Biointerfaces. 100, 116–125.

Nipro, 2017. Cellulose triacetate dialyzer/hollow-fiber - FB-U - Nipro. http://www.medicalexpo.com/prod/nipro/product-94053-579043.html

Nissenson, A.R., Fine, R.E., 2016. *Handbook of Dialysis Therapy E-Book*. Elsevier Health Sciences.

Nubé, M.J., Grooteman, M.P., Blankestijn, P.J. (2015). *Hemodiafiltration: Theory, Technology and Clinical Practice*. Springer.

Qin, H., Nie, S., Cheng, C., Ran, F., He, C., Ma, L., Yin, Z., Zhao, C., 2014. Insights into the surface property and blood compatibility of polyethersulfone/polyvinylpyrrolidone composite membranes: Toward high-performance hemodialyzer. Polymers for Advanced Technologies. 25(8), 851–860.

Qin, H., Sun, C., He, C., Wang, D., Cheng, C., Nie, S., Sun, S., Zhao, C., 2014. High efficient protocol for the modification of polyethersulfone membranes with anticoagulant and antifouling properties via in situ cross-linked copolymerization. Journal of Membrane Science. 468, 172–183.

Ran, F., Nie, S., Zhao, W., Li, J., Su, B., Sun, S., Zhao, C., 2011. Biocompatibility of modified polyethersulfone membranes by blending an amphiphilic triblock co-polymer of poly (vinyl pyrrolidone)–b-poly (methyl methacrylate)–b-poly (vinyl pyrrolidone). Acta Biomaterialia. 7(9), 3370–3381.

Ran, F., Niu, X., Song, H., Zhao, W., Nie, S., Wang, L., . . . Zhao, C., 2014. Toward a highly hemocompatible membrane for blood purification via a physical blend of miscible comb-like amphiphilic copolymers. Biomaterials Science. 2(4), 538–547.

Roy, A., Dadhich, P., Dhara, S., De, S., 2015. In vitro cytocompatibility and blood compatibility of polysulfone blend, surface-modified polysulfone and polyacrylonitrile membranes for hemodialysis. RSC Advances. 5(10), 7023–7034.

Roy, A., De, S., 2017. State-of-the-art materials and spinning technology for hemodialyzer membranes. Separation & Purification Reviews. 46(3), 216–240.

Saito, A., 2011. Definition of high-performance membranes–From the clinical point of view. In A. Saito, H. Kawanishi, A.C. Yamashita, M. Mineshima (Eds.), *High-Performance Membrane Dialyzers* (Vol. 173, pp. 1–10). Karger Publishers.

Sakai, K., Matsuda, M., 2011. Solute removal efficiency and biocompatibility of the high-performance membrane–from engineering points of view. In A. Saito, H. Kawanishi, A.C. Yamashita, M. Mineshima (Eds.), *High-Performance Membrane Dialyzers* (Vol. 173, pp. 11–22). Karger Publishers.

Saran, R., Robinson, B., Abbott, K., Agodoa, L., Ayanian, J., Bragg-Gresham, J., . . . Eggers, P., 2017. US renal data system 2016 annual data report: Epidemiology of Kidney Disease in the United States. American Journal of Kidney Diseases: The Official Journal of the National Kidney Foundation. 69(3S1), A7.

Senthilkumar, S., Rajesh, S., Jayalakshmi, A., Mohan, D., 2013. Biocompatibility and separation performance of carboxylated poly (ether–imide) incorporated polyacrylonitrile membranes. Separation and Purification Technology. 107, 297–309.

Slinin, Y., Greer, N., Ishani, A., MacDonald, R., Olson, C., Rutks, I., Wilt, T.J., 2015. Timing of dialysis initiation, duration and frequency of hemodialysis sessions, and membrane flux: A systematic review for a KDOQI clinical practice guideline. American Journal of Kidney Diseases. 66(5), 823–836.

Sunohara, T., Masuda, T., 2011. Cellulose triacetate as a high-performance membrane. In A. Saito, H. Kawanishi, A.C. Yamashita, M. Mineshima (Eds.), *High-Performance Membrane Dialyzers* (Vol. 173, pp. 156–163). Karger Publishers.

Sunohara, T., Masuda, T., 2017. Fundamental characteristics of the newly developed ATA™ membrane dialyzer. In T. Sunohara, T. Masuda (Eds.), *Scientific Aspects of Dialysis Therapy* (Vol. 189, pp. 215–221). Karger Publishers.

Takouli, L., Hadjiyannakos, D., Metaxaki, P., Sideris, V., Filiopoulos, V., Anogiati, A., Vlassopoulos, D., 2010. Vitamin E-coated cellulose acetate dialysis membrane: Long-term effect on inflammation and oxidative stress. Renal Failure. 32(3), 287–293.

Tang, M., Xue, J., Yan, K., Xiang, T., Sun, S., Zhao, C., 2012. Heparin-like surface modification of polyethersulfone membrane and its biocompatibility. Journal of Colloid and Interface Science. 386(1), 428–440.

Tian, X., Qiu, Y.-R., 2017. 2-methoxyethylacrylate modified polysulfone membrane and its blood compatibility. Archives of Biochemistry and Biophysics. 631, 49–57.

Toray, 2017. Products | Toray Medical Co., Ltd. http://www.toray-medical.com/en/products/index.html

Urbani, A., Lupisella, S., Sirolli, V., Bucci, S., Amoroso, L., Pavone, B., Pieroni, L., Sacchetta, P., Bonomini, M., 2012. Proteomic analysis of protein adsorption capacity of different haemodialysis membranes. Molecular BioSystems. 8(4), 1029–1039.

Vandentorren, S., Zeman, F., Morin, L., Sarter, H., Bidondo, M.-L., Oleko, A., Leridon, H., 2011. Bisphenol-A and phthalates contamination of urine samples by catheters in the Elfe pilot study: Implications for large-scale biomonitoring studies. Environmental Research. 111(6), 761–764.

Verma, S. K., Modi, A., Singh, A. K., Teotia, R., Bellare, J., 2017. Improved hemodialysis with hemocompatible polyethersulfone hollow fiber membranes: In vitro performance. Journal of Biomedical Materials Research Part B: Applied Biomaterials. 106, 1286–1298.

Wang, L., Cai, Y., Jing, Y., Zhu, B., Zhu, L., Xu, Y., 2014. Route to hemocompatible polyethersulfone membranes via surface aminolysis and heparinization. Journal of Colloid and Interface Science. 422, 38–44.

Wei, X., Fei, Y., Shi, Y., Chen, J., Lv, B., Chen, Y., Xiang, H., 2016. Hemocompatibility and ultrafiltration performance of PAN membranes surface-modified by hyperbranched polyesters. Polymers for Advanced Technologies. 27(12), 1569–1576.

Xie, Y., Li, S.-S., Jiang, X., Xiang, T., Wang, R., Zhao, C.-S., 2015. Zwitterionic glycosyl modified polyethersulfone membranes with enhanced anti-fouling property and blood compatibility. Journal of Colloid and Interface Science. 443, 36–44.

Xue, J., Zhao, W., Nie, S., Sun, S., Zhao, C., 2013. Blood compatibility of polyethersulfone membrane by blending a sulfated derivative of chitosan. Carbohydrate Polymers. 95(1), 64–71.

Yamashita, A.C., Sakurai, K., 2015. Dialysis Membranes—Physicochemical Structures and Features. In H. Suzuki (Ed.), *Updates in Hemodialysis*. InTech.

Yamashita, A.C., Tomisawa, N., 2009. Importance of membrane materials for blood purification devices in critical care. Transfusion and Apheresis Science. 40(1), 23–31.

Yamazaki, K., Matsuda, M., Yamamoto, K.-I., Yakushiji, T., Sakai, K., 2011. Internal and surface structure characterization of cellulose triacetate hollow-fiber dialysis membranes. Journal of Membrane Science. 368(1), 34–40.

Yin, Z., Su, B., Nie, S., Wang, D., Sun, S., Zhao, C., 2012. Poly (vinylpyrrolidone-co-acrylonitrile-co-vinylpyrrolidone) modified polyethersulfone hollow fiber membranes with improved blood compatibility. Fibers and Polymers. 13(3), 269–276.

Young-Hyman, D.L., 2013. National institute of diabetes and digestive and kidney diseases. *Encyclopedia of Behavioral Medicine* (pp. 1290–1291). Springer.

Yu, X., Shen, L., Zhu, Y., Li, X., Yang, Y., Wang, X., Zhu, M., Hsiao, B.S., 2017. High performance thin-film nanofibrous composite hemodialysis membranes with efficient middle-molecule uremic toxin removal. Journal of Membrane Science. 523, 173–184.

Zhu, L., Song, H., Wang, J., Xue, L., 2017. Polysulfone hemodiafiltration membranes with enhanced anti-fouling and hemocompatibility modified by poly (vinyl pyrrolidone) via in situ cross-linked polymerization. Materials Science and Engineering: C. 74, 159–166.

20 Development of Redox-Responsive Membranes for Biomedical Applications

Weifeng Zhao and Changsheng Zhao

CONTENTS

20.1 INTRODUCTION

Redox reaction, so-called oxidation-reduction reaction, is one kind of reaction in which the electrons are transferred between two chemical species. The redox reaction is the chemical reaction in which the oxidation number of chemical species changes by gaining or losing electrons. Redox reactions are very common and crucial to many basic functions of life, such as combustion, respiration, photosynthesis, and rusting. The main technique to understand the energetics of biological electron-transfer processes is redox potentiometry, which provides the description of biological oxidation-reduction systems, including photosynthesis, metabolism and drug detoxification (Dutton, 1978).

A redox component can be represented as follows:

$$A_{reduced} = A_{oxidized} + ne^{-} \tag{20.1}$$

where n is the number of electrons (e^{-}). Such an equation represents only half the complete redox reaction. Actually, redox reactions include two parts, a reduced part and an oxidized part. The complete equation is:

$$xA_{red} + yB_{ox} = mA_{ox} + nB_{red} \tag{20.2}$$

the ratios of A and B (x/y and m/n) should be adjusted appropriately to the account for the number of electrons exchanged. The equilibrium constant ($K_{eq} = [A_{ox}]\,[B_{red}]/[A_{red}]\,[B_{ox}]$) of Eq. (20.2) depends on the electrons of the A and B couples. The standard free energy of the reaction (i.e., $\Delta G°$) is commonly represented as $\Delta G° = -RT\mathit{ln}K_{eq}$.

$$\Delta G = \Delta G° + RT\mathit{ln}[A_{ox}]\,[B_{red}]\,/\,([A_{red}]\,[B_{ox}]) \tag{20.3}$$

The two species that exchange electrons in an oxidation-reduction reaction are given a certain definition. The molecule accepting electrons is the oxidizing agent: it causes the oxidation of other species. In contrast, the species donating electrons is the reducing agent: when the reaction occurs, it reduces the other species.

As an advanced material for membranes, redox-responsive agents for synthesis and modification of the membrane are a rapidly growing field, and this chapter is intended to serve as a guide to the latest research and references, offering summaries of the current state-of-the-art breakthroughs of redox-responsive membranes. Readers can be expected to understand that the entire spectrum of redox-responsive membranes covers fundamental research, cutting-edge technologies, and basic applications. Looking ahead to the topics covered in the chapter, we begin by giving a broad overview of redox-responsive agents. Section 20.3 covers basic fundamental research on the redox-responsive membranes and their applications in controlled drug release, biosensors, as well as separation.

20.2 REDOX-RESPONSIVE AGENTS

The main molecules involving in redox reaction are the oxidizing agents and the reducing agents.

20.2.1 Oxidizing Agents

Non-metallic elements. Non-metallic elements in redox reaction, such as chlorine, bromine, iodine, oxygen, ozone, etc., are easy to achieve electron and tuned into the corresponding anion. For example, when bromine reacts with silver, it generates bromine ions. At the same time, the silver ion is combined and turned into silver bromide, making it easily detectable.

The peroxide molecules. Examples of molecules that have -O-O- structure are hydrogen peroxide (H_2O_2), sodium peroxide (Na_2O_2), and barium peroxide (BaO_2). Since the -O-O- structure is very unstable, the peroxide in the reaction is very easy to release oxygen for oxidizing other substances. For instance, H_2O_2 is an unstable compound that breaks down easily and frequently. But the H_2O_2 solution is relatively stable. In biomedical fields, H_2O_2 is used to disinfect wounds, especially in the oral cavity. It can also be used to wipe off old oil paintings, so that the oil painting can regain its luster. This is because the lead in the paint is exposed to the effects of hydrogen sulphide in the air and turns into black lead sulphide. Meanwhile the black sulphide reacts with an overload of hydrogen and becomes white with lead sulfate.

$$PbS + 4H_2O_2 = PbSO_4 + 4H_2O \tag{20.4}$$

Compounds with the highest valence. Compounds with the highest valence of the element include certain oxygen-containing acids (e.g., sulfuric acid (H_2SO_4) and nitric acid (HNO_3)) and oxygen-containing salts (e.g., potassium permanganate ($KMnO_4$), potassium dichromate (K_2CrO_7), and potassium chlorate ($KClO_3$)). The main elements in these compounds have the highest valence number, so when they participate in the reaction, the electron tends to act as an oxidant. Here are a few common examples:

$$2KClO_3 \xrightarrow{\triangle} 2KCl + 3O_2\uparrow \tag{20.5}$$

$$2KMnO_4 \xrightarrow{\triangle} K_2MnO_4 + MnO_2 + O_2\uparrow \tag{20.6}$$

It is important to note that the resulting products are not only dependent on the oxidant itself but are also determined by the concentration of the solution, the strength of the reducing agent, and the nature of the medium during the reaction (Marcus, 1957).

20.2.2 Reducing Agents

The metal elements. The large active metals lose electrons easily, so they are all reductants. Sodium amalgam and aluminum, which are commonly used in laboratories and in production, are good reductants. For example, the aluminum agent shows the nature of the strong reducing agent. Hot aluminum is a mixture of trife and aluminum powder. The valence number of iron in Fe_3O_4 is +3 and +2, so its formula can be written as $Fe_2O_3 \bullet FeO$. When it reacts with aluminum at a high temperature, aluminum takes all the oxygen in its components and turns into a free iron. In this reaction, the trivalent iron and the divalent iron get the electron to be reduced and so is the oxidizing agent, and the electron is oxidized, the reducing agent.

$$3(Fe_2O_3 \bullet FeO) + 8Al \rightarrow 9Fe + 4Al_2O_3 \text{ (high temperature)} \tag{20.7}$$

The materials that easily make electronic "deviation". In the redox reaction, this kind of material can often atomically covalently bond with other elements, making the electrons "deviate" from the outer atoms. For example, hydrogen, carbon, and carbon monoxide have been involved in the redox reaction, producing compounds of carbon monoxide and carbon dioxide. The main reaction of the blast furnace iron making is in this type of reaction. In this reaction, the trivalent iron gets the electron and is the oxidizing agent.

$$Fe_2O_3 + 3CO \rightarrow 2Fe + 3CO_2 \uparrow \text{(high temperature)} \quad (20.8)$$

The element compounds with low valency. These compounds include ferrous salt, stannous salt, sulfur dioxide, sulfite, etc., which can lose electrons and turn into iron, tin salt, sulfate, etc. For example, the tin chloride solution is mixed with the iron chloride solution, and the tin chloride is then oxidized to tin chloride, while the iron chloride is reduced to ferrous chloride.

$$2FeCl_3 + SnCl_2 = 2FeCl_2 + SnCl_4$$

Various acid anion. Various acid anions include hydroiodic acid sulfate (H_2S), hydrobromic acid (HBr), hydrochloric acid (HCl), etc. These compounds always lose electrons into anions in redox reaction. For example, hydrogen sulfuric acid and hydrogen iodide must be oxidized in the air for a long time to become to sulfur and iodine.

Some organic compounds. Oxalate $H_2C_2O_4$ is often used as a reducing agent; for example, using it with potassium permanganate solution can turn seven valent numbers of Mn to bivalent Mn.

$$H_2C_2O_4 + 2KMnO_4 + 3H_2SO_4 = K_2SO_4 + 2MnSO_4 + 8H_2O + 10CO_2\uparrow \quad (20.9)$$

Some oxidizers can only be used as reducing agent if they react with oxidative strength. For example, H_2O_2 is normally used as an oxidant (in an acidic solution with ferrous sulfate), but when it reacts with potassium permanganate, it is oxidized to generate water and give off oxygen. Here, hydrogen peroxide is a reducing agent.

$$H_2O_2 + 2FeSO_4 + H_2SO_4 = Fe_2(SO_4)_3 + 2H_2O \quad (20.10)$$

$$5H_2O_2 + 2KMnO_4 + 3H_2SO_4 = K_2SO_4 + 2MnSO_4 + 8H_2O + 5O_2\uparrow \quad (20.11)$$

Commonly used oxidants and reducing agents are summarized in Table 20.1.

TABLE 20.1
Commonly Used Oxidants and Reducing Agents

Oxidants		**Reducing Agents**	
Non-metallic elements	O_3, Cl_2, Br_2, I_2	Metal elements	K, Na, Ca, Mg, Al
Peroxide molecules	H_2O_2, Na_2O_2, BaO_2	Element compounds with low valency	CO, H_2
The highest valence of the elements	MnO_2, $KMnO_4$, $K_2Cr_2O_7$, HNO_3, H_2SO_4	Various acid anions	H_2S, NH_3, HCl
		Some organic compounds	SO_2, H_2SO_3, $FeCl_2$, $SnCl_2$

20.3 REDOX-RESPONSIVE MEMBRANES FOR BIOMEDICAL APPLICATIONS

Redox-responsive systems have attracted great attention in past two decades, since a few redox reactions are carried out in living systems and natural environment. Large reports on redox-responsive materials have been produced. These materials have great potential to be used in controlled drug release (Chen et al., 2016; Zhang et al., 2014a), biomedical sensors (Puzzo et al., 2009), and membrane separation (Elbert et al., 2014).

20.3.1 Controlled Drug Release

In the area of controlled cargo release for biomedical applications, stimuli-responsive drug release systems capable of reacting to their microenvironment are required (Che and van Hest, 2016; Freudenberg et al., 2009). The change of redox potential has been widely applied to trigger release of a cargo to the microenvironment (Dai et al., 2011). The intracellullar microenvironment displayed reductive potential for relatively high concentrations (2–10 mM) of glutathione (GSH) compared to extra-cellular fluids (2–20 μM). Additionally, a material internalized into the endo-/lysosomal system is thought to face a reducing environment, mainly regulated by gamma-interferon-induced lysosomal thiol reductase in the presence of cysteine.

Taking advantage of these mechanisms, many researchers have reported disulfide bond based reductive-responsive nanocontainers for controlled release of various payloads (Liu et al. 2015; Yan et al., 2011; Zhang et al., 2017a). Wilke et al. (2017) prepared a vesicle composed of an amphiphilic β-cyclodextrin derivative, which could selectively bind guest molecules via host-guest recognition, with a redox-responsive disulfide crosslinked polymer shell anchored on the vesicle surface (Figure 20.1). In the reductive intracellular microenvironment, these disulfide crosslinks could be split inducing a diminished stabilization from the shell for cargo release (de Vries et al., 2017). Under the reduction of tris(2-carboxyethyl) phosphine hydrochloride (TCEP), polymer shelled vesicles exhibited a rapid release of carboxyfluorescein (CF) ($t_{1/2} = 2$ h at 400 mm TCEP), and the release profiles were highly dependent on the concentration of the reducing agent. Yan et al. (2010) synthesized thiolated poly(methacrylic acid) to form disulfide crosslinked hollow hydrogel capsules via layer by layer assemble

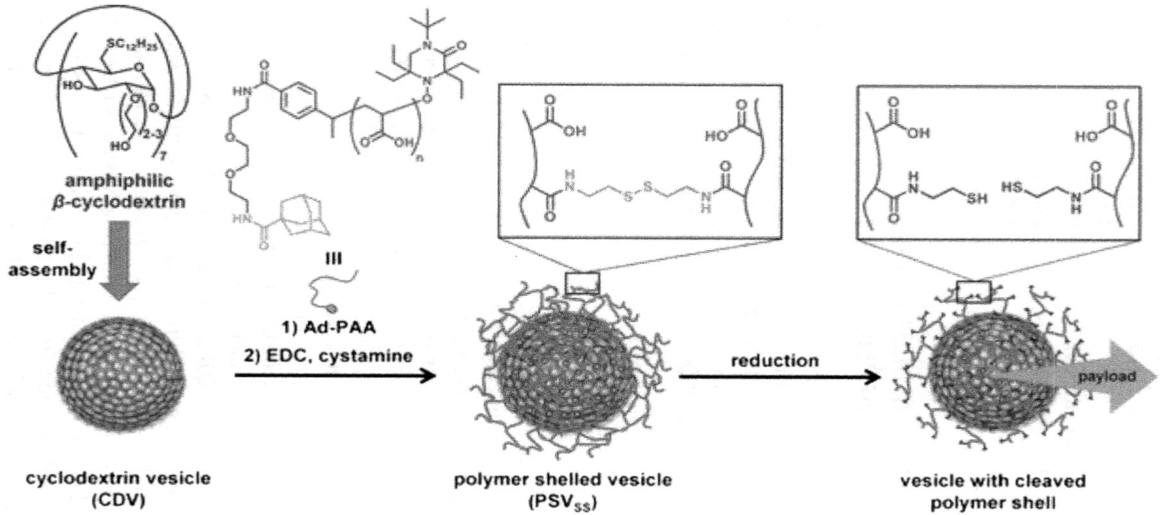

FIGURE 20.1 Stepwise preparation of a redox-responsive nanocontainer and the redox-triggered release of a hydrophilic payload. (From de Vries, W.C. et al., Angewandte Chemie-International Edition. 56(32), 9603–9607, 2017.)

method followed by subsequent cross-linking of the thiol groups using oxidizing agent. The capsules are stable in oxidizing conditions, such as bloodstream, but the disulfide bonds could be broken in reducing environments such as the cytoplasm of the cell, so it can be used for drug delivery to cancer cells (Yan et al., 2010).

Besides the disulfide based reduction-responsive drug delivery systems, oxidation-responsive drug delivery systems have also been widely studied (Napoli et al., 2004b; Power-Billard et al., 2004) To address the oxidation environment present in inflammation sites, Hubbell et al. (2004) fabricated oxidation-responsive vesicles in pure water from poly(ethylene glycol)-*b*-poly(propylene sulfide)-*b*-poly(ethylene glycol) (PEO-*b*-PPS-*b*-PEO) triblock copolymer (PPS stands for poly(propylene sulfide). The hydrophobic PPS segments could be converted into hydrophilic poly(propylene sulfoxide) and poly(propylene sulfone) upon exposure to an oxidative environment, which destabilized the vesicle membrane and led to the release of the loaded cargo (Napoli et al., 2004a).

Ferrocene-modified copolymers were also popular oxidation-responsive materials owing to reversible oxidation and reduction of ferrocene units by electrochemical and chemical means and were widely used as the membrane of nanocontainers (Kakizawa et al., 2001; Saleem et al., 2017) For example, Shi et al. (2016) prepared multi-compartment vesicles containing on-off switchable pores in the vesicular membrane from ferrocene-containing triblock terpolymer. The incompatible solvophobic poly(benzyl methacrylate) (PBzMA) and poly(4-vinylbenzyl ferrocenecarboxylate) (PVFC) parts form the porous phase-segregated membrane and the solvophilic poly(2-(dimethylamino) ethyl methacrylate) part is present at both the inner and outer sides of the membrane (Figure 20.2). The porous membranes are redox-responsive,

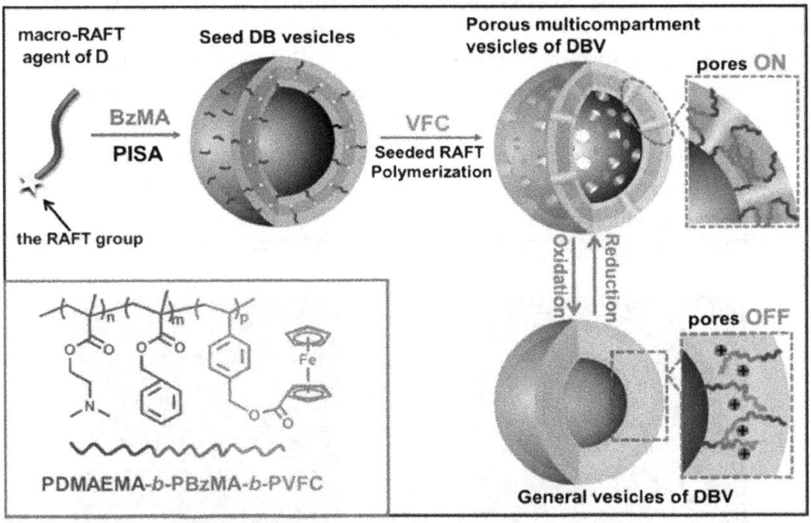

FIGURE 20.2 Synthesis of the porous multicompartment vesicles of the PDMAEMA-*b*-PBzMA-*b*-PVFC triblock polymer by seeded reversible addition-fragmentation chain transfer (RAFT) polymerization and the schematic on-off switch of the membrane pores of the multicompartment vesicles through oxidation/reduction. (From Shi, P. et al., ACS Macro Letters. 5(1), 88–93, 2016.)

and these pores can be on-off switched through a redox-stimuli to control the releasing behaviors (Shi et al. 2016). That is, the reduced vesicles showed a porous phase-segregated membrane. In contrast, the oxidized vesicles exibited a uniform membrane.

20.3.2 Biosensors

Unregulated generation of reactive oxygen species (ROS) will cause oxidative damage to proteins, lipids, and nucleic acids, which is connected to serious human diseases such as acute and chronic inflammation, cancers, diabetes, and neurodegenerative disorders (Chu et al., 2016; Miller et al., 2007; Narayanaswamy et al., 2016). Therefore, precise determination of ROS and real-time monitoring of the redox homeostasis changing should have wide physiological and pathological perspectives. A few techniques and tools have been employed to design redox-responsive biosensors that can detect transient ROS burst and relevant redox events in living systems (Lou et al., 2014; Manjare et al., 2013). Depending on the category of the ROS, the designed biosensors can also be classified as $ONOO^-$ sensitive biosensors, HbrO-sensitive biosensors, H_2O_2 sensitive biosensors, or O_2 sensitive biosensors.

20.3.2.1 $ONOO^-$ Sensitive Biosensors

The peroxynitrite ($ONOO^-$), a highly ROS, plays an important role in living biological systems. Great efforts have been devoted to design $ONOO^-$ biosensors with high sensitivity, fast response, and great specificity towards $ONOO^-$ against other ROS both in vitro and in vivo (Nascimento et al., 2011). For example, by using the organoselenium compound of monoselenides (a major glutathione peroxidase mimic), an on-off switch near Infrared (NIR) biosensor Cy-Pse was designed containing a 4-(phenylselanyl)aniline fluorescent modulator. As a result, biosensors with specific and sensitive responses to $ONOO^-$ and a NIR fluorescent heptamethine cyanine (Cy) dye have been achieved. The fluorescence was switched on upon detection of $ONOO^-$ with an increase factor of 23.3 fold but was switched off in response to the main in vivo antioxidants glutathione (GSH) and L-cysteine (Yu et al., 2011). Tang et al. (2011) fabricated a biosensor BzSe-Cy based on the organoselenium compound of monoselenides, which contained the heptamethine cyanine dye as the NIR fluorophore and the divalent selenium as a redox-responsive group (Figure 20.3). It emitted

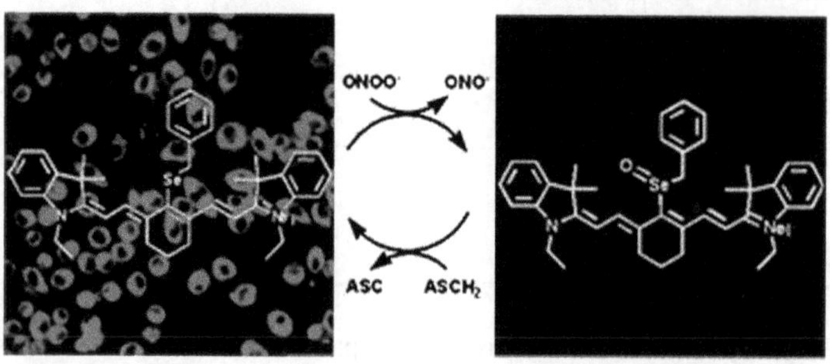

FIGURE 20.3 A near-infrared reversible fluorescent probe for peroxynitrite and imaging of redox cycles in living cells. (From Xu, K. et al., Chemical Communications. 47(33), 9468–9470, 2011.)

NIR light with a maximum emission around 795 nm and its fluorescence emission was switched "off" and "on" when it responded to $ONOO^-$ and the antioxidant ascorbate ($ASCH_2$), respectively (Xu et al., 2011).

20.3.2.2 HBrO-Sensitive Biosensors

In vivo excessive production of hypobromous acid (HBrO) can cause tissue damage and thus leads to many diseases like asthma, cardiovascular diseases, arthritis, and cancers. Based on the redox reaction cycles between HbrO and ascorbic acid and between HbrO and H_2S, HBrO biosensors only responding to HBrO were designed. For example, Han et al. (2012) designed the Cy-TemOH biosensor, where a heptamethine cyanine platform acted as the fluorophore and the integrated 4-hydroxylamino-2,2,6,6-tetramethylpiperidine-*N*-oxyl (TemOH) moiety acted as the fluorescent modulator that responded to the redox changes of HbrO (Figure 20.4) (Yu et al., 2012b). That is to say, by using this catalytic redox cycle, the NIR fluorescent probes containing a hydroxylamine moiety could be used to detect intracellular HOBr level.

20.3.2.3 H_2O_2-Sensitive Biosensor

H_2O_2 is an important ROS species. In vivo excessive concentration of hydrogen peroxide can cause oxidative stress that connects to neurodegenerative and cardiovascular diseases, cancers, etc. To design a biosensor responding only to H_2O_2 at physiologically relevant levels and allowing for real-time monitoring of its multiple redox cycles in living cells, tissues and living animals, Yu et al. (2012) integrated a dopamine unit into a cyanine scaffold via central substitution to fabricate DA-Cy biosensor. Upon detection of H_2O_2, its fluorescence intensity decreased and switched off, but the fluorescence was gradually switched on with the GSH reduction (Yu et al., 2012a). Xu et al. (2013) combined the ebselen (Eb) moiety (a modulator with a unique response to H_2O_2/GSH) with the heptamethine cyanine fluorophore platform to prepare H_2O_2-biosensor (Figure 20.5). The fluorescence intensity decreased rapidly with the GSH reduction, and the intensity recovered and increased rapidly upon the detection of H_2O_2 (Xu et al., 2013).

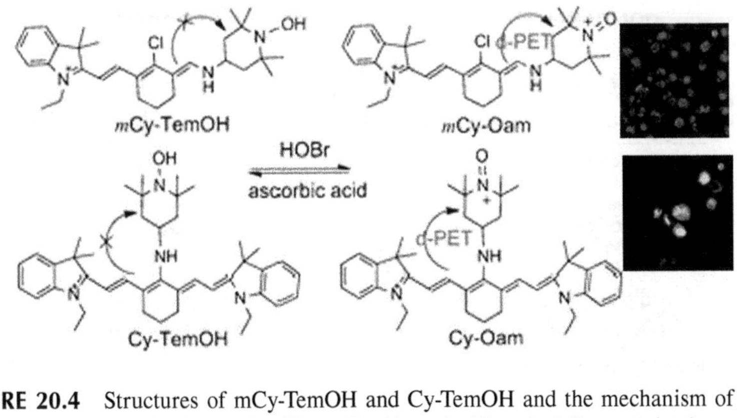

FIGURE 20.4 Structures of mCy-TemOH and Cy-TemOH and the mechanism of HOBr/ascorbic acid induced redox cycle. (From Yu, F. et al., Chemical Communications. 48(62), 7735–7737, 2012.)

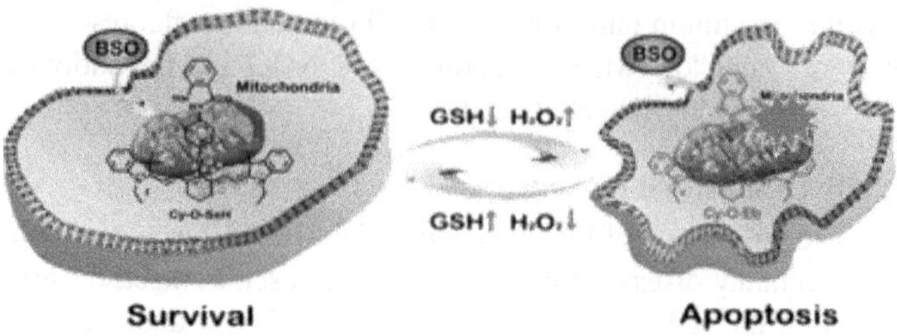

FIGURE 20.5 A near-infrared reversible fluorescent probe for real-time imaging of redox status changes in vivo. (From Xu, K. et al., Chemical Communications. 47(33), 9468–9470, 2011; Xu, K. et al., Chemical Science. 4(3), 1079–1086, 2013.)

20.3.2.4 O_2-Sensitive Biosensor

Diseases such as cardiovascular diseases, tumors, and stroke all related to hypoxia. Takahashi et al. (2012) developed the reversible biosensor RhyCy5 to monitor in real time the repeated hypoxia-normoxia cycles in living cells. The sensing mechanism of RHyCy5 is associated with the redox behavior of QSY-21. Under hypoxia, QSY-21 can be one-electron reduced to its radical form QSY-21-radical, and this reduced radical form undergoes rapid oxidation when exposed to air (Figure 20.6). Such rapid oxidation and the large separation between Cy5 and QSY-21 in RHyCy5 caused the decreased fluorescence of the biosensor from hypoxia to normoxia, instead of the fluorescence quenching of the Cy5 fluorophores. The strong fluorescence under hypoxia is due to no Förster resonance energy transfer (FRET) in RHyCy5-radical (Takahashi et al., 2012).

20.3.3 Separation Membranes

Stimuli-responsive materials, which exhibit large and sharp changes in response to small variations of external parameters such as temperature, pH, and ionic strength, have been widely incorporated into porous polyelectrolyte membranes to fabricate stimuli-responsive membranes (Cheng et al., 2011; Luo et al., 2014; Wandera et al.,

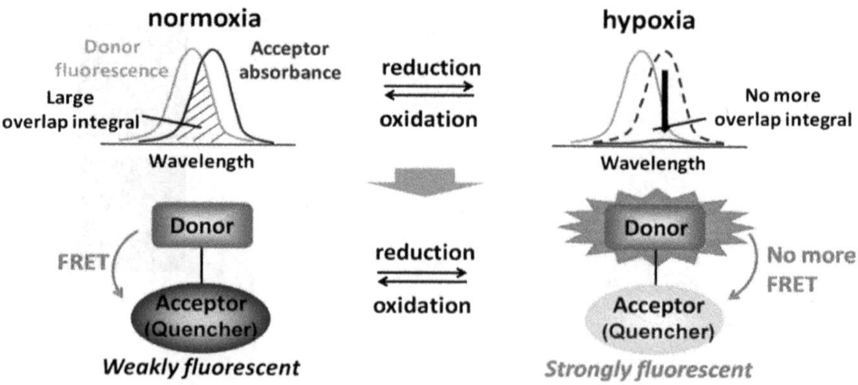

FIGURE 20.6 Design strategy for a reversible fluorescence probe for hypoxia. (From Takahashi, S. et al., Journal of the American Chemical Society. 134(48), 19588–19591, 2012.)

2011). These smart membranes are capable of reversibly changing the pore size or the surface wettability after exposure to certain stimulation (Bera et al., 2015; Zhang et al., 2017b). Besides these traditional smart membranes, redox-responsive membranes have recently emerged as a promising smart membrane for various applications (Elbert et al., 2014). For these redox-responsive membranes, the functional polymers with ferrocene groups as redox active groups were usually coated or grafted onto the porous membrane platforms. By oxidizing the ferrocene moieties chemically, the polymer chain conformation usually undergoes a remarkable transition, leading to the transition of membrane surface from hydrophobic to hydrophilic or a change of pore size.

Elbert et al. (2014) grafted poly(2-(methacryloyloxy)ethyl ferrocenecarboxylate) (PFcMA) and polyvinylferrocene (PVFc) onto an ordered mesoporous silica thin film, and the modified membrane showed great effect on ionic permselectivity by a redox-controlled membrane gating process (Elbert et al., 2014). Mesoporous silica films were obtained after sol-gel synthesis and evaporation induced self assembly of non-ionic block copolymer template Pluronic-F127. After the modification, 150-300-nm-thick mesoporous silica films with the pore size about 10 nm were formed, confirmed by transmission electron microscopy (TEM) images (Figure 20.7).

Zhang et al. (2014b) coated poly(ferrocenylsilane) onto a porous membrane. The membrane was oxidized and reduced by 10 mM $Fe(ClO_4)_3$ and 10 mM ascorbic acid, leading to the switching of permeability originating from the changes in interconnectivity between pores (Figure 20.8). Compared to traditional electrochemical oxidation and reduction, the chemical treatment provided a fast change of the redox state due to high diffusion rates in the porous membrane. In the oxidized form, the membranes had a visibly higher density of openings. In contrast, a higher density of closed cells was observed in the reduced state of the porous membranes as confirmed by scanning electron microscopy (SEM) images.

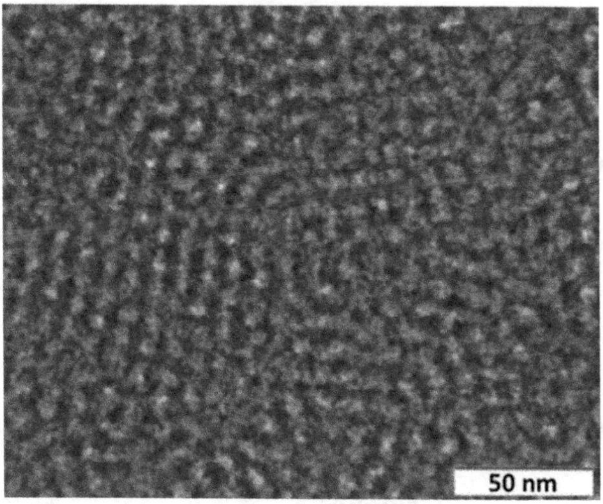

FIGURE 20.7 TEM image of a calcinated mesoporous silica film surface templated with Pluronic F127. The porous structure is clearly visible. (From Elbert, J. et al., Advanced Functional Materials. 24(11), 1591–1601, 2014.)

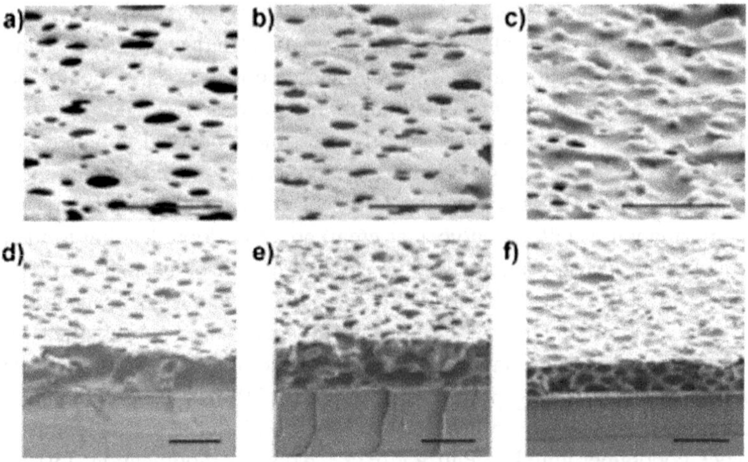

FIGURE 20.8 Surface and cross-sectional SEM images of complex membrane: untreated membrane (a and d), after oxidation by 10 mM $Fe(ClO_4)_3$ aqueous solution (b and e), and then following reduction by 10 mM ascorbic acid in aqueous solution (c and f). Scale bar: 1 μm.

Shi et al. (2015) synthesized ferrocene and β-cyclodextrin grafted PES. The two functionalized PES were blended with pristine PES to form a membrane, and the two functional PES interacted with each other by host-guest interactions (Figure 20.9). By cyclic oxidating and reducing process, the permeability of the membrane could be significantly changed (Shi et al., 2015) For the surface morphology of the membranes (Figure 20.10), the pristine PES membrane (M-0) showed a smooth surface while the modified PES membrane (M-5) showed a rough surface, which was

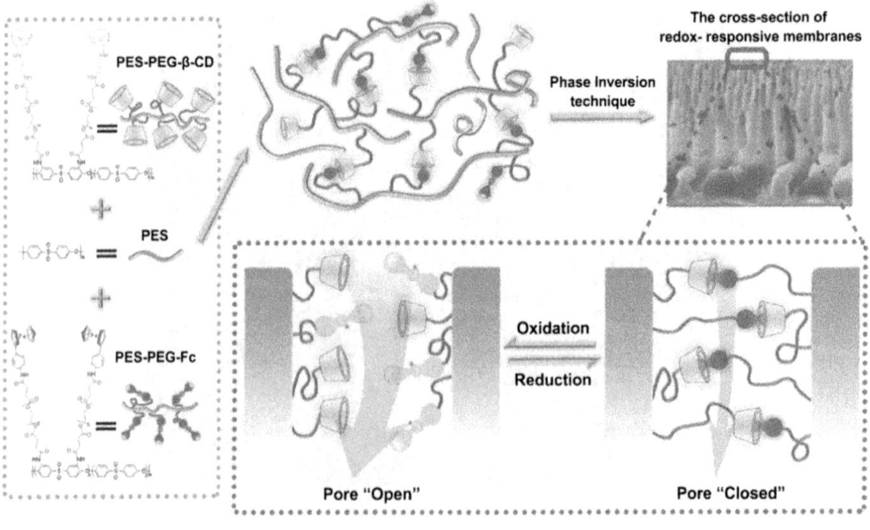

FIGURE 20.9 The modified membranes with redox-responsive hydraulic permeability. (From Shi, W. et al., Journal of Membrane Science. 480, 139–152, 2015.)

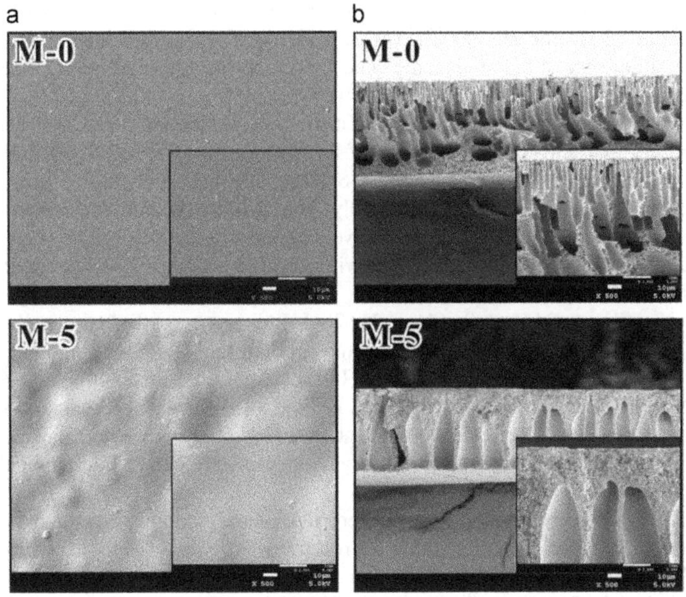

FIGURE 20.10 SEM of the membranes (a) surface and (b) cross-sections. (Magnification: 500× and 2000×) (From Shi, W. et al., Journal of Membrane Science. 480, 139–152, 2015.)

caused by the phase separation during the membrane preparation. A typical structure for membranes prepared by a phase inversion method was observed in the cross-sectional views of the membranes: a skin layer was found on the top surfaces, followed by a finger-like structure. After the modification of the PES membranes by the PES-PEG-Fc/PES-PEG-β-CD, the finger-like structure was gradually suppressed.

20.4 CONCLUSION

"Redox-responsive" membranes have been gaining increasing attention due to their potential to be controlled by the external stimulus of redox-sensitive agents. However, the applications of these advanced materials have a long way to go. Clearly having a redox-responsive membrane that can also be sensitive to other stimuli, such as temperature, light, as well as magnetic fields, is best as an intelligent membrane. Furthermore, the response times of these redox-responsive membranes are usually slow and need to be improved. Despite the above-mentioned challenges, the cutting edge research and the enthusiasm of scientists in redox-responsive membranes remain undaunted.

REFERENCES

Bera, A., Kumar, C.U., Parui, P., Jewrajka, S.K., 2015. Stimuli responsive and low fouling ultrafiltration membranes from blends of polyvinylidene fluoride and designed library of amphiphilic poly(methyl methacrylate) containing copolymers. Journal of Membrane Science. 481, 137–147.

Che, H., van Hest, J.C., 2016. Stimuli-responsive polymersomes and nanoreactors. Journal of Materials Chemistry B. 4(27), 4632–4647.

Chen, Q., Qi, R., Chen, X., Yang, X., Huang, X., Xiao, H., Wang, X., Dong, W., 2016. Polymeric nanostructure compiled with multifunctional components to exert tumor-targeted delivery of antiangiogenic gene for tumor growth suppression. ACS Applied Materials & Interfaces. 8(37), 24404–24414.

Cheng, C., Ma, L., Wu, D., Ren, J., Zhao, W., Xue, J., Sun, S., Zhao, C., 2011. Remarkable pH-sensitivity and anti-fouling property of terpolymer blended polyethersulfone hollow fiber membranes. Journal of Membrane Science. 378(1–2), 369–381.

Chu, T.-S., Lu, R., Liu, B.-T., 2016. Reversibly monitoring oxidation and reduction events in living biological systems: Recent development of redox-responsive reversible NIR biosensors and their applications in in vitro/in vivo fluorescence imaging. Biosensors & Bioelectronics. 86, 643–655.

Dai, J., Lin, S., Cheng, D., Zou, S., Shuai, X., 2011. Interlayer-crosslinked micelle with partially hydrated core showing reduction and pH dual sensitivity for pinpointed intracellular drug release. Angewandte Chemie International Edition. 50(40), 9404–9408.

de Vries, W.C., Grill, D., Tesch, M., Ricker, A., Nuesse, H., Klingauf, J., Studer, A., Gerke, V., Ravoo, B.J., 2017. Reversible stabilization of vesicles: Redox-responsive polymer nanocontainers for intracellular delivery. Angewandte Chemie-International Edition. 56(32), 9603–9607.

Dutton, P.L., 1978. Redox potentiometry: Determination of midpoint potentials of oxidation-reduction components of biological electron-transfer systems. Methods in Enzymology. 54, 411–435.

Elbert, J., Krohm, F., Ruettiger, C., Kienle, S., Didzoleit, H., Balzer, B.N., Hugel, T., Stuehn, B., Gallei, M., Brunsen, A., 2014. Polymer-modified mesoporous silica thin films for redox-mediated selective membrane gating. Advanced Functional Materials. 24(11), 1591–1601.

Freudenberg, U., Hermann, A., Welzel, P.B., Stirl, K., Schwarz, S.C., Grimmer, M., Zieris, A., Panyanuwat, W., Zschoche, S., Meinhold, D., Storch, A., Werner, C., 2009. A star-PEG-heparin hydrogel platform to aid cell replacement therapies for neurodegenerative diseases. Biomaterials. 30(28), 5049–5060.

Kakizawa, Y., Sakai, H., Yamaguchi, A., Kondo, Y., Yoshino, N., Abe, M., 2001. Electrochemical control of vesicle formation with a double-tailed cationic surfactant bearing ferrocenyl moieties. Langmuir. 17(26), 8044–8048.

Liu, S.-T., Tuan-Mu, H.-Y., Hu, J.-J., Jan, J.-S., 2015. Genipin cross-linked PEG-block-poly(L-lysine)/disulfide-based polymer complex micelles as fluorescent probes and pH-/redox-responsive drug vehicles. RSC Advances. 5(106), 87098–87107.

Lou, Z., Yang, S., Li, P., Zhou, P., Han, K., 2014. Experimental and theoretical study on the sensing mechanism of a fluorescence probe for hypochloric acid: A Se··· N nonbonding interaction modulated twisting process. Physical Chemistry Chemical Physics. 16(8), 3749–3756.

Luo, T., Lin, S., Xie, R., Ju, X.-J., Liu, Z., Wang, W., Mou, C.-L., Zhao, C., Chen, Q., Chu, L.-Y., 2014. pH-responsive poly(ether sulfone) composite membranes blended with amphiphilic polystyrene-block-poly(acrylic acid) copolymers. Journal of Membrane Science. 450, 162–173.

Manjare, S.T., Kim, S., Heo, W.D., Churchill, D.G., 2013. Selective and sensitive superoxide detection with a new diselenide-based molecular probe in living breast cancer cells. Organic Letters. 16(2), 410–412.

Marcus, R.A., 1957. Theory of oxidation-reduction reactions involving electron transfer. 2. Applications to data on the rates of isotopic exchange reactions. Journal of Chemical Physics. 26(4), 867–871.

Miller, E.W., Bian, S.X., Chang, C.J., 2007. A fluorescent sensor for imaging reversible redox cycles in living cells. Journal of the American Chemical Society. 129(12), 3458–3459.

Napoli, A., Boerakker, M.J., Tirelli, N., Nolte, R.J., Sommerdijk, N.A., Hubbell, J.A., 2004a. Glucose-oxidase based self-destructing polymeric vesicles. Langmuir. 20(9), 3487–3491.

Napoli, A., Valentini, M., Tirelli, N., Müller, M., Hubbell, J.A., 2004b. Oxidation-responsive polymeric vesicles. Nature Materials. 3(3), 183.

Narayanaswamy, N., Narra, S., Nair, R.R., Saini, D.K., Kondaiah, P., Govindaraju, T., 2016. Stimuli-responsive colorimetric and NIR fluorescence combination probe for selective reporting of cellular hydrogen peroxide. Chemical Science. 7(4), 2832–2841.

Nascimento, V., Alberto, E.E., Tondo, D.W., Dambrowski, D., Detty, M.R., Nome, F., Braga, A.L., 2011. GPx-Like activity of selenides and selenoxides: Experimental evidence for the involvement of hydroxy perhydroxy selenane as the active species. Journal of the American Chemical Society. 134(1), 138–141.

Power-Billard, K.N., Spontak, R.J., Manners, I., 2004. Redox-active organometallic vesicles: aqueous self-assembly of a diblock copolymer with a hydrophilic polyferrocenylsilane polyelectrolyte block. Angewandte Chemie International Edition. 43(10), 1260–1264.

Puzzo, D.P., Arsenault, A.C., Manners, I., Ozin, G.A., 2009. Electroactive inverse opal: A single material for all colors. Angewandte Chemie-International Edition. 48(5), 943–947.

Saleem, M., Wang, L., Yu, H., Zain ul Abdin, Akram, M., Ullah, R.S., 2017. Synthesis of amphiphilic block copolymers containing ferrocene-boronic acid and their micellization, redox-responsive properties and glucose sensing. Colloid and Polymer Science. 295(6), 995–1006.

Shi, P., Qu, Y., Liu, C., Khan, H., Sun, P., Zhang, W., 2016. Redox-responsive multicompartment vesicles of ferrocene-containing triblock terpolymer exhibiting On Off switchable pores. ACS Macro Letters. 5(1), 88–93.

Shi, W., Zhang, L., Deng, J., Wang, D., Sun, S., Zhao, W., Zhao, C., 2015. Redox-responsive polymeric membranes via supermolecular host-guest interactions. Journal of Membrane Science. 480, 139–152.

Takahashi, S., Piao, W., Matsumura, Y., Komatsu, T., Ueno, T., Terai, T., Kamachi, T., Kohno, M., Nagano, T., Hanaoka, K., 2012. Reversible off–on fluorescence probe for hypoxia and imaging of hypoxia–normoxia cycles in live cells. Journal of the American Chemical Society. 134(48), 19588–19591.

Wandera, D., Wickramasinghe, S.R., Husson, S.M., 2011. Modification and characterization of ultrafiltration membranes for treatment of produced water. Journal of Membrane Science. 373(1–2), 178–188.

Xu, K., Chen, H., Tian, J., Ding, B., Xie, Y., Qiang, M., Tang, B., 2011. A near-infrared reversible fluorescent probe for peroxynitrite and imaging of redox cycles in living cells. Chemical Communications. 47(33), 9468–9470.

Xu, K., Qiang, M., Gao, W., Su, R., Li, N., Gao, Y., Xie, Y., Kong, F., Tang, B., 2013. A near-infrared reversible fluorescent probe for real-time imaging of redox status changes in vivo. Chemical Science. 4(3), 1079–1086.

Yan, Y., Johnston, A.P.R., Dodds, S.J., Kamphuis, M.M.J., Ferguson, C., Parton, R.G., Nice, E.C., Heath, J.K., Caruso, F., 2010. Uptake and intracellular fate of disulfide-bonded polymer hydrogel capsules for doxorubicin delivery to colorectal cancer cells. ACS Nano. 4(5), 2928–2936.

Yan, Y., Wang, Y., Heath, J.K., Nice, E.C., Caruso, F., 2011. Cellular association and cargo release of redox-responsive polymer capsules mediated by exofacial thiols. Advanced Materials. 23(34), 3916–3921.

Yu, F., Li, P., Li, G., Zhao, G., Chu, T., Han, K., 2011. A near-IR reversible fluorescent probe modulated by selenium for monitoring peroxynitrite and imaging in living cells. Journal of the American Chemical Society. 133(29), 11030–11033.

Yu, F., Li, P., Song, P., Wang, B., Zhao, J., Han, K., 2012a. Facilitative functionalization of cyanine dye by an on–off–on fluorescent switch for imaging of H_2O_2 oxidative stress and thiols reducing repair in cells and tissues. Chemical Communications. 48(41), 4980–4982.

Yu, F., Song, P., Li, P., Wang, B., Han, K., 2012b. Development of reversible fluorescence probes based on redox oxoammonium cation for hypobromous acid detection in living cells. Chemical Communications. 48(62), 7735–7737.

Zhang, B., Luo, Z., Liu, J., Ding, X., Li, J., Cai, K., 2014a. Cytochrome c end-capped mesoporous silica nanoparticles as redox-responsive drug delivery vehicles for liver tumor-targeted triplex therapy in vitro and in vivo. Journal of Controlled Release. 192, 192–201.

Zhang, F., Gong, S., Wu, J., Li, H., Oupicky, D., Sun, M., 2017a. CXCR4-targeted and redox responsive dextrin nanogel for metastatic breast cancer therapy. Biomacromolecules. 18(6), 1793–1802.

Zhang, K., Feng, X., Sui, X., Hempenius, M.A., Vancso, G.J., 2014b. Breathing pores on command: Redox-responsive spongy membranes from poly(ferrocenylsilane)s. Angewandte Chemie-International Edition. 53(50), 13789–13793.

Zhang, X., Zhou, J., Wei, R., Zhao, W., Sun, S., Zhao, C., 2017b. Design of anion species/strength responsive membranes via in-situ cross-linked copolymerization of ionic liquids. Journal of Membrane Science. 535, 158–167.

Index

A

B

C

D

E

F

G

H

M

N

O

P

Q

R

S

V

W

X

Y

Z

Milton Keynes UK
Ingram Content Group UK Ltd.
UKHW021932071024
449327UK00022B/1772

9 780367 656980